Yaochu Jin (Ed.)

Multi-Objective Machine Learning

Studies in Computational Intelligence, Volume 16

Editor-in-chief
Prof. Janusz Kacprzyk
Systems Research Institute
Polish Academy of Sciences
ul. Newelska 6
01-447 Warsaw
Poland
E-mail: kacprzyk@ibspan.waw.pl

Further volumes of this series can be found on our homepage: springer.com

Vol. 1. Tetsuya Hoya
Artificial Mind System – Kernel Memory Approach, 2005
ISBN 3-540-26072-2

Vol. 2. Saman K. Halgamuge, Lipo Wang (Eds.)
Computational Intelligence for Modelling and Prediction, 2005
ISBN 3-540-26071-4

Vol. 3. Bożena Kostek
Perception-Based Data Processing in Acoustics, 2005
ISBN 3-540-25729-2

Vol. 4. Saman K. Halgamuge, Lipo Wang (Eds.)
Classification and Clustering for Knowledge Discovery, 2005
ISBN 3-540-26073-0

Vol. 5. Da Ruan, Guoqing Chen, Etienne E. Kerre, Geert Wets (Eds.)
Intelligent Data Mining, 2005
ISBN 3-540-26256-3

Vol. 6. Tsau Young Lin, Setsuo Ohsuga, Churn-Jung Liau, Xiaohua Hu, Shusaku Tsumoto (Eds.)
Foundations of Data Mining and Knowledge Discovery, 2005
ISBN 3-540-26257-1

Vol. 7. Bruno Apolloni, Ashish Ghosh, Ferda Alpaslan, Lakhmi C. Jain, Srikanta Patnaik (Eds.)
Machine Learning and Robot Perception, 2005
ISBN 3-540-26549-X

Vol. 8. Srikanta Patnaik, Lakhmi C. Jain, Spyros G. Tzafestas, Germano Resconi, Amit Konar (Eds.)
Innovations in Robot Mobility and Control, 2006
ISBN 3-540-26892-8

Vol. 9. Tsau Young Lin, Setsuo Ohsuga, Churn-Jung Liau, Xiaohua Hu (Eds.)
Foundations and Novel Approaches in Data Mining, 2005
ISBN 3-540-28315-3

Vol. 10. Andrzej P. Wierzbicki, Yoshiteru Nakamori
Creative Space, 2005
ISBN 3-540-28458-3

Vol. 11. Antoni Ligęza
Logical Foundations for Rule-Based Systems, 2006
ISBN 3-540-29117-2

Vol. 12. Jonathan Lawry
Modelling and Reasoning with Vague Concepts, 2006
ISBN 0-387-29056-7

Vol. 13. Nadia Nedjah, Ajith Abraham, Luiza de Macedo Mourelle (Eds.)
Genetic Systems Programming, 2006
ISBN 3-540-29849-5

Vol. 14. Spiros Sirmakessis (Ed.)
Adaptive and Personalized Semantic Web, 2006
ISBN 3-540-30605-6

Vol. 15. Lei Zhi Chen, Sing Kiong Nguang, Xiao Dong Chen
Modelling and Optimization of Biotechnological Processes, 2006
ISBN 3-540-30634-X

Vol. 16. Yaochu Jin (Ed.)
Multi-Objective Machine Learning, 2006
ISBN 3-540-30676-5

Yaochu Jin
(Ed.)

Multi-Objective Machine Learning

Dr. Yaochu Jin
Honda Research Institute
Europe GmbH
Carl-Legien-Str. 30
63073 Offenbach
Germany
E-mail: yaochu.jin@honda-ri.de

ISSN print edition: 1860-949X
ISSN electronic edition: 1860-9503
ISBN 978-3-642-06796-9
e-ISBN 978-3-540-33019-6

Springer is a part of Springer Science+Business Media
springer.com

Softcover reprint of the hardcover 1st edition 2006

Cover design: *design & production* GmbH, Heidelberg

To Fanhong, Robert and Zewei

Preface

Feature selection and model selection are two major elements in machine learning. Both feature selection and model selection are inherently multi-objective optimization problems where more than one objective has to be optimized. For example in feature selection, minimization of the number of features and maximization of feature quality are two common objectives that are likely conflicting with each other. It is also widely realized that one has to deal with the trade-off between approximation or classification accuracy and model complexity in model selection.

Traditional machine learning algorithms try to satisfy multiple objectives by combining the objectives into a scalar cost function. A good example is the training of neural networks, where the main target is to minimize a cost function that accounts for the approximation or classification error on given training data. However, reducing the training error often leads to overfitting, which means that the error on unseen data will become very large, though the neural network performs perfectly on the training data. To improve the generalization capability of neural networks, i.e., to improve their ability to perform well on unseen data, a regularization term, e.g., the complexity of neural networks weighted by a hyper-parameter (regularization coefficient) has to be included in the cost function. One major challenge to implement the regularization technique is how to choose the regularization coefficient appropriately, which is non-trivial for most machine learning problems.

Several other examples exist in machine and human learning where a trade-off between conflicting objectives has to be taken into account. For example in object recognition, a learning system should learn as many details of a given object as possible; on the other hand, it should also be able to abstract the general features of different objects. Another example is the stability and plasticity dilemma, where the learning systems have to trade off between learning new information without forgetting old information. From the viewpoint of multi-objective optimization, there is no single learning model that can satisfy different objectives at the same time. In this sense, Pareto-based multi-objective optimization is the only way to deal with the conflicting objectives in machine learning. However, this seemingly straightforward idea has not been implemented in machine learning until late 1990's. Liu and Kadirkamanathan [1] considered three criteria in designing neural networks for system identification. They used a genetic algorithm to minimize the maximum of the three normalized objectives. Similar work has been presented in [2]. Kottathra and Attikiouzel [3] employed a branch-and-bound search algorithm to determine the structure of neural networks by trading off the mean square error and the number of hidden nodes. The trade-off between sum of squared error and the norm of weights of neural networks was reported in [4], whereas the trade-off between training error and test error has been considered in [5]. An interesting work on Pareto-based neural network learning is reported by Kupinski and Anastasio [6], where a multi-objective genetic algorithm is implemented to generate the receiver operating characteristics curve of neural network classifiers. In generating fuzzy systems, Ishibuchi et al [7] used a multi-objective genetic algorithm to minimize the classification error and the number of rules. Gomez-Skarmeta et al suggested the idea of using Pareto-based multi-objective genetic algorithms to optimize multiple objectives in fuzzy modeling [8], though no simulation results were provided. A genetic algorithm is used to minimize approximation error, complexity, sensitivity to noise and continuity of rules [9].

With the boom of the research on evolutionary multi-objective optimization, Pareto-optimality based multi-objective machine learning has gained new impetus. Compared to the early works, not only more sophisticated multi-objective algorithms are used, but new research areas are also being opened up. For example, it is found that one is able to determine the number of clusters by analyzing the shape of the Pareto front obtained by a multi-objective clustering method [10], see Chapter 2 for more detailed results. In [11], it is found that interpretable rules can be extracted from the simple, Pareto-optimal neural networks. Further research reveals that neural networks with good generalization capability can also be identified by analyzing the Pareto front, as reported in Chapter 13 in this book. Our most recent work suggests that multi-objective machine learning provides an efficient approach to addressing catastrophic forgetting [12]. Besides, Pareto-based multi-objective learning has shown particularly powerful in generating ensembles, support vector machines (SVMs), and interpretable fuzzy systems.

This edited book presents a collection of most representative research work on multi-objective machine learning. The book is structured into five parts. Part I discusses multi-objective feature extraction and selection, such as rough set based feature selection, clustering and cluster validation, supervised and unsupervised feature selection for ensemble based handwritten digit recognition, and edge detection in image processing. In the second part, multi-objective model selection is presented for improving the performance of single objective learning algorithms in generating various machine learning models, including linear and nonlinear regression models, multi-layer perceptrons (MLPs), radial-basis-function networks (RBFNs), support vector machines

(SVMs), decision trees, and learning classifier systems. Multi-objective model selection for creating interpretable models is described in Part III. Generating interpretable learning models plays an important rule in data mining and knowledge extraction, where the preference is put on gaining insights into unknown systems. From the work presented, the reader can see that how understandable symbolic or fuzzy rules can be extracted from trained neural networks, or from data directly within the framework of multi-objective optimization. The merit of multi-objective optimization is fully demonstrated in Part IV, where techniques for generating ensembles of machine learning models are concerned. Diverse member of neural networks or fuzzy systems can be generated by trading off between training and test errors, between accuracy and complexity, or between accuracy and diversity. Compared to single objective based ensemble generation methods, diversity is imposed more explicitly in multi-objective learning so that structural or functional diversity of ensemble members can be guaranteed. To conclude the book, Part V presents a number of successful applications of multi-objective machine learning, such as multi-class receiver operating curve analysis, mobile robot navigation, docking maneuver of automated ground vehicles, information retrieval and object detection.

I am confident that by reading this book, the reader is able to bring home a complete view of the emerging research area and to gain hands-on experience of a variety of multi-objective machine learning approaches. Furthermore, I do hope that this book, which is the first book dedicated to multi-objective machine learning to the best of my knowledge, will inspire more creative ideas to further promote the research in this research area.

I would like to thank all contributors who prepared excellent chapters for this book. Many thanks go to Prof. Janusz Kacprzyk for his interest in this book. I would also like to thank Dr. Thomas Ditzinger and Ms. Heather King at Springer for their kind assistance. I am most grateful to Mr. Tomohiko Kawanabe, Prof. Dr. Edgar Körner, Dr. Bernhard Sendhoff and Mr. Andreas Richter at the Honda Research Institute Europe for their full understanding and kind support.

Offenbach am Main
October 2005 *Yaochu Jin*

References

[1] G.P. Liu and V. Kadirkamanathan. Learning with multi-objective criteria. In *IEE Conference on Artificial Neural Networks*, pages 53–58, 1995.

[2] S. Park, D. Nam, and C. H. Park. Design of a neural controller using multi-objective optimization for nonminimum phase systems. In *Proceedings of 1999 IEEE Int. Conf. on Fuzzy Systems*, volume I, pages 533–537, 1999.

[3] K. Kottathra and Y. Attikiouzel. A novel multicriteria optimization algorithm for the structure determination of multilayer feedforward neural networks. *Journal of Network and Computer Applications*, 19:135–147, 1996.
[4] R. de A. Teixeira, A.P. Braga, R. H.C. Takahashi, and R.R. Saldanha. Improving generalization of MLP with multi-objective optimization. *Neurocomputing*, 35:189–194, 2000.
[5] H.A. Abbass. A memetic Pareto approach to artificial neural networks. In *Proceedings of the 14th Australian Joint Conference on Artificial Intelligence*, pages 1–12, 2001.
[6] M.A. Kupinski and M Anastasio. Multiobjective genetic optimization of diagnostic classifiers with implementations for generating receiver operating characteristic curves. *IEEE Transactions on Medical Imaging*, 18(8):675–685, 1999.
[7] H. Ishibuchi, T. Murata, and B. Turksen. Single-objective and two-objective genetic algorithms for selecting linguistic rules for pattern classification problems. *Fuzzy Sets and Systems*, 89:135–150, 1997.
[8] A. Gomez-Skameta, F. Jimenez, and J. Ibanez. Pareto-optimality in fuzzy modeling. In *Proc. of the 6th European Congress on Intelligent Techniques and Soft Computing*, pages 694–700, 1998.
[9] T. Suzuhi, T. Furuhashi, S. Matsushita, and H. Tsutsui. Efficient fuzzy modeling under multiple criteria by using genetic algorithm. In *IEEE Conf. on Systems, Man and Cybernetics*, volume 5, pages 314–319, 1999.
[10] J. Handl and J. Knowles. Exploiting the trade-off – the benefits of multiple objectives in data clustering. In *Evolutionary Multi-Criteria Optimization*, LNCS 3410, pages 547–560. Springer, 2005.
[11] Y. Jin, B. Sendhoff, and E. Körner. Evolutionary multi-objective optimization for simultaneous generation of signal-type and symbol-type representations. In *Evolutionary Multi-Criteria Optimization*, LNCS 3410, pages 752–766. Springer, 2005.
[12] Y. Jin and B. Sendhoff. Avoiding catastrophic forgetting via multi-objective learning. Congress of Evolutionary Computation. 2006

Contents

Part I

Multi-Objective Clustering, Feature Extraction and Feature Selection

1

Feature Selection Using Rough Sets

Mohua Banerjee[1], Sushmita Mitra[2], and Ashish Anand[1]

[1] Indian Institute of Technology, Kanpur, India
{mohua,aanand}@iitk.ac.in
[2] Indian Statistical Institute, Kolkata, India
sushmita@isical.ac.in

Summary. Feature selection refers to the selection of input attributes that are most predictive of a given outcome. This is a problem encountered in many areas such as machine learning, signal processing, and recently bioinformatics/computational biology. Feature selection is one of the most important and challenging tasks, when it comes to dealing with large datasets with tens or hundreds of thousands of variables. Areas of web-mining and gene expression array analysis provide examples, where selection of interesting and useful features determines the performance of subsequent analysis. The intrinsic nature of noise, uncertainty, incompleteness of data makes extraction of hidden and useful information very difficult. Capability of handling imprecision, inexactness and noise, has attracted researchers to use rough sets for feature selection. This article provides an overview on recent literature in this direction.

1.1 Introduction

Feature selection techniques aim at reducing the number of irrelevant and redundant variables in the dataset. Unlike other dimensionality reduction methods, feature selection preserves the original features after reduction and selection. Benefit of feature selection is many fold: it improves subsequent analysis by removing the noisy data and outliers, makes faster and more cost-effective post-analysis, makes data visualization easier and provides a better understanding of the underlying process that generated the data.

Here, we will consider an example which will serve us as an illustration throughout the chapter. Consider gene selection from microarray data. In this problem, the features are expression levels of genes corresponding to the abundance of mRNA in a sample (e.g. particular time point of development or treatment), for a number of patients and replicates. A typical analysis task is to find genes which are differentially expressed in different cases or can classify different classes with high accuracy. Usually very few data samples are available altogether for testing and training. But, the number of features (genes) ranges from 10,000 to 15,000.

M. Banerjee et al.: *Feature Selection Using Rough Setsx*, Studies in Computational Intelligence (SCI) **16**, 3–20 (2006)
www.springerlink.com

Rough set theory (RST) [13, 14] was developed by Pawlak as a tool to deal with inexact and incomplete data. Over the years, RST has become a topic of great interest to researchers and has been applied to many domains, in particular to knowledge databases. This success is due in part to the following aspects of the theory:

- only the facts hidden in data are analyzed;
- no additional information about the data is required;
- minimal knowledge representation is obtained.

Consider an information system consisting of a domain U of objects / observations and a set A of attributes/features. A induces a partition (classification) of U by A, by grouping together objects having identical attribute values. But the whole set A may not always be necessary to define the classification/partition of U. Many of the attributes may be redundant, and we may find *minimal* subsets of attributes which give the same classification as the whole set A. These subsets are called *reducts* in RST, and correspond to the *minimal feature sets* that are *necessary* and *sufficient* to represent a *correct* decision about classification. Thus RST provides a methodology for addressing the problem of relevant feature selection that could be applied, e.g. to the case of microarray data described earlier.

The task of finding reducts is reported to be NP-hard [15]. The high complexity of this problem has motivated investigators to apply various approximation techniques to find near-optimal solutions. There are some studies reported in literature, e.g., [17, 3], where genetic algorithms [9] have been applied to find reducts.

Genetic algorithms (GAs) provide an efficient search technique in a large solution space, based on the theory of evolution. A population of chromosomes is made to evolve over generations by optimizing a fitness function, which provides a quantitative measure of the fitness of individuals in the pool. When there are two or more conflicting characteristics to be optimized, often the single-objective GA requires an appropriate formulation of the single fitness function in terms of an additive combination of the different criteria involved. In such cases *multi-objective* GAs (MOGAs) [7] provide an alternative, more efficient, approach to search for optimal solutions.

In this article, we present various attempts of using GA's (both single- and multi-objective) in order to obtain reducts, and hence provide some solution to the challenging task of feature selection. The rest of the chapter is organized as follows. Section 1.2 introduces the preliminaries of rough sets and genetic algorithms. Section 1.3 deals with feature selection and the role of rough sets. Section 1.4 and Section 1.5 describe recent literature on single- and multi-objective feature selection using rough sets. Section 1.7 concludes the chapter.

1.2 Preliminaries

In this section we discuss the preliminaries of rough sets and genetic algorithms, with emphasis on multi-objective GAs. The issues relevant to this chapter are explained briefly. For detailed discussion, pointers to references are given at the appropriate places.

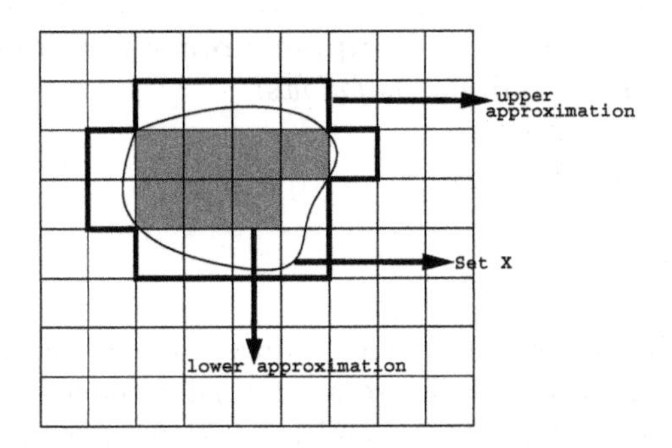

Fig. 1.1. Lower and upper approximations of a rough set

1.2.1 Rough sets

The theory of rough sets deals with uncertainty that arises from granularity in the domain of discourse, the latter represented formally by an *indiscernibility* relation (typically an equivalence) on the domain. The intention is to approximate a *rough* (imprecise) concept in the domain of discourse by a pair of *exact* concepts, that are determined by the indiscernibility relation. These exact concepts are called the lower and upper approximations of the rough concept. The lower approximation is the set of objects *definitely* belonging to the rough concept, whereas the upper approximation is the set of objects *possibly* belonging to the same. Fig. 1.1 illustrates a rough set with its approximations. The small squares represent equivalence classes induced by the indiscernibility relation on the domain. Lower approximation of the set X is shown as the shaded region and upper approximation consists of all the elements inside the thick line. The formal definitions of the above notions and others required for the present work are given below.

Definition 1. *An* **Information System** $\mathcal{A} = (U, A)$ *consists of a non-empty, finite set* U *of* objects *(cases, observations, etc.) and a non-empty, finite set* A *of* attributes a *(features, variables), such that* $a : U \rightarrow V_a$, *where* V_a *is a* value set. *Often, the attribute set* A *consists of two parts* C *and* D, *called* condition *and* decision *attributes respectively. In that case the information system* $\mathcal{A}$ *is called a* **decision table**. *Decision tables are termed* consistent, *whenever objects* x, y *are such that for each condition attribute* a, $a(x) = a(y)$, *then* $d(x) = d(y)$, *for any* $d \in D$.

Definition 2. *Let $B \subseteq A$. A* **B-indiscernibility relation** *$IND(B)$ is defined as*

$$IND(B) = \{(x,y) \in U : a(x) = a(y),\ \forall a \in B\}. \tag{1.1}$$

It is clear that $IND(B)$ partitions the universe U into equivalence classes

$$[x_i]_B = \{x_j \in U : (x_i, x_j) \in IND(B)\},\ x_i \in U. \tag{1.2}$$

Definition 3. *The* **B-lower** *and* **B-upper approximations** *of a given set $X(\subseteq U)$ are defined, respectively, as follows:*
$\underline{B}X = \{x \in U : [x]_B \subseteq X\}$,
$\overline{B}X = \{x \in U : [x]_B \cap X \neq \phi\}$.

Reducts and Core

Reducts are the basic attributes which induce same partition on universe U as the whole set of attributes. These are formally defined below for both a general information system (U, A), and a decision table $(U, C \cup D)$. In an information system, there may exist many reducts.

For a given information system, $\mathcal{A} = (U, A)$, an attribute $b \in B \subseteq A$ is *dispensable* if $IND(B) = IND(B - \{b\})$, otherwise b is said to be *indispensable* in B. If all attributes in B are indispensable, then B is called *independent* in A. Attribute set $B \subseteq A$ is called *reduct*, if B is independent in A and $IND(B) = IND(A)$.

In a decision table $\mathcal{A} = (U, C \cup D)$, one is interested in eliminating redundant *condition* attributes, and *relative* (D)-reducts are computed.

Let $B \subseteq C$, and consider the B-*positive region* of D, *viz.*, $POS_B(D) = \bigcup_{[x]_D} \underline{B}[x]_D$. An attribute $b \in B(\subseteq C)$ is **D-dispensable** in B if $POS_B(D) = POS_{B\setminus\{b\}}(D)$, otherwise b is **D-indispensable** in B. B is said to be **D-independent** in $\mathcal{A}$, if every attribute from B is D-indispensable in B.

Definition 4. *$B(\subseteq C)$ is called a* **D-reduct** *in $\mathcal{A}$, if B is D-independent in $\mathcal{A}$ and $POS_C(D) = POS_B(D)$.*

If a consistent decision table has a single decision attribute d, then $U = POS_C(d) = POS_B(D)$, for any d-reduct B.

The *core* is the set of essential attributes of any information system. Mathematically, $core(A) = \bigcap reduct(A)$, *i.e.*, the set consists of those attributes, which are members of all reducts.

Discernibility Matrix

D-reducts can be computed with the help of D-discernibility matrices [15]. Let $U = \{x_1, \cdots, x_m\}$. A **D-discernibility matrix** $M_D(\mathcal{A})$ is defined as an $m \times m$ matrix of the information system $\mathcal{A}$ with the (i, j)th entry c_{ij} given by:

$$c_{ij} = \{a \in C : a(x_i) \neq a(x_j), \text{and } (x_i, x_j) \notin IND(D)\}, \quad i, j \in \{1, \cdots, m\}. \tag{1.3}$$

A variant of the discernibility matrix, *viz.*, *distinction table* [17] is generally used in many applications to enable faster computation.

Definition 5. *A* **distinction table** *is a binary matrix with dimensions* $\frac{(m^2-m)}{2} \times N$*, where* N *is the number of attributes in* A*. An entry* $b((k,j),i)$ *of the matrix corresponds to the attribute* a_i *and pair of objects* (x_k, x_j)*, and is given by*

$$b((k,j),i) = \begin{cases} 1 \text{ if } a_i(x_k) \neq a_i(x_j), \\ 0 \text{ if } a_i(x_k) = a_i(x_j). \end{cases} \tag{1.4}$$

The presence of a '1' signifies the ability of the attribute a_i to discern between the pair of objects (x_k, x_j).

1.2.2 Genetic Algorithms

Genetic algorithms [9] are heuristic techniques applied to solve complex search and optimization problems. They are motivated by the principles of natural genetics and natural selection. Unlike classical optimization methods, GAs deal with a population of solutions/individuals. With the basic genetic/evolutionary operators, like selection, crossover and mutation, new solutions are generated. A population of chromosomes, representing solutions, is made to evolve over generations by optimizing a fitness function, which provides a quantitative measure of the fitness of individuals in the pool. Selection operator selects better solutions to participate into crossover. Crossover operator is responsible for creating new solutions from the old ones. Mutation also creates new solutions, but only in the vicinity of old solutions. Mutation operator plays a great role in case of multi-modal problems.

When there are two or more conflicting objectives to be optimized, often weighted sum of objectives are taken to convert them as single-objective problem. In such cases multi-objective GAs provide an alternative, more efficient approach to searching for optimal solutions.

Multi-Objective GAs

As the name suggests, a multi-objective optimization problem deals with more than one objective function. In contrast to single-objective problems, multiple objective problems give rise to a set of optimal solutions, known as Pareto-optimal solution [5]. Over the past decade, a number of multi-objective genetic algorithms have been suggested. The basic advantage of multi-objective GAs over classical optimization methods is their ability to find multiple Pareto-optimal solutions in one single simulation run. Detailed discussion about multi-objective genetic algorithms can be found in [7]. Here, we will discuss the main features of Non-dominated Sorting Genetic Algorithm, viz. NSGA-II [8], which is one of the frequently used multi-objective genetic algorithms. This has been used in the studies of multi-objective feature selection using rough sets, as discussed in Section 1.5.

Among the different multi-objective algorithms, it has been observed that NSGA-II has the features required for a good multi-objective GA. It uses the concept of non-domination to select the better individuals, when they are compared with respect to all objectives. NSGA-II uses crowding distance to find the population density near each individual. To get an estimate of the density around the solution i, average distance of two solutions on either side of solution i along each of the objectives is taken. Crowding distance metric is defined in [8] such that the solution which resides in less crowded region will get higher value of crowding distance. Thereby NSGA-II tries to maintain the diversity among the non-dominated solutions. The algorithm assumes each solution in the population has two characteristics:

- a non-domination rank r_i;
- a local crowding distance d_i.

By using crowded tournament selection operator, NSGA-II not only tries to converge to Pareto-front but also tries to have diverse solution on the front. Crowded tournament selection operation is described below.

Definition 6. *It is said that solution i wins tournament with another solution j if any one of the following is true:*
(i) solution i has better rank i.e. $r_i < r_j$;
(ii) both the solutions are in the same front, i.e. $r_i = r_j$ but solution i is less densely located in the search space, i.e. $d_i > d_j$.

The NSGA-II algorithm can be summarized as follows.

1. Initialize the population;
2. Calculate the fitness;
3. Rank the population using the dominance criteria;
4. Calculate the crowding distance;
5. Do selection using crowding selection operator;
6. Do crossover and mutation to generate children population;
7. Combine parent and children population and do non-dominated sorting;
8. Replace the parent population by the best members of the combined population. Initially, members of lower fronts replace the parent population. When it is not possible to accommodate all the members of a particular front, that front is sorted according to the crowding distance. The number of individuals selected on the basis of higher crowding distance, is that which makes size of the new parent population same as size of the old one.

1.3 Feature Selection and Rough Sets

Feature selection plays an important role in data selection and preparation for subsequent analysis. It reduces the dimensionality of a feature space, and removes redundant, irrelevant, or noisy data. It enhances the immediate effects for any application by speeding up subsequent mining algorithms, improving data quality and thereby performance of such algorithms, and increasing the comprehensibility of their output. In this section we highlight the basics of feature selection followed by the role of rough sets in this direction.

1.3.1 Feature Selection

It is a process that selects a minimum subset of M features from an original set of N features ($M \leq N$), so that the feature space is optimally reduced according to an evaluation criterion. Finding the best feature subset is often intractable or NP-hard.

Feature selection typically involves the following steps:

- Subset generation: For N features, the total number of candidate subsets is 2^N. This makes an exhaustive search through the feature space infeasible, even with moderate value of N. Often heuristic and non-deterministic strategies are found to be more practical.
- Subset evaluation: Each generated subset needs to be evaluated by a criterion, and compared with the previous best subset.
- Stopping criterion: The algorithm may stop when either of the following holds.
  - A pre-defined number of features are selected,
  - a pre-defined number of iterations are completed,
  - when addition or deletion of any feature does not produce a better subset, or
  - an optimal subset is obtained according to the evaluation criterion.
- Validation: The selected best feature subset needs to be validated with different tests.

Search is a key issue in feature selection, involving search starting point, search direction, and search strategy. One also needs to measure the goodness of the generated feature subset. Feature selection can be supervised as well as unsupervised, depending on class information availability in data. The algorithms are typically categorized under filter and wrapper models [18], with different emphasis on dimensionality reduction or accuracy enhancement.

1.3.2 Role of Rough Sets

Rough sets provide a useful tool for feature selection. We explain its role with reference to the bioinformatics domain. A basic issue addressed in many practical applications, such as the gene expression analysis example discussed in

Section 1.1, is that the whole set of attributes/features is not always necessary to define an underlying partition/classification. Many of the features may be superfluous, and minimal subsets of attributes may give the same classification as whole set of attributes. For example, only few genes in microarray gene expression study are supposed to define the underlying process and hence working with all genes only reduces the quality and significance of analysis. In rough set terminology, these minimal subsets of features are just the *reducts*, and correspond to the minimal feature sets that are necessary and sufficient to represent underlying classification.

The high complexity of the reduct finding problem has motivated investigators to apply various approximation techniques to find near optimal solutions. There are some studies reported in literature, e.g., [17, 3], where genetic algorithms (GAs) have been applied to find reducts. Each of the studies in [17, 3] employs a single-objective function to obtain reducts.

The essential properties of reducts are:

- to classify among all elements of the universe with the same accuracy as the starting attribute (feature) set, and
- to be of small cardinality.

A close observation reveals that these two characteristics are of a conflicting nature. Hence the determination of reducts is better represented as a bi-objective problem. The idea was first presented in [1], and a preliminary study was conducted. Incorporating some modifications in this proposal, [2] investigates the multi-objective feature selection criteria for classification of cancer microarray data.

We will first discuss single-objective feature selection approach using rough sets in Section 1.4, and then pass on to the multi-objective approach in Section 1.5.

1.4 Single-objective Feature Selection Approach

Over the past few years, there has been a good amount of study in effectively applying GAs to find minimal reducts. First we will discuss algorithms proposed by Wroblewski [17].

Wroblewski's Algorithms

Wroblewski has proposed three heuristic-based approaches for finding minimal reducts. While the first approach is based on classical GAs, the other two are permutation-based greedy approaches.

Method 1

Solutions are represented by binary strings of length N, where N is the number of attributes (features). In the bit representation '1' means that the attribute is present and '0' means that it is not. The following fitness function is considered for each individual:

$$F_1(\boldsymbol{\nu}) = \frac{N - L_{\boldsymbol{\nu}}}{N} + \frac{C_{\boldsymbol{\nu}}}{(m^2 - m)/2}, \tag{1.5}$$

where $\boldsymbol{\nu}$ is a reduct candidate, N is the number of available attributes, $L_{\boldsymbol{\nu}}$ is the number of 1's in $\boldsymbol{\nu}$, $C_{\boldsymbol{\nu}}$ is the number of object combinations that $\boldsymbol{\nu}$ can discern, and m is the number of objects.

First part of the fitness function gives the candidate credit for containing less attributes (few 1's) and the second part of the function determines the extent to which the candidate can discern among objects.

When the fitness function is calculated for each individual, the selection process begins. A particular selection operator 'Roulette Wheel' is used. One-point crossover is used with crossover probability $P_c = 0.7$. Probability P_m of mutation on a single position of individual is taken as 0.05. Mutation of one position means replacement of '1' by '0' or '0' by '1'. Complexity of the algorithm is governed by that of fitness calculation, and it can be shown that the latter complexity is $O(Nm^2)$.

Method 2

This method uses greedy algorithms to generate the reducts. Here the aim is to find the proper order of attributes. We can describe this method as follows:

Step 1 : Generate an initial set of random permutations of attributes $\tau(a_1, \ldots, a_N)$, each of them representing an ordered list of attributes, i.e., $(b_1,\ldots,b_N) = \tau(a_1,\ldots,a_N)$.
Step 2 : For each ordered list, start with empty reduct $R = \phi$ and set $i \leftarrow 0$.
Step 3 : Check whether R is a reduct. If R is a reduct, Stop.
Step 4 : Else, add one more element from the ordered list of attributes, i.e. define $R := R \cup b_{i+1}$.
Step 5 : Go to step 3.

The result of this algorithm will be either a reduct or a set of attributes containing a reduct as a subset. GAs help to find reducts of different order. *Genetic Operators*: Different permutations represent different individuals. The fitness function of an individual $\boldsymbol{\nu}$ is defined as:

$$F(\tau) = \frac{1}{L_{\boldsymbol{\nu}}}, \tag{1.6}$$

where $L_{\boldsymbol{\nu}}$ is the length of the subset R found by the greedy algorithm.

The same selection methods are used as in method 1. But different mutation and crossover operators are used, with an interchange of two randomly chosen attributes being done in mutation with some probability. Although one can choose any order-based crossover method, Wroblewski [17] has suggested the use of PMX (Partially Mapped Crossover [9]).

Method 3

This method again uses greedy algorithms to generate reducts. We can describe this method as follows:

Step 1 : Generate an initial set of random permutations of attributes $\tau(a_1, \ldots, a_N)$, each of which represents an ordered list of attributes, i.e., $(b_1, \ldots, b_N) = \tau(a_1, \ldots, a_N)$.
Step 2 : For each ordered list, define reduct R as the whole set of attributes.
Step 3 : Set $i \leftarrow 1$ and let $R := R - b_i$.
Step 4 : Check whether R is a reduct. If it is not, then undo step 3 and $i \leftarrow i + 1$. Go back to step 3.

All genetic operators are chosen as in method 2. The result of this algorithm will always be a reduct, the proof of which is discussed in [17]. However, a disadvantage of this method is its high complexity.

'Rough Enough' Approach to Calculating Reducts

Bjorvand [3] has proposed another variant of finding reducts using GAs. In his approach, a different fitness function is used. The notations used are the same as those for the previous methods.

$$F_1(\nu) = \begin{cases} (\frac{N-L_\nu}{N} + \frac{C_\nu}{(m^2-m)/2})^2 & \text{if } C_\nu < (m^2-m)/2 \\ (\frac{N-L_\nu}{N} + (\frac{C_\nu}{(m^2-m)/2} + \frac{1}{2}) \times \frac{1}{2})^2 & \text{if } C_\nu = (m^2-m)/2. \end{cases}$$

Bjorvand argues that by squaring the fitness values it becomes easy to separate the different values. In case of total coverings (candidate is possibly a reduct), the second part of fitness values is added and then multiplied by 1/2 to avoid getting low fitness values as compared to the candidates almost covering all objects and also having a low number of attributes. Instead of constant mutation rate, Bjorvand uses adaptive mutation. If there are many individuals with same fitness value, higher mutation rate is chosen to avoid premature convergence or getting stuck at local minima. To give more preference to finding shorter reducts, higher mutation probability is chosen for mutation from 1 to 0 than for the reverse direction.

Other Approaches

In literature, there are some more approximation approaches to calculate reducts. Among these algorithms are Johnson's algorithm [12] and hitting set approach by Vinterbo *et. al* [16]. Johnson's algorithm is a greedy approach to find a single reduct. In the hitting set approach, non-empty elements of the discernibility matrix are chosen as elements of a multiset ℓ. The minimal hitting sets of ℓ are exactly the reducts. Since, finding minimal hitting sets is again an NP-hard problem [16], GAs are used for finding the approximate hitting sets (reducts).

Here the fitness function again has two parts, and a weighted sum of the two parts are taken. The following fitness function is defined for each candidate solution $\boldsymbol{\nu}$:

$$F(\boldsymbol{\nu}) = (1-\alpha) \times \frac{cost(A) - cost(\boldsymbol{\nu})}{cost(A)} + \alpha \times min\{\varepsilon, \frac{|[S \in \ell | S \cap \boldsymbol{\nu} \neq \phi]|}{|\ell|}\}. \quad (1.7)$$

In the above equation, α lies between 0 and 1, A is the set containing elements of the discernibility matrix. Candidate solutions $\boldsymbol{\nu}(\subset A)$ are found through evolutionary search algorithms. The parameter ε signifies a minimal value for the hitting fraction. First term of the above equation rewards the shorter element and the second term tries to ensure that hitting sets get reward. Cost function in above definition specifies the cost of an attribute subset. Means of defining the cost function are discussed in Rosetta [12]. One can trivially define a cost function as "the cardinality of the candidate $\boldsymbol{\nu}$, $|\boldsymbol{\nu}|$". Rosetta describes all the required parameters in brief, and detailed description of the fitness function and all parameters can be found in [16].

1.5 Multi-objective Feature Selection Approach

All the algorithms discussed above concentrate more on finding the minimal reducts and thus use variations of different single fitness functions. In many applications, such as gene expression analysis, a user may not like to just have a minimal set of genes, but explore a range of different sets of features and the relations among them. Multi-objective criterion has been used successfully in many engineering problems, as well as in feature selection algorithms [11]. Here, the basic idea is to give freedom to the user to choose features from a wide spectrum of trade-off features, which will be useful for them. As discussed in Section 1.3.2, a reduct exhibits a conflicting nature of having small cardinality and ability to discern among all objects. Combining this conflicting nature of reducts with MOGAs may give the desired set of trade-off solutions. This motivated the work in [1], to use multi-objective fitness functions for finding reducts of all lengths.

In this section we first discuss the multi-objective reduct finding algorithm proposed initially in [1]. This is followed by a discussion on a modification and implementation of this proposal, for gene expression classification problem, in [2].

1.5.1 Finding Reducts Using MOGA – I

In [1] the fitness function proposed by Wroblewski (eqn. (1.5)) was split into two parts, to exploit the basic properties of reducts as two conflicting objectives. The two fitness functions F_1 and F_2 are as follows:

$$F_1(\boldsymbol{\nu}) = \frac{N - L_{\boldsymbol{\nu}}}{N}, \tag{1.8}$$

$$F_2(\boldsymbol{\nu}) = \frac{C_{\boldsymbol{\nu}}}{(m^2 - m)/2}. \tag{1.9}$$

Hence, in this case, the first fitness function gives the solution credit for containing less attributes (few 1's) and the second fitness function determines the extent to which the solution can discern among objects.

Non-domination sorting brings out the difference between the proposed algorithm and the earlier algorithms. The algorithm makes sure that

- the true reducts come to the best non-domination front, and
- two different candidates also come into the same non-domination front, if one is not the superset of the other and the two can discern between the same number of objects.

For two solutions i and j, non-domination procedure is outlined as follows:

if $F_2^i = F_2^j$ and $F_1^i \neq F_1^j$
  if one (i, say) is superset of other (j)
    then put i at inferior domination level
  else put both in same domination level
else
  do regular domination checking with respect to two fitness values.

Remark 1. If the two solutions discern the same number of pair of objects, then their non-domination level is determined by the first objective. In this way, we make sure that candidates with different cardinality can come to the same non-domination level, if they do not violate the superset criteria, i.e. one solution is not a superset of the other. Explicit checking of superset is intended to ensure that only true reducts come into the best non-domination level.

The representation scheme of solutions discussed in Section 1.4, and the usual crowding binary tournament selection as suggested in Section 1.2.2 are used. The complete algorithm can be summarized in the following steps:

Step 1 : A random population of size n is generated.

Step 2 : The two fitness values for each individual is calculated.
Step 3 : Non-domination sorting is performed, to identify different fronts.
Step 4 : Crowding sort is performed to get a wide spread of the solution.
Step 5 : Offspring solution is created using crowded tournament selection, crossover and mutation operators.
Step 6 : Steps 2 to 5 are repeated for a pre-specified number of generations.

An advantage of the multi-objective approach can be shown by taking an example. Consider two solutions (a, b) and (c, d, e) giving the same classification. Then the earlier approach will give less preference to the second solution, whereas the proposed algorithm puts both solutions in the same non-domination level. Thus the probability of selecting reducts of larger cardinalities is the same as that of smaller cardinalities. The explicit check of superset also increases the probability of getting only true reducts.

This algorithm was implemented on some simple data sets. However, there are complexity problems when faced with large data.

1.5.2 Finding Reducts Using MOGA – II

For a decision table $\mathcal{A}$ with N condition attributes and a single decision attribute d, the problem of finding a d-reduct is equivalent to finding a minimal subset of columns $R(\subseteq \{1, 2, \cdots, N\})$ in the distinction table, satisfying

$$\forall (k, j) \exists i \in R : b((k, j), i) = 1, \text{whenever } d(x_k) \neq d(x_j).$$

So, in effect, the distinction table consists of N columns, and rows corresponding to only those object pairs (x_k, x_j) such that $d(x_k) \neq d(x_j)$. We call this shortened distinction table, a *d-distinction table.* Note that, as $\mathcal{A}$ is taken to be consistent, there is no row with all 0 entries in a d-distinction table.

In [2], NSGA-II is modified to effectively handle large datasets. We focus on two-class problems. An initial redundancy reduction is done to generate a reduced attribute value table $\mathcal{A}_r$. From this we form the d-distinction table consisting of N columns, with rows corresponding to only those object pairs (x_k, x_j) such that $d(x_k) \neq d(x_j)$. As object pairs corresponding to the same class do not constitute a row of the d-distinction table, there is a considerable reduction in its size, thereby leading to a decrease in computational cost.

The modified feature selection algorithm is implemented on microarray data consisting of three different cancer samples, as summarized in Table 1.1 [2]. After the initial redundancy reduction, the feature sets are reduced:

- Colon dataset: 1102 attributes for the normal and cancer classes,
- Lymphoma dataset: 1867 attributes for normal and malignant lymphocyte cells, and
- Leukemia dataset: 3783 attributes for classes ALL and AML.

The algorithm is run on the d-distinction table, with different population sizes, to generate reducts upon convergence. Fitness functions of eqns. (1.8)-(1.9),

Table 1.1. Usage details of the two-class microarray data

Data used	# Attributes	Classes	# Samples
Colon	2000	Colon cancer	40
		Normal	22
Lymphoma	4026	Other type	54
		B-cell lymphoma	42
Leukemia	7129	ALL	47
		AML	25

adapted to the case of two-class problems, are used. Results indicate convergence to 8, 2, 2 attributes respectively, for the *minimal* reduct on the three sets of two-class microarray gene expression data after 15,000 generations.

On the other hand, feature selection (without rough sets) in microarray gene expression analysis has also been reported in literature [10, 4, 6]. Huang [10] uses a probabilistic neural network for feature selection, based on correlation with class distinction. In case of *Leukemia* data, they report a reduction to a ten-genes set. For *Colon* data, a ten-genes set is generated. Chu *et al.* [6] employ a t-test based feature selection with a fuzzy neural network. A five-genes set is generated for *Lymphoma* data.

1.6 Example

In this section we will explain single-objective and multi-objective based feature selection through an example. For illustration, a sample data viz. *Cleveland* data set is taken from Rosetta [12]. The data has 14 attributes and it does not contain any decision variable. We have removed all objects with missing values and hence there are 297 objects. A part of the data with 10 attributes and 3 objects, is shown in Table 1.2.

Table 1.2. Part of Cleveland data, taken from Rosetta

age	sex	cp	trestbps	chol	fbs	restecg	thalach	exang	oldpeak
63	M	Typical angina	145	233	T	LV hypertrophy	150	N	2.3
67	M	Asymptomatic	160	286	F	LV hypertrophy	108	Y	1.5
67	M	Asymptomatic	120	229	F	LV hypertrophy	129	Y	2.6
37	M	Non-anginal pain	130	250	F	Normal	187	N	3.5

To illustrate the single-objective feature selection approach, we have used Wroblewski's algorithm implemented in Rosetta, to find reducts. For multi-objective feature selection approach, the algorithm proposed in [1] is used on the same data set.

1.6.1 Illustration for the Single-objective Approach

Wroblewski's algorithm searches for reducts using GA until either it has exhausted the search space, or a pre-defined number of reducts has been found. Three different parameters can be chosen to control the thoroughness and speed of the algorithm.

Rosetta gives an option of finding reducts based on a complete set of objects or a subset of objects. In this example, we have chosen discernibility based on all objects. In case of decision table, one can select *modulo decision* to avoid discerning between objects belonging to the same decision class. Thus the resultant distinction table consists of object pairs belonging to different classes only. Rosetta also provides options to users for selecting parameters such as *number of reducts*, *seed* to start with different random populations, and *calculation speed* to choose one of the different versions of Wroblewski's algorithm discussed in Section 1.4. We have chosen *normal calculation speed* and *number of reducts = 50*. It may be remarked that other approaches (e.g. Vinterbo's method [16]) for finding reducts have also been implemented in Rosetta.

Results

On running Wroblewski's algorithm implemented in Rosetta, we obtain 25 reducts. Table 1.3 summarizes the results. All reducts were tested for reduct membership [15] and found to be true reducts.

Table 1.3. Results of Rosetta

Reduct Length	# Reducts
3	10
4	10
5	2
6	3

1.6.2 Illustration for the Multi-objective Approach

Solutions or chromosomes are binary strings of 1 and 0, of length equal to the total number of attributes. 1 indicates that the particular attribute is present. For example, 10001000000001 means that 1st, 5th and 14th attributes are present.

A random population of size n is generated. Each individual is nothing but a binary string, as just explained. Fitness functions are calculated and non-domination ranking is done, as discussed in Section 1.5.1. For illustration, let us take 4 individuals with the following strings and assume that all 4 individuals can discern among all the objects, i.e. second fitness function of all individuals is 1.0.

Individual 1: 10001000100000
Individual 2: 01001000010000
Individual 3: 11001000010000
Individual 4: 01001001000100

Since individual 3 is a superset of individual 2, individual 2 dominates it. But individuals 1 and 2 are non-dominated, and are put in the same front. Though individual 4 has cardinality four, which is more than the cardinality of individual 1 and individual 2, it is still kept in the same non-dominated front as individuals 1 and 2.

Offspring solutions are created using crowding selection, crossover and mutation operators.

Results

The following parameters are used to run the multi-objective algorithm.
Population Size = 50
Number of Generations = 500
Crossover Probability = 0.6
Mutation Probability = 0.08
Table 1.4 summarizes the results. Again, all reducts were tested for reduct membership [15] and found to be true reducts. A comparison with the results obtained using the single-objective algorithm indicates a greater effectiveness of the multi-objective approach.

Table 1.4. Results of multi-objective implementation

Reduct Length	# Reducts
3	10
4	17
5	14
6	7
7	2

1.7 Conclusion

In this article we have provided a study on the use of rough sets for feature selection. Handling of high-dimensional data requires a judicious selection of attributes. Feature selection is hence very important for such data analysis. *Reducts* in rough set theory, prove to be relevant for this task. However, reduct computation is a hard problem. It is found that evolutionary algorithms, particularly multi-objective GA, is useful in computing optimal reducts. Application to microarray data is described. An illustrative example is also provided, to explain the single- and multi-objective approaches.

Identifying the essential features amongst the non-redundant ones, also appears to be important in feature selection. The notion of *core* (the common part of all reducts) of an information system, could be relevant in this direction, and it may be a worthwhile future endeavor to conduct an investigation into its role.

Acknowledgment

Ashish Anand is financially supported by DBT project no. DBT/BSBE/20030360. We thank the referees for their suggestions.

References

[1] A. Anand. Representation and learning of inexact information using rough set theory. Master's thesis, Department of Mathematics, Indian Institute of Technology, Kanpur, India, 2002.
[2] M. Banerjee, S. Mitra, and H. Banka. Evolutionary-rough feature selection in gene expression data. *IEEE Transactions on Systems, Man, and Cybernetics, Part C: Applications and Reviews*, 2005. Accepted.
[3] A. T. Bjorvand. 'Rough Enough' – A system supporting the rough sets approach. In *Proceedings of the Sixth Scandinavian Conference on Artificial Intelligence*, pages 290–291, Helsinki, Finland, 1997.
[4] L. Cao, H. P. Lee, C. K. Seng, and Q. Gu. Saliency analysis of support vector machines for gene selection in tissue classification. *Neural Computing and Applications*, 11:244–249, 2003.
[5] V. Chankong and Y. Y. Haimes. *Multiobjective Decision Making Theory and Methodology.* North-Holland, 1983.
[6] F. Chu, W. Xie, and L. Wang. Gene selection and cancer classification using a fuzzy neural network. In *Proceedings of 2004 Annual Meeting of the North American Fuzzy Information Processing Society (NAFIPS 2004)*, volume 2, pages 555–559, 2004.
[7] K. Deb. *Multi-Objective Optimization using Evolutionary Algorithms.* John Wiley & Sons, London, 2001.

[8] K. Deb, S. Agarwal, A. Pratap, and T. Meyarivan. A fast and elitist multi-objective genetic algorithm: NSGA-II. *IEEE Transactions on Evolutionary Computation*, 6:182–197, 2002.
[9] D.E. Goldberg. *Genetic Algorithms for Search, Optimization, and Machine Learning.* Addison-Wesley, Reading, 1989.
[10] C. -J. Huang. Class prediction of cancer using probabilistic neural networks and relative correlation metric. *Applied Artificial Intelligence*, 18:117–128, 2004.
[11] R. Jensen. *Combining rough and fuzzy sets for feature selection.* PhD thesis, School of Informatics, University of Edinburgh, 2004.
[12] J. Komorowski, A. Øhrn, and A. Skowron. The rosetta rough set software system. In W. Klasgen and J. Zytkow, editors, *Handbook of Data Mining and Knowledge Discovery*, chapter D.2.3. Oxford University Press, 2002.
[13] Z. Pawlak. Rough sets. *International J. Comp & Inf. Sc.*, 1982.
[14] Z. Pawlak. *Rough Sets, Theoretical Aspects of Reasoning about Data.* Kluwer Academic Publishers, Dordrecht, 1991.
[15] A. Skowron and C. Rauszer. The discernibility matrices and functions in information systems. In R. Slowinski, editor, *Handbook of Applications and Advances of the Rough Set Theory*, pages 331–362. Kluwer Academic Publishers, Dordrecht, 1992.
[16] S. Vinterbo and A. Øhrn. Minimal approximate hitting sets and rule templates. *International Journal of Approximate Reasoning*, pages 123–143, 2000.
[17] J. Wroblewski. Finding minimal reducts using genetic algorithms. In *Second Annual Joint Conference on Information Sciences*, pages 186–189, 1995.
[18] L. Yu and H. Liu. Efficient feature selection via analysis of relevance and redundancy. *Journal of Machine Learning Research*, 5:1205–1224, 2004.

2

Multi-Objective Clustering and Cluster Validation

Julia Handl and Joshua Knowles

University of Manchester
jhandl@postgrad.manchester.ac.uk, jknowles@manchester.ac.uk

Summary. This chapter is concerned with unsupervised classification, that is, the analysis of data sets for which no (or very little) training data is available. The main goals in this data-driven type of analysis are the discovery of a data set's underlying structure, and the identification of groups (or clusters) of homogeneous data items — a process commonly referred to as cluster analysis.

Clustering relies on the use of certain criteria that attempt to capture those aspects that humans perceive as the properties of a good clustering solution. A variety of such criteria exist, many of which are partially complementary or even conflicting, and may favour different types of solutions. Like many other machine learning problems, clustering can therefore be considered as an intrinsically multiobjective optimization problem.

In this chapter, we consider two steps in the clustering process that may benefit from the use of multiple objectives. First, we consider the generation of clustering solutions and show that the use of two complementary clustering objectives results in an improved and more robust performance vis-a-vis single-objective clustering algorithms. Second, we consider the problem of model selection, that is, the choice of the best clustering solution out of a set of different alternatives, and investigate whether a multiobjective approach may also be beneficial vis-a-vis more traditional validation criteria.

2.1 Unsupervised Classification

The increasing volume of data arising in the fields of document retrieval and bioinformatics are two prominent examples of a trend in a wide range of different research areas. Novel technologies (such as the Internet in document retrieval, microarray experiments in bioinformatics, physical simulations in scientific computing and many more) give rise to large 'warehouses' of data, which can only be handled and processed by means of computers. The efficient analysis and the generation of new knowledge from these masses of data requires the extensive use of data-driven inductive approaches. Data-driven techniques are also referred to as unsupervised techniques, and stand in contrast to supervised techniques, which require the presence of training data,

J. Handl and J. Knowles: *Multi-Objective Clustering and Cluster Validation*, Studies in Computational Intelligence (SCI) **16**, 21–47 (2006)
www.springerlink.com

that is, a (sufficiently large) set of data samples for which the correct classification is known. In the absence of training data, algorithms rely on the presence of distinct structure in the data, that is, it must be hoped that a distance measure or a reduced feature space can be identified under which related data items cluster together in data space. This type of method is also referred to as clustering or unsupervised classification [7, 10, 17, 19].

2.1.1 Clustering

Informally, clustering is concerned with the division of data into homogeneous subgroups, subject to the following two aims: data items within one cluster should be similar to each other, while those within different clusters should be dissimilar. Formally, the *clustering problem* can be defined as an optimization problem[1] [3]:

Definition 1.1: The clustering problem

INSTANCE: A finite set X, a distance measure $\delta(i,j) \in \Re_0^+$ for $i,j \in X$, a positive integer K, and a criterion function $J(C, \delta(\cdot,\cdot))$ on a K-partition $C = \{C_1, \ldots, C_K\}$ of X and measure $\delta(\cdot,\cdot)$.

OPTIMIZATION: Find the partition of X into disjoint sets $C_1, \ldots, C_K$ that maximises the expression $J(C, \delta(\cdot,\cdot))$.

The number $R(K,N)$ of possible solutions for the division of a data set of size N into K partitions is given by the Stirling number of the second kind:

$$R(K,N) = \frac{1}{K!} \sum_{i=1}^{K} (-1)^{K-i} \binom{K}{i} (i)^N \approx \frac{K^N}{K!}$$

Hence, even with a fixed number of partitions K, the search space for the clustering problem grows exponentially and cannot be scanned exhaustively even for medium sized problems. Indeed, the clustering problem is known to be NP-hard in many of its forms [3].

Definitions of the clustering problem vary in the optimization criterion J and the distance function $\delta(\cdot,\cdot)$ used. Unfortunately, choosing an optimization criterion, that is, grasping the intuitive notion of a cluster by means of an explicit definition, is one of the fundamental dilemmas in clustering [9, 20]. While there are several valid properties that may be ascribed to a good clustering, these may be partially conflicting and/or inappropriate in a particular context (see Figure 2.1). Yet, most existing clustering methods attempt, explicitly or otherwise, to optimize just one such property — and through this

[1] Without loss of generality, we here assume maximization.

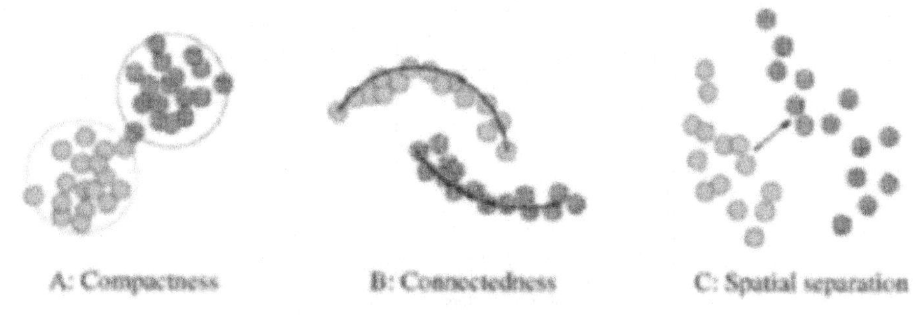

Fig. 2.1. There are several valid properties that may be ascribed to a good partitioning, but these are partly in conflict and are generally difficult to express in terms of objective functions. Despite this, existing clustering criteria/algorithms do fit broadly into three fundamental categories:

(1) Compactness. This concept is generally implemented by keeping intra-cluster variation small. This category includes algorithms like K-means [21], average link agglomerative clustering [32] or model-based clustering approaches [22]. The resulting methods tend to be very effective for spherical or well-separated clusters, but they may fail to detect more complicated cluster structures [7, 10, 17, 19].

(2) Connectedness. This is a more local concept of clustering based on the idea that neighbouring data items should share the same cluster. Algorithms implementing this principle are density-based methods [1, 8] and methods such as single link agglomerative clustering [32]. They are well-suited for the detection of arbitrarily shaped clusters, but can lack robustness when there is little spatial separation between clusters.

(3) Spatial separation. Spatial separation on its own is a criterion that gives little guidance during the clustering process and can easily lead to trivial solutions. It is therefore usually combined with other objectives, most notably measures of compactness or balance of cluster sizes. The resulting clustering objectives can be tackled by general-purpose meta-heuristics (such as simulated annealing, tabu search and evolutionary algorithms [2, 25]).

Examples of data sets exhibiting compactness, connectedness and spatial separation, respectively, are shown above. Clearly, connectedness and spatial separation are related (albeit opposite) concepts. In principle, the cluster structure in the data sets B and C can be identified by a clustering algorithm based on either connectedness or on spatial separation, but not by one based on compactness.

choice they make a priori assumptions on the type of clusters that can later be identified. This confinement to a particular clustering criterion is one of the reasons for the fundamental discrepancies observable between the solutions produced by different algorithms and will cause a clustering method to fail in a context where the criterion employed is inappropriate.

In practical data-mining scenarios, researchers attempt to circumvent this problem through the application and comparison of multiple clustering algorithms [16], or through the a posteriori combination of different clustering results by means of ensemble methods [28, 31].

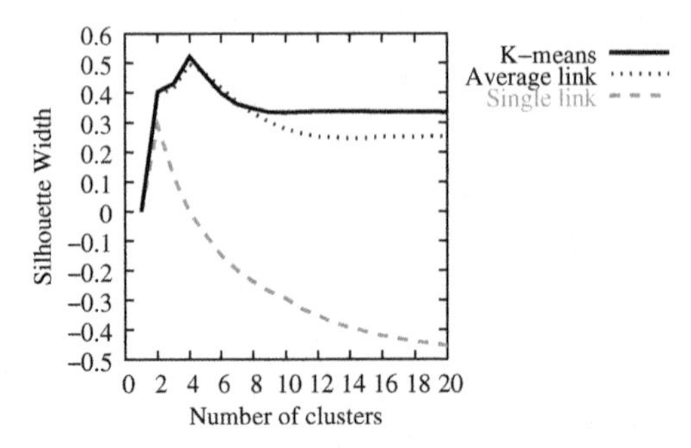

Fig. 2.2. If the structure of a data set and the best number of clusters is not known, a typical approach is to run several clustering algorithms for a range of numbers of clusters. Here, K-means, average link and single link agglomerative clustering have been run on a four-cluster data set (Square3) for $K \in [1, 20]$. The solutions are evaluated under a validation measure, here, the Silhouette Width [26], which takes both cluster compactness and cluster separation into account. For each algorithm, clear local maxima within the corresponding curve (here, at $K = 4$ for K-means and average link and at $K = 2$ for single link), are considered as good solutions. The algorithm resulting in the highest validation values (here, K-means) may be selected as the most suitable clustering method. This way of choosing one partitioning from a set of different partitionings is frequently referred to as model selection.

2.1.2 Model Selection

Determining the most appropriate number of clusters K for a given data set is a second important challenge in clustering. Automated approaches to the determination of the number of clusters frequently work by selecting the most appropriate clustering from a range of partitionings with different Ks, a process often referred to as model selection (see Figure 2.2). Here, the partitioning that is most appropriate is again defined by means of a clustering criterion. Formally, the process of model selection considered in this chapter can be defined as follows:[2]

Definition 1.2: Model selection

INSTANCE: A finite set X, a distance measure $\delta(i, j) \in \Re_0^+$ for $i, j \in X$, a set of M partitionings $\{P_1, P_2, \ldots, P_M\}$, and a criterion function $J(P, \delta(\cdot, \cdot))$ on a partition P of X and measure $\delta(\cdot, \cdot)$.

OPTIMIZATION: Find the partition $P \in \{P_1, P_2, \ldots, P_M\}$, for which the expression $J(P, \delta(\cdot, \cdot))$ is maximal.

[2] Without loss of generality, we here assume maximization.

Again, fundamentally different clustering objectives can be used in this process, and the final result may vary strongly depending on the specific objective employed. There is a consensus that good partitionings tend to perform well under multiple complementary clustering criteria and the best clustering is therefore commonly determined using linear or non-linear combinations of different clustering criteria.

2.1.3 Scope of This Work

Evidently, the choice of an appropriate clustering criterion is of major importance in both clustering and model selection. Unfortunately, it is also clear that a secure choice requires knowledge of the underlying data distribution, which is, in most cases, not known a priori.

In this chapter, we contend that the framework of Pareto optimization can provide a principled way to deal with this issue. Multiobjective Pareto optimization allows the simultaneous optimization of a set of complementary clustering objectives, which abolishes the need for multiple runs of different clustering algorithms, and is more general than the fixed linear or non-linear combination of individual objectives (see Figure 2.3). We therefore expect the direct optimization of multiple objectives to be beneficial both in the context of clustering and model selection.

In the following, we provide evidence to support this hypothesis in several steps. In Section 2.2, we describe a multiobjective clustering algorithm, MOCK, which simultaneously optimizes two different complementary clustering objectives. Experimental results show that the algorithm has a significantly improved classification performance compared to traditional single-objective clustering methods, and is highly robust over a range of different data properties. In Section 2.3, we consider model selection and discuss how the output of MOCK can be directly used to assess the quality of individual clustering results. Experimental results demonstrate the high performance of the resulting multiobjective validation approach, and discuss performance differences with respect to single-objective validation techniques. Finally, Section 2.4 concludes.

2.2 Multiobjective Clustering

In order to develop a clustering algorithm that simultaneously considers several complementary aspects of clustering quality, we embrace the framework of Pareto optimization. Specifically, we employ a multiobjective evolutionary algorithm (MOEA) to optimize several clustering objectives, and to obtain a set of trade-off solutions, which represent a good approximation to the Pareto front. The resulting multiobjective clustering algorithm is named "MultiObjective Clustering with automatic K-determination" (MOCK, [13, 14, 15]).

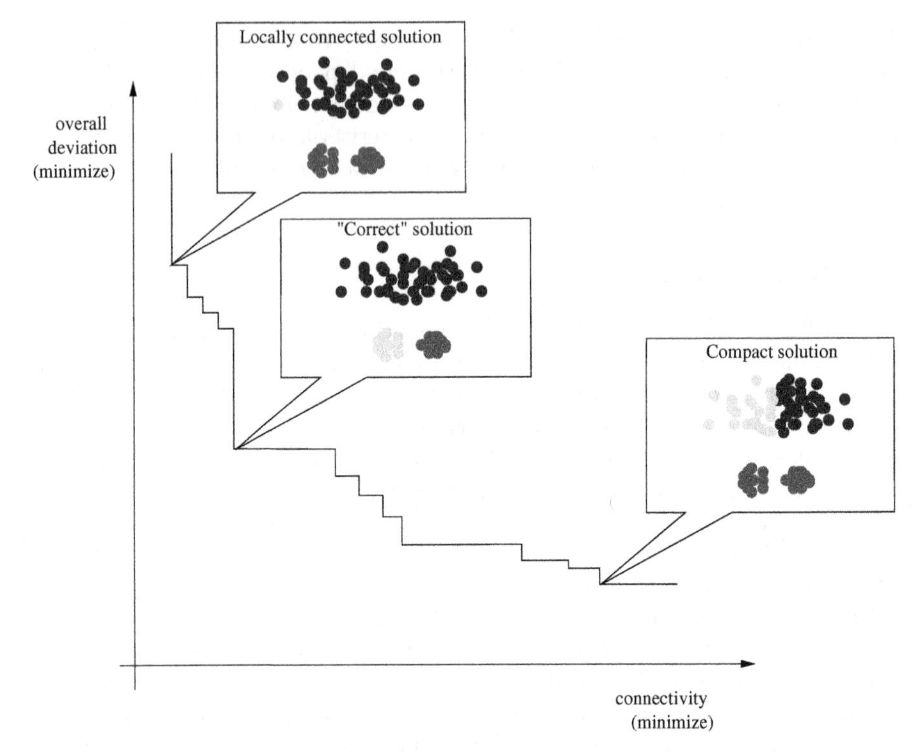

Fig. 2.3. Clustering can be considered as a multiobjective optimization problem. Instead of finding a single clustering solution, the aim in a multiobjective clustering scenario is to find an approximation to the Pareto front, that is, the set of partitionings that are Pareto optimal with respect to the objectives optimized. Here, two objectives are to be minimized and the approximation set is represented by the corresponding attainment surface, which is the boundary in the objective space separating those points that are dominated by or equal to at least one of the data points in the approximation set from those that no data point in the approximation set dominates or equals. The figure highlights three solutions within this approximation set. The solutions to the top left tend to be locally connected but not compact, whereas those to the bottom right tend to be highly compact. The correct solution corresponds to a compromise between the two objectives, and can therefore only be found through the optimization of both objectives. For sake of clarity, the approximation set in this example only contains solutions for $K = 3$. More generally, the number of clusters can also be kept dynamic — in this case an approximation set is obtained in which the number of clusters varies along the Pareto front.

2.2.1 PESA-II

MOCK is based on an existing MOEA, PESA-II. PESA-II [5, 6] is a well-known algorithm in the evolutionary multiobjective optimization literature, and has been used in comparison studies by several researchers. A high-level description is given in Algorithm 1. Two populations of solutions are maintained: an internal population, IP of fixed size, and an external population EP, of non-fixed but limited size. The purpose of EP is to *exploit* good solutions: to this end it implements elitism by maintaining a large and diverse set of nondominated solutions. The internal population's job is to *explore* new solutions, and achieves this by the standard EA processes of reproduction and variation (i.e., recombination and mutation). Selection occurs at the interface between the two populations, primarily in the update of EP.

The solutions in EP are stored in 'niches', implemented as a hypergrid in the objective space. A tally of the number of solutions that occupy each niche is kept and this is used to encourage solutions to cover the whole objective space, rather than bunch together in one region. To this end, nondominated solutions that try to enter a full EP can only do so if they occupy a less crowded niche than some other solution (lines 36 and 37 of Algorithm 1). Moreover, when the internal population of each generation is constructed from EP (lines 9–12), they are selected uniformly from among the populated niches — thus highly populated niches do not contribute more solutions than less populated ones.

An important advantage of PESA-II is that this niching policy uses an adaptive range equalization and normalization of the objective function values. This means that difficult parameter tuning is avoided, and objective functions that have very different ranges can be readily used. PESA-II can also handle any number of objective functions. For further details on PESA-II, the reader is referred to [4].

2.2.2 Details of MOCK

The application of PESA-II to the clustering problem requires the choice of

- two or more objective functions,
- a suitable genetic encoding of a partitioning,
- one or more genetic variation operators (e.g. mutation and/or crossover),
- and an effective initialization scheme.

These choices are non-trivial and are crucial for the performance and particularly the scalability of the algorithm: many encodings work well for data sets with a few hundred data points, but their performance breaks down rapidly for larger data sets. The design of an effective EA for clustering requires a close harmonization of the encoding, the operators and the objective functions, which permits to narrow down the search space and guide the search

Algorithm 1 PESA-II (high-level pseudocode)

```
1: procedure PESA-II(ipszize, epmaxsize, p_m ∈ [0,1], p_c ∈ [0,1], #gens)
2:     IP := ∅; EP := ∅
3:     for each i in 1 to ipsize do                      /* INITIALIZATION */
4:         s_i := initialize_solution(i)
5:         evaluate(s_i)
6:         UpdateEP(EP, s_i, epmaxsize)      /* Procedure defined in line 30 */
7:     end for
8:     for gen in 1 to #gens do                               /* MAIN LOOP */
9:         for each i in 1 to ipsize do
10:            select a populated niche n uniformly at random from EP
11:            select a solution s_i uniformly at random from n
12:            IP := IP ∪ {s_i}
13:        end for
14:        i := 0
15:        while i < ipsize do
16:            if random deviate R(0,1) < p_c then
17:                s_i, s_{i+1} := crossover(s_i, s_{i+1})
18:            end if
19:            s_i := mutate(s_i, p_m); s_{i+1} := mutate(s_{i+1}, p_m)
20:            i := i + 2
21:        end while
22:        for each i in 1 to ipsize do
23:            evaluate(s_i)
24:            UpdateEP(EP, s_i, epmaxsize)
25:        end for
26:        IP := ∅
27:    end for
28:    return EP, a set of nondominated solutions
29: end procedure

30: procedure UpdateEP(EP, s_i, epmaxsize)/* Update EP with solution s_i */
31:    if ∃s ∈ EP, s_i dominates s then
32:        EP := EP ∪ {s_i} \ {s ∈ EP, s_i dominates s}
33:    else if s_i is nondominated in EP then
34:        if EP < epmaxsize then
35:            EP := EP ∪ {s_i}
36:        else if ∃s ∈ EP, s_i is in a less crowded niche than s then
37:            EP := EP ∪ {s_i} \ {s, a solution from a most-crowded niche}
38:        end if
39:    end if
40:    update all niche counts
41: end procedure
```

effectively. Following extensive experiments using a range of different encodings, operators and objective functions, an effective combination of encoding, operators and objective functions was derived, the details of which are explained in the following.

Objective Functions

For the clustering objectives, we are interested in selecting optimization criteria that reflect fundamentally different aspects of a good clustering solution. From the groups identified in Figure 2.1, we therefore select two types of complementary objectives: one based on compactness, the other one based on connectedness of clusters. We refrain from using a third objective based on spatial separation, as the concept of spatial separation is intrinsic (opposite) to that of connectedness of clusters.

In order to express cluster compactness we compute the *overall deviation* of a partitioning. This is simply computed as the overall summed distances between data items and their corresponding cluster centre:

$$Dev(C) = \sum_{C_k \in C} \sum_{i \in C_k} \delta(i, \mu_k),$$

where C is the set of all clusters, μ_k is the centroid of cluster C_k and $\delta(.,.)$ is the chosen distance function. As an objective, overall deviation should be minimized. The criterion is strongly biased towards spherically shaped clusters.

As an objective reflecting cluster connectedness, we use a measure, connectivity, which evaluates the degree to which neighbouring data-points have been placed in the same cluster. It is computed as

$$Conn(C) = \sum_{i=1}^{N} \left(\sum_{j=1}^{L} x_{i,nn_{ij}} \right), \text{where } x_{r,s} = \begin{cases} \frac{1}{j} \text{ if } \nexists C_k : r \in C_k \land s \in C_k \\ 0 \text{ otherwise,} \end{cases}$$

nn_{ij} is the jth nearest neighbour of datum i, and L is a parameter determining the number of neighbours that contribute to the connectivity measure. As an objective, connectivity should be minimized. This criterion captures local densities — it can therefore detect arbitrarily shaped clusters, but is not robust toward overlapping clusters.

Overall deviation (like our mutation operator, see below) requires the one-off computation of the nearest neighbour lists of all data items in the initialization phase. Subsequently, both objectives, overall deviation and connectedness, can be efficiently computed in linear time.

An important aspect in the choice of these objective functions is their potential to balance each other's tendency to increase or decrease the number

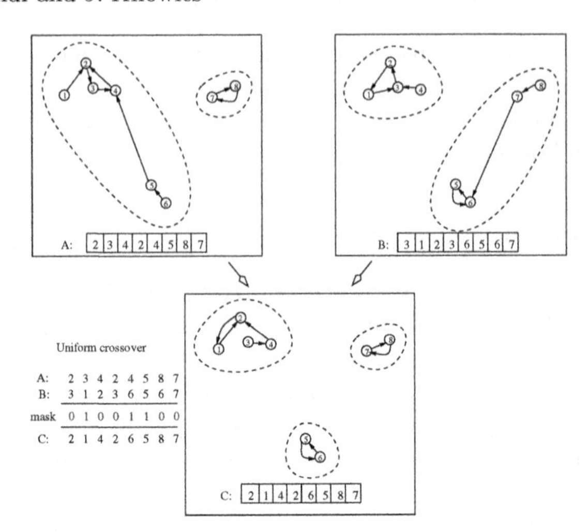

Fig. 2.4. Two parent partitionings, their graph structure, and their respective genotypes, A and B are shown. A standard uniform crossover of the genotypes yields the child C, which has inherited much of its structure from its parents, but differs from both of them.

of clusters. While the objective value associated with overall deviation necessarily improves with an increasing number of clusters, the opposite is the case for connectivity. The interaction of the two is important in order to explore sensible parts of the solution space, and not to converge to trivial solutions (which would be N singleton clusters for overall deviation and only one cluster for connectivity).

Encoding

For the encoding, we employ the locus-based adjacency representation proposed in [23]. In this graph-based representation, each individual g consists of N genes $g_1, \ldots, g_N$, where N is the size of the clustered data set, and each gene g_i can take allele values j in the range $\{1, \ldots, N\}$. Thus, a value of j assigned to the ith gene, is then interpreted as a link between data items i and j: in the resulting clustering solution they will be in the same cluster. The decoding of this representation requires the identification of all subgraphs. All data items belonging to the same subgraph are then assigned to one cluster. Note that, using a simple backtracking scheme, this decoding step can be done in linear time [13].

The locus-based adjacency encoding scheme has several major advantages for our application. Most importantly, there is no need to fix the number of clusters in advance, as it is automatically determined in the decoding step. Hence, we can evolve and compare solutions with different numbers of clusters in just one run of the GA. Furthermore, the representation is well-suited for use with standard crossover-operators such as uniform, one-point or two-point crossover. In more traditional encodings for clustering these straightforward crossover operators are usually highly disruptive and therefore detrimental for the clustering process. In a link-based encoding, in contrast, they effortlessly implement merging and splitting operations on individual clusters, while maintaining the remainder of the partitioning (see Figure 2.4).

Uniform Crossover

We choose the uniform crossover [29] in favour of one- or two-point because it is unbiased with respect to the ordering of genes and can generate any combination of alleles from the two parents (in a single crossover event) [33].

Neighbourhood-biased Mutation Operator

While the encoding results in a very large search space with N^N possible combinations, a suitable mutation operator can be employed to significantly reduce the size of the search space. We use a restricted *nearest neighbour mutation* where each data item can only be linked to one of its L nearest neighbours. Hence, $g_i \in \{nn_{i1}, \ldots, nn_{iL}\}$, where nn_{il} denotes the lth nearest neighbour of data item i. This reduces the extent of the search space to just L^N. Note that the nearest neighbour list can be precomputed in the initialization phase of the algorithm.

The properties of the encoding can additionally be employed to bias the mutation probabilities of individual genes. Intuitively, 'longer links' in the encoding are expected to be less favourable, that is, a link $i \rightarrow j$ with $j = nn_{il}$ (item j is the lth nearest neighbour of item i) may be preferred over a link $i \rightarrow j^*$ with $j^* = nn_{ik}$ (item j^* is the kth nearest neighbour of item i) if $l < k$. This can be used to bias the mutation probability of individual links $i \rightarrow j$, which we now define as

$$p_m = \frac{1}{N} + \left(\frac{l}{N}\right)^2 ,$$

where $j = nn_{il}$ and N is the size of the data set.

Initialization

Our initialization routine is based on the observation that different clustering algorithms tend to perform better (find better approximations) in different regions of the Pareto front [14]. In particular, algorithms consistent with connectivity tend to generate close to optimal solutions in those regions of the Pareto front where connectivity is low, whereas algorithms based on compactness perform well in the regions where overall deviation has significantly

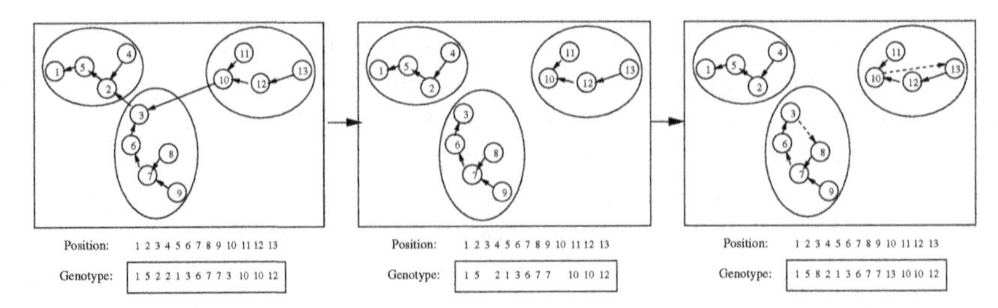

Fig. 2.5. Construction of an MST-similar solution from a given K-means solution. Starting from the original MST solution, all links that cross cluster boundaries (defined by the three-cluster K-means solution indicated here by the ellipses) are removed. A missing link emanating from data item i is then replaced by a (randomly determined) link to one of its L nearest neighbours.

decreased. We therefore use an initialization based on two different single-objective algorithms, in order to obtain a good initial spread of solutions and a close initial approximation to the Pareto front.

Solutions performing well under connectivity are generated using minimum spanning trees (MSTs). The use of MSTs has the advantage that the links present in a given MST can be directly translated as the encoding of individual solutions. Solutions performing well under overall deviation are generated using the K-means algorithm. The translation of these flat (that is, non-hierarchical) partitionings to the encoding of individual solutions is more involved, as described below.

Generation of Interesting MST Solutions

For a given data set, we first compute the complete MST using Prim's algorithm [34]. As the complete MST corresponds to a one-cluster solution, we are then interested in obtaining a range of good solutions with different numbers of clusters. Simply removing the largest links from this MST does not yield the desired results: in the absence of spatially separated clusters the method tends to isolate outliers so that many of the solutions generated are highly similar.

In order to avoid this effect, we employ a definition of interestingness that distinguishes between 'uninteresting' links whose removal leads to the separation of outliers, and 'interesting' links whose removal leads to the discovery of real cluster structures.

Definition 1.3: A link $i \rightarrow j$ is considered as *interesting*, iff $i = nn_{jl} \wedge j = nn_{ik} \wedge l > L \wedge k > L$, where L is a parameter. Its *degree of interestingness* is $d = \min(l, k)$. Informally, this means, that a link between two items i and j is considered interesting if neither of them is a part of the other item's set of L nearest neighbours.

Definition 1.4: A clustering solution C is considered as *interesting*, if it can be deduced from the full MST through the removal of interesting links only.

For a given data set, a set of interesting MST-derived solutions can then be constructed as follows. In a first step, all I interesting links from the MST are detected and are sorted by their degree of interestingness. Using this sorted list, a set of clustering solutions is then constructed: for $n \in [0, \min(I, 0.5 \times \text{fsize})]$, where fsize is the total number of initial solutions, clustering solution C_n is generated by removing the first n interesting links. The missing links are then replaced by a link to a randomly chosen neighbour j with $j = nn_{il} \wedge l \leq L$.

Generation of K-means Solutions

Next, we consider the generation of K-means solutions. We start by running the K-means algorithm (for 10 iterations) for different numbers of clusters $K \in [2, fsize - (\min(I, 0.5 \times fsize) + 1)]$. The resulting partitionings are then converted to MST-based genotypes as illustrated in Figure 2.5. The preservation of a high degree of MST information (within clusters) at this stage is crucial for the quick convergence of the algorithm. Note that the numbers of clusters obtained as the final phenotypes are not pre-defined and can increase or decrease depending on the structure of the underlying MST.

2.2.3 Experimental Setup

MOCK generates a set of different clustering solutions, which correspond to different trade-offs between the two objectives and to different numbers of clusters. While this provides a large set of solutions to choose from (typically more than one solution for each K), this experimental section primarily focuses on MOCK's 'peak' performance at generating high-quality clustering solutions. In order to assess this, we evaluate the quality of the best solution present in the final Pareto set. In order to put MOCK's results into perspective, we compare to three single-objective algorithms, which are run for a range of different numbers of clusters $K \in [1, 50]$. Again, only the best solution within this range is selected (note that, on many data sets, the number of clusters of the best solution may be quite different from the known 'correct' number of clusters, e.g. because of the isolation of outliers).

We compare these algorithms on a range of data sets exhibiting different difficult data properties including cluster overlap (the Square series), unequally sized clusters (the Sizes series) and elongated clusters of arbitrary shape (the Long, Smile and Triangle series). See Figure 2.6 for an illustration of some of these data sets, and [12] for a detailed description.

Clustering quality is objectively evaluated using the Adjusted Rand Index, an external measure of clustering quality. External indices evaluate a partitioning by comparing to a 'gold standard', that is, the true class labels.

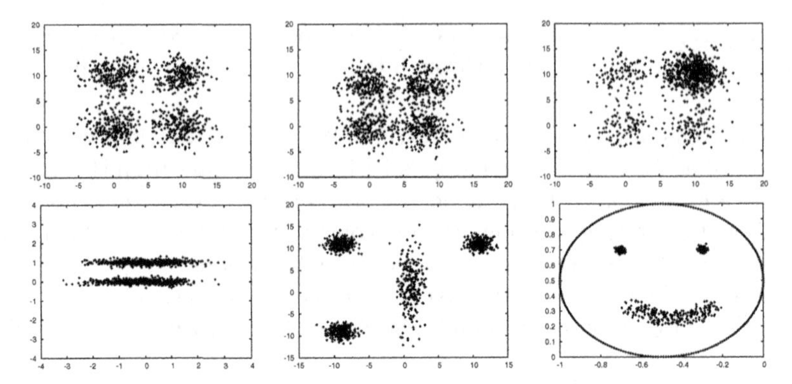

Fig. 2.6. Examples of data sets exhibiting different properties. The clusters in these data sets (with the exception of two clusters in Smile) are described by Normal Distributions, and, in every run, a new instance is sampled from these distributions (hence, all instances differ). Top row (from left to right): Square1, Square3 and Sizes3. Bottom row (from left to right): Long1, Triangle1 and Smile1.

Compared to other external indices, the Adjusted Rand Index has the strong advantage that it takes into account biases introduced due to the distribution of class sizes and differences in the number of clusters. This is of particular importance, as we compare clustering results across a range of different numbers of clusters.

Implementation details for the individual algorithms and the Adjusted Rand Index are given in the Appendix.

2.2.4 Experimental Results

Experimental results confirm that MOCK is indeed robust towards the range of different data properties studied, and it clearly outperforms traditional single-objective clustering algorithms in this respect.

Table 2.1 shows the results of the comparison of MOCK to K-means, average link and single link. The results demonstrate that all single-objective algorithms perform well for certain subsets of the data, but that their performance breaks down dramatically for those data sets violating the assumptions made by the employed clustering criterion. As can be expected, K-means and agglomerative clustering perform very well for spherically shaped clusters but fail to detect clusters of elongated shapes. Single link detects well-separated clusters of arbitrary shape, but fails for 'noisy' data sets, such as the Sizes and Square series. Here, frequently, the failure of an algorithm becomes evident not only through a low Adjusted Rand Index, but also through a dramatic increase in the number of clusters of the best solution.

MOCK, in contrast, shows a consistently good performance across the entire range of data properties. It is the best performer on ten out of fourteen

Table 2.1. Number of clusters and quality of the best solution (in terms of the Adjusted Rand Index) generated by MOCK, K-means, average link and single link on a number of artificial data sets exhibiting different cluster properties (averages over 50 runs). The best and second best performer are highlighted in bold and italic face respectively.

Problem		MOCK		K-means		Average link		Single link	
Name	K	K	Rand	K	Rand	K	Rand	K	Rand
Square1	4	4.12	*0.96342*	4	**0.96505**	4.16	0.93701	37.76	0.82555
Square2	4	4.2	*0.92661*	4	**0.93574**	4.24	0.89713	40.12	0.45803
Square3	4	4.42	*0.86767*	4	**0.88436**	4.12	0.82932	41.32	0.10803
Sizes1	4	4.24	*0.96465*	4	**0.96663**	4.32	0.94117	36.22	0.84291
Sizes2	4	4.16	**0.97109**	4	*0.96532*	4.28	0.95103	35.64	0.82880
Sizes3	4	4.14	**0.97575**	4	*0.96336*	4.26	0.96008	34.94	0.87194
Long1	2	2	**0.99984**	5	0.35544	7.64	0.47172	3.48	*0.99570*
Long2	2	2	**0.99992**	4.88	0.33936	7.84	0.45719	3.42	*0.99595*
Long3	2	2.02	**0.99965**	4.42	0.21742	7.32	0.37855	3.32	*0.99612*
Smile1	4	4	**1**	30.48	0.74404	11.98	0.80083	4	**1**
Smile2	4	4	**1**	27.4	0.55456	10.54	0.75607	4	**1**
Smile3	4	4	**1**	33.48	0.34918	11.22	0.38256	4	**1**
Triangle1	4	4	**1**	4	0.95800	4.82	0.99259	4.14	*0.99979*
Triangle2	4	4	**1**	4	0.88607	5.02	*0.95435*	24.4	0.94976

of the data sets. On the remaining four data sets it is the second best performer with only little difference to the best performer K-means. The number of clusters associated with the best solution is usually very close to the real number of clusters in the data set.

For additional experimental results, including a comparison to an advanced ensemble technique, and results on complex high-dimensional and real test data with large numbers of clusters, the reader is referred to [12, 15]. All of these experiments confirm the high performance of the multiobjective clustering approach.

2.3 Multiobjective Model Selection

After dealing with the generation of clustering solutions in the previous section, we now turn our interest to model selection, that is, the identification of one or several promising solutions from a large set of given clustering solutions. Model selection is particularly relevant for multiobjective clustering, as the algorithm does not return a single solution, but a set of solutions representing an approximation to the Pareto front. The individual partitionings in this approximation set correspond to different trade-offs between the two objectives but also consist of different numbers of clusters. While this may be a very useful feature under certain circumstances (e.g. human experts may find it preferable to have the opportunity to choose from a set of clustering solutions, to have the opportunity to analyze several alternative solutions and bring to bear any specialized domain expertise available), other applications may require the automatic selection of just one 'best' solution. In this section, we therefore introduce a method for identifying the most promising clustering solutions in the candidate set. We will also show how this methodology, originally developed for MOCK, can also be applied to analyze the output of other clustering algorithms.

2.3.1 Related Work

The approach we have developed is inspired by Tibshirani et al's Gap statistic [30], a statistical method to determine the number of clusters in a data set. The Gap statistic is based on the expectation that the most suitable number of clusters shows in a significant 'knee' when plotting the performance of a clustering algorithm (in terms of a selected internal validation measure) as a function of the number of clusters. As internal validation measures are generally biased by the number of clusters (they show an increasing/decreasing trend that is solely due to a change in the number of clusters), the 'knee' can be best identified in a normalized plot, that is, a performance plot that takes out the bias resulting purely from a change in the number of clusters. Tibshirani et al realize this by generating a number of reference partitionings for random data. From the normalized performance curve they then identify the smallest number of clusters for which the gain in performance is not higher than would be expected for random data.

2.3.2 Proposed Approach

Several aspects of this idea can be carried over to our case of two objectives. Intuitively, we equally expect the structure of the data to be reflected in the shape of the Pareto front. From the two objectives employed, overall deviation decreases with an increasing number of clusters, whereas connectivity decreases. Hence, generally, when considering two solutions with $K = k$ and $K = k+1$ respectively (where $k \in [1, N]$, and N is the size of the data set), we can say that we gain an improvement in overall deviation $\delta_{overall\_deviation}$ at the cost of a degradation in connectivity $\delta_{connectivity}$. For a number of clusters smaller than the true number K, we expect the ratio $R = \frac{\delta_{overall\_deviation}}{\delta_{connectivity}}$ to be large: the separation of two clusters will trigger a great decrease in overall deviation, with only a small increase in connectivity. When we surpass the correct number of clusters this ratio will diminish: the decrease in overall deviation will be less significant but come at a high cost in terms of connectivity (because a true cluster is being split). Using this knowledge, let us consider a plot of the Pareto front. Due to the natural bias of both measures, the solutions are approximately ordered by the number of clusters they contain:

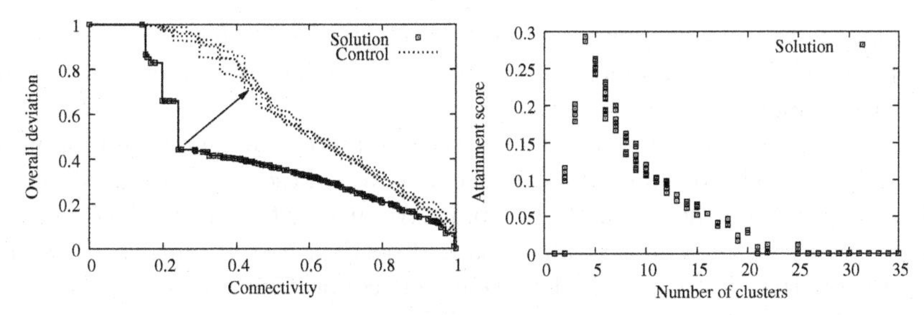

Fig. 2.7. (left) Solution and control reference fronts for a run of MOCK on the *Square1* data set. The solution with the largest minimum distance to the reference fronts is indicated by the arrow and corresponds to the desired $K = 4$ cluster solution. (right) Attainment scores for the *Square1* data set. Plot of the scores as a function of K. The global maximum at $K = 4$ is clearly visible.

K gradually increases from left to right. The distinct change in R occurring for the correct number of clusters can therefore be seen as a 'knee'. In order to help us correctly determine this knee, we can again use random reference data distributions. Clustering a number of such distributions using MOCK, provides us with a set of 'reference fronts' (see Figure 2.7).

Control Data

The reference distributions are obtained using a Poisson model in eigen-space as suggested by Sarle [27]. Specifically, a principal component analysis is applied to the covariance matrix of the original data. The eigenvectors and eigenvalues obtained are then used for the definition of a uniform distribution: the data is generated within a hyperbox in eigenspace, where each side of the hyperbox is proportional in length to the size of the eigenvalue corresponding to this dimension. The resulting data is then back-transformed to the original data space.

Unfortunately, a normalization of the original 'solution front' using the 'reference fronts' is not as straightforward as the normalization of the performance curve for the Gap statistic. This is because both solution and control fronts contain not just one, but a set of solutions for every value of K, and it is therefore not clear how individual points in the solution front should be normalized. We overcome this problem by a heuristic approach, described next.

Alignment of Solution and Control Fronts

Given both solution and reference fronts, we set $K_{min} = 1$ and identify K_{max}, the highest number of clusters shared by all fronts. Subsequently, we restrict the analysis to solutions with a number of clusters $K \in [K_{min}, K_{max}]$. Solution points that are dominated by any reference point are also excluded from further consideration.

For each front, we then determine the minimum and maximum value of both overall deviation and connectivity, and use these to scale all objective values to lie within the region $[0, 1] \times [0, 1]$. We further transform the objective values by taking the square root of each objective, a step motivated by the observation that both overall deviation and connectivity show a non-linear development with respect to K. Overall deviation decreases very rapidly for the first few K, while changes for higher number of clusters are far less marked. Connectivity, in contrast rises very quickly for higher numbers of clusters, while initial changes in the degree of connectivity are rather small. This results in an uneven sampling of the range of objective values, which — in a plot of the Pareto front — shows in a high density of points at the tails, with fewer solution points in the centre. Taking the square root of the objective values is an attempt to reduce this 'squeezing' effect, and give a higher degree of emphasis to small (but distinct) changes in the objectives. By this means, the algorithm becomes more precise at identifying solutions situated in all parts of the Pareto front, in particular those at the tails which may correspond e.g. to partitionings with elongated cluster shapes or a high number of clusters.

Attainment Scores

For both solution and reference fronts, we subsequently compute the attainment surfaces [11]. The attainment surface of a Pareto front is uniquely defined by the points in the front. It is the boundary in the objective space separating those points that are dominated by or equal to at least one of the data points in the front from those that no data point in the front dominates or equals (see Figure 2.7). For each point in the solution front we then compute its distance to the attainment surfaces of each of the reference fronts, and we refer to this distance as the 'attainment score'. For a given solution point p, we compute its attainment score as the Euclidean distance between p and the closest point on the the reference attainment surface.

Finally, we plot the attainment scores as a function of the number of clusters K (see Figure 2.7). All solutions corresponding to the local optima in the resulting plot are considered as promising solutions. The global maximum in this plot may be considered as a hypothesis as to the best solution.

2.3.3 Experimental Setup

In this experimental section, we investigate the performance of MOCK's attainment method as a general tool for model selection. To make the investigation more meaningful, the performance of the attainment method is compared to an existing cluster validation technique, the Silhouette Width, which is highly-regarded in the clustering literature as a means to model selection.

To get a broad and detailed picture of how both methods perform, they are applied to the clustering results from four different clustering algorithms. Specifically, we run MOCK and three single-objective algorithms (the latter run for a range of different numbers of clusters $K \in [1, 50]$), and do this across the full range of data sets introduced in Section 2.2.3. On each data set, and for each algorithm, the two validation techniques are then used to perform model selection, that is, to select one or more partitionings from those generated by the clustering algorithm, as candidate 'best' solutions. The selected solutions are then objectively evaluated using the Adjusted Rand Index, an external measure of clustering quality. To isolate more clearly the performance of the validation techniques from the performance of the clustering algorithms, we also give the evaluation of the overall best solution returned by the clustering algorithm. In the following, we briefly describe the Silhouette Width (see also the Appendix) and state how the two validation techniques are applied to our data.

The Silhouette Width is a popular validation technique commonly used for model selection in cluster analyses. Notably, it can be considered a (fixed) nonlinear combination of two objectives, as it takes both intra-cluster distances and inter-cluster distances into account. The procedure for its use in model selection is as follows: for each $K \in [1, 50]$, the Silhouette Width of the corresponding clustering solution is computed. In a plot of the Silhouette Width as a function of the number of clusters, the global maximum and all local maxima are identified. The global maximum is considered the estimated best solution. All local maxima are considered as alternative, potentially interesting solutions.

Similarly, MOCK's attainment method can be adapted for model selection on the clustering results of any single-objective algorithm. For this purpose, clustering results for a range of $K \in [1, 50]$ are obtained both on the original data set and random control data (which is generated in the same way as described in Section 2.3.2). The resulting sequences are then subjected to the procedure detailed in Section 2.3.2, in order to obtain attainment scores for all solutions. In a plot of the attainment score as a function of the number of clusters, the global maximum and all local maxima are identified. The global maximum is considered the guess for the best solution. All local maxima are considered as alternative, potentially interesting solutions.

2.3.4 Experimental Results

The Silhouette Width assumes compact, spherically shaped clusters. We would therefore expect it to perform well for data sets exhibiting this type of structure, but to perform poorly for data containing arbitrarily shaped, elongated clusters. In contrast, we would expect MOCK's attainment method to perform more uniformly well for data of different structures.

Table 2.2, Table 2.3, Table 2.4 and Table 2.5 show the results obtained for model selection on the results returned by the four different algorithms.

Table 2.2. Results for MOCK. The overall best solution found (in terms of the Adjusted Rand Index), and the solutions identified using the Silhouette Width and the attainment score are evaluated using the Adjusted Rand Index (Rand) and the number of clusters (K). Here, Max A-score' and 'Max S-score' refer to the solution corresponding to the global maximum in the plot of the attainment score and the Silhouette Width, respectively. 'Best local A-score' and 'Best local S-score' refer to the best (in terms of the Adjusted Rand Index) out of the set of solutions corresponding to local maxima in the plot of the attainment score and the Silhouette Width, respectively. Values presented are averages over 50 sample data sets.

Problem	Best Solution		Max A-score		Max S-score		Best local A-score		Best local S-score	
	K	Rand	K	Rand	K	Rand	K	Rand	K	Rand
Square1	4.12	0.96342	4.02	0.95919	4	**0.96184**	4.02	0.95919	4	**0.96184**
Square2	4.2	0.92661	4.12	0.91546	4	**0.92438**	4.24	0.91790	4	**0.92438**
Square3	4.42	0.86767	4.08	0.85653	4	**0.86251**	4.08	0.85653	4	**0.86251**
Sizes1	4.24	0.96465	4.02	0.96113	4	**0.96314**	4.02	0.96113	4	**0.96314**
Sizes2	4.16	0.97110	4.06	0.96724	4	**0.96868**	4.1	0.96727	4	**0.96868**
Sizes3	4.14	0.97575	3.92	0.9411	4	**0.97394**	4.04	0.97109	4	**0.97394**
Long1	2	0.99984	2.02	**0.99745**	7.68	0.32103	2.02	**0.99745**	6.48	0.36782
Long2	2	0.99992	2.04	**0.99748**	7.76	0.31094	2.04	**0.99748**	6.44	0.37415
Long3	2.02	0.99965	2.44	**0.93692**	7	0.21439	2.2	**0.96878**	3.72	0.38018
Smile1	4	1	4	**1**	14.24	0.72711	4	**1**	10.56	0.77839
Smile2	4	1	4	**1**	13	0.55059	4	**1**	10.5	0.63436
Smile3	4	1	4	**1**	15.96	0.32214	4	**1**	11.48	0.39329
Triangle1	4	1	4	**1**	4.02	0.99680	4	**1**	4.02	0.99680
Triangle2	4.04	0.98652	4.04	**0.97692**	4	0.97165	4.04	**0.97692**	4	0.97165

We have seen before, that MOCK is the only one of the three algorithms that perform robustly across the entire range of data sets. The results in Table 2.2 indicate that this advantage can be largely maintained when using the attainment score for model selection, but is strongly reduced, if the Silhouette Width is used. While the Silhouette Width performs (as expected) very well for model selection on the Square and the Sizes data sets, its performance breaks down drastically for those data with elongated cluster shapes.

In Table 2.3, no advantage can be observed for MOCK's attainment score method. This is intuitively plausible, as the K-means algorithm only generates spherically shaped, compact clusters. All of the partitionings generated by the algorithm are therefore in concordance with the assumptions made by the Silhouette Width, and the Silhouette Width can therefore be expected to perform well. In such a scenario, the additional flexibility of the attainment score is not rewarded, but instead results in the introduction of noise.

In Table 2.4 and Table 2.5 a slight advantage of the attainment method can be observed. In general, it seems to cope better with the data sets containing elongated data sets, and, for single link, its estimate of the best clustering solution is generally of much higher quality.

Overall, it seems that the utility of multiobjective cluster validation strongly depends on the algorithm used for the generation of the clustering solutions. In the context of multiobjective clustering, where the full extent of

Table 2.3. Results for K-means. The overall best solution found (in terms of the Adjusted Rand Index), and the solutions identified using the Silhouette Width and the attainment score are evaluated using the Adjusted Rand Index (Rand) and the number of clusters (K). Here, Max A-score' and 'Max S-score' refer to the solution corresponding to the global maximum in the plot of the attainment score and the Silhouette Width, respectively. 'Best local A-score' and 'Best local S-score' refer to the best (in terms of the Adjusted Rand Index) out of the set of solutions corresponding to local maxima in the plot of the attainment score and the Silhouette Width, respectively. Values presented are averages over 50 sample data sets.

Problem	Best Solution		Max A-score		Max S-score		Best local A-score		Best local S-score	
	K	Rand	K	Rand	K	Rand	K	Rand	K	Rand
Square1	4	0.96500	4	**0.96500**	4	**0.96500**	4	**0.96500**	4	**0.96500**
Square2	4	0.935684	4	**0.93568**	4	**0.93568**	4	**0.93568**	4	**0.93568**
Square3	4	0.88432	4.02	0.88259	4	**0.88432**	4.02	0.88259	4	**0.88432**
Sizes1	4	0.96648	4	**0.96648**	4	**0.96648**	4	**0.96648**	4	**0.96648**
Sizes2	4	0.96532	4	**0.96532**	4	**0.96532**	4	**0.96532**	4	**0.96532**
Sizes3	4	0.96336	4	**0.96336**	4	**0.96336**	4	**0.96336**	4	**0.96336**
Long1	5.04	0.35500	12.04	0.20836	8.42	**0.27396**	8.38	0.27602	7.62	**0.29356**
Long2	4.92	0.33803	12.48	0.18235	8.32	**0.25925**	8.78	0.24886	7.02	**0.29083**
Long3	4.46	0.21694	13.86	0.08793	6.64	0.18584	11.36	0.11241	4.54	0.21598
Smile1	30.6	0.74397	9.66	0.65296	41.58	**0.73858**	14.8	0.69363	30.18	**0.74356**
Smile2	25.22	0.55347	12.58	0.53044	35.2	**0.53460**	7.64	0.53681	24.94	**0.55340**
Smile3	34.02	0.34886	13.1	0.29340	43.02	**0.33693**	13.84	0.29585	32.28	**0.34815**
Triangle1	4	0.95805	5.34	0.90414	4	**0.95805**	5.3	0.90565	4	**0.95805**
Triangle2	4	0.88575	5.16	0.77339	4	**0.88575**	5.12	0.77700	4	**0.88575**

the true Pareto front is approximated, a clear advantage to the multiobjective approach can be observed. In contrast, for traditional clustering algorithms such as K-means, which are based on very specific cluster models, the use of an analogous validation technique based on identical assumptions may be preferable.

2.4 Conclusion

In this chapter we have identified two subtasks in clustering, in which the use of a multiobjective framework may be beneficial: (1) the actual process of clustering, that is, the generation of clustering solutions; and (2) model selection, that is, the selection of the best clustering out of a set of given partitionings. Both processes require the definition of a clustering criterion, and the algorithms' capability to deal with different data properties highly depends on this choice. We have argued that that the simultaneous consideration of several objectives may be a way to obtain a more consistent performance (across a range of data properties) for both subtasks. This idea has been illustrated through the development of a multiobjective clustering algorithm and a multiobjective scheme for model selection. The resulting algorithms have been evaluated across a range of different data sets, and results have been compared to those

Table 2.4. Results for average link. The overall best solution found (in terms of the Adjusted Rand Index), and the solutions identified using the Silhouette Width and the attainment score are evaluated using the Adjusted Rand Index (Rand) and the number of clusters (K). Here, Max A-score' and 'Max S-score' refer to the solution corresponding to the global maximum in the plot of the attainment score and the Silhouette Width, respectively. 'Best local A-score' and 'Best local S-score' refer to the best (in terms of the Adjusted Rand Index) out of the set of solutions corresponding to local maxima in the plot of the attainment score and the Silhouette Width, respectively. Values presented are averages over 50 sample data sets.

Problem	Best Solution		Max A-score		Max S-score		Best local A-score		Best local S-score	
	K	Rand	K	Rand	K	Rand	K	Rand	K	Rand
Square1	4.16	0.93701	4	**0.93671**	4	**0.93671**	4	**0.93671**	4	**0.93671**
Square2	4.24	0.89713	4	**0.89661**	4	**0.89661**	4.1	**0.89664**	4	0.89661
Square3	4.12	0.82932	4.02	0.82147	4.06	**0.82921**	4.06	**0.82921**	4.06	**0.82921**
Sizes1	4.32	0.94117	4	**0.94079**	4	**0.94079**	4	**0.94079**	4	**0.94079**
Sizes2	4.28	0.95103	4	**0.95089**	4	**0.95089**	4	**0.95089**	4	**0.95089**
Sizes3	4.26	0.96008	4	**0.95990**	4	**0.95990**	4.08	**0.95990**	4	0.95990
Long1	7.64	0.47172	14.38	**0.21911**	7.78	0.17631	11.92	0.28632	9.3	**0.37684**
Long2	7.84	0.45719	15.4	**0.18893**	6.7	0.12544	12.94	0.24133	9.3	**0.36842**
Long3	7.32	0.37855	16.16	**0.08473**	5.84	0.082558	15.98	0.09140	10.24	**0.25754**
Smile1	11.98	0.80083	8.9	0.68408	17.42	**0.72330**	11.78	**0.74898**	17.1	0.72458
Smile2	10.54	0.75607	4.82	**0.55600**	18.62	0.52529	11.08	**0.69273**	16.64	0.54517
Smile3	11.22	0.38256	10.28	0.30292	18	**0.30426**	12.2	**0.37947**	17.1	0.31562
Triangle1	4.82	0.99259	4.8	**0.9886**	4	0.98493	4.8	**0.98863**	4	0.98493
Triangle2	5.02	0.954345	4.64	0.935819	4.06	**0.94234**	4.58	**0.94283**	4.06	0.94234

Table 2.5. Results for single link. The overall best solution found (in terms of the Adjusted Rand Index), and the solutions identified using the Silhouette Width and the attainment score are evaluated using the Adjusted Rand Index (Rand) and the number of clusters (K). Here, Max A-score' and 'Max S-score' refer to the solution corresponding to the global maximum in the plot of the attainment score and the Silhouette Width, respectively. 'Best local A-score' and 'Best local S-score' refer to the best (in terms of the Adjusted Rand Index) out of the set of solutions corresponding to local maxima in the plot of the attainment score and the Silhouette Width, respectively. Values presented are averages over 50 sample data sets.

Problem	Best Solution		Max A-score		Max S-score		Best local A-score		Best local S-score	
	K	Rand	K	Rand	K	Rand	K	Rand	K	Rand
Square1	37.76	0.82555	14.54	**0.32341**	2.72	0.03742	14.24	0.33951	38.3	**0.82400**
Square2	40.12	0.45803	11.56	**0.08107**	2	8.02944e-07	10.46	0.08108	40.88	**0.45713**
Square3	41.32	0.10803	11.84	**0.00568**	2	1.07145e-06	12.36	0.00570	43.08	**0.10770**
Sizes1	36.22	0.84291	14.4	**0.38400**	3.06	0.06278	15.5	0.40410	36.28	**0.84184**
Sizes2	35.64	0.82880	13.72	**0.42990**	2.18	0.02102	15.92	0.47064	35.38	**0.82803**
Sizes3	34.94	0.87194	12.58	**0.49804**	2.7	0.042279	15.04	0.53626	35.74	**0.87094**
Long1	3.48	0.99570	4.42	**0.97150**	2.02	0.30000	4.32	0.97182	8.7	**0.97394**
Long2	3.42	0.99595	4.58	**0.99073**	2.06	0.20012	4	**0.99366**	9.52	0.97057
Long3	3.32	0.99612	4.98	**0.98881**	2.06	0.26364	4.46	**0.99138**	7.58	0.97434
Smile1	4	1	4	**1**	4	**1**	4	**1**	4	**1**
Smile2	4	1	4.04	0.99992	4	**1**	4.04	0.99992	4	**1**
Smile3	4	1	4.08	**0.99991**	3.6	0.88220	4.08	0.99991	4	**1**
Triangle1	4.14	0.99979	4.14	**0.99979**	4.14	**0.99979**	4.14	**0.99979**	4.14	**0.99979**
Triangle2	24.4	0.94976	11.12	**0.63951**	5.46	0.29776	13.7	0.72440	24.68	**0.94832**

obtained using traditional single-objective approaches. Experimental results indicate some general advantages to the multiobjective approach.

Acknowledgements

Julia Handl gratefully acknowledges support of a scholarship from the Gottlieb Daimler- and Karl Benz-Foundation, Germany. Joshua Knowles is supported by a David Phillips Fellowship from the Biotechnology and Biological Sciences Research Council (BBSRC), UK.

References

[1] M. Ankerst, M. Breunig, H.-P. Kriegel, and J. Sander. OPTICS: Ordering points to identify clustering structure. In *Proceedings of the 1999 International Conference on Management of Data*, pages 49–60. ACM Press, 1999.
[2] S. Bandyopadhyay and U. Manlik. Nonparametric genetic clustering: comparison of validity indices. *IEEE Transactions on Systems, Man and Cybernetics*, 31:120–125, 2001.
[3] J. Bilmes, A. Vahdat, W. Hsu, and E.-J. Im. Empirical observations of probabilistic heuristics for the clustering problem. Technical Report TR-97-018, International Computer Science Institute, University of California, Berkeley, CA, 1997.
[4] D. W. Corne, Nick R. Jerram, Joshua D. Knowles, and Martin J. Oates. PESA-II: Region-based selection in evolutionary multiobjective optimization. In *Proceedings of the Genetic and Evolutionary Computation Conference*, pages 283–290. Morgan Kaufmann Publishers, 2001.
[5] D. W. Corne, J. D. Knowles, and M. J. Oates. The Pareto envelope-based selection algorithm for multiobjectice optimization. In *Proceedings of the Fifth Conference on Parallel Problem Solving from Nature*, pages 839–848, 2000.
[6] D. W. Corne, J. D. Knowles, and M. J. Oates. PESA-II: region-based selection in evolutionary multiobjective optimization. In *Proceedings of the Genetic and Evolutionary Computation Conference*, pages 283–290, 2001.
[7] R. O. Duda, P. E. Hart, and D. G. Stork. *Pattern Classification, Second edition.* John Wiley and Son Ltd, 2001.
[8] M. Ester, H. P. Kriegel, and J. Sander. A density-based algorithm for discovering clusters in large spatial databases with noise. In *Proceedings of the Second International Conference on Knowledge Discovery and Data-Mining*, pages 226–231. AIII Press, 1996.
[9] V. Estivill-Castro. Why so many clustering algorithms: A position paper. *ACM SIGKDD Explorations Newsletter Archive*, 4:65–75, 2002.
[10] B. S. Everitt. *Cluster Analysis.* Edward Arnold, 1993.
[11] C. M. Fonseca and P. J. Fleming. On the performance assessment and comparison of stochastic multiobjective optimizers. In *Proceedings of the Fourth International Conference on Parallel Problem Solving from Nature*, pages 584–593. Springer-Verlag,, 1996.

[12] J. Handl and J. Knowles. Evolutionary multiobjective clustering. In *Proceedings of the Eighth International Conference on Parallel Problem Solving from Nature*, pages 1081–1091. Springer-Verlag, 2004.
[13] J. Handl and J. Knowles. Multiobjective clustering with automatic determination of the number of clusters. Technical Report TR-COMPSYSBIO-2004-02, UMIST, Manchester, UK, 2004.
[14] J. Handl and J. Knowles. Exploiting the trade-off: the benefits of multiple objectives in data clustering. In *Proceedings of the Third International Conference on Evolutionary Multicriterion Optimization*, pages 547–560. Springer-Verlag, 2005.
[15] J. Handl and J. Knowles. Improvements to the scalability of multiobjective clustering. In *IEEE Congress on Evolutionary Computation*, pages 632–639. IEEE Press, 2005.
[16] J. Handl, J. Knowles, and D. B. Kell. Computational cluster validation in post-genomic data analysis. *Bioinformatics*, 21:3201–3212, 2005.
[17] T. Hastie, R. Tibshirani, and J. Friedman. *The elements of statistical learning: data mining, inference and prediction.* Springer-Verlag, 2001.
[18] A. Hubert. Comparing partitions. *Journal of Classification*, 2:193–198, 1985.
[19] A. K. Jain, M. N. Murty, and P. J. Flynn. Data clustering: a review. *ACM Computing Surveys*, 31:264–323, 1999.
[20] J. Kleinberg. An impossibility theorem for clustering. In *Proceedings of the 15th Conference on Neural Information Processing Systems.* The Internet, 2002.
[21] L. MacQueen. Some methods for classification and analysis of multivariate observations. In *Proceedings of the Fifth Berkeley Symposium on Mathematical Statistics and Probability*, pages 281–297. University of California Press, 1967.
[22] G. McLachlan and T. Krishman. *The EM Algorithm and Extensions.* John Wiley and Son Ltd, 1997.
[23] Y.-J. Park and M.-S. Song. A genetic algorithm for clustering problems. In *Proceedings of the Third Annual Conference on Genetic Programming*, pages 568–575, Madison, WI, 1998. Morgan Kaufmann.
[24] J. M. Pena, J. A. Lozana, and P. Larranaga. An empirical comparison of four initialization methods for the k-means algorithm. *Pattern Recognition Letters*, 20:1027–1040, 1999.
[25] V. J. Rayward-Smith, I. H. Osman, C. R. Reeves, and G. D. Smith. *Modern Heuristic Search Methods.* John Wiley and Son Ltd, 1996.
[26] P. J. Rousseeuw. Silhouettes: a graphical aid to the interpretation and validation of cluster analysis. *Journal of Computational and Applied Mathematics*, 20:53–65, 1987.
[27] W. S. Sarle. Cubic clustering criterion. Technical report, SAS Technical Report A-108, Cary, NC: SAS Institute Inc, 1983.
[28] A. Strehl and J. Ghosh. Cluster ensembles — a knowledge reuse framework for combining multiple partitions. *Journal on Machine Learning Research*, 3:583–617, 2002.
[29] G. Syswerda. Uniform crossover in genetic algorithms. In *Proceedings of the Third International Conference on Genetic Algorithms*, pages 2–9. Morgan Kaufmann Publishers, 1989.
[30] R. Tibshirani, G. Walther, and T. Hastie. Estimating the number of clusters in a dataset via the Gap statistic. *Journal of the Royal Statistical Society: Series B (Statistical Methodology)*, 63:411–423, 2001.

[31] A. Topchy, A. K. Jain, and W. Punch. Clustering ensembles: Models of consensus and weak partitions. Submitted to IEEE Transactions on Pattern Analysis and Machine Intelligence, 2004.

[32] E. Vorhees. *The effectiveness and efficiency of agglomerative hierarchical clustering in document retrieval.* PhD thesis, Department of Computer Science, Cornell University, 1985.

[33] D. Whitley. A genetic algorithm tutorial. *Statistics and Computing*, 4:65–85, 1994.

[34] R. J. Wilson and J. J. Watkins. *Graphs: An Introductory Approach: A First Course in Discrete Mathematics.* John Wiley and Sons, 1990.

2.5 Appendix

2.5.1 Evaluation Functions

Silhouette Width

The Silhouette Width [26] for a partitioning is computed as the average Silhouette value over all data items. The Silhouette value for an individual data item i, which reflects the confidence in this particular cluster assignment, is computed as

$$S(i) = \frac{b_i - a_i}{\max(b_i, a_i)},$$

where a_i denotes the average distance between i and all data items in the same cluster, and b_i denotes the average distance between i and all data items in the closest other cluster (which is defined as the one yielding the minimal b_i).

The Silhouette Width return values in the interval $[-1, 1]$ and is to be maximized.

Adjusted Rand Index

In all experiments, clustering quality is objectively evaluated using the Adjusted Rand Index, an external measure of clustering quality, which is a generalization of the Rand Index.

The Rand indices are based on counting the number of pair-wise co-assignments of data items. The Adjusted Rand Index additionally introduces a statistically induced normalization in order to yield values close to 0 for random partitions. Using a representation based on contingency tables, the Adjusted Rand Index [18] is given as

$$R(U,V) = \frac{\sum_{lk}\binom{n_{lk}}{2} - \left[\sum_l\binom{n_{l.}}{2}\cdot\sum_k\binom{n_{.k}}{2}\right]/\binom{n}{2}}{\frac{1}{2}\left[\sum_l\binom{n_{l.}}{2} + \sum_k\binom{n_{.k}}{2}\right] - \left[\sum_l\binom{n_{l.}}{2}\cdot\sum_k\binom{n_{.k}}{2}\right]/\binom{n}{2}},$$

where n_{lk} denotes the number of data items that have been assigned to both cluster l and cluster k.

The Adjusted Rand Index returns values in the interval $[0, 1]$ and is to be maximized.

2.5.2 Algorithms

K-means

Starting from a random partitioning, the K-means algorithm repeatedly (i) computes the current cluster centres (that is, the average vector of each cluster in data space) and (ii) reassigns each data item to the cluster whose centre is closest to it. It terminates when no more reassignments take place. By this

Table 2.6. Parameter settings for MOCK, where N is data set size.

Parameter	*setting*
Number of generations	500
External population size	1000
Internal population size	10
#(Initial solutions) $fsize$	100
Initialization	Minimum spanning tree and K-means ($L = 20$)
Mutation type	L nearest neighbours ($L = 20$)
Mutation rate p_m	$p_m = \frac{1}{N} + (\frac{l}{N})^2$
Recombination	Uniform crossover
Recombination rate p_c	0.7
Objective functions	Overall deviation and connectivity ($L = 20$)
#(Reference distributions)	3

means, the intra-cluster variance, that is, the sum of squares of the differences between data items and their associated cluster centres, is locally minimized.

Our implementation of the K-means algorithm is based on the batch version of K-means, that is, cluster centres are only recomputed after the reassignment of all data items. To reduce suboptimal solutions K-means is run repeatedly (100 times) using random initialisation (which is known to be an effective initialization method [24]) and only the best result in terms of intra-cluster variance is returned.

Hierarchical Clustering

In general, agglomerative clustering algorithms start with the finest partitioning possible (that is, singletons) and, in each iteration, merge the two least distant clusters. They terminate when the target number of clusters has been obtained. Alternatively, the entire dendrogram can be generated and be cut at a later point.

Single link and average link agglomerative clustering only differ in the linkage metric used. For the linkage metric of average link, the distance between two clusters C_i and C_j is computed as the average dissimilarity between all possible pairs of data elements i and j with $i \in C_i$ and $j \in C_j$. For the linkage metric of single link, the distance between two clusters C_i and C_j is computed as the smallest dissimilarity between all possible pairs of data elements i and j with $i \in C_i$ and $j \in C_j$.

Parameter Settings for MOCK

Parameter settings for MOCK are given in Table 2.6 and are kept constant over all experiments.

3

Feature Selection for Ensembles Using the Multi-Objective Optimization Approach

Luiz S. Oliveira[1], Marisa Morita[2], and Robert Sabourin[3]

[1] Pontifical Catholic University of Paraná, Curitiba, PR, Brazil
soares@ppgia.pucpr.br
[2] HSBC Bank Brazil, Curitiba, PR, Brazil
marisa.e.morita@hsbc.com.br
[3] École de Technologie Supérieure, Montreal, Canada
robert.sabourin@etsmtl.ca

Summary. Feature selection for ensembles has shown to be an effective strategy for ensemble creation due to its ability of producing good subsets of features, which make the classifiers of the ensemble disagree on difficult cases. In this paper we present an ensemble feature selection approach based on a hierarchical multi-objective genetic algorithm. The underpinning paradigm is the "overproduce and choose". The algorithm operates in two levels. Firstly, it performs feature selection in order to generate a set of classifiers and then it chooses the best team of classifiers. In order to show its robustness, the method is evaluated in two different contexts: supervised and unsupervised feature selection. In the former, we have considered the problem of handwritten digit recognition and used three different feature sets and multi-layer perceptron neural networks as classifiers. In the latter, we took into account the problem of handwritten month word recognition and used three different feature sets and hidden Markov models as classifiers. Experiments and comparisons with classical methods, such as Bagging and Boosting, demonstrated that the proposed methodology brings compelling improvements when classifiers have to work with very low error rates.

3.1 Introduction

Ensemble of classifiers has been widely used to reduce model uncertainty and improve generalization performance. Developing techniques for generating candidate ensemble members is a very important direction of ensemble of classifiers research. It has been demonstrated that a good ensemble is one where the individual classifiers in the ensemble are both accurate and make their errors on different parts of the input space (there is no gain in combining identical classifiers) [11, 17, 31]. In other words, an ideal ensemble consists of good classifiers (not necessarily excellent) that disagree as much as possible on difficult cases.

L.S. Oliveira et al.: *Feature Selection for Ensembles Using the Multi-Objective Optimization Approach*, Studies in Computational Intelligence (SCI) **16**, 49–74 (2006)
www.springerlink.com

The literature has shown that varying the feature subsets used by each member of the ensemble should help to promote this necessary diversity [12, 31, 24, 35]. Traditional feature selection algorithms aim at finding the best trade-off between features and generalization. On the other hand, ensemble feature selection has the additional goal of finding a set of feature sets that will promote disagreement among the component members of the ensemble. The Random Subspace Method (RMS) proposed by Ho in [12] was one early algorithm that constructs an ensemble by varying the subset of features. More recently some strategies based on genetic algorithms (GAs) have been proposed [31]. All these strategies claim better results than those produced by traditional methods for creating ensembles such as Bagging and Boosting. In spite of the good results brought by GA-based methods, they still can be improved in some aspects, e.g., avoiding classical methods such as the weighted sum to combine multiple objective functions. It is well known that when dealing with this kind of combination, one should deal with problems such as scaling and sensitivity towards the weights.

It has been demonstrated that feature selection through multi-objective genetic algorithm (MOGA) is a very powerful tool for finding a set of good classifiers, since GA is quite effective in rapid global search of large, non-linear and poorly understood spaces [30]. Besides, it can overcome problems such as scaling and sensitivity towards the weights. Kudo and Sklansky [18] have compared several algorithms for feature selection and concluded that GAs are suitable when dealing with large-scale feature selection (number of features is over 50). This is the case of most of the problems in handwriting recognition, which is the test problem in this work.

In this light, we propose an ensemble feature selection approach based on a hierarchical MOGA. The underlying paradigm is the "overproduce and choose" [32, 9]. The algorithm operates in two levels. The former is devoted to the generation of a set of good classifiers by minimizing two criteria: error rate and number of features. The latter combines these classifiers in order to find an ensemble by maximizing the following two criteria: accuracy of the ensemble and a measure of diversity.

Recently, the issue of using diversity to build ensemble of classifiers has been widely discussed. Several works have demonstrated that there is a weak correlation between diversity and ensemble performance [23]. In light of this, some authors have claimed that diversity brings no benefits in building ensemble of classifiers [33], on the other hand, others suggest that the study of diversity in classifier combination might be one of the lines for further exploration [19].

In spite of the weak correlation between diversity and performance, we argue that diversity might be useful to build ensembles of classifiers. We demonstrated through experimentation that using diversity jointly with performance to guide selection can avoid overfitting during the search. In order to show robustness of the proposed methodology, it was evaluated in two different contexts: supervised and unsupervised feature selection. In the former, we have considered the problem of handwritten digit recognition and used three different feature sets and multi-layer perceptron (MLP) neural networks as classifiers. In such a case, the classification accuracy is supplied by the MLPs in conjunction with the sensitivity analysis. This approach makes it feasible to deal with huge databases in order to better represent the pattern recognition problem during the fitness evaluation. In the latter, we took into account the problem of handwritten month word recognition and used three different feature sets and hidden Markov models (HMM) as classifiers. We demonstrate that it is feasible to find compact clusters and complementary high-level representations (codebooks) in subspaces without using the recognition results of the system. Experiments and comparisons with classical methods, such as Bagging and Boosting, demonstrated that the proposed methodology brings compelling improvements when classifiers have to work with very low error rates.

The remainder of this paper is organized as follows. Section 3.2 presents a brief review about the methods for ensemble creation. Section 3.3 provides a overview of the strategy. Section 3.4 introduces briefly the the multi-objective genetic algorithm we are using in this work. Section 3.5 describes the classifiers and feature sets for both supervised and unsupervised contexts. Section 3.6 introduces how we have implemented both levels of the proposed methodology and Section 3.7 reports the experimental results. Finally, Section 3.8 discusses the reported results and Section 3.9 concludes the paper.

3.2 Related Works

Assuming the architecture of the ensemble as the main criterion, we can distinguish among serial, parallel, and hierarchical schemes, and if the classifiers of the ensemble are selected or not by the ensemble algorithm we can divide them into selection-oriented and combiner-oriented methods [20]. Here we are more interested in the first class, which try to improve the overall accuracy of the ensemble by directly boosting the accuracy and the diversity of the experts of the ensemble. Basically, they can be divided into resampling methods and feature selection methods.

Resampling techniques can be used to generate different hypotheses. For instance, bootstrapping techniques [6] may be used to generate different training sets and a learning algorithm can be applied to the obtained subsets of data in order to produce multiple hypotheses. These techniques are effective especially with unstable learning algorithms, which are algorithms very sensitive to small changes in the training data. In bagging [1] the ensemble is formed by making bootstrap replicates of the training sets, and then multiple generated hypotheses are used to get an aggregated predictor. The aggregation can be performed by averaging the outputs in regression or by majority or weighted voting in classification problems.

While in bagging the samples are drawn with replacement using a uniform probability distribution, in boosting methods [7] the learning algorithm is called at each iteration using a different distribution or weighting over the training examples. This technique places the highest weight on the examples most often misclassified by the previous base learner: in this manner the classifiers of the ensemble focus their attention on the hardest examples. Then the boosting algorithm combines the base rules taking a weighted majority vote of the base rules.

The second class of methods regards those strategies based on feature selection. The concept behind these approaches consists in reducing the number of input features of the classifiers, a simple method to fight the effects of the classical curse of dimensionality problem. For instance, the random subspace method [12, 35] relies on a pseudorandom procedure to select a small number of dimensions from a given feature space. In each pass, such a selection is made and a subspace is fixed. All samples are projected to this subspace, and a classifier is constructed using the projected training samples. In the classification a sample of an unknown class is projected to the same subspace and classified using the corresponding classifier. In the same vein of the random subspace method lies the input decimation method [37], which reduces the correlation among the errors of the base classifiers, by decoupling the classifiers by training them with different subsets of the input features. It differs from the random subspace as for each class the correlation between each feature and the output of the class is explicitly computed, and the classifier is trained only on the most correlated subset of features.

Recently, several authors have been investigated GA to design ensemble of classifiers. Kuncheva and Jain [21] suggest two simple ways to use genetic algorithm to design an ensemble of classifiers. They present two versions of their algorithm. The former uses just disjoint feature subsets while the latter considers (possibly) overlapping feature subsets. The fitness function employed is the accuracy of the ensemble, however, no measure of diversity is considered. A more elaborate method, also based on GA, was proposed by Optiz [31]. In his work, he stresses the importance of a diversity measure by including it in the fitness calculation. The drawback of this method is that the objective functions are combined through the weighted sum. It is well known that when dealing with this kind of combination, one should deal with problems such as scaling and sensitivity towards the weights. More recently Gunter and Bunke [10] have applied feature selection in conjunction with floating search to create ensembles of classifiers for the field of handwriting recognition. They used handwritten words and HMMs as classifiers to evaluate their algorithm. The feature set was composed of nine discrete features, which makes simpler the feature selection process. A drawback of this method is that one must set a priori the number of classifiers in the ensemble.

3.3 Methodology Overview

In this section we outline the hierarchical approach proposed. As stated before, it is based on an "overproduce and choose" paradigm where the first level generates several classifiers by conducting feature selection and the second one chooses the best ensemble among such classifiers. Figure 3.1 depicts the proposed methodology. Firstly, we carry out feature selection by using a MOGA. It gets as inputs a trained classifier and its respective data set. Since the algorithm aims at minimizing two criteria during the search[1], it will produce at the end a 2-dimensional Pareto-optimal front, which contains a set of classifiers (trade-offs between the criteria being optimized). The final step of this first level consists in training such classifiers.

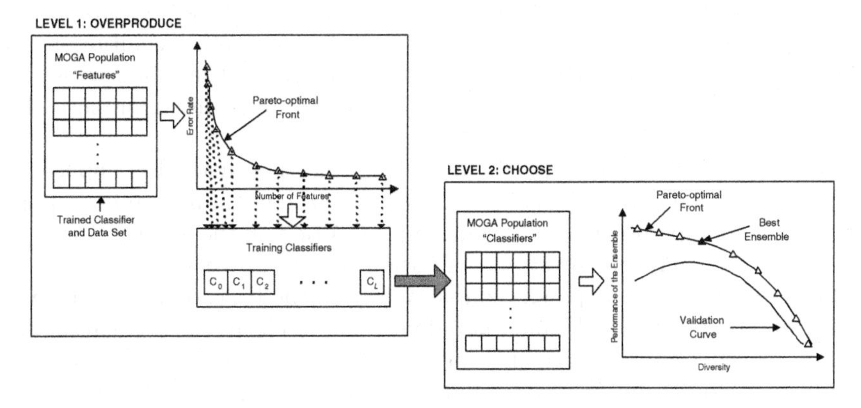

Fig. 3.1. An overview of the proposed methodology.

Once the set of classifiers have been trained, the second level is suggested to pick the members of the team which are most diverse and accurate. Let $A = C_1, C_2, \ldots, C_L$ be a set of L classifiers extracted from the Pareto-optimal and B a chromosome of size L of the population. The relationship between A and B is straightforward, i.e., the gene i of the chromosome B is represented by the classifier C_i from A. Thus, if a chromosome has all bits selected, all classifiers of A will be included in the ensemble. Therefore, the algorithm will produce a 2-dimensional Pareto-optimal front which is composed of several ensembles (trade-offs between accuracy and diversity). In order to choose the best one, we use a validation set, which points out the most diverse and accurate team among all. Later in this paper, we will discuss the issue of using diversity to choose the best ensemble.

[1] Error rate and number of features in the case of supervised feature selection and a clustering index and the number of features in the case of unsupervised feature selection.

In both cases, MOGAs are based on bit representation, one-point crossover, and bit-flip mutation. In our experiments, MOGA used is a modified version of the Non-dominated Sorting Genetic Algorithm (NSGA) [4] with elitism.

3.4 Multi-Objective Genetic Algorithm

Since the concept of multi-objective genetic algorithm (MOGA) will be explored in the remaining of this paper, this section briefly introduces it.

A general multi-objective optimization problem consists of a number of objectives and is associated with a number of inequality and equality constraints. Solutions to a multi-objective optimization problem can be expressed mathematically in terms of nondominated points, i.e., a solution is dominant over another only if it has superior performance in all criteria. A solution is said to be Pareto-optimal if it cannot be dominated by any other solution available in the search space. In our experiments, the algorithm adopted is the Non-dominated Sorting Genetic Algorithm (NSGA) with elitism proposed by Srinivas and Deb in [4, 34].

The idea behind NSGA is that a ranking selection method is applied to emphasize good points and a niche method is used to maintain stable subpopulations of good points. It varies from simple GA only in the way the selection operator works. The crossover and mutation remain as usual. Before the selection is performed, the population is ranked on the basis of an individual's nondomination. The nondominated individuals present in the population are first identified from the current population. Then, all these individuals are assumed to constitute the first nondominated front in the population and assigned a large dummy fitness value. The same fitness value is assigned to give an equal reproductive potential to all these nondominated individuals. In order to maintain the diversity in the population, these classified individuals are made to share their dummy fitness values. Sharing is achieved by performing selection operation using degraded fitness values obtained by dividing the original fitness value of an individual by a quantity proportional to the number of individuals around it. After sharing, these nondominated individuals are ignored temporarily to process the rest of population in the same way to identify individuals for the second nondominated front. These new set of points are then assigned a new dummy fitness value which is kept smaller than the minimum shared dummy fitness of the previous front. This process is continued until the entire population is classified into several fronts.

Thereafter, the population is reproduced according to the dummy fitness values. A stochastic remainder proportionate selection is adopted here. Since individuals in the first front have the maximum fitness value, they get more copies than the rest of the population. The efficiency of NSGA lies in the way multiple objectives are reduced to a dummy fitness function using nondominated sorting procedures. More details about NSGA can be found in [4].

3.5 Classifiers and Feature Sets

As stated before, we have carried out experiments in both supervised and unsupervised contexts. The remaining of this section describes the feature sets and classifiers we have used.

3.5.1 Supervised Context

To evaluate the proposed methodology in the supervised context, we have used three base classifiers trained to recognize handwritten digits of NIST SD19. Such classifiers were trained with three well-known feature sets: Concavities and Contour (CC*sc*), Distances (DDD*sc*), and Edge Maps (EM*sc*) [29].

All classifiers here are MLPs trained with the gradient descent applied to a sum-of-squares error function. The transfer function employed is the familiar sigmoid function. In order to monitor the generalization performance during learning and terminate the algorithm when there is no longer an improvement, we have used the method of cross-validation. Such a method takes into account a validation set, which is not used for learning, to measure the generalization performance of the network. During learning, the performance of the network on the training set will continue to improve, but its performance on the validation set will only improve to a point, where the network starts to overfit the training set, that the learning algorithm is terminated. All networks have one hidden layer where the units of input and output are fully connected with units of the hidden layer, where the number of hidden units were determined empirically (see Table 3.1). The learning rate and the momentum term were set at high values in the beginning to make the weights quickly fit the long ravines in the weight space, then these parameters were reduced several times according to the number of iterations to make the weights fit the sharp curvatures.

Among the different strategies of rejection we have tested, the one proposed by Fumera et al [8] provided the better error-reject trade-off for our experiments. Basically, this technique suggests the use of multiple reject thresholds for the different data classes $(T_0, \ldots, T_n)$ to obtain the optimal decision and reject regions. In order to define such thresholds we have developed an iterative algorithm, which takes into account a decreasing function of the threshold variables $R(T_0, \ldots, T_n)$ and a fixed error rate T_{error}. We start from all threshold values equal to 1, i.e., the error rate equal to 0 since all images are rejected. Then, at each step, the algorithm decreases the value of one of the thresholds in order to increase the accuracy until the error rate exceeds T_{error}.

The training (TRDB*sc*) and validation (VLDB1*sc*) sets are composed of 195,000 and 28,000 samples from hsf_0123 series respectively while the test set (TSDB*sc*) is composed of 30,089 samples from the hsf_7. We consider also a second validation set (VLDB2*sc*), which is composed of 30,000 samples of hsf_7. This data is used to select the best ensemble of classifiers. Figure 3.2 shows the performance on the test set of all classifiers for error rates varying

Table 3.1. Description and performance of the classifiers on TSDB (zero-rejection level).

Feature Set	Number. of Features	Units in the Hidden Layer	Rec. Rate (%)
CC*sc*	132	80	99.13
DDD*sc*	96	60	98.17
EM*sc*	125	70	97.04

from 0.10 to 0.50%, while Table 3.1 reports the performance of all classifiers at zero-rejection level. The curve depicted in Figure 3.2 is much more meaningful when dealing with real applications since they describe the recognition rate in relation to a specific error rate, including implicitly a corresponding reject rate. This rate also allows us to compute the reliability of the system for a given error rate. It can be done by using Equation 3.1.

$$\text{Reliability} = \frac{\text{Rec.Rate}}{\text{Rec.Rate} \quad + \quad \text{Error Rate}} \times 100 \tag{3.1}$$

Figure 3.2 corroborates that recognition of handwritten digits is still an open problem when very low error rates are required. Consider for example our best classifier, which reaches 99.13% at zero-rejection level on the test set. If we allow an error rate of 0.1%, i.e., just one error in 1,000, the recognition rate of such classifier drops from 99.13% to 91.83%. This means that we have to reject 8.07% to get 0.1% of error (Figure 3.2). We will demonstrate that the ensemble of classifiers can significantly improve the performance of the classifiers for low error rates.

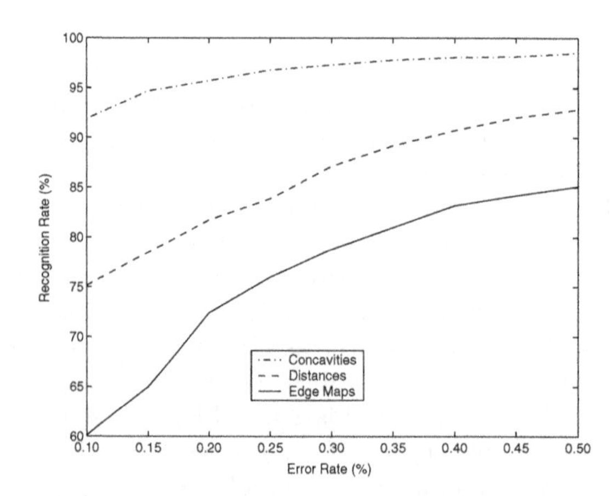

Fig. 3.2. Performance of the classifiers on the test set for error rates varying from 0.10 to 0.50%.

3.5.2 Unsupervised Context

To evaluate the proposed methodology in unsupervised context we have used three HMM-based classifiers trained to recognize handwritten Brazilian month words ("Janeiro", "Fevereiro", "Março", "Abril", "Maio", "Junho", "Julho", "Agosto", "Setembro", "Outubro", "Novembro", "Dezembro"). The training (TRDBuc), validation (VLDB1uc), and testing (TSDBuc) sets are composed of 1,200, 400, and 400 samples, respectively. In order to increase the training and validation sets, we have also considered 8,300 and 1,900 word images, respectively, extracted from the legal amount database. This is possible because we are considering character models. We consider also a second validation set (VLDB2uc) of 500 handwritten Brazilian month words [14]. Such data is used to select the best ensemble of classifiers.

Given a discrete HMM-based approach, each word image is transformed as a whole into a sequence of observations by the successive application of preprocessing, segmentation, and feature extraction. Preprocessing consists of correcting the average character slant. The segmentation algorithm uses the upper contour minima and some heuristics to split the date image into a sequence of segments (graphemes), each of which consists of a correctly segmented, an under-segmented, or an over-segmented character. A detailed description of the preprocessing and segmentation stages is given in [28].

The word models are formed by the concatenation of appropriate elementary HMMs, which are built at letter and space levels. The topology of space model consists of two states linked by two transitions that encode a space or no space. Two topologies of letter models were chosen based on the output of our grapheme-based segmentation algorithm which may produce a correct segmentation of a letter, a letter under-segmentation or a letter over-segmentation into two, three, or four graphemes depending on each letter. In order to cope with these configurations of segmentations, we have designed topologies with three different paths leading from the initial state to the final state.

Considering uppercase and lowercase letters, we need 42 models since the legal amount alphabet is reduced to 21 letter classes and we are not considering the unused ones. Thus, regarding the two topologies, we have 84 HMMs which are trained using the Baum-Welch algorithm with the Cross-Validation procedure.

Since no information on recognition is available on the writing style (uppercase, lowercase), the word model consists of two letter HMMs in parallel and four space HMMs linked by four transitions: two uppercase-letters (UU), two lowercase-letters (LL), one uppercase letter followed by one lowercase-letter (UL), and one lowercase letter followed by one uppercase-letter (LU). The probabilities of these transitions are estimated by their frequency of occurrence in the training set. In the same manner, the probabilities of beginning a word by an uppercase-letter (0U) or a lowercase letter (0L) are also estimated in the training set. This architecture handles the problem related to the mixed handwritten words detecting implicitly the writing style during recognition using the Backtracking of the Viterbi algorithm.

The feature set that feeds the first classifier is a mixture of concavity and contour features (CCuc) [29]. In this case, each grapheme is divided into two equal zones (horizontal) where for each region a concavity and contour feature vector of 17 components is extracted. Therefore, the final feature vector has 34 components. The other two classifiers make use of a feature set based on distances [28]. The former uses the same zoning discussed before (two equal zones), but in this case, for each region a vector of 16 components is extracted. This leads to a final feature vector of 32 components (DDD$32_u c$). For the latter we have tried a different zoning. The grapheme is divided into four zones using the reference baselines, hence, we have a final feature vector composed of 64 components (DDD$64_u c$). Table 3.2 reports the performance of all classifiers on the test set at zero-rejection level. Figure 3.3 shows the performance of all classifiers for error rates varying from 1% to 4%. The strategy for rejection used in this case is the one discussed previously. We have chosen higher error rates in this case due to the size of the database we are dealing with.

Table 3.2. Performance of the classifiers on the test set.

Feature Set	Number of Features	Codebook Size	Rec Rate (%)
CCuc	34	80	86.1
DDD32uc	32	40	73.0
DDD64uc	64	60	64.5

It can be observed from Table 3.3 that the recognition rates with error fixed at 1% are very poor, hence, the number of rejected patterns is very high. We will see in the next sections that the proposed methodology can improve these results considerably.

3.6 Implementation

This section introduces how we have implemented both levels of the proposed methodology. First we discuss the supervised context and then the unsupervised.

3.6.1 Supervised Context

Supervised Feature Subset Selection

The feature selection algorithm used in here was introduced in [30]. To make this paper self-contained, a brief description is included in this section.

Regarding feature selection algorithms, they can be classified into two categories based on whether or not feature selection is performed independently of the learning algorithm used to construct the classifier. If feature selection is done independently of the learning algorithm, the technique is said to follow a filter approach. Otherwise, it is said to follow a wrapper approach [13]. While the filter approach is generally computationally more efficient than the wrapper approach, its major drawback is that an optimal selection of features may not be independent of the inductive and representational biases of the learning algorithm that is used to construct the classifier. On the other hand, the wrapper approach involves the computational overhead of evaluating candidate feature subsets by executing a given learning algorithm on the database using each feature subset under consideration.

As stated elsewhere, the idea of using feature selection is to promote diversity among the classifiers. To tackle such a task we have to optimize two objective functions: minimization of the number of features and minimization of the error rate of the classifier. Computing the first one is simple, i.e., the number of selected features. The problem lies in computing the second one, i.e., the error rate supplied by the classifier. Regarding a wrapper approach, in each generation, evaluation of a chromosome (a feature subset) requires training the corresponding neural network and computing its accuracy. This evaluation has to be performed for each of the chromosomes in the population. Since such a strategy is not feasible due to the limits imposed by the learning time of the huge training set considered in this work, we have adopted the strategy proposed by Moody and Utans in [26], who use the sensitivity of the network to estimate the relationship between the input features and the network performance.

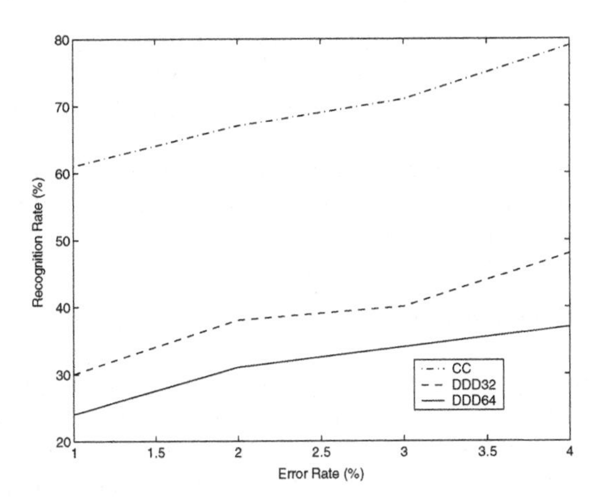

Fig. 3.3. Performance of the classifiers on the test set for error rates varying from 1 to 4%.

The sensitivity of the network model to variable β is defined as:

$$S_\beta = \frac{1}{N}\sum_{j=1}^{N} ASE(\bar{x}_\beta) - ASE(x_\beta) \tag{3.2}$$

with

$$\bar{x}_\beta = \frac{1}{N}\sum_{j=1}^{N} x_{\beta_j} \tag{3.3}$$

where x_{β_j} is the βth input variable of the jth exemplar. S_β measures the effect on the training ASE (average square error) of replacing the βth input x_β by its average $\bar{x}_\beta$. Moody and Utans show that when variables with small sensitivity values with respect to the network outputs are removed, they do not influence the final classification. So, in order to evaluate a given feature subset we replace the unselected features by their averages. In this way, we avoid training the neural network and hence turn the wrapper approach feasible for our problem. We call this strategy modified-wrapper. Such a scheme has been employed also by Yuan et al in [38], and it makes it feasible to deal with huge databases in order to better represent the pattern recognition problem during the fitness evaluation[2]. Moreover it can accommodate multiple criteria such as the number of features and the accuracy of the classifier, and generate the Pareto-optimal front in the first run of the algorithm. Figure 3.4 shows the evolution of the population in the objective plane and its respective Pareto-optimal front.

It can be observed in Figure 3.4b that the Pareto-optimal front is composed of several different classifiers. In order to get a better insight about them, they were classified into 3 different groups: weak, medium, and strong. It can be observed that among all those classifiers there are very good ones. To find out which classifiers of the Pareto-optimal front compose the best ensemble, we carried out a second level of search. Once we did not train the models during the search (the training step is replaced by the sensitivity analysis), the final step of feature selection consists of training the solutions provided by the Pareto-optimal front (3.1).

Choosing the Best Ensemble

As defined in Section 3.3 each gene of the chromosome is represented by a classifier produced in the previous level. Therefore, if a chromosome has all bits selected, all classifiers of will compose the team. In order to find the best ensemble of classifiers, i.e., the most diverse set of classifiers that brings a good generalization, we have used two objective functions during this level

[2] If small databases are considered, then a full-wrapper could replace the proposed modified-wrapper.

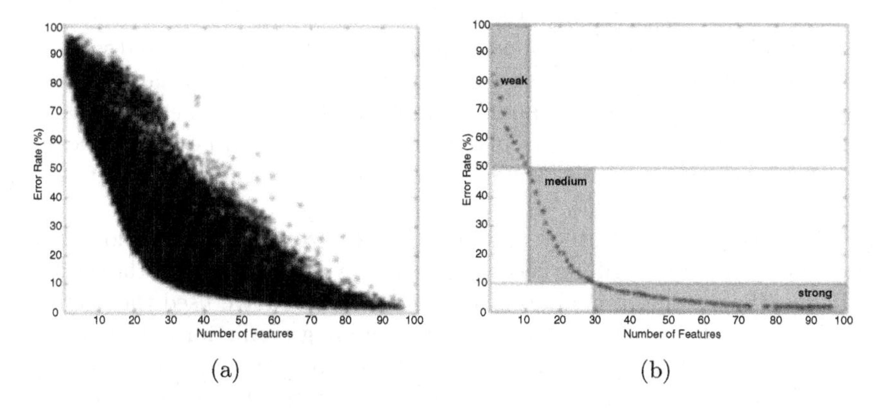

Fig. 3.4. Supervised feature selection using a Pareto-based approach (a) Evolution of the population in the objective plane, (b) Pareto-optimal front and its different classes of classifiers.

of the search, namely, maximization of the recognition rate of the ensemble and maximization of a measure of diversity. We have tried different measures such as overlap, entropy [22], and ambiguity [17]. The results achieved with ambiguity and entropy were very similar. In this work we have used ambiguity as diversity measure. The ambiguity is defined as follows:

$$a_i(x_k) = [V_i(x_k) - \overline{V}(x_k)]^2 \tag{3.4}$$

where a_i is the ambiguity of the ith classifier on the example x_k, randomly drawn from an unknown distribution, while V_i and $\overline{V}$ are the ith classifier and the ensemble predictions, respectively. In other words, it is simply the variance of ensemble around the mean, and it measures the disagreement among the classifiers on input x. Thus the contribution to diversity of an ensemble member i as measured on a set of M samples is:

$$A_i = \frac{1}{M}\sum_{k=1}^{M} a_i(x_k) \tag{3.5}$$

and the ambiguity of the ensemble is

$$\overline{A} = \frac{1}{N}\sum A_i \tag{3.6}$$

where N is the number of classifiers. So, if the classifiers implement the same functions, the ambiguity $\overline{A}$ will be low, otherwise it will be high. In this scenario the error from the ensemble is

$$E = \overline{E} - \overline{A} \tag{3.7}$$

where $\overline{E}$ is the average errors of the single classifiers and $\overline{A}$ is the ambiguity of the ensemble. Equation 3.7 expresses the trade-off between bias and variance in the ensemble, but in a different way than the common bias-variance relation in which the averages are over possible training sets instead of ensemble averages. If the ensemble is strongly biased the ambiguity will be small, because the classifiers implement very similar functions and thus agree in inputs even outside the training set [17].

At this level of the strategy we want to maximize the generalization of the ensemble, therefore, it will be necessary to use a way of combining the outputs of all classifiers to get a final decision. To do this, we have used the average, which is a simple and effective scheme of combining predictions of the neural networks [16]. Other combination rules such as product, min, and max have been tested but the simple average has produced slightly better results. In order to evaluate the objective functions during the search described above we have used the validation set VLDB1*sc*.

3.6.2 Unsupervised Context

Unsupervised Feature Subset Selection

A lot of work done in the field of handwritten word recognition takes into account discrete HMMs as classifiers, which have to be fed with a sequence of discrete values (symbols). This means that before using a continuous feature vector, we must convert it to discrete values. A common way to do that is through clustering. The problem is that for the most of real-life situations we do not know the best number of clusters, what makes it necessary to explore different numbers of clusters using traditional clustering methods such as the K-means algorithm and its variants. In this light, clustering can become a trial-and-error work. Besides, its result may not be very promising especially when the number of clusters is large and not easy to estimate.

Unsupervised feature selection emerges as a clever solution to this problem. The literature contains several studies on feature selection for supervised learning, but only recently, the feature selection for unsupervised learning has been investigated [5, 15]. The objective in unsupervised feature selection is to search for a subset of features that best uncovers "natural" groupings (clusters) from data according to some criterion. In this way, we can avoid the manual process of clustering and find the most discriminative features in the same time. Hence, we will have at the end a more compact and robust high-level representation (symbols).

In the above context, unsupervised feature selection also presents a multi-criterion optimization function, where the objective is to find compact and well separated hyper-spherical clusters in the feature subspaces. Differently of the supervised feature selection, here the criteria optimized by the algorithm are a validity index and the number of features. [27].

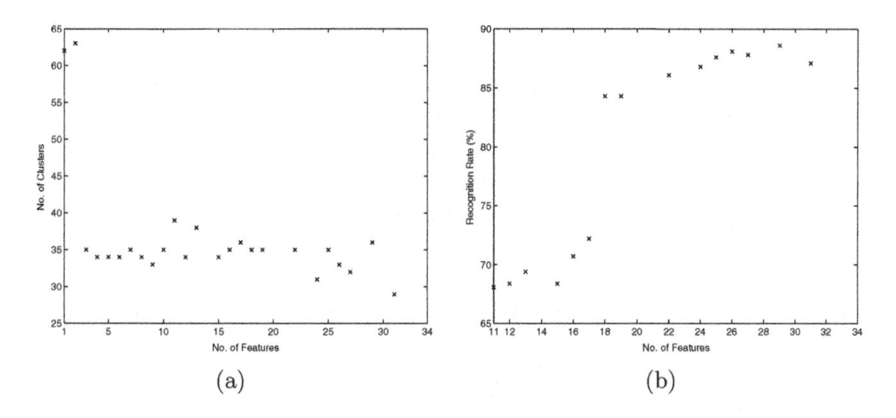

Fig. 3.5. (a) Relationship between the number of clusters and the number of features and (b) Relationship between the recognition rate and the number of features.

In order to measure the quality of clusters during the clustering process, we have used the Davies-Bouldin (DB)-index [3] over 80,000 feature vectors extracted from the training set of 9,500 words. To make such an index suitable for our problem, it must be normalized by the number of selected features. This is due to the fact that it is based on geometric distance metrics and therefore, it is not directly applicable here because it is biased by the dimensionality of the space, which is variable in feature selection problems.

We have noticed that the value of DB index decreases as the number of features increases. We have correlated this effect with the normalization of DB-index by the number of features. In order to compensate this, we have considered as second objective the minimization of the number of features. In this case, one feature must be set at least. Figure 3.5 depicts the relationship between the number of clusters and number of features and the relationship between the recognition rate on the validation set and the number of features.

Like in the supervised context, here we also divided the classifiers of the Pareto into classes. In this case, we have realized that those classifiers with very few features are not selected to compose the ensemble, and therefore, just the classifiers with more than 10 features were used into the second level of search. In Section 3.7.2 we discuss this issue in more detail. The way of choosing the best ensemble is exactly the same as introduced in Section 3.6.1.

3.7 Experimental Results

All experiments in this work were based on a single-population master-slave MOGA. In this strategy, one master node executes the genetic operators (selection, crossover and mutation), and the evaluation of fitness is distributed among several slave processors. We have used a Beowulf cluster with 17 (one master and 16 slaves) PCs (1.1Ghz CPU, 512Mb RAM) to execute our experiments.

The following parameter settings were employed in both levels: population size = 128, number of generations = 1000, probability of crossover = 0.8, probability of mutation = $1/L$ (where L is the length of the chromosome), and niche distance (σ_{share}) = [0.25,0.45]. The length of the chromosome in the first level is the number of components in the feature set (see Table 3.1), while in the second level is the number of classifiers picked from the Pareto-optimal front in the previous level.

In order to define the probabilities of crossover and mutation, we have used the one-max problem, which is probably the most frequently-used test function in research on genetic algorithms because of its simplicity [2]. This function measures the fitness of an individual as the number of bits set to one on the chromosome. We have used a standard genetic algorithm with a single-point crossover and the maximum generations of 1000. The fixed crossover and mutation rates are used in a run, and the combination of the crossover rates 0.0, 0.4, 0.6, 0.8 and 1.0 and the mutation rates of $0.1/L$, $1/L$ and $10/L$, where L is the length of the chromosome. The best results were achieved with $P_c = 0.8$ and $P_m = 1/L$. Such results confirmed the values reported by Miki et al in [25]. The parameter σ_{share} was tuned empirically.

3.7.1 Experiments in the Supervised Context

Once all parameters have been defined, the first step, as described in Section 3.6.1, consists of performing feature selection for a given feature set. As depicted in Figure 3.4, this procedure produces quite a large number of classifiers, which should be trained for use in the second level. After some experiments, we found out that the second level always chooses "strong" classifiers to compose the ensemble. Thus, in order to speed up the training process and the second level of search as well, we decide to train and use in the second level just the "strong" classifiers. This decision was made after we realized that in our experiments the "weak" and "medium" classifiers did not cooperate with the ensemble at all. To train such classifiers, the same databases reported in Section 3.5.1 were used. Table 3.3 summarizes the "strong" classifiers produced by the first level for the three feature sets we have considered.

Table 3.3. Summary of the classifiers produced by the first level.

Feature Set	No. of Classifiers	Range of Features	Range of Rec. Rates (%)
CC*sc*	81	24-125	90.5 - 99.1
DDD*sc*	54	30-84	90.6 - 98.1
EM*sc*	78	35-113	90.5 - 97.0

Considering for example the feature set CC_{sc}, the first level of the algorithm provided 81 "strong" classifiers which have the number of features ranging from 24 to 125 and recognition rates ranging from 90.5% to 99.1% on $TSDB_{sc}$. This shows the great diversity of the classifiers produced by the feature selection method. Based on the classifiers reported in Table 3.3 we define four sets of base classifiers as follows: $S_1 = \{CCsc_0, \ldots, CCsc_{80}\}$, $S_2 = \{DDDsc_0, \ldots, DDDsc_{53}\}$, $S_3 = \{EMsc_0, \ldots, EMsc_{77}\}$, and $S_4 = \{S_1 \bigcup S_2 \bigcup S_3\}$. All these sets could be seen as ensembles, but in this work we reserve the word ensemble to characterize the results yielded by the second-level of the algorithm. In order to assess the objective functions of the second-level of the algorithm (generalization of the ensemble and diversity) we have used the validation set (VLDB1sc).

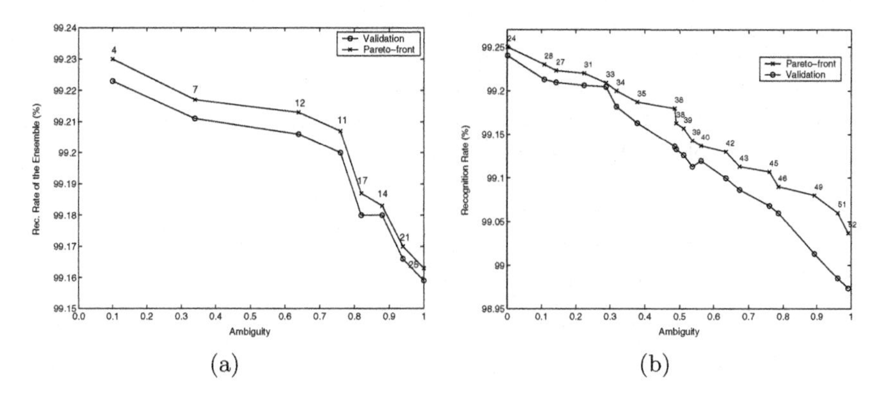

Fig. 3.6. The Pareto-optimal front produced by the second-level MOGA: (a) S_1 and (b) S_4

Like the first level, the second one also generates a set of possible solutions which are the trade-offs between the generalization of the ensemble and its diversity. Thus the problem now lies in choosing the most accurate ensemble among all. Due to the limited space we have, Figure 3.6 only depicts the variety of ensembles yielded by the second-level of the algorithm for S_1 and S_4. The number over each point stands for the number of classifiers in the ensemble. In order to decide which ensemble to choose we validate the Pareto-optimal front using VLDB2sc, which was not used so far. Since we are aiming at performance, the direct choice will be the ensemble that provides better generalization on $VLDB2_{sc}$. Table 3.4 summarizes the best ensembles produced for the four sets of base classifiers and their performance at zero-rejection level on the test set. For facility, we reproduce in this table the results of the original classifiers.

We can notice from Table 3.4 that the ensembles and base classifiers have very similar performance at zero-rejection level. On the other hand, Figure

Table 3.4. Performance of the ensembles on the test set.

Feature Set	Number of Classifiers	Rec. Rate (%) zero-rejection level	Rec. Rate (%) Original Classifiers
S_1	4	99.22	99.13
S_2	4	98.18	98.17
S_3	7	97.10	97.04
S_4	24	99.25	

3.7 shows that the ensembles respond better for error rates fixed at very low levels than single classifiers. The most expressive result was achieved for the ensemble S_3, which attains a reasonable performance at zero-rejection level but performs very poorly at low error rates. In such a case, the ensemble of classifiers brought an improvement of about 8%. We have noticed that the ensemble reduces the high outputs of some outliers so that the threshold used for rejection can be reduced and consequently the number of samples rejected is reduced. Thus, aiming for a small error rate we have to consider the important role of the ensemble.

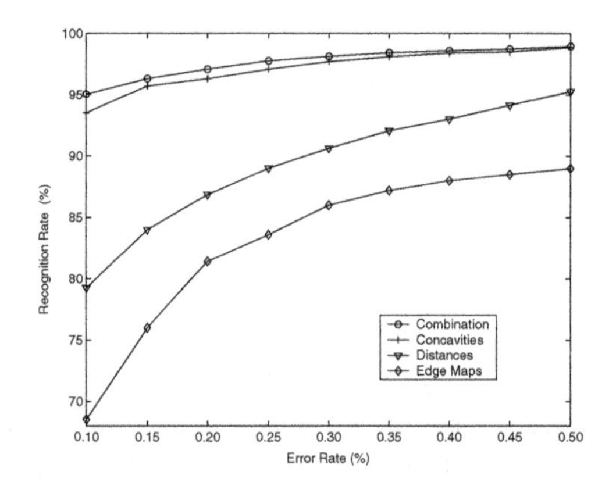

Fig. 3.7. Improvements yielded by the ensembles.

Regarding the ensemble S_4, we can notice that it achieves a performance similar to S_1 at zero-rejection level (see Table 3.4). Besides, it is composed of 24 classifiers, against four of S_1. The fact worths noting though, is the performance of S_4 at low error rates. For the error rate fixed at 1% it reached 95.0% against 93.5% of S_1. S_4 is composed of 14, 6, and 4 classifiers from S_1, S_2, and S_3, respectively. This emphasizes the ability of the algorithm in finding good ensembles when more original classifiers are available.

3.7.2 Experiments in the Unsupervised Context

The experiments in the unsupervised context follow the same vein of the supervised one. As discussed in Section 3.6.2, the main difference lies in the way the feature selection is carried out. In spite of that, we can observe that the number of classifiers produced during unsupervised feature selection is quite large as well. In light of this, we have applied the same strategy of dividing the classifiers into groups (see Figure 3.5). After some experiments, we found out that the second level always chooses "strong" classifiers to compose the ensemble. Thus, in order to speed up the training process and the second level of search as well, we decide to train and use in the second level just "strong" classifiers. To train such classifiers, the same databases reported in Section 3.5.2 were considered. Table 3.5 summarizes the "strong" classifiers (after training) produced by the first level for the three feature sets we have considered.

Table 3.5. Summary of the classifiers produced by the first level.

Feature Set	Number of Classifiers	Range of Features	Range of Codebook	Range of Rec. Rates (%)
CC*uc*	15	10-32	29-39	68.1 - 88.6
DDD32*uc*	21	10-31	20-30	71.7 - 78.0
DDD64*uc*	50	10-64	52-80	60.6 - 78.2

Considering for example the feature set CC*uc*, the first level of the algorithm provided 15 "strong" classifiers which have the number of features ranging from 10 to 32 and recognition rates ranging from 68.1% to 88.6% on VLDB1*uc* . This shows the great diversity of the classifiers produced by the feature selection method. Based on the classifiers reported in Table 3.5 we define four sets of base classifiers as follows: $F_1 = \{CCuc_0, \ldots, CCuc_{14}\}$, $F_2 = \{DDD32uc_0, \ldots, DDD32uc_{20}\}$, $F_3 = \{DDD64uc_0, \ldots, DDD64uc_{49}\}$, and $F_4 = \{F_1 \bigcup F_2 \bigcup F_3\}$.

Again, due to the limited space we have, Figure 3.8 only depicts the variety of ensembles yielded by the second-level of the algorithm for F_2 and F_4. The number over each point stands for the number of classifiers in the ensemble. Like in the previous experiments, the second validation set (VLDB2*uc*) was used to select the best ensemble. After selecting the best ensemble the final step is to assess them on the test set. Table 3.6 summarizes the performance of the ensembles on the test set. For the sake of comparison, we reproduce in Table 3.6 the results presented in Table 3.2.

Figure 3.8b shows the performance of the ensembles generated with all base classifiers available, i.e., Ensemble F_4. Like in the previous experiments (supervised context), the result achieved by the ensemble F_4 shows the ability of the algorithm in finding good ensembles when more base classifiers are

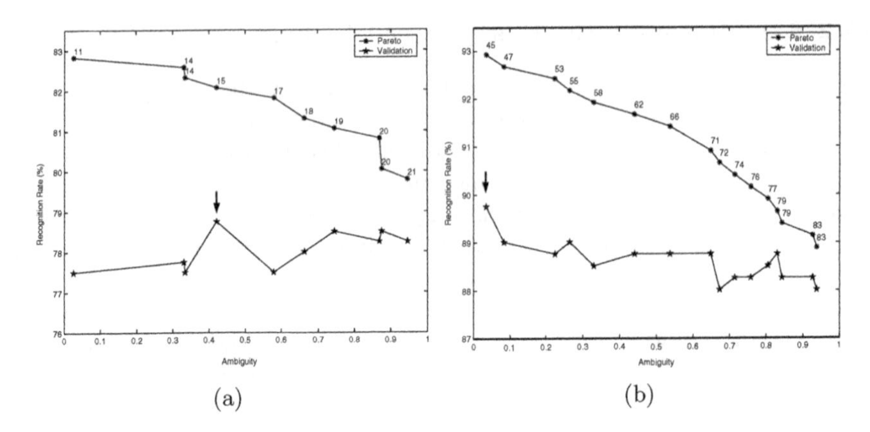

Fig. 3.8. The Pareto-optimal front (and validation curves where the best solutions are highlighted with an arrow) produced by the second-level MOGA: (a) F_2 and (b) F_4.

Table 3.6. Comparison between ensembles and original classifiers.

Base Classifiers	Number of Classifiers	Rec. Rate (%)	Original Feature Set	Rec. Rate (%)
F_1	10	89.2	CC	86.1
F_2	15	80.2	DDD32	73.0
F_3	36	80.7	DDD64	64.5
F_4	45	90.2		

considered. The ensemble F_4 is composed of 9, 11, and 25 classifiers from F_1, F_2, and F_3, respectively.

In light of this, we decided to introduce a new feature set, which, based on our experience, has a good discrimination power when combined with other features such as concavities. This feature set, which we call "global features", is composed of primitives such as ascenders, descenders, and loops. The combination of these primitives plus a primitive that determines whether a grapheme does not contain ascender, descender, and loop produces a 20-symbol alphabet. For more details, see Ref. [28]. In order to train the classifier with this feature set, we have used the same databases described in Section 3.5.2. The recognition rates at zero-rejection level are 86.1% and 87.2% on validation and testing sets, respectively. This performance compares with the CC*uc* classifier.

Since we have a new base classifier, our sets of base classifiers must be modified to cope with it. Thus, $F_{1G} = \{F_1 \bigcup G\}$, $F_{2G} = \{F_2 \bigcup G\}$, $F_{3G} = \{F_3 \bigcup G\}$, and $F_{4G} = \{F_1 \bigcup F_2 \bigcup F_3 \bigcup G\}$. In such cases, G stands for the classifier trained with global features. Table 3.7 summarizes the ensembles found using these new sets of base classifiers. It is worthy of remark the reduction of the size of the teams. This shows the ability of the algorithm

Table 3.7. Performance of the ensembles with global features.

Base Classifiers	Number of Classifiers	Rec. Rate (%) Testing
F_{1G}	2	92.2
F_{2G}	2	89.7
F_{3G}	7	85.5
F_{4G}	23	92.0

in finding not just diverse but also uncorrelated classifiers to compose the ensemble [36]. Besides, it corroborates to our claim that the classifier G when combined with other features bring an improvement to the performance.

In Figure 3.9 we compare the error-reject trade-offs for some ensembles reported in Table 3.7. Like the results at zero-rejection level, the improvement observed here also are quite impressive. Table 3.7 shows that F_{1G} and F_{4G} reach similar results on the test set at zero-rejection level, however, F_{1G} contains just two classifiers against 23 of F_{4G}. On the other hand, the latter features a slightly better error-reject trade-off in the long run (Figure 3.9b).

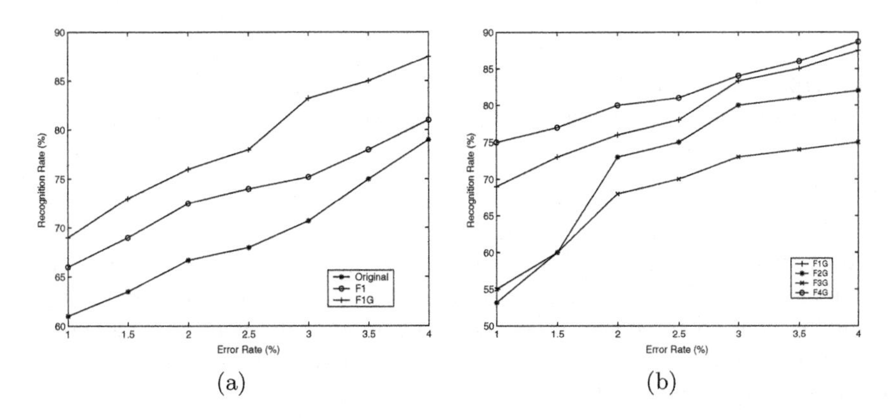

Fig. 3.9. Improvements yielded by the ensembles: (a) F_1 and (b) Comparison among all ensembles.

Based on the experiments reported so far we can affirm that the unsupervised feature selection is a good strategy to generate diverse classifiers. This is made very clear in the experiments regarding the feature set DDD64. In such a case, the original classifier has a poor performance (about 65% on the test set), but when it is used to generate the set of base classifiers, the second-level MOGA was able to produce a good ensemble by maximizing the performance and the ambiguity measure. Such an ensemble of classifiers brought an improvement of about 15% in the recognition rate at zero-rejection level.

3.8 Discussion

The results obtained here attest that the proposed strategy is able to generate a set of good classifiers in both supervised and unsupervised contexts. To better evaluate our results, we have used two traditional ensemble methods (Bagging and Boosting) in the supervised context. Figure 3.10 reports the results. As we can see, the proposed methodology achieved better results, especially when considering very low error rates.

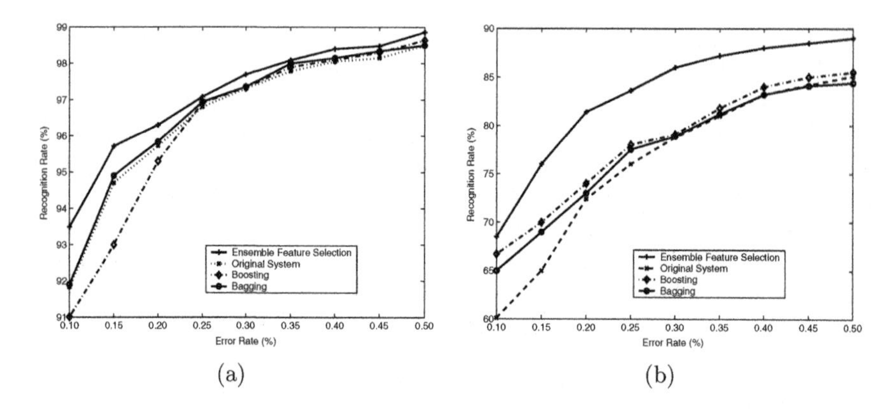

Fig. 3.10. Comparison among feature selection for ensembles, bagging, and boosting for the two feature sets used in the supervised context:(a) CC_{sc} and (b) EM_{sc}

Diversity is an issue that deserves some attention when discussing ensemble of classifiers. As we have mentioned before, some authors advocated that diversity does not help at all. In our experiments, most of the time, the best ensembles of the Pareto-optimal also were the best for the unseen data. This could lead one to agree that diversity is not important when building ensembles, since even using a validation set the selected team is always the most accurate and with less diversity.

However, if we look carefully the results, we will observe that there are cases where the validation curve does not have the same shape of the Pareto-optimal. In such cases diversity is very useful to avoid selecting overfitted solutions.

One can argue that using a single GA and considering the entire final population, perhaps the similar solutions found in the Pareto-optimal produced by the MOGA will be there. To show that it does not happen, we have carried out some experiments with a single GA where the fitness function was the maximization of the ensemble´s accuracy. Since a single-objective optimization algorithm searches for an optimum solution, it is natural to expect that it will converge towards the fittest solution, hence, the diversity of solutions presented in the Pareto-optimal is not present in the final population of the single genetic algorithm.

To illustrate that, we present the results we got using a GA to find ensemble in F_2 (unsupervised context). The parameters used here are the same we have used for the MOGA (Section 3.7). Figure 3.11a plots all the classifiers found in the final population of the genetic algorithm. For the sake of comparison we reproduce Figure 3.8a in Figure 3.11b. As we can see, the population is very homogeneous and it converged, as expected, towards the most accurate ensemble.

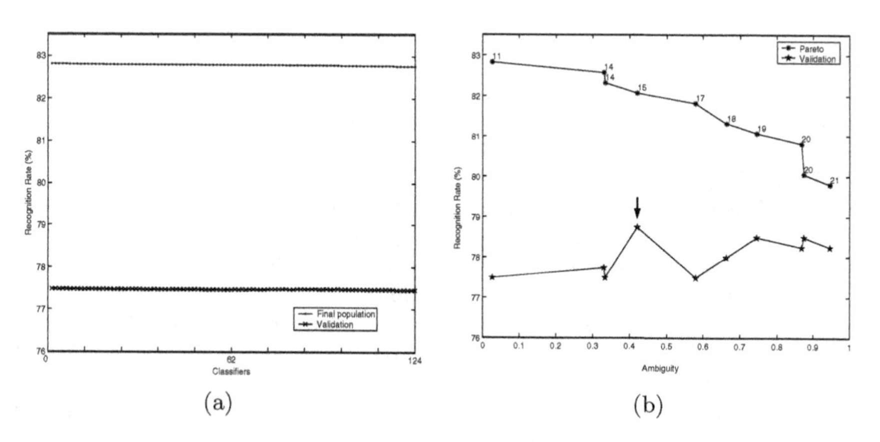

Fig. 3.11. Benefits of using diversity: (a) population (classifiers) of the final generation of the GA and (b) classifiers found by the MOGA.

Some attempts in this direction were made by Optiz [31]. He combined accuracy and diversity through the weighted-sum approach. As stated somewhere, when dealing with this kind of combination, one should deal with problems such as scaling and sensitivity towards the weights. We believe that our strategy offers a clever way to find the ensemble using genetic algorithms.

3.9 Conclusion

We have described a methodology for ensemble creation underpinned on the paradigm "overproduce and choose". It takes two levels of search where the first level overproduces a set of classifiers by performing feature selection while the second one chooses the best team of classifiers.

The feasibility of the strategy was demonstrated through comprehensive experiments carried out in the context of handwriting recognition. The idea of generating classifiers through feature selection was proved to be successful in both supervised and unsupervised contexts. The results attained in both situations and using different feature sets and base classifiers demonstrated the efficiency of the proposed strategy by finding powerful ensembles, which succeed in improving the recognition rates for classifiers working with a very low error rates. Such results compare favorably to traditional ensemble methods such as Bagging and Boosting.

Finally we have addressed the issue of using diversity to build ensembles. As we have seen, using diversity jointly with the accuracy of the ensemble as selection criterion might be very helpful to avoid choosing overfitted solutions. Our results certainly bring some contribution to the field, but this still is an open problem.

References

[1] L. Breiman. Stacked regressions. *Machine Learning*, 24(1):49–64, 1996.

[2] E. Cantu-Paz. *Efficient and Accurate Parallel Genetic Algorithms.* Kluwer Academic Publishers, 2000.

[3] D. L. Davies and D. W. Bouldin. A cluster separation measure. *IEEE Trans. on Pattern Analysis and Machine Intelligence*, 1(224-227):550–554, 1979.

[4] K. Deb. *Multi-Objective Optimization using Evolutionary Algorithms.* John Wiley and Sons Ltd, 2nd edition, April 2002.

[5] J. G. Dy and C. E. Brodley. Feature subset selection and order identification for unsupervised learning. In *Proc. 17th International Conference on Machine Learning*, 2000.

[6] B. Efron and Tibshirani R. *An introduction to the Bootstrap.* Chapman and Hall, 1993.

[7] Y. Freund and R. Schapire. Experiments with a new boosting algorithm. In *Proc. of* 13th *International Conference on Machine Learning*, pages 148–156, Bary-Italy, 1996.

[8] G. Fumera, F. Roli, and G. Giacinto. Reject option with multiple thresholds. *Pattern Recognition*, 33(12):2099–2101, 2000.

[9] G. Giacinto and F. Roli. Design of effective neural network ensemble for image classification purposes. *Image Vision and Computing Journal*, 9-10:697–705, 2001.

[10] S. Gunter and H. Bunke. Creation of classifier ensembles for handwritten word recogntion using feature selection algorithms. In *Proc. of* 8th *IWFHR*, pages 183–188, Niagara-on-the-Lake, Canada, 2002.

[11] S. Hashem. Optimal linear combinations of neural networks. *Neural Networks*, 10(4):599–614, 1997.

[12] T. K. Ho. The random subspace method for constructing decision forests. *IEEE Trans. on Pattern Analysis and Machine Intelligence*, 20(8):832–844, 1998.

[13] G. John, R. Kohavi, and K. Pfleger. Irrelevant features and the subset selection problems. In *Proc. of* 11th *International Conference on Machine Learning*, pages 121–129, 1994.

[14] J. J. Oliveira Jr., J. M. Carvalho, C. O. A. Freitas, and R. Sabourin. Evaluating NN and HMM classifiers for handwritten word recognition. In *Proceedings of the 15th Brazilian Symposium on Computer Graphics and Image Processing*, pages 210–217. IEEE Computer Society, 2002.

[15] Y. S. Kim, W. N. Street, and F. Menczer. Feature selection in unsupervised learning via evolutionary search. In *Proc. 6th ACM SIGKDD International Conference on Knowledge Discovery and Data Mining*, pages 365–369, 2000.
[16] J. Kittler, M. Hatef, R. Duin, and J. Matas. On combining classifiers. *IEEE Trans. on Pattern Analysis and Machine Intelligence*, 20(3):226–239, 1998.
[17] A. Krogh and J. Vedelsby. Neural networks ensembles, cross validation, and active learning. In G.Tesauro et al, editor, *Advances in Neural Information Processing Systems 7*, pages 231–238. MIT Press, 1995.
[18] M. Kudo and J. Sklansky. Comparision of algorithms that select features for pattern classifiers. *Pattern Recognition*, 33(1):25–41, 2000.
[19] L. Kuncheva. That elusive diversity in classifier ensembles. In *Proc. of ibPRIA, LNCS 2652*, pages 1126–1138, Mallorca, Spain, 2003.
[20] L. Kuncheva, J. C. Bezdek, and R. P. W. Duin. Decision templates for multiple classifier fusion: An experimental comparison. *Pattern Recognition*, 34(2):299–314, 2001.
[21] L. Kuncheva and L. C. Jain. Designing classifier fusion systems by genetic algorithms. *IEEE Trans. on Evolutionary Computation*, 4(4):327–336, 2000.
[22] L. I. Kuncheva and C. J. Whitaker. Ten measures of diversity in classifier ensembles:limits for two classifiers. In *Proc. of IEE Workshop on Intelligent Sensor Processing*, pages 1–10, 2001.
[23] L. I. Kuncheva and C. J. Whitaker. Measures of diversity in classifier ensembles. *Machine Learning*, 51:181–207, 2003.
[24] M. Last, H. Bunke, and A. Kandel. A feature-based serial approach to classifier combination. *Pattern Analysis and Applications*, 5:385–398, 2002.
[25] M. Miki, T. Hiroyasu, K. Kaneko, and K. Hatanaka. A parallel genetic algorithm with distributed environment scheme. In *Proc. of International Conference on System, Man, and Cybernetics*, volume 1, pages 695–700, 1999.
[26] J. Moody and J. Utans. Principled architecture selection for neural networks: Application to corporate bond rating prediction. In J. Moody, S. J. Hanson, and R. P. Lippmann, editors, *Advances in Neural Information Processing Systems 4*. Morgan Kaufmann, 1991.
[27] M. Morita, R. Sabourin, F. Bortolozzi, and C. Y. Suen. Unsupervised feature selection using multi-objective genetic algorithms for handwritten word recognition. In *Proceedings of the 7th International Conference on Document Analysis and Recognition*, pages 666–670. IEEE Computer Society, 2003.
[28] M. Morita, R. Sabourin, F. Bortolozzi, and Suen C. Y. Segmentation and recognition of handwritten dates: An hmm-mlp hybrid approach. *International Journal on Document Analysis and Recognition*, 6:248–262, 2003.
[29] L. S. Oliveira, R. Sabourin, F. Bortolozzi, and C. Y. Suen. Automatic recognition of handwritten numerical strings: A recognition and verification strategy. *IEEE Trans. on Pattern Analysis and Machine Intelligence*, 24(11):1438–1454, 2002.
[30] L. S. Oliveira, R. Sabourin, F. Bortolozzi, and C. Y. Suen. A methodology for feature selection using multi-objective genetic algorithms for handwritten digit string recognition. *International Journal of Pattern Recognition and Artificial Intelligence*, 17(6):903–930, 2003.
[31] D. W. Optiz. Feature selection for ensembles. In *Proc. of 16th International Conference on Artificial Intelligence*, pages 379–384, 1999.

[32] D. Partridge and W. B. Yates. Engineering multiversion neural-net systems. *Neural Computation*, 8(4):869–893, 1996.
[33] D. Ruta. Multilayer selection-fusion model for pattern classification. In *Proceedings of the IASTED Artificial Intelligence and Application Conference*, Insbruck, Austria, 2004.
[34] N. Srinivas and K. Deb. Multiobjective optimization using nondominated sorting in genetic algorithms. *Evolutionary Computation*, 2(3):221–248, 1995.
[35] A. Tsymbal, S. Puuronen, and D. W. Patterson. Ensemble feature selection with the simple Bayesian classification. *Information Fusion*, 4:87–100, 2003.
[36] K. Tumer and J. Ghosh. Error correlation and error reduction in ensemble classifiers. *Connection Science*, 8(3-4):385–404, 1996.
[37] K. Tumer and N. C. Oza. Input decimated ensembles. *Pattern Analysis and Applications*, 6:65–77, 2003.
[38] H. Yuan, S. S. Tseng, W. Gangshan, and Z. Fuyan. A two-phase feature selection method using both filter and wrapper. In *Proc. of IEEE International Conference on Systems, Man, and Cybernetics*, volume 2, pages 132–136, 1999.

4

Feature Extraction Using Multi-Objective Genetic Programming

Yang Zhang and Peter I Rockett

Department of Electronic and Electrical Engineering, University of Sheffield, Sheffield, S1 3JD, UK
yang.zhang,p.rockett@shef.ac.uk

Summary. A generic, optimal feature extraction method using multi-objective genetic programming (MOGP) is presented. This methodology has been applied to the well-known edge detection problem in image processing and detailed comparisons made with the Canny edge detector. We show that the superior performance from MOGP in terms of minimizing the misclassification is due to its effective *optimal* feature extraction. Furthermore, to compare different evolutionary approaches, two popular techniques - PCGA and SPGA - have been extended to genetic programming as PCGP and SPGP, and applied to five datasets from the UCI database. Both of these evolutionary approaches provide comparable misclassification errors within the present framework but PCGP produces more compact transformations.

4.1 Introduction

4.1.1 Feature Extraction

Over recent decades, the effort made on designing ever-more sophisticated classifiers has almost eclipsed the importance of feature extraction, typically the pre-processing step in a pattern classification system - see Figure 4.1. Indeed many elegant results have been obtained in the area of classification since the 1970s. Nonetheless, feature extraction maintains a key position in the field since it is well-known that feature extraction can enhance the performance of a pattern classifier via appropriate pre-processing of the raw measurement data [2].

To utilize the potential information in a dataset to its maximum extent has always been highly desirable and subset selection, dimensionality reduction and transformation of features (feature extraction) have all been applied to patterns in the past before they are labeled by a classifier. Often though, the feature selection and/or extraction stages are omitted or are implicit in the recognition paradigm - a multi-layer perceptron (MLP) is a good example, where a distinct feature selection/extraction stages are not readily identifiable. In many application domains, such as medical diagnosis or credit scoring,

Y. Zhang and P. I Rockett: *Feature Extraction Using Multi-Objective Genetic Programming*, Studies in Computational Intelligence (SCI) **16**, 75–99 (2006)
www.springerlink.com

however, the *interpretability* of the features used in classification may contain much important information.

A generic and domain-independent methodology for generating feature extraction stages has hitherto been an unattained goal in pattern recognition. Normally, feature extraction approaches are hand-crafted using domain-specific knowledge and optimality is hard to guarantee - in fact the subject of optimality is rarely even addressed. Indeed, much of image processing research, for example, has been devising what are really feature extraction algorithms to detect edges, corners, *etc.* Ideally, we require some measure of class separability in the transformed decision space to be maximized but with hand-crafted methods, this is usually hard to assess.

Most importantly, no generic and domain-independent methodology exists to *automate* the process of creating or searching for good feature extractors for classification tasks where domain knowledge either does not exist or is incomplete.

Generally speaking, there are two main streams of feature extractions: linear and non-linear. The former is exemplified by principal component analysis (PCA) and related methods and are mainly applied to reduce the dimensionality of the original input space by projecting the data down into the sub-space with the greatest amount of data variability - this does not necessarily equate to optimal *class* separability. To obtain the optimal (possibly non-linear) transformation $\mathbf{x} \rightarrow \mathbf{y}$ from input vector, $\mathbf{x}$ to the decision space vector, $\mathbf{y}$ where:

$$\mathbf{y} = f(\mathbf{x}) \tag{4.1}$$

is a more challenging task. Popular supervised learning network systems such as the multi-layer perceptrons, radial basis functions and LVQs, coupled with powerful training algorithms, can provide model-free or semi-parametric methods to design non-linear mappings between the input and classification spaces from a set of training examples. In such paradigms, the three processing steps shown in Figure 4.1 are merged into a single, indivisible stage. Nonetheless, optimality is hard to guarantee with such methods. For example, with multi-layer perceptrons, critical quantities such the *best* number of hidden neurons are often determined empirically using cross-validation or other techniques.

Following the above insight that the feature extraction pre-processing stage is a mapping from input space to a decision space which can be represented as a *sequence* of transformations, we seek the mapping which maximizes the class separability in decision space and hence the classification performance. Thus the problem reduces to one of finding an optimal sequence of operations, subject to some criterion.

4.1.2 Genetic Programming

Genetic programming (GP) is an evolutionary problem-solving method which has been extensively used to evolve programs, or sequences of operations [1].

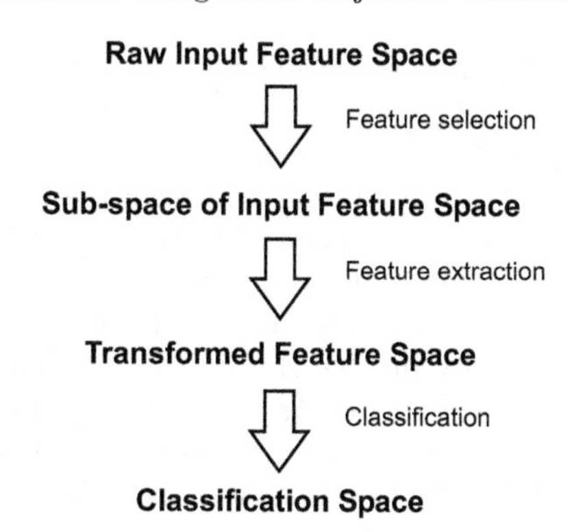

Fig. 4.1. Prototypical pattern recognition system

Typically, a prospective solution in GP is represented as a parse tree which can be interpreted straightforwardly as a sequence of operations. Indeed, GP has been used before to design feature extraction.

Koza [1] has evolved character detectors using genetic programming while Tackett [9] evolved a symbolic expression for image classification based on image features. Bot [4] has used GP to evolve new features in a decision space, adding these one-at-a-time to a k-nearest neighbor (k-NN) classifier until the newly added feature fails to improve the classification performance by more than a predefined amount; in fact, Bot's method is a greedy algorithm and therefore sub-optimal.

Pipelined image processing operations to transform multi-spectral input synthetic aperture radar (SAR) image planes into a new set of image planes were evolved by Harvey *et al.* [29]. A conventional supervised classifier was employed to classify the transformed features. Training data were used to derive a Fisher linear discriminant and GP was applied to find a threshold to reduce the output to a binary image. The discriminability, however, is constrained in the discriminant-finding phase and the GP only used as a one-dimensional search tool to find a threshold.

Sherrah *et al.* [14] proposed an Evolutionary Pre-Processor (EPrep) system which used GP to evolve a good feature (or feature vector) by minimizing misclassification error. Generalized linear machine, k-nearest neighbor (k-NN) and maximum likelihood classifiers were selected randomly and trained in conjunction with the search for feature extractors. The misclassification error over the training set was used as a raw fitness for the individuals in the evolutionary population. This approach not only has a large search space in which to work, but also depends on the classifiers in an opaque way such that there is a potential risk that the evolved pre-processing can be excellent but the classifier can be ill-matched to the task in hand, giving a poor overall performance or *vice versa*.

Kotani *et al.* [23] used GP to determine the polynomial combination of raw features to pass to a k-NN classifier. Krawiec [33] proposed a new method of using GP to design a fixed-length decision-space vector which protects 'useful' blocks during the evolution. Unfortunately, the protection mechanism actually contributes to the over-fitting which is evident in his experiments.

Significantly, all previous work on GP feature extraction has used a single objective comprising either a raw misclassification score or an aggregation of this with a tree complexity measure - see also [25]. The single objective function used in this previous work has a number of shortcomings; indeed, many real-world problems naturally involve the simultaneous optimization of multiple, often conflicting objectives. As far as genetic algorithms are concerned, many methods have been proposed to apply multi-objective optimization: *e.g.* SPEA-II[7, 17], MOGA [19], PCGA [20]. Also see [39] for a detailed review of multi-objective evolutionary algorithms.

4.1.3 Multi-objective Optimization

As we have noted, there are often competing multiple objectives in real world design problems. For a pattern classification system, the obvious design objective is to minimize the misclassification error (over some finite size training set). Unfortunately, unless specific measures are taken to prevent it, the trees in a GP optimization tend to carry-on growing in complexity, a phenomenon known as *bloat.* This results in excessive computational demands, typically accompanied by a degradation in the generalization performance of the trained classifiers. Effectively, bloat is an *over-parameterization* of the problem. To counter this, various heuristic techniques have been tried to suppress bloat, for example, adding a further objective to penalize more complex individuals. Such strategies stem from Occam's Razor - for a given error, the simpler individual is always preferred. Ekárt and Németh [18] have shown that embedding tree complexity within a multi-objective GP tends to prevent bloat by exerting selective pressure in favor of trees of smaller size [7]. Thus a multi-objective framework based on Pareto optimality [16] is presented here.

Both Strength Pareto GP (SPGP) and Pareto Converging GP (PCGP) are applied in the present framework to find the Pareto set for our multi-objective optimization problem - identifying the (near-)optimal sequence of transformations which map input patterns to decision space with maximized separability between the classes. As far as the authors are aware, this is the first report of the comparison between these two evolutionary techniques (generational *vs.* steady-state) applied to multi-objective genetic programming.

The rest of this chapter is organized as follows: Our generic framework for evolving optimal feature extractors will be presented in Section 4.2 In Section 4.3, we will describe the principal application domain to which the methodology will be applied - the edge detection problem in image processing. In Section 4.4, quantitative comparisons are made with the 'gold-standard' Canny edge detector to investigate the effectiveness of the present method on the USF real world dataset [24]. In addition, comparisons between two multi-objective evolutionary strategies - SPGP and PCGP - have been made on five datasets from UCI Machine Learning database [8]. We offer conclusions and thoughts on future work in Section 4.5.

4.2 Methodology

There are two ways of applying evolutionary optimization techniques in the pattern classification domain - one could be regarded as putting the whole system inside the evolutionary learning loop to directly output class labels. This is a tempting route due to its 'simplicity'. Unfortunately, it not as simple as it first appears since the search space is very large, comprising the space of all feature extractors *and* the space of all classifiers. The other option - and the one we favor - is that treating the feature extraction as a stage distinct from classification. Our target is to invest all available computational effort into evolving the feature extraction and to utilize established knowledge from the relatively well-understood area of classifier design. Consequently, we adopt the approach of evolving the optimal feature extraction and performing the classification task using a standard, simple and fast-to-train classifier. The "fast-to-train" requirement on the classifier comes from the fact that we still have to use classification performance as an individual's fitness and thus the classifier has to be trained anew within the evolutionary loop for every fitness evaluation. In addition, evolving a distinct feature extraction stage retains the possibility of human interpretation of the evolved meta-features and therefore the use of our method as a data exploration technique.

Our target is to obtain a generic, domain-independent method to fully automate the process of producing optimal feature transformations. Hence we make no assumptions about the statistical distributions of the original data, nor indeed about the dimensionality of the input space.

4.2.1 Multi-objective Genetic Programming

In order to implement the method, multi-objective genetic programming (MOGP) is used as the search tool driven by Pareto optimality to search for the optimal sequence of transformations. If the optimization is effective, features in the transformed space should not only be much easier to separate than in the original pattern space but should also be *optimal* in the decision space with the classifier employed. In the context of *optimality*, we conjecture that our method should yield a classification performance which is at least as good as the best of the set of all classifiers on a particular dataset [6].

The two evolutionary strategies mentioned above - SPGP and PCGP - are both implemented here on five UCI datasets. The SPGP approach uses a generational strategy with tournament selection to maintain two sets of individuals during evolution, one representing the current population and the other containing the current approximation to the Pareto set. Ranking is done by calculating the strength, or fitness, of each individual in both sets - when calculating an individual's strength, we use the method proposed in SPEA-II [7]. In our modified binary tournament selection method, two individuals from the union of the population and the non-dominated set are randomly selected. If both trees have been drawn from the same set we compare the normalized fitness to determine the winner; if not, the raw fitness vector is used to decide which should be chosen.

For the steady-state PCGP strategy, we followed the method designed in [20] as PCGA with straightforward modifications for use with GP trees. In comparing the two evolutionary strategies, we are aiming to focus on the issue of the search convergence. The Pareto Converging Genetic Algorithm (PCGA) ensures population advancement towards the Pareto-front by naturally sampling the solution space. The motivation for this approach is aimed particularly at solving difficult (real-world) problems with local minima. For detailed information see [20].

We have used a fairly standard chromosomal tree structure for the MOGP implementation. There are two types of nodes in each individual - function nodes and terminal nodes. Details of node types are summarized in Table 4.1.

Non-destructive, depth-dependent crossover [13] and a depth-dependent mutation operator were used to avoid the breaking of building blocks. Here a set of candidate sub-trees is chosen to mutate based on their depth in the tree using the depth-fair operator [13]. Then, one particular sub-tree from this set is selected biased in its complexity (*i.e.* the number of nodes) using roulette wheel selection; that is, we favor mutating more complex trees. Detailed information can be found in [5].

The stopping criterion for SPGP is any of the following being is met:

- The maximum number of generations is exceeded (500 in present case)
- The evolution process stops as adjudged by the Bayes error of the best individual failing to improve for $0.04 \times$ the maximum number of generations (the 0.04 value is an empirical value determined by experiment)
- The misclassification error $= 0$.

For PCGP, the termination condition was set simply as exceeding the maximum number of generations, 2000 in present work. The population size for SPGP in all experiments presented here was 500 while a population of only 200 individuals was used for PCGP. Note that when *fully* converged, all the non-dominated individuals in the PCGP population comprise the Pareto-front. For more detailed information see [20].

4.2.2 Multiple Objectives

Within the multi-objective framework, our three-dimensional fitness vector of objectives comprises: tree complexity, misclassification error and Bayes error, as follows:

Table 4.1. MOGP Settings

Terminal set	Input pattern vector elements
Constant	10 floating point numbers 0.0, ..., 1.0
Raw fitness vector	Bayes error, misclassification error, number of nodes
Standardized fitness	Strength-based fitness
Original population	Half full-sized trees, half random trees
Original tree depth	5
Probabilities	0.3 mutation, 0.7 crossover

Tree Complexity Measurement

As pointed-out above, tree bloat in GP can produce trees with extremely small classification errors over the training set but a very poor error estimated over an independent validation set. This is clearly an example of the familiar over-fitting phenomenon and grounded on the principle of Occam's Razor, we use the node count of a tree as a straightforward measure of tree complexity. Within the concept of Pareto optimality in which all objectives are weighted equally, this complexity measure exerts a selective pressure which favors small trees.

Misclassification Error

In addition to the tree complexity measure, the fraction of misclassified patterns counted over the training set is used as a second objective [33]. Since the GP tree chromosomes used here naturally lead to an n-to-1 mapping into the one-dimensional decision space, a simple threshold in this decision space is adapted during the determination of fitness value to obtain the minimum misclassification error. This means we are trying to evolve an optimal feature extractor *conditioned* on a thresholding classifier operating in the decision space.

Under Pareto optimality [16], all objectives are treated as equally important since we are aiming to explore the trade-off between the complexity and the misclassification error (over the training set). With the aid of the two competing objectives, we are aiming to maximize the class separability in the decision space using the simplest possible mapping. Nonetheless, during our early experiments, we found that learning is often very slow and sometimes the optimization fails to converge at all. After detailed investigation into the searching process, we found that in the initial stages, when all the randomly-created individuals possessed roughly the same (very high) misclassification error, there was insufficient selective pressure to pick individuals with slightly more promise than the irredeemably poor performers. Thus the search stagnated. Consequently, the third objective of Bayes error was investigated as an additional measure of inter-class separability.

Bayes Error

The search performance of all evolutionary algorithms is critically dependent on the use of appropriate fitness functions. Our motive for choosing the Bayes error as an objective is because it is the fundamental lower bound on classification performance, independent of the class distributions and classifier. In a two class problem, we map the n-dimensional input pattern space into the 1D decision space forming two class-conditioned probability density functions (PDFs) in the decision space and the overlapping region(s) of these two class-conditioned PDFs can be used to estimate the Bayes error with a simple histogramming procedure.

If the Bayes error is used as a direct replacement for the misclassification error, the optimization converges rapidly - clearly the Bayes error is exerts more sensitive selective pressure in the the situation where the population is far from convergence. Unfortunately, the subsequently estimated validation error was disappointingly high. Further investigation revealed that in minimizing the overlap between two class-conditioned PDFs, the GP often achieved this goal by producing two PDFs with non-coincident, *'comb'-like* features as illustrated in Figure 4.2. The Bayes error is indeed small when calculated over the *training set* but this does not generalize to an independent validation set. Clearly what was desired were two well-separated PDFs although the GP was meeting this goal in an unintended and unhelpful way - such opportunistic behavior has been observed previously in evolutionary algorithms. As a consequence, we employed both misclassification error *and* the Bayes error estimate in a three-dimensional fitness vector.

The Bayes error allows the evolutionary search to make rapid initial progress after which the misclassification error objective appears provide selective pressure which separates the transformed distributions. We have also observed that on those occasions when the optimization has got temporarily stuck in a local minimum, it is the Bayes error which improves first and appears to 'lead' algorithm out of the local minimum.

4.3 Application Domain

4.3.1 Edge Detection in Image Processing

Edge detection is an important image processing technique in many applications, such as object segmentation, shape recognition, *etc.* and so we chose this well-understood problem as the starting point to demonstrate our methodology. Edge detection is a well-researched field which should be suitable for assessing the effectiveness of the evolved feature extraction method in comparison to an established and conventional method *e.g.* the Canny edge detector; the Canny algorithm is widely held to be a 'gold standard' among edge detection algorithms [21, 22, 24]. Harris [34] used GP with a single objective

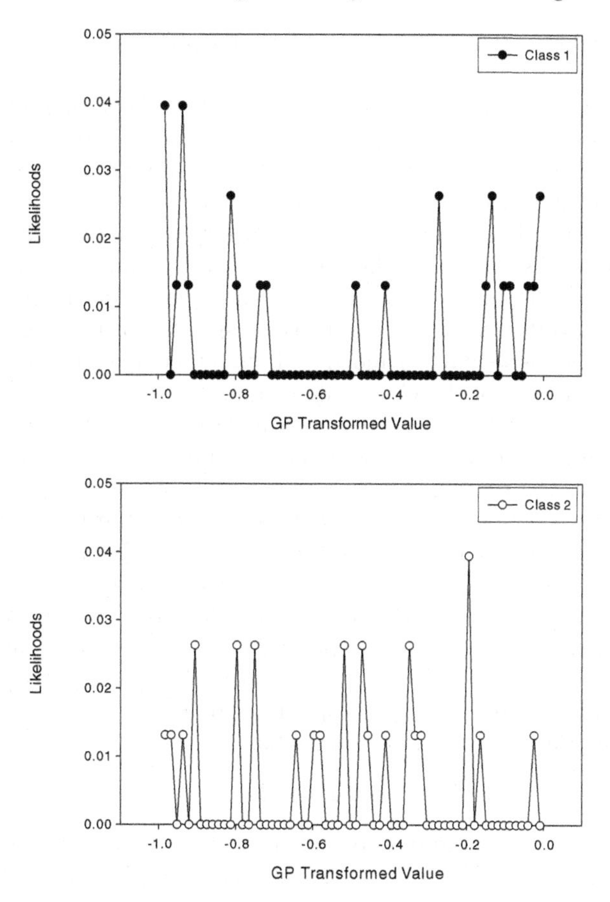

Fig. 4.2. Example of *'comb'-like* class-conditioned densities evolved using the Bayes error metric alone

to evolve an 'optimal' edge detector but terminated the evolution when he obtained a performance comparable to the Canny detector See also [10, 11].

We used a synthetic dataset for training on the edge detection task because Chen *et al.* [30] concluded that hand-labeling real image training set did not adequately sample the pattern space, leading to deficient learning. We have followed a very similar approach of synthesizing a training set from a physically realistic model of the edge imaging process. Three distinct types of patterns are identified: edges, non-obvious-non edges and uniform patches. Further details can be found in [30, 31, 32]. We have employed an image patch size of 13×13 in the generated training set - probably larger than is needed - deliberately to investigate whether the most useful features would be selected by the objective of minimizing the tree size, and by implication, the number of raw input features used. The training set comprised 10,000 samples and the realistic figure of 0.05 is chosen for the prior probability of the edge class. See also [6].

The Canny algorithm includes a significant number of post-processing steps, such as non-maximal suppression (NMS) which appear to be responsible in large measure, if not completely, for its superiority over other conventional edge detectors [32]. The NMS step is a heuristic sequential and spatially-localized classification stage, which serves to reduce the fraction of false positives (*FPs*) with a slight attendant sacrifice in the fraction of true positives (*TPs*). Our interest here is a domain-independent methodology and a step such as NMS is very much a heuristic specific to the image processing domain; nonetheless we have included NMS in our comparisons since it is held to be an integral part of the Canny algorithm.

Whereas the our GP detector has no adjustable parameters (once trained), the edge labelling threshold is a user-defined tuning parameter in the Canny algorithm. Thus fair comparison is not completely straightforward. The principled basis we have chosen for comparison between the GP-generated and Canny edge detectors is the operating points which minimize the Bayes risk for both detectors. The operating point which naturally emerges from our GP algorithm is that which minimizes the misclassification error over the training set which was constructed with edge prior of 0.05. For the Canny algorithm we can locate the corresponding decision threshold by plotting the Bayes risk *versus* threshold to locate the optimal operating point (*i.e.* the minimum Bayes risk) [15]. The Bayes risk can be written as:

$$R = P \times (1 - TP) + (1 - P) \times FP \tag{4.2}$$

where, TP and FP are the fractions of true and false positives, respectively. P is the prior of edge. Generally, both TP and FP will vary with threshold and other operating conditions. Through locating the minimum of the risk-threshold plot we can identify the 'optimal' operating point of the detector at the given prior.

Note we have used the assumption of equal costs: In fact, cost ratios are always subjectively chosen and vary from application to application. For example, in medical image processing, the cost of a false negative may be unacceptably high and so a suitable cost would be used, biasing the classifier operating point. In other applications, false negatives resulting in line fragmentation, may be tolerable in order to keep processing times below some limit. Here we adopt a neutral position of using a cost ratio of unity since we have no basis for regarding one sort of error as more or less important than any other.

Detailed comparison using synthetic datasets between the MOGP edge detector and the Canny algorithm had been made elsewhere [5]. In order to investigate the labeling performance on real image data, we have applied the GP-generated edge detector to images taken from the ground-truth labeled USF dataset [24] and drawn comparison with the Canny edge detector with and without NMS; the results are presented in Section 4.4.

4.3.2 UCI data

In addition to the edge detection task, we have applied our method to five other datasets from the UCI Machine Learning databases [35] in Section 4.4.2 where we make comparison between the two evolutionary strategies - SPGP and PCGP - to investigate the relative merits of generational and steady-state evolutionary techniques. The datasets used in the current work are:

- BUPA Liver Disorders (BUPA): To predict whether a patient has a liver disorder. There are two classes, six numerical attributes and 345 records.
- Wisconsin Diagnostic Breast Cancer (WDBC): This dataset has been discussed before by Mangasarian *et al.* [26]. 569 examples with thirty numerical attributes.
- Pima Indians Diabetes (PID): All records with missing attributes were removed. This dataset comprises 532 complete examples with seven attributes.
- Wisconsin Breast Cancer (WBC): Sixteen instances with missing values were removed; 683 out of original 699 instances have been used here. Each record comprises ten attributes. This dataset has been used previously in [27].
- Thyroid (THY): This dataset includes 7200 instances with 21 attributes (15 are binary, 6 are continuous) from 3 classes. It has been reconfigured as a two-class problem with 166 instances from class1 and the remaining, 7034 instances from non-class1. This dataset has been discussed by [36].

For convenience, the details of the datasets used in the current work are summarized in Table 4.2.

Table 4.2. Five UCI Datasets

Name	Number of Features	Size and Distributions
BUPA	6	345 = 200 (Benign) + 145 (Malignant)
WDBC	30	569 = 357 (Benign) + 212 (Malignant)
PID	7	532 = 355 + 177 (Diabetic)
WBC	10	699 = 458 (Benign) + 241 (Malignant)
THY	21	7200 = 166 (Class1) + 7034 (Others)

4.4 Results

4.4.1 Comparisons on USF Datasets

In order to examine the performance of the GP edge detector as well as make fair comparison with the Canny algorithm, we have assessed the performance of both detectors with and without the NMS post-processing step.

As pointed out in our previous work [6], fair comparisons turn-out to be somewhat harder than first appear. We are trying to compare the GP edge detector (with or without NMS) to the Canny edge detector (with or without NMS) at the detector operating points which minimize the Bayes risk. The principal complication here is that the USF dataset has been subjectively censored since the non-obvious non-edge patterns [24],[32] - which are the patterns most likely to be confused by any classifier - have been hand-labeled as belonging to a distinct "don't care" class. Hence we have used the following methodology: Quantifying the performance of the Canny detector over the USF images is straightforward. We have used the same optimal threshold as determined over the synthetic data [5],[6] which assumes an edge prior of 0.05 - the GP detector assumes this same prior therefore neither detector is comparatively disadvantaged. The labeling performance is summarized in Table 4.3 for each of the USF images shown in Figs 4.3(a) - 4.6(a).

Table 4.3. Canny [TP, FP] Operating Points for USF Test Images

Figure	Edge Prior	Without NMS		With NMS	
		TP	FP	TP	FP
4.3	0.087	0.0003	0.0001	0.3821	0.0466
4.4	0.080	0.0204	0.0003	0.4326	0.0695
4.5	0.211	0.0142	0.0063	0.3886	0.0061
4.6	0.066	0.0048	0.0003	0.3580	0.0049

Table 4.4. GP [TP, FP] Operating Points for USF Test Images

Figure	Edge Prior	Without NMS		With NMS	
		TP	FP	TP	FP
4.3	0.087	0.5045	0.0278	0.3657	0.0083
4.4	0.080	0.4151	0.0298	0.3388	0.0052
4.5	0.211	0.6228	0.0129	0.4433	0.00036
4.6	0.066	0.5581	0.0246	0.3415	0.0047

It is again straightforward to obtain the labeling performance of the GP detector *without* NMS and these results are shown in Table 4.4. To determine the GP performance *with* NMS we have devised a special method of carrying-out non-maximal suppression on the output of the GP detector. First, we estimate the orientation of an edgel using the familiar difference-of-boxes operator. We then quantize the edge direction into one of the eight principal directions of the compass and examine the decision variable responses of the GP detector for the three pixels centered on the edge and (approximately) normal to the edge direction. The 'distance' of a given edge pixel's response from the decision threshold can be taken as a measure of its edge strength and we perform non-maximal suppression on this quantity. The results are again summarized in Table 4.4.

It is apparent from Table 4.3 that the Canny algorithm without NMS performs very poorly, an observation consistent with the preceding results on synthetic edge data reported in [6]. In particular, the *TP* values are very low. Consistent to the conclusion made in [32] that there is no clear minimum risk operating point for the Canny algorithm without NMS. Hence the success of this algorithm owes very little to the feature extraction (pre-processing) step, a finding which will be a surprise to many in the image processing community.

Table 4.5. Bayes Risk Comparisons for USF Test Images

Figure	GP		Canny	
	Without NMS	With NMS	Without NMS	With NMS
4.3	0.06848	0.06276	0.0871	0.09630
4.4	0.07420	0.05768	0.0786	0.10933
4.5	0.08977	0.11774	0.2129	0.13381
4.6	0.05214	0.04785	0.06596	0.04694

With NMS, the Canny algorithm has a much higher *TP* fraction although the *FP* fraction also increases. This is again consistent with the result reported in [32] that after the NMS step, there is a clearer optimal operating point in terms of minimization of Bayes risk. Indeed, after NMS the (increased) *FP* fraction becomes the principal contributor to the Bayes risk.

The received wisdom in the image processing community would suggest that the difference between the detector output with and without NMS can be explained by the thinning of otherwise thick edges - this turns-out not to be the case since NMS significantly changes the ROC of the Canny edge detector and hence its optimal operating point.

From the comparisons we can see that before NMS, GP gives larger *TP* values while after NMS, the two algorithms give comparable *TP* values. The Canny algorithm, however, gives much bigger *FP* values. Comparisons of the Bayes risk figures are shown in Table 4.5 from which it can be seen that the risk values of the Canny algorithm are higher than for the GP, with or without NMS. The only exception to this is the image in Fig. 4.6 where, after NMS, the GP has a slightly higher risk. The reason seems to be that the noise level in this image is much lower than the others, allowing the Canny algorithm to yield a higher *TP* value while having an *FP* value roughly equal to the GP detector. In fact, it is not statistically significantly different using t-test under 95% confidence level. The final labeled images in Fig 4.6 (e, f) for Canny and GP (with NMS) look much more similar than for any of the other images from the USF dataset.

In the USF dataset, a number of regions have been labeled as "don't care" - the white regions in Fig. 4.3b, 4.4b, 4.5b, and 4.6b are the distinct "don't care" class. For the edge detection problem we have concluded that it is these non-obvious non-edge (NONE) patterns labeled as "don't care"in the USF images which make the classification task difficult. This is consistent with the observation of Konishi *et al.* [22] that the USF images are easier to label than the Sowerby dataset which these authors also considered and which has no "don't care" regions.

Whole image labeling results for four, typical USF images using the Canny and GP detectors (with and without NMS) are shown in Fig. 4.3 to Fig. 4.6. In these figures, (a) denotes the original image from the USF dataset, (b) shows the ground truth data, (c) shows the labeling results from the Canny detector without NMS and (d) shows images labeled with the GP feature extractor; (e) shows images from the Canny edge detector with NMS, (f) shows the output from the GP feature extractor with the gradient-direction based NMS.

In these labeling results, it is striking that Fig. 4.3(c) contains only six labeled pixels in the top right hand corner of the image and the same poor performance in other corresponding (c) figures are consistent with the results from Table 4.3. These results further confirm our conclusion that the Canny edge detector's performance is not due to the feature extraction stage but to the sophisticated post-processing steps coupled with subjectively - and implicitly - set cost ratios.

Discussion

The evolution described here, driven by multiple objectives is able to generate separated class-conditioned distributions in the 1D decision space. In contrast to the hand-crafted, heuristic post-processing steps of the Canny edge detector [21], we concentrate our computational resource on optimizing the feature extraction stage to yield greater separability in the mapped feature space. Further, compared to the Canny detector which is based on extensive domain knowledge - see Canny's original work on deducing the 'optimal' filter kernel [21] - we did not supply *any* domain-dependent knowledge to MOGP apart from the carefully constructed training set. This lends evidence to support our conjecture that our method is able to automatically produce (near-)optimal feature extraction stages of a classification system.

Although feature selection was not explicitly intended in our approach, we have deliberately employed an overly large image patch (13 × 13) to investigate how the GP selects input features within this patch given that one of our fitness objectives is to minimize tree size and, therefore, the number of leaf nodes (*i.e.* individual pixel intensity values). A histogram of the number of times each pixel was used by the trees in a typical converged Pareto set is shown in Fig. 4.7, where the central pixel of the 13 × 13 patch has the row and column indices of (0, 0). This figure illustrates that the MOGP optimization has a strong tendency to select pixels from around the center of the image patch, a fact which is intuitively pleasing because most of the edge/non-edge discriminatory information can be considered to come from this region. We reiterate that we have not embedded any domain knowledge in this optimization. Thus we believe that feature selection is occuring as a beneficial by-product of the feature extraction process due to the way we have incorporated parsimony into the optimization.

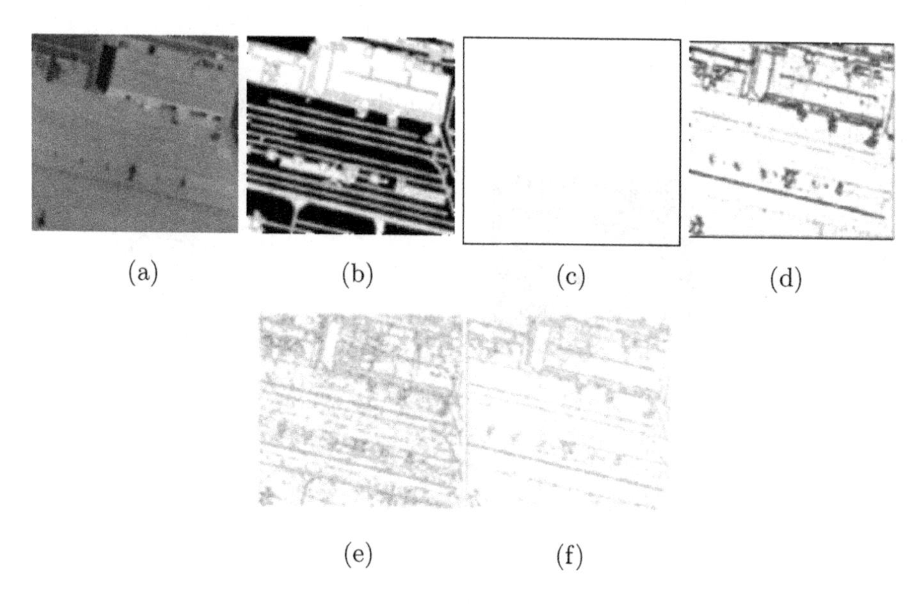

(a) (b) (c) (d)

(e) (f)

Fig. 4.3. a to f illustrate the comparisons from the Canny edge detector and GP, details refer to the text.

4.4.2 Comparison of Generational and Steady-state Evolutionary Algorithms on UCI Datasets

As mentioned in the introduction, evidence to substantiate the generic property of our method has been demonstrated in previous work [6] where we have shown that our methodology, implemented using the SPGP algorithm, exhibited either superior or equivalent performance to nine conventional classifiers. Here we report the results of a preliminary investigation into the efficacy and performance of different evolutionary techniques within the proposed framework. SPGP and PCGP have been compared on five typical problems from the

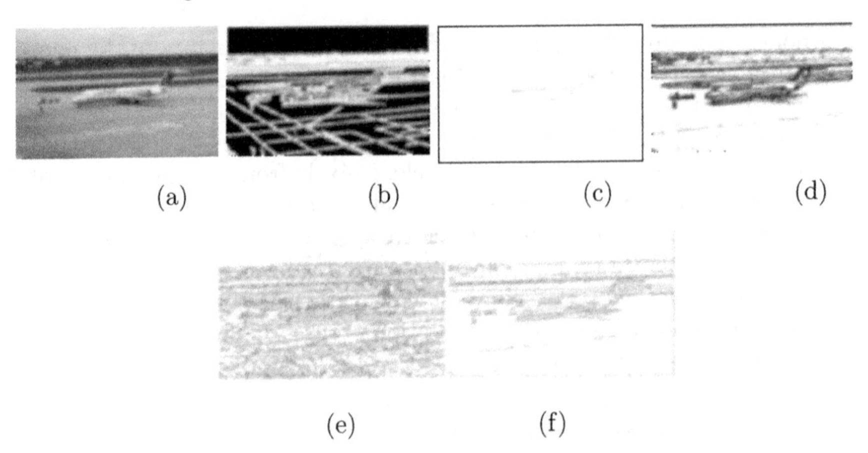

Fig. 4.4. a to f illustrate the comparisons from the Canny edge detector and GP, details refer to the text.

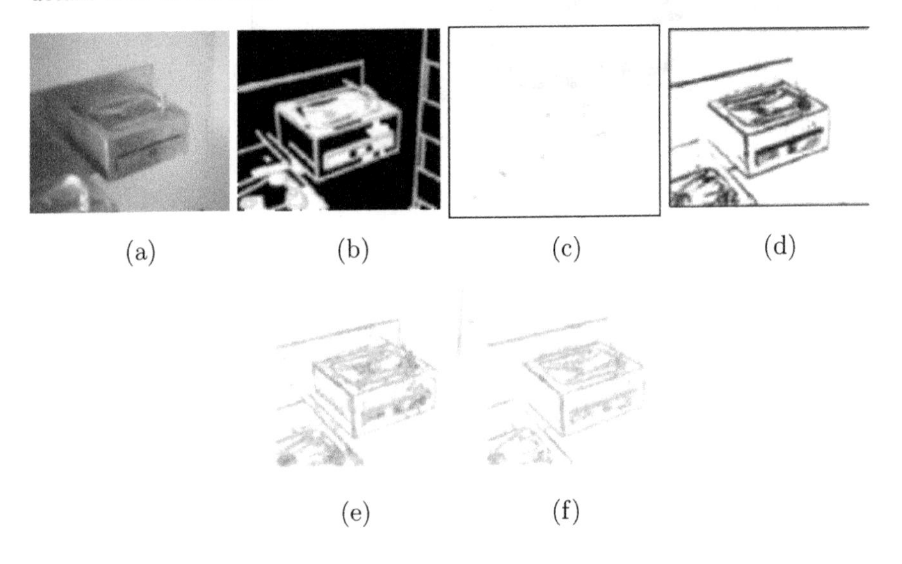

Fig. 4.5. a to f illustrate the comparisons from the Canny edge detector and GP, details refer to the text.

UCI database [35]. As far as we are aware, there has been little detailed analysis and comparison of multi-objective evolutionary strategies on real world GP problems.

We have applied SPGP and PCGP to each of the problems listed in Table 4.2. The GP settings are the same as those listed in Table 4.1 except that the terminal nodes are now the elements in the pattern vectors of the five UCI datasets rather than image pixel values. For the generational SPGP algorithm,

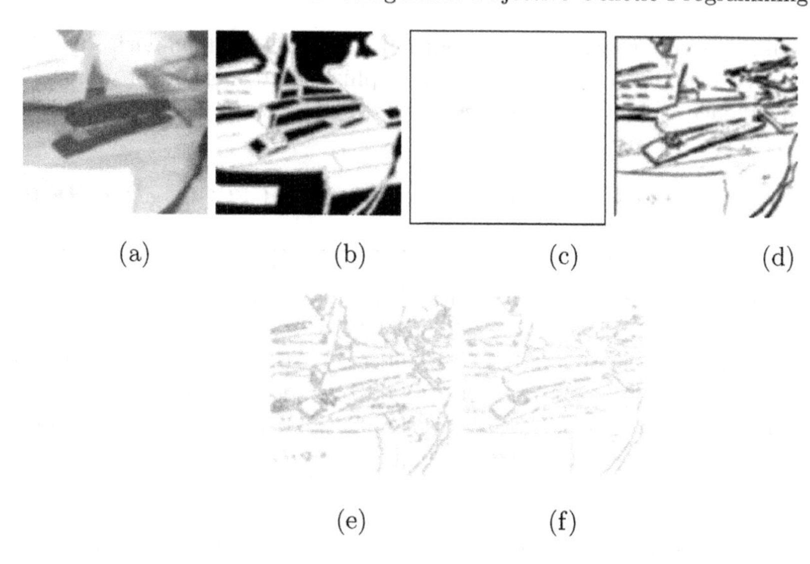

Fig. 4.6. a to f illustrate the comparisons from the Canny edge detector and GP, details refer to the text.

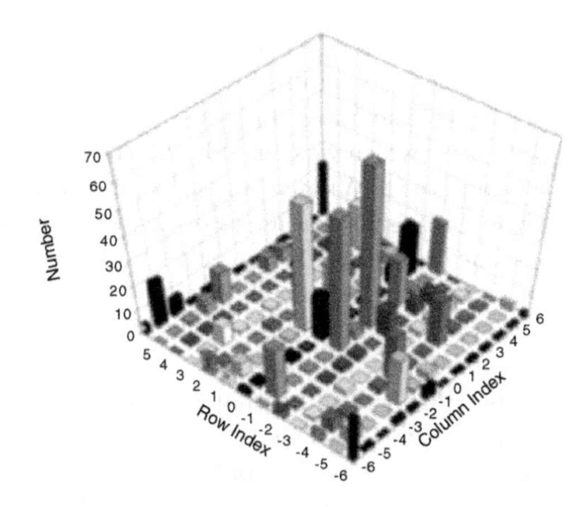

Fig. 4.7. Histogram of numbers of pixels used in the GP trees in a typical Pareto set, relative to the 13×13 input patch.

we used a population size of 500 and a maximum of 500 generations; the three possible stopping criteria listed in Table 4.1 are reused. For the steady-state PCGP algorithm, only 200 individuals were used in the population, while for simplicity, the stopping criterion used was to run for a fixed number of 2000 generations. This means that PCGP is performing around one-tenth the number of fitness evaluations of the SPGP algorithm and is thus comparatively disadvantaged; this was an intentional feature of this preliminary study to determine if approximately equivalent solutions could be obtained with PCGP but with much less computing resource. A sophisticated method of gauging actual convergence of PCGA have been discussed by Kumar & Rockett [20].

Table 4.6. Mean Error Comparisons Between SPGP and PCGP on Five UCI Datasets

Datasets	SPGP			PCGP		
	BE	ME	Mean Nodes	BE	ME	Mean Nodes
BUPA	0.185	0.272	58.928 ± 54.636	0.219	0.278	7.830 ± 5.242
PID	0.160	0.203	32.633 ± 40.133	0.173	0.200	3.870 ± 1.401
WBC	0.019	0.025	24.071 ± 46.320	0.014	0.022	6.280 ± 2.647
WDBC	0.025	0.028	36.941 ± 43.417	0.028	0.028	7.785 ± 5.343
THY	0.0044	0.0061	37.964 ± 43.299	0.0044	0.0058	11.220 ± 5.492

The mean error comparisons between the two evolutionary algorithms over 10 repetitions are summarized in Table 4.6, where ME stands for misclassification error from optimal thresholding in the decision space after feature extraction. BE, the corresponding Bayes error estimates are listed to give an indication of the degree to which the misclassification error approaches its fundamental lower bound. Table 4.6 contains only data from the *non-dominated* solutions from each of the evolutionary paradigms. We have applied Alpaydin's F-test [28] to these data from which we conclude that none of these differences in error is statistically significant. What *is* a notable and significant difference between these two algorithms is the numbers of mean nodes required to obtain 'identical' misclassification errors. The mean numbers of nodes for the steady-state PCGP approach are very much smaller than for the generational SPGP algorithm. We have observed that some inactive/redundant sub-trees exist in solutions evolved by SPGP [6] and the fact that we generate significantly smaller trees using PCGP implies that PCGP is much more effective in controlling tree bloat than SPGP. This result saves a lot of computational effort during evolution as well as having a practical implication.

To further explore the differences between the 'quality' of the solutions produced by these two evolutionary paradigms, we have plotted the misclassification error *versus* the number of nodes for the final solutions produced by the two algorithms on each of the datasets in Figures 4.8 to 4.12. Although the SPGP results shown are this algorithm's best approximation to the Pareto front, for PCGP we have plotted the *whole* population which in practice, contains a number of *dominated* solutions. Our aim here is to compare the coverage and sampling of the Pareto front produced by the two algorithms, albeit at an intermediate stage for PCGP. Note that these figures do *not* depict Pareto-optimal sets *per se* since these plots are a two-dimensional representation of a three-dimensional Pareto front. This in part explains why a number of solutions appear at first glance to be dominated (in the Pareto

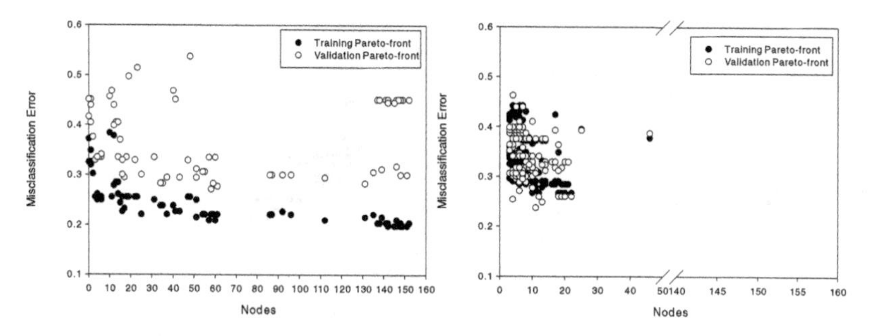

Fig. 4.8. Misclassification error *vs.* Number of nodes for the members of the Pareto sets generated by SPGP (left) and PCGP (right) on the BUPA dataset

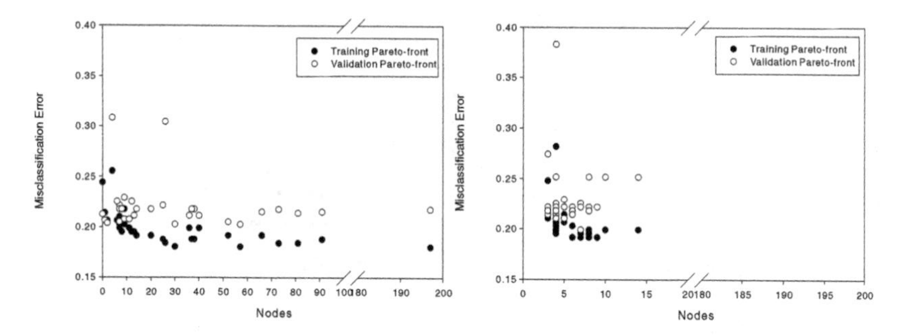

Fig. 4.9. Misclassification error *vs.* Number of nodes for the members of the Pareto sets generated by SPGP (left) and PCGP (right) on the PID dataset

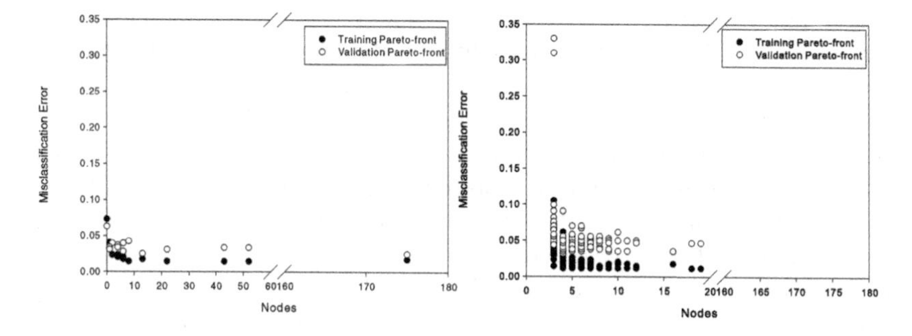

Fig. 4.10. Misclassification error *vs.* Number of nodes for the members of the Pareto sets generated by SPGP (left) and PCGP (right) on the WBC dataset

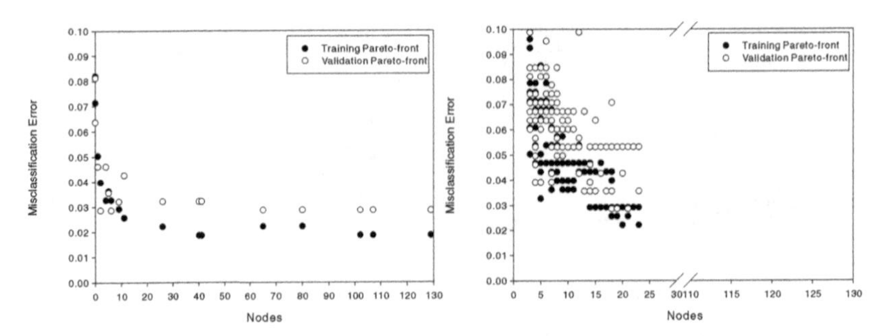

Fig. 4.11. Misclassification error *vs.* Number of nodes for the members of the Pareto sets generated by SPGP (left) and PCGP (right) on the WDBC dataset

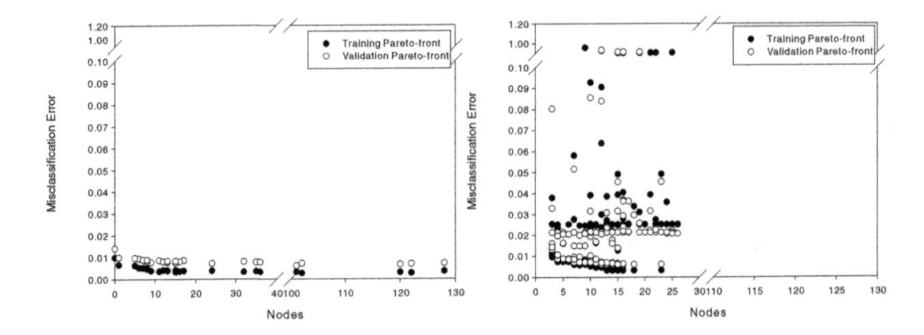

Fig. 4.12. Misclassification error *vs.* Number of nodes for the members of the Pareto sets generated by SPGP and PCGP on the THY dataset

sense) whereas in fact, they are actually non-dominated if one takes the third (Bayes error) objective into account. (Indeed, many of the PCGP solutions actually are dominated within their population.)

From Figures 4.8 to 4.12, it is apparent that PCGP produces a solution set which is much more strongly centered on lower node numbers. This preliminary comparison of PCGP is thus extremely promising since even in the state of not being fully converged and only having had around one-tenth the computing resources invested in it, PCGP is able to produce solutions which are significantly smaller than those from SPGP but with indistinguishable error rates.

Tree Interpretation

By way of example, we show a typical GP-generated tree in Figure 4.13. This individual was a non-dominated population member obtained from a PCGP run on the THY dataset; we present the tree as it was generated without pruning. The leaf nodes in Figure 4.13 are labeled as Xn, for $\in [1...N]$, where N is the number of raw attributes in the input pattern vector; for the THY dataset here, $N = 21$ (see Table 4.2). All the function nodes in the graph are elementary functions except *pow2* which calculates the child value raised to the power of two and *ifelse* which returns the second child value if the first child value is 0, otherwise it returns the third child value. Actually, in the example tree shown in Figure 4.13, the sub-tree containing *ifelse* could be simplified as $X19$. Although it is trivial to remove this obvious redundancy by hand (or even as a non-evolutionary post-processing stage), it appears that even the PCGP evolutionary strategy has the propensity to generate inactive code.

The value returned from the tree root - $-(minus)$ - is the transformed feature value in the one-dimensional decision space in which the separability between classes is maximized. Each input pattern will be fed into the tree and will return a scalar value as the extracted feature. Again, the set of raw input features has been selected during the optimization process which seeks to minimize the number of tree nodes. Surprisingly, with this dataset only two original elements can be used to extract the *optimal* new feature to obtain the desirable classification performance.

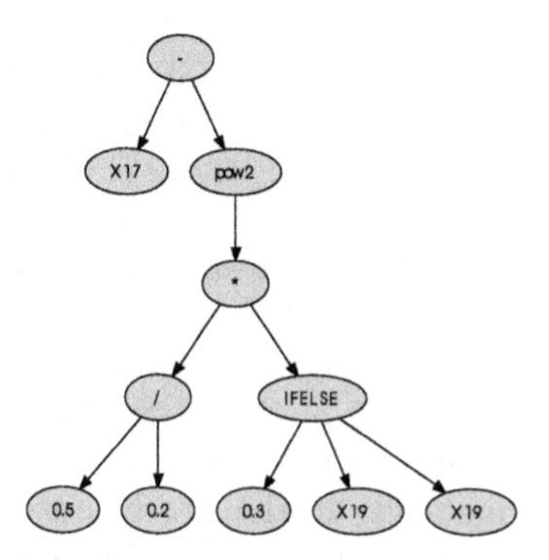

Fig. 4.13. One typical evolved GP feature extractor on THY dataset

4.5 Conclusions and Future Research Work

In the present work we conjecture that efficient feature extraction can yield a classification system, the performance of which is *at least* as good as the best of all available classifiers on any given problem. In essence, the evolutionary algorithm is 'inventing' the optimal classifier for a given labeling problem/dataset. It is, of course, entirely feasible that the genetic optimization is re-inventing an existing classifier rather than devising a novel algorithm but this is of little practical consequence. When confronted with a new dataset, the common approach among pattern recognition practitioners is to empirically try a range of classification paradigms since there is no principled method for predicting which classifier will perform best on a given problem; there is always the chance that better performance could be obtained with another, but untried classifier. Our conjecture that multi-objective evolutionary optimization produces the best possible classifier effectively eliminates the risk from this trial-and-select approach. Needless to say, although the results presented here and in previous work [6] support our conjecture, by definition, we cannot offer a definitive proof.

Also of considerable importance is the fact that we do require any domain knowledge of the application. The feature extraction transformation sequence is identified automatically, driven solely by the optimality criteria of the multi-objective algorithm. We have presented the application to edge detection, for example, in which our method yields superior results to the very well-established and mature Canny algorithm. The generic property of the presented method enables it to be straightforwardly applied to other application domains.

Although the principle of evolutionary optimization is straightforward, exact *implementation* is still an open research issue. Here we present the preliminary results of applying two different evolutionary paradigms, one generational (SPGP) and one steady-state (PCGP). The differences between the misclassification errors attained with the two types of genetic search are not statistically significant although the *complexity* of the generated feature transformation sequences is markedly different. Even where the PCGP algorithm has not fully converged, we find that the steady-state method produces much smaller trees which implies that PCGP is more responsive to the objective trying to minimize tree size. The concentration of the solutions at low node numbers also means that the sampling of the Pareto front is better with PCGP. (Previous comparisons between PCGA and other multi-objective GAs imply a similar, fundamental superiority in the steady-state approach [20].) Faster convergence and better control of the tree-bloating make the PCGP approach very promising for practical applications.

We have demonstrated the use of multi-objective genetic programming (MOGP) to evolve an "optimal" feature extractor which transforms the input patterns into a decision space such that pattern separability in this decision space in is maximized. In the present work we have projected the input pattern to a *one-dimensional* decision space since this transformation naturally arises from a genetic programming tree although potentially, superior classification performance could be obtained by projecting into a multi-dimensional decision space [14] - this is currently an area of active research.

Finally, as with all multi-objective optimizations, what arises from our method is a family of equivalent solutions - the Pareto set - which presents the system designer with the trade-off surface between classification performance and the complexity of the feature extractor. Exactly which pre-processing solution is selected for the final application will depend on the *generalization* properties of the solutions - invariably one would like the solution which is best able to predict class of as-yet unseen patterns with the greatest accuracy. Predicting generalization performance is very much a major and continuing issue in statistical pattern recognition - ultimately we would like to formulate a measure of classifier generalization which potentially, could be optimized along with the other multiple objectives. This remains an area of open research.

References

[1] J.R. Koza. Genetic Programming II, Automatic Discovery of Reusable Programs. The MIT Press, Cambridge, Massachusetts, 1994

[2] R. O. Duda and P. E. Hart and D. G. Stork. Pattern Classification (2nd Edition), Wiley-Interscience, 2000

[3] D. Addison, S. Wermter and G. Arevian. A comparison of feature extraction and selection techniques. In: *Proceedings of the International Conference on Artificial Neural Networks*, Istanbul, Turkey, Supplementary Proceedings, pp. 212–215, 2003

[4] M. C. J. Bot. Feature extraction for the k-Nearest neighbor classifier with genetic programming. In: *Genetic Programming, Proceedings of EuroGP'2001*, Lake Como, Italy, pp. 256-267, 2001

[5] Y. Zhang and P.I. Rockett. Evolving optimal feature extraction using multi-objective genetic programming: A methodology and preliminary study on edge detection. In: *GECCO 2005*, pp. 795-802, 2005

[6] Y. Zhang and P.I. Rockett. A Generic Optimal Feature Extraction Method using Multiobjective Genetic Programming: Methodology and Applications. *IEEE Trans. Systems, Man, and Cybernetics*, 2005 (submitted)

[7] S. Bleuler, M.Brack, L. Thiele, and E.Zitzler. Multiobjective genetic programming: Reducing bloat using SPEA2. In: *Congress on Evolutionary Computation (CEC 2001)*, pp. 536-543, 2001

[8] Z.J. Huang, M. Pei, E. Goodman, Y. Huang, and G. Liu. Genetic algorithm optimized feature transformation - A comparison with different classifiers. In: *GECCO 2003*, LNCS 2724, pp. 2121-2133, 2003

[9] W.A. Tackett. Genetic programming for feature discovery and image discrimination. In: *Proceedings of the Fifth International Conference on Genetic Algorithms*, Morgan Kaufmann, pp. 303-309, 1993

[10] M. Ebner and A. Zell. Evolving a task specific image operator. In: *Joint Proceedings of the First European Workshops on Evolutionary Image Analysis, Signal Processing and Telecommunications (EvoIASP'99 and EuroEcTel'99)*, Göteborg, Sweden, Springer-Verlag, pp. 74-89, 1999

[11] M. Ebner. On the evolution of interest operators using genetic programming. In: *Late Breaking Papers at EuroGP'98: the First European Workshop on Genetic Programming*, Paris, France, pp. 6-10, 1998

[12] M.D. Heath, S. Sarkar, T. Sanocki, and K.W. Bowyer. Comparison of edge detectors: A methodology and initial study. In: *Computer Vision and Pattern Recognition, Proceedings CVPR '96*, pp. 143-148, 1996

[13] T. Ito, I. Iba, and S. Sato. Non-destructive depth-dependent crossover for genetic programming. In: *Proceedings of the First European Workshop on Genetic Programming*, LNCS, Paris, pp. 14-15, 1998

[14] J.R. Sherrah, R.E. Bogner, and A. Bouzerdoum. The evolutionary preprocessor: Automatic feature extraction for supervised classification using genetic programming. In: *Genetic Programming 1997: Proceedings of the Second Annual Conference*. Stanford University, CA, USA. pp. 304-312, 1997
[15] T. Kanungo and R.M. Haralick. Receiver operating characteristic curves and optimal Bayesian operating points. In: *International Conference on Image Processing - Proceedings*, vol.3, pp. 256-259, Washington, DC., 1995
[16] C.A.C. Coello. An updated survey of evolutionary multiobjective optimization techniques: State of the art and future trends. In: *Congress on Evolutionary Computation*, pp. 3-13, Washington, D.C., 1999
[17] E. Zitzler and L. Thiele. An evolutionary algorithm for multiobjective optimization: The strength Pareto approach. *Technical Report*, 43, Computer Engineering and Communication Networks Lab (TIK), Swiss Federal Institute of Technology (ETH), Zurich, Switzerland, 1998
[18] A. Ekárt and S.Z. Németh. Selection based on the Pareto nondomination criterion for controlling code growth in genetic programming. *Genetic Programming and Evolvable Machines*, vol. 2, pp. 61-73, 2001
[19] C.M. Fonseca and P.J. Fleming. Multiobjective optimization and multiple constraint handling with evolutionary algorithms -Part I: A unified formulation. *IEEE Transactions on Systems, Man and Cybernetics-Part A: Systems and Humans*, vol. 28, pp.26-37, 1998
[20] R. Kumar and P.I. Rockett. Improved sampling of the Pareto-Front in multiobjective genetic optimization by Steady-State evolution: A Pareto converging genetic algorithm. *Evolutionary Computation*, vol.10, no. 3, pp. 283-314, 2002
[21] J. Canny. A computational approach to edge detection. *IEEE Trans. Pattern Analysis and Machine Intelligence*, vol. 8, no. 6, pp. 679-698, 1986
[22] S. Konishi, A.L. Yuille, J.M. Coughlan, and S.C. Zhu. Statistical edge detection: Learning and evaluating edge cues. *IEEE Trans. Pattern Analysis and Machine Intelligence*, vol. 25, no. 1, pp. 57-74, 2003
[23] M. Kotani, M. Nakai, and K. Akazawa. Feature extraction using evolutionary computation. In: *Proceedings of the Congress of Evolutionary Computation*, IEEE Press, pp. 1230-1236, 1999
[24] K. Bowyer, C. Kranenburg, and S. Dougherty. Edge detector evaluation using empirical ROC curves. *Computer Vision and Image Understanding*, vol.84, no.1, pp. 77-103, 2001
[25] D. P. Muni, N. R. Pal, and J. Das. A novel approach to design classifiers using genetic programming. *IEEE Transactions on Evolutionary Computation*, vol. 8, no. 2, pp.183-196, 2004
[26] O.L. Mangasarian, W.N. Street and W.H. Wolberg. Breast cancer diagnosis and prognosis via linear programming. *Operations Research*, vol. 43, no. 4, pp. 570-577, 1995
[27] O. L. Mangasarian and W. H. Wolberg. Cancer diagnosis via linear programming. *SIAM News*, vol. 23, pp. 1-18, 1990
[28] E. Alpaydin. Combined 5 2 cv F-test for comparing supervised classification learning algorithms. *Neural Computation*, vol. 11, no. 8, pp. 1885-1892, 1999
[29] N.R. Harvey, S.P. Brumby, S. Perkins, J.J. Szymanski, J. Theiler, J.J. Bloch, R.B. Porter, M. Galassi and A.C. Young. Image feature extraction: GENIE vs conventional supervised classification techniques. *IEEE Transactions on Geoscience and Remote Sensing*, vol. 40, no. 2, pp. 393-404, 2002

[30] W.C. Chen, N.A. Thacker and P.I. Rockett. An adaptive step edge model for self-consistent training of a neural network for probabilistic edge labeling. *IEE Proceedings - Vision, Image and Signal Processing*, vol. 143, no.1, pp. 41-50, 1996

[31] P.I. Rockett. Performance assessment of feature detection algorithms: A methodology and case study on corner detectors. *IEEE Transactions on Image Processing*, vol.12, no.11. pp. 1668-1676, 2003

[32] Y. Zhang and P.I. Rockett. The Bayesian operating point of the Canny edge detector. *IEEE Trans. Image Processing*, 2005 (submitted)

[33] K. Krawiec. Genetic programming-based construction of features for machine learning and knowledge discovery tasks. *Genetic Programming and Evolvable Machines*, vol.3, no.4, pp.329-343, 2002

[34] C. Harris. An investigation into the application of genetic programming techniques to signal analysis and feature detection. PhD. thesis, Dept. Comp. Science., Univ. College of London, Sep. 1997.

[35] C.L. Blake and C.J. Merz. UCI Repository of machine learning databases [http://www.ics.uci.edu/ mlearn/MLRepository.html]. Irvine, CA: University of California, Department of Information and Computer Science, 1998

[36] W. Schiffmann, M. Joost, and R. Werner. Synthesis and performance analysis of multilayer neural network architectures. Technical Report 16/1992, University of Koblenz, Institute für Physics, 1992.

Part II

Multi-Objective Learning for Accuracy Improvement

5

Regression Error Characteristic Optimisation of Non-Linear Models

Jonathan E. Fieldsend

School of Engineering, Computer Science and Mathematics,
University of Exeter, Exeter, UK, EX4 4QF
J.E.Fieldsend@exeter.ac.uk

Summary. In this chapter recent research in the area of multi-objective optimisation of regression models is presented and combined. Evolutionary multi-objective optimisation techniques are described for training a population of regression models to optimise the recently defined Regression Error Characteristic Curves (REC). A method which meaningfully compares across regressors and against benchmark models (i.e. 'random walk' and maximum *a posteriori* approaches) for varying error rates. Through bootstrapping training data, degrees of confident out-performance are also highlighted.

This approach is then extending to encapsulate the complexity of the model as a third objective to minimise. Results are shown for a number of data sets, using multi-layer perceptron neural networks.

5.1 Introduction

When forecasting a time series there are often different measurements of the *quality* of the signal prediction. These are numerous and often problem specific (a wide range of which are described in [1]). Recent advances in regressor comparison has been concerned not solely with the chosen error measure itself, but also with the distributional properties of a regressor's error – this has been formulated in a methodology called Regression Error Characteristic (REC) curve, introduced by Bi and Bennett [2]. This curve traces out the proportion of residual errors of a model which lie below a certain error threshold, allowing the comparison of models given different error property preferences.

This chapter proceeds with the introduction of evolutionary computation techniques as a process for generating REC curves for regressor families, instead of for individual regressors, allowing the visualisation of the potential prediction properties for an entire class of method for a problem.

This approach is then augmented with the simultaneous optimisation of model complexity, with a similar framework to that introduced by Fieldsend et al. [13]. The highlighting of regions of the REC curve on which we can

J.E. Fieldsend: *Regression Error Characteristic Optimisation of Non-Linear Models*, Studies in Computational Intelligence (SCI) **16**, 103–123 (2006)
www.springerlink.com

confidently outperform the maximum *a posteriori* (MAP) trained model is also introduced, through bootstrapping the optimised REC curve.

The chapter proceeds as follows: REC curves are formally defined and their properties discussed in Section 5.2. A general model for multi-objective REC optimisation is introduced in Section 5.3, and the regression models to be optimised and compared are presented in Section 5.4.3. Empirical results are presented on a range of problems in Section 5.4, and the methodology is extended to include complexity trade-off in Section 5.5. The chapter ends with a final discussion of results in Section 5.6.

5.2 Regression Error Characteristic Curves

Receiver Operating Characteristic (ROC) curves have proved useful for a number of years as a method to compare classifiers when the costs of misclassification are *a priori* unknown. In the binary classification case it plots the rates of correct classification of one class against the misclassification rates of the other class, typically derived through changing the threshold/cost of a particular parameterised classifier. This formulation allows the user to see what range of classifications they can obtain from a model, and also allows them to compare models where misclassification costs are unknown by using such measures as the area under the curve (AUC) and the Gini coefficient. A more in depth discussion of ROC curves can be found in the chapter on ROC optimisation in this book.

Inspired by this, Bi and Bennett [2] developed the Regression Error Characteristic curve methodology to represent the properties of regression models. In a regression problem, the task is to generate a good estimate of a signal y_i (called the dependent variable), from a transformation of one or more input signals $\mathbf{x}_i$ (called independent variables), such that

$$\hat{y}_i = f(\mathbf{x}_i, \mathbf{u}). \tag{5.1}$$

$\hat{y}_i$ is the regression model prediction of y_i, given the data $\mathbf{x}_i$ and the model parameters $\mathbf{u}$. The optimisation process of regression models typically takes the form of varying $\mathbf{u}$ in order to make $\hat{y}_i$ as close to y_i as possible, where *closeness* is calculated through some error term (like Euclidean distance or absolute error). This error is calculated for all the n training data points used, $\boldsymbol{\xi} = error(\hat{\mathbf{y}}, \mathbf{y})$, and the average error, $\bar{\xi}$ typically used as the scaler evaluation of the parameters $\mathbf{u}$.

In REC, instead of dealing purely with the average error of a regressor, the entire error distribution is of interest. The proportion of points forecast below a certain error threshold are plotted against the error threshold, for a range of error thresholds (from zero to the maximum obtained error for a null regressor on a single point). This effectively traces out an estimate of the cumulative distribution function of the error experienced by a regressor, and

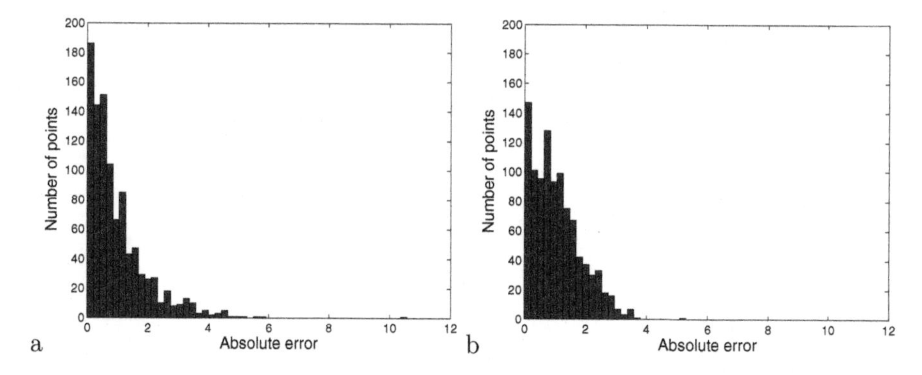

Fig. 5.1. Distribution of residual errors of two regression models, *a:* A and *b:* B.

can be created by ordering $\boldsymbol{\xi}$ in ascending order, and plotting this against the element index divided by n.

There are many useful properties of this representation, for instance the area over the curve (AOC) is a good estimate of the expected error of the model. However, probably the most useful contribution of the REC formulation is the easy visualisation of error information about a model across the range of data used to train or test it. This gives information to the user which may lead them to select a model other than the one with the lowest average error.

5.2.1 Illustration

Let us start with a toy example where we have two regression models available to us, A and B, which are predicting some time series (for instance the demand of a product in the next month). If model A experiences, on average, an absolute error of 0.99 and model B an absolute error of 1.06, then without any additional information one would typically choose to use model A as it exhibits a lower mean error. Knowing the distributional properties of this error, however, may lead to a different decision.

Figure 5.1 shows an (illustrative) error distribution of the absolute errors of models A and B, Figure 5.2 in turn traces out the REC curves for A (solid line) and B (dashed line). This shows that although model A has a lower average error than model B, its largest errors (the top 15%) are proportionally bigger than that of model B – meaning it makes more extreme errors than model B. Given that the cost of extreme errors to the user of the forecast may be proportionally greater than small errors (small under predictions of demand

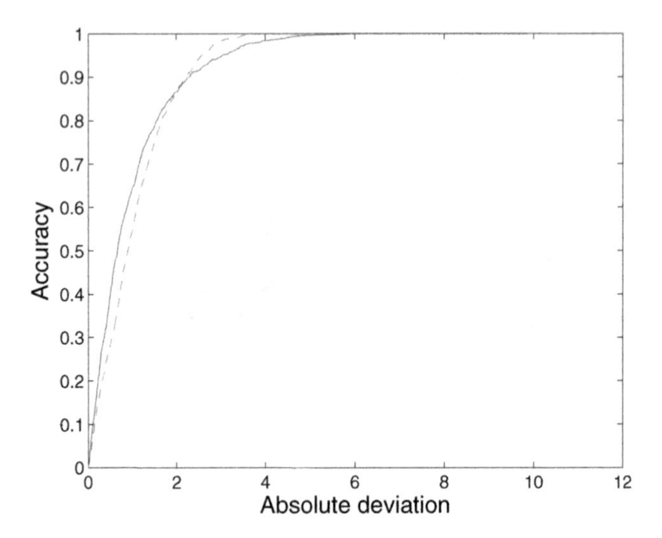

Fig. 5.2. REC curves of two regression models, A (solid) and B (dashed).

can be taken up by inventoried goods, large under predictions may result in turning customers away), B may actually be preferred.[1]

The final choice of model will depend upon the preferences and costs of the user of the model which may be difficult to incorporate in an optimisation algorithm when there is no *a priori* knowledge of the *shape* of the REC curves possible (see Das and Dennis [5]).

Bi and Bennett [2] used REC to compare different regressors trained to minimise the average error, and so were effectively interested in those models which minimised the AOC. When we are interested in a particular region of the REC curve (distributional property of our residual errors), minimising the AOC of a single model will not necessarily lead us to the best model given our preferences. However minimising the AOC of a *set* of models can.

Using the previous illustration, if we merge the two REC curves of models A and B, taking only the portions which are in front, we can create an REC curve which illustrates the possible error/accuracy combinations given the available models. The illustration in Figure 5.3 shows this composite REC curve along with the REC curves of two new models, C and D. As the REC of model C lies completely below the composite REC curve, we can see that for any possible error/accuracy combination models A and/or B are better than model C. Model D however is slightly in front of the composite REC for a small range of accuracy, and so would be useful to retain and offer to the end user as a possible model.

[1] It should be noted in the original work by Bi and Bennett they recommended ranking models by the AOC – i.e. an estimate of the mean expected error. However this assumes proportional costs of error, which in many situations is not the case.

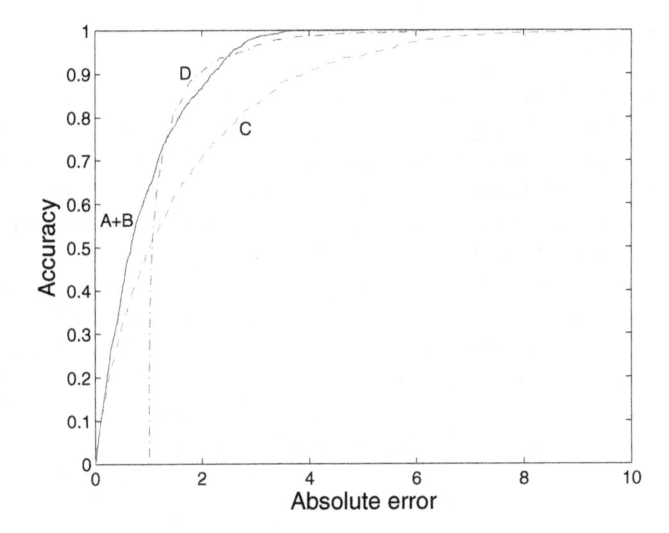

Fig. 5.3. Composite REC of models A (solid) and B (dashed), and REC curves of models C and D.

5.3 Multi-Objective Evolutionary Computation for REC

The generation of a composite REC curve to describe the possible error/accuracy combinations for a problem (given a model family or families) is easily cast in terms of multi-objective optimisation problem. However the casting itself can be in two different forms, which affect both the representation and the complexity of the optimisation process.

5.3.1 Casting as A Two-Objective Problem

The obvious representation of the REC optimisation problem is as a 2-objective problem, the first objective (error threshold) to be minimised and the second objective (accuracy) to be maximised. In this case a single parameterisation, $\mathbf{u}$, results in n error/accuracy pairs, which need to be compared to the current best estimate of the composite REC curve. Any values on the current best estimate which have higher error threshold for the same accuracy as $\mathbf{u}$ need to be removed as they are *dominated* and replaced by relevant the pair(s) from the evaluation of $\mathbf{u}$. Formulated in this fashion, $\mathcal{O}(n \log n)$ domination comparisons are needed for each parameter vector compared to the current best estimate of the Pareto front/composite REC curve.

We can instead however represent the problem as an n objective problem, which actually turns out to be faster.

5.3.2 Casting as An n-Objective Problem

On calculating the error, $\boldsymbol{\xi}$, of two parameterisations $\mathbf{u}$ and $\mathbf{v}$, and arranging them in ascending order, we can use these n dimensional errors as the fitness vectors for $\mathbf{u}$ and $\mathbf{v}$. The accuracy term (the second objective in the previous formulation) always takes on the value of the index (over n) of the elements of $\boldsymbol{\xi}$, so each element of the ordered $\boldsymbol{\xi}$ for $\mathbf{u}$ is coupled with the corresponding element of the ordered $\boldsymbol{\xi}$ of $\mathbf{v}$. If all the error thresholds at each index for the regressor with decision vector (parameters) $\mathbf{u}$ are no larger than the error thresholds at the corresponding index for regressor $\mathbf{v}$ and at least one threshold is lower, then the regressor parameterised by $\mathbf{u}$ is said to *strictly dominate* that parameterised by $\mathbf{v}$ ($\mathbf{u} \succ \mathbf{v}$).

Traditionally, a set F of decision vectors is said to be *non-dominated* if no member of the set is dominated by any other member:

$$\mathbf{u} \not\prec \mathbf{v} \quad \forall \mathbf{u}, \mathbf{v} \in F, \tag{5.2}$$

and this formulation is used to store the archive of the best estimate of the *Pareto front* found by an optimisation process. In the REC optimisation situation, because the set F itself traces out the composite REC, our decision to add or remove an element to/from F is not quite so straightforward as that in (5.2). If we define $REC(F)$ as the composite curve generated by the elements in F, then we actually want to maintain F such that:

$$\mathbf{u} \not\prec REC(F \setminus \mathbf{u}) \quad \forall \mathbf{u} \in F. \tag{5.3}$$

For this formulation a single domination comparison is needed for each parameter vector compared to the current best estimate of the Pareto front/ composite REC curve (albeit a comparison across n objectives as opposed to 2). Additionally we know that at most n unique model parameterisations will describe the REC composite curve.

Interestingly the points returned by casting the problem as a 2-objective or as an n-objective problem are slightly different – the n objective optimisation of the curve returning an attainment surface representation of the REC curve (along the error axis) [35, 29], whereas the 2-objective optimisation returns a strictly non-dominated front representation (and therefore may have less than n elements, as illustrated in Figure 5.4

5.4 Empirical Illustration

In this section a simple multi-objective evolutionary algorithm (MOEA) is introduced to show the generation of REC curves for a number of different well-known autoregressive problems from the literature, namely the Santa Fe competition suite of problems (see Weigend and Gershenfeld [32]). Note that

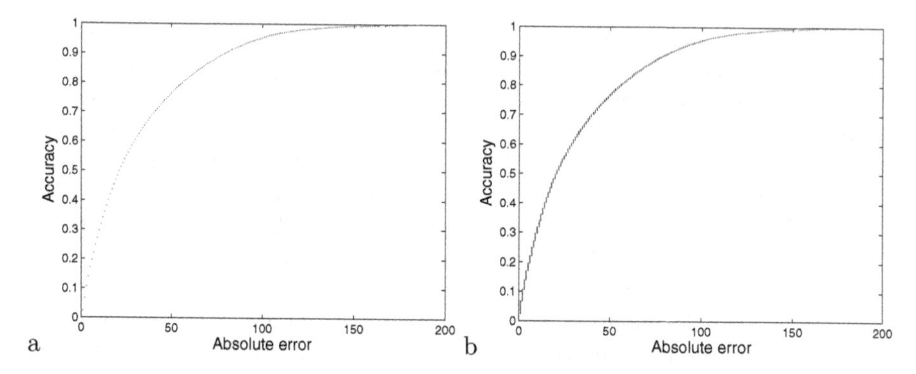

Fig. 5.4. Composite REC curves from casting the REC optimisation problem as a *a:* 2-objective or a *b:* n-objective one. Note that all the points in *(a)*, are included in *(b)*.

Algorithm 2 REC optimising MOEA.

1:	$F :=$ initialise()	Initial front estimate
2:	$n := 0$	
3:	`while` $n < N$:	Loop
4:	$\mathbf{u} :=$ evolve(select(F))	Copy and evolve from F
5:	REC($\mathbf{u}$)	Evaluate
6:	**if** $\mathbf{u} \not\preceq REC(F)$	If non-dominated
7:	$F := F \cup \mathbf{u}$	Insert in archive
8:	$F := \{\mathbf{v} \in F \mid \mathbf{v} \not\prec REC(F \setminus \mathbf{v})\}$	Remove any dominated
9:	**end**	
10:	$n := n + 1$	
11:	**end**	

any recent MOEA from the literature could equally be used [17, 20, 36, 4, 31, 6, 7] – however they would need augmentation to compensate for the problem specific archive update shown in (5.3).

5.4.1 The Optimiser

The MOEA used here is based on a simple (1+1)-evolutionary algorithm, a model which has been used extensively in the literature [20, 21, 15, 9, 14, 12, 23, 10, 22]. An overview is provided in Algorithm 2. The process commences with the generation of an initial estimate of the REC curve for the problem (Algorithm 2, line 1). This can typically be provided by the generation of random parameterisations of the model and/or the optimisation of the model with traditional scaler optimisers concerned with the mean error (i.e. back-propagation or scaled conjugate gradient for a neural network (NN) model [27]). These decision vector(s) are stored in F. The algorithm continues by iterating through a number of generations (line 3), and at each generation creating a new model parameterisation, $\mathbf{u}$, through mutation and/or crossover of elements of F (line 4). $\mathbf{u}$ is compared at each generation to F (line 6). If it is non-dominated by the composite front defined by F, it is inserted into F (line 7), and any elements in F which no longer contribute to the composite front are removed (line 8).

An equivalent formulation would be to simply update F to minimise its AOC, and remove any elements that do not contribute to its minimisation – effectively a set based scalar optimisation. However as later in this chapter the additional imposition of a complexity minimisation objective will be introduced, the nominally multi-objective formulation will be adhered to here.

The evolve(select(F)) methods (line 4) either generates a new solution by single point crossover of two members from F and perturbing the weights, or by copying a single solution from F and perturbing its weights. Crossover probability was 0.5, weight perturbation probability 0.8 and weight perturbation multiplier 0.01 (perturbation values themselves were drawn from a Laplacian distribution). Parameters used are encoded as real values and crossover occurred at the transform unit level. Different representations and crossover/perturbation/selection methods could equally be used, and an excellent review of those concerned with NN training can be found in Yao [34].

5.4.2 Data

The data used in this chapter to show the properties of REC optimisation is the Santa Fe competition data [32], which is a suite of autoregressive and multi-variate data problems exhibiting different properties. The final series, Series F, is not used here as this exhibits the missing data property, which this chapter is not concerned with.

The other 5 series are:

- Series A: Laser generated data.
- Series B: Three sets of physiological data, spaced by 0.5 second intervals. The first set is the heart rate, the second is the chest volume (respiration force), and the third is the blood oxygen concentration (measured by ear oximetry).
- Series C: Tickwise bids for the exchange rate from Swiss Francs to US dollars. Recorded by a currency trading group for 12 days.
- Series D: computer generated time series.
- Series E: Astrophysical data. A set of measurements of the light curve (time variation of the intensity) of the variable white dwarf star PG1159-035 during March 1989, at 10 second intervals.

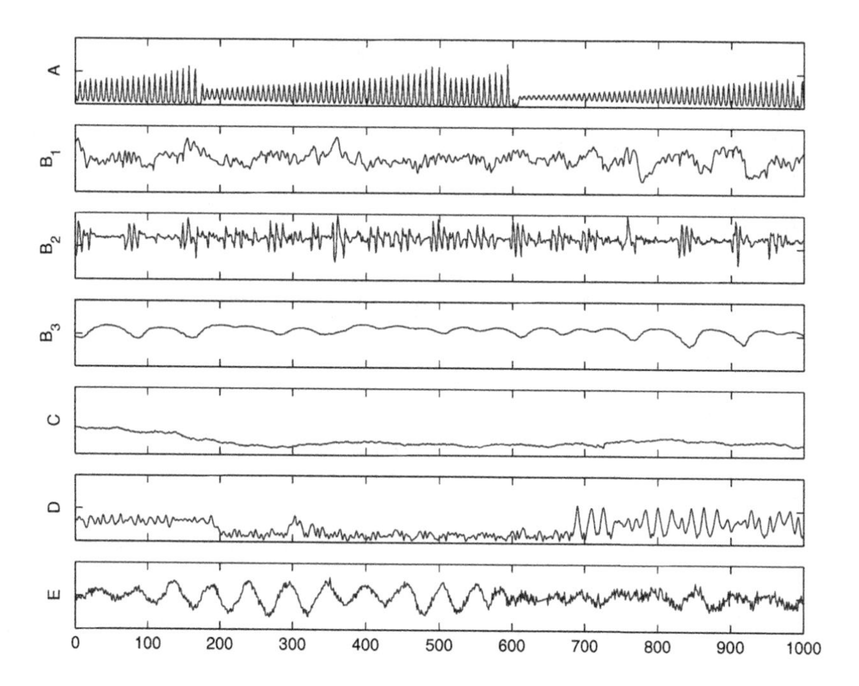

Fig. 5.5. Santa Fe competition series. 1000 sequential points of each used for training.

All may be obtained from `http://www-psych.stanford.edu/~andreas/Time-Series/SantaFe.html`. Figure 5.5 shows the first 1000 samples of each of the series (except for series B, where the second 1000 is shown). As can be seen, the series exhibit varying degrees of oscillation and rapid change. In the prediction tasks used here, all series are standardised (the mean of the training data subtracted and divided through by the variance of the training data), however the error reported is based on the prediction transformed back into the original range.

Series A, C, D and E are autoregressive problems (i.e. past values of the same series are used to predict future values). One step ahead predictions are made, and the number of lags (past steps) to use is determined by observing the auto-correlation between steps (shown in Figure 5.6). Series B is a multivariate problem, and is formulated here as a problem of predicted one of the series at time t given the values of the two other series at t.

On inspection of the autocorrelations, 40 lags are used for series A and D, 10 lags for series E. Series C is highly correlated even with very large lags (see Figure 5.6). 5 lags were chosen for this series, but the correlation levels and *a priori* knowledge of the exchange rate market would indict results outperforming a random walk would be surprising.

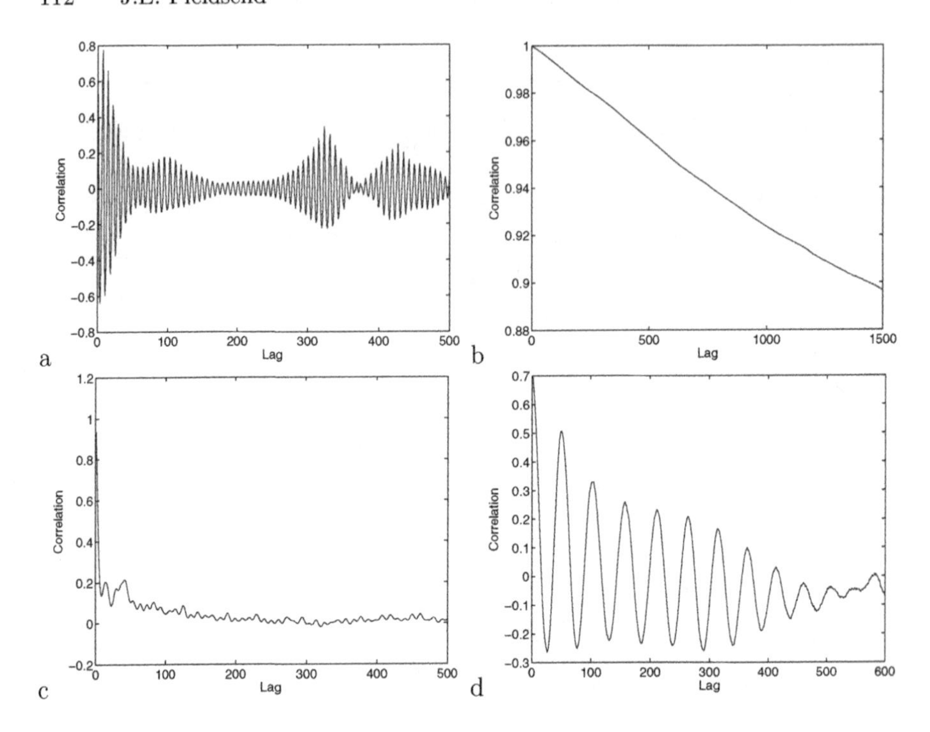

Fig. 5.6. Autocorrelation of series *a:* A, *b:* C, *c:* D and *d:* E.

5.4.3 Non-linear Models

In the traditional linear regression model the functional transformation takes the form of

$$\hat{y} = u_1x_1 + u_2x_2 + \ldots + u_{m-1}x_p + u_m \tag{5.4}$$

with correspondingly $m = p+1$ model parameters to fit, where p is the number of independent variables.

Here however we shall use the non-linear multi-layer perceptrons (MLP) neural network regression model. In an MLP, the functional transformation takes the form of a number of parallel and sequential functional transformations. With k parallel transformation functions (known as the hidden layer), and a single hidden layer, they can be represented in their regression form as:

$$\hat{y} = f_1(\mathbf{x}, u_1, \ldots, u_l) + f_2(\mathbf{x}, u_{l+1}, \ldots, u_{2l}) + \ldots + f_k(\mathbf{x}, u_{(k-1)l}, \ldots, u_{kl}) + u_m \tag{5.5}$$

In the case of the MLP transfer, these units take the form of a hyperbolic tangent:

$$f(\mathbf{x}, \mathbf{q}) = \tanh\left(q_{p+1} + \sum_{i=1}^{p} q_i x_i\right) q_{p+2} \tag{5.6}$$

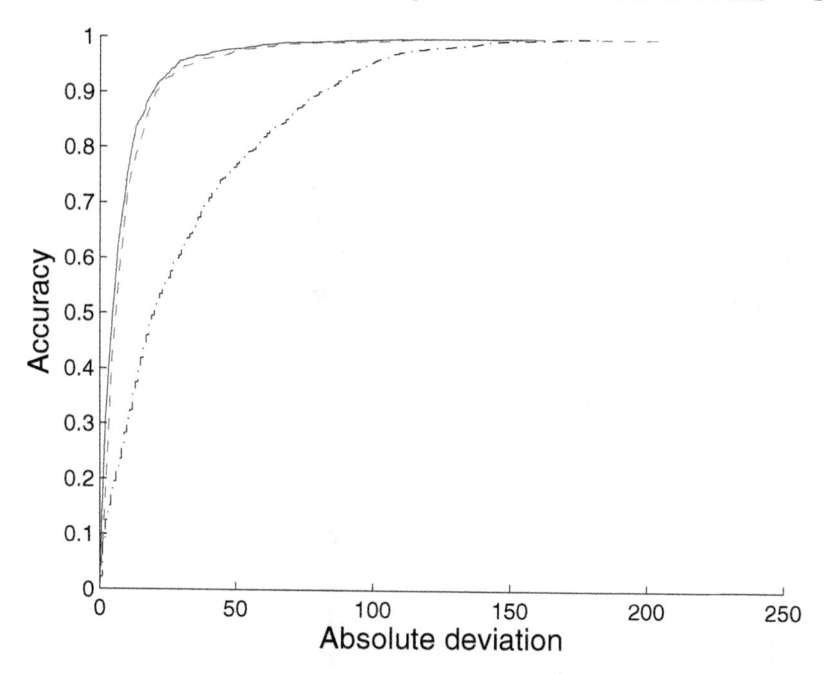

Fig. 5.7. REC curves of Santa Fe competition series 'A' for a 5 hidden unit MLP. Solid line evolved composite REC, dashed line optimised REC curve of scalar best model and dot-dashed line REC of null model.

The first p elements of $\mathbf{q}$ are the weights between the inputs to the hidden unit. The $p+1$th element is the unit bias and the $p+2$th element is the weight between the unit and the output. It therefore has $m = k(p+2)+1$ parameters.

5.4.4 Results

The first example results are shown here for a 5 hidden unit multi-layer perceptron neural network. The model is initially trained using a scaled conjugate gradient algorithm [27], with acts as an initial F. As series A is highly oscillatory the first difference is used as the dependant variable, and the final prediction reconstructed from this. The null model, which is typically the mean of the data, in this formulation is therefore the more appropriate random walk model (which predicts that $\hat{y}_t = y_{t-1}$) – as differencing the data gives a mean of zero.

Figure 5.7 shows the composite (evolved) REC curve after 20000 generations, the REC curve of a single model optimised with the scaled conjugate gradient algorithm for 1000 epochs, and the null model. It should be noted that the training of the single neural network and the subsequent run of the MOEA took approximately the same computation time. The composite REC

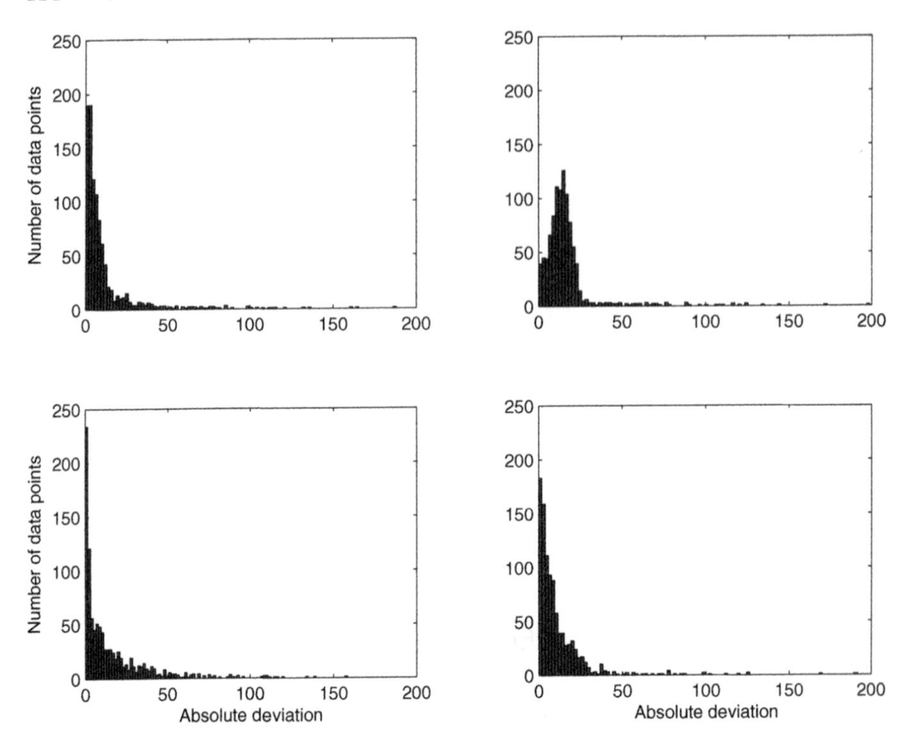

Fig. 5.8. Histograms of models on evolved composite REC. Bottom left, lowest threshold at 0.1 level. Top left lowest threshold at 0.5 level. Top right lowest threshold at 0.9 level. Bottom right histogram of lowest mean error model.

curve is only slightly in front of the single AOC minimising model, however it does completely dominate it. Both curves are well in front of the null model – implying there is indeed information in the series which enables a degree of prediction beyond the most simple formulation.

Figure 5.8 gives the error histograms of three different points on the composite REC curve, to better illustrate the qualitative difference between the errors made by regressors on different points on the composite REC curve, the bottom right histogram shows that of the single minimising AOC Model (trained with the scaled conjugate gradient algorithm – the maximum *a posteriori* model). The bottom left histogram (corresponding to the model with the lowest threshold at the 0.1 level) can be seen to exhibit the greatest number of points with very low absolute error. Conversely, although the mean error of the histogram in the top right is pushed higher than the other four models shown, it exhibits fewer very high errors than the others. Figure 5.9 shows the actual errors corresponding to the histograms provided in Figure 5.8, which shows where these errors are being made.

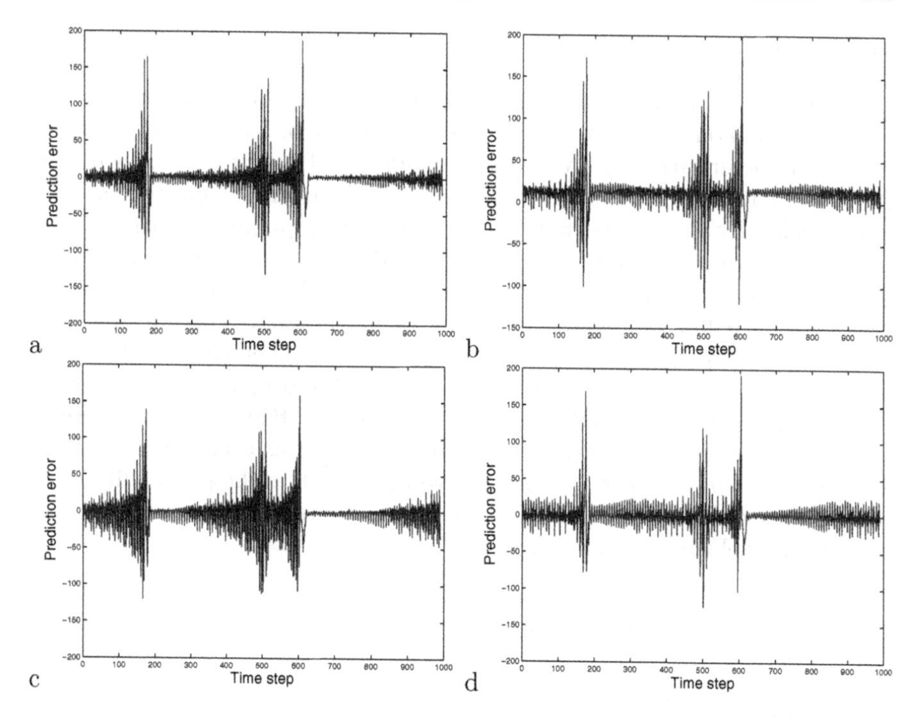

Fig. 5.9. Errors of models on evolved composite REC. *a:* lowest threshold at 0.5 level. *b:* lowest threshold at 0.9 level. *c:*, lowest threshold at 0.1 level. *d:* histogram of lowest mean error model.

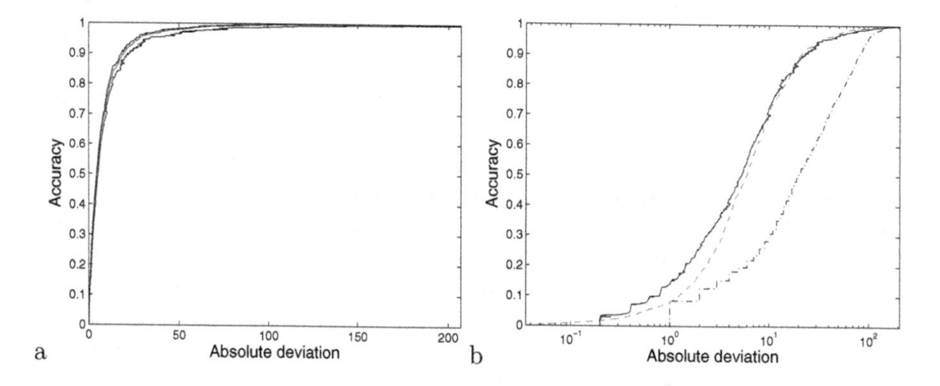

Fig. 5.10. *a:* Probability contours of REC for MLP on series A, 5% and 95% level contours shown around composite REC curve. *b:* 95% level composite REC contour and REC curve of MAP model (dashed) and null model (dot-dashed). Error in log scale to aid visualisation.

Uncertainty

In [2] the REC curve presented are the means of a number of different runs (on cross validation data), meaning they were an average of a number of different model parameterisations on a number of different datasets – as each model was trained in turn on the different data. What we are interested in here more specifically is the expected variation of a *single* parameterisation, as in the end we have to decide upon a single model. We can do this effectively by bootstrapping our training data [8], and noting the variation in errors to generate a probability of operating in a particular objective space region [11].

More formally, by bootstrapping the data we are generating a data set of the same size, which is statistically equivalent to the original. If we generate p bootstrap replications, and evaluate our composite REC curve on these p data sets, we are provided with p error thresholds for each accuracy level. This gives us an $n \times p$ set of error values Ξ. We can calculate the probability that the regressor defining the REC curve point at accuracy level, i, will, have a lower error level than a value $\tilde{e}$ as:

$$p(\Xi_i < \tilde{e}) = \frac{1}{p} \sum_{j=1}^{p} I(\Xi_{i,j} < \tilde{e}) \tag{5.7}$$

where $I(\cdot)$ is the indicator function.

Figure 5.10a shows the probability contours for composite REC front shown in Figure 5.7, at the 5% and 95% levels, created from 200 bootstrap resamples of the data. Figure 5.10b in turn shows the 95% REC contour and the REC curve of the single model trained using the conjugate gradient algorithm. From this we can say not only (from Figure 5.7) that the single MAP curve lies completely in front of that of the single MAP model, but that we are confident (at the 95% level) that it will lie in front of the single REC model for accuracy levels from 0.02 up to 0.85 on statistically equivalent data.

Figure 5.11 shows these fronts for the other four test problems (once more using an MLP with 5 hidden units). Again the 95% composite REC contour (solid line), the REC of the MAP model (dashed line) and the null model (dot-dashed line) are plotted. (The null model is again set as the random walk model – which is a far better fit than the mean allocation model suggested in [2].) From these we can see that we are confident (at the 95% level) of the composite REC models outperforming the single MAP model on the accuracy range 0.01-0.5 for series B, 0.00-0.10 for series D and 0.00-0.30 on series E. In the case of series C, the REC of the MAP model (and for most of the null model) lies in front of the 95% composite REC contour, implying there is little or no information in the series beyond a random walk prediction.

Until this point we have been concerned with the uncertainty over the error preferences for a regression model (leading to the optimisation of REC curves), and the uncertainty over the variability of the data used (leading to the use of probability contours). It is also very likely that, although we may

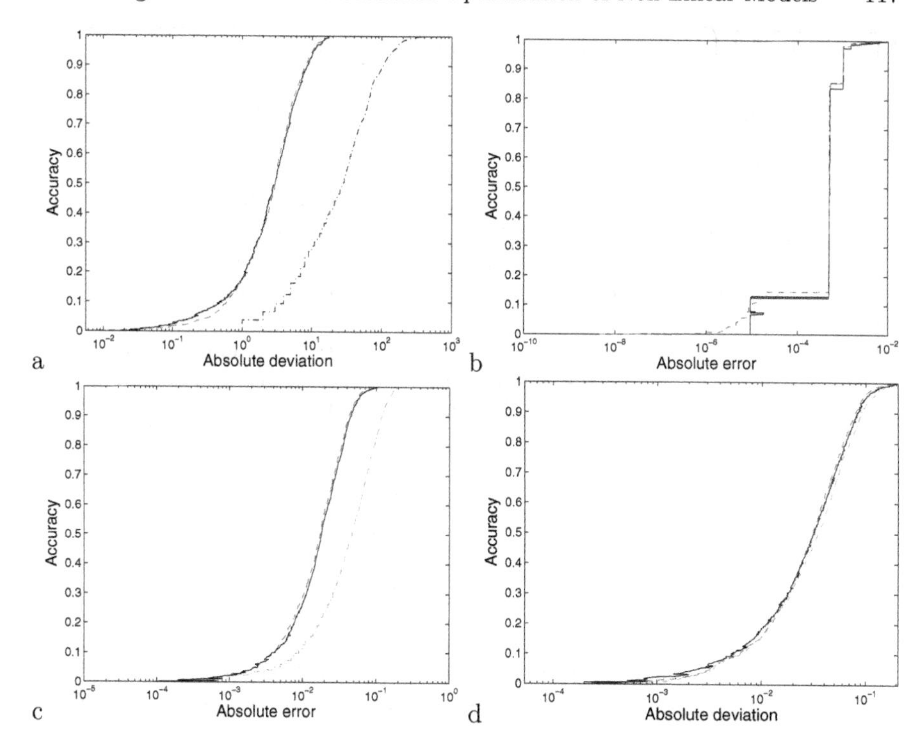

Fig. 5.11. 95% level composite REC contour and REC curve of MAP model (dashed) and null model (dot-dashed) for series *a:* B, *b:* C, *c:* D and *d:* E. Error in log scale to aid visualisation.

have a preferred regression model type, we may not know how complex a model we should use – i.e., in the case of NNs, how many hidden units, how many inputs, what level of connectivity? In the final section of this chapter the previous optimising model will be extended so that a set of REC curves can be returned, in one optimising process, which describe the error/accuracy trade-off possibilities for a range of model complexities.

5.5 Complexity as An Additional Objective

An additional problem which is manifest when training regression models is how to specify a model that is sufficient to provide 'good' results to the task at hand without any *a priori* knowledge of how complex the function you wish to emulate actually is. Too simple a model and the performance will be worse than is actually realisable, too complex a model and one runs the risk of 'overfitting' the model and promoting misleading confidence on the actual error properties of your chosen technique. Depending upon the regressor used, various methods to tackle this problem are routinely in use. In the

Algorithm 3 REC and complexity optimising MOEA.

1:	$F :=$ initialise()	Initial front estimate
2:	$n := 0$	
3:	`while` $n < N$:	Loop
4:	$\quad \mathbf{u} :=$ evolve(select(F))	Copy and evolve from F
5:	$\quad$ REC($\mathbf{u}$)	Evaluate
6:	$\quad \tilde{F}^- := F \setminus \mathbf{v} \in F$ where $\lvert\mathbf{v}\rvert \leq \lvert\mathbf{u}\rvert$	Lower and equal complexity
7:	$\quad$ `if` $\mathbf{u} \npreceq REC(\tilde{F})$	If non-dominated
8:	$\qquad F := F \cup \mathbf{u}$	Insert in archive
9:	$\qquad \tilde{F}^+ := F \setminus \mathbf{v} \in F$ where $\lvert\mathbf{v}\rvert \geq \lvert\mathbf{u}\rvert$	Higher and equal complexity
10:	$\qquad \tilde{F}^- := F \setminus \mathbf{v} \in F$ where $\lvert\mathbf{v}\rvert < \lvert\mathbf{u}\rvert$	Lower complexity
11:	$\qquad \tilde{F}^+ := \{\mathbf{v} \in \tilde{F}^+ \lvert \mathbf{v} \nprec REC(\tilde{F}^+ \setminus \mathbf{v})\}$	Remove any dominated
12:	$\qquad F := \tilde{F}^+ \cup \tilde{F}^-$	
13:	$\quad$ `end`	
14:	$\quad n := n + 1$	
15:	`end`	

neural network domain these take the guise of weight decay regularisation [28, 3], pruning [24], complexity loss functions [33] and cross validation topology selection [30].

A multi-objective formulation of this problem, which recognised the implicit assumptions about the interaction of model complexity and accuracy that penalisation methods make (e.g. [25]), was recently proposed by the author [16], and Jin et al. [19]. This casts the problem of accuracy and complexity as an explicit trade-off, which could be traced out and visualised without imposing any complexity costs *a priori.*

This method can be applied to the REC optimisation to trace out estimated optimal REC curves for different levels of model complexity. Here complexity shall be cast in terms of the number of model parameters – the larger the parameterisation of a model from a family, the more complex. In the linear regression case this is simply the number of coefficients. In the MLP case this is the number of weights and biases.

5.5.1 Changes to the Optimiser

The evolve(select(F)) methods (line 4) of Algorithm 3 is adjusted in this extension to allow the generation and evolutionary interaction of parameterisations of differing complexity (dimensionality). The existing crossover allows the interaction of models with different dimensionality – in addition weight deletion is also incorporated with a probability of 0.1.

F now contains a set of models generating a composite REC curve for each complexity level, as such its update is also modified from Algorithm 2.

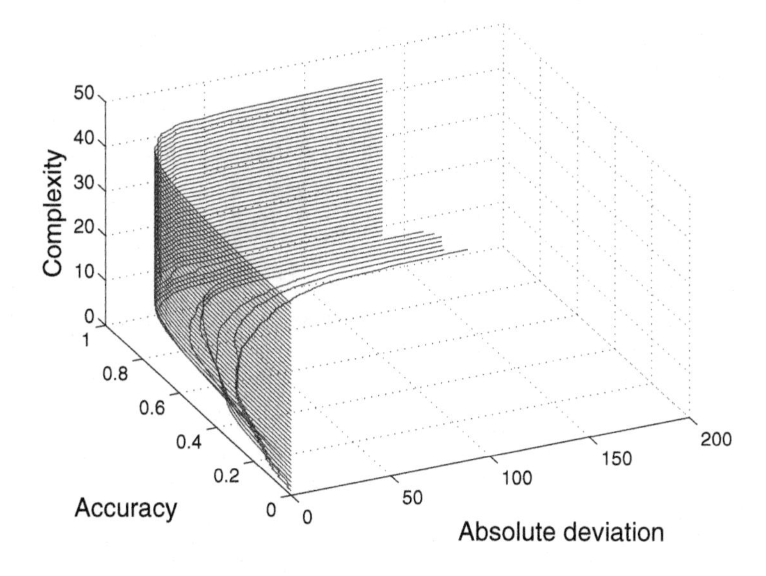

Fig. 5.12. REC/complexity surface for problem A.

Line 6 of Algorithm 3 shows the selection from F of those members with equal or lower complexity to the new parameterisation $\mathbf{u}$ into $\tilde{F}^{-}$. $\mathbf{u}$ is then compared to the composite REC curve defined by the members of $\tilde{F}^{-}$, if it is non-dominated, then it is inserted into F (line 8) and those members of F with and equal or higher complexity than $\mathbf{u}$ are then compared to $\mathbf{u}$, and any dominated members removed (line 11).

Although the method in [19] uses a MOEA with a constrained archive, Jin et al. also report the occurrence of degrading estimates of the Pareto front as the search progressed (the elite archive containing points dominated by some solutions found earlier in the search). This problem with using constrained archives in multi-objective search has been highlighted previously in [9, 12]. As such the methodology of [16] is maintained here and an unconstrained elite archive of non-dominated solutions is maintained. Although the computational cost is higher with this level of maintenance, there are data structures available to make this process more efficient [9, 26, 12, 18], and we have the additional benefit in this application that we know there is a limit on the archive size (as discussed earlier in Section 5.3).

Apart from these alterations the algorithm is run as previously, this time for 50000 generations.

5.5.2 Empirical Results

Figure 5.12 shows the composite REC curves for various MLP complexities for the Series A problem previously defined. As can be see, the REC curve rapidly approaches a stable position with only 10 weights in the network.

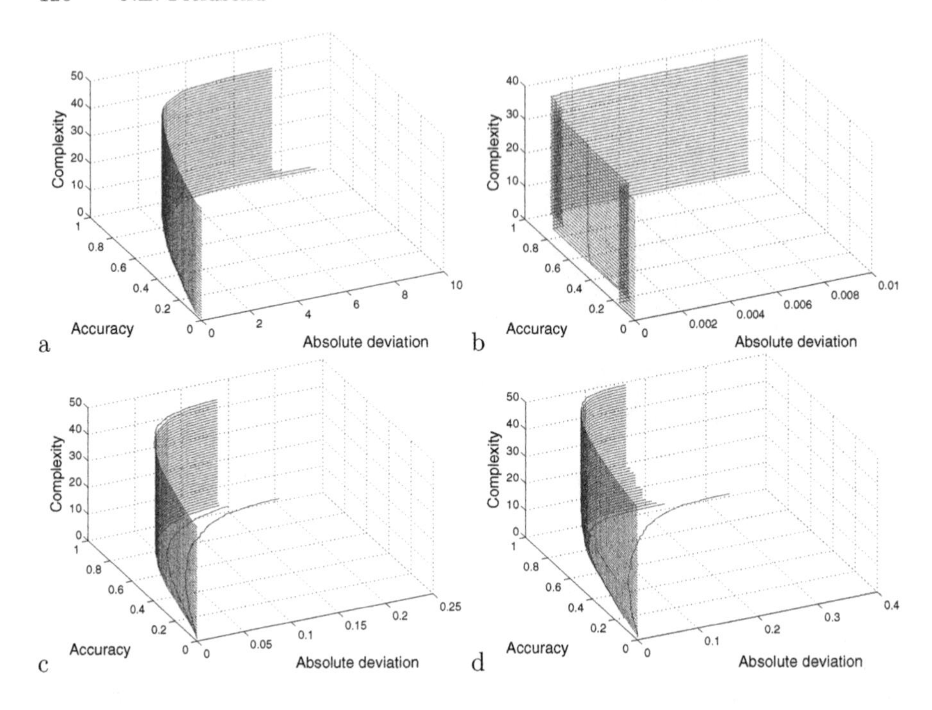

Fig. 5.13. REC curves of series *a:* B, *b:* C, *c:* D and *d:* E, for varying complexities.

MLPs with less than 10 weights are seen to suffer from larger absolute deviations at the limit of accuracy, as well as higher deviations for given accuracy levels in the range. Beyond 10 weights the absolute deviations at the limit of accuracy are constant, and only very minimal improvements in deviation for given accuracy are observed; inferring all that is left to model is noise, or is not predictable given the inputs.

Figure 5.13 in turn provides these plots for the other 4 problems previously described. Test problem B can also be adequately described with only 10 weights whereas the REC curves of test problem C are relatively unchanged across all complexities. There are very small adjustments throughout the range of complexities shown for problem D, although good results can be obtained with relatively low complexity. On the other hand 20 weights are needed for problem E before the improvements to the REC curve become relatively small for higher complexity levels.

5.6 Conclusion

This chapter has introduced the use of MOEAs to optimise a composite REC curve, which describes the (estimated) best possible error/accuracy trade off a regression model can produce for a particular problem.

REC specific properties where also highlighted – the hard limit of the number of different parameterisations possible, the casting as a 2 objective, n objective or even scalar problem, and the different computational complexities of these different formulations.

The optimisation method was extended by the use of bootstrapping to show the probability of performance, on statistically equivalent data. The problem itself was then expanded so that the complexity of the model was also optimised, producing a REC surface, which makes it easy to identify the minimum level of complexity needed to achieve specific error/accuracy combinations for a particular model family on a particular problem.

It is also useful to note that the use of bootstrapping is equally applicable for examining the significance of performance of a single model parameterisation, across accuracy ranges, to a baseline model.

Acknowledgements

The author gratefully acknowledges support from the EPRSC grant number GR/R24357/01 during the writing of this chapter. The NETLAB toolbox for MATLAB, available from `http://www.ncrg.aston.ac.uk/netlab/index.php`, was used as the basis of the NNs used here.

References

[1] J.S. Armstrong and F. Collopy. Error measures for generalizing about forecasting methods: Empirical comparisons. *International Journal of Forecasting*, 8(1):69–80, 1992.
[2] J. Bi and K.P. Bennett. Regression Error Characteristic Curves. In *Proceedings of the Twentieth International Conference on Machine Learning (ICML-2003)*, pages 43–50, Washington DC, 2003.
[3] C.M. Bishop. *Neural Networks for Pattern Recognition.* Oxford University Press, 1998.
[4] C.A. Coello Coello. A Comprehensive Survey of Evolutionary-Based Multiobjective Optimization Techniques. *Knowledge and Information Systems. An International Journal*, 1(3):269–308, 1999.
[5] I. Das and J.Dennis. A closer look at drawbacks of minimizing weighted sums of objectives for pareto set generation in multicriteria optimization problems. *Structural Optimization*, 14(1):63–69, 1997.
[6] K. Deb. *Multi-Objective Optimization Using Evolutionary Algorithms.* Wiley, Chichester, 2001.
[7] K. Deb, A. Pratap, S. Agarwal, and T. Meyarivan. Fast and elitist multiobjective genetic algorithm: NSGA–II,. *IEEE Transactions on Evolutionary Computation*, 6(2):182–197, 2002.
[8] B. Efron and R.J. Tibshirani. *An Introduction to the Bootstrap.* Number 57 in Monographs on Statistics and Probability. Chapman & Hall, New York, 1993.

[9] R.M. Everson, J.E. Fieldsend, and S. Singh. Full Elite Sets for Multi-Objective Optimisation. In I.C. Parmee, editor, *Adaptive Computing in Design and Manufacture V*, pages 343–354. Springer, 2002.
[10] J.E. Fieldsend and R.M. Everson. ROC Optimisation of Safety Related Systems. In J. Hernández-Orallo, C. Ferri, N. Lachiche, and P. Flach, editors, *Proceedings of ROCAI 2004, part of the 16th European Conference on Artificial Intelligence (ECAI)*, pages 37–44, Valencia, Spain, 2004.
[11] J.E. Fieldsend and R.M. Everson. Multi-objective Optimisation in the Presence of Uncertainty. In *Proceedings of the IEEE Congress on Evolutionary Computation (CEC'05)*, 2005. Forthcoming.
[12] J.E. Fieldsend, R.M. Everson, and S. Singh. Using Unconstrained Elite Archives for Multi-Objective Optimisation. *IEEE Transactions on Evolutionary Computation*, 7(3):305–323, 2003.
[13] J.E. Fieldsend, J. Matatko, and M. Peng. Cardinality constrained portfolio optimisation. In Z.R. Yang, R. Everson, and H. Yin, editors, *Proceedings of the Fifth International Conference on Intelligent Data Engineering and Automated Learning (IDEAL'04)*, number 3177 in Lecture Notes in Computer Science, pages 788–793. Springer, 2004.
[14] J.E. Fieldsend and S. Singh. A Multi-Objective Algorithm based upon Particle Swarm Optimisation, an Efficient Data Structure and Turbulence. In *Proceedings of UK Workshop on Computational Intelligence (UKCI'02)*, pages 37–44, Birmingham, UK, Sept. 2-4, 2002.
[15] J.E. Fieldsend and S. Singh. Pareto Multi-Objective Non-Linear Regression Modelling to Aid CAPM Analogous Forecasting. In *Proceedings of the 2002 IEEE International Joint Conference on Neural Networks*, pages 388–393, Hawaii, May 12-17, 2002. IEEE Press.
[16] J.E. Fieldsend and S. Singh. Optimizing forecast model complexity using multi-objective evolutionary algorithms. In C.A.C Coello and G.B. Lamont, editors, *Applications of Multi-Objective Evolutionary Algorithms*, pages 675–700. World Scientific, 2005.
[17] C.M. Fonseca and P.J. Fleming. An Overview of Evolutionary Algorithms in Multiobjective Optimization. *Evolutionary Computation*, 3(1):1–16, 1995.
[18] M.T. Jensen. Reducing the run-time complexity of multiobjective EAs: The NSGA-II and other algorithms. *IEEE Transactions on Evolutionary Computation*, 7(5):503–515, 2003.
[19] Y. Jin, T. Okabe, and B. Sendhoff. Neural network regularization and ensembling using multi-objective evolutionary algorithms. In *Proceedings of the IEEE Congress on Evolutionary Computation (CEC'04)*, pages 1–8. IEEE Press, 2004.
[20] J. Knowles and D. Corne. The pareto archived evolution strategy: A new baseline algorithm for pareto multiobjective optimisation. In *Proceedings of the 1999 Congress on Evolutionary Computation*, pages 98–105, Piscataway, NJ, 1999. IEEE Service Center.
[21] J.D. Knowles and D. Corne. Approximating the Nondominated Front Using the Pareto Archived Evolution Strategy. *Evolutionary Computation*, 8(2):149–172, 2000.
[22] M. Laumanns, L. Thiele, and E. Zitzler. Running Time Analysis of Multiobjective Evolutionary Algorithms on Pseudo-Boolean Functions. *IEEE Transactions on Evolutionary Computation*, 8(2):170–182, 2004.

[23] M. Laumanns, L. Thiele, E. Zitzler, E. Welzl, and K. Deb. Running Time Analysis of Multi-objective Evolutionary Algorithms on a Simple Discrete Optimization Problem. In J.J. Merelo Guervós, P. Adamidis, H-G Beyer, J-L Fernández-Villacañas, and H-P Schwefel, editors, *Parallel Problem Solving from Nature—PPSN VII*, Lecture Notes in Computer Science, pages 44–53. Springer-Verlag, 2002 2002.
[24] Y. LeCun, J. Denker, S. Solla, R. E. Howard, and L. D. Jackel. Optimal brain damage. In D. S. Touretzky, editor, *Advances in Neural Information Processing Systems II*, pages 598–605, San Mateo, CA, 1990. Morgan Kauffman.
[25] Y. Liu and X. Yao. Towards designing neural network ensembles by evolution. *Lecture Notes in Computer Science*, 1498:623–632, 1998.
[26] S. Mostaghim, J. Teich, and A. Tyagi. Comparison of Data Structures for Storing Pareto-sets in MOEAs. In *Congess on Evolutionary Computation (CEC'2002)*, volume 1, pages 843–848, Piscataway, New Jersey, May 2002. IEEE Press.
[27] I.T. Nabney. *Netlab: Algorithms for Pattern Recognition.* Springer, 2002.
[28] Y. Raviv and N. Intrator. Bootstrapping with noise: An effective regularization technique. *Connection Science*, 8:356–372, 1996.
[29] K.I. Smith, R.M. Everson, and J.E. Fieldsend. Dominance Measures for Multi-Objective Simulated Annealing. In *Proceedings of the IEEE Congress on Evolutionary Computation (CEC'04)*, pages 23–30. IEEE Press, 2004.
[30] J. Utans and J. Moody. Selecting neural network architectures via the prediction risk: application to corporate bond rating prediction. In *Proc. of the First Int. Conf on AI Applications on Wall Street*, pages 35–41, Los Alamos,CA, 1991. IEEE Computer Society Press.
[31] D. Van Veldhuizen and G. Lamont. Multiobjective Evolutionary Algorithms: Analyzing the State-of-the-Art. *Evolutionary Computation*, 8(2):125–147, 2000.
[32] A. S. Weigend and N. A. Gershenfeld, editors. *Time Series Prediction: Forecasting the Future and Understanding the Past.* Addison-Wesley, Reading, MA, 1994.
[33] D. Wolpert. On bias plus variance. *Neural Computation*, 9(6):1211–1243, 1997.
[34] X. Yao. Evolving Artificial Neural Networks. *Proceedings of the IEEE*, 87(9):1423–1447, 1999.
[35] E. Zitzler. *Evolutionary Algorithms for Multiobjective Optimization: Methods and Applications.* PhD thesis, Swiss Federal Institute of Technology Zurich (ETH), 1999. Diss ETH No. 13398.
[36] E. Zitzler and L. Thiele. Multiobjective Evolutionary Algorithms: A Comparative Case Study and the Strength Pareto Approach. *IEEE Transactions on Evolutionary Computation*, 3(4):257–271, 1999.

6

Regularization for Parameter Identification Using Multi-Objective Optimization

Tomonari Furukawa[1], Chen Jian Ken Lee[2] and John G. Michopoulos[3]

[1] ARC Centre of Excellence in Autonomous Systems, School of Mechanical and Manufacturing Engineering, The University of New South Wales, NSW 2052 Australia
t.furukawa@unsw.edu.au

[2] School of Science for Open and Environmental Systems, Keio University, 3-14-1 Hiyoshi, Kohoku-ku, Yokohama 223-8522 Japan
ken@noguchi.sd.keio.ac.jp

[3] Naval Research Laboratory, Center of Computational Material Science Special Projects Group, Code 6390 Computational Multiphysics Systems Lab., Washington DC 20375 USA
john.michopoulos@nrl.navy.mil

Summary. Regularization is a technique used in finding a stable solution when a parameter identification problem is exposed to considerable errors. However a significant difficulty associated with it is that the solution depends upon the choice of the value assigned onto the weighting regularization parameter participating in the corresponding formulation. This chapter initially and briefly describes the weighted regularization method. It continues by introducing a weightless regularization approach that reduces the parameter identification problem to multi-objective optimization. Subsequently, a gradient-based multi-objective optimization method with Lagrange multipliers, is presented. Comparative numerical results with explicitly defined objective functions demonstrate that the technique can search for appropriate solutions more efficiently than other existing techniques. Finally, the technique was successfully applied for the parameter identification of a material model[1].

6.1 Introduction

Inverse modeling is used in application areas where the system model is derived from its measured response and its corresponding stimulus [23]. In engineering applications, this is achieved via the identification of a continuous parameter set associated with a continuous deterministic mathematical model, based on a set of measured data related to the system's behavior. When the

[1] This work is supported in part by the ARC Centre of Excellence program, funded by the Australian Research Council (ARC) and the New South Wales State Government.

T. Furukawa et al.: *Regularization for Parameter Identification Using Multi-Objective Optimization*, Studies in Computational Intelligence (SCI) **16**, 125–149 (2006)
www.springerlink.com

model is complicated, the parameter identification problem is often converted into an optimization problem where an objective function is formed by the measured behavioral data and the analytically predicted data generated from the partial model is minimized. Various optimization methods have been used, depending upon the characteristics of the objective function [25, 30]. A difficulty of this approach is finding a solution when measurement data and/or the model contain large errors [3] because the objective function becomes too complicated to be solved by conventional optimization methods.

One of the approaches capable of overcoming this difficulty adds a regularization term [4] to the objective function, which normally consists of a function multiplied by a weighting factor that plays the role of the corresponding regularization parameter. This term makes the resulting functional smooth, so that conventional calculus-based optimization techniques can obtain an appropriate parameter set in a stable fashion. Nevertheless, the obtained solution depends upon the value of the weighting factor, and most of the research reports leave the determination of its values open for further studies, while they only demonstrate results based on a limited number of *a priori* selected values [27, 18].

However, some techniques for finding the best value of the weighting factor have been proposed by several researchers. Conventional techniques include the Morozov discrepancy principle [28], which obtains a regularization parameter based on minimization of an error criterion, and the generalized cross validation [14], which is derived from a statistical method. Later, Kitagawa [19, 20] proposed a technique based on the sensitivity of the regularization term with respect to the regularization parameter. Reginska [31] considered the maximizer of the L-curve, defined by Hansen [15], as the optimal parameter. Kubo, et al. [24] also proposed a technique using Singular Value Decomposition, while Zhang and Zhu [33] developed a multi-time-step method for inverse problems involving systems consisting of partial differential equations. A comparative study of some of the techniques can be found in [21, 22]. Although improvements have been demonstrated by these techniques, the fundamental question remains whether the automatic determination of a single solution through the selection of the best weighting factor is necessary. To find the best regularization parameter, all these techniques result in introducing another parameter that in turn influences the solution and therefore no conclusive result can ever be produced.

Meanwhile, multi-objective optimization methods have been proposed to solve multi-objective design optimization problems [13, 8, 9, 5]. These methods allow the design parameters to be optimized without weighting factors that depend on design criteria such as weight and energy consumption. Since the solution of this vector functional formulation is henceforth represented as a space, namely the *solution space*, rather than a point, the corresponding methods attempt to find a set of admissible solutions (points) in the solution space. Because of the multiple character of the solution and the possible complexity of the associated objective functions, the corresponding solution methods have been mostly based on the evolutionary algorithms (EAs) [2]. They execute robust searches from multiple search points for single objective optimization. Many EAs are however excessively robust at the expense of efficiency in contrast to the conventional calculus-based methods that are not inefficient for parameter identification problems involving a continuous deterministic objective function with continuous search space [16]. Moreover, since these algorithms find only a fixed number of solutions in the solution space, the solutions are sparse and are not well distributed in the solution space.

In this chapter, a framework for solving a regularized parameter identification problem without weighting factors [12] is presented initially. According to this technique, regularization terms are each formulated as another objective function, and the multi-objective optimization problem is solved by a multi-objective optimization method. Furthermore, a multi-objective gradient-based method with Lagrange multipliers (MOGM-LM) is introduced as an optimization method to find solutions for this class of problems efficiently [26]. The method is also formulated such that its solutions can configure the solution space to be derived.

The next section deals with the overview of the parameter identification discipline. The proposed weightless regularized identification approach and a general framework for multi-objective optimization are presented in Section 6.3 whereas MOGM-LM implemented in the framework is formulated in Sec. 6.4. Section 6.5 presents numerical results demonstrating the superiority of the proposed technique to conventional techniques and its applicability for parameter identification. In the first three subsections, the performance of the proposed technique is tested with explicitly defined objective functions, and the last subsection deals with the parameter identification of a viscoplastic material for practical use. The final section summarizes conclusions.

6.2 Parameter Identification

6.2.1 Formulation

We assume the existence of a set of stimulus and response experimental data $[\mathbf{u}_i^*, \mathbf{v}_i^*]$, respectively, where $\mathbf{u}_i^* \in \mathcal{U}$ and $\mathbf{v}_i^* \in \mathcal{V}$. The corresponding model $\hat{\mathbf{v}}$ is defined in terms of parameters $\mathbf{x} \in \mathcal{X}$. The experimental data can be related to the model by

$$\hat{\mathbf{v}}\left(\mathbf{u}_i^*; \mathbf{x}\right) + \mathbf{e}_i = \mathbf{v}_i^* \tag{6.1}$$

where $\mathbf{e}_i$ represents the sum of the model errors and measurement errors:

$$\mathbf{e}_i = \mathbf{e}_i^{\text{mod}} + \mathbf{e}_i^{\text{exp}}. \tag{6.2}$$

The parameter identification process is typically defined as the activity of identifying a continuous vector $\mathcal{X} \subseteq \Re^n$, provided a set of continuous experimental data, $\mathcal{U}, \mathcal{V} \subseteq \Re^n$ is available. In order to determine it numerically, the parameter identification problem is often converted into a minimization of a continuous functional:

$$f(\mathbf{x}) \to \min_{\mathbf{x}} \tag{6.3}$$

where $f : \mathcal{X} \to \Re$. The parameter set minimizing such an objective function is found within the bounds:

$$\mathbf{x}_{\min} \leq \mathbf{x} \leq \mathbf{x}_{\max} \tag{6.4}$$

where $[\mathbf{x}_{\min}, \mathbf{x}_{\max}] = \mathcal{X}$ and may be further subject to equality constraints

$$h_i(\mathbf{x}) = 0, \forall i \in \{1, \ldots, m_e\} \tag{6.5}$$

as well as inequality constraints

$$g_i(\mathbf{x}) \leq 0, \forall i \{m_e + 1, \ldots, m\}. \tag{6.6}$$

In this formulation, the solution of the identification problem is said to exist if there is at least one minimum within the search range satisfying constraints (6.4)-(6.6), and the solution is said to be unique if there is only one minimum within the range.

As an example approach for determining an objective functional (6.3), one may consider the popular method of least squares, where the objective function of which is often represented as:

$$f(\mathbf{x}) = \sum_{i=1}^{n} \left\| \hat{\mathbf{v}}(\mathbf{u}_i^*; \mathbf{x}) - \mathbf{v}_i^* \right\|^2. \tag{6.7}$$

This equation clearly indicates that the form of the objective function depends upon the modeled data that depend on the model of the system itself, and the measured data. The difficulty of the parameter identification is therefore that the objective function can become complex if the model and measured data contain considerable errors. It is more apparent when the number of measured data is small.

On the other hand, the majority of optimization methods can consistently find a global minimum only if the objective function is nearly convex. Otherwise, the solution may diverge or vibrate depending on the *a priori* selected initial search point. An approach for overcoming this problem is to introduce an additional term to the objective function in order to make it near-convex. This gives rise to the need for regularization described in the next section.

6.2.2 Regularization

If an objective function is complex, even a small change of the parameters may lead to a significant change to the value of the objective function. The regularization contributes to the stabilization of such a function. In the Tikhonov regularization method [32], (arguably the most popular regularization technique), the objective function is reformulated as

$$\Pi(\mathbf{x}) = f(\mathbf{x}) + \omega \Lambda(\mathbf{x}) \tag{6.8}$$

where ω and $\Lambda(\mathbf{x})$ are known as the *Tikhonov regularization parameter* and the *Tikhonov regularization function*, respectively. If the solution is known to be adjacent to $\mathbf{x}^*$, the regularization term may be given by

$$\Lambda(\mathbf{x}) = \|\mathbf{K}(\mathbf{x} - \mathbf{x}^*)\|^2 \tag{6.9}$$

where $\mathbf{K}$ is a weighting matrix. The matrix is most often set simply to the unity matrix unless some prior information is available about the system.

A solution of the parameter identification problem is thus obtained by first specifying ω and subsequently minimizing the function with an optimization method. In other words, the specification of ω becomes the fundamental issue in the regularized parameter identification as mentioned in Section 6.1. The next section will describe how to derive a solution which is not influenced by the regularization parameter, thus avoiding the uncertainty introduced by the unknown character of ω.

6.3 Weightless Regularization

6.3.1 Problem Formulation

Instead of attempting to find solutions that depend on the weighting factors we focus into finding solutions that do not depend on the weighting factors by entirely removing them from our formulation. This results into a multi-objective formulation. With the unity weighting matrix of $\mathbf{K}$, Tikhonov regularization parameter ω is the only weighting factor, and the objective of the problem is thus expressed as

$$\mathbf{f}(\mathbf{x})^\top = \left[f(\mathbf{x}), \|\mathbf{x} - \mathbf{x}^*\|^2\right] \rightarrow \min_{\mathbf{x}} \tag{6.10}$$

where $\mathbf{f}(\mathbf{x}) : \mathcal{X} \rightarrow \Re^2$. If the weighting matrix is diagonal,

$$\mathbf{f}(\mathbf{x})^\top = \left[f(\mathbf{x}), \|x_1 - x_1^*\|^2, \ldots, \|x_n - x_n^*\|^2\right] \tag{6.11}$$

where $\mathbf{f}(\mathbf{x}) : \mathcal{X} \rightarrow \Re^{1+n}$. The regularized parameter identification formulated as a multi-objective optimization problem is conclusively characterized as that:

- the problem is multi-objective,
- the objective functions $\mathbf{f}(\mathbf{x})$ are continuous but can be complex such as non-convex,
- the problem is subject to equality constraints (6.5) and inequality constraints (6.6),
- the search space is continuous.

6.3.2 Problem Solution

For the sake of generality, we consider a problem where m objective functions, $f_k : \mathcal{X} \rightarrow \Re, \forall k \in \{1, ..., m\}$ are defined as

$$\mathbf{f}(\mathbf{x})^\top = [f_1(\mathbf{x}), \ldots, f_m(\mathbf{x})] \rightarrow \min_{\mathbf{x}}. \tag{6.12}$$

The solution of this multi-objective optimization problem becomes a space rather than a point, but there is no analytical technique that derives the solution space. Numerical techniques can configure the solution space in an approximate manner by finding points that satisfy the *Pareto-optimality*, which is the requirement for points to be in the solution space:

Pareto-optimality: A decision vector $\mathbf{x}_i \in \mathcal{X}$ is said to be Pareto-optimal if and only if there is no vector $\mathbf{x}_j \in \mathcal{X}$ for which $f_k(\mathbf{x}_i), \forall k \in \{1, \ldots, m\}$, dominates $f_k(\mathbf{x}_j)$, i.e., there is no vector $\mathbf{x}_i$ such that

$$f_k(\mathbf{x}_j) \leq f_k(\mathbf{x}_i), \forall k \in \{1, \ldots, m\}. \tag{6.13}$$

To find a good approximate solution space, the numerical technique must find Pareto-optimal solutions which are well distributed to configure the solution space.

6.3.3 A General Framework for Searching Pareto-Optimal Solutions

Figure 6.1 shows the flowchart of the framework of the multi-objective optimization proposed in this chapter. In order to find multiple solutions, the multi-objective optimization searches with p multiple points, i.e.,

$$X(K) = \{\mathbf{x}_1^K, ..., \mathbf{x}_p^K\} \in (\Re^n)^p, \tag{6.14}$$

where $\mathbf{x}_i^K$ is the ith search point at K^{th} iteration. The initial population, $X(0)$, is generated randomly within a specified range $[\mathbf{x}_{\min}, \mathbf{x}_{\max}]$. Each objective function value $f_j(\mathbf{x}_i^K)$ is then calculated with each parameter set $\mathbf{x}_i^K$, finally yielding

$$F(K) = \{\mathbf{f}(\mathbf{x}_1^K), ...\mathbf{f}(\mathbf{x}_p^K)\}. \tag{6.15}$$

Unlike multi-objective EAs, two scalar criteria are evaluated for each search point in the proposed framework. One is the rank in Pareto-optimality expressed as usual

$$\Theta(K) = \{\theta(\mathbf{x}_1^K), ..., \theta(\mathbf{x}_p^K)\}, \tag{6.16}$$

where $\theta : \Re^n \rightarrow \aleph$, and the other is a positive real-valued scalar objective function or a fitness, which is derived by taking into account the rank:

$$\Phi(K) = \{\phi(\mathbf{x}_1^K), ..., \phi(\mathbf{x}_p^K)\}, \tag{6.17}$$

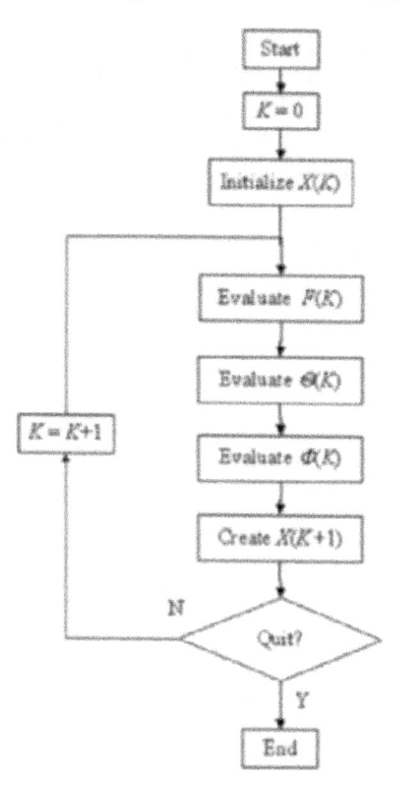

Fig. 6.1. Flowchart of multi-objective evolutionary algorithms.

where $\phi : \Re^n \rightarrow \Re^+$. Whilst the rank is evaluated to check the degree of Pareto-optimality of each search point, the fitness is used to create the next search point $\mathbf{x}_i^{K+1}$. Since the creation depends upon the search methods to be used, the next population is written in canonical form as

$$X(K+1) = s(X(K), \Phi(K), \nabla\Phi(K), \nabla^2\Phi(K)), \tag{6.18}$$

where s is the search operator.

Once the iterative computation has been enabled, we want to find effective Pareto-optimal solutions as many as possible such that the solution space can be configured. Another technique proposed here is a *Pareto pooling strategy* where the set of Pareto-optimal solutions created in the past are pooled as $P(K)$ besides the population of search points $X(K)$.

The process of the Pareto pooling technique is as follows. The whole Pareto-optimal solutions obtained in the first iteration are stored, i.e., $P(0) = X(0)$. From the second iteration, the newly created Pareto-optimal solutions in the optimization loop, $X(K+1)$, are compared to the stored Pareto-optimal solutions $P(K)$, and the new set of Pareto-optimal solutions $P(K+1)$ is saved in the storage as illustrated in Fig. 6.2. Some Pareto-optimal solutions may be identical or very close to an existing point. The storage of such solutions is simply a waste of memory, so that they are discarded if they are closer than the resolution specified *a priori*. The creations of the new population and the Pareto-optimal solutions are repeated until a terminal condition is satisfied.

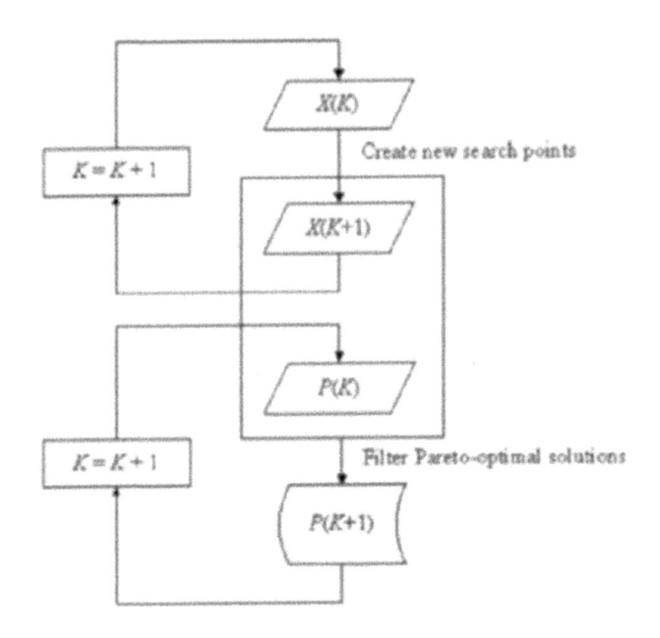

Fig. 6.2. Creation of Pareto-optimal solutions.

6.3.4 Selection of a Single Solution by Centre-of-Gravity Method (CoGM)

Figure 6.3 illustrates Pareto-optimal solutions where two objective functions $\mathbf{f} = [f_1, f_2]^\top$ are minimized to identify three parameters $\mathbf{x} = [x_1, x_2, x_3]^\top$. As two-dimensional function space and three-dimensional parameter space are still easy to visualize, one may incorporate human knowledge into computational knowledge-based techniques such as expert systems and fuzzy logic [17] for automatic selection of a single solution. However, if the numbers of objective functions and parameters are large, the knowledge to be constructed exponentially increases, and such techniques are no longer practically possible. In this case, one prominent way is to select the solution residing in the center of solution space since this solution is robust. Here, we propose a technique where the closest solution to the center-of-gravity is chosen as the solution. Let the Pareto-optimal solutions finally obtained be $\bar{\mathbf{x}}^i, \forall i \in \{1, ..., q\}$. If each solution is evaluated in a scalar manner, i.e., $\varphi(\bar{\mathbf{x}}^i)$, the center-of-gravity is in general given by

$$\bar{\mathbf{x}} = \frac{\sum_{i=1}^{q} \bar{\mathbf{x}}^i \cdot \varphi(\bar{\mathbf{x}}^i)}{\sum_{i=1}^{q} \varphi(\bar{\mathbf{x}}^i)}. \tag{6.19}$$

Since the Pareto-optimal solutions must be evaluated equally, we can consider all the Pareto-optimal solutions possess the same scalar value, i.e., $\varphi(\bar{\mathbf{x}}^1) = \cdots = \varphi(\bar{\mathbf{x}}^q)$. No matter what the value is, the center-of-gravity results in the form:

$$\bar{\mathbf{x}} = \frac{\sum_{i=1}^{q} \bar{\mathbf{x}}^i}{q}. \tag{6.20}$$

The effectiveness of the CoGM has not been proved theoretically, but it is highly acceptable, as it has been commonly used in fuzzy logic[17] to find a solution from the solution space described by fuzzy sets.

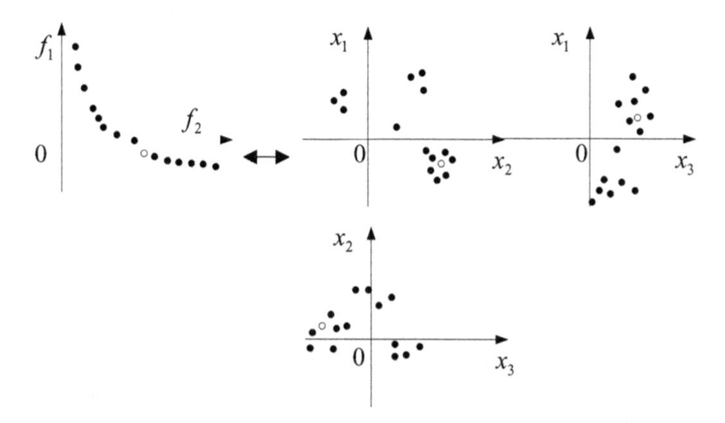

Fig. 6.3. Process of deriving a single solution.

6.4 Multi-objective Gradient-based Method with Lagrange Multipliers

6.4.1 Evaluation of Functions

Rank function

Figure 6.4 depicts the process to rank the search points and accordingly derive $\Theta(K)$ in Eq. (6.16). The process is based purely on an elimination rule.

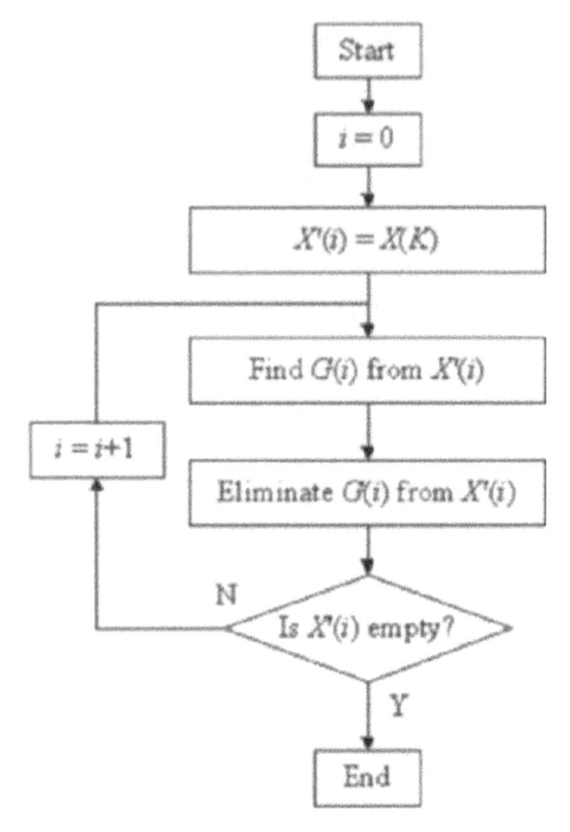

Fig. 6.4. Ranking process.

According to the rule, every objective function at every search point $f_j(\mathbf{x}_i^K)$, $\forall i \in \{1, ..., p\}$, $\forall j \in \{1, ..., m\}$, is first calculated, and the Pareto-optimal set in the population is ranked No. 1, i.e., $\theta(\mathbf{x}_i^K) = 1$ if the search point $\mathbf{x}_i^K$ is in the Pareto-optimal set. The group of search points ranked No. 1 is denoted as $G(1)$ in the figure. The points with rank No. 1 are then eliminated from the population, and the Pareto-optimal set in the current population is ranked No. 2, $\theta(\mathbf{x}_i^K) = 2$. Ranking is continued in the same fashion until all the points are ranked [13].

Real-valued Function

The evaluation of real-valued function of each search point starts with finding the maximum and minimum values of each objective function among the population:

$$(f_{\max})_j(K) = \max\{f_j(\mathbf{x}_i^K) | \forall i \in \{1, ..., p\}\} \tag{6.21}$$

$$(f_{\min})_j(K) = \min\{f_j(\mathbf{x}_i^K) | \forall i \in \{1, ..., p\}\}. \tag{6.22}$$

If we temporarily define the real-valued function as

$$\phi'_j(\mathbf{x}_i^K) = \frac{f_j(\mathbf{x}_i^K) - (f_{\min})_j(K)}{(f_{\max})_j(K) - (f_{\min})_j(K)}, \tag{6.23}$$

the following normalized conditions can be obtained,

$$0 \leq \phi'_j(\mathbf{x}_i^K) \leq 1, \tag{6.24}$$

and this allows the value of each function to be treated in the same scale. The real values of points with the same rank have to be identical, and the true value of each objective function is thus defined as:

$$\phi_j(\mathbf{x}_i^K) = \max\{\phi'_j(\mathbf{x}_l^K) | \theta(\mathbf{x}_l^K) = \theta(\mathbf{x}_i^K), l \neq i, \forall l \in \{1, ..., p\}\}. \tag{6.25}$$

The value of each search point can be conclusively calculated as

$$\phi(\mathbf{x}_i^K) = \sum_{j=i}^{m} w_j^K \phi_j(\mathbf{x}_i^K) \tag{6.26}$$

where $w_j^K = \text{rand}(0,1)$ contributes to a wide distribution of resultant Pareto-optimal solutions – $\text{rand}(0,1)$ refers to a random number between 0 and 1 following a uniform distribtion. The real value will appear within the range:

$$0 \leq \phi(\mathbf{x}_i^K) \leq m. \tag{6.27}$$

6.4.2 Search Method

The introduction of real-valued function $\Phi(k)$ makes gradient-based methods implemented easily without additional formulations. With $\phi(\mathbf{x}_i^K)$ calculated for each $\mathbf{x}_i^K$, the next state for a search point is given by

$$\mathbf{x}_i^{K+1} = \mathbf{x}_i^K + \Delta\mathbf{x}_i^K \tag{6.28}$$

where the step $\Delta\mathbf{x}_i^K$ of the search point determined by

$$\Delta\mathbf{x}_i^K = \alpha_K \mathbf{d}(\mathbf{x}_i^K, \phi(\mathbf{x}_i^K), \nabla\phi(\mathbf{x}_i^K), \nabla^2\phi(\mathbf{x}_i^K)). \tag{6.29}$$

In Eq. (6.29), α_K is the search step length iteratively searched as a subproblem by the Armijo rule [29], whereas $\mathbf{d}$ maps the direction of the search step. The Armijo rule is governed by three parameters ξ_1, ξ_2 and τ.

Various existing techniques can be used to find the direction of the search step. The most basic technique may be the Steepest Descent (SD) method:

$$\mathbf{d}_{SD}(\mathbf{x}_i^K, \nabla\phi(\mathbf{x}_i^K)) = -\nabla\phi(\mathbf{x}_i^K) \tag{6.30}$$

The efficient and most popular technique is the Quasi-Newton (QN) method, which is described as:

$$\mathbf{d}_{QN}(\mathbf{x}_i^K, \nabla\phi(\mathbf{x}_i^K)) = -{\mathbf{A}_K}^{-1}\nabla\phi(\mathbf{x}_i^K) \tag{6.31}$$

where $\mathbf{A}_K \approx \nabla^2\phi(\mathbf{x}_i^K)$ is the approximated Hessian of the function defined in (6.26).

6.4.3 Constraint Handling with Lagrange Multipliers

Equality constraints

The introduction of real-valued function $\Phi(K)$ also contributes to handling constraints using the standard technique of Lagrange multiplier method [1]. To handle equality constraints with real-valued function $\Phi(K)$, Lagrangian function can be defined as

$$L(\mathbf{x}, \lambda) \equiv \phi(\mathbf{x}) + \mathbf{h}(\mathbf{x})^\top \lambda \tag{6.32}$$

where λ is a set of Lagrange multipliers for equality constraints. Suppose the solution is denoted by $[\mathbf{x}^*, \lambda^*]$. The optimality conditions in this case, known as Kuhn-Tucker conditions, are described as

$$\nabla_{\mathbf{x}} L(\mathbf{x}^*, \lambda^*) = \nabla_{\mathbf{x}} \phi(\mathbf{x}^*) + \nabla_{\mathbf{x}} \mathbf{h}(\mathbf{x}^*)^\top \lambda^* = 0, \tag{6.33}$$

$$\nabla_\lambda L(\mathbf{x}^*, \lambda^*) = \mathbf{h}(\mathbf{x}^*) = 0. \tag{6.34}$$

Let the parameters and Lagrange Multipliers be updated iteratively with $\mathbf{x}_{K+1} = \mathbf{x}_K + \Delta\mathbf{x}_K$ and $\lambda_{K+1} = \lambda_K + \Delta\lambda_K$ respectively. The approximation of the first-order Taylor expansion yields the following equations:

$$\begin{aligned} \nabla_{\mathbf{x}} L(\mathbf{x}_{K+1}, \lambda_{K+1}) \equiv \nabla_{\mathbf{x}} L(\mathbf{x}_K, \lambda_K) &+ \nabla_{\mathbf{xx}} L(\mathbf{x}_K, \lambda_K) \Delta\mathbf{x}_K \\ &+ \nabla_{\mathbf{x}\lambda} L(\mathbf{x}_K, \lambda_K) \Delta\lambda_K, \end{aligned} \tag{6.35}$$

$$\begin{aligned} \nabla_\lambda L(\mathbf{x}_{K+1}, \lambda_{K+1}) \equiv \nabla_\lambda L(\mathbf{x}_K, \lambda_K) &+ \nabla_{\lambda\mathbf{x}} L(\mathbf{x}_K, \lambda_K) \Delta\mathbf{x}_K \\ &+ \nabla_{\lambda\lambda} L(\mathbf{x}_K, \lambda_K) \Delta\lambda_K. \end{aligned}$$

The substitution of

$$\nabla_{\mathbf{x}} L(\mathbf{x}_K, \lambda_K) = \nabla_{\mathbf{x}} \phi(\mathbf{x}_K) + \nabla_{\mathbf{x}} \mathbf{h}(\mathbf{x}_K)^\top \lambda_K = 0, \tag{6.36}$$

$$\nabla_{\mathbf{x}\lambda} L(\mathbf{x}_K, \lambda_K) = \nabla_{\lambda\mathbf{x}} L(\mathbf{x}_K, \lambda_K) = \nabla_{\mathbf{x}} \mathbf{h}(\mathbf{x}_K)^\top, \tag{6.37}$$

$$\nabla_{\lambda\lambda} L(\mathbf{x}_K, \lambda_K) = 0 \tag{6.38}$$

rewrites Eqs. (6.35) and (6.36) as

$$\begin{aligned} \nabla_{\mathbf{x}} L(\mathbf{x}_{K+1}, \lambda_{K+1}) = \nabla_{\mathbf{x}} \phi(\mathbf{x}_K) &+ \nabla_{\mathbf{x}} \mathbf{h}(\mathbf{x}_K)^\top \lambda_K + \nabla_{\mathbf{xx}} L(\mathbf{x}_K, \lambda_K) \Delta\mathbf{x}_K \\ &+ \nabla_{\mathbf{x}} \mathbf{h}(\mathbf{x}_K)^\top \Delta\lambda_K, \end{aligned} \tag{6.39}$$

$$\nabla_\lambda L(\mathbf{x}_{K+1}, \lambda_{K+1}) = \mathbf{h}(\mathbf{x}_K) + \nabla_{\mathbf{x}} \mathbf{h}(\mathbf{x}_K)^\top \Delta\mathbf{x}_K. \tag{6.40}$$

It remains to determine $[\Delta\mathbf{x}_K, \Delta\lambda_K]$ such that the optimality conditions are satisfied at $[\mathbf{x}_{K+1}, \lambda_{K+1}]$. The resulting equation in matrix-vector form is:

$$\begin{bmatrix} \nabla_{\mathbf{xx}} L(\mathbf{x}_K, \lambda_K) & \nabla_{\mathbf{x}} \mathbf{h}(\mathbf{x}_K) \\ \nabla_{\mathbf{x}} \mathbf{h}(\mathbf{x}_K)^\top & 0 \end{bmatrix} \begin{bmatrix} \Delta\mathbf{x}_K \\ \lambda_{K+1} \end{bmatrix} = \begin{bmatrix} -\nabla_{\mathbf{x}} \phi(\mathbf{x}_K) \\ -\mathbf{h}(\mathbf{x}_K) \end{bmatrix}. \tag{6.41}$$

An inversion of the matrix-vector Eq. (6.41) results in equations for evaluating $[\Delta \mathbf{x}_K, \lambda_{K+1}]$:

$$\begin{bmatrix} \Delta \mathbf{x}_K \\ \lambda_{K+1} \end{bmatrix} = \begin{bmatrix} \nabla_{\mathbf{xx}} L(\mathbf{x}_K, \lambda_K) & \nabla_{\mathbf{x}} \mathbf{h}(\mathbf{x}_K) \\ \nabla_{\mathbf{x}} \mathbf{h}(\mathbf{x}_K)^\top & 0 \end{bmatrix}^{-1} \begin{bmatrix} -\nabla_{\mathbf{x}} \phi(\mathbf{x}_K) \\ -\mathbf{h}(\mathbf{x}_K) \end{bmatrix}. \tag{6.42}$$

It is important to note that $\nabla_{\mathbf{xx}} L(\mathbf{x}_K, \lambda_K)$ is the Hessian of the Lagrange equation (6.32). The computation of the Hessian can be performed by the QN method.

Inequality constraints

Parameter identification may also be subject to inequality constraints. Applying the theory of Lagrange multipliers for equality constraints, a simple selection algorithm is proposed for inequality constraints. The equality constraint $\mathbf{h}(\mathbf{x})$ in Eq. (6.32) is replaced with the inequality constraint $\mathbf{g}(\mathbf{x})$.

$$L(\mathbf{x}, \lambda) \equiv \mathbf{f}(\mathbf{x}) + \mathbf{g}(\mathbf{x})^\top \lambda. \tag{6.43}$$

As it is evident from Eq. (6.42), a positive value of $\mathbf{g}(\mathbf{x}_K)$ would lead to an increase in $\Delta \mathbf{x}_K$. When $\mathbf{g}(\mathbf{x}_K) < 0$, then the constraint is satisfied. The mentioned selection algorithm is exemplified by

$$\begin{array}{ll} \text{if } (\mathbf{g}(\mathbf{x}_K) < 0) & \mathbf{g}(\mathbf{x}_K) = 0 \\ \text{else if } (\mathbf{g}(\mathbf{x}_K) < 1) & \mathbf{g}(\mathbf{x}_K) = -\sqrt{\mathbf{g}(\mathbf{x}_K)} \\ \text{else} & \mathbf{g}(\mathbf{x}_K) = -\mathbf{g}(\mathbf{x}_K)^2 \end{array}$$

These relations indicate that, when the constraint is satisfied, there does not exist a need for further search. However, when the constraint is violated, the search is greatly improved by square-rooting the value of $\mathbf{g}(\mathbf{x}_K)$ if $0 < \mathbf{g}(\mathbf{x}_K) < 1$ and squaring the value of $\mathbf{g}(\mathbf{x}_K)$ if $\mathbf{g}(\mathbf{x}_K) > 1$. This process has several advantages including keeping in check the ideal solution as explained in the next section.

Equality and inequality constraints

With both constraints involved, the Lagrangian function becomes

$$L(\mathbf{x}, \lambda, \mu) \equiv \mathbf{f}(\mathbf{x}) + \mathbf{h}(\mathbf{x})^\top \lambda + \mathbf{g}(\mathbf{x})^\top \mu \tag{6.44}$$

where λ and μ are both Lagrange multipliers for equality and inequality constraints respectively. The solution is now denoted by $[\mathbf{x}^*, \lambda^*, \mu^*]$. Every iteration gives $[\Delta \mathbf{x}_K, \Delta \lambda_{K+1}, \Delta \mu_{K+1}]$ by

$$\begin{bmatrix} \Delta \mathbf{x}_K \\ \lambda_{K+1} \\ \mu_{K+1} \end{bmatrix} = \begin{bmatrix} \nabla_{\mathbf{xx}} L(\mathbf{x}_K, \lambda_K) & \nabla_{\mathbf{x}} \mathbf{h}(\mathbf{x}_K) & \nabla_{\mathbf{x}} \mathbf{g}(\mathbf{x}_K) \\ \nabla_{\mathbf{x}} \mathbf{h}(\mathbf{x}_K)^\top & \mathbf{0} & \mathbf{0} \\ \nabla_{\mathbf{x}} \mathbf{g}(\mathbf{x}_K)^\top & \mathbf{0} & \mathbf{0} \end{bmatrix}^{-1} \begin{bmatrix} -\nabla_{\mathbf{x}} \mathbf{f}(\mathbf{x}_K) \\ -\mathbf{h}(\mathbf{x}_K) \\ -\mathbf{g}'(\mathbf{x}_K) \end{bmatrix} \tag{6.45}$$

where $\mathbf{g}'(\mathbf{x}_K)$ is described by the inequality selection algorithm described earlier.

6.5 Numerical Examples

6.5.1 Regularized Parameter Identification in Two-dimensional Parameter Space

In order to confirm its suitability for finding Pareto-optimal solutions and the increase of solutions over iterations, MOGM-LM was first used to identify two parameters by minimizing a simple objective function where the exact set of solutions is known and can be seen visually in two-dimensional space. In this example, the function is given by a simple quadratic function:

$$f_1(\mathbf{x}) = \sum_{i=1}^{n} x_i^2, \tag{6.46}$$

where $n = 2$, the set of parameters $\mathbf{x} \in \Re^2$ is subject to inequality constraint (6.4) with $\mathbf{x}_{\min}^{\top} = [-50, -50]$ and $\mathbf{x}_{\max}^{\top} = [50, 50]$. The solution of the problem is clearly $\mathbf{x}^* = [0, 0]$, but, to apply regularization, it is assumed that the solution is known to be adjacent to [2,4]. A Tikhonov regularization term is also added as another objective function:

$$f_2(\mathbf{x}) = \sum_{i=1}^{n} (x_i - z_i)^2, \tag{6.47}$$

where $\mathbf{z}^{\top} = [2, 4] \in \Re^2$. The problem therefore becomes to minimize functions (6.46) and (6.47). The exact solution for this problem can be determined analytically and is given by the space

$$X^* = \{\mathbf{x} | \mathbf{x} = r\mathbf{z}, \forall r \in [0, 1]\} \tag{6.48}$$

We can evaluate the performance of the proposed technique by comparing the computed Pareto-optimal solutions to the exact solution space. Values of major parameters for MOGM-LM used to solve the problem are listed in Table 6.1. ξ_1, ξ_2 and τ are parameters used to find the step length α_K.

Table 6.1. Parameters for MOGM-LM

Parameter	second
p	10
ξ_1	0.4
ξ_2	0.5
τ	0.7

Figures 6.5(a)-(d) show the computed Pareto-optimal set in $f_1 - f_2$ space (left) and $x_1 - x_2$ space (right) at 5th, 15th, 25th and 50th iterations respectively together with the exact solution space. The results indicate that some

good approximate solutions appear after 5 iterations and well distributed solutions have been already obtained in the 15th iteration. The number of Pareto-optimal solutions with respect to the number of iterations can be seen in Figure 6.6. One can see that the number of solutions increases rapidly and saturates around the 15th iteration. The increase is due to the Pareto-pooling technique, and this contributes to visualizing the solution space. The saturation and good distribution of the solutions, on the other hand, are controlled by the input resolution and avoids delaying the optimization process by pooling too many Pareto-optimal solutions. The final single solution selected by the CoGM was [1.021, 2.008], which is very close to the center of gravity of the exact solution space.

To demonstrate the effectiveness of the proposed method, the same problem was solved using conventional MOEAs [12]. Figure 6.7 shows the resulting Pareto-optimal solutions in $f_1 - f_2$ space (left) and $x_1 - x_2$ space (right) at the 50th, 250th and 1000th iteration. Due to the robust but inefficient search of evolutionary algorithms, the result at the 50th iteration is poor with MOEAs, underscoring the efficacy of MOGM-LM compared to MOEAs. In MOEAs, the result at the 250th iteration is still significantly inaccurate, but, because of the premature convergence, the result continues to be inaccurate at the 1000th iteration.

6.5.2 Regularized Parameter Identification within a General Parameter Space

To see the capability of MOGM-LM in general multi-dimensional parameter space, the second example deals with five parameters ($n = 5$) where the parameter space is constrained by inequality (6.4) with $\mathbf{x}_{\text{min}}^{\top} =$ [-50,-50,-50, -50,-50] and $\mathbf{x}_{\text{max}}^{\top} =$ [50,50,50,50,50]. The objective function and Tikhonov regularization term to be minimized are given by functions (6.46) and (6.47), respectively, where $\mathbf{z}^{\top} = [2, 3, 4, 5, 6]$. The exact solution space is therefore given by Eq. (6.48). Again, parameters listed in Table 6.1 were used for MOGM-LM.

Figures 6.8(a)-(d) show the Pareto-optimal solutions computed in $f_1 - f_2$ space and $x_1 - x_3$ space at the 5, 15, 25 and 50th iterations together with the exact solution. Although the search space is much larger than the previous example, the Pareto-optimal set which well describes the exact solution was already found after 15 iterations. It is shown that the result at the 25th iteration of this five-dimensional problem is much better than the result of the two-dimensional problem by MOEAs at the 1000th iteration.

The Pareto-optimal solutions of the two-objective optimization problem with functions (6.46) and (6.47) obtained by the MOGM-LM contains a solution of a single-objective optimization problem with one of the functions. In order to evaluate the performance of MOGM-LM to find a solution of a single-objective optimization problem through multi-objective optimization, the single-objective optimization problem with function (6.46) was solved using the most popular single-objective optimization method of Sequential

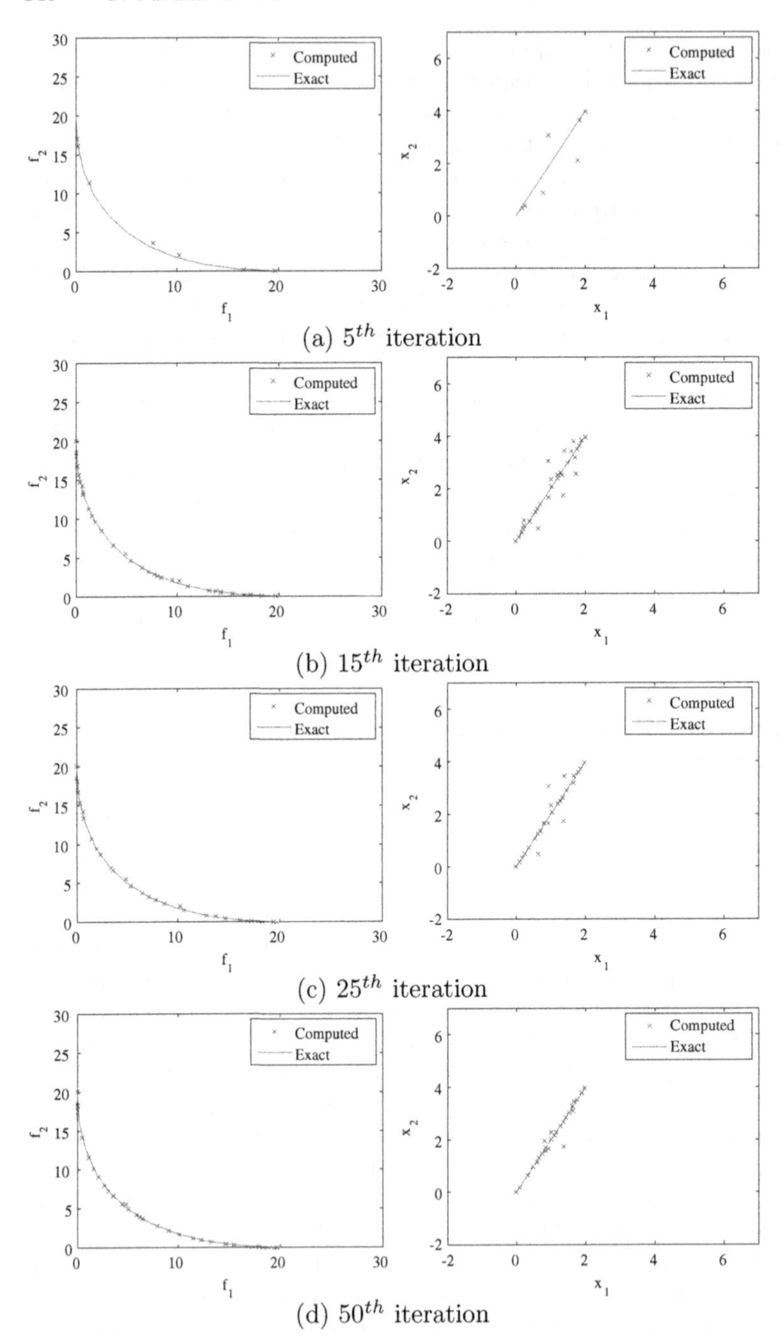

Fig. 6.5. Pareto-optimal set for Example I.

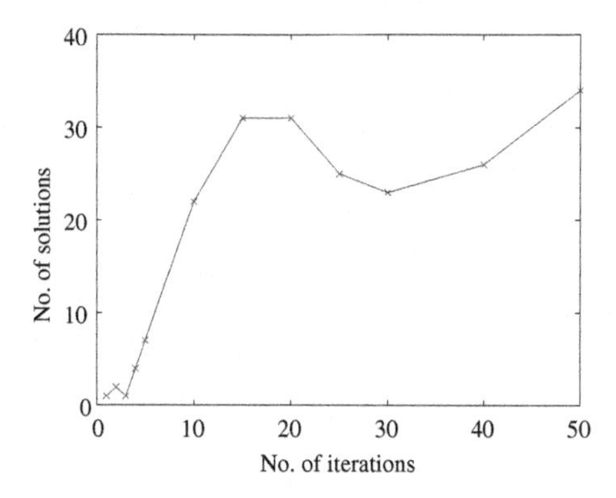

Fig. 6.6. Number of computed solutions vs number of iterations.

Quadratic Programming (SQP), and the result was compared to the Pareto-optimal solution by MOGM-LM. The Pareto-optimal solution used for comparison was the one that produced the minimal value of function (6.46) when both the functions (6.46) and (6.47) were minimized.

Figure 6.9 shows the minimal values with respect to iterations by both SQP and MOGM-LM. The figure shows the superiority of SQP to MOGM-LM. This is due to the random selection of MOGM-LM in Eq. (6.26), which does not take place in the single-objective optimization by SQP. However, one may more importantly conclude that the solution by MOGM-LM is also close to the exact solution, the mean square error being in the order of 10^{-2}. As MOGM-LM can find many other Pareto-optimal solutions, the effectiveness of the MOGM-LM is clearer from this result.

6.5.3 Regularized Parameter Identification with a Multimodal Objective Function

Having the appropriate performance of the proposed technique for identification with a simple objective function been demonstrated, the identification with an objective function, which is more likely in real applications, was investigated. In this Example III, the objective function has an additional term to Eq. (6.46) and is given by

$$f_1(\mathbf{x}) = 10n + \sum_{i=1}^{n} x_i^2 - 10\cos x_i, \tag{6.49}$$

where $n = 5$. The cosine term clearly makes the function multimodal with a number of local minima, so that the function has been used as a good example for a multimodal continuous function [7]. Again, Eq. (6.47) was used as the Tikhonov regularization term, and Table 6.1 as MOGM-LM parameters.

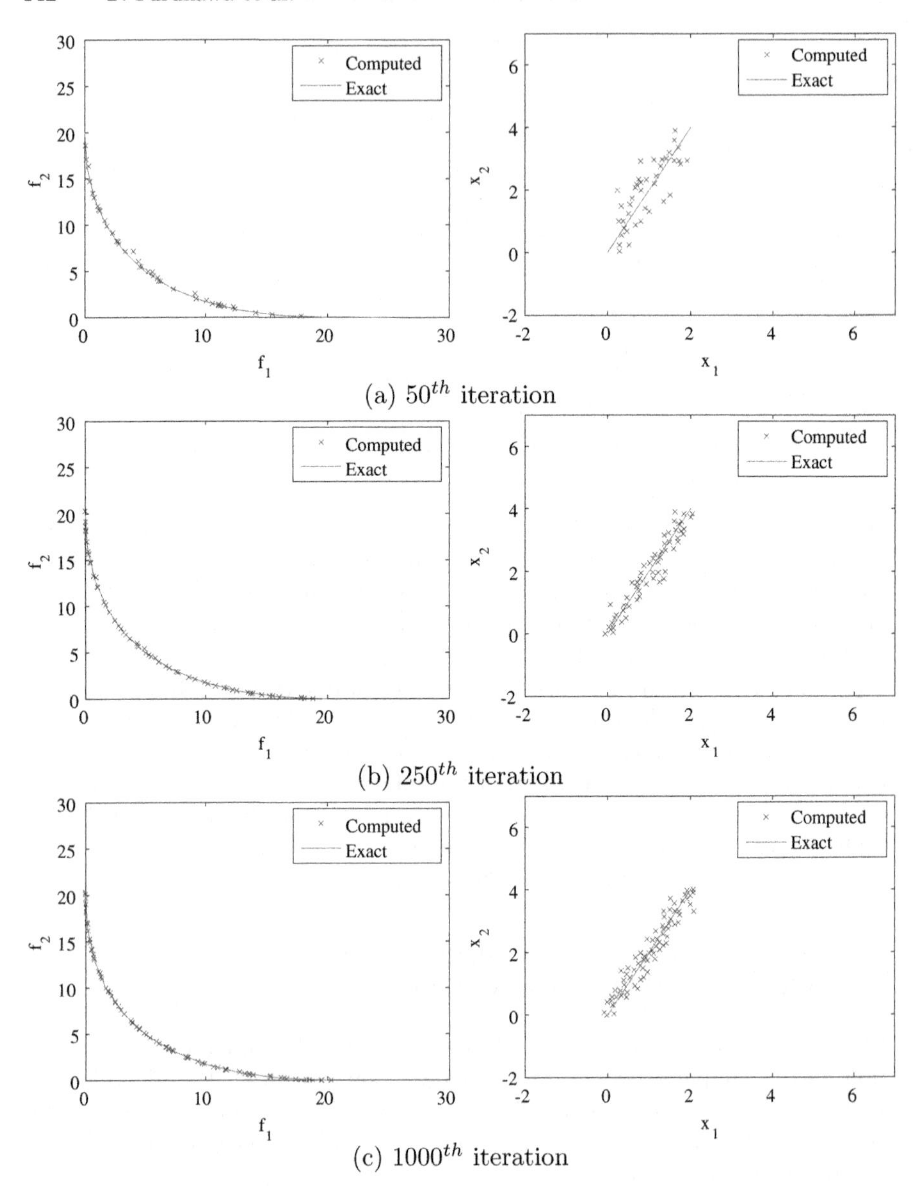

Fig. 6.7. Pareto-optimal set by MOEAs for Example I.

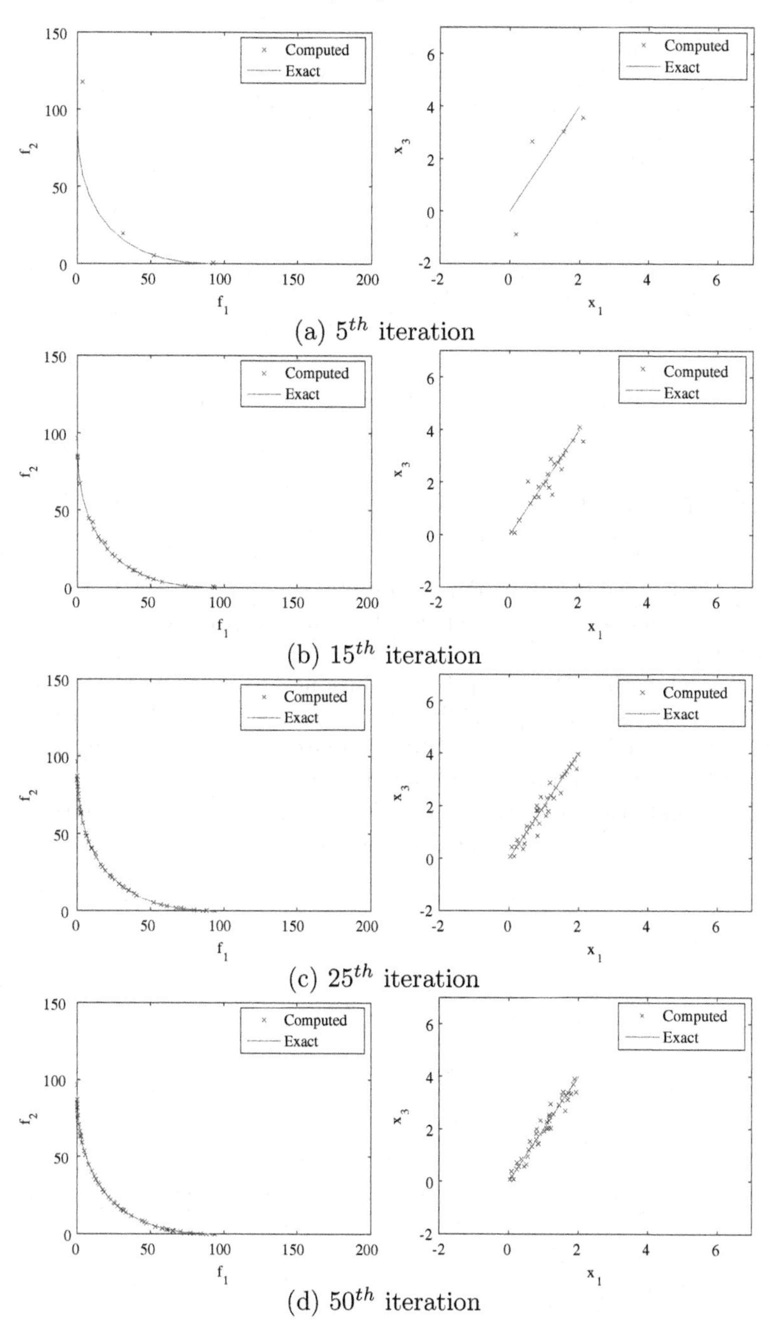

Fig. 6.8. Pareto-optimal set for Example II.

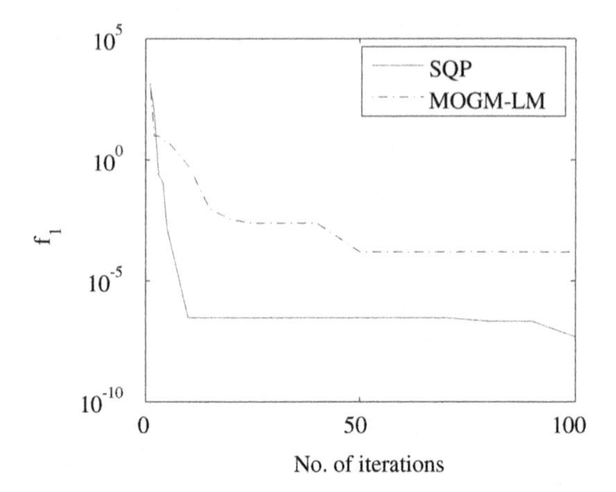

Fig. 6.9. Minimal value of objective function.

Figure 6.10 shows the resultant Pareto-optimal solutions at 50th iteration. Due to the complexity of f_1, it is seen that the Pareto-optimal solutions are spread over the parameter space. The appropriateness of the solutions can be verified by observing that Pareto-optimal solutions are found near the two distinct exact solutions $[0, 0]$ and $[2, 4]$. The coarse distribution of solutions with small f_1 is yielded by its complexity compared to f_2. The solution of the optimization of such a multimodal objective function by SQP diverges or vibrates. The probabilistic formulation enables MOGM-LM to find solutions with such a function and visualize the solution space.

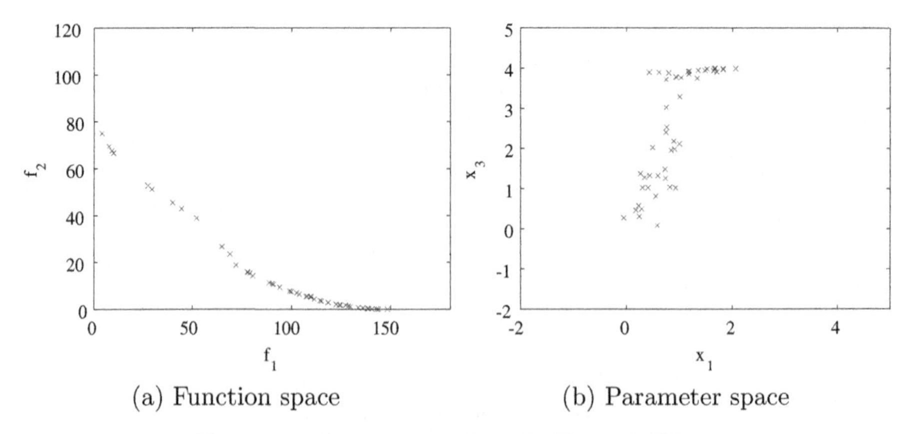

(a) Function space (b) Parameter space

Fig. 6.10. Pareto-optimal set for Example III.

6.5.4 Regularized Parameter Identification of Viscoplastic Material Models

Finally, the proposed technique was applied to a practical parameter identification problem of a material model. In the mechanical tests of material, stress-strain data can be derived as experimental data. Let stress and strain be represented by σ and ϵ, the problem in the robust least square formulation is given by

$$f_1(\mathbf{x}) = \sum_i \|\hat{\sigma}(\epsilon_i^*; \mathbf{x}) - \sigma_i^*\|^2 \tag{6.50}$$

where $[\epsilon_i^*, \sigma_i^*]$ are a set of experimental stress-strain data and $\sigma = \hat{\sigma}(\epsilon; \mathbf{x})$ is a material model having $\mathbf{x}$ as material parameters. Measurement errors in the mechanical tests are relatively small, but the difficulty of solving this problem is created by the complex description of material model. The material model used in the numerical example is Chaboche model [6], which can describe the major material behaviors of viscosity and cyclic plasticity accurately and is thus used to model a variety of metallic materials. The model under stationary temperature and uniaxial load conditions is of the form:

$$\dot{\epsilon}^{v_p} = \left\langle \frac{|\sigma - \chi|}{K} \right\rangle \operatorname{sgn}(\sigma - \chi) \tag{6.51}$$

$$\dot{\chi} = H\dot{\epsilon}^{v_p} - D|\dot{\epsilon}^{v_p}| \tag{6.52}$$

$$\dot{R} = h|\dot{\epsilon}^{v_p}| - d|\dot{\epsilon}^{v_p}| \tag{6.53}$$

where state variables $[\epsilon^{v_p}, \chi, R]$ are the viscoplastic strain, kinematic hardening and isotropic hardening, $[K, n, H, D, h, d]$ are inelastic material parameters, and $\langle . \rangle$ is McCauley bracket [10, 11]. The stress-strain relationship cannot be explicitly written as described in Eq. (6.50), but, given strain ϵ as a control input and the initial condition of state variables $[\epsilon^{v_p}|_{t=0}, \chi|_{t=0}, R|_{t=0}] = [\epsilon_0^{v_p}, \chi_0, R_0]$, the viscoplastic strain with respect to time can be derived iteratively, and the stress can be ultimately calculated using

$$\sigma = E(\epsilon - \epsilon^{vp}) \tag{6.54}$$

where E is the elastic modulus. In the model, parameters often unknown are inelastic material parameters $[K, n, H, D, h, d]$ plus the initial condition of isotropic hardening variable R_0. To facilitate the analysis of identification, the numerical example uses pseudo-experimental data of cyclic plasticity shown in Fig. 6.11, created from Chaboche model with a set of parameters described in Table 6.2. We will hence identify only parameters which influence cyclic plasticity; i.e., $\mathbf{x} = [R_0, K, H]$, assuming that the others are exactly known. Material models are formulated by considering the effect of each parameter, and therefore we often have a coarse estimate on the values of the material parameters $\mathbf{x} = [R_0, K, H]$. This gives Tikhonov regularization term expressed

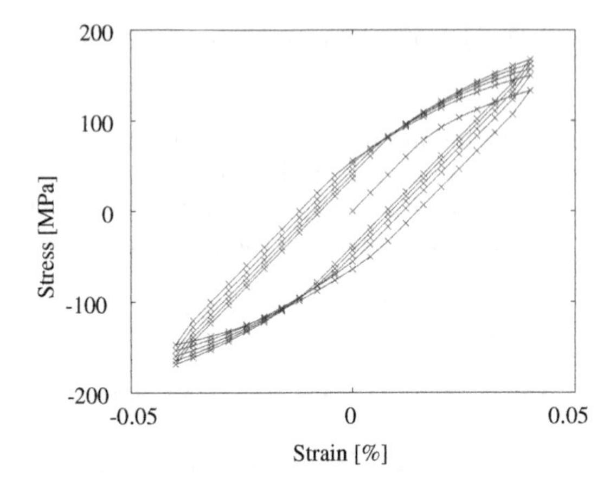

Fig. 6.11. Pseudo-experimental data of cyclic plasticity.

Table 6.2. Parameters for Chaboche model

Parameter	R_0	K	n	H	D	h	d
Exact	50	100	3	5000	100	300	0.6
Known	-	-	3	-	100	300	0.6
Coarse estimate	45	95	-	4800	-	-	-

by Eq. (6.47) as the second objective function f_2. The parameters used for optimization were again those in Table 6.1.

Figure 6.12(a) depicts the Pareto-optimal solutions in function space obtained from the parameter identification after 20 iterations. A total of 234 solutions are obtained, and it is easily seen in the figure that the solutions are well distributed. Respectively shown in Figs. 6.12(a), (b) and (c) are the solutions in $R_0 - K$, $K - H$ and $H - R_0$ parameter spaces. All the figures show that the solutions are distributed along the straight line linking the exact values and the initial coarse estimates. This result indicates that the proposed technique could find appropriate Pareto-optimal solutions for this problem.

6.6 Conclusions

A weightless regularized identification technique and a multi-objective optimization method of MOGM-LM, which can search for solutions efficiently for this class of problems, have been proposed. The use of multi-objective method allows for the derivation of the whole solution set of the problem rather than a single solution to be derived by one optimization. The user can select a single solution later by applying CoGM. After the Pareto-optimality of solutions derived by MOGM-LM was confirmed with a simple example, the

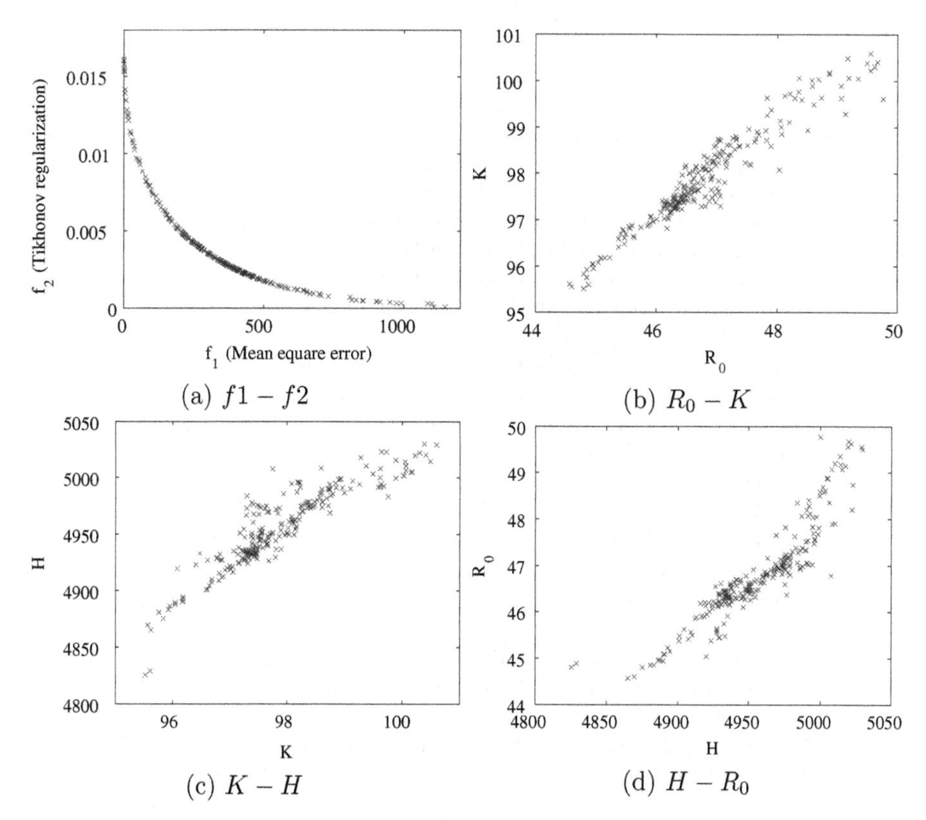

Fig. 6.12. Pareto-optimal set for material parameter identification.

proposed technique was applied to identification problems including material parameter identification, and the technique could find appropriate solutions in all the problems. The searching capability of the technique was also compared to that of a single-objective optimization method, and its superiority has been demonstrated. Conclusively, the overall effectiveness of the proposed technique for parameter identification has been confirmed.

One of the issues that have not been discussed thoroughly is the selection of a final solution. The CoGM described in this chapter is one of the many possible techniques. Because the solution of an inverse problem is never known, many other techniques can be thought of. One of the firm steps to take is to incorporate prior knowledge and solve the problem stochastically. Other issues include the improvement of the technique for high-dimensional problems since Chaboche model described in this chapter, for instance, contains 23 parameters in its multiaxial formulation.

References

[1] M. Aoki. Introduction to Optimizaion Techniques; fundamentals and applications of nonlinear programming. Macmillian, 1971

[2] T. Bäck, H.- P. Schwefel. *Evol Comp*, 1(1):1-23, 1993

[3] Y. Bard. Nonlinear Parameter Estimation. Academic Press, New York, 1974

[4] J. Baumeister. Stable Solution of Inverse Problems. Vieweg, Braunschweig, 1987

[5] C.A. Coello. *Int J Knowl Info Sys*, 1(1):1-25, 1999

[6] J.L. Chaboche. *Int J Plast*, 5:247-254, 1989

[7] K. De Jong. An Analysis of the Behaviour of a Class of Genetic Adaptive Systems. Ph.D Thesis, University of Michigan, 1975

[8] C.M. Fonseca, P.J. Fleming PJ. Genetic Algorithms for Multi-objective Optimisation: Formulation, Discussion and Generalisation. In: Forrest S (ed)., *Proceedings of the Fifth International Conference on Genetic Algorithms*, Morgan Kaufmann, San Mateo, CA, 416-423, 1993

[9] C.M. Fonseca, P.J. Fleming. *Evol Comp*, 3(1):1-16, 1995

[10] T. Furukawa, G. Yagawa. *Int J Numer Meth Eng*, 40:1071-1090, 1997

[11] T. Furukawa, G. Yagawa. *Int J Numer Meth Eng*, 43:195-219, 1998

[12] T. Furukawa. *Int J Numer Meth Eng*, 52:219-238, 2001

[13] D. Goldberg. Genetic Algorithms in Search, Optimization and Machine Learning. Addison-Wesley, Reading, MA, 1989

[14] C.W. Groetsche. The Theory of Tikhonov Regularization for Fredholm Integral Equation of the First Kind. Pitman, Boston, 1984

[15] P.C. Hansen. *SIAM Review* 34(4):561-580, 1992

[16] F. Hoffmeister, T. Bäck. Genetic Algorithms and Evolution Strategies: Similarities and Differences. *Technical Report*, University of Dortmund, Germany, Sys-1/92, 1992

[17] J.H. Holland. Adaptation in Natural and Artificial Systems. The University of Michigan Press, Michigan, 1975

[18] Y. Honjo, N. Kudo. Matching Objective and Subjective Information in Geotechnical Inverse Analysis Based on Entropy Minimization. In: Tanaka M, Dulikravich GS (eds). *Inverse Problems in Engineering Mechanics*, Elsevier Science, 263-271, 1998

[19] T. Kitagawa. *Jap J Appl Math*, 4:371-379, 1987

[20] T. Kitagawa. *J Info Proc*, 11:263-270, 1988

[21] T. Kitagawa. Methods in Estimating the Optimal Regularization Parameters. In: Yamaguti M (ed), *Inverse Problems in Mathematical Engineering*, Springer Verlag, 37-42, 1991

[22] T. Kitagawa. A Comparison between Two Classes of the Method for the Optimal Regularization. In: Kubo S (ed) *Inverse Problems*, Atlanta Technology Publications, 25-35, 1992

[23] S. Kubo. Inverse Problems, Baifu-kan (in Japanese), 1993

[24] S. Kubo, T. Takahashi, K. Ohji. Characterization of the Tikhonov Regularization for Numerical Analysis of Inverse Boundary Value Problems by Using the Singular Value Decomposition. In: Tanaka M, Dulikravich GS (eds), *Inverse Problems in Engineering Mechanics*, 337-344, 1998

[25] H.P. Kunzi, H.G. Tzschach, C.A. Zehnder. Numerical Methods of Mathematical Optimization. Academic Press, New York, 1971

[26] C.J.K. Lee, T. Furukawa, S. Yoshimura. *Int J Numer Meth Eng*, 2005 (in print)

[27] R. Mahnken, E. Stein. Gradient-based Methods for Parameter Identification of Viscoplastic Materials. In: Bui T, Tanaka M (eds), *Inverse Problems in Engineering Mechanics*, 137-144, 1994
[28] V.A. Morozov. Methods for Solving Incorrectly Posed Problems. Springer-Verlag, New York, 1984
[29] P.Y. Papalambros, D.J. Wilde. Principles of Optimal Design. Cambridge University Press, 2000
[30] W.H. Press, B.P. Flannery, S.A. Teukolsky, W.T. Vetterling. Numerical Recipes in C. Cambridge University Press, 1988
[31] T. Reginska. *SIAM J Sci Comp*, 17:740-749, 1996
[32] A.N. Tikhonov, V.Y. Arsenin. Solutions to Ill-posed Problems. John Willy Sons, New York, 1977
[33] X. Zhang, J. Zhu. An Efficient Numerical Algorithm with Adaptive Regularization for Parameter Estimations. In: Tanaka M, Dulikravich GS (eds), *Inverse Problems in Engineering Mechanics*, Elsevier Science 299-308, 1998

7

Multi-Objective Algorithms for Neural Networks Learning

Antônio Pádua Braga[1], Ricardo H. C. Takahashi[1], Marcelo Azevedo Costa[1], and Roselito de Albuquerque Teixeira[2]

[1] Federal University of Minas Gerais
apbraga@cpdee.ufmg.br
[2] Eastern University Centre of Minas Gerais
roselito@unilestemg.br

Summary. Most supervised learning algorithms for Artificial Neural Networks (ANN)aim at minimizing the sum of the squared error of the training data [12, 11, 5, 10]. It is well known that learning algorithms that are based only on error minimization do not guarantee good generalization performance models. In addition to the training set error, some other network-related parameters should be adapted in the learning phase in order to control generalization performance. The need for more than a single objective function paves the way for treating the supervised learning problem with multi-objective optimization techniques. Although the learning problem is multi-objective by nature, only recently it has been given a formal multi-objective optimization treatment [16]. The problem has been treated from different points of view along the last two decades.

In this chapter, an approach that explicitly considers the two objectives of minimizing the squared error and the norm of the weight vectors is discussed. The learning task is carried on by minimizing both objectives simultaneously, using vector optimization methods. This leads to a set of solutions that is called the Pareto-optimal set [2], from which the best network for modeling the data is selected. This method is named MOBJ (for Multi-OBJective training).

7.1 Introduction

The different approaches to tackle the supervised learning problem usually include error minimization and some sort of network complexity control, where *complexity* can be *structural* or *apparent*. Structural complexity is associated to the number of network parameters (weights) and apparent complexity is associated to the network response, regardless to its size. Thus, a large structural complexity network may have a small apparent complexity if it behaves like a lower order model. If the network is over-sized it is possible to control its apparent complexity so that it behaves properly.

A.P. Braga et al.: *Multi-Objective Algorithms for Neural Networks Learning*, Studies in Computational Intelligence (SCI) **16**, 151–171 (2006)
www.springerlink.com

Structural complexity control can be accomplished by shrinking (pruning) [9, 4, 8] or growing (constructive) methods [24, 25], whereas apparent complexity control can be achieved by cross-validation [26], smoothing (regularization) [17] and by a restricted search into the set of possible solutions. The latter includes the multi-objective approach that will be described in this chapter [16].

The concepts of flexibility and rigidity are also embodied by the notion of complexity. The larger the network complexity the higher its flexibility to fit the data. On the other extreme, the lower its complexity, the larger its rigidity to adapt itself to the data set. A model that is too rigid tends to concentrate its responses into a limited region, whereas a flexible one spans its possible solutions into a wider area of the solutions space. These concepts are also related to the bias-variance dilemma [27], since a flexible model has a large variance and a rigid one is biased. The essence of our problem is therefore to obtain a proper balance between error and complexity (structural or apparent). Restrictions to the apparent complexity can be obtained by limiting the value of the norm of the weight vectors for a given network solution.

The MOBJ training approach considers two simultaneous objectives: minimizing the squared error and the norm of the weight vectors. using vector optimization methods. This leads to a set of solutions that is called the Pareto-optimal set [2], from which the best network for modeling the data is selected.

The step of finding the Pareto-optimal set can be interpreted as a way for reducing the search space to a one-dimensional set of candidate solutions, from which the best one is to be chosen. This one-dimensional set exactly follows a trade-off direction between flexibility and rigidity, which means that it can be used for reaching a suitable compromise solution [16].

7.2 Multi-objective Learning Approach

The proposed MOBJ method is composed of the following steps:

Vector optimization step: From the training data, find the Pareto-optimal solutions;

Decision step: Using a set of validation data, choose the optimal solution from the Pareto-optimal set.

The main principle that is behind these two steps is: some data is employed in order to adjust the model parameters (the training data). These data cannot be re-used for the purpose of choosing a model that does not over-fit the data (fitting the noise too), from a set of candidate models, since the models fit the data by construction. Therefore, other data (the validation data) must be employed in the model selection step. This structure of the problem of ANN learning imposes this two-step procedure, that corresponds to the scheme of multi-objective optimization with *a posteriori* decision. The first step when using multi-objective optimization, in an *a posteriori* decision scheme, is to obtain the *Pareto-optimal set* [2], which contains the set of efficient solutions $\mathcal{X}^*$. Let $f_1(x)$ designate the sum of squared errors, and let $f_2(x)$ designate the norm of the weighting vector. The set $\mathcal{X}^*$ is defined as:

$$\begin{aligned}\mathcal{X}^* = \{x \mid \ \nexists \bar{x} \neq x \text{ such that does not occur:} \\ f_1(\bar{x}) \leq f_1(x) \text{ and } f_2(\bar{x}) < f_2(x) \\ f_1(\bar{x}) < f_1(x) \text{ and } f_2(\bar{x}) \leq f_2(x)\}\end{aligned} \tag{7.1}$$

The second step aims at selecting the most appropriate solution within the Pareto-optimal set.

Several algorithms are known in the literature to deal with this multi-objective approach [2].

7.2.1 Vector Optimization Step

A variation of the ϵ-constraint problem, proposed by Takahashi *et al.* [14], was adopted. The algorithm intrinsicaly avoids the generation of non-feasible solutions, what increases its efficiency. The Pareto set is obtained by first obtaining its two extremes, which are formed by the underfitted and overfitted solutions. Figure 7.1 shows the two extremes of the Pareto set, denoted by $\mathbf{f}_1^*$ and $\mathbf{f}_2^*$, which correspond to the underfitted and overfitted solutions. $\mathbf{f}_1^*$ is obtained by training a network with a standard training algorithm, such as Backpropagation [12]. $\mathbf{f}_2^*$ is trivial, since it is obtained by making all network weights equal to zero. Intermediate solutions obtained by the multiobjective (MOBJ) algorithm are called Pareto-optimal (Fig. 7.1). Solutions belonging to the Pareto-optimal set cannot be improved considering both objective functions simultaneously. This is in fact the definition of the Pareto-optimal set, which is used here to obtain a good compromise between the two conflicting objectives: error and norm.

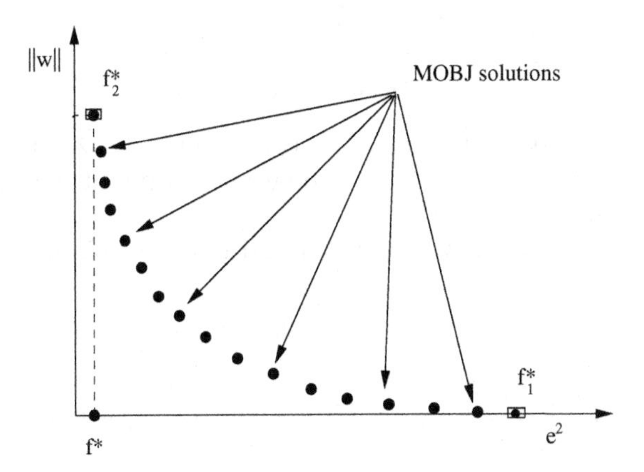

Fig. 7.1. Pareto-optimal set

The next step aims at selecting the most appropriate solution within the Pareto-optimal set. Several algorithms are described in the literature to deal with this multi-objective problem [2]. In this work, a variation of the ϵ-constraint problem, called relaxation method [14] was adopted. Its formulation is presented next.

- $\mathbf{f}^* \in \Re^m$ is the objective vector corresponding to the "utopian solution" [2] of the problem;
- ϕ_i^* is the ith objective optimum;
- ϕ_i is the value of the ith objective in any point;
- $\mathbf{f}_i^* \in \Re^m$; $i = 1, \ldots, m$ is the vector formed by optimal solutions of the individual objective i and values corresponding to the other objective functions.
- $\mathcal{C}$ is the cone generated by vectors $(\mathbf{f}_i^* - \mathbf{f}^*)$, with origin in $\mathbf{f}^*$;
- $\mathbf{v}_k \in \mathcal{C}$ is a vector constructed according to Eq. (7.2), which performs a convex combination of the individual objective vectors.

$$\mathbf{v}_k = \mathbf{f}^* + \gamma_k(\mathbf{f}_1^* - \mathbf{f}^*) + (1 - \gamma_k)(\mathbf{f}_2^* - \mathbf{f}^*), \tag{7.2}$$

for $0 \leq \gamma_k \leq 1$.

Equation (7.2) results always in a vector within the cone (in the objective space) of feasible solutions. For every γ_k there is a vector $\mathbf{v}_k$ within this cone that results in a Pareto solution (a vector on the Pareto set). In the simulations presented in this chapter, γ_k is initialized with zero and then incremented to one, so that the Pareto set can be generated from subsequent vectors $\mathbf{v}_k$.

The objectives "sum of squared errors" and "norm of the weight vector", considered in the optimization problem, are described by Eq. (7.3) and (7.4), respectively. The ANN output is denoted by $\mathbf{y}(\mathbf{w}, \mathbf{x}_j)$.

$$f_1(\mathbf{w}) = \frac{1}{N} \sum_{j=1}^{N} (\mathbf{d}_j - \mathbf{y}(\mathbf{w}, \mathbf{x}_j))^2 \tag{7.3}$$

$$f_2(\mathbf{w}) = \|\mathbf{w}\| \tag{7.4}$$

where $\mathbf{w}$ is the ANN weight vector, N is the training set size, $\mathbf{d}_j$ and $\mathbf{y}(\mathbf{w}, \mathbf{x}_j)$ are, respectively, the desired output and current output at iteration j and $\mathbf{x}_j$ is the input pattern.

Equation (7.5)–(7.9) show how $\mathbf{f}_1^*$, $\mathbf{f}_2^*$ and $\mathbf{f}^*$ can be obtained.

$$\mathbf{w}_1^* = \arg\min f_1 \tag{7.5}$$

$$\mathbf{f}_1^* = \begin{bmatrix} \phi_1^* \\ \phi_2 \end{bmatrix}, \quad \phi_1^* = f_1(\mathbf{w}_1^*) \text{ and } \phi_2 = f_2(\mathbf{w}_1^*) \tag{7.6}$$

$$\mathbf{w}_2^* = \arg\min f_2 \tag{7.7}$$

$$\mathbf{f}_2^* = \begin{bmatrix} \phi_1 \\ \phi_2^* \end{bmatrix}, \quad \phi_1 = f_1(\mathbf{w}_2^*) \text{ and } \phi_2^* = f_2(\mathbf{w}_2^*) \tag{7.8}$$

The vector $\mathbf{w}_2^*$ can be obtained easily and the vector $\mathbf{w}_1^*$ can be obtained by the Backpropagation algorithm. The utopian solution denoted by $\mathbf{f}^*$ is a vector formed by two elements according to the Eq. 7.9.

1. ϕ_1^*, minimum value of the objective function f_1;
2. ϕ_2^*, minimum value of the objective function f_2.

$$\mathbf{f}^* = \begin{bmatrix} \phi_1^* \\ \phi_2^* \end{bmatrix} \tag{7.9}$$

The multi-objective problem can be redefined now as a single-objective one, by considering the multiple objectives as constraints to the optimization algorithm. The problem can be solved by a constrained optimization method such as the "ellipsoidal algorithm" [13]. The multi-objective problem can now be described by Eq. (7.10) and (7.11).

$$\mathbf{w}^* = arg_w \min_{\mathbf{w},\eta} \eta \tag{7.10}$$

$$\text{subject to}: \ \mathbf{f}_i(\mathbf{w}) \leq \mathbf{f}^* + \eta \mathbf{v}_k \tag{7.11}$$

where η is the auxiliary variable.

Substituting Eq. (7.3) and (7.4) into 7.11 leads to the constrained problem described by Eq. (7.12) and (7.13):

$$\mathbf{w}^* = arg_w \min_{\mathbf{w},\eta} \eta \tag{7.12}$$

subject to:

$$\begin{cases} \frac{1}{N} \sum_{j=1}^{N} (\mathbf{d}_j - \mathbf{y}(\mathbf{w}, \mathbf{x}_j))^2 - \phi_1^* - \eta v_{k_1} \leq 0 \\ \|\mathbf{w}\| - \phi_2^* - \eta v_{k_2} \leq 0 \end{cases} \tag{7.13}$$

As can be observed in the graph of Fig. 7.1, the two extremes of the Pareto set are the two opposite solutions to the minimization problem. At one extreme, the solution $\mathbf{f}_1^*$ yields small mean square error with large norm weight vectors, what would result in poor generalization (overfitting). At the other extreme, the solution $\mathbf{f}_2^*$ would result in minimum (zero) norm with large error (underfitting). The well balanced solution is picked up on the Pareto set, between the two extremes, via the decision step.

7.2.2 Decision Step

After the Pareto-optimal solutions being found, the best ANN among them must be chosen. This procedure can be performed via a very simple algorithm that relies on the comparison of the solutions through validation data:

1. Simulate the response of all candidate ANN's for a set of validation data points;
2. Compute the sum of squared errors of each candidate ANN, for these data points;
3. Choose the ANN with smallest error.

7.2.3 A Note on the Two-step Algorithm

The MOBJ algorithm, as presented above, has used a simple schematic procedure that has been divided in two separate steps: the vector optimization and the decision. In fact, a more efficient algorithm can be built, that iteratively takes two points of the Pareto-set and makes one decision step. This algorithm would approximate arbitrarily the best ANN, with possibly less computational effort. See detais in [19].

7.2.4 The Sliding Mode Approach

The original multi-objective algorithm [16] fits a neural network model by simultaneous optimization of two cost functions, the sum of squared error and the norm of the weights vector.

The Multi-Objective Sliding Mode (SMC-MOBJ) algorithm applies sliding mode control to the network training. Pre-established values for error and norm are set as training targets. The theoretical setting of the gains aggregates robustness to the convergence and guarantee its convergence even from adverse initial conditions. The algorithm allows to reach any solution within the solution space as long as it is previously selected. A particular case occurs when a null error is defined for arbitrary values of norm which represents Pareto solutions targets. Once the Pareto-optimal candidates are found the decision step is done with validation data as described in section 7.2.2.

The SMC-MOBJ algorithm uses two sliding surfaces to lead the error and norm cost functions to a pre-establish coordinate $(E_t, ||\mathbf{w}_t||)$ into the cost functions space. The first surface is defined as the difference between the actual network error function at the k-iteration and the target error ($S_{E(k)} = E_{(k)} - E_t$). The second surface is the difference between the actual and the target norm functions ($S_{||\mathbf{w}(k)||} = ||\mathbf{w}_{(k)}||^2 - ||\mathbf{w}_t||^2$). The final weight update equation is based on two descent gradients related to the error and norm functions. Both are controlled by its respective sliding surface signals leading the networks to the target point. Equation 7.14 defines the weight update formula according to the SMC-MOBJ proposition.

$$\Delta w_{ji(k)} = -\alpha.sgn\left(S_{E(k)}\right).\frac{\partial E_{(k)}}{\partial w_{ji(k)}} - \beta.sgn\left(S_{||\mathbf{w}(k)||}\right).w_{ji(k)} \tag{7.14}$$

where $S_{E(k)}$ e $S_{||\mathbf{w}(k)||}$ are the sliding surfaces, α, β the respectives gains calculated with sliding model theory and $sgn(.)$, the sign function.

$$sgn(S) = \begin{cases} -1 & \text{if } S < 0 \\ 0 & \text{if } S = 0 \\ +1 & \text{if } S > 0 \end{cases}$$

Multiple solutions can be generated from different targets or trajectories into the space. Approximations to the Pareto set are achieved setting targets with null error cost function ($E_t = 0$) and pre-established norm values or through arbitrary trajectories that cross the Pareto boundary as shown in Figure 7.2.

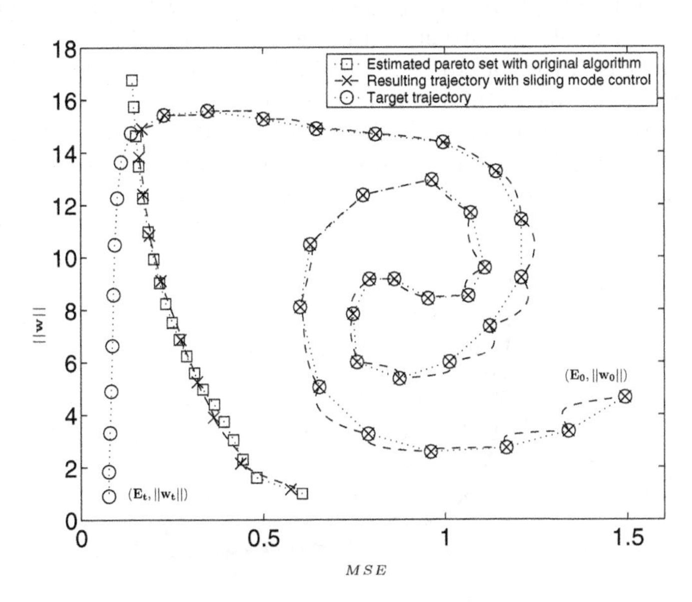

Fig. 7.2. Arbitrary trajectory in the plane of objectives

7.3 Constructive and Pruning Methods with Multi-objective Learning

The Multi-objective algorithm for training neural networks [16] is based on the idea of picking up a network solution within the restricted set of the Optimal Pareto set. The solutions generated by the MOBJ algorithm can be tuned to be oversized and yet with good generalization performance. The final solution represents a network whose amplitude weights were properly set during training. Therefore, the norm of the weights is optimal since it represents a solution with minimal validation error and the network is suitable for pruning techniques. If networks are known to be oversized in advance, their structure can be shrinked. Another approach consists of a constructive algorithm based on the MOBJ algorithm. In this case the Pareto shape can be seen as a threshold for hidden layer growing.

Original pruning algorithms for MOBJ training [22] are based on: linearization of the hidden nodes, random sampling of weights, similarity of hidden nodes responses and, finally, a mixture of some of these methods.

7.3.1 Network Growing

This constructive algorithm is based on the idea of gradually increasing the number of nodes of a MLP and, for each new neural network, a Pareto set is generated by one of the current multi-objective approaches. As the number of nodes increases, the Pareto set shape in the space of objectives shifts left, converging to a stable form, as can be seen in the sketch presented in Figure 7.3. The Pareto set shape is nearly the same regardless of network complexity. Based on this principle, training stops as training set error stabilizes.

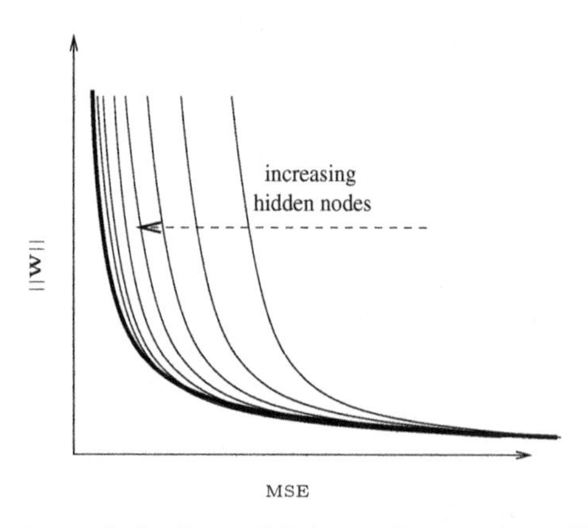

Fig. 7.3. Pareto set behaviour with increasing number of hidden nodes

The multi-objective growing algorithm searches for a solution with improved generalization performance within the three-dimensional space defined by the following functions:

1. Error function ($E_{(\mathbf{w})}$)
2. Norm function ($||\mathbf{w}||$)
3. Number of Hidden nodes (H)

In this case, the optimization problem is described according to Equation 7.15:

$$\mathbf{w}^* = arg \min_{\mathbf{w} \in \Re^{N(H)}} \begin{cases} E_{(\mathbf{w})} \\ ||\mathbf{w}|| \\ H \end{cases} \tag{7.15}$$

7.3.2 Linearization of Hidden Nodes

A non-linear node with hiperbolic tangent activation function may have a linear output response. This means that, for a limited range of the outputs, $f(-\psi) < f(x) < f(+\psi)$, the sigmoidal activation function could be replaced by a linear one. For every network node, the output range is calculated and, if it is within the pre-established range it is substituted by a linear function within that range. The effect of linearizing hidden nodes is that the mapping between network input and output can be performed by a single linear transformation, simplifying network structure. The new network has a mixed structure with non-linear and linear hidden nodes. Equation 7.16 expresses a MLP's output equation for a two layer perceptron network with H hidden nodes, p outputs, N inputs and L linearized hidden nodes ($L \leq H$).

$$y_p = f_p \left\{ \sum_{i=1, i \neq t}^{H} w2_{ip}.f_i \left[\sum_{j=1}^{N} (x_j w1_{ji}) + b1_i \right] + \sum_{j=1}^{N} (x_j.wln_{jp}) + b2'_p \right\} \tag{7.16}$$

where $wln_{jp} = \sum_{t=1}^{L} w1_{jt}.w2_{tp}$, $b2'_p = b2_p + \sum_{t=1}^{L} b1_t.w2_{tp}$, f_p and f_i are the activation functions.

7.3.3 Similar Nodes Response Simplification

This method works by identifying pairs of hidden nodes whose responses are highly correlated. The nodes could be replaced by a single one. In order to measure the similarity between two hidden nodes r and t, the norm of their responses to the training set input vectors is calculated, as shown in Equation 7.17.

$$||\mathbf{f}_r^H - \mathbf{f}_t^H|| \approx 0 \tag{7.17}$$

The algorithm consists of finding the pair of hidden nodes with minimum distance between their outputs for the whole training set. The norms of the weight vectors that connect each one of the two nodes to the output layer is then computed and the one with the smallest norm is chosen to be pruned and the remaining output weights are updated with the sum of the hidden nodes's output weights. The stop criterion is based on the validation error. Pruning stops when it starts to increase.

7.3.4 Pruning Randomly Selected Weights

This method consists of pruning randomly selected weights. If the extraction results on validation error decrease, the node is pruned, otherwise it is maintained. For large MLPs, the performance may decrease, depending on the number of weights to be pruned, that is a user-defined parameter. Due to its simplicity, this method has a good balance between algorithm complexity and network performance.

7.3.5 Mixed Method

A combination of the previously described methods is used to simplify network structures. The execution order of the methods is not restricted but the following order is suggested:
1. Pruning randomly selected weigths;
2. Similar nodes response;
3. Linearization.

The methods are tested in the sequence above and, for each algorithm, weights and nodes are pruned. The final topology may have linear connections between inputs and outputs and also reduced number of hidden nodes with prunned connections.

7.4 Classification Problem

In order to test the algorithm's performance in classification task, two examples were selected. The first one consists of data sampled from two gaussian distributions with mean $\mu_1 = (2, 2)$ (class A) and $\mu_2 = (4, 4)$ (class B) and variance $\sigma^2 = 1.5^2$. The training set has 80 input-output patterns with 40 examples from each class. A network with over-estimated size with topology 2–50–1 was used in order allow complexity control by the learning algorithms.

The ANNs topology 2–50–1 trained by standard Backpropagation (BP), Weight Decay (WD), SVM [3] algorithms and the proposed multi-objective approach (MOBJ). The decay constant to the weight decay algorithm was chosen equal to 0.0004. For the SVM algorithm, the chosen kernel was RBF with variance equal to 6 and the limit of the Lagrange multipliers was 10. Several trials were carried on in order to obtain appropriate WD and and SVM solutions.

The MOBJ algorithm was executed in order to generate 20 Pareto-optimal solutions. Figure 7.4 shows the twenty generated solutions, from which the final MOBJ solution is chosen.

Figure 7.5 shows the solutions generated by all methods in the objective space and the Pareto-optimal set generated by the MOBJ algorithm. Note that BP and WD solutions are far from the Pareto-optimal set. It means that these solutions are not Pareto-optimal solutions and therefore they could still

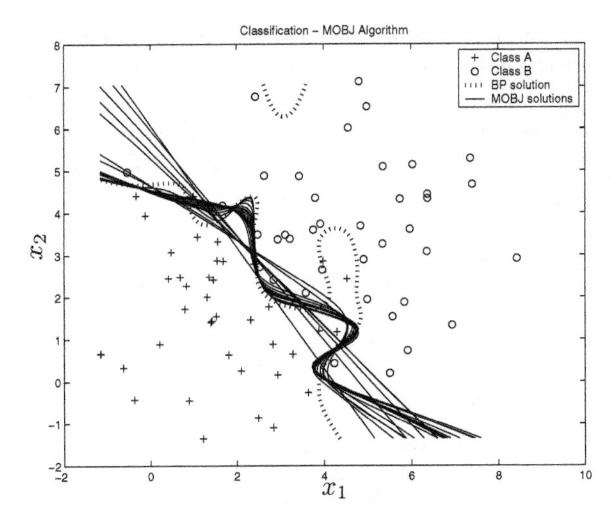

Fig. 7.4. MOBJ classification solutions

be minimized considering both objectives. It is also evident that the MOBJ solution has the smaller norm value.

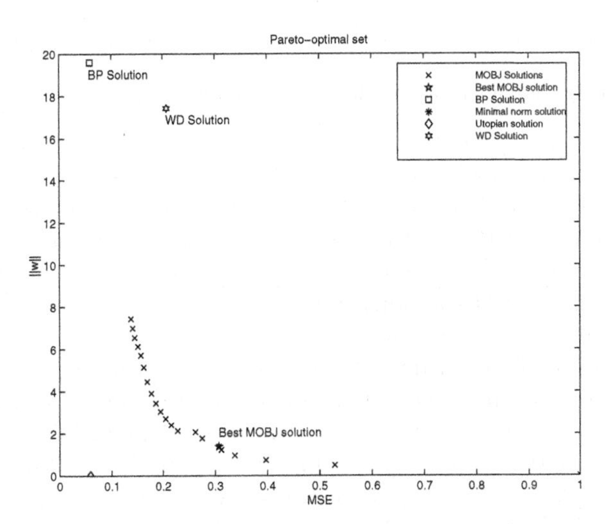

Fig. 7.5. Solutions within the norm versus MSE space

Figure 7.6 shows also the decision regions obtained by a 2-50-1 MLP trained with BP, WD, SVM and MOBJ algorithms. Overfitting can be clearly observed in the standard Backpropagation response, since it tends to separate every single component of each class. Several trials were carried on in order to choose an appropriate decay constant for the WD solution. The one presented in the graph was the best one obtained. The MOBJ and SVM solutions are quite similar but the generation of the SVM solution demanded selecting network and training parameters, such as the limit of the Lagrange multipliers that was not obtained in a straight way. The SVM norm comparison was not provided due to the respective network structure (RBF) that does not have input weights.

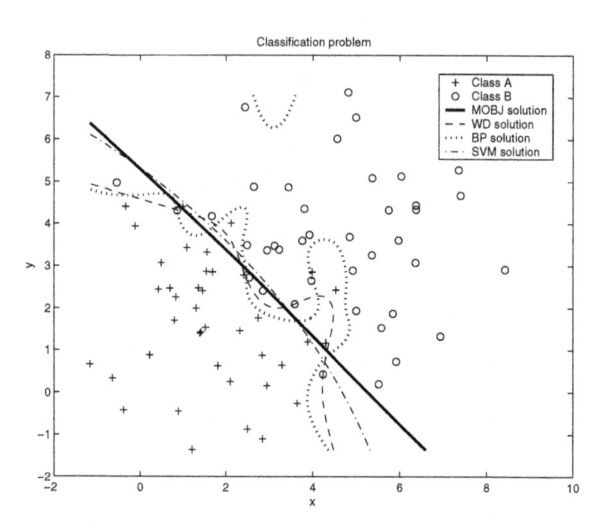

Fig. 7.6. Classification problem. BP, WD, ES and MOBJ solutions

An important property of the multi-objective approach is the characteristic of the decision surfaces in classification problems. Since it reduces the norm of the weight vector, the MOBJ surfaces are smoother than those of other methods which avoid overfitting. Figure 7.7 and Figure 7.8 show the back-propagation and MOBJ solutions respectively, where the smoothness of the MOBJ solution can be observed.

A second example is called the *Chess-Board* classification problem which is a two class problem where the data is spread into the bi-dimensional space as a $4x4$ chess board. The classes present a narrow overlapping in the boundaries. The data distribution is shown in Figure 7.9 and consists of 960 patterns.

ANNs with 2-50-1 topology were trained with BP, WD, OBD (Optimal Brain Damage), ES, CV and MOBJ. The SVM approach was also evaluated. Figure 7.10 shows the solutions as well as the approximated Pareto set.

In this example the CV, ES, and WD solutions were very close to the Pareto-set. Although in the previous example only the SVM solution had this behavior, it is not guaranteed that any of them will have similar responses as the MOBJ approach. Mainly because their generalization control is sensitive to their parameters, initial weights and number of iterations differently from the MOBJ algorithm which is insensitive to initialization process and parameters.

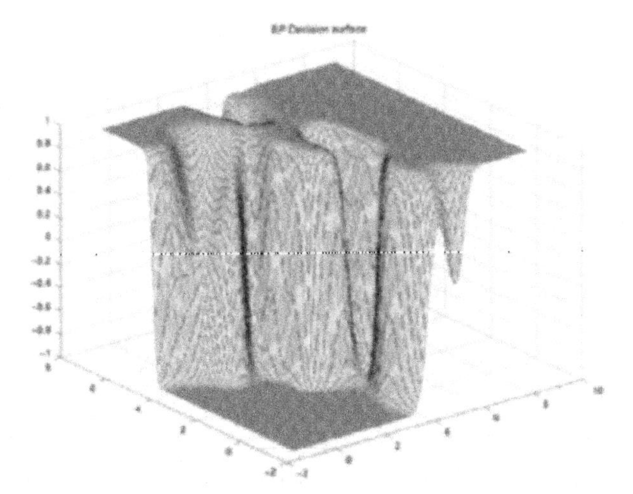

Fig. 7.7. BP decision surface

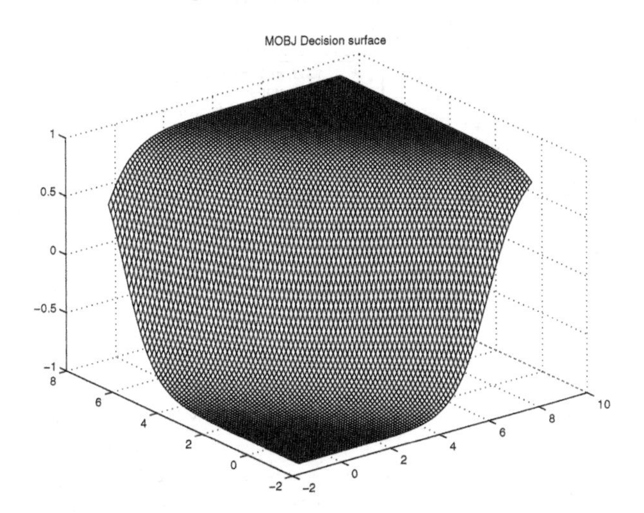

Fig. 7.8. MOBJ decision surface

Figures 11(a) and 11(b) show the surface decision generated with the MOBJ and BP algorithms. The BP surface is rougher than the MOBJ surface which is smoother. The BP algorithm aims at minimizing the MSE which overfits the data in contrast with the MOBJ that takes into account the generalization degree measured as the validation error. In addition to the validation set, the MOBJ approach starts with an underfitted network which is gradually fitted to the data according to the Pareto's principle [18].

Table 7.1 displays the percentage of patterns correctly classified for 100 sets generated from the original set (960 patterns) which was randomly divided into training and validation sets. For multiple tests, all the algorithm

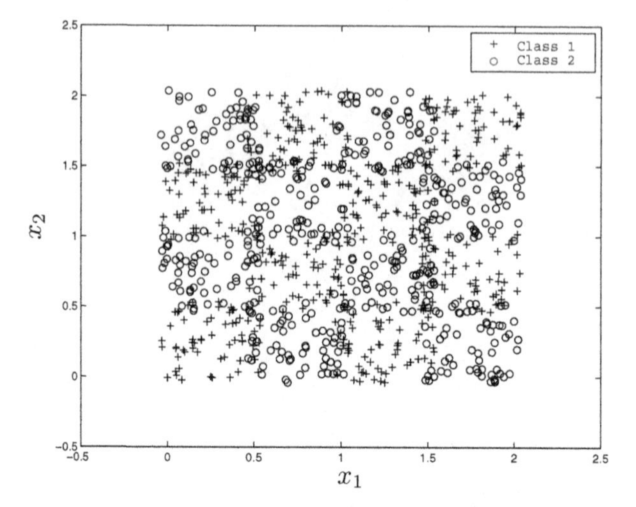

Fig. 7.9. The Chess Board data set

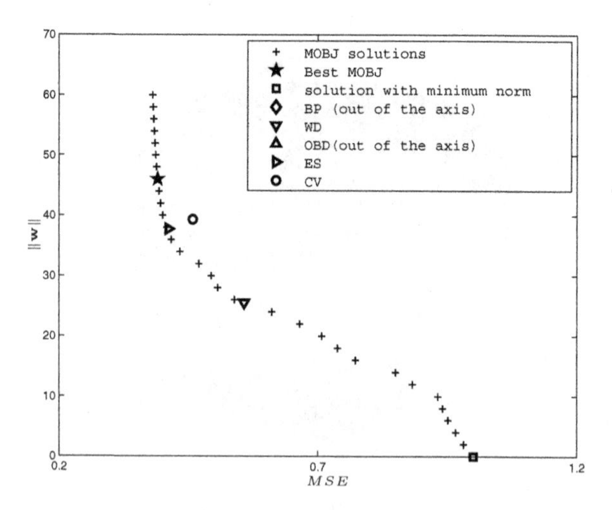

Fig. 7.10. The *Chess Board* solutions

performed good classifications with closer means. The MOBJ approach had a slightly better performance.

7.5 Regression Problem

A MLP with topology 1-50-1 was trained by BP, WD, SVM and MOBJ algorithms. The networks were trained with the 40 noisy patterns sampled from the function described by Eq. (7.18), which was generated by adding a zero-

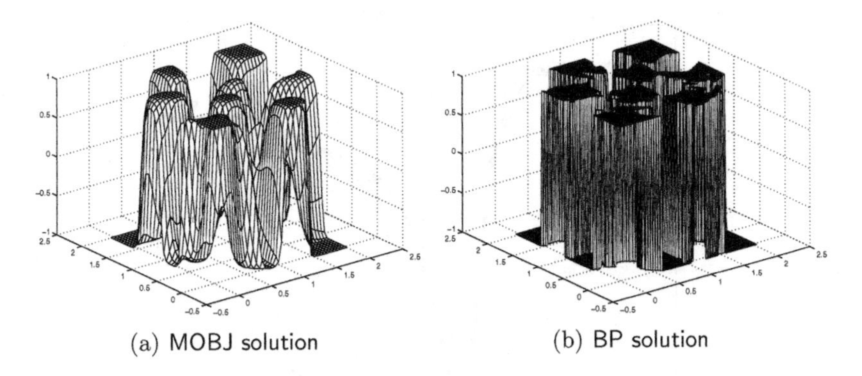

(a) MOBJ solution (b) BP solution

Fig. 7.11. Decision surface for BP and MOBJ algorithms

Table 7.1. Proportions of patterns correctly identified for 100 simulations sets

Algorithms	Percentage	σ
Backpropagation	75.9635%	±1.3103%
Weight decay	77.8958%	±1.2566%
Optimal Brain Damage	77.2031%	±1.2293%
Early stopping	77.7312%	±1.3177%
Cross-validation	77.7604%	±1.3257%
SVM	77.7583%	±1.3525%
MOBJ	77.9667%	±1.2863%

mean normally distributed random noise with variance of 0.15^2 to the original curve.

The decay constant to the weight decay algorithm has been chosen equals to 0.00015. For the SVM algorithm, the chose kernel was RBF with variance equals to 0.8 and the chosen limit of Lagrange multipliers was 1. The weight decay and SVM parameters were not obtained easily once many attempts were necessary.

$$f(x) = \frac{(x-2)(2x+1)}{(1+x^2)} \tag{7.18}$$

The MOBJ algorithm was executed to generate 20 Pareto-optimal solutions. Figure 7.12 shows the twenty generated solutions. Figure 7.13 shows each algorithm solution. As can be observed, the BP, WD and SVM solutions presented some degrees of overfitting. To carry on this approximation is specially difficult for some algorithms, since the target function has both steep and smooth regions.

In the design of the SVM solution, when we changed the parameters in order to model the smooth region, the steep one was not well approximated and, on the other hand, when we tried to model the steep region, overfitting

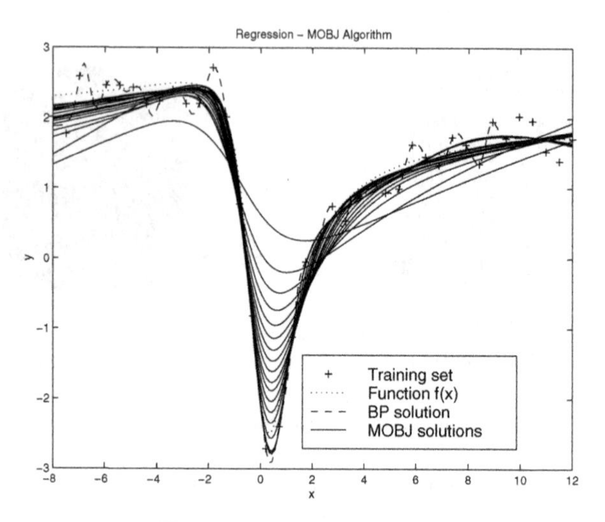

Fig. 7.12. MOBJ solutions

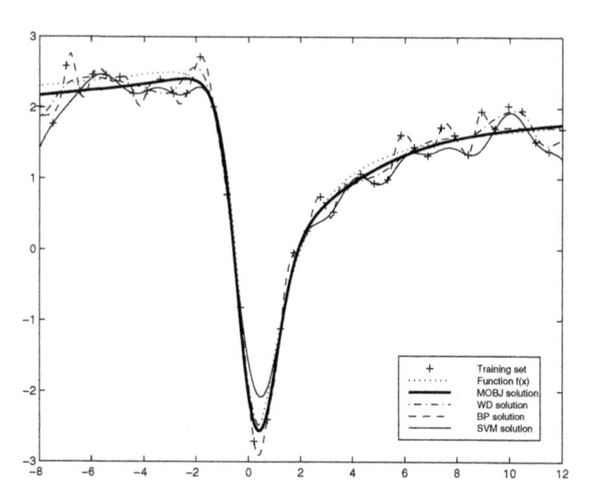

Fig. 7.13. BP, WD, SVM and MOBJ solution

occurred in the smooth region. The MOBJ algorithm reached a solution with high generalization capacity independently of the user choices.

Although the MOBJ algorithm works by reducing norm, the reduction is not carried on with the same magnitude for all the network weights. This differs from weight decay, for example, that applies the same decay constant for all weights, what penalizes mainly the large ones. Figure 7.14 shows that in the multi-objective approach it is possible to reduce the norm of some weight vectors while other vectors have their norm increased. This characteristic of the MOBJ algorithm yields more flexibility to neural networks, since the resulting model can be smooth in some regions and steep in other ones.

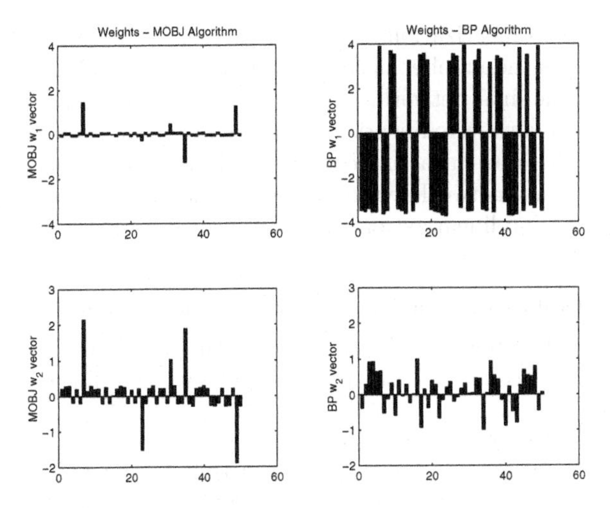

Fig. 7.14. MOBJ and BP weight vectors

7.6 Pruning Problem

Four real based data sets picked-up from *PROBEN* [23] consisting of three classification and one function approximation problems were used to test the Pruning methods with Multi-objective learning. The data sets *cancer*, *card* and *gene* were used for classification and *building* data set were used for function approximation. The data sets were divided into training and validation sets with 60% and 40% of the whole data, respectively.

Table 7.2 presents the results for oversized network with the MOBJ algorithm, without pruning. The networks are the best generalization solutions obtained without reducing the network number of weights and nodes. These initial solutions were used further as a reference for the pruning methods. The pruning methods are capable to reduce these initial over-sized networks without loss in validation performance.

Table 7.2. Initial results with the Multi-Objective algorithm

Data set	Training Error	Validation Error	Oversized Topology	$\|\|\mathbf{w}\|\|$
cancer	98.09%	98.57%	9-15-2	14.263
card	88.89%	88.04%	51-15-2	4.413
gene	98.85%	89.53%	120-15-3	8.831
building	0.0196	0.0228	14-15-3	6.992

For classification problems the training and validation error are the percentage of patterns correctly assigned and represents the mean squared error for the function approximation data set.

Results for the Linearization method are presented in Table 7.3. The Final Topology presents the number of linear and non-linear nodes between inputs and outputs of the final network. For the *card* data set, all the hidden nodes of the original network were linearized, what reduced the network into a single layer network with non-linear nodes. The final network replaced the original two-layers non-linear network, without loss in generalization performance. For the *gene* data set, no linearization was possible.

Table 7.3. Results for Linearization of hidden nodes

Data Set	Training Error	Validation Error	Initial Topology	Final Topology	
				non-linear	linear
cancer	96.42%	98.57%	9-15-2	9-2-2	9-2
card	88.89%	88.04%	51-15-2	-	51-2
gene	98.85%	89.53%	120-15-3	120-15-3	-
building	0.0234	0.0224	14-15-3	14-7-3	14-3

Similar nodes response identification results are presented in Table 7.4. The network for the *cancer* data set was reduced to a 9-6-2 topology, what represented a reduction of 9 nodes in the hidden layer. For the *card* data set the reduction was of 11 nodes in the hidden layer. There was no reduction for the *gene* data set and reduction only one node for the *building* data set.

Table 7.4. Similar nodes response identification results

Data Set	Training Error	Validation Error	Initial Topology	Final Topology
cancer	96.18%	98.57%	9-15-2	9-6-2
card	88.89%	88.04%	51-15-2	51-4-2
gene	98.85%	89.53%	120-15-3	120-15-3
building	0.0189	0.0227	14-15-3	14-14-3

Table 7.5 shows the results for the Randomly Selected Weights method. Topologies with pruned weights are indicated with an asterisk (*). The result for the *cancer* data set reduced the network to a similar structure to the one obtained by linearization and Method 3. In addition to the final 9-4-2 topology, weights were pruned throughout the network. The results for the *card* data set showed that some input variables are not relevant for solving the problem, since the number of inputs was reduced from 51 to 14. The final network was also reduced in the hidden layer to 9 nodes, in contrast with the 15 nodes of the original network. A similar result was obtained for the *gene* data set, that was reduced from 120 inputs to only 44. The number of hidden nodes was reduced to 12. The network for the *building* data set had also a significant reduction in size and number of inputs (inputs were reduced to 4).

Table 7.5. Randomly selected weights

Data Set	Training Error	Validation Error	Final Topology
cancer	96.18%	99.64%	9-4-2*
card	85.99%	88.41%	14-9-2*
gene	90.08%	90.31%	44-12-3*
building	0.0257	0.0165	4-8-3*

As shown in Table 7.6, the best results occurs when the methods are mixed. This can be justified by the fact that it can take the best features of each one. For the *cancer* data set, 13 hidden nodes were pruned; the final network has also 3 linear hidden nodes. There was an impressive reduction in the *card* data set. The problem could be solved with a single hidden layer with 8 input variables. The *gene* data set problem had the best reduction in the number of inputs of all methods: the original 120 variables was reduced to only 20. The *building* data set had also significant reduction in the number of inputs.

Table 7.6. Results for arranged pruning methods

Data	Training	Validation	Initial	Final Topology	
Set	Error	Error	Topology	non-linear	linear
cancer	96.18%	99.64%	9-15-2	9-2-2*	3-2
card	85.27%	88.77%	51-15-2	-	8-2
gene	91.18%	91.26%	120-15-3	20-8-3*	1-1
building	0.0257	0.0164	14-15-3	4-7-3*	1-2

7.7 Conclusions

The multi-objective algorithms described are able to obtain good generalization solutions for regression and classification problems. The algorithms do not demand external user parameters in order to generate the Pareto-optimal set and to choose the best solution. The solution is obtained by a restricted search over the space of objectives norm and error. Current research aim at describing new decision strategies and at extending MOBJ concepts to other machine learning approaches.

References

[1] B. Boser, I. Guyon, and V. Vapnik. A training algorithm for optimal margin classifiers. *Fifth Annual Workshop on Computational Learning Theory*, pages 144–152, 1992.

[2] V. Chankong and Y. Y. Haimes. *Multiobjective Decision Making: Theory and Methodology*, volume 8. North-Holland (Elsevier), New York, 1983.

[3] C. Cortes and V. Vapnik. Support vector networks. *Machine Learning*, 20:273–279, 1995.

[4] Yann Le Cun, John S. Denker, and Sara A. Solla. Optimal brain damage. In *Advances in Neural Information Processing Systems 2*, pages 598–605, 1990.

[5] S. E. Fahlman. Faster-learning variations on back-propagation: an empirical study. In D. Touretzky, G. Hinton, and T. Sejnowski, editors, *Proceedings of the 1988 Connectionist Models Summer School, Pittsburg*, pages 38–51, San Mateo, CA, 1988. Morgan Kaufmann.

[6] S. R. Gunn. Support vector machines for classification and regression. Technical report, Image Speech and Intelligent Systems Research Group, University of Southampton, 1997.

[7] S. Haykin. *Neural Networks: A Comprehensive Foundation.* Prentice Hall, 1999.

[8] Ehud D. Karnin. A simple procedure for pruning back-propagation trained neural networks. *IEEE Transactions on Neural Networks*, 1(2):239–242, 1990.

[9] M. C. Mozer and P. Smolensky. Skeletonization: A technique for trimming the fat from a network via relevance assessment. *Advances in Neural Information Processing*, vol. 1, pages 107–115, 1989.

[10] Gustavo G. Parma, Antonio P. Braga, and Benjamim R. Menezes. Sliding mode algorithm for training multi-layer neural networks. *IEE Electronics Letters*, 38(1):97–98, January 1998.

[11] Martin Riedmiller and Heinrich Braun. A direct adaptive method for faster backpropagation learning: The RPROP algorithm. In *Proc. of the IEEE Intl. Conf. on Neural Networks*, pages 586–591, San Francisco, CA, April 1993.

[12] D. E. Rumelhart, G. E. Hinton, and R. J. Williams. Learning representations by back-propagating errors. *Nature*, 323:533–536, 1986.

[13] N. Z. Shor. Cut-off method with space extension in convex programming problems. *Cybernetics*, 12:94–96, 1977.

[14] R. H. C. Takahashi, P. L. D. Peres, and P. A. V. Ferreira. H2/h-infinity multiobjective pid design. *IEEE Control Systems Magazine*, 17(5):37–47, June 1997.

[15] V. Vapnik. *The Nature of Statistical Learning Theory.* Springer-Verlag, 1995.

[16] R. A. Teixeira, A. P. Braga, R. H. C. Takahashi, and R. R. Saldanha. Improving generalization of mlps with multi-objective optimization. *Neurocomputing*, 35(1–4):189–194, 2000.

[17] G. A. Hinton. Connectionist learning procedures. *Artificial Intelligence*, 40:185-234, 1989.

[18] V. Pareto. Cours D'Economie Politique. Rouse, Lausanne, 1896. vols. I and II.

[19] R. A. Teixeira, A. P. Braga, R. H. C. Takahashi, and R. R. Saldanha. Utilização de seção áurea no cálculo de soluções eficientes para treinamento de redes neurais artificiais através de otimização multi-objetivo. *8th Brazilian Symposium on Neural Networks*, November 2004.

[20] U. Itkis. Control Systems of Variable Structure. Keter Publishing House Jerusalem LTD, 1976.

[21] M. A. Costa, A. P. Braga, B. R. de Menezes, G. G. Parma, and R. A. Teixeira. Training neural networks with a multi-objective sliding mode control algorithm. *Neurocomputing*, 51:467-473, 2003.

[22] M. A. Costa, A. P. Braga and B. R. de Menezes. Improving neural networks generalization with new constructive and pruning methods. *Journal of Intelligent & Fuzzy Systems*, 10:1-9, 2003.

[23] C.L. Blake and C.J. Merz. {UCI} Repository of machine learning databases. *University of California, Irvine, Dept. of Information and Computer Sciences*, `http://www.ics.uci.edu/~mlearn/MLRepository.html`, 1998.

[24] S. E. Fahlman and C. Lebiere, *The cascade-correlation learning architecture*, Morgan Kaufmann, In Advances in Neural Information Processing Systems 2 (D. S. Touretzky, Editor), 1990

[25] Jean-Pierre Nadal, Study of a growth algorithm for a feedforward network, *International Journal of Neural Systems*, 1(1):55-59, 1989.

[26] Ron Kohavi, A Study of Cross-Validation and Bootstrap for Accuracy Estimation and Model Selection, `http://citeseer.ist.psu.edu/105046.html`, 1995.

[27] S. Geman and E. Bienenstock and R. Doursat. Neural Networks and the Bias/Variance Dilemma, *Neural Computation*, 4(1):1-58, 1992.

8

Generating Support Vector Machines Using Multi-Objective Optimization and Goal Programming

Hirotaka Nakayama[1] and Yeboon Yun[2]

[1] Konan University, Dept. of Information Science and Systems Engineering
8-9-1 Okamoto, Higashinada, Kobe 658-8501, Japan
nakayama@konan-u.ac.jp

[2] Kagawa University, Kagawa 761-0396, Japan
yun@eng.kagawa-u.ac.jp

Summary. Support Vector Machine (SVM) is gaining much popularity as one of effective methods for machine learning in recent years. In pattern classification problems with two class sets, it generalizes linear classifiers into high dimensional feature spaces through nonlinear mappings defined implicitly by kernels in the Hilbert space so that it may produce nonlinear classifiers in the original data space. Linear classifiers then are optimized to give the maximal margin separation between the classes. This task is performed by solving some type of mathematical programming such as quadratic programming (QP) or linear programming (LP). On the other hand, from a viewpoint of mathematical programming for machine learning, the idea of maximal margin separation was employed in the multi-surface method (MSM) suggested by Mangasarian in 1960's. Also, linear classifiers using goal programming were developed extensively in 1980's. This chapter introduces a new family of SVM using multi-objective programming and goal programming (MOP/GP) techniques, and discusses its effectiveness throughout several numerical experiments.

8.1 Introduction

For convenience, we consider pattern classification problems. Let X be a space of conditional attributes. For binary classification problems, the value of $+1$ or -1 is assigned to each pattern $\boldsymbol{x}_i \in X$ according to its class $\mathcal{A}$ or $\mathcal{B}$. The aim of machine learning is to predict which class newly observed patterns belong to on the basis of the given training data set $(\boldsymbol{x}_i, y_i)$ $(i = 1, \ldots, \ell)$, where $y_i = +1$ or -1. This is performed by finding a discriminant function $f(\boldsymbol{x})$ such that $f(\boldsymbol{x}) \geqq 0$ for $\boldsymbol{x} \in \mathcal{A}$ and $f(\boldsymbol{x}) < 0$ for $\boldsymbol{x} \in \mathcal{B}$. Linear discriminant functions, in particular, can be expressed by the following linear form

$$f(\boldsymbol{x}) = \boldsymbol{w}^T \boldsymbol{x} + b$$

H. Nakayama and Y. Yun: *Generating Support Vector Machines Using Multi-Objective Optimization and Goal Programming*, Studies in Computational Intelligence (SCI) **16**, 173–198 (2006)
www.springerlink.com

with the property

$$\begin{aligned} \boldsymbol{w}^T \boldsymbol{x} + b &\geqq 0 \quad \text{for} \quad \boldsymbol{x} \in \mathcal{A} \\ \boldsymbol{w}^T \boldsymbol{x} + b &< 0 \quad \text{for} \quad \boldsymbol{x} \in \mathcal{B}. \end{aligned}$$

For such a pattern classification problem, artificial neural networks have been widely applied. However, the back propagation method is reduced to nonlinear optimization with multiple local optima, and hence difficult to apply to large scale problems. Another drawback in the back propagation method is in the fact that it is difficult to change the structure adaptively according to the change of environment in incremental learning. Recently, Support Vector Machine (SVM, for short) is attracting interest of researchers, in particular, people who are engaged in mathematical programming, because it is reduced to quadratic programming (QP) or linear programming (LP). One of main features in SVM is that it is a linear classifier with maximal margin on the feature space through nonlinear mappings defined implicitly by kernels in the Hilbert space.

The idea of maximal margin in linear classifiers is intuitive, and its reasoning in connection with perceptrons was given in early 1960's (e.g., Novikoff [17]). The maximal margin is effectively applied for discrimination analysis using mathematical programming, e.g., MSM (Multi-Surface Method) by Mangasarian [11]. Later, linear classifiers with maximal margin were formulated as linear goal programming, and extensively studied through 1980's to the beginning of 1990's. The pioneering work was given by Freed-Glover [9], and a good survey can be seen in Erenguc-Koehler *et al.* [8]. This chapter discusses SVMs using techniques of multi-objective programming (MOP) and goal programming (GP), and proposes several extensions of SVM along MOP/GP.

8.2 Support Vector Machine

Support vector machine (SVM) was developed by Vapnik *et al.* [6], [22] (see also Cristianini and Shawe-Taylor [7], Schölkopf-Smola [20]) and its main features are

1) SVM maps the original data set into a high dimensional feature space by nonlinear mapping implicitly defined by kernels in the Hilbert space,

2) SVM finds linear classifiers with maximal margin on the feature space,

3) SVM provides an evaluation of the generalization ability using VC dimension.

Namely, in cases where training data set X is not linearly separable, we map the original data set X to a feature space Z by some nonlinear map ϕ.

Increasing the dimension of the feature space, it is expected that the mapped data set becomes linearly separable. We try to find linear classifiers with maximal margin in the feature space. Letting $\boldsymbol{z}_i = \phi(\boldsymbol{x}_i)$, the separating hyperplane with maximal margin can be given by solving the following problem with the normalization $\boldsymbol{w}^T\boldsymbol{z} + b = \pm 1$ at points with the minimum interior deviation:

$$\begin{array}{lll} \text{minimize} & ||\boldsymbol{w}|| & (\text{SVM}_{hard})_P \\ \text{subject to} & y_i\left(\boldsymbol{w}^T\boldsymbol{z}_i + b\right) \geqq 1, \ i = 1, \ldots, \ell. & \end{array}$$

Several kinds of norm are possible. When $||\boldsymbol{w}||_2$ is used, the problem is reduced to quadratic programming, while the problem with $||\boldsymbol{w}||_1$ or $||\boldsymbol{w}||_\infty$ is reduced to linear programming (see, e.g., [12]).

Dual problem of $(\text{SVM}_{hard})_P$ with $\frac{1}{2}||\boldsymbol{w}||_2^2$ is

$$\begin{array}{lll} \text{maximize} & \displaystyle\sum_{i=1}^{\ell} \alpha_i - \frac{1}{2}\sum_{i,j=1}^{\ell} \alpha_i\alpha_j y_i y_j \phi(\boldsymbol{x}_i)^T\phi(\boldsymbol{x}_j) & (\text{SVM}_{hard})_D \\ \text{subject to} & \displaystyle\sum_{i=1}^{\ell} \alpha_i y_i = 0, & \\ & \alpha_i \geqq 0, \ i = 1, \ldots, \ell. & \end{array}$$

Using the kernel function $K(\boldsymbol{x}, \boldsymbol{x}') = \phi(\boldsymbol{x})^T\phi(\boldsymbol{x}')$, the problem $(\text{SVM}_{hard})_D$ can be reformulated as follows:

$$\begin{array}{lll} \text{maximize} & \displaystyle\sum_{i=1}^{\ell} \alpha_i - \frac{1}{2}\sum_{i,j=1}^{\ell} \alpha_i\alpha_j y_i y_j K(\boldsymbol{x}_i, \boldsymbol{x}_j) & (\text{SVM}_{hard}) \\ \text{subject to} & \displaystyle\sum_{i=1}^{\ell} \alpha_i y_i = 0, & \\ & \alpha_i \geqq 0, \ i = 1, \ldots, \ell. & \end{array}$$

Several kinds of kernel functions have been suggested: among them, q-polynomial

$$K(\boldsymbol{x}, \boldsymbol{x}') = (\boldsymbol{x}^T\boldsymbol{x}' + 1)^q$$

and Gaussian

$$K(\boldsymbol{x}, \boldsymbol{x}') = \exp\left(-\frac{||\boldsymbol{x} - \boldsymbol{x}'||^2}{r^2}\right)$$

are most popularly used.

8.3 Review of Multi-objective Programming and Goal Programming

Multi-objective programming (MOP) problems are formulated as follows:

$$\text{(MOP)} \qquad \text{Maximize} \quad g(\boldsymbol{x}) \equiv (g_1(\boldsymbol{x}),\ g_2(\boldsymbol{x}), \dots,\ g_p(\boldsymbol{x}))$$
$$\text{over} \quad \boldsymbol{x} \in X.$$

The constraint set X may be given by

$$c_j(\boldsymbol{x}) \leqq 0, \qquad j = 1, \dots, m,$$

and/or a subset of R^n itself. For the problem (MOP), Pareto solutions are candidates of final decision ($\hat{\boldsymbol{x}}$ is said *Pareto optimal*, if there is no better solution $\boldsymbol{x} \in X$ other than $\hat{\boldsymbol{x}}$).

In general, there may be many Pareto solutions. The final decision is made among them taking the total balance over all criteria into account. This is a problem of value judgment of decision maker (in abbreviation, DM). The totally balancing over criteria is usually called *trade-off*. It is important to help DM to trade-off easily in practical decisin making problems.

There have been developed several kinds of methods for multi-objective programming (see, e.g., Steuer [21], Chankong-Haims [4], Sawaragi-Nakayama-Tanino [18], Nakayama [15], Miettinen [14]). Among them, interactive multi-objective programming methods, which were developed remarkably in 1980's, have been observed to be effective in various fields of practial problems. Those methods search a solution in an interactive way with DM while eliciting information on his/her value judgment.

On the other hand, Goal Programming (GP) was developed by Charnes-Cooper [5] much earlier than interactive programming methods. The idea was originated from getting rid of no feasible solution in usual mathematical programming. Namely, many constraints should be regarded as "goal" to be attained, and we try to find a solution which attains those goals as much as possible.

For example, suppose that we want to make

$$g_i(\boldsymbol{x}) \geqq \overline{g}_i, \quad i = 1, \dots, p.$$

Introducing the degree of overattainment (or surplus, or interior deviation) η_i and the degree of unattainment (or slackness, or exterior deviation) ξ_i, we have the following goal programming formulation:

$$\begin{aligned} &\text{minimize} && \sum_{i=1}^{p} h_i \xi_i && (\text{GP}_0) \\ &\text{subject to} && g_i(\boldsymbol{x}) - \overline{g}_i = \eta_i - \xi_i, \\ &&& \xi_i,\ \eta_i \geqq 0,\ i = 1, \dots, p \\ &&& \boldsymbol{x} \in X. \end{aligned}$$

where h_i $(i = 1, \ldots, p)$ are positive weighting parameters which are given by DMs.

It should be noted that in order for η_i and ξ_i in the above formulation to have the meaning of the degree of overattainment and the degree of unattainment, respectively, the relation $\xi_i \cdot \eta_i = 0$ has to be satisfied. The above formulation assures this property due to the following lemma (Lemma 7.3.1 of [18]):

Lemma 1. *Let* $\boldsymbol{\xi}$ *and* $\boldsymbol{\eta}$ *be vectors of* R^p. *Then consider the following problem:*

$$\begin{array}{ll} \textit{minimize} & P(\boldsymbol{\xi}, \boldsymbol{\eta}) \\ \textit{subject to} & g_i(\boldsymbol{x}) - \overline{g}_i = \eta_i - \xi_i, \\ & \xi_i,\ \eta_i \geqq 0,\ i = 1, \ldots, p, \\ & \boldsymbol{x} \in X. \end{array}$$

Suppose that the function P *is monotononically increasing with respect to elements of* $\boldsymbol{\xi}$ *and* $\boldsymbol{\eta}$ *and strictly monotonically increasing with respect to at least either* ξ_i *or* η_i *for each* i $(i = 1, \ldots, p)$. *Then, the solution* $\hat{\boldsymbol{\xi}}$ *and* $\hat{\boldsymbol{\eta}}$ *to the preceding problem satisfy*

$$\hat{\xi}_i \hat{\eta}_i = 0, \quad i = 1, \ldots, p.$$

In the original formulation of goal programming, once a solution which attains every goal, no efforts are made for further improvement. Therefore, the obtained solution by goal programming is not necessarily Pareto optimal. This is due to the fact that the idea of goal programming is based on "satisficing" rather than "optimization".

In order to overcome this difficulty, we can put the degree of overattainment in the objective function in (GP_0) as follows:.

$$\begin{array}{ll} \text{minimize} & \displaystyle\sum_{i=1}^{p} h_i \xi_i - \sum_{i=1}^{p} k_i \eta_i \\ \text{subject to} & g_i(\boldsymbol{x}) - \overline{g}_i = \eta_i - \xi_i, \\ & \xi_i,\ \eta_i \geqq 0,\ i = 1, \ldots, p, \\ & \boldsymbol{x} \in X. \end{array} \qquad (\mathrm{GP}_1)$$

Note that if the relation $h_i > k_i$ for each $i = 1, \ldots, p$ holds, then the relation $\xi_i \eta_i = 0$ for each $i = 1, \ldots, p$ is satisfied at the solution. This follows in a similar fashion to Lemma 1 by considering

$$\sum_{i=1}^{p} h_i \xi_i - \sum_{i=1}^{p} k_i \eta_i = \sum_{i=1}^{p} k_i (\xi_i - \eta_i) + \sum_{i=1}^{p} (h_i - k_i) \xi_i.$$

Moreover, if $k_i = h_i$ for each $i = 1, \ldots, p$, then by substituting the right hand side of the equality constraints of (GP_1) into the objective function we have

$$\begin{aligned} &\text{maximize} && \sum_{i=1}^{p} h_i(g_i(\boldsymbol{x}) - \overline{g}_i) && (\text{MOP/GP}_0) \\ &\text{subject to} && \boldsymbol{x} \in X. \end{aligned}$$

Since the term of $-\overline{g}_i$ does not affect to maximizing the objective function, it can be removed. Namely the formulation (MOP/GP$_0$) is reduced to the usual scalarization using the linearly weighted sum in multi-objective programming.

However, the scalarization of linearly weighted sum has another drawbacks: e.g., it can not yield solutions on nonconvex parts of the Pareto frontier. To overcome this, the formulation of improvement of the worst level of objective function as much as possibel is applied as follows:

$$\begin{aligned} &\text{maximize} && \eta && (\text{MOP/GP}_1) \\ &\text{subject to} && g_i(\boldsymbol{x}) - \overline{g}_i \geqq \eta, \quad i = 1, \ldots, p, \\ &&& \boldsymbol{x} \in X. \end{aligned}$$

The solution to (MOP/GP$_1$) is guaranteed to be weakly Pareto optimal. Further discussion on scalarization functions can be seen in the literatures ([21], [4], [18], [15], [14]).

8.4 MOP/GP Approaches to Pattern Classification

In 1981, Freed-Glover suggested to get just a hyperplane separating two classes with as few misclassified data as possible by using goal programming [9] (see also [8]). Let ξ_i denote the exterior deviation which is a deviation from the hyperplane of a point $\boldsymbol{x}_i$ improperly classified. Similarly, let η_i denote the interior deviation which is a deviation from the hyperplane of a point $\boldsymbol{x}_i$ properly classified. Some of main objectives in this approach are as follows:

i) Minimize the maximum exterior deviation (decrease errors as much as possible)

ii) Maximize the minimum interior deviation (i.e., maximize the margin)

iii) Maximize the weighted sum of interior deviation

iv) Minimize the weighted sum of exterior deviation

Although many models have been suggested, the one considering iii) and iv) above may be given by the following linear goal programming:

$$\begin{array}{lll} \text{minimize} & \displaystyle\sum_{i=1}^{\ell}(h_i\xi_i - k_i\eta_i) & \text{(GP)} \\ \text{subject to} & y_i(\boldsymbol{x}_i^T\boldsymbol{w} + b) = \eta_i - \xi_i, & \\ & \xi_i,\ \eta_i \geqq 0,\ i = 1,\ldots,\ell, & \end{array}$$

where since $y_i = +1$ or -1 according to $\boldsymbol{x}_i \in \mathcal{A}$ or $\boldsymbol{x}_i \in \mathcal{B}$, two equations $\boldsymbol{x}_i^T\boldsymbol{w} + b = \eta_i - \xi_i$ for $\boldsymbol{x}_i \in \mathcal{A}$ and $\boldsymbol{x}_i^T\boldsymbol{w} + b = -\eta_i + \xi_i$ for $\boldsymbol{x}_i \in \mathcal{B}$ can be reduced to the following one equation

$$y_i(\boldsymbol{x}_i^T\boldsymbol{w} + b) = \eta_i - \xi_i.$$

Here, h_i and k_i are positive constants. As was stated in the preceding section, if $h_i > k_i$ for $i = 1,\ldots,\ell$, then we have $\xi_i\eta_i = 0$ for every $i = 1,\ldots,\ell$ at the solution to (GP). Hence then, ξ_i and η_i are assured to have the meaning of the exterior deviation and the interior deviation respectively at the solution.

It should be noted that the above formulation may yield some unacceptable solutions such as $\boldsymbol{w} = 0$ and unbounded solution. In the goal programming approach to linear classifiers, therefore, some appropriate normality condition must be imposed on $\boldsymbol{w}$ in order to provide a bounded nontrivial optimal solution. One of such normality conditions is $||\boldsymbol{w}|| = 1$.

If the classification problem is linearly separable, then using the normalization $||\boldsymbol{w}|| = 1$, the separating hyperplane $H : \boldsymbol{w}^T\boldsymbol{x} + b = 0$ with maximal margin can be given by solving the following problem [3]:

$$\begin{array}{lll} \text{maximize} & \eta & (\text{MOP/GP}_2) \\ \text{subject to} & y_i(\boldsymbol{x}_i^T\boldsymbol{w} + b) \geqq \eta,\ i = 1,\ldots,\ell, & \\ & ||\boldsymbol{w}|| = 1. & \end{array}$$

However, this normality condition makes the problem to be of nonlinear optimization. Instead of maximizing the minimum interior deviation in (MOP/GP$_2$), we can use the following equivalent formulation with the normalization $\boldsymbol{x}^T\boldsymbol{w} + b = \pm 1$ at points with the minimum interior deviation [13]:

$$\begin{array}{lll} \text{minimize} & ||\boldsymbol{w}|| & (\text{MOP/GP}_2') \\ \text{subject to} & y_i\left(\boldsymbol{x}_i^T\boldsymbol{w} + b\right) \geqq \eta,\ i = 1,\ldots,\ell, & \\ & \eta = 1. & \end{array}$$

This formulation is the same as the one used in SVM.

8.5 Soft Margin SVM

Separating two sets $\mathcal{A}$ and $\mathcal{B}$ completely is called the hard margin method, which tends to make overlearning. This implies the hard margin method is easily affected by noise. In order to overcome this difficulty, the soft margin method is introduced. The soft margin method allows some slight error which is represented by slack variables (exterior deviation) ξ_i $(i = 1, \ldots, \ell)$. Using the trade-off parameter C between minimizing $||\boldsymbol{w}||$ and minimizing $\sum_{i=1}^{\ell} \xi_i$, we have the following formulation for the soft margin method:

$$\begin{array}{lll} \text{minimize} & \frac{1}{2}||\boldsymbol{w}||_2^2 + C\sum_{i=1}^{\ell} \xi_i & (\text{SVM}_{soft})_P \\ \text{subject to} & y_i \left(\boldsymbol{w}^T \boldsymbol{z}_i + b\right) \geqq 1 - \xi_i, & \\ & \xi_i \geqq 0, \ \ i = 1, \ldots, \ell. & \end{array}$$

Using a kernel function in the dual problem yields

$$\begin{array}{lll} \text{maximize} & \sum_{i=1}^{\ell} \alpha_i - \frac{1}{2} \sum_{i,j=1}^{\ell} \alpha_i \alpha_j y_i y_j K(\boldsymbol{x}_i, \boldsymbol{x}_j) & (\text{SVM}_{soft}) \\ \text{subject to} & \sum_{i=1}^{\ell} \alpha_i y_i = 0, & \\ & 0 \leqq \alpha_i \leqq C, \ i = 1, \ldots, \ell. & \end{array}$$

It can be seen that the idea of soft margin method is the same as the goal programming approach to linear classifiers. This idea was used in an extension of MSM by Benett [2]. Not only exterior deviations but also interior deviations can be considered in SVM. Such MOP/GP approaches to SVM are discussed by the authors and their coresearchers [1], [16], [23]. When applying GP approaches, it was pointed out in Section 3 that we need some normality condition in order to avoid unacceptable solutions.

Glover suggested the following necessary and sufficient condition for avoiding unacceptable solutions [10]:

$$\left(-l_{\mathcal{A}} \sum_{i \in I_{\mathcal{B}}} \boldsymbol{x}_i + l_{\mathcal{B}} \sum_{i \in I_{\mathcal{A}}} \boldsymbol{x}_i \right)^T \boldsymbol{w} = 1, \tag{8.1}$$

where $l_{\mathcal{A}}$ and $l_{\mathcal{B}}$ denote the number of data for the category $\mathcal{A}$ and $\mathcal{B}$, respectively. Geometrically, the normalization (8.1) means that the distance between two hyperplanes passing through centers of data respectively for $\mathcal{A}$ and $\mathcal{B}$ is scaled by $l_{\mathcal{A}} l_{\mathcal{B}}$.

Lately, taking into account the objectives (ii) and (iv) of goal programming stated in the previous section, Schölkopf et al. [19] suggested ν-support vector algorithm:

$$\begin{aligned} \text{minimize} \quad & \frac{1}{2}\|\boldsymbol{w}\|_2^2 - \nu\rho + \frac{1}{\ell}\sum_{i=1}^{\ell}\xi_i & (\nu\text{–SVM})_P \\ \text{subject to} \quad & y_i\left(\boldsymbol{w}^T\boldsymbol{z}_i + b\right) \geqq \rho - \xi_i, \\ & \rho \geqq 0,\ \ \xi_i \geqq 0,\ \ i = 1, \ldots, \ell. \end{aligned}$$

where $0 \leqq \nu \leqq 1$ is a parameter.

Compared with the existing soft margin algorithm, one of the differences is that the parameter C for slack variables does not appear, and another difference is that the new variable ρ appears in the above formulation. The problem $(\nu\text{–SVM})_P$ maximizes the variable ρ which corresponds to the minimum interior deviation (i.e., the minimum distance between the separating hyperplane and correctly classified points).

The Lagrangian dual problem to the problem $(\nu\text{–SVM})_P$ is as follows:

$$\begin{aligned} \text{maximize} \quad & -\frac{1}{2}\sum_{i,j=1}^{\ell} y_i y_j \alpha_i \alpha_j K\left(\boldsymbol{x}_i, \boldsymbol{x}_j\right) & (\nu\text{–SVM}) \\ \text{subject to} \quad & \sum_{i=1}^{\ell} y_i \alpha_i = 0, \\ & \sum_{i=1}^{\ell} \alpha_i \geqq \nu, \\ & 0 \leqq \alpha_i \leqq \frac{1}{\ell},\ \ i = 1, \ldots, \ell. \end{aligned}$$

8.6 Extensions of SVM by MOP/GP

In this section, we propose various algorithms of SVM considering both slack variables for misclassified data points (i.e., exterior deviations) and surplus variables for correctly classified data points (i.e., interior deviations).

8.6.1 Total Margin Algorithm

In order to minimize the slackness and to maximize the surplus, we have the following optimization problem:

$$\begin{aligned} \text{minimize} \quad & \frac{1}{2}\|\boldsymbol{w}\|_2^2 + C_1\sum_{i=1}^{\ell}\xi_i - C_2\sum_{i=1}^{\ell}\eta_i & (\text{SVM}_{total})_P \\ \text{subject to} \quad & y_i\left(\boldsymbol{w}^T\boldsymbol{z}_i + b\right) \geqq 1 - \xi_i + \eta_i, \\ & \xi_i \geqq 0,\ \ \eta_i \geqq 0,\ \ i = 1, \ldots, \ell, \end{aligned}$$

where C_1 and C_2 are chosen in such a way that $C_1 > C_2$ which ensures that at least one of ξ_i and η_i becomes zero. The Lagrangian function for the problem $(\mathrm{SVM}_{total})_P$ is

$$\begin{aligned} L(\boldsymbol{w}, b, \boldsymbol{\xi}, \boldsymbol{\eta}, \boldsymbol{\alpha}, \boldsymbol{\beta}, \boldsymbol{\gamma}) = & \frac{1}{2}\|\boldsymbol{w}\|_2^2 + C_1 \sum_{i=1}^{\ell} \xi_i - C_2 \sum_{i=1}^{\ell} \eta_i \\ & - \sum_{i=1}^{\ell} \alpha_i \left[y_i \left(\boldsymbol{w}^T \boldsymbol{z}_i + b \right) - 1 + \xi_i - \eta_i \right] \\ & - \sum_{i=1}^{\ell} \beta_i \xi_i - \sum_{i=1}^{\ell} \gamma_i \eta_i, \end{aligned}$$

where $\alpha_i \geqq 0, \ \ \beta_i \geqq 0$ and $\gamma_i \geqq 0$.

Differentiating the Lagrangian function with respect to $\boldsymbol{w}$, b, $\boldsymbol{\xi}$ and $\boldsymbol{\eta}$ yields the following conditions:

$$\begin{aligned} \frac{\partial L(\boldsymbol{w}, b, \boldsymbol{\xi}, \boldsymbol{\eta}, \boldsymbol{\alpha}, \boldsymbol{\beta}, \boldsymbol{\gamma})}{\partial \boldsymbol{w}} &= \boldsymbol{w} - \sum_{i=1}^{\ell} \alpha_i y_i \boldsymbol{z}_i = \boldsymbol{0}, \\ \frac{\partial L(\boldsymbol{w}, b, \boldsymbol{\xi}, \boldsymbol{\eta}, \boldsymbol{\alpha}, \boldsymbol{\beta}, \boldsymbol{\gamma})}{\partial \xi_i} &= C_1 - \alpha_i - \beta_i = 0, \\ \frac{\partial L(\boldsymbol{w}, b, \boldsymbol{\xi}, \boldsymbol{\eta}, \boldsymbol{\alpha}, \boldsymbol{\beta}, \boldsymbol{\gamma})}{\partial \eta_i} &= -C_2 + \alpha_i - \gamma_i = 0, \\ \frac{\partial L(\boldsymbol{w}, b, \boldsymbol{\xi}, \boldsymbol{\eta}, \boldsymbol{\alpha}, \boldsymbol{\beta}, \boldsymbol{\gamma})}{\partial b} &= \sum_{i=1}^{\ell} \alpha_i y_i = 0. \end{aligned}$$

Substituting the above stationary conditions into the Lagrangian function L and using kernel representation, we obtain the following dual optimization problem:

$$\begin{array}{lll} \text{maximize} & \displaystyle\sum_{i=1}^{\ell} \alpha_i - \frac{1}{2} \sum_{i,j=1}^{\ell} y_i y_j \alpha_i \alpha_j K\left(\boldsymbol{x}_i, \boldsymbol{x}_j\right) & (\mathrm{SVM}_{total}) \\ \text{subject to} & \displaystyle\sum_{i=1}^{\ell} y_i \alpha_i = 0, & \\ & C_2 \leqq \alpha_i \leqq C_1, \ \ i = 1, \ldots, \ell. & \end{array}$$

Let $\boldsymbol{\alpha}^*$ be the optimal solution to the problem (SVM_{total}). Then, the discrimination function can be written by

$$f(\phi(\boldsymbol{x})) = \sum_{i=1}^{\ell} \alpha_i^* y_i K\left(\boldsymbol{x}, \boldsymbol{x}_i\right) + b.$$

The offset b is given as follows: Let n_+ be the number of $\boldsymbol{x}_j$ with $C_2 < \alpha_j^* < C_1$ and $y_j = +1$, and let n_- be the number of $\boldsymbol{x}_j$ with $C_2 < \alpha_j^* < C_1$ and $y_j = -1$, respectively. From the Karush-Kuhn-Tucker complementarity conditions, if $C_2 < \alpha_j^* < C_1$, then $\beta_j > 0$ and $\gamma_j > 0$. This implies that $\xi_j = \eta_j = 0$. Then,

$$b^* = \frac{1}{n_+ + n_-}\left((n_+ - n_-) - \sum_{j=1}^{n_+ + n_-} \sum_{i=1}^{\ell} y_i \alpha_i^* K(\boldsymbol{x}_i, \boldsymbol{x}_j) \right).$$

8.6.2 μ–SVM

Minimizing the worst slackness and maximizing the sum of surplus, we have a reverse formulation of ν–SVM. We introduce a new variable σ which represents the maximal distance between the separating hyperplane and misclassified data points. Thus, the following problem is obtained:

$$\begin{array}{lll} \text{minimize} & \dfrac{1}{2}\|\boldsymbol{w}\|_2^2 + \mu\sigma - \dfrac{1}{\ell}\displaystyle\sum_{i=1}^{\ell} \eta_i & (\mu\text{–SVM})_P \\ \text{subject to} & y_i\left(\boldsymbol{w}^T \boldsymbol{z}_i + b\right) \geqq \eta_i - \sigma, & \\ & \sigma \geqq 0, \ \ \eta_i \geqq 0, \ \ i = 1, \ldots, \ell, & \end{array}$$

where μ is a parameter which reflects the trade-off between σ and the sum of η_i.

The Lagrangian function for the problem $(\mu\text{–SVM})_P$ is

$$\begin{aligned} L(\boldsymbol{w}, b, \boldsymbol{\eta}, \sigma, \boldsymbol{\alpha}, \boldsymbol{\beta}, \gamma) = & \frac{1}{2}\|\boldsymbol{w}\|_2^2 + \mu\sigma - \frac{1}{\ell}\sum_{i=1}^{\ell} \eta_i \\ & - \sum_{i=1}^{\ell} \alpha_i \left[y_i\left(\boldsymbol{w}^T \boldsymbol{z}_i + b\right) - \eta_i + \sigma \right] - \sum_{i=1}^{\ell} \beta_i \eta_i - \gamma\sigma, \end{aligned}$$

where $\alpha_i \geqq 0$, $\beta_i \geqq 0$ and $\gamma \geqq 0$.

Differentiating the Lagrangian function with respect to $\boldsymbol{w}$, b, $\boldsymbol{\eta}$ and σ yields the following conditions:

$$\begin{aligned} \frac{\partial L(\boldsymbol{w}, b, \boldsymbol{\eta}, \sigma, \boldsymbol{\alpha}, \boldsymbol{\beta}, \gamma)}{\partial \boldsymbol{w}} &= \boldsymbol{w} - \sum_{i=1}^{\ell} \alpha_i y_i \boldsymbol{z}_i = \boldsymbol{0}, \\ \frac{\partial L(\boldsymbol{w}, b, \boldsymbol{\eta}, \sigma, \boldsymbol{\alpha}, \boldsymbol{\beta}, \gamma)}{\partial \eta_i} &= -\frac{1}{\ell} + \alpha_i - \beta_i = 0, \\ \frac{\partial L(\boldsymbol{w}, b, \boldsymbol{\eta}, \sigma, \boldsymbol{\alpha}, \boldsymbol{\beta}, \gamma)}{\partial \sigma} &= \mu - \sum_{i=1}^{\ell} \alpha_i - \gamma = 0, \\ \frac{\partial L(\boldsymbol{w}, b, \boldsymbol{\eta}, \sigma, \boldsymbol{\alpha}, \boldsymbol{\beta}, \gamma)}{\partial b} &= \sum_{i=1}^{\ell} \alpha_i y_i = 0. \end{aligned}$$

Substituting the above stationary conditions into the Lagrangian function L, we obtain the following dual optimization problem:

$$\begin{aligned} &\text{maximize} && -\frac{1}{2}\sum_{i,j=1}^{\ell} \alpha_i \alpha_j y_i y_j K\left(\boldsymbol{x}_i, \boldsymbol{x}_j\right) && (\mu\text{–SVM}) \\ &\text{subject to} && \sum_{i=1}^{\ell} \alpha_i y_i = 0, \\ &&& \sum_{i=1}^{\ell} \alpha_i \leqq \mu, \\ &&& \alpha_i \geqq \frac{1}{\ell}, \quad i = 1, \ldots, \ell. \end{aligned}$$

Let $\boldsymbol{\alpha}^*$ be the optimal solution to the problem (μ–SVM). To compute the offset b, we take the set $\mathcal{A}$ of $\boldsymbol{x}_j$ which is the same size n with $\frac{1}{\ell} < \alpha_j^*$. From the Karush-Kuhn-Tucker complementarity conditions, if $\frac{1}{\ell} < \alpha_j^*$, then $\beta_j > 0$ which implies $\eta_j = 0$. Thus,

$$b^* = -\frac{1}{2n} \sum_{\boldsymbol{x}_j \in \mathcal{A}} \sum_{i=1}^{\ell} \alpha_i^* y_i K\left(\boldsymbol{x}_i, \boldsymbol{x}_j\right).$$

8.6.3 μ–ν–SVM

Applying SVM$_{total}$ and μ–SVM, all training points become support vectors due to the second constraint of the problem (SVM$_{total}$) and the third constraint of the problem (μ–SVM). In other words, the algorithms (SVM$_{total}$) and (μ–SVM) lack in the sparsity of support vectors. In order to overcome this problem in (SVM$_{total}$) and (μ–SVM), we suggest the following formulation, which combines the ideas of ν–SVM and μ–SVM:

$$\begin{aligned} &\text{minimize} && \frac{1}{2}\|\boldsymbol{w}\|_2^2 - \nu\rho + \mu\sigma && (\mu - \nu\text{–SVM})_P \\ &\text{subject to} && y_i\left(\boldsymbol{w}^T \boldsymbol{z}_i + b\right) \geqq \rho - \sigma, \quad i = 1, \ldots, \ell, \\ &&& \rho \geqq 0, \quad \sigma \geqq 0, \end{aligned}$$

where ν and μ are parameters.

The Lagrangian function to the problem (μ–ν–SVM)$_P$ is

$$\begin{aligned} L(\boldsymbol{w}, b, \rho, \sigma, \boldsymbol{\alpha}, \beta, \gamma) = & \frac{1}{2}\|\boldsymbol{w}\|_2^2 - \nu\rho + \mu\sigma \\ & - \sum_{i=1}^{\ell} \alpha_i \left[y_i \left(\boldsymbol{w}^T \boldsymbol{z}_i + b\right) - \rho + \sigma \right] - \beta\rho - \gamma\sigma, \end{aligned}$$

where $\alpha_i \geqq 0, \ \beta \geqq 0$ and $\gamma \geqq 0$.

Differentiating Lagrangian function with respect to $\boldsymbol{w}$, b, ρ and σ yields the four conditions

$$\frac{\partial L(\boldsymbol{w}, b, \rho, \sigma, \boldsymbol{\alpha}, \beta, \gamma)}{\partial \boldsymbol{w}} = \boldsymbol{w} - \sum_{i=1}^{\ell} \alpha_i y_i \boldsymbol{z}_i = \boldsymbol{0},$$

$$\frac{\partial L(\boldsymbol{w}, b, \rho, \sigma, \boldsymbol{\alpha}, \beta, \gamma)}{\partial \rho} = -\nu + \sum_{i=1}^{\ell} \alpha_i - \beta = 0,$$

$$\frac{\partial L(\boldsymbol{w}, b, \rho, \sigma, \boldsymbol{\alpha}, \beta, \gamma)}{\partial \sigma} = \mu - \sum_{i=1}^{\ell} \alpha_i - \gamma = 0,$$

$$\frac{\partial L(\boldsymbol{w}, b, \rho, \sigma, \boldsymbol{\alpha}, \beta, \gamma)}{\partial b} = \sum_{i=1}^{\ell} \alpha_i y_i = 0.$$

Substituting the above stationary conditions into the Lagrangian function L and using kernel representation, we obtain the following dual optimization problem:

$$\begin{array}{lll} \text{maximize} & -\frac{1}{2} \sum_{i,j=1}^{\ell} \alpha_i \alpha_j y_i y_j K(\boldsymbol{x}_i, \boldsymbol{x}_j) & (\mu - \nu\text{–SVM}) \\ \text{subject to} & \sum_{i=1}^{\ell} \alpha_i y_i = 0, & \\ & \nu \leqq \sum_{i=1}^{\ell} \alpha_i \leqq \mu, & \\ & \alpha_i \geqq 0, \;\; i = 1, \ldots, \ell. & \end{array}$$

Letting $\boldsymbol{\alpha}^*$ be the optimal solution to the problem (μ–ν–SVM), the offset b^* can be chosen easily for any i satisfying $\alpha_i^* > 0$. Otherwise, b^* can be obtained by the similar way with the decision of the b^* in the other algorithms.

8.7 Numerical Examples

In order to investigate the performance of our proposed method, we compare the results for four data sets in the following: (The data can be downloaded from `http://www.ics.uci.edu/˜ mlearn/MLSummary.html`)

I. MONK's Problem (all data sets with 7 attributes)
   a) case 1
      i. training : 124 instances ($\mathcal{A}$: 62 instances, $\mathcal{B}$: 62 instances)
      ii. test : 432 instances ($\mathcal{A}$: 216 instances, $\mathcal{B}$: 216 instances)

b) case 2
   i. training : 169 instances ($\mathcal{A}$: 64 instances, $\mathcal{B}$: 105 instances)
   ii. test : 432 instances ($\mathcal{A}$: 142 instances, $\mathcal{B}$: 290 instances)

c) case 3
   i. training : 122 instances ($\mathcal{A}$: 60 instances, $\mathcal{B}$: 62 instances)
   ii. test : 432 instances ($\mathcal{A}$: 228 instances, $\mathcal{B}$: 204 instances)

II. Cleveland heart-disease from Long Beach and Cleveland Clinic Foundation : 303 instances ($\mathcal{A}$: 164 instances, $\mathcal{B}$: 139 instances) with 14 attributes

III. BUPA liver disorders from BUPA Medical Research Ltd. : 345 instances ($\mathcal{A}$: 200 instances, $\mathcal{B}$: 145 instances) with 7 attributes

IV. PIMA Indians diabetes database : 768 instances ($\mathcal{A}$: 268 instances, $\mathcal{B}$: 500 instances) with 9 attributes

In the following numerical experiments, QP solver of MATLAB was used for solving QP problems in SVM formulations; Gaussian kernels with $r = 1.0$ were used with the data normalization for each sample $\boldsymbol{x}_i$

$$\tilde{x}_{ki} = \frac{x_{ki} - \mu_k}{\sigma_k}$$

where μ_k and σ_k are the mean value and the standard deviation of k-th component of given the sample data $\{\boldsymbol{x}_1, \ldots, \boldsymbol{x}_p\}$, respectively. For parameters in applying GP model, we set $h_1 = h_2 = \cdots = h_\ell = C_1$ and $k_1 = k_2 = \cdots = k_\ell = C_2$.

For the dataset I, we followed both the training data and the test data as in the benchmark of the WEB site. Tables 8.1–8.6 compare the classification rates by using the existing algorithms (GP), (SVM_{soft}) and (ν−SVM) with the proposed algorithms (SVM_{total}), (μ−SVM) and (μ−ν−SVM), respectively.

For the datasets II and III, we adopt the 'cross validation test' method which makes 10 trials for randomly selected training data of 70% from the original data set and the test data of the rest 30%. Tables 8.7–8.18 compare the average (AVE) and the standard deviation (STDV) of classification rates by using the existing algorithms (GP), (SVM_{soft}) and (ν−SVM) with the proposed algorithms (SVM_{total}), (μ−SVM) and (μ−ν−SVM), respectively.

For the dataset IV, there is an unbalance between the number of elements of two classes: $\mathcal{A}$ (tested positive for diabetes) has 268 elements, while $\mathcal{B}$ (tested non-positive for diabetes) 500 elements. We selected randomly 70% from the whole data set as the training samples, and set the rest 30% as the test samples. We compared the results by (GP), (SVM_{soft}) and (SVM_{total} ν−SVM) with the proposed algorithms (SVM_{total}), (μ−SVM) and (μ−ν−SVM) as seen in Tables 8.19–8.24, respectively.

Table 8.25 shows the rate of support vectors in terms of percentage for each problem and each method.

Throughout our numerical experiments, it has been observed that even though the result depends on the value of parameters, the family of SVM using MOP/GP such as ν−SVM, SVM_{total}, μ−SVM and $\mu - \nu$−SVM show a relatively good performance in comparison with the simple SVM_{soft}. Sometimes unbalanced data sets cause a difficulty in predicting the category with fewer samples. In our experiments, MONK (case2) and PIMA diabetes are of this kind. It can be seen in those problems that the classification ability for the class with fewer samples is much sensitive to the value of C in SVM_{soft}. In other words, we have to select the appropriate value of C in SVM_{soft} carefully in order to attain some reasonable classification rate for unbalanced data sets. SVM_{total} and $\mu - \nu$–SVM, however, have advantage over SVM_{soft} in classification rate of the class with fewer elements. In addition, the data set of MONK seems not to be linearly separated. In this example, therefore, SVMs using MOP/GP show much better performance than the mere GP.

Table 8.1. Classification Rate by GP for MONK's Problem

	C_1	1			10			100			
	C_2	0.001	0.01	0.1	0.01	0.1	1	0.1	1	10	average
Training	case 1	73.39	73.39	73.39	73.39	73.39	71.77	73.39	73.39	73.39	73.21
	case 2	63.31	63.31	63.31	63.31	63.31	63.91	63.31	63.31	65.09	63.57
	$\mathcal{A}$	53.13	53.13	45.31	51.56	51.56	48.44	51.56	51.56	45.31	50.17
	$\mathcal{B}$	69.52	69.52	74.29	70.48	70.48	73.33	70.48	70.48	77.14	71.75
	case 3	88.52	88.52	88.52	88.52	88.52	88.52	88.52	88.52	88.52	88.52
Test	case 1	66.67	66.67	66.67	66.67	66.67	65.97	66.67	66.67	66.67	66.59
	case 2	58.33	58.33	58.33	59.03	59.03	59.26	59.03	59.03	61.11	59.05
	$\mathcal{A}$	39.44	39.44	35.92	40.14	40.14	37.32	40.14	40.14	35.92	38.73
	$\mathcal{B}$	67.59	67.59	70.69	68.28	68.28	70.00	68.28	68.28	73.45	69.16
	case 3	88.89	88.89	88.89	88.89	88.89	88.89	88.89	88.89	88.89	88.89

Table 8.2. Classification Rate by SVM_{soft} for MONK's Problem

	C	0.1	1	10	100	average
Training	case 1	87.90	95.16	100	100	95.77
	case 2	62.13	85.80	100	100	86.98
	$\mathcal{A}$	0.00	64.06	100	100	66.02
	$\mathcal{B}$	100	99.05	100	100	40.84
	case 3	81.15	99.18	100	100	95.08
Test	case 1	78.94	83.80	92.36	92.36	86.86
	case 2	67.13	70.14	79.63	80.09	74.25
	$\mathcal{A}$	0.00	40.14	82.39	83.10	51.41
	$\mathcal{B}$	100	84.83	78.28	78.62	85.43
	case 3	69.44	95.83	91.67	91.67	87.15

Table 8.3. Classification Rate by ν−SVM for MONK's Problem

	ν	0.1	0.2	0.3	0.4	0.5	0.6	0.7	average
Training	case 1	100	100	100	99.19	98.39	94.35	91.94	97.70
	case 2	100	100	100	98.82	98.82	95.27	88.17	97.30
	$\mathcal{A}$	100	100	100	96.88	96.88	89.06	70.31	93.30
	$\mathcal{B}$	100	100	100	100	100	99.05	99.05	99.73
	case 3	100	99.18	99.18	99.18	97.54	95.90	94.26	97.89
Test	case 1	92.36	92.13	91.20	88.43	87.04	84.03	80.56	87.96
	case 2	80.09	80.09	79.40	78.70	77.78	74.31	71.06	77.35
	$\mathcal{A}$	83.10	83.10	82.39	80.28	73.94	60.56	45.07	72.64
	$\mathcal{B}$	78.62	78.62	77.93	77.93	79.66	81.03	83.79	79.66
	case 3	91.67	94.44	95.14	96.06	95.60	93.52	92.13	94.08

Table 8.4. Classification Rate by SVM_{total} for MONK's Problem

	C_1	1			10			100			
	C_2	0.001	0.01	0.1	0.01	0.1	1	0.1	1	10	average
Training	case 1	95.16	95.16	95.97	100	100	100	100	100	90.38	97.40
	case 2	86.98	87.57	88.76	100	100	100	100	100	80.47	93.75
	$\mathcal{A}$	70.31	71.88	76.56	100	100	100	100	100	100	90.97
	$\mathcal{B}$	97.14	97.14	96.19	100	100	100	100	100	68.57	95.45
	case 3	99.18	99.18	99.18	100	100	100	100	100	95.0	99.18
Test	case 1	84.49	84.26	84.03	92.59	92.59	86.57	92.59	86.57	79.40	87.01
	case 2	69.68	69.91	70.83	77.78	78.01	78.01	77.78	78.01	69.91	74.43
	$\mathcal{A}$	47.18	47.89	50.70	86.62	87.32	89.44	87.32	89.44	85.92	74.65
	$\mathcal{B}$	80.69	80.69	80.69	73.45	73.45	72.41	73.10	72.41	62.07	74.33
	case 3	95.83	95.83	96.06	91.90	91.90	91.90	91.90	91.90	90.51	93.08

Table 8.5. Classification Rate by μ−SVM for MONK's Problem

	μ	1.3	1.4	1.5	1.6	1.7	1.8	1.9	2	average
Training	case 1	90.32	90.32	90.32	90.32	90.32	90.32	90.32	90.32	90.32
	case 2	71.01	71.01	71.01	71.01	71.01	71.01	71.01	71.01	71.01
	$\mathcal{A}$	100	100	100	100	100	100	100	100	100
	$\mathcal{B}$	53.33	53.33	53.33	53.33	53.33	53.33	53.33	53.33	53.33
	case 3	100	100	100	100	100	100	100	100	100
Test	case 1	77.46	77.46	77.46	77.46	77.46	77.46	77.46	77.46	77.46
	case 2	62.73	62.73	62.73	62.73	62.73	62.73	62.73	62.73	62.73
	$\mathcal{A}$	97.18	97.18	97.18	97.18	97.18	97.18	97.18	97.18	97.18
	$\mathcal{B}$	45.86	45.86	45.86	45.86	45.86	45.86	45.86	45.86	45.86
	case 3	93.52	93.52	93.52	93.52	93.52	93.52	93.52	93.52	93.52

Table 8.6. Classification Rate by $\mu-\nu$–SVM for MONK's Problem

	μ	1			10			100			
	ν	0.001	0.01	0.1	0.01	0.1	1	0.1	1	10	average
Training	case 1	100	100	100	100	100	100	100	100	100	100
	case 2	100	100	100	100	100	100	100	100	100	100
	$\mathcal{A}$	100	100	100	100	100	100	100	100	100	100
	$\mathcal{B}$	100	100	100	100	100	100	100	100	100	100
	case 3	100	100	100	100	100	100	100	100	100	100
Test	case 1	95.37	93.06	92.59	92.59	92.59	92.36	92.59	92.36	92.36	92.88
	case 2	80.56	75.69	75.46	75.46	75.46	80.09	75.46	80.09	80.09	77.60
	$\mathcal{A}$	95.77	92.96	92.96	92.96	92.96	83.10	92.96	83.10	83.10	89.98
	$\mathcal{B}$	73.10	67.24	66.90	66.90	66.90	78.62	66.90	78.62	78.62	71.53
	case 3	93.98	93.52	93.52	93.52	93.52	91.67	93.52	91.67	91.67	92.95

Table 8.7. Classification Rate by GP for Cleveland Heart-disease

c_1	1																	
c_2	0.001						0.01						0.1					
	training			test			training			test			training			test		
	rate	$\mathcal{A}$	$\mathcal{B}$	rate	$\mathcal{A}$	$\mathcal{B}$	rate	$\mathcal{A}$	$\mathcal{B}$	rate	$\mathcal{A}$	$\mathcal{B}$	rate	$\mathcal{A}$	$\mathcal{B}$	rate	$\mathcal{A}$	$\mathcal{B}$
AVE	87.93	88.80	86.88	79.11	80.05	78.07	88.03	88.98	86.88	79.11	80.05	78.07	88.22	89.50	86.68	78.89	80.30	77.35
STD	1.07	1.21	2.05	3.61	5.73	5.14	1.03	1.22	2.05	3.61	5.73	5.14	0.93	1.47	1.94	3.02	5.24	4.48
c_1	10																	
c_2	0.01						0.1						1					
AVE	88.12	89.16	86.88	78.56	78.86	78.44	88.12	89.16	86.88	78.56	78.86	78.44	88.26	89.77	86.47	79.22	80.29	78.20
STD	1.41	1.56	2.17	2.77	5.32	5.45	1.41	1.56	2.17	2.77	5.32	5.45	1.07	1.44	2.23	3.26	5.77	5.28
c_1	100																	
c_2	0.1						1						10					
AVE	88.12	89.16	86.88	78.56	78.86	78.44	88.12	89.16	86.88	78.56	78.86	78.44	88.22	89.77	86.38	79.00	80.12	77.92
STD	1.41	1.56	2.17	2.77	5.32	5.45	1.41	1.56	2.17	2.77	5.32	5.45	1.08	1.30	1.95	3.08	5.54	5.84

Table 8.8. Classification Rate by SVM_{soft} for Cleveland Heart-disease

C	0.01						0.1						1.0					
	training			test			training			test			training			test		
	rate	$\mathcal{A}$	$\mathcal{B}$	rate	$\mathcal{A}$	$\mathcal{B}$	rate	$\mathcal{A}$	$\mathcal{B}$	rate	$\mathcal{A}$	$\mathcal{B}$	rate	$\mathcal{A}$	$\mathcal{B}$	rate	$\mathcal{A}$	$\mathcal{B}$
AVE	53.05	90.00	10.00	53.89	90.00	10.00	53.05	90.00	10.00	53.89	90.00	10.00	99.72	100	99.40	73.89	75.19	74.30
STD	1.67	30.00	30.00	7.36	30.00	30.00	1.67	30.00	30.00	7.36	30.00	30.00	0.23	0.00	0.49	4.31	11.86	16.65
C	10						100											
AVE	100	100	100	74.56	74.49	76.28	100	100	100	74.56	74.49	76.28						
STD	0.00	0.00	0.00	3.93	10.80	14.14	0.00	0.00	0.00	3.93	10.80	14.14						

Table 8.9. Classification Rate by ν−SVM for Cleveland Heart-disease

ν	0.1						0.2						0.3					
	training			test			training			test			training			test		
	rate	$\mathcal{A}$	$\mathcal{B}$	rate	$\mathcal{A}$	$\mathcal{B}$	rate	$\mathcal{A}$	$\mathcal{B}$	rate	$\mathcal{A}$	$\mathcal{B}$	rate	$\mathcal{A}$	$\mathcal{B}$	rate	$\mathcal{A}$	$\mathcal{B}$
AVE	100	100	100	74.56	74.49	76.28	100	100	100	74.56	74.49	76.28	100	100	100	74.56	74.49	76.28
STD	0.00	0.00	0.00	3.93	10.80	14.14	0.00	0.00	0.00	3.93	10.80	14.14	0.00	0.00	0.00	3.93	10.80	14.14
ν	0.4						0.5						0.6					
AVE	100	100	100	74.67	74.87	76.04	100	100	100	74.33	74.45	75.72	99.91	100	99.80	74.33	74.64	75.48
STD	0.00	0.00	0.00	4.18	10.74	14.59	0.00	0.00	0.00	3.83	10.66	14.52	0.19	0.00	0.40	3.99	10.75	14.65
ν	0.7						0.8											
AVE	99.86	100	99.70	74.67	75.05	75.76	99.72	100	99.40	73.78	75.02	74.30						
STD	0.22	0.00	0.46	4.38	10.75	15.40	0.23	0.00	0.49	4.59	12.21	16.65						

Table 8.10. Classification Rate by SVM_{total} for Cleveland Heart-disease

C_1	1																	
C_2	0.0001						0.001						0.01					
	training			test			training			test			training			test		
	rate	$\mathcal{A}$	$\mathcal{B}$	rate	$\mathcal{A}$	$\mathcal{B}$	rate	$\mathcal{A}$	$\mathcal{B}$	rate	$\mathcal{A}$	$\mathcal{B}$	rate	$\mathcal{A}$	$\mathcal{B}$	rate	$\mathcal{A}$	$\mathcal{B}$
AVE	99.72	100	99.40	74.44	74.80	75.97	99.72	100	99.40	74.11	74.23	75.97	99.72	100	99.40	74.11	74.23	75.97
STD	0.23	0.00	0.49	4.99	11.80	17.23	0.23	0.00	0.49	4.87	11.99	17.23	0.23	0.00	0.49	4.87	11.99	17.23
C_1	10																	
C_2	0.001						0.01						0.1					
AVE	100	100	100	74.22	73.26	77.00	100	100	100	74.22	73.26	77.00	100	100	100	74.22	73.05	77.25
STD	0.00	0.00	0.00	3.47	10.62	13.93	0.00	0.00	0.00	3.47	10.62	13.93	0.00	0.00	0.00	3.47	10.56	14.14
C_1	100																	
C_2	0.01						0.1						1					
AVE	100	100	100	74.22	73.26	77.00	100	100	100	74.22	73.05	77.25	100	100	100	72.56	57.91	91.63
STD	0.00	0.00	0.00	3.47	10.62	13.93	0.00	0.00	0.00	3.47	10.56	14.14	0.00	0.00	0.00	3.37	5.67	5.76

Table 8.11. Classification Rate by μ−SVM for Cleveland Heart-disease

μ	1.2						···						1.5					
	training			test			training			test			training			test		
	rate	$\mathcal{A}$	$\mathcal{B}$	rate	$\mathcal{A}$	$\mathcal{B}$	rate	$\mathcal{A}$	$\mathcal{B}$	rate	$\mathcal{A}$	$\mathcal{B}$	rate	$\mathcal{A}$	$\mathcal{B}$	rate	$\mathcal{A}$	$\mathcal{B}$
AVE	99.81	99.67	100.00	81.00	82.25	79.72	···			···			99.81	99.67	100.00	81.00	82.25	79.72
STD	0.33	0.59	0.00	2.19	3.33	4.02	···			···			0.33	0.59	0.00	2.19	3.33	4.02
μ	1.6						···						2.0					
AVE	99.81	99.67	100.00	81.00	82.25	79.72	···			···			99.81	99.67	100.00	81.00	82.25	79.72
STD	0.33	0.59	0.00	2.19	3.33	4.02	···			···			0.33	0.59	0.00	2.19	3.33	4.02

Table 8.12. Classification Rate by $\mu - \nu-$SVM for Cleveland Heart-disease

μ	1																	
ν	0.0001						0.001						0.01					
	training			test			training			test			training			test		
	rate	$\mathcal{A}$	$\mathcal{B}$	rate	$\mathcal{A}$	$\mathcal{B}$	rate	$\mathcal{A}$	$\mathcal{B}$	rate	$\mathcal{A}$	$\mathcal{B}$	rate	$\mathcal{A}$	$\mathcal{B}$	rate	$\mathcal{A}$	$\mathcal{B}$
AVE	100	100	100	94.67	95.50	93.30	100	100	100	85.00	86.66	82.69	100	100	100	80.44	82.56	77.61
STD	0.00	0.00	0.00	1.71	2.74	3.49	0.00	0.00	0.00	2.91	4.30	4.90	0.00	0.00	0.00	2.73	5.00	4.14
μ	10																	
ν	0.001						0.01						0.1					
AVE	100	100	100	85.00	86.66	82.69	100	100	100	80.44	82.56	77.61	100	100	100	79.11	80.93	76.59
STD	0.00	0.00	0.00	2.91	4.30	4.90	0.00	0.00	0.00	2.73	5.00	4.14	0.00	0.00	0.00	2.80	4.82	4.22
μ	100																	
ν	0.01						0.1						1					
AVE	100	100	100	80.44	82.56	77.61	100	100	100	79.11	80.93	76.59	100	100	100	74.56	74.49	76.28
STD	0.00	0.00	0.00	2.73	5.00	4.14	0.00	0.00	0.00	2.80	4.82	4.22	0.00	0.00	0.00	3.93	10.80	14.14

Table 8.13. Classification Rate by GP for Liver Disorders

c_1	1																	
c_2	0.001						0.01						0.1					
	training			test			training			test			training			test		
	rate	$\mathcal{A}$	$\mathcal{B}$	rate	$\mathcal{A}$	$\mathcal{B}$	rate	$\mathcal{A}$	$\mathcal{B}$	rate	$\mathcal{A}$	$\mathcal{B}$	rate	$\mathcal{A}$	$\mathcal{B}$	rate	$\mathcal{A}$	$\mathcal{B}$
AVE	71.32	75.48	65.57	69.71	73.17	64.92	71.32	75.62	65.38	69.71	73.50	64.39	72.31	77.61	64.97	70.10	74.89	63.51
STD	1.42	1.45	1.64	2.67	5.38	3.38	1.46	1.67	1.44	2.70	5.26	3.61	1.83	2.23	1.92	2.81	5.02	4.27
c_1	10																	
c_2	0.01						0.1						1					
AVE	71.36	75.48	65.67	69.71	73.17	64.92	71.45	75.62	65.67	69.81	73.34	64.92	72.31	77.75	64.78	70.10	75.06	63.30
STD	1.37	1.45	1.58	2.67	5.38	3.38	1.44	1.61	1.58	2.62	5.36	3.38	1.80	2.23	1.87	2.97	5.17	4.56
c_1	100																	
c_2	0.1						1						10					
AVE	71.36	75.48	65.67	69.71	73.17	64.92	71.45	75.62	65.67	69.81	73.34	64.92	72.31	77.75	64.78	70.10	75.06	63.30
STD	1.37	1.45	1.58	2.67	5.38	3.38	1.44	1.61	1.58	2.62	5.36	3.38	1.80	2.23	1.87	2.97	5.17	4.56

Table 8.14. Classification Rate by SVM_{soft} for Liver Disorders

C	0.01						0.1						1					
	training			test			training			test			training			test		
	rate	$\mathcal{A}$	$\mathcal{B}$	rate	$\mathcal{A}$	$\mathcal{B}$	rate	$\mathcal{A}$	$\mathcal{B}$	rate	$\mathcal{A}$	$\mathcal{B}$	rate	$\mathcal{A}$	$\mathcal{B}$	rate	$\mathcal{A}$	$\mathcal{B}$
AVE	58.02	100	0.00	57.86	100	0.00	58.02	100	0.00	57.86	100	0.00	86.69	93.79	76.89	70.10	85.05	49.79
STD	1.32	0.00	0.00	3.11	0.00	0.00	1.32	0.00	0.00	3.11	0.00	0.00	1.42	1.16	3.16	3.93	4.83	5.89
C	10						100											
AVE	95.29	96.29	93.91	66.12	73.70	56.22	99.46	99.36	99.61	63.20	69.54	54.92						
STD	1.34	0.80	2.81	3.72	2.39	8.67	0.32	0.49	0.48	4.20	5.31	7.59						

Table 8.15. Classification Rate by ν−SVM for Liver Disorders

ν	0.1						0.2						0.3					
	training			test			training			test			training			test		
	rate	$\mathcal{A}$	$\mathcal{B}$	rate	$\mathcal{A}$	$\mathcal{B}$	rate	$\mathcal{A}$	$\mathcal{B}$	rate	$\mathcal{A}$	$\mathcal{B}$	rate	$\mathcal{A}$	$\mathcal{B}$	rate	$\mathcal{A}$	$\mathcal{B}$
AVE	99.46	99.36	99.61	63.59	70.04	55.12	98.26	98.29	98.22	65.34	72.72	55.73	96.07	96.50	95.46	65.05	71.90	56.21
STD	0.32	0.49	0.48	3.34	5.20	6.55	0.40	0.49	1.15	3.01	4.61	8.44	0.89	1.01	1.75	3.07	2.95	8.74
ν	0.4						0.5						0.6					
AVE	93.47	95.51	90.64	67.86	76.06	57.12	91.98	95.15	87.63	68.64	78.52	55.44	90.12	94.45	84.15	69.22	80.86	53.54
STD	0.80	0.68	1.73	4.17	3.11	8.25	0.95	0.95	1.69	3.61	3.64	7.69	1.22	1.62	2.30	3.86	3.82	8.07
ν	0.7						0.8											
AVE	87.81	93.80	79.48	69.42	83.09	50.94	99.72	100	99.40	73.78	75.02	74.30						
STD	1.19	1.23	2.97	3.08	4.17	6.41	0.23	0.00	0.49	4.59	12.21	16.65						

Table 8.16. Classification Rate by SVM_{total} for Liver Disorders

C_1	1																	
C_2	0.0001						0.001						0.01					
	training			test			training			test			training			test		
	rate	$\mathcal{A}$	$\mathcal{B}$	rate	$\mathcal{A}$	$\mathcal{B}$	rate	$\mathcal{A}$	$\mathcal{B}$	rate	$\mathcal{A}$	$\mathcal{B}$	rate	$\mathcal{A}$	$\mathcal{B}$	rate	$\mathcal{A}$	$\mathcal{B}$
AVE	82.73	90.66	71.66	65.53	74.71	53.00	86.74	92.94	78.17	69.51	82.72	51.56	86.74	92.94	78.17	69.42	82.55	51.56
STD	1.81	2.71	4.91	5.11	6.74	7.08	1.33	1.20	2.89	3.17	3.99	5.34	1.29	1.20	2.72	3.17	4.18	5.34
C_1	10																	
C_2	0.001						0.01						0.1					
AVE	95.25	95.86	94.40	65.83	71.85	57.97	95.25	95.86	94.40	65.83	71.85	57.97	95.29	95.86	94.50	65.53	71.35	57.97
STD	1.17	0.91	2.51	3.09	2.99	7.16	1.17	0.91	2.51	3.09	2.99	7.16	1.19	0.91	2.50	3.02	3.18	7.16
C_1	100																	
C_2	0.01						0.1						1					
AVE	99.42	99.29	99.61	61.84	64.54	58.20	99.42	99.29	99.61	61.84	64.76	58.00	99.46	99.29	99.70	63.11	66.45	58.76
STD	0.27	0.44	0.48	4.43	6.13	6.65	0.27	0.44	0.48	4.47	5.96	6.50	0.26	0.44	0.46	3.53	5.86	6.85

Table 8.17. Classification Rate by μ−SVM for Liver Disorders

μ	1.2						1.3						1.4					
	training			test			training			test			training			test		
	rate	$\mathcal{A}$	$\mathcal{B}$	rate	$\mathcal{A}$	$\mathcal{B}$	rate	$\mathcal{A}$	$\mathcal{B}$	rate	$\mathcal{A}$	$\mathcal{B}$	rate	$\mathcal{A}$	$\mathcal{B}$	rate	$\mathcal{A}$	$\mathcal{B}$
AVE	69.75	93.11	37.37	65.34	92.87	27.54	73.93	90.99	50.32	69.03	90.26	39.75	74.92	90.64	53.14	68.83	88.96	41.10
STD	1.79	3.19	7.83	4.27	4.75	6.06	1.04	1.70	3.22	4.61	3.71	5.73	1.83	1.85	5.37	3.65	3.39	4.35
μ	1.5						1.6						1.7					
AVE	78.22	91.71	59.57	68.64	88.59	41.11	87.89	95.09	77.87	68.45	87.78	41.77	92.19	97.00	85.49	69.13	88.12	42.93
STD	7.85	3.13	15.24	3.63	2.46	4.27	12.85	5.18	23.69	3.58	2.93	5.17	12.62	4.84	23.61	3.29	3.27	5.21
μ	1.8						...						2.0					
AVE	100	100	100	69.22	87.64	43.87	...			...			100	100	100	69.22	87.64	43.87
STD	0.00	0.00	0.00	3.58	3.30	5.06	...			...			0.00	0.00	0.00	3.58	3.30	5.06

Table 8.18. Classification Rate by $\mu - \nu$−SVM for Liver Disorders

μ	1																	
ν	0.0001						0.001						0.01					
	training			test			training			test			training			test		
	rate	$\mathcal{A}$	$\mathcal{B}$	rate	$\mathcal{A}$	$\mathcal{B}$	rate	$\mathcal{A}$	$\mathcal{B}$	rate	$\mathcal{A}$	$\mathcal{B}$	rate	$\mathcal{A}$	$\mathcal{B}$	rate	$\mathcal{A}$	$\mathcal{B}$
AVE	97.81	97.25	98.58	93.69	93.58	93.85	100	100	100	72.52	76.50	66.84	100	100	100	63.20	66.74	58.24
STD	3.15	5.43	3.33	2.50	5.75	6.19	0.00	0.00	0.00	4.45	5.64	5.31	0.00	0.00	0.00	4.71	5.39	5.30
μ	10																	
ν	0.001						0.01						0.1					
AVE	100	100	100	72.52	76.50	66.84	100	100	100	63.20	66.74	58.24	100	100	100	62.14	67.87	54.53
STD	0.00	0.00	0.00	4.45	5.64	5.31	0.00	0.00	0.00	4.71	5.39	5.30	0.00	0.00	0.00	4.23	4.48	8.78
μ	100																	
ν	0.01						0.1						1					
AVE	100	100	100	63.20	66.74	58.24	100	100	100	62.14	67.87	54.53	100	100	100	62.14	67.87	54.53
STD	0.00	0.00	0.00	4.71	5.39	5.30	0.00	0.00	0.00	4.23	4.48	8.78	0.00	0.00	0.00	4.23	4.48	8.78

Table 8.19. Classification Rate by GP for PIMA

C_1	1																	
C_2	0.001						0.01						0.1					
	training			test			training			test			training			test		
	rate	$\mathcal{A}$	$\mathcal{B}$	rate	$\mathcal{A}$	$\mathcal{B}$	rate	$\mathcal{A}$	$\mathcal{B}$	rate	$\mathcal{A}$	$\mathcal{B}$	rate	$\mathcal{A}$	$\mathcal{B}$	rate	$\mathcal{A}$	$\mathcal{B}$
AVE	78.10	67.75	83.49	77.35	66.40	83.67	78.18	67.24	83.85	77.57	65.82	84.36	78.72	63.34	86.70	76.87	60.93	86.06
STD	1.19	1.58	1.02	2.92	5.37	2.05	1.14	1.79	1.02	2.87	5.34	1.96	1.37	2.32	1.07	3.00	5.92	1.74
C_1	10																	
C_2	0.01						0.1						1					
AVE	78.14	68.02	83.40	77.43	66.08	84.01	78.14	67.41	83.71	77.74	66.18	84.43	78.74	63.77	86.51	77.00	61.55	85.93
STD	1.29	1.83	1.07	2.86	5.15	1.92	1.13	1.55	1.08	2.71	5.09	1.91	1.38	2.00	1.25	3.41	6.82	1.94
C_1	100																	
C_2	0.1						1						10					
AVE	78.12	67.97	83.40	77.48	66.19	84.01	78.12	67.41	83.68	77.78	66.18	84.49	78.79	63.94	86.51	77.00	61.55	85.93
STD	1.27	1.76	1.07	2.82	5.00	1.92	1.15	1.55	1.11	2.75	5.09	1.92	1.39	1.86	1.31	3.51	7.07	1.94

Table 8.20. Classification Rate by SVM_{soft} for PIMA

C	0.01						0.1						1					
	training			test			training			test			training			test		
	rate	$\mathcal{A}$	$\mathcal{B}$	rate	$\mathcal{A}$	$\mathcal{B}$	rate	$\mathcal{A}$	$\mathcal{B}$	rate	$\mathcal{A}$	$\mathcal{B}$	rate	$\mathcal{A}$	$\mathcal{B}$	rate	$\mathcal{A}$	$\mathcal{B}$
AVE	65.61	0.00	100	63.91	0.00	100	65.61	0.00	100	63.91	0.00	100	91.47	80.71	97.08	74.17	52.51	86.40
STD	1.03	0.00	0.00	2.41	0.00	0.00	1.03	0.00	0.00	2.41	0.00	0.00	0.70	2.26	0.57	1.82	3.02	2.25
C	10						100											
AVE	99.44	98.43	99.97	69.83	54.29	78.60	100	100	100	68.91	54.45	77.09						
STD	0.20	0.63	0.08	1.63	4.24	2.05	0.00	0.00	0.00	2.07	3.90	2.71						

Table 8.21. Classification Rate by ν–SVM for PIMA

ν	0.1						0.2						0.3					
	training			test			training			test			training			test		
	rate	$\mathcal{A}$	$\mathcal{B}$	rate	$\mathcal{A}$	$\mathcal{B}$	rate	$\mathcal{A}$	$\mathcal{B}$	rate	$\mathcal{A}$	$\mathcal{B}$	rate	$\mathcal{A}$	$\mathcal{B}$	rate	$\mathcal{A}$	$\mathcal{B}$
AVE	99.91	99.73	100	69.04	54.33	77.34	99.33	98.16	99.94	70.04	54.62	78.74	97.66	94.69	99.20	71.61	54.51	81.24
STD	0.12	0.36	0.00	1.88	3.81	2.37	0.25	0.66	0.11	1.80	4.25	2.23	0.46	0.69	0.46	1.42	4.24	1.59
ν	0.4						0.5						0.6					
AVE	94.93	88.52	98.27	72.65	53.40	83.57	93.03	84.18	97.64	73.43	52.66	85.18	90.65	78.75	96.85	74.09	51.75	86.76
STD	0.58	1.82	0.53	2.30	5.27	2.29	0.46	1.92	0.65	2.25	3.92	2.56	0.86	3.12	0.79	2.08	3.49	2.47

Table 8.22. Classification Rate by SVM_{total} for PIMA

C_1	1																	
C_2	0.0001						0.001						0.01					
	training			test			training			test			training			test		
	rate	$\mathcal{A}$	$\mathcal{B}$	rate	$\mathcal{A}$	$\mathcal{B}$	rate	$\mathcal{A}$	$\mathcal{B}$	rate	$\mathcal{A}$	$\mathcal{B}$	rate	$\mathcal{A}$	$\mathcal{B}$	rate	$\mathcal{A}$	$\mathcal{B}$
AVE	68.77	87.51	58.95	65.22	85.36	53.85	91.62	82.29	96.48	73.91	57.85	82.99	91.67	82.51	96.45	73.57	58.19	82.23
STD	1.29	1.31	1.84	2.01	2.97	3.02	0.72	2.23	0.92	2.39	2.78	2.45	0.74	2.14	0.94	2.44	2.75	2.51
C_1	10																	
C_2	0.001						0.01						0.1					
AVE	99.48	98.70	99.89	69.17	63.45	72.36	99.48	98.70	99.89	69.17	63.69	72.22	99.44	98.86	99.75	68.65	66.35	69.89
STD	0.20	0.37	0.19	2.40	2.84	3.14	0.20	0.37	0.19	2.47	2.78	3.18	0.28	0.52	0.23	2.64	3.41	3.31
C_1	100																	
C_2	0.01						0.1						1					
AVE	100	100	100	67.35	64.60	68.89	100	100	100	67.35	67.23	67.37	70.13	100	54.52	62.22	91.43	45.59
STD	0.00	0.00	0.00	2.51	4.05	3.14	0.00	0.00	0.00	2.54	3.57	2.94	3.36	0.00	4.68	2.98	2.62	5.69

Table 8.23. Classification Rate by μ–SVM for PIMA

μ	1.4						1.5						1.6					
	training			test			training			test			training			test		
	rate	$\mathcal{A}$	$\mathcal{B}$	rate	$\mathcal{A}$	$\mathcal{B}$	rate	$\mathcal{A}$	$\mathcal{B}$	rate	$\mathcal{A}$	$\mathcal{B}$	rate	$\mathcal{A}$	$\mathcal{B}$	rate	$\mathcal{A}$	$\mathcal{B}$
AVE	78.01	46.68	94.62	66.91	26.41	89.14	93.92	85.70	97.86	67.87	30.92	88.14	100	1000	100	68.00	31.32	88.06
STD	2.13	7.25	1.20	3.13	5.78	2.84	9.81	23.91	3.69	4.70	6.78	2.47	0.00	0.00	0.00	4.41	6.13	2.40
μ	1.7						...						2.0					
AVE	100	100	100	68.00	31.32	88.06	...			...			100	100	100	68.00	31.32	88.06
STD	0.00	0.00	0.00	4.41	6.13	2.40	...			...			0.00	0.00	0.00	4.41	6.13	2.40

Table 8.24. Classification Rate by $\mu - \nu$–SVM for PIMA

μ	1																	
ν	0.0001						0.001						0.01					
	training			test			training			test			training			test		
	rate	$\mathcal{A}$	$\mathcal{B}$	rate	$\mathcal{A}$	$\mathcal{B}$	rate	$\mathcal{A}$	$\mathcal{B}$	rate	$\mathcal{A}$	$\mathcal{B}$	rate	$\mathcal{A}$	$\mathcal{B}$	rate	$\mathcal{A}$	$\mathcal{B}$
AVE	98.05	99.95	97.02	89.57	94.29	87.13	100	100	100	73.00	64.93	77.64	100	100	100	69.17	60.46	74.17
STD	2.01	0.16	3.12	3.20	3.99	5.43	0.00	0.00	0.00	2.14	4.26	2.77	0.00	0.00	0.00	2.25	4.46	3.44
μ	10																	
ν	0.001						0.01						0.1					
AVE	100	100	100	73.00	64.93	77.64	100	100	100	69.17	60.46	74.17	100	100	100	69.09	56.07	76.49
STD	0.00	0.00	0.00	2.14	4.26	2.77	0.00	0.00	0.00	2.25	4.46	3.44	0.00	0.00	0.00	1.72	5.06	3.63
μ	100																	
ν	0.01						0.1						1					
AVE	100	100	100	69.17	60.46	74.17	100	100	100	69.09	56.07	76.49	100	100	100	68.91	54.45	77.09
STD	0.00	0.00	0.00	2.25	4.46	3.44	0.00	0.00	0.00	1.72	5.06	3.63	0.00	0.00	0.00	2.07	3.90	2.71

Table 8.25. Rates of Support Vectors (unit : %)

		SVM_{soft}	ν−SVM	SVM_{total}	μ−SVM	$\mu-\nu$−SVM
MONK	AVE	74.60	76.69	100	100	62.83
(case 1)	STD	15.19	13.65	0	0	0.23
MONK	AVE	76.90	73.10	100	100	69.00
(case 2)	STD	7.43	4.78	0	0	0.85
MONK	AVE	70.70	74.23	100	100	56.11
(case 3)	STD	17.40	12.53	0	0	1.41
Cleveland	AVE	97.40	96.97	100	100	96.83
Heart-disease	STD	0.75	0.70	0	0	0.70
Liver Disorders	AVE	81.59	75.37	100	100	59.02
	STD	2.13	2.18	0	0	3.54
PIMA	AVE	72.53	71.35	100	100	64.26
	STD	1.74	1.71	0	0	1.98

8.8 Concluding Remarks

In this chapter, we introduced various SVM algorithms using MOP/GP. The authors have given a generalization error bound, and proved that the error bound can be decreased by minimizing slack variables and maximizing surplus variables [23]. As a total, $\mu-\nu-$SVM shows relatively good performance in our experiences. However, SVM_{total} and $\mu-$SVM among the proposed algorithms are inferior to the standard SVM algorithms in terms of sparsity of support vectors. This means that those methods cause some difficulty in computation for large scale data sets. It is observed in our experience, moreover, that some values of μ yield unacceptable solutions in $\mu-$SVM algorithm. However, $\mu-\nu-$SVM overcomes the lack of sparsity of support vectors, and does not cause so much difficulty in computation even for large scale data sets. For regression problems, moreover, $\mu-\nu-$SVM minimizing the exterior deviation is akin to function approximation using the Tchebyshev error, which is widely applied to many real problems. This is another point for which $\mu-\nu-$SVM is promising. The details on regression by $\mu-\nu-$SVM will be discussed elsewhere.

Acknowledgement

This research was supported by JSPS.KAKENHI13680540.

References

[1] Asada, T. and Nakayama, H. (2003) SVM using Multi Objective Linear Programming and Goal Programming, in T. Tanino, T. Tanaka and M. Inuiguchi (eds.), *Multi-objective Programming and Goal Programming*, 93-98

[2] Bennett, K.P. and Mangasarian, O.L. (1992) Robust Linear Programming Discrimination of Two Linearly Inseparable Sets, *Optimization Methods and Software*, **1**, 23-34

[3] Cavalier, T.M., Ignizio, J.P. and Soyster, A.L., (1989) Discriminant Analysis via Mathematical Programming: Certain Problems and their Causes, *Computers and Operations Research*, **16**, 353-362

[4] Chankong, V. and Haimes, Y.Y., (1983) *Multiobjective Decision Making Theory and Methodlogy* , Elsevier Science Publsihing

[5] Charnes, A. and Cooper W.W., (1961) *Management Models and Industrial Applications of Linear Programming* , vol. 1, Wiley

[6] Cortes, C. and Vapnik, V., (1995) Support Vector Networks, *Machine Learning*, **20**, pp. 273–297

[7] Cristianini, N. and Shawe-Taylor, J., (2000) *An Introduction to Support Vector Machines and Other Kernel-based Learning Methods*, Cambridge University Press

[8] Erenguc, S.S. and Koehler, G.J., (1990) Survey of Mathematical Programming Models and Experimental Results for Linear Discriminant Analysis, *Managerial and Decision Economics*, **11**, 215-225

[9] Freed, N. and Glover, F., (1981) Simple but Powerful Goal Programming Models for Discriminant Problems, *European J. of Operational Research*, **7**, 44-60

[10] Glover, F. (1990) Improved Linear Programming Models for Discriminant Analysis, *Decision Sciences*, **21**, 771-785

[11] Mangasarian, O.L., (1968) Multisurface Method of Pattern Separation, *IEEE Transact. on Information Theory*, **IT-14**, 801-807

[12] Mangasarian, O.L., (1999) *Arbitrary-Norm Separating Plane*, Operations Research Letters **23**

[13] Marcotte, P. and Savard, G., (1992) Novel Approaches to the Discrimination Problem, *ZOR–Methods and Models of Operations Research*, **36**, pp.517-545

[14] Miettinen, K. M., (1999) *Nonlinear Multiobjective Optimization* , Kluwer Academic Publishers

[15] Nakayama, H., (1995) Aspiration Level Approach to Interactive Multi-objective Programming and its Applications, *Advances in Multicriteria Analysis*, ed. by P.M. Pardalos, Y. Siskos and C. Zopounidis, Kluwer Academic Publishers, pp. 147-174

[16] Nakayama, H. and Asada, T., (2001) Support Vector Machines formulated as Multi Objective Linear Programming, *Proc. of ICOTA2001*, **3**, pp.1171-1178

[17] Novikoff, A.B., (1962) On the Convergence Proofs on Perceptrons, In*Symposium on the Mathematical Theory of Automata*, vol. 12, pp. 615–622, Polytechnic Institute of Brooklyn

[18] Sawaragi, Y., Nakayama, H. and Tanino, T., (1994) *Theory of Multiobjective Optimization*, Academic Press

[19] Schölkopf, B. and Smola, A.J., (1998) New Support Vector Algorithms, *NeuroCOLT2 Technical report Series*, NC2-TR-1998-031

[20] B.Schölkopf, and A.J.Smola, (2002) *Learning with Kernels: Support Vector Machines, Regularization, Optimization, and Beyond-*, MIT Press

[21] Steuer, R.E., (1986) *Multiple Criteria Optimization: Theory, Computation, and Application* , Wiley

[22] Vapnik, V.N., (1998) *Statistical Learning Theory*, John Wiley & Sons, New York

[23] Yoon, M., Yun, Y.B. and Nakayama, H., (2003) A Role of Total Margin in Support Vector Machines, *Proc. IJCNN'03*, 2049-2053

9

Multi-Objective Optimization of Support Vector Machines

Thorsten Suttorp and Christian Igel

Institut für Neuroinformatik
Ruhr-Universität Bochum
44780 Bochum, Germany
thorsten.suttorp@neuroinformatik.rub.de
christian.igel@neuroinformatik.rub.de

Summary. Designing supervised learning systems is in general a multi-objective optimization problem. It requires finding appropriate trade-offs between several objectives, for example between model complexity and accuracy or sensitivity and specificity. We consider the adaptation of kernel and regularization parameters of support vector machines (SVMs) by means of multi-objective evolutionary optimization. Support vector machines are reviewed from the multi-objective perspective, and different encodings and model selection criteria are described. The optimization of split modified radius-margin model selection criteria is demonstrated on benchmark problems. The MOO approach to SVM design is evaluated on a real-world pattern recognition task, namely the real-time detection of pedestrians in infrared images for driver assistance systems. Here the three objectives are the minimization of the false positive rate, the false negative rate, and the number of support vectors to reduce the computational complexity.

9.1 Introduction

The design of supervised learning systems for classification requires finding a suitable trade-off between several objectives, especially between model complexity and accuracy on a set of noisy training examples ($\rightarrow$ bias vs. variance, capacity vs. empirical risk). In many applications, it is further advisable to consider sensitivity and specificity (i.e., true positive and true negative rate) separately. For example in medical diagnosis, a high false alarm rate may be tolerated if the sensitivity is high. The computational complexity of a solution can be an additional design objective, in particular under real-time constraints.

This multi-objective design problem is usually tackled by aggregating the objectives into a scalar function and applying standard methods to the resulting single-objective task. However, such an approach can only lead to satisfactory solutions if the aggregation (e.g., a linear weighting of empirical

T. Suttorp and C. Igel: *Multi-Objective Optimization of Support Vector Machines*, Studies in Computational Intelligence (SCI) **16**, 199–220 (2006)
www.springerlink.com

error and regularization term) matches the problem. A better way is to apply "true" multi-objective optimization (MOO) to approximate the set of Pareto-optimal trade-offs and to choose a final solution afterwards from this set. A solution is Pareto-optimal if it cannot be improved in any objective without getting worse in at least one other objective [11, 14, 40].

We consider MOO of support vector machines (SVMs), which mark the state-of-the-art in machine learning for binary classification in the case of moderate problem dimensionality in terms of the number of training patterns [42, 43, 45]. First, we briefly introduce SVMs from the perspective of MOO. In Section 9.3 we discuss MOO model selection for SVMs. We review model selection criteria, optimization methods with an emphasis on evolutionary MOO, and kernel encodings. Section 9.4 summarizes results on MOO of SVMs considering model selection criteria based on radius-margin bounds [25]. In Section 9.5 we present a real-world application of the proposed methods: Pedestrian detection for driver assistance systems is a difficult classification task, which can be approached using SVMs [30, 32, 36]. Here fast classifiers with a small false alarm rate are needed. We therefore propose MOO to minimize the false positive rate, the false negative rate, and the complexity of the classifier.

9.2 Support Vector Machines

Support vector machines are learning machines based on two key elements: a general purpose learning algorithm and a problem specific kernel that computes the inner product of input data points in a feature space. In this section, we concisely summarize SVMs and illustrate some of the underlying concepts. For an introduction to SVMs we refer to the standard literature [12, 42, 45].

9.2.1 General SVM Learning

We start with a general formulation of binary classification. Let

$$S = ((\mathbf{x}_1, y_1), \ldots, (\mathbf{x}_\ell, y_\ell)),$$

be the set of training examples, where $y_i \in \{-1, 1\}$ is the label associated with input pattern $\mathbf{x}_i \in X$. The task is to estimate a function f from a given class of functions that correctly classifies unseen examples $(\mathbf{x}, y)$ by the calculation of $\text{sign}(\text{f}(\mathbf{x}))$. The only assumption that is made is that the training data as well as the unseen examples are generated independently by the same, but unknown probability distribution $\mathcal{D}$.

The main idea of SVMs is to map the input vectors to a feature space $\mathcal{H}$, where the transformed data is classified by a linear function f. The transformation $\Phi : X \to \mathcal{H}$ is implicitly done by a kernel $k : X \times X \to \mathbb{R}$, which computes an inner product in the feature space and thereby defines the reproducing kernel Hilbert space (RKHS) $\mathcal{H}$. The kernel matrix $\mathbf{K} = (K_{ij})_{i,j=1}^{\ell}$ has the entries $K_{ij} = \langle \Phi(\mathbf{x}_i), \Phi(\mathbf{x}_j) \rangle$. The kernel has to be positive semi-definite, that is, $\mathbf{v}^T \mathbf{K} \mathbf{v} \geq 0$ for all $\mathbf{v} \in \mathbb{R}^\ell$ and all S.

The best function f for classification is the one that minimizes the generalization error, that is, the probability of misclassifying unseen examples $P_{\mathcal{D}}(\text{sign}(f(\mathbf{x})) \neq y)$. Because the example's underlying distribution $\mathcal{D}$ is unknown, a direct minimization is not possible. Thus, upper bounds on the generalization error from statistical learning theory are studied that hold with a probability of $1 - \delta,\ \ \delta \in (0, 1)$.

We follow the way of [43] for the derivation of SVM learning and give an upper bound that directly incorporates the concepts of margin and slack variables. The margin of an example $(\mathbf{x}_i, y_i)$ with respect to a function $f : X \to \mathbb{R}$ is defined by $y_i f(\mathbf{x}_i)$. If a function f and a desired margin γ are given, the example's slack variable $\xi_i(\gamma, f) = \max(0, \gamma - y_i f(\mathbf{x}_i))$ measures how much the example fails to meet the margin (Figure 9.1 and 9.2). It holds [43]:

Theorem 1. *Let* $\gamma > 0$ *and* $f \in \{f_{\mathbf{w}} : X \to \mathbb{R},\ \ f_{\mathbf{w}}(\mathbf{x}) = \mathbf{w} \cdot \Phi(\mathbf{x}),\ \ \|\mathbf{w}\| < 1\}$ *a linear function in a kernel-defined RKHS with norm at most* 1. *Let*

$$S = \{(\mathbf{x}_1, y_1), \ldots, (\mathbf{x}_\ell, y_\ell)\}$$

be drawn independently according to a probability distribution $\mathcal{D}$ *and fix* $\delta \in (0, 1)$. *Then with probability at least* $1 - \delta$ *over samples of size* ℓ *we have*

$$P_{\mathcal{D}}(y \neq \text{sign}(f(\mathbf{x}))) \leq \frac{1}{\ell\gamma} \sum_{i=1}^{\ell} \xi_i + \frac{4}{\ell\gamma} \sqrt{\text{tr}(\mathbf{K})} + 3\sqrt{\frac{\ln(2/\delta)}{2\ell}} \ ,$$

where $\mathbf{K}$ *is the kernel matrix for the training set* S.

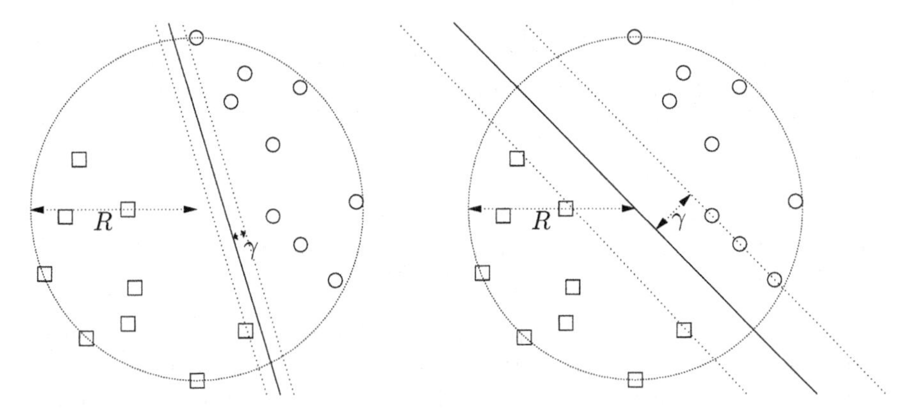

Fig. 9.1. Two linear decision boundaries separating circles from squares in some feature space. In the right plot, the separating hyperplane maximizes the margin γ, in the left not. The radius of the smallest ball in feature space containing all training examples is denoted by R.

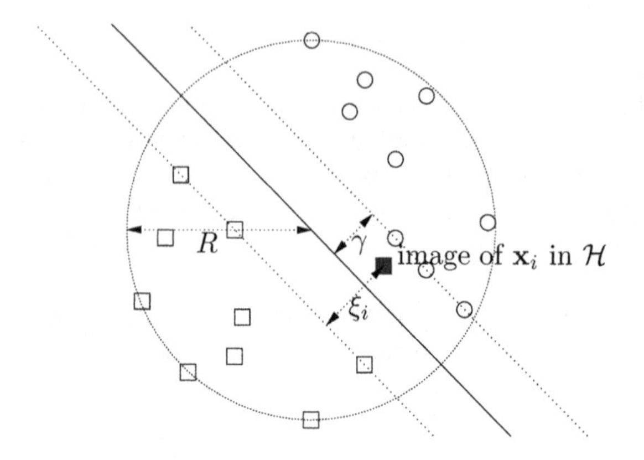

Fig. 9.2. The concept of slack variables.

The upper bound of Theorem 1 gives a way of controlling the generalization error $P_{\mathcal{D}}(y \neq \text{sign}(f(\mathbf{x})))$. It states that the described learning problem has a multi-objective character with two objectives, namely the margin γ and the sum of the slack variables $\sum_{i=1}^{\ell} \xi_i$.

This motivates the following definition of SVM learning[1]:

$$\mathcal{P}_{\text{SVM}} = \begin{cases} \max & \gamma \\ \min & \sum_{i=1}^{\ell} \xi_i \\ \text{subject to} & y_i((\mathbf{w} \cdot \Phi(\mathbf{x}_i)) + b) \geq \gamma - \xi_i \ , \\ & \xi_i \geq 0, \ \ i = 1, \ldots, \ell \text{ and } \|\mathbf{w}\|^2 = 1 \ . \end{cases}$$

In this formulation γ represents the *geometric margin* due to the fixation of $\|\mathbf{w}\|^2 = 1$. For the solution of $\mathcal{P}_{\text{SVM}}$ all training patterns $(\mathbf{x}_i, y_i)$ with $\xi_i = 0$ have a distance of at least γ to the hyperplane.

The more traditional formulation of SVM learning that will be used throughout this chapter is slightly different. It is a scaled version of $\mathcal{P}_{\text{SVM}}$, but finally provides the same classifier:

$$\mathcal{P}'_{\text{SVM}} = \begin{cases} \min & (\mathbf{w} \cdot \mathbf{w}) \\ \min & \sum_{i=1}^{\ell} \xi_i \\ \text{subject to} & y_i((\mathbf{w} \cdot \Phi(\mathbf{x}_i)) + b) \geq 1 - \xi_i \ , \\ & \xi_i \geq 0, \ \ i = 1, \ldots, \ell \ . \end{cases} \tag{9.1}$$

There are more possible formulations of SVM learning. For example, when considering the kernel as part of the SVM learning process, as it is done in [6], it becomes necessary to incorporate the term $\text{tr}(\mathbf{K})$ of Theorem 1 into the optimization problem.

[1] We do not take into account that the bound of Theorem 1 has to be adapted in the case $b \neq 0$.

9.2.2 Classic C-SVM Learning

Until now we have only considered multi-objective formulations of SVM learning. In order to obtain the classic single-objective C-SVM formulation the weighted sum method is applied to (9.1). The factor C determines the trade-off between the margin γ and the sum of the slack variables $\sum_{i=1}^{\ell} \xi_i$:

$$\mathcal{P}_{C\text{-SVM}} = \begin{cases} \min & \frac{1}{2}(\mathbf{w}\cdot\mathbf{w}) + C\sum_{i=1}^{\ell}\xi_i \\ \text{subject to} & y_i((\mathbf{w}\cdot\Phi(\mathbf{x}_i)) + b) \geq 1 - \xi_i \ , \\ & \xi_i \geq 0, \ i = 1,\dots,\ell \ . \end{cases}$$

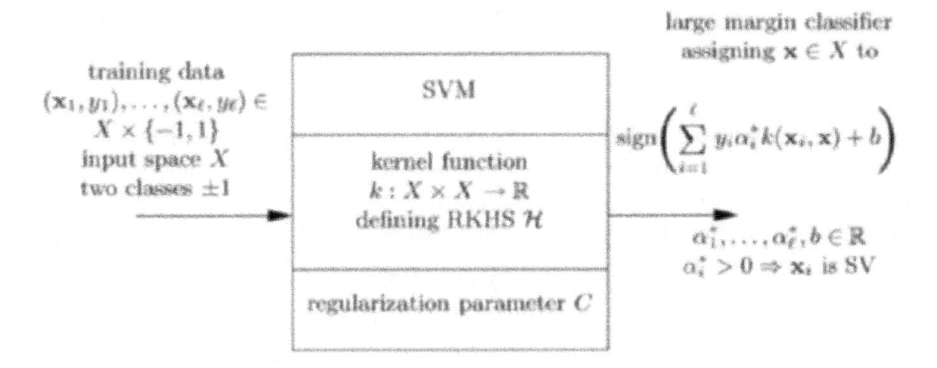

Fig. 9.3. Classification by a soft margin SVM. The learning algorithm is fully specified by the kernel function k and the regularization parameter C. Given training data, it generates the coefficients of a decision function.

This optimization problem $\mathcal{P}_{C\text{-SVM}}$ defines the soft margin L_1-SVM schematically shown in Figure 9.3. It can be solved by Lagrangian methods. The resulting classification function becomes

$$\text{sign}(f(\mathbf{x})) \quad \text{with} \quad f(\mathbf{x}) = \sum_{i=1}^{\ell} y_i\alpha_i^* k(\mathbf{x}_i, \mathbf{x}) + b \ .$$

The coefficients α_i^* are the solution of the following quadratic optimization problem

$$\begin{aligned} \text{maximize} \quad & W(\boldsymbol{\alpha}) = \sum_{i=1}^{\ell} \alpha_i - \frac{1}{2}\sum_{i,j=1}^{\ell} y_i y_j \alpha_i, \alpha_j k(\mathbf{x}_i, \mathbf{x}_j) \\ \text{subject to} \quad & \sum_{i=1}^{\ell} \alpha_i y_i = 0 \ , \\ & 0 \leq \alpha_i \leq C, \quad i = 1,\dots,\ell \ . \end{aligned} \tag{9.2}$$

The optimal value for b can then be computed based on the solution $\boldsymbol{\alpha}^*$. The vectors $\mathbf{x}_i$ with $\alpha_i^* > 0$ are called support vectors. The number of support vectors is denoted by $\#\text{SV}$. The regularization parameter C controls the trade-off between maximizing the margin

$$\gamma^* = \left(\sum_{i,j=1}^{\ell} y_i y_j \alpha_i^* \alpha_j^* k(\mathbf{x}_i, \mathbf{x}_j) \right)^{-1/2}$$

and minimizing the L_1-norm of the final margin slack vector $\boldsymbol{\xi}^*$ of the training data, where

$$\xi_i^* = \max \left(0, 1 - y_i \left(\sum_{j=1}^{\ell} y_j \alpha_j^* k(\mathbf{x}_j, \mathbf{x}_j) + b \right) \right) .$$

In the following we give an extension of the classic C-SVM. It is especially important for practical applications, where the case of highly unbalanced data appears very frequently. To realize a different weighting for wrongly classified positive and negative training examples different cost-factors C_+ and C_- are introduced [31] that change the optimization problem $\mathcal{P}_{C\text{-SVM}}$ to

$$\mathcal{P}_{\tilde{C}\text{-SVM}} = \begin{cases} \min & \frac{1}{2}(\mathbf{w} \cdot \mathbf{w}) + C_+ \sum_{i \in I^+} \xi_i + C_- \sum_{i \in I^-} \xi_i \\ \text{subject to} & y_i((\mathbf{w} \cdot \Phi(\mathbf{x}_i)) + b) \geq 1 - \xi_i \ , \\ & \xi_i \geq 0, \ \ i = 1, \dots, \ell \ , \end{cases}$$

where $I^+ = \{i \in \{1, \dots, \ell\} \mid y_i = 1\}$ and $I^- = \{i \in \{1, \dots, \ell\} \mid y_i = -1\}$. The quadratic optimization problem remains unchanged, except for constraint (9.2) that has be adapted to

$$0 \leq \alpha_i \leq C_-, \ \ i \in I^- \ ,$$

$$0 \leq \alpha_j \leq C_+, \ \ j \in I^+ \ .$$

9.3 Model Selection for SVMs

So far we have considered the inherent multi-objective nature of SVM training. For the remainder of this chapter, we focus on a different multi-objective design problem in the context of C-SVMs. We consider model selection of SVMs, subsuming hyperparameter adaptation and feature selection with respect to different model selection criteria, which are discussed in this section.

Choosing the right kernel for an SVM is crucial for its training accuracy and generalization capabilities as well as the complexity of the resulting classifier. When a parameterized family of kernel functions is considered, kernel adaptation reduces to finding an appropriate parameter vector. These parameters together with the regularization parameter C are called hyperparameters of the SVM.

In the following, we first discuss optimization methods used for SVM model selection with an emphasis on evolutionary multi-objective optimization. Then different model selection criteria are briefly reviewed. Section 9.3.3 deals with appropriate encodings for Gaussian kernels and for feature selection.

9.3.1 Optimization Methods for Model Selection

In practice, the standard method to determine the hyperparameters is grid-search. In simple grid-search the hyperparameters are varied with a fixed step-size through a wide range of values and the performance of every combination is measured. Because of its computational complexity, grid-search is only suitable for the adjustment of very few parameters. Further, the choice of the discretization of the search space may be crucial.

Perhaps the most elaborate techniques for choosing hyperparameters are gradient-based approaches [9, 10, 21, 22, 28]. When applicable, these methods are highly efficient. However, they have some drawbacks and limitations. The most important one is that the score function for assessing the performance of the hyperparameters (or at least an accurate approximation of this function) has to be differentiable with respect to all hyperparameters, which excludes reasonable measures such as the number of support vectors. In some approaches, the computation of the gradient is only exact in the hard-margin case (i.e., for separable data / L_2-SVMs) when the model is consistent with the training data. Further, as the objective functions are indeed multi-modal, the performance of gradient-based heuristics may strongly depend on the initialization—the algorithms are prone to getting stuck in sub-optimal local optima. Evolutionary methods partly overcome these problems.

Evolutionary Algorithms

Evolutionary algorithms (EAs) are a class of iterative, direct, randomized global optimization techniques based on principles of neo-Darwinian evolution theory. In canonical EAs, a set of individuals forming the parent population is maintained, where each individual has a genotype that encodes a candidate solution for the optimization problem at hand. The fitness of an individual is equal to the objective function value at the point in the search space it represents. In each iteration of the algorithm, new individuals, the offspring, are generated by partially stochastic variations of parent individuals. After the fitness of each offspring has been computed, a selection mechanism that prefers individuals with better fitness chooses the new parent population from the current parents and the offspring. This loop of variation and selection is repeated until a termination criterion is met.

In [19, 39], single-objective evolution strategies were proposed for adapting SVM hyperparameters. A single-objective genetic algorithm for SVM feature selection (see below) was used in [17, 20, 27, 29], where in [20] additionally the (discretized) regularization parameter was adapted.

Multi-objective Optimization

Training accuracy, generalization capability, and complexity of the SVM (measured by the number of support vectors) are multiple, probably conflicting objectives. Therefore, it can be beneficial to treat model selection as a multi-objective optimization (MOO) problem.

Consider an optimization problem with M objectives $f_1, \ldots, f_M : X \to \mathbb{R}$ to be minimized. The elements of X can be partially ordered using the concept of Pareto dominance. A solution $\mathbf{x} \in X$ dominates a solution $\mathbf{x}'$ and we write $\mathbf{x} \prec \mathbf{x}'$ if and only if $\exists m \in \{1, \ldots, M\} : f_m(\mathbf{x}) < f_m(\mathbf{x}')$ and $\nexists m \in \{1, \ldots, M\} : f_m(\mathbf{x}) > f_m(\mathbf{x}')$. The elements of the (Pareto) set $\{\mathbf{x} \,|\, \nexists \mathbf{x}' \in X : \mathbf{x}' \prec \mathbf{x}\}$ are called Pareto-optimal. Without any further information, no Pareto-optimal solution can be said to be superior to another element of the Pareto set. The goal of MOO is to find in a single trial a diverse set of Pareto-optimal solutions, which provide insights into the trade-offs between the objectives. When approaching a MOO problem by linearly aggregating all objectives into a scalar function, each weighting of the objectives yields only a limited subset of Pareto-optimal solutions. That is, various trials with different aggregations become necessary—but when the Pareto front (the image of the Pareto set in the m-dimensional objective space) is not convex, even this inefficient procedure does not help (cf. [14, 40]).

Evolutionary multi-objective algorithms have become the method of choice for MOO [11, 14]. Applications of evolutionary MOO to model selection for neural networks can be found in [1, 2, 26, 46], for SVMs in [25, 33, 44].

9.3.2 Model Selection Criteria

In the following, we list performance indices that have been considered for SVM model selection. They can be used alone or in linear combination for single-objective optimization. In MOO a subset of these criteria can be used as different objectives.

Accuracy on Sample Data

The most straightforward way to evaluate a model is to consider its classification performance on sample data. One can always compute the empirical risk given by the error on the training data. To estimate the generalization performance of an SVM, one monitors its accuracy on data not used for training. In the simplest case, the available data is split into a training and validation set, the first one is used for building the SVM and the second for assessing the performance of the classifier. In L-fold cross-validation (CV) the available data is partitioned into L disjoint sets $D_1, \ldots, D_L$ of (approximately) equal size. For given hyperparameters, the SVM is trained L times. In the ith iteration, all data but the patterns in D_i are used to train the SVM and afterwards the performance on the ith validation data set D_i is determined.

At last, the errors observed on the L validation data sets are averaged yielding the L-fold CV error. In addition, the average empirical risk observed in the L iterations can be computed, a quantity we call L-fold CV training error. The ℓ-fold CV (training) error is called the leave-one-out (training) error. The L-fold CV error is an unbiased estimate of the expected generalization error of the SVM trained with $\lfloor \ell - \ell/L \rfloor$ i.i.d. patterns. Although the bias is low, the variance may not be, in particular for large L. Therefore, and for reasons of computational complexity, moderate choices of L (e.g., 5 or 10) are usually preferred [24].

It can be reasonable to split the classification performance into false negative and false positive rate and consider sensitivity and specificity as two separate objectives of different importance. This topic is discussed in detail in Section 9.5.

Number of Input Features

Often the input space X can be decomposed into $X = X_1 \times \cdots \times X_m$. The goal of feature selection is then to determine a subset of indices (feature dimensions) $\{i_1, \ldots, i_{m'}\} \subset \{1, \ldots, m\}$ that yields classifiers with good performance when trained on the reduced input space $X' = X_{i_1} \times \cdots \times X_{i_{m'}}$. By detecting a set of highly discriminative features and ignoring non-discriminative, redundant, or even deteriorating feature dimensions, the SVM may give better classification performance than when trained on the complete space X. By considering only a subset of feature dimensions, the computational complexity of the resulting classifier decreases. Therefore reducing the number of feature dimensions is a common objective.

Feature selection for SVMs is often done using single-objective [17, 20, 27, 29] or multi-objective [33, 44] evolutionary computing. For example, in [44] evolutionary MOO of SVMs was used to design classifiers for protein fold prediction. The three objective functions to be minimized were the number of features, the CV error, and the CV training error. The features were selected out of 125 protein properties such as the frequencies of the amino acids, polarity, and van der Waals volume. In another bioinformatics scenario [33] dealing with the classification of gene expression data using different types of SVMs, the subset of genes, the leave-one-out error, and the leave-one-out training error were minimized using evolutionary MOO.

Modified Radius-Margin Bounds

Bounds on the generalization error derived using statistical learning theory (see Section 9.2.1) can be (ab)used as criteria for model selection.[2] In the

[2] When used for model selection in the described way, the assumptions of the underlying theorems from statistical learning theory are violated and the term "bound" is misleading.

following, we consider radius-margin bounds for L_1-SVMs as used for example in [25] and Section 9.4.

Let R denote the radius of the smallest ball in feature space containing all ℓ training examples given by

$$R = \sqrt{\sum_{i=1}^{\ell} \beta_i^* K(\mathbf{x}_i, \mathbf{x}_i) - \sum_{i,j=1}^{\ell} \beta_i^* \beta_j^* K(\mathbf{x}_i, \mathbf{x}_j)} \ ,$$

where $\boldsymbol{\beta}^*$ is the solution vector of the quadratic optimization problem

$$\begin{aligned} \underset{\boldsymbol{\beta}}{\text{maximize}} \quad & \sum_{i=1}^{\ell} \beta_i K(\mathbf{x}_i, \mathbf{x}_i) - \sum_{i,j=1}^{\ell} \beta_i \beta_j K(\mathbf{x}_i, \mathbf{x}_j) \\ \text{subject to} \quad & \sum_{i=1}^{\ell} \beta_i = 1 \\ & \beta_i \geq 0 \quad , \quad i = 1, \ldots, \ell \ , \end{aligned}$$

see [41]. The modified radius-margin bound

$$T_{\mathrm{DM}} = (2R)^2 \sum_{i=1}^{\ell} \alpha_i^* + \sum_{i=1}^{\ell} \xi_i^* \ ,$$

was considered for model selection of L_1-SVMs in [16]. In practice, this expression did not lead to satisfactory results [10, 16]. Therefore, in [10] it was suggested to use

$$T_{\mathrm{RM}} = R^2 \sum_{i=1}^{\ell} \alpha_i^* + \sum_{i=1}^{\ell} \xi_i^* \ ,$$

based on heuristic considerations and it was shown empirically that T_{RM} leads to better models than T_{DM}.[3] Both criteria can be viewed as two different aggregations of the following two objectives

$$f_1 = R^2 \sum_{i=1}^{\ell} \alpha_i^* \text{ and } f_2 = \sum_{i=1}^{\ell} \xi_i^* \tag{9.3}$$

penalizing model complexity and training errors, respectively. For example, a highly complex SVM classifier that very accurately fits the training data has high f_1 and small f_2.

[3] Also for L_2-SVMs it was shown empirically that theoretically better founded weightings of such objectives (e.g., corresponding to tighter bounds) need not correspond to better model selection criteria [10].

Number of Support Vectors

There are good reasons to prefer SVMs with few support vectors (SVs). In the hard-margin case, the number of SVs (#SV) is an upper bound on the expected number of errors made by the leave-one-out procedure (e.g., see [9, 45]). Further, the space and time complexity of the SVM classifier scales with the number of SVs.

For example, in [25] the number of SVs was optimized in combination with the empirical risk, see also Section 9.4.

9.3.3 Encoding in Evolutionary Model Selection

For feature selection a binary encoding is appropriate. Here, the genotypes are n-dimensional bit vectors $(b_1, \ldots, b_n)^T = \{0,1\}^n$, indicating that the ith feature dimension is used or not depending on $b_i = 1$ or $b_i = 0$, respectively [33, 44].

When a parameterized family of kernel functions is considered, the kernel parameters can be encoded more or less directly. In the following, we focus on the encoding of Gaussian kernels.

The most frequently used kernels are Gaussian functions. General Gaussian kernels have the form

$$k_{\mathbf{A}}(\mathbf{x}, \mathbf{z}) = \exp\left(-(\mathbf{x} - \mathbf{z})^T \mathbf{A} (\mathbf{x} - \mathbf{z})\right)$$

for $\mathbf{x}, \mathbf{z} \in \mathbb{R}^n$ and $\mathbf{A} \in M$, where $M := \{\mathbf{B} \in \mathbb{R}^{n \times n} \mid \forall x \neq 0 : x^T \mathbf{B} x > 0 \wedge \mathbf{B} = \mathbf{B}^T\}$ is the set of positive definite symmetric $n \times n$ matrices.

When adapting Gaussian kernels, the questions of how to ensure that the optimization algorithm only generates positive definite matrices arises. This can be realized by an appropriate parameterization of $\mathbf{A}$. Often the search is restricted to $k_{\gamma \mathbf{I}}$, where $\mathbf{I}$ is the unit matrix and $\gamma > 0$ is the only adjustable parameter. However, allowing more flexibility has proven to be beneficial (e.g., see [9, 19, 21]). It is straightforward to allow for independent scaling factors weighting the input components and consider $k_{\mathbf{D}}$, where $\mathbf{D}$ is a diagonal matrix with arbitrary positive entries. This parameterization is used in most of the experiments described in Sections 9.4 and 9.5. However, only by dropping the restriction to diagonal matrices one can achieve invariance against linear transformations of the input space. To allow for arbitrary covariance matrices for the Gaussian kernel, that is, for scaling and rotation of the search space, we use a parameterization of M mapping $\mathbb{R}^{n(n+1)/2}$ to M such that all modifications of the parameters by some optimization algorithm always result in feasible kernels. In [19], a parameterization of M is used which was inspired by the encoding of covariance matrices for mutative self-adaptation in evolution strategies. We make use of the fact that for any symmetric and positive definite $n \times n$ matrix $\mathbf{A}$ there exists an orthogonal $n \times n$ matrix $\mathbf{T}$ and a diagonal $n \times n$ matrix $\mathbf{D}$ with positive entries such that $\mathbf{A} = \mathbf{T}^T \mathbf{D} \mathbf{T}$ and

$$\mathbf{T} = \prod_{i=1}^{n-1} \prod_{j=i+1}^{n} \mathbf{R}(\alpha_{i,j}) \ ,$$

as proven in [38]. The $n \times n$ matrices $\mathbf{R}(\alpha_{i,j})$ are elementary rotation matrices. These are equal to the unit matrix except for $[\mathbf{R}(\alpha_{i,j})]_{ii} = [\mathbf{R}(\alpha_{i,j})]_{jj} = \cos \alpha_{ij}$ and $[\mathbf{R}(\alpha_{i,j})]_{ji} = -[\mathbf{R}(\alpha_{i,j})]_{ij} = \sin \alpha_{ij}$. However, this is not a canonical representation. It is not invariant under reordering the axes of the coordinate system, that is, applying the rotations in a different order (as discussed in the context of evolution strategies in [23]). The natural injective parameterization is to use the exponential map

$$\exp : \mathfrak{m} \to M \ , \qquad \mathbf{A} \mapsto \sum_{i=0}^{\infty} \frac{\mathbf{A}^i}{i!} \ ,$$

where $\mathfrak{m} := \{\mathbf{A} \in \mathbb{R}^{n \times n} \mid \mathbf{A} = \mathbf{A}^T\}$ is the vector space of symmetric $n \times n$ matrices, see [21]. However, also the simpler, but non-injective function $\mathfrak{m} \to M$ mapping $\mathbf{A} \mapsto \mathbf{A}\mathbf{A}^T$ should work.

9.4 Experiments on Benchmark Data

In this section, we summarize results from evolutionary MOO obtained in [25]. In that study, L_1-SVMs with Gaussian kernels were considered and the two objectives given in (9.3) were optimized.

The evaluation was based on four common medical benchmark datasets *breast-cancer*, *diabetes*, *heart*, and *thyroid* with input dimensions n equal to 9, 8, 13, and 5, and ℓ equal to 200, 468, 170, and 140. The data originally from the UCI Benchmark Repository [7] were preprocessed and partitioned as in [37]. The first of the splits into training and external test set D_{train} and D_{extern} was considered.

Figure 9.4 shows the results of optimizing $k_{\gamma\mathbf{I}}$ (see Section 9.3.3) using the objectives (9.3). For each f_1 value of a solution the corresponding f_2, T_{RM}, T_{DM}, and the percentage of wrongly classified patterns in the test data set $100 \cdot \text{CE}(D_{\text{extern}})$ are given. For *diabetes*, *heart*, and *thyroid*, the solutions lie on typical convex Pareto fronts; in the *breast-cancer* example the convex front looks piecewise linear.

Assuming convergence to the Pareto-optimal set, the results of a single MOO trial are sufficient to determine the outcome of single-objective optimization of any (positive) linear weighting of the objectives. Thus, we can directly determine and compare the solutions that minimizing T_{RM} and T_{DM} would suggest.

The experiments confirm the findings in [10] that the heuristic bound T_{RM} is better suited for model selection than T_{DM}. When looking at $\text{CE}(D_{\text{extern}})$ and the minima of T_{RM} and T_{DM}, we can conclude that T_{DM} puts too much emphasis on the "radius-margin part" yielding worse classification results on

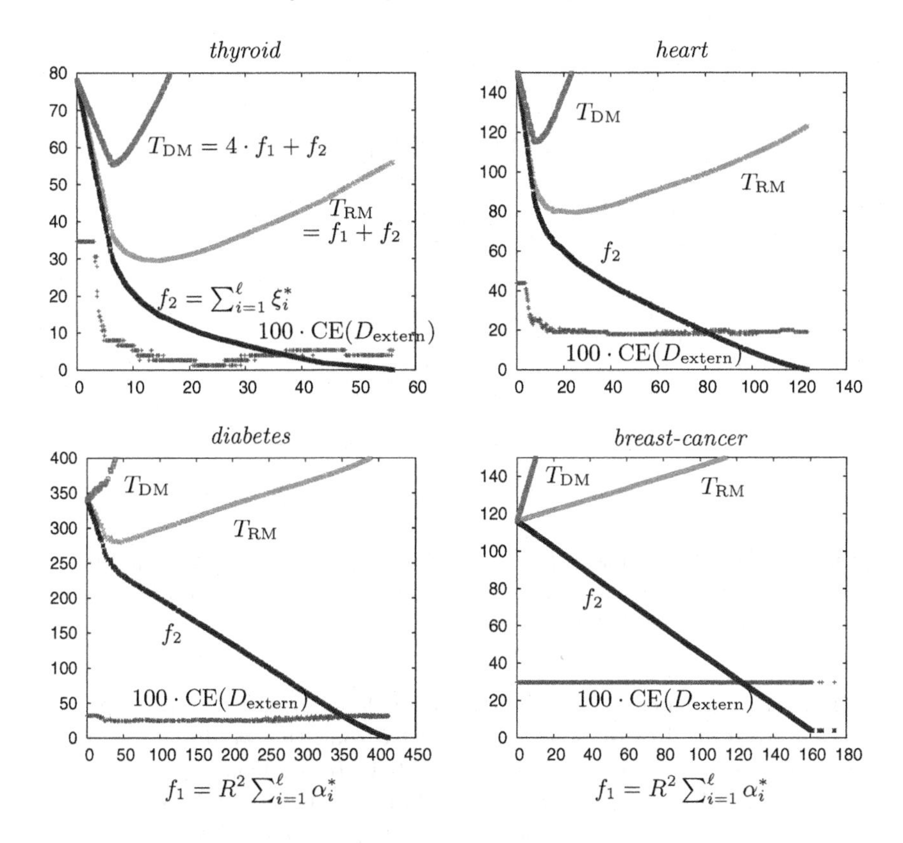

Fig. 9.4. Pareto fronts (i.e., (f_1, f_2) of non-dominated solutions) after 1500 fitness evaluations, see [25] for details. For every solution the values of T_{RM}, T_{DM}, and $100 \cdot \mathrm{CE}(D_{\mathrm{extern}})$ are plotted against the corresponding f_1 value, where $\mathrm{CE}(D_{\mathrm{extern}})$ is the proportion of wrongly classified patterns in the test data set. Projecting the minimum of T_{RM} (for T_{DM} proceed analogously) along the y-axis on the Pareto front gives the (f_1, f_2) pair suggested by the model selection criterion T_{RM}—this would also be the outcome of single-objective optimization using T_{RM}. Projecting an (f_1, f_2) pair along the y-axis on $100 \cdot \mathrm{CE}(D_{\mathrm{extern}})$ yields the corresponding error on an external test set.

the external test set (except for *breast-cancer* where there is no difference on D_{extern}). The *heart* and *thyroid* results suggest that even more weight should be given to the slack variables (i.e., the performance on the training set) than in T_{RM}.

In the MOO approach, degenerated solutions resulting from a not appropriate weighting of objectives (which we indeed observed—without the chance to change the trade-off afterwards—in single-objective optimization of SVMs) become obvious and can be excluded. For example, one would probably not pick the solution suggested by T_{DM} in the *diabetes* benchmark. A typical

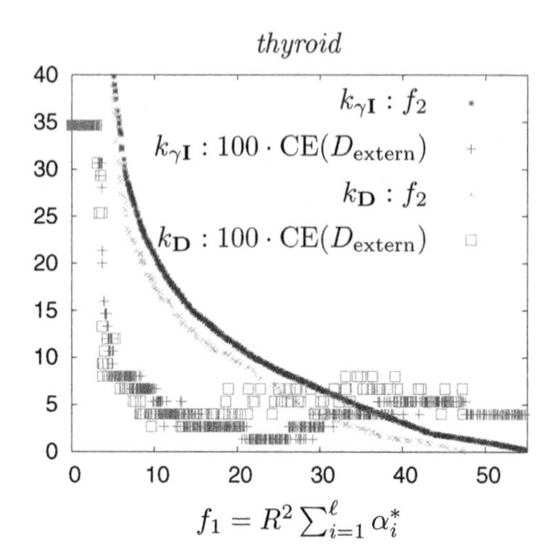

Fig. 9.5. Pareto fronts after optimizing $k_{\gamma\mathbf{I}}$ and $k_{\mathbf{D}}$ for objectives (9.3) and *thyroid* data after 1500 fitness evaluations [25]. For both kernel parameterizations, f_2 and $100 \cdot \mathrm{CE}(D_{\mathrm{extern}})$ are plotted against f_1.

MOO heuristic is to choose a solution that belongs to the "interesting" part of the Pareto front. In case of a typical convex front, this would be the area of highest "curvature" (the "knee", see Figure 9.4). In our benchmark problems, this leads to results on a par with T_{RM} and much better than T_{DM} (except for *breast-cancer*, where the test errors of all optimized trade-offs were the same). Therefore, this heuristic combined with T_{RM} (derived from the MOO results) is an alternative for model selection based on modified radius margin bounds.

Adapting the scaling of the kernel (i.e., optimizing $k_{\mathbf{D}}$) sometimes led to better objective values compared to $k_{\gamma\mathbf{I}}$, see Figure 9.5 for an example, but not necessarily to better generalization performance.

9.5 Real-world Application: Pedestrian Detection

In this section, we consider MOO of SVM classifiers for online pedestrian detection in infrared images for driver assistance systems. This is a challenging real-world task with strict real-time constraints requiring highly optimized classifiers and a considerate adjustment of sensitivity and specificity. Instead of optimizing a single SVM and varying the bias parameter to get a ROC (receiver operating characteristic) curve [30, 35], we apply MOO to decrease the false positive rate, the false negative rate, as well as the number of support vectors. Reducing the latter directly corresponds to decreasing the capacity and the computational complexity of the classifier. We automatically select the kernel parameters, the regularization parameter, and the weighting of positive and negative examples during training. Gaussian kernel functions with individual scaling parameters for each component of the input are adapted. As neither gradient-based optimization methods nor grid-search techniques are applicable, we solve the problem using the real-valued non-dominated sorting genetic algorithm NSGA-II [14, 15].

9.5.1 Pedestrian Detection

Robust object detection systems are a key technology for the next generation of driver assistance systems. They make major contributions to the environment representation of the ego-vehicle, which serves a basis for different high-level driver assistance applications. Besides vehicle detection the early detection of pedestrians is of great interest since it is one important step towards avoiding dangerous situations. In this section we focus on the special case of the detection of pedestrians in a single frame. This is an extremely difficult problem, because of the large variety of human appearances, as pedestrians are standing or walking, carrying bags, wearing hats, etc. Another reason making pedestrian detection very difficult is that pedestrians usually appear in urban environment with complex background (e.g., containing buildings, cars, traffic signs, and traffic lights).

Most of the past work in detecting pedestrians was done using visual cameras. These approaches use a lot of different techniques so we can name only a few. In [48] segmentation was done by means of stereo vision and classification by the use of neural networks. Classification with SVMs that are working on wavelet features was suggested in [34]. A shape-based method for classification was applied in [4]. In [13] a hybrid approach for pedestrian detection was presented, which evaluates the leg-motion and tracks the upper part of the pedestrian.

Recently some pedestrian detection systems have been developed that are working with infrared images, where the color depends on the heat of the object. The advantage of infrared based systems is that they are almost independent on the lighting conditions, so that night-vision is possible. A shape-based method for the classification of pedestrians in infrared images was developed by [3, 5] and an SVM-based one was suggested in [47].

9.5.2 Pedestrian Detection System

In this section we give a description of our pedestrian detection system that is working with infrared images. We keep it rather short because our focus mainly lies on the classification task.

The task of detecting pedestrians is usually divided into two steps, namely the segmentation of candidate regions for pedestrians and the classification of the segmented regions (Figure 9.6). In our system the segmentation of candidate regions for pedestrians is based on horizontal gradient information, which is used to find vertical structures in the image. If the candidate region

Fig. 9.6. Results of our pedestrian detection system on an infrared image; the left picture shows the result of the segmentation step, which provides candidate regions for pedestrians; the image on the right shows the regions that has been labeled as pedestrian.

is at least of size 10×20 pixels a feature vector is calculated and classified using an SVM.

The calculation of the feature vectors is based on contour points and the corresponding discretized angles, which are obtained using a Canny filter [8]. To make the approach scaling-invariant we put a 4×8 grid on the candidate region and determine the histograms of eight different angles for each of these fields. In a last step the resulting 256-dimensional feature vector is normalized to the range $[-1, 1]^{256}$.

9.5.3 Model Selection

In practice the common way for assigning a performance to a classifier is to analyze its ROC curve. This analysis visualizes the trade-off between the two partially conflicting objectives false negative and false positive rate and allows for the selection of a problem specific solution. A third objective for SVM model-selection, which is especially important for real-time tasks like pedestrian detection, is the number of support vectors, because it directly determines the computational complexity of the classifier.

We use an EA for the tuning of much more parameters than would be possible with grid-search, thus making a better adaptation to the given problem possible. Concretely we tune the parameters C_+, C_-, and $\mathbf{D}$, that is, independent scaling factors for each component of the feature vector (see Sections 9.2.2 and 9.3.3).

For the optimization we generated four datasets D_{train}, D_{val}, D_{test}, and D_{extern}, whose use will become apparent in the discussion of the optimization algorithm. Each of the datasets consists of candidate regions (256-dimensional feature vectors) that are manually labeled pedestrian or non-pedestrian. The candidate regions are obtained by our segmentation algorithm to ensure that the datasets are realistic in that way that all usually appearing critical cases are contained. Furthermore the segmentation algorithm provides much more non-pedestrians than pedestrians and therefore negative and positive examples in the data are highly unbalanced. The datasets are obtained from different image sequences, which have been captured on the same day to ensure similar environmental conditions, but no candidate region from the same sequence is in the same dataset.

For optimization we use the NSGA-II, where the fitness of an individual is determined on dataset D_{val} with an SVM that has been trained on dataset D_{train}. This training is done using the individual's corresponding SVM parameterization. To avoid overfitting we keep an external archive of non-dominated solutions, which have been evaluated on the validation set D_{test} for every individual that has been created by the optimization process. The dataset D_{extern} is used for the final evaluation (cf. [46]).

For the application of the NSGA-II we choose a population size of 50 and create the initial parent population by randomly selecting non-dominated solutions from a 3D-grid-search on the parameters C_+, C_-, and one global scaling factor γ, that is $\mathbf{D} = \gamma\mathbf{I}$. The other parameters of the NSGA-II are chosen like in [15] ($p_c = 0.9$, $p_m = 1/n$, $\eta_c = 20$, $\eta_m = 20$). We carried out 10 optimization trials, each of them lasting for 250 generations.

9.5.4 Results

In this section we give a short overview about the results of the MOO of the pedestrian detection system.

The progress of one optimization trial is exemplary shown in Figure 9.7. It illustrates the Pareto-optimal solutions in the objective space that are contained in the external archive after the first and after the 250th generation. The solutions in the archive after the first generation roughly correspond to the solutions that have been found by 3D-grid search. The solutions after the

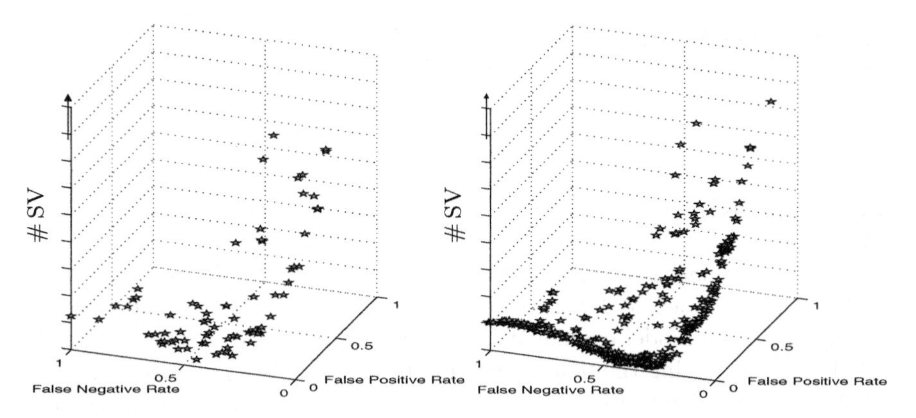

Fig. 9.7. Pareto-optimal solutions that are contained in the external archive after the first (left plot) and after the 250th generation (right plot).

250th generation have obviously improved and clearly reveal the trade-off between the three objectives, thereby allowing for a problem-specific choice of an SVM.

For assessing the performance of a stochastic optimization algorithm it is not sufficient to evaluate a single optimization trial. A possibility for visualizing the outcome of a series of optimization trials are the so-called summary attainment surfaces [18] that provide the points in objective space that have been attained in a certain fraction of all trials.

We give the summary attainment curve for the two objectives true positive and false positive rate, which are the objectives of the ROC curve. Figure 9.8 shows the points that have been attained by all, 50%, and the best of our optimization trials.

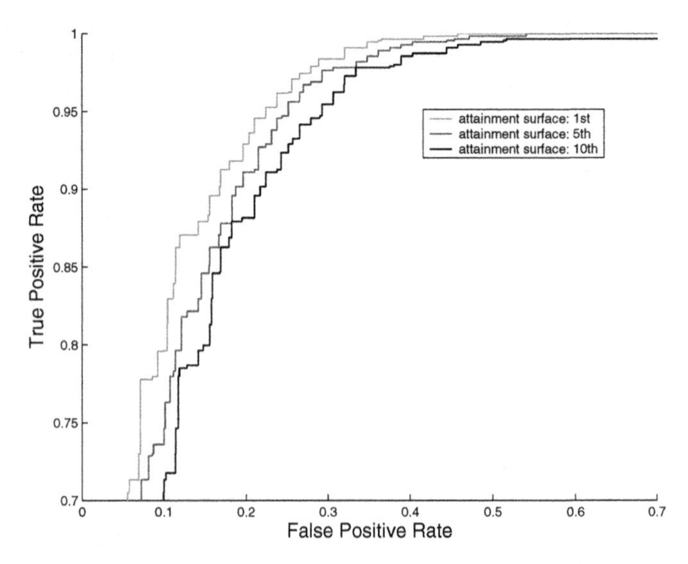

Fig. 9.8. Summary attainment curves for the objectives true positive rate and false positive rate ("ROC curves").

9.6 Conclusions

Designing classifiers is a multi-objective optimization (MOO) problem. The application of "true" MOO algorithms allows for visualizing trade-offs, for example between model complexity and learning accuracy or sensitivity and specificity, for guiding the model selection process.

We considered evolutionary MOO of support vector machines (SVMs). This approach can adapt multiple hyperparameters of SVMs based on conflicting, not differentiable criteria.

When optimizing the norm of the slack variables and the radius-margin quotient as two objectives, it turned out that standard MOO heuristics based on the curvature of the Pareto front led to comparable models as corresponding single-objective criteria proposed in the literature. In benchmark problems it appears that the latter should put more emphasis on minimizing the slack variables.

We demonstrated MOO of SVMs for the detection of pedestrians in infrared images for driver assistance systems. Here the three objectives are the false positive rate, the false negative rate, and the number of support vectors. The Pareto front of the first two objectives can be viewed as a ROC curve where each point corresponds to a learning machine optimized for that particular trade-off between sensitivity and specificity. The third objective reduces the model complexity in order to meet real-time constraints.

Acknowledgments

We thank Tobias Glasmachers for proofreading and Aalzen Wiegersma for providing the thoroughly preprocessed pedestrian image data. We acknowledge support from BMW Group Research and Technology.

References

[1] Hussein A. Abbass. An evolutionary artificial neural networks approach for breast cancer diagnosis. *Artificial Intelligence in Medicine*, 25(3):265–281, 2002.
[2] Hussein A. Abbass. Speeding up backpropagation using multiobjective evolutionary algorithms. *Neural Computation*, 15(11):2705–2726, 2003.
[3] Massimo Bertozzi, Alberto Broggi, Marcello Carletti, Alessandra Fascioli, Thorsten Graf, Paolo Grisleri, and Michael Meinecke. IR pedestrian detection for advanced driver assistance systems. In *Proceedings of the 25th Pattern Recognition Symposium*, volume 2781 of *LNCS*, pages 582–590. Springer-Verlag, 2003.
[4] Massimo Bertozzi, Alberto Broggi, Alessandra Fascioli, and Massimiliano Sechi. Shape-based pedestrian detection. In *Proceedings of the IEEE Intelligent Vehicles Symposium 2000*, pages 215–220, 2000.
[5] Massimo Bertozzi, Alberto Broggi, Thorsten Graf, Paolo Grisleri, and Michael Meinecke. Pedestrian detection in infrared images. In *Proceedings of the IEEE Intelligent Vehicles Symposium 2003*, pages 662–667, 2003.
[6] J. Bi. Multi-objective programming in SVMs. In Tom Fawcett and Nina Mishra, editors, *Machine Learning, Proceedings of the 20th International Conference (ICML 2003)*, pages 35–42. AAAI Press, 2003.
[7] C.L. Blake and C.J. Merz. UCI repository of machine learning databases, 1998.
[8] J Canny. A computational approach to edge detection. *IEEE Transactions on Pattern Analysis and Machine Intelligence*, 8(6):679–698, 1986.
[9] O. Chapelle, V. Vapnik, O. Bousquet, and S. Mukherjee. Choosing multiple parameters for support vector machines. *Machine Learning*, 46(1):131–159, 2002.

[10] K.-M. Chung, W.-C. Kao, C.-L. Sun, and C.-J. Lin. Radius margin bounds for support vector machines with RBF kernel. *Neural Computation*, 15(11):2643–2681, 2003.

[11] Carlos A. Coello Coello, David A. Van Veldhuizen, and Gary B. Lamont. *Evolutionary Algorithms for Solving Multi-Objective Problems.* Kluwer Academic Publishers, 2002.

[12] Nello Cristianini and John Shawe-Taylor. *An Introduction to Support Vector Machines and other kernel-based learning methods.* Cambridge University Press, 2000.

[13] C. Curio, J. Edelbrunner, T. Kalinke, C. Tzomakas, and W. von Seelen. Walking pedestrian recognition. *IEEE Transactions on Intelligent Transportation Systems*, 1(3):155–163, 2000.

[14] Kalyanmoy Deb. *Multi-Objective Optimization Using Evolutionary Algorithms.* Wiley, 2001.

[15] Kalyanmoy Deb, Samir Agrawal, Amrit Pratap, and T. Meyarivan. A fast and elitist multiobjective genetic algorithm: NSGA-II. *IEEE Transactions on Evolutionary Computation*, 6(2):182–197, 2002.

[16] K. Duan, S. S. Keerthi, and A.N. Poo. Evaluation of simple performance measures for tuning SVM hyperparameters. *Neurocomputing*, 51:41–59, 2003.

[17] Damian R. Eads, Daniel Hill, Sean Davis, Simon J. Perkins, Junshui Ma, Reid B. Porter, and James P. Theiler. Genetic algorithms and support vector machines for time series classification. In Bruno Bosacchi, David B. Fogel, and James C. Bezdek, editors, *Applications and Science of Neural Networks, Fuzzy Systems, and Evolutionary Computation V.*, volume 4787 of *Proceedings of the SPIE*, pages 74–85, 2002.

[18] C. M. Fonseca, J. D. Knowles, L. Thiele, and E. Zitzler. A tutorial on the performance assessment of stochastic multiobjective optimizers. Presented at the Third International Conference on Evolutionary Multi-Criterion Optimization (EMO 2005), 2005.

[19] Frauke Friedrichs and Christian Igel. Evolutionary tuning of multiple SVM parameters. *Neurocomputing*, 64(C):107–117, 2005.

[20] H. Fröhlich, O. Chapelle, and B. Schölkopf. Feature selection for support vector machines using genetic algorithms. *International Journal on Artificial Intelligence Tools*, 13(4):791–800, 2004.

[21] Tobias Glasmachers and Christian Igel. Gradient-based adaptation of general gaussian kernels. *Neural Computation*, 17(10):2099–2105, 2005.

[22] Carl Gold and Peter Sollich. Model selection for support vector machine classification. *Neurocomputing*, 55(1-2):221–249, 2003.

[23] Nikolaus Hansen. Invariance, self-adaptation and correlated mutations and evolution strategies. In *Proceedings of the 6th International Conference on Parallel Problem Solving from Nature (PPSN VI)*, volume 1917 of *LNCS*, pages 355–364. Springer-Verlag, 2000.

[24] Trevor Hastie, Robert Tibshirani, and Jerome Friedman. *The Elements of Statistical Learning Data Mining, Inference, and Prediction.* Springer Series in Statistics. Springer-Verlag, 2001.

[25] C. Igel. Multi-objective model selection for support vector machines. In C. A. Coello Coello, E. Zitzler, and A. Hernandez Aguirre, editors, *Proceedings of the Third International Conference on Evolutionary Multi-Criterion Optimization (EMO 2005)*, volume 3410 of *LNCS*, pages 534–546. Springer-Verlag, 2005.

[26] Yaochu Jin, Tatsuya Okabe, and Bernhard Sendhoff. Neural network regularization and ensembling using multi-objective evolutionary algorithms. In *Congress on Evolutionary Computation (CEC'04)*, pages 1–8. IEEE Press, 2004.

[27] Kees Jong, Elena Marchiori, and Aad van der Vaart. Analysis of proteomic pattern data for cancer detection. In G. R. Raidl, S. Cagnoni, J. Branke, D. W. Corne, R. Drechsler, Y. Jin, C. G. Johnson, P. Machado, E. Marchiori, F. Rothlauf, G. D. Smith, and G. Squillero, editors, *Applications of Evolutionary Computing*, volume 3005 of *LNCS*, pages 41–51. Springer-Verlag, 2004.

[28] S. S. Keerthi. Efficient tuning of SVM hyperparameters using radius/margin bound and iterative algorithms. *IEEE Transactions on Neural Networks*, 13(5):1225–1229, 2002.

[29] M. T. Miller, A. K. Jerebko, J. D. Malley, and R. M. Summers. Feature selection for computer-aided polyp detection using genetic algorithms. In Anne V. Clough and Amir A. Amini, editors, *Medical Imaging 2003: Physiology and Function: Methods, Systems, and Applications*, volume 5031 of *Proceedings of the SPIE*, pages 102–110, 2003.

[30] Anuj Mohan, Constantine Papageorgiou, and Thomas Poggio. Example-based object detection in images by components. *IEEE Transactions on Pattern Analysis and Machine Intelligence*, 23(4):349–361, 2001.

[31] Katharina Morik, Peter Brockhausen, and Thorsten Joachims. Combining statistical learning with a knowledge-based approach - a case study in intensive care monitoring. In *Proceedings of the 16th International Conference on Machine Learning*, pages 268–277. Morgan Kaufmann, 1999.

[32] M. Oren, C. P. Papageorgiou, P. Sinha, E. Osuna, and T. Poggio. Pedestrian detection using wavelet templates. In *Proceedings of the IEEE Conference on Computer Vision and Pattern Recognition*, pages 193–199, 1997.

[33] S. Pang and N. Kasabov. Inductive vs. transductive inference, global vs. local models: SVM, TSVM, and SVMT for gene expression classification problems. In *International Joint Conference on Neual Networks (IJCNN)*, volume 2, pages 1197–1202. IEEE Press, 2004.

[34] C. Papageorgiou, T. Evgeniou, and T. Poggio. A trainable pedestrian detection system. In *Proceedings of the IEEE International Conference on Intelligent Vehicles Symposium 1998*, pages 241–246, 1998.

[35] C. P. Papageorgiou. A trainable system for object detection in images and video sequences. Technical Report AITR-1685, Massachusetts Institute of Technology, Artificial Intelligene Laboratory, 2000.

[36] Constantine Papageorgiou and Tomaso Poggio. A trainable system for object detection. *International Journal of Computer Vision*, 38(1):15–33, 2000.

[37] G. Rätsch, T. Onoda, and K.-R. Müller. Soft margins for AdaBoost. *Machine Learning*, 42(3):287–320, 2001.

[38] Günther Rudolph. On correlated mutations in evolution strategies. In R. Männer and B. Manderick, editors, *Parallel Problem Solving from Nature 2 (PPSN II)*, pages 105–114. Elsevier, 1992.

[39] Thomas Philip Runarsson and Sven Sigurdsson. Asynchronous parallel evolutionary model selection for support vector machines. *Neural Information Processing – Letters and Reviews*, 3(3):59–68, 2004.

[40] Y. Sawaragi, H. Nakayama, and T. Tanino. *Theory of Multiobjective Optimization*, volume 176 of *Mathematics in Science and Engineering*. Academic Press, 1985.

[41] B. Schölkopf, C. J. C. Burges, and V. Vapnik. Extracting support data for a given task. In U. M. Fayyad and R. Uthurusamy, editors, *Proceedings of the First International Conference on Knowledge Discovery & Data Mining*, pages 252–257, Menlo Park, CA, 1995. AAAI Press.

[42] B. Schölkopf and A. J. Smola. *Learning with Kernels: Support Vector Machines, Regularization, Optimization, and Beyond.* MIT Press, 2002.

[43] John Shawe-Taylor and Nello Cristianini. *Kernel Methods for Pattern Analysis.* Cambridge University Press, 2004.

[44] S. Y. M. Shi, P. N. Suganthan, and K. Deb. Multi-class protein fold recognition using multi-objective evolutionary algorithms. In *IEEE Symposium on Computational Intelligence in Bioinformatics and Computational Biology*, pages 61–66, 2004.

[45] Vladimir N. Vapnik. *The Nature of Statistical Learning Theory.* Springer-Verlag, 1995.

[46] S. Wiegand, C. Igel, and U. Handmann. Evolutionary multi-objective optimization of neural networks for face detection. *International Journal of Computational Intelligence and Applications*, 4(3):237–253, 2004. Special issue on Neurocomputing and Hybrid Methods for Evolving Intelligence.

[47] F. Xu, X. Liu, and K. Fujimura. Pedestrian detection and tracking with night vision. *IEEE Transactions on Intelligent Transportation Systems*, 6(1):63–71, 2005.

[48] L. Zhao and C. Thorpe. Stereo- and neural network-based pedestrian detection. In *Proceedings of the IEEE International Conference on Intelligent Transportation Systems'99*, pages 298–303, 1999.

10

Multi-Objective Evolutionary Algorithm for Radial Basis Function Neural Network Design

Gary G. Yen

School of Electrical and Computer Engineering
Oklahoma State University
Stillwater, OK 74078-5032, USA
gyen@okstate.edu

Summary. In this chapter, we present a multiobjective evolutionary algorithm based design procedure for radial-basis function neural networks. A Hierarchical Rank Density Genetic Algorithm (HRDGA) is proposed to evolve the neural network's topology and parameters simultaneously. Compared with traditional genetic algorithm based designs for neural networks, the hierarchical approach addresses several deficiencies highlighted in literature. In addition, the rank-density based fitness assignment technique is used to optimize the performance and topology of the evolved neural network to tradeoff between the training performance and network complexity. Instead of producing a single *optimal* solution, HRDGA provides a set of near-optimal neural networks to the designers so that they can have more flexibility for the final decision-making based on certain preferences. In terms of searching for a near-complete set of candidate networks with high performances, the networks designed by the proposed algorithm prove to be competitive, or even superior, to three state-of-the-art designs for radial-basis function neural networks to predict Mackey-Glass chaotic time series.

10.1 Introduction

Neural Networks (NN's) and Genetic Algorithms (GA's) represent two emerging technologies inspired by biologically motivated computational paradigms. NN's are derived from the information-processing framework of a human brain to emulate the learning behavior of *an individual*, while GA's are motivated by the theory of evolution to evolve *a whole population* toward better fitness. Although these two technologies seem quite different in the time period of action, number of involved individuals, and the process scheme, their similar dynamic behaviors stimulate research on whether a synergistic combination of these two technologies may provide more problem solving power than either alone [22].

There has been an extensive analysis of different classes of neural networks possessing various architectures and training algorithms. Without a proven

G.G. Yen: *Multi-Objective Evolutionary Algorithm for Radial Basis Function Neural Network Design*, Studies in Computational Intelligence (SCI) **16**, 221–239 (2006)
www.springerlink.com

guideline, the design of an optimal neural network for a given problem is an *ad hoc* process. Given a sufficient number of neurons, more than one neural network structure (i.e., with different weighting coefficients and numbers of neurons) can be trained to solve a given problem within an error bound if given sufficient training time. The decision of "which network is the best" is often decided by which network will better meet the user's needs for a given problem. It is known that the performance of neural networks is sensitive to the number of neurons. Too few neurons can result in underfitting problems (poor approximation), while too many neurons may contribute to overfitting problems. Obviously, achieving a better network performance and simplifying the network topology are two competing objectives. This has promoted research on how to identify an optimal and efficient neural network structure. AIC (Akaike Information Criterion) [19] and PMDL (Predictive Minimum Description Length) [8] are two well-adopted approaches. However, AIC can be inconsistent and has a tendency to overfit a model, while PMDL only succeeded in relatively simple neural network structures and seemed very difficult to extend to a complex NN structure optimization problem. Moreover, all of these approaches tend to produce a single neural network for each run, offering the designers no alternative choices.

Since the 1990's, evolutionary algorithms have been successfully applied to the design of network topologies and the choice of learning parameters [1, 2, 18, 17, 15]. They reported some encouraging results that are comparable with conventional neural network design approaches. However, multiobjective trade-off characteristic of the neural network design has not been well studied and applied in the real world applications. In this chapter, we propose a Hierarchical Rank Density Genetic Algorithm (HRDGA) for neural network design in order to evolve a set of near-optimal neural networks. Without loss of generality, we will restrict our discussions to the radial basis function neural network. In HRDGA, each chromosome is a candidate neural network and is coded by three different gene segments– high level segments have control genes that can determine the status (activated or deactivated) of genes in lower level segments. Hidden layers and neurons are added or deleted by this "on/off" scheme to achieve an optimal structure through a survival of the fittest evolution. Meanwhile, weights and biases are evolved along with the neural network topology. Treating the neural network design as a bi-objective optimization problem, a new rank-density based fitness assignment technique is developed to evaluate the structure complexity and the performance of the evolved neural network. More importantly, instead of a single network, HRDGA produces a set of near-optimal candidate networks with different trade-off traits from which the designers or decision makers can make flexible choices based on their preferences.

The remainder of this chapter is organized as follows. Section 10.2 discusses the neural network design dilemma and the difficulty of finding a single *optimal* neural network. Section 10.3 reviews various approaches to applying genetic algorithms for neural network design and introduces the proposed hierarchical structure, genetic operators, and multi-fitness measures of the proposed genetic algorithm. Section 10.4 applies hierarchical genotype representation to a radial-basis function neural network design. Section 10.5 introduces the proposed rank-density fitness assignment technique for multiobjective genetic algorithms and describes HRDGA parameters and design flowchart. Section 10.6 presents a feasible study on the Mackey-Glass chaotic time series prediction using HRDGA evolved neural networks. A time series with chaotic character is trained and the performance is compared with those of the k-nearest neighbors, generalized regression, and orthogonal least square training algorithms. Finally, Section 10.7 provides some concluding remarks along with pertinent observations.

10.2 Neural Network Design Dilemma

To generate a neural network that possesses the practical applicability, several essential conditions need to be considered.

1. A training algorithm that can search for the optimal parameters (i.e., weights and biases) for the specified network structure and training task.
2. A rule or algorithm that can determine the network complexity and ensure it to be sufficient for solving the given training problem.
3. A metric or measure to evaluate the reliability and generalization of the produced neural network.

The design of an optimal neural network involves all of these three problems. As given in [6], the ultimate goal of the construction of a neural network with the input-output relation $\mathbf{y} = f_{NS}(\mathbf{x}, \omega)$ is the minimization of the expectation of a cost function $g_T(f_{NS}(\mathbf{X}, \omega), \mathbf{Y})$ as:

$$E[g_T(f_{NS}(\mathbf{X}, \omega, \mathbf{Y})] = \int \int g_T(f_{NS}(\mathbf{x}, \omega), \mathbf{y}) f_{x,y}(\mathbf{x}, \mathbf{y}) d\mathbf{x} d\mathbf{y}, \tag{10.1}$$

where $f_{\mathbf{x},\mathbf{y}}(\mathbf{x}, \mathbf{y})$ denotes the joint *pdf* that depends on the input vector $\mathbf{x}$ and the target output vector $\mathbf{y}$. $\mathbf{X}$ and $\mathbf{Y}$ are spaces spanned by all individual training samples, $\mathbf{x}$ and $\mathbf{y}$. Given a network structure NS, a family of input-output relations $F_{NS} = \{f_{NS}(\mathbf{x}, \omega)\}$, parameterizd by ω, consisting of all network functions that may be formed with different choices of the weights can be assigned. The structure NS' is said to be *dominated* by NS" if $F_{NS'} \subset F_{NS"}$. In order to choose the optimal neural network, two problems have to be solved.

1. Determination of the network function $f^*_{NS}(\mathbf{x})$ (i.e., the determination of the respective weights) that gives the minimal cost value within the family F_{NS}:

$$f_{NS}^*(\mathbf{x}) = f_{NS}(\mathbf{x}, \omega^*) = \arg\min_\omega \mathbf{E}[g_L(f_{NS}(\mathbf{X}, \omega), \mathbf{Y})], \qquad (10.2)$$

where $g_L(\cdot,\cdot)$ denotes the cost function measuring the performance over the training set.

2. Determination of the network structure NS^* that realizes the minima cost value within a set of structures $\{NS\}$:

$$NS^* = \arg\min_{NS \in F_{NS}} \mathbf{E}[g_T(f_{NS}^*(\mathbf{X}), \mathbf{Y})]. \qquad (10.3)$$

Obviously, the solutions of both tasks need not result into a unique network. In [6], if several structures $NS_1^*, NS_2^*, \cdots$ meet the criterion as shown in Equation (10.3), the one with the minimal number of hidden neurons is defined as an *optimal.* However, as a neural network can only tune the weights by the given training data sets, and these data sets are always finite, there will be a trade-off between NN learning capability and the number of the hidden neurons. A network with insufficient neurons might not be able to approximate well enough the functional relationship between input and target output. On the other hand, if the number of neurons is excessive, the realized network function will depend greatly on the resulting realization of the given limited training set. This trade-off characteristic implies that a single *optimal* neural network is very difficult to find as extracting $f_{NS}^*(\mathbf{x})$ from F_{NS} by using a finite training data set is a difficult task, if not impossible [9]. Therefore, instead of trying to obtain a single *optimal* neural network, finding a set of *near-optimal* networks with different network structures seems more feasible. Each individual in this neural network set may provide different training and test performances for different training and test data sets. Moreover, the idea of providing "a set of" candidate networks to the decision makers can offer more flexibilities in selecting an appropriate network judged by their own preferences. For this reason, genetic algorithms and multiobjective optimization techniques can be introduced in neural network design problems to evolve network topology along with parameters and present a set of alternative network candidates.

10.3 Neural Network Design with Genetic Algorithm

In the literature of applying genetic algorithms to assist neural networks design, several approaches have been proposed for different objectives. These approaches can be categorized into four different areas.

10.3.1 Data Preparation

GA's were primarily used to help NN design by pre-processing data. Kelly and Davis used a genetic algorithm to find the rotation of a data set and scaling factors for each attribute to improve the performance of a KNN classifier [13]. Chang and Lippmann used a genetic algorithm to reduce the dimensionality of a feature set for a KNN classifier in a speech recognition task [3].

10.3.2 Evolving Network Parameters

Belew and his colleagues used a genetic algorithm to identify a good initial weighting configuration for a back-propagation network [1]. On the other hand, Bruce and Timothy used a genetic algorithm to evolve the centers and widths for a radial basis function neural network [2]. Genetic algorithms are used to evolve the weights or biases in a fixed topology neural network. The structure, number of layers and number of neurons, is pre-determined based upon some heuristic judgments.

10.3.3 Evolving Network Topology

This is the most targeted area with which genetic algorithm can be used in neural network design. Miller *et. al.*, used a genetic algorithm to evolve an optimally connected matrix to form a neural network [16], but this method can only be used for simple problems. Whiltley *et. al.*, used a genetic algorithm to find which links could be eliminated in order to achieve a specific learning objective [21]. Lucas proposed a GA based adaptive neural architecture selection method to evolve a back-propagation neural network [15], and Davila applied GA's schema theory to aid the design of genetic coding for NN topology optimization. Three main problems exist in current topology design research, namely network feasibility, one genotype mapping multiple phenotypes and one phenotype mapping different genotypes [23].

10.3.4 Evolving NN Structures together with Weights and Biases

Dasgupta and McGregor proposed an sGA (Structure Genetic Algorithm) to evolve neural networks [5]. But in their work, only a XOR problem and a 4×4 encoder/decoder problem were tested, which is relatively simple. Since an N-neuron neural network must be expressed as a chromosome with a bit string of length N^2, a complex phenotype will map to a much more complex genotype. As a result, using sGA to evolve a large neural network is computationally expensive, if not impossible. Zhang and Cho proposed Bayesian evolutionary algorithms to evolve the structures and parameters of neural trees, which are then used to predict a time series [24]. However, both of these algorithms use the connection matrix and from-to units, which had been shown to easily produce "one phenotype mapping different genotypes" problem.

To avoid this problem, a hierarchical genotype representation is adopted in this study. HGA (Hierarchical Genetic Algorithm) was first proposed by Ke for fuzzy controller design [12]. They used two layer genes to evolve membership functions for a fuzzy logic design. Based on this idea, Yen and Lu designed an HGA Neural Network (HGA-NN) [23]. In the HGA-NN, a three-layer HGA is used to evolve a Multi-layer Perceptron (MLP) neural network. The chromosome structure (genotype) is shown in Figure 10.1(a). As shown

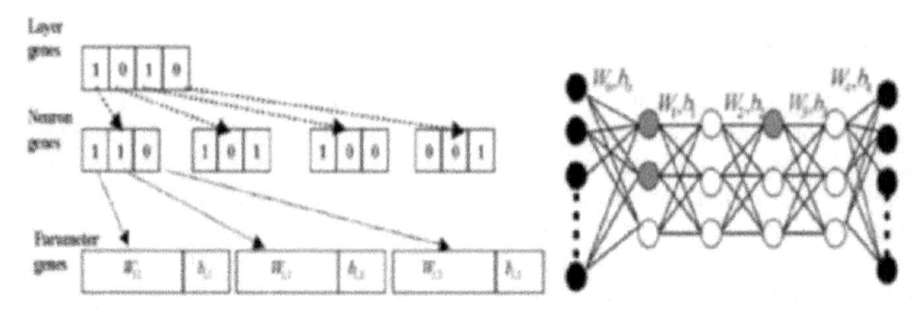

Fig. 10.1. Genotype structure of an individual MLP neural network (left) and the corresponding phenotype with layered neural network topology (right).

in Figure 10.1(a), each candidate chromosome corresponding to a neural network implementation is assumed to have at most four hidden layers (shown in the high-level *layer genes*), where the first and the third hidden layers are activated (as indicated with binary bits 1) and the second and the fourth hidden layers are deactivated (with binary bits 0). Additionally, we assume at most three neurons in each hidden layer as shown in the space available in neuron gene corresponding to each element in layer gene.

The mid-level *neuron genes* indicate that two out of three neurons in the first hidden layer are activated, while only one neuron in the third hidden layer is activated. Since the second and the fourth layers are deactivated, their neurons are not used. The low-level *parameter genes* are then used to represent the weighting and bias parameters of each corresponding neuron activated. The active status of one control gene determines whether the parameters of the next level controlled by this gene will be activated or not. As an example, a genetic chromosome (genotype) shown in Figure 10.1(a) corresponds to an individual neural network (phenotype) with two hidden layers and two and one neuron in each layer in Figure 10.1(b). By using this hierarchical genotype, the problem of "one phenotype mapping different genotypes" can be prevented.

10.4 HGA Evolved Radial-Basis Function NN

In a similar spirit, HGA is tailored in this chapter to evolve an RBF (Radial-Basis Function) neural network. A radial-basis function can be formed as:

$$f(\mathbf{x}) = \sum_{i=1}^{m} \omega_i \exp(-||\mathbf{x} - \mathbf{c}_i||^2), \tag{10.4}$$

where $\mathbf{c}_i$denotes the center of the ith localized function, ω_i is the weighting coefficient connecting the ith Gaussian neuron to the output neuron, and m is the number of Gaussian neurons in the hidden layer. Without loss of generality, we choose the variance as unity for each Gaussian neuron.

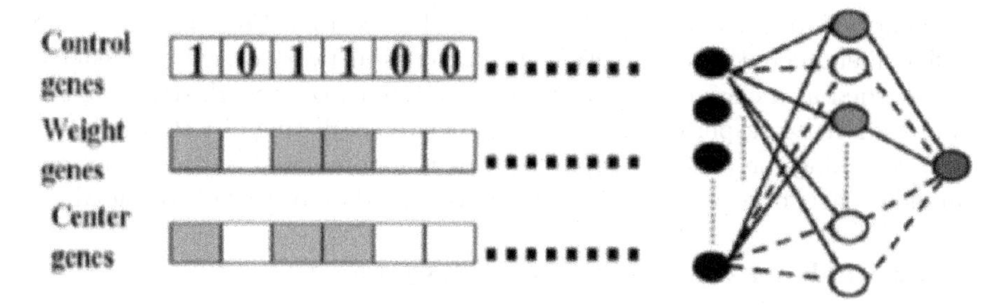

Fig. 10.2. Genotype (left) and phenotype (right) of HGA based RBF neural network.

In HGA based RBF neural network design, genes in the genotype are hierarchically structured into three layers: control genes, weight genes, and center genes. The lengths of these three layers of genes are the same and specified by the user. The value of each control gene (0 or 1) determines the activation status (off or on) of the corresponding weight gene and center gene. On the other hand, the weight genes and center genes are represented by real values. Control genes and weight genes are randomly initialized and the center genes are randomly selected from given training data samples. Figure 10.2 shows the genotype and phenotype of a HGA based RBF neural network, where the first, third and fifth hidden neurons are activated. As a result, their corresponding weight and center parameters are used.

10.5 Multiobjective Genetic Algorithm

As discussed in Section 10.3, neural network design problems have a multiobjective trade-off characteristic in terms of optimizing network topology and performance. Therefore, multiobjective genetic algorithm is applied in NN design procedure.

10.5.1 Multiobjective Optimization Problems

Multiobjective Optimization (MO) is a very important research topic, because most real world problems have not only a multiobjective nature, but also many open issues to be answered qualitatively and quantitatively. In many optimization problems, there is not even a universally accepted definition of "optimum" as in single-objective optimization [10], because the solution to a MO problem is generally not a single point. It consists of a family of non-dominated points, a so-called Pareto front [7], which describes the trade-off among contradicted objectives. The Pareto front yields many candidate solutions— non-dominated points, from which we can choose the desired one under different trade-off conditions. In most cases, the Pareto front is on the boundary of the feasible range as shown in Figure 10.3. Considering the NN design dilemma outlined in Section 10.2, a neural network design problem can be regarded as a class of MO problems as minimizing network structure and improving network performance, which are two conflicting objectives. Therefore, searching for a near-complete set of non-dominated and near-optimal candidate networks as the design solutions (i.e., Pareto front) is our goal.

10.5.2 Rank-density Based Fitness Assignment

Since the 1980's, several Multiobjective Genetic Algorithms (MOGAs) have been proposed and applied in multiobjective optimization problems [25]. These algorithms all have almost the same purpose— searching for a uniformly distributed and near-optimal Pareto front for a given MO problem. However, this ultimate goal is far from been accomplished by the existing MOGAs described in literature due to trade-off decisions between homogenously distributing the computational resources and GA's strong tendencies to restrict searching efforts (i.e., genetic drift).

In this chapter, we propose a new rank-density based fitness assignment technique in a multiobjective genetic algorithm to assist neural network design. Compared to traditional fitness assignment methods, the proposed rank-density based technique possesses the following characteristics of a) simplifying the problem domain by converting high-dimensional multiple objectives into two objectives to minimize the individual rank value and population density value, b) searching for and keeping better-approximated Pareto points by diffusion and elitism schemes, and c) preventing harmful individuals by introducing a "forbidden region" concept. Three essential techniques were applied in this technique.

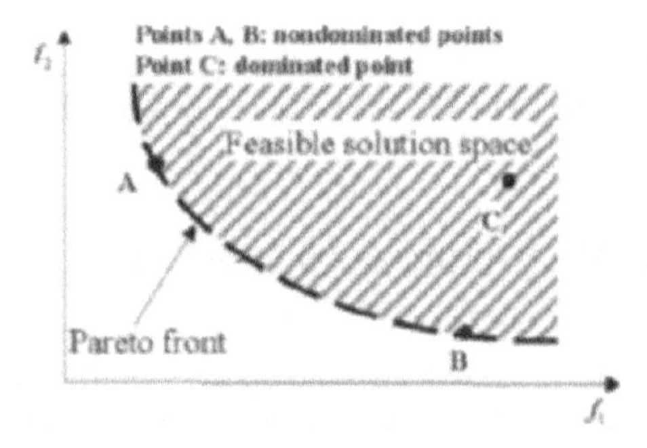

Fig. 10.3. Graphical illustration of the Pareto optimality.

Automatic Accumulated Ranking Strategy (AARS)

In HRDGA, an Automatic Accumulated Ranking Strategy (AARS) is applied to calculate the Pareto rank value, which represents the dominated relationship among individuals. In AARS, an individual's rank value is defined as the summation of the rank values of the individuals that dominate it. Assume at generation t, individual y is dominated by $p^{(t)}$ individuals $y_1, y_2, \cdots, y_{p^{(t)}}$, whose rank values are already known as $\text{rank}(y_1, t)$, $\text{rank}(y_2, t)$, $\cdots$, $\text{rank}(y_{p^{(t)}}, t)$. Its rank value can be computed by

$$\text{rank}(y, t) = 1 + \sum_{j=1}^{p^{(t)}} \text{rank}(y_j, t). \tag{10.5}$$

Therefore, by AARS, all the non-dominated individuals are still assigned rank value 1, while dominated ones are penalized to reduce the population density and redundancy.

Adaptive Density Value Calculation

To maintain the diversity of the obtained Pareto front, HRDGA adopts an adaptive cell density evaluation scheme as shown in Figure 10.4. The cell width in each objective dimension can be computed as :

$$d_i = \frac{\max_{x \in X} f_i(\mathbf{x}) - \min_{x \in X} f_i(\mathbf{x})}{K_i}, \quad i = 1, \cdots, n, \tag{10.6}$$

where d_i is the width of the cell in the ith dimension, K_i denotes the number of cells designated for the ith dimension (i.e., in Figure 10.4, $K_1 = 12$ and $K_2 = 8$), and X denotes the decision vector space. As the maximum and minimum fitness values will change with different generations, the cell size will vary from generation to generation to maintain the accuracy of the density calculation. The density value of an individual is defined as the number of the individuals located in the same cell.

Rank-density Based Fitness Assignment

Because rank and density values represent fitness and population diversity, respectively, the new rank-density fitness formulation can convert any multiobjective optimization problem into a bi-objective optimization problem. Here, population rank and density values are designated as the two fitness values for GA to minimize. Before fitness evaluation, the entire population is divided into two subpopulations with equal sizes; each subpopulation is filled with individuals that are randomly chosen from the current population according to rank and density value, respectively. Afterwards, the entire population is shuffled, and crossover and mutation are then performed.

For crossover, the parent selection and replacement schemes are borrowed from Cellular GA [14] to explore the new search area by "diffusion." For each subpopulation, a fixed number of parents are randomly selected for crossover. Then, each selected parent performs crossover with the best individual (the one with the lowest rank value) within the same cell and the nearest neighboring cells that contain individuals. If one offspring produces better fitness (a lower rank value or a lower population density value) than its corresponding parent, it replaces its parent. The replacement scheme of the mutation operation is analogous.

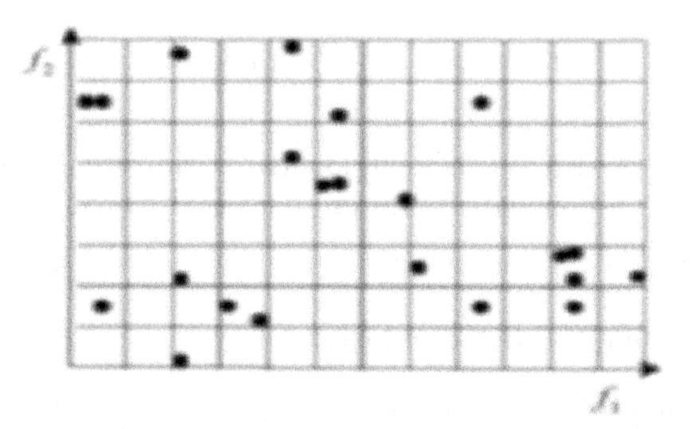

Fig. 10.4. Density map and density grid.

Meanwhile, we take the minimization of the population density value as one of the objectives. It is expected that the entire population will move toward an opposite direction to the Pareto front where the population density value is being minimized. Although moving away from the true Pareto front can reduce population density value, obviously, these individuals are harmful to the population to converge to the Pareto front. To prevent "harmful" offspring surviving and affecting the evolutionary direction and speed, a *forbidden region* concept is proposed in the replacement scheme for the density subpopulation, thereby preventing the "backward" effect. The *forbidden region* includes all the cells dominated by the selected parent. The offspring located in the forbidden region will not survive in the next generation, and thus the selected parent will not be replaced. As shown in Figure 10.5, suppose our goal is to minimize objectives f_1 and f_2, and a resulting offspring of the selected parent $p$is located in the forbidden region. This offspring will be eliminated even if it reduces the population density, because this kind of offspring has the tendency to push the entire population away from the desired evolutionary direction.

Finally, the simple elitism scheme [7] is also applied as the bookkeeping for storing the Pareto individuals obtained in each generation. These individuals are compared to achieve the final Pareto front after the evolution process has stopped.

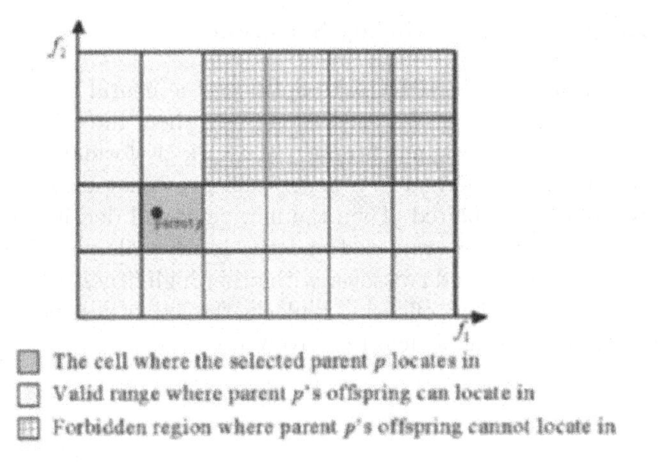

Fig. 10.5. Illustration of the valid range and the forbidden region.

10.5.3 HRDGA for NN Design

To assist RBF network design, HRDGA is applied to carry out the fitness evaluation and mating selection schemes. The HRDGA operators are designed as followed.

Chromosome Representation

In HRDGA, each individual (chromosome) represents a candidate neural network. The control genes are binary bits (0 or 1). For the weight and center genes, real values are adopted as the gene representation to reduce the length of the chromosome. The population size is fixed and chosen *ad hoc* by the difficulty of the problem to be solved.

Crossover and Mutation

We used one-point crossover in the control gene segments and two-point crossover in the other two gene segments. The crossover points were randomly selected and the crossover rates were chosen to be 0.8, 0.7, and 0.7 for the control, weight, and center genes, respectively. One-point mutation was applied in each segment. In the control gene segment, common binary value mutation was adopted. In the weight and center gene segments, real value mutation was performed by adding aGaussian$(0,1)$, which denotes a Gaussian function with zero mean and unit variance. The mutation rates were set to be 0.1, 0.05, and 0.05 for the control, weight, and center genes, respectively.

Fitness Evaluations and Mating Selection

Since we are trying to use HRDGA to optimize the neural network topology along with its performance, we need to convert them into the rank-density domain. Therefore, the original fitness— network performance and number of neurons—of each individual in a generation is evaluated and ranked, and the density value is calculated. Then the new rank and density fitness values of each individual will be evaluated and the individuals with higher fitness measures will reproduce and crossover with other high fitness individuals with a certain probability. Their offspring replaces the low fitness parents forming a new generation. Mating is then iteratively processed.

Stopping Criteria

When the desired number of generations is met, the evolutionary process stops.

10.6 Mackey-Glass Chaotic Time Series Prediction

Since the proposed HRDGA is designed to evolve the neural network topology together with its best performance, it proves useful in solving complex problems such as time series prediction or pattern classification. For a feasibility check, we use the HRDGA assisted NN design to predict the Mackey-Glass chaotic time series.

10.6.1 Mackey-Glass Time Series

The Mackey-Glass time series is a continuous time-delay data differential equation:

$$\frac{d(x(t))}{d(t)} = \frac{a\, x(t-\tau)}{(1+x^c(t-\tau))} - b\, x(t). \tag{10.7}$$

The chaotic behavior of the Mackey-Glass time series is determined by the delay parameter τ. Some examples are listed in Table 10.1. Larger values of τ produce more chaotic dynamics which are much more difficult to predict. Here we assign $a = 0.2$, $b = 0.1$ and $c = 10$ for Equation (10.7). In this study, we used HRDGA evolved neural networks to predict a chaotic Mackey-Glass time series with $\tau = 150$. The network is set to predict $x(t+6)$ based on $x(t)$, $x(t-6)$, $x(t-12)$, and $x(t-18)$.

In the proposed HRDGA, 150 initial center genes are selected, 150 control genes and 150 weight genes are initially generated as well. Population size was set to be 400. For comparison, we applied three other center selection methods—KNN (K-Nearest Neighbor) [11], GRNN (Generalized Regression

Table 10.1. Characteristics of Mackey-Glass time series

delay parameter τ	Chaotic characteristics
$\tau < 4.53$	A stable fixed point attractor
$4.53 < \tau < 13.3$	A stable limit cycle attractor
$13.3 < \tau < 16.8$	Period limit cycle doubles
$\tau > 16.8$	Chaotic attractor characterized by τ

Neural Network) [20], and OLS (Orthogonal Least Square Error) [4] methods on the same time series prediction problem. For KNN and GRNN types of networks, 70 networks are generated with the neuron numbers increasing from 11 to 80 with the step equals to one. Each of these networks will be trained by KNN and GRNN methods, respectively, and the stopping criterion is the same with the one used in HRDGA. For the OLS method, the selection of the tolerance parameter ρ determines the trade-off between the performance and complexity of the network. Ideally, ρ should be larger than, but very close to, the ratio $\sigma_\varepsilon^2/\sigma_d^2$, where σ_ε^2 is the variance of the residuals, and σ_d^2 is the variance of the desired output. A smaller ρ value will produce a neural network with more neuron number, whereas a larger ρ value generally results in a network with less number of neurons. Therefore, by using different ρ values, we generated a group of neural networks with various training performances and numbers of hidden neurons. For the given Mackey-Glass time series prediction problem, we selected 100 different ρvalues, which are from 0.01 to 0.4 with the step size of 0.01. The stopping criteria for KNN, GRNN, and OLS algorithms is either the epochs exceed 5,000, or the training Sum Square Error (SSE) between two sequential generations is smaller than 0.01. For HRDGA, the stopping generation is set to be 5,000. We used the first 250 seconds of the data as the training data set, and then the data from 250 – 499, 500 – 749, 750 – 999, and 1,000 – 1,249 seconds were used as the corresponding test data sets to be predicted by four different approaches. Each approach runs 30 times with different parameter initializations to obtain the average results. Figure 10.6(a) shows the resulting average training SSEs of neural networks with different number of hidden neurons by four training approaches. Figure 10.6(b) shows the approximated Pareto fronts (i.e., non-dominated sets) by the selected four approaches. Figure 10.7(a) shows the average test SSEs of the resulting networks by using the first test data set for each approach, and Figure 10.7(b) shows their corresponding Pareto fronts. Furthermore, Figures 10.8, 10.9 and 10.10 show the same types of results by using the second, third, and fourth test data, respectively.

Table 10.2 shows the best training and test performances and their corresponding numbers of hidden neurons. From Figures 10.6- 10.10, we can see, comparing to KNN and GRNN, HRDGA and OLS algorithms have much smaller training and test errors for the same network structures. KNN trained networks produce the worst performances, because the RBF centers of the

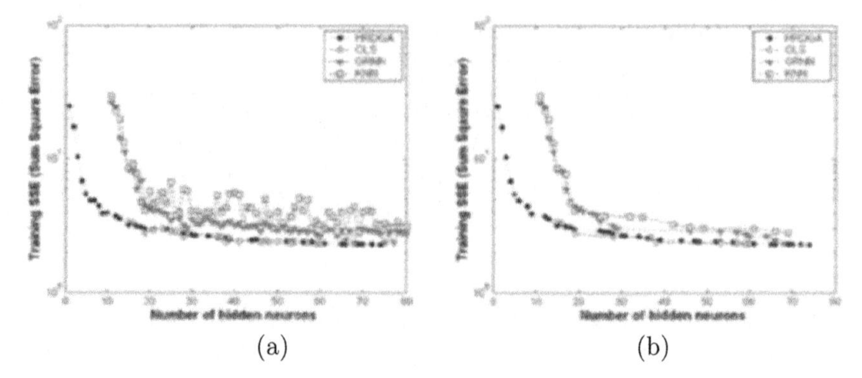

(a) (b)

Fig. 10.6. (a) Training performances for the resulting neural networks with different number of hidden neurons and (b) The corresponding Pareto fronts (non-dominated sets).

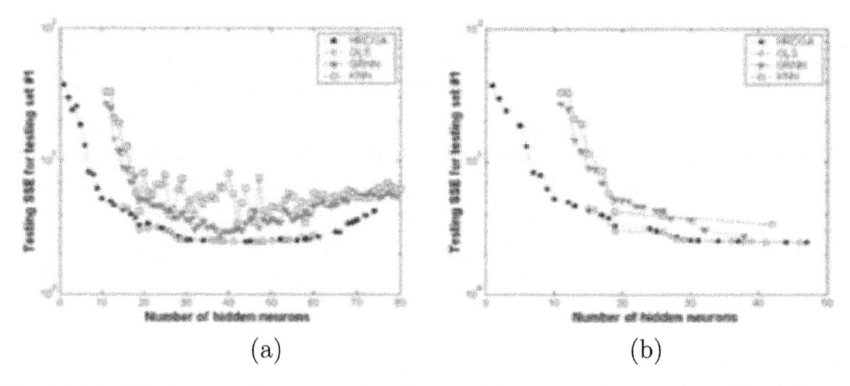

(a) (b)

Fig. 10.7. (a) Test performances for the resulting neural networks with different number of hidden neurons and by using test set #1 (b) The corresponding Pareto fronts (non-dominated sets).

Table 10.2. Comparison of best performance (SSE) and structure (number of neurons) between KNN, OLS, GRNN and HRDGA

	Training set		**Test set #1**		**Test set #2**		**Test set #3**		**Test set #4**	
	SSE	no.	SSE	no.	SSE	no.	SSE	no.	SSE	no.
KNN	2.8339	69	3.3693	42	3.4520	42	4.8586	48	4.8074	19
GRNN	2.3382	68	2.7720	38	3.0711	43	2.9644	40	3.2348	37
OLS	2.3329	60	2.4601	46	2.5856	50	2.5369	37	2.7199	54
HRDGA	2.2901	74	2.4633	47	2.5534	52	2.5226	48	2.7216	58

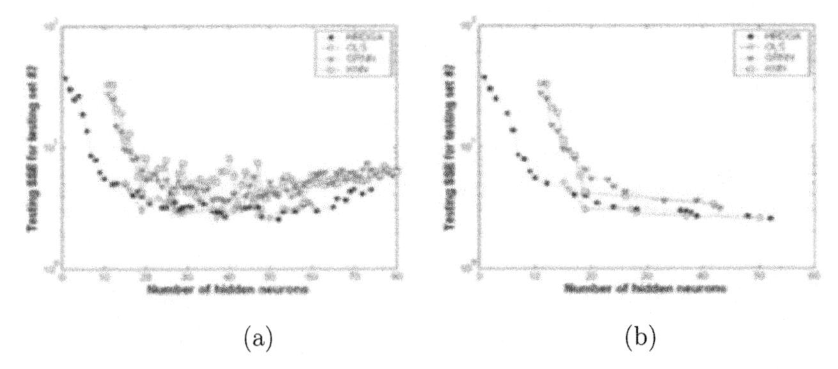

(a) (b)

Fig. 10.8. (a) Test performances for the resulting neural networks with different number of hidden neurons and by using test set #2 (b) The corresponding Pareto fronts (non-dominated sets).

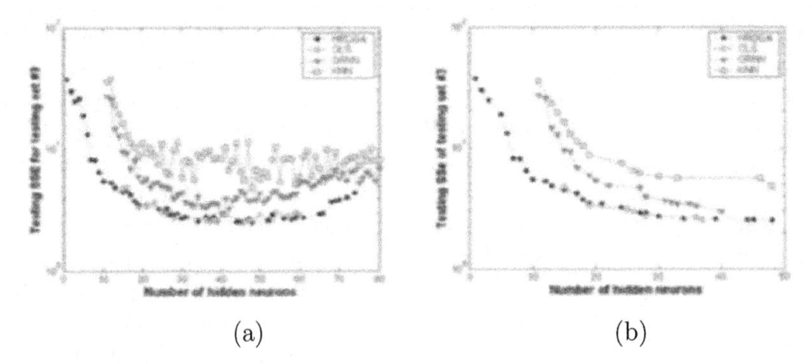

(a) (b)

Fig. 10.9. (a) Test performances for the resulting neural networks with different number of hidden neurons and by using test set #3 (b) The corresponding Pareto fronts (non-dominated sets).

KNN algorithm are randomly selected, which make KNN to achieve only a "local optimum" solution. Since GA always seeks "global optimum", and the orthogonal result is near optimal, the performances of OLS are comparable to HRDGA.

Moreover, from Figure 10.6, we can see that when the network complexity increases, the training error decreases. This phenomenon can be observed from the results by all of the selected training approaches. However, this phenomenon is only partially maintained for the relationship between the test performances and the network complexity. Before the number of hidden neurons reaches a certain threshold, the test error still decreases as the network complexity increases. After that, the test error has the tendency to fluctuate even when the number of hidden neurons increases. This occurrence can be considered as that the resulting networks are overfitted. The network with the

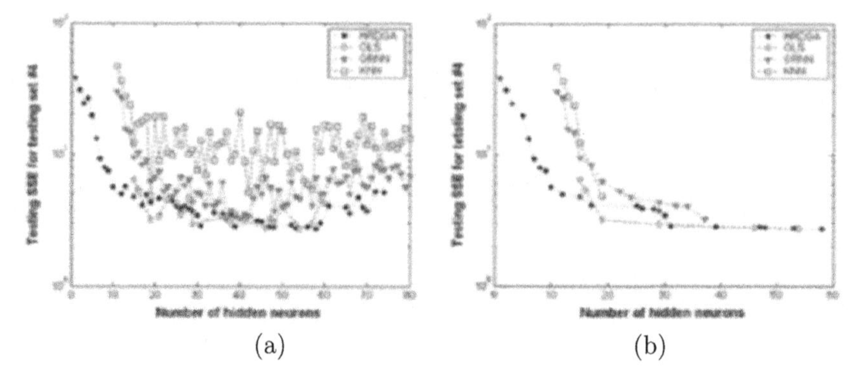

(a) (b)

Fig. 10.10. (a) Test performances for the resulting neural networks with different number of hidden neurons and by using test set #4 (b) The corresponding Pareto fronts (non-dominated sets).

best test performance before overfitting occurs is called the *optimal* network and is judged as the final single solution by conventional NN design algorithms. However, from Figures 10.6– 10.10 and Table 10.1, it is very difficult to find a single *optimal* network that can offer the best performances for all the test data sets, since these data sets possess different traits. Therefore, instead of searching for a single *optimal* neural network, an algorithm that can result in a near-complete set of near-optimal networks can be a more reasonable and applicable option. This is the essential reason that multiobjective genetic algorithms can be justified for this type of neural network design problems.

From the simulation results, althoughK KNN and GRNN approaches did not provide better training and test results comparing to the other two approaches, they have the advantage that the designer can control the network complexity by increasing or decreasing the neuron numbers at will. On the other hand, although the OLS algorithm always provides near-optimal network solutions with good training and test performance, it also has serious problem to generate a set of network solutions in that the designers cannot manage the network structure directly. The trade-off characteristic between network performance and complexity totally depends on the value of tolerance parameterρ. Same ρ value means completely different trade-off features for different NN design problems. In addition, as shown in Figure 10.11, the relationship between ρ value and network topology is a nonlinear, many-to-one mapping, which may cause a redundant computation effort in order to generate a near-complete neural network solution set. Compared with the other three training approaches, HRDGA does not have problems in designing trade-off parameters, because it treats each objective equally and independently, and its population diversity preserving techniques help it build a near-uniformly distributed non-dominated solution set.

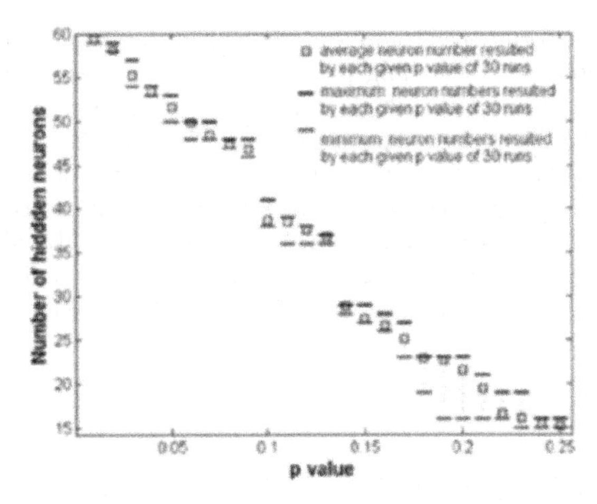

Fig. 10.11. Relationship between ρ values and network complexity.

Therefore, comparing to the other three traditional training approaches, the proposed HRDGA algorithm offers several benefits for the neural network design problems in terms of:

1. providing a set of candidate solutions, which is caused by GA's population-based optimization capability and the definition of Pareto optimality;
2. presenting competitive or even superior individuals with high training and test performances. This is resulted from GA's feature of seeking "global optimum" and HRDGAs' Pareto ranking technique; and
3. offering a near-complete, non-dominated set, and long-extended Pareto front, which is originated from HRDGA's population diversity keeping design that can be found in AARS, density preserving technique, and the concept of "forbidden region".

10.7 Conclusions

In this study, we propose a multiobjective genetic algorithm based design procedure for the radial-basis function neural network. A Hierarchical Rank Density Genetic Algorithm (HRDGA) is developed to evolve both the neural network's topology and parameters simultaneously. Instead of producing a single solution, HRDGA provides a set of near-optimal neural networks from the perspective of Pareto optimality to the designers so that they can have more flexibility for the final decision-making based on certain preferences. From the results presented above, HRDGA shows potential in estimating neural network topology and weighting parameters for complex problems when a heuristic estimation of the neural network structure is not readily available. For the given Mackey–Glass chaotic time series prediction, HRDGA shows competitive, or even superior performances comparing with the other three selected training algorithms in terms of searching for a set of non-dominated, near-complete neural network solutions with high training and test performances. While we considered radial-basis function neural networks, the proposed hierarchical genetic algorithm may be easily extended to the designs of other neural networks (i.e., feed-forward, feedback, or self-organized). In addition, as some of the traditional neural network training approaches (i.e., OLS algorithm) also provide competitive results, a hybrid algorithm that synergistically integrates traditional training method with the proposed algorithm can be a promising future work.

References

[1] R.K. Belew, J. McInerney, N.N. Schraudolph. Evolving networks: Using genetic algorithms with connectionist learning, *CSE Technical Report*, University of California at Dan Diego, CS90-174, 1990

[2] A.W. Bruce, D.C. Timothy. Cooperative-competitive genetic evolution of radial basis function centers and widths for time series prediction. *IEEE Transactions Neural Networks*, 7:869-880, 1996

[3] E.J. Chang, R.P. Lippmann. Using genetic algorithms to improve pattern classification performance. *Neural Information Processing Systems*, 797–803, 1991

[4] S. Chen, C.F. Cowan, P.M. Grant. Orthogonal least square learning algorithm for radial basis function networks. *IEEE Trans. Neural Networks*, 2:302–309, 1991

[5] D. Dasgupta, D.R. McGregor. Designing application-specific neural networks using the structured genetic algorithm. In: *Proc. Int. Workshop on Combinations of Genetic Algorithms and Neural Networks*, pp.87–96, 1992

[6] A. Doering, M. Galicki, H. Witte. Structure optimization of neural networks with the A*-Algorithm. *IEEE Trans. Neural Networks*, 8:1434–1445, 1997

[7] C.M. Fonseca, P.J. Fleming. An overview of evolutionary algorithms in multiobjective optimization. *Evolutionary Computation*, 3:1–16, 1995

[8] X.M. Gao, S.J. Ovaska, Z.O. Hartimo. Speech signal restoration using an optimal neural network structure. In: *Proc. of the IEEE Int. Conf. Neural Networks*, pp.1841–1846, 1996

[9] S. Geman, E. Bienenstock , R. Dousat. Neural networks and the bias/variance dilemma. *Neural Computation*, 2:303–314, 1989

[10] C.L. Hwang, A.S.M. Masud. Multiple objective decision making—methods and applications. Springer, Berlin

[11] T. Kaylani, S. Dasgupta. A new method for initializing radial basis function classifiers. In: *Proc. IEEE Int. Conf. Systems, Man, and Cybernetics*, pp.2584–2587, 1994

[12] T.Y. Ke, K.S. Tang, K.F. Man, P.C. Luk. Hierarchical genetic fuzzy controller for a solar power plant. In: *Proc. IEEE Int. Symp. Industrial Electronics*, pp.584–588, 1998

[13] J.D. Kelly, L. Davis. Hybridizing the genetic algorithm and the K-nearest neighbors classification algorithm. In: *Proc. Intl. Conf. Genetic Algorithms*, pp.377–383, 1991

[14] T. Krink, R.K. Ursem. Parameter control using agent based patchwork model. In: *Proc. IEEE Congress on Evolutionary Computation*, pp.77–83, 2000

[15] S. Lucas. Specifying intrinsically adaptive architectures. In: *Proc. 1st IEEE Symp. Combination of Evolutionary Computation and Neural Networks*, pp.224–231, 2000

[16] G.F. Miller, P.M. Todd, S.U. Hedge. Designing neural networks using genetic algorithms. In: *Proc. 3rd Int. Conf. Genetic Algorithms*, pp.379–384, 1989

[17] S.W. Moon, S.G. Kong. Block-based neural network. *IEEE Transactions on Neural Networks*, 12:307–317, 2001

[18] A.K. Morales. Non-standard norms in genetically trained neural networks. In: *Proc. IEEE Symp. Combination of Evolutionary Computation and Neural Networks*, pp.43–51, 2000

[19] N. Murata, S. Yoshizawa, S. Amari. Network information criterion – determining the number of hidden units for an artificial neural network model. *IEEE Transactions on Neural Networks*, 5:865–872, 1994

[20] P.D. Wasserman. Advanced Method in Neural Computing. VNR, New York, 1993

[21] D. Whitley, T. Starkweather, C. Bogart. Genetic algorithms and neural networks: optimizing connections and connectivity. *Parallel Computting*, 14:347–361, 1990

[22] X. Yao. Evolving artificial neural network. *International Journal of Neural Systems*. 4:203–222, 1993

[23] G.G. Yen, H. Lu. Hierarchical genetic algorithm based neural network design. In: *Proc. IEEE Symp. Combination of Evolutionary Computation and Neural Networks*, pp.168–175, 2000

[24] B. Zhang, D. Cho. Evolving neural trees for time series prediction using Bayesian evolutionary algorithms. In: *Proc. IEEE Symp. Combination of Evolutionary Computation and Neural Networks*, 17–23, 2000

[25] E. Zitzler, L. Thiele. Multiobjective evolutionary algorithms: A comparative case study and the strength Pareto approach. *IEEE Trans. Evolutionary Computation*, 3:257–271, 1999

11

Minimizing Structural Risk on Decision Tree Classification

DaeEun Kim

Cognitive Robotics
Max Planck Institute for Human Cognitive & Brain Sciences
Munich, 80799, Germany
daeeun@cbs.mpg.de

Summary. Tree induction algorithms use heuristic information to obtain decision tree classification. However, there has been little research on how many rules are appropriate for a given set of data, that is, how we can find the best structure leading to desirable generalization performance. In this chapter, an evolutionary multi-objective optimization approach with genetic programming will be applied to the data classification problem in order to find the minimum error rate or the best pattern classifier for each size of decision trees. As a result, we can evaluate the classification performance under various structural complexity of decision trees. Following structural risk minimization suggested by Vapnik, we can determine a desirable number of rules with the best generalization performance. The suggested method is compared with C4.5 application for machine learning data.

11.1 Introduction

The recognition of patterns and the discovery of decision rules from data examples is one of the challenging problems in machine learning. When data points with numerical attributes are involved, the continuous-valued attributes should be discretized with threshold values. Decision tree induction algorithms such as C4.5 build decision trees by recursively partitioning the input attribute space [24]. Thus, a conjunctive rule is obtained by following the tree traversal from the root node to each leaf node. Each internal node in the decision tree has a splitting criterion or threshold for continuous-valued attributes to partition a part of the input space, and each leaf represents a class depending on the conditions of its parent nodes.

The creation of decision trees often relies on heuristic information such as information gain measurement. Yet how many nodes are appropriate for a given set of data has been an open question. Mitchell [22] showed the curve of the accuracy rate of decision trees with respect to the number of nodes over the independent test examples. There exists a peak point of the accuracy rate in a certain size of decision trees; a larger size of decision trees can

D.E. Kim: *Minimizing Structural Risk on Decision Tree Classification*, Studies in Computational Intelligence (SCI) **16**, 241–260 (2006)
www.springerlink.com

increase its classification performance on the training samples but reduces the accuracy over the test samples which have not been seen before. This problem is related to the overfitting problem to increase the generalization error[1]. Many techniques such as tree growing with stopping criterion, tree prunning or bagging [25, 21, 24, 5] have been studied to reduce the generalization error. However, the methods are dependent upon a heuristic information or measure to estimate the generalization error, and they do not explore every size of trees.

An evolutionary approach to decision trees has been studied to obtain optimal classification performance [16, 15, 4], since the decision tree based on heuristics is not optimal in structure and performance. Freitas et al. [15] have shown evolutionary multi-objective optimization to obtain both the minimum error rate and minimum size of trees. Their method was based on the information gain measurement; it followed the C4.5 splitting method and selected the attributes with genetic algorithms. They were able to reduce the size of decision trees, but had higher test error rates than C4.5 in some data sets. Recently a genetic programming approach with evolutionary multi-objective optimization (EMO) was applied to decision trees [4, 3]. A new representation of decision trees for genetic programming was introduced [3], where the structure of decision trees is similar to linear regression trees [6]. Two objectives, tree size and accuracy rate in data classification, were considered in the method. The method succeeded in reducing both error rates and size of decision trees in some data sets. However, searching for the best structure of decision trees has not been considered in their works.

It has been shown that EMO is very effective for optimization of multi-objectives or constraints in continuous range [27, 7]. Also EMO is a useful tool even when the best performance for each discrete genotype or structure should be determined [19, 17]. The EMO approach was used to minimize the training error and the tree size for decision tree classification [3, 8]. Also other works using fitness and size or complexity as objectives have been reported [2, 20]. Yet there has been no effort so far to find what is the best structure of decision trees to have the minimal generalization error. Vapnik [26] showed an analytic study to reduce the generalization error, and he suggested the structural risk minimization to find the best structure. It can be achieved by exploring the empirical error (training error) and generalization error (test error) for various structure complexity. We will follow the approach and the tree size will be the parameter to control the structure.

In this work, the EMO with genetic programming for two objectives, the tree size and the training error, is first used to obtain the Pareto-optimal solutions, that is, the minimum training error rate for each size of trees. Then the best tree for each size will be examined to see the generalization performance for a given set of test data. By observing the distribution of the test error rates over the size of trees, we can pinpoint the best structure to minimize the generalization error. In our EMO approach, a special elitism strategy

[1] This is also called test error in this chapter.

for discrete structures is applied. Genetic programming evolves decision trees with variable thresholds and attributes, and an incremental evolution from small to large structures with Pareto ranking is used. The suggested method provides the accuracy rate of classification for each size of trees as well as the best structure of decision trees. The approach will be compared with the tree induction algorithm C4.5. A preliminary study of the approach was published in [18].

11.2 Method

11.2.1 Decision Tree Classification

Decision tree learning is a popular induction algorithm. A decision tree classifies data samples by a set of decision classifiers; the classifiers are located in the internal nodes of the tree and for each instance, the tree traversal depending on the decision of the classifiers from the root node to some leaf node determines the corresponding class. The decision trees can be easily represented as a set of decision rules (if-then rules) to assist the interpretation. Inductive learning methods create such decision trees, often based on heustic information or statistical probability.

One of the most popular learning algorithm is to construct decision trees from the root node to leaf nodes with a top-down greedy method [25, 24]. Each data attribute (with appropriate threshold if the attribute is continuous-valued) is evaluated using the information theory with entropy. This evaluation decides which attribute or what threshold of the selected attribute classifies well a given set of instances. It has been shown that information gain measure efficiently selects one of attribute vectors and its thresholds [23].

Let Y be the set of examples and let C be the set of k classes. Let $p(C_i, Y)$ be the probability of the examples in Y that belong to class C_i. The split probability $p(Y_j)$ in a continuous-valued attribute among m partitions is given as the probability of the examples that belong to the partition Y_j when the range of the attribute is divided into several regions.

Then the information gain of an attribute A over a collection of instances Y is defined as

$$Gain(Y, A) = \frac{E(Y) - \sum_{i=1}^{m} \frac{|Y_i|}{|Y|} E(Y_i)}{\sum_{j=1}^{m} p(Y_j) \log p(Y_j)}$$

where $E(Y) = \sum_{i=1}^{k} p(C_i, Y) \log p(C_i, Y)$ is an entropy function, A has m partitions, and Y_j is one of m partitions for the attribute A.

For a given set of instances, each attribute A has its own threshold τ_i that produces the greatest information gain. This threshold is automatically selected by information gain measurement, and the threshold τ_i is one of the best cut for the attribute A to make good decision boundaries. One decision

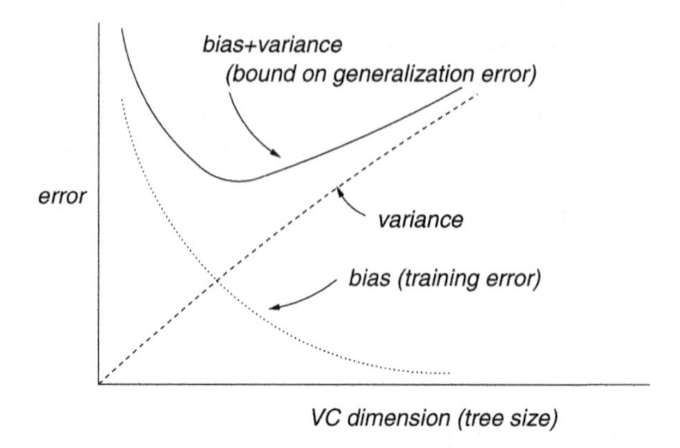

Fig. 11.1. Relationship between training error and generalization error

boundary divides the parameter space into two non-overlapping subsets, depending on τ_i; it has a dichotomy $A > \tau_i$ and $A \leq \tau_i$. However, multiple intervals of the attribute can improve the classification [10].

With the above process of information gain measurement, the decision tree algorithm finds the best attribute and its threshold. More sophisticated algorithms to improve the classification have been developed, but the style of tree induction is not much changed. In our experiments, C4.5 [24] will be used for inductive tree classication.

11.2.2 Structural Risk Minimization

According to the statistical estimation theory by Vapnik [26], while the complexity of a model over a given set of data increases, learning algorithms such as decision trees and neural networks can reduce the approximation error called bias but increase the variance of the model. Much research is concerned with reducing the generalization error which is a combination of bias and variance terms [12, 14, 13]. The generalization error is the rate of errors caused by the model when the model is tested on samples which have not been seen before. Vapnik showed the general bounds for the variance and the generalization error. The generalization error varies with respect to a control parameter of the learning algorithm to model a given set of data; Vapnik [26] mentioned this control parameter as VC (Vapnik-Chervonenkis) dimension. The VC dimension is a measure of the capacity of a set of classification functions.

The number of decision-tree nodes in induction trees can be a control parameter to be related to the generalization error, because increasing the number of nodes can decrease the bias and increases the variance in classification problem. Fig. 11.1 shows the relationship between the training error and generalization error. Here, we are interested in finding a tree structure with minimal generalization error at the expense of increase in training error.

In structural risk minimization, a hierarchical space of structures is enumerated and then the function to minimize the empirical risk (training error, bias) for each structure space is found. Among a collection of those functions, we can choose the best model function to minimize the generalization error. The number of leaf nodes (rules) in decision trees corresponds to the VC-dimension that Vapnik mentioned in the structural risk minimization [26].

The structure of decision trees can be specified by the number of leaf nodes. Thus, we define a set of pattern classifiers as follows:

$$S_k = \{F(x, \beta) | \beta \in D_k\}$$

where x is a set of input-output vectors, $F(x, \beta)$ is a pattern classifier with parameter vector β, D_k is a set of decision trees with k terminal nodes and S_k is a set of pattern classifiers formed by decision trees with k terminal nodes. Then we have

$$S_1 \subset S_2 \subset \cdots \subset S_n.$$

From the method of structural risk minimization [26], we can easily set the VC dimension into the number of leaf nodes and the VC dimension of each pattern classifier is finite. In this chapter, the training error for each set of pattern classifiers, S_k, for $k = 2, ..., n$, is minimized with the EMO method and then the generalization error for the selected pattern classifier is identified. The best pattern classifier or the best structure of pattern classifiers is the one with the minimum generalization error; in the experiments, 10-fold cross validation will be used to estimate the generalization error.

When we wish to have a desirable set of rules over a given set of data, we do not have a prior knowledge about what is the best number of rules to minimize the generalization error. Thus, a two-phase algorithm with the EMO method can be applied to general classification problems. First, we can apply the EMO method to the whole training instances and obtain a set of rules for each size of trees. Then the method of finding the best structure with 10-fold cross validation or other validation process can be applied to the training instances. From this information, we can decide the best VC-dimension, or best size of trees among a collection of rule sets for the original data set. As a result, we can obtain a desirable set of rules to avoid the overfitting problem.

11.2.3 Evolutionary Multiobjective Optimization

We use evolutionary multiobjective optimization to obtain Pareto-optimal solutions which have minimal training error for each size of trees. In the suggested evolutionary approach, a decision tree is encoded as a genotype chromosome; each internal node specifies one attribute for training instances and its threshold. The terminal node defines a class, depending on the conjunctive conditions of its parent nodes through the tree traversal from the root node to the leaf node. Unlike many genetic programming approaches, the current method encodes only a binary tree classification; the only one function set is a comparison operator for a variable and its threshold, and the terminal set consists of classes determined by decision rules.

The genetic pool in the evolutionary computation handles decision trees as chromosomes. The chromosome size (tree size) is proportional to the number of leaf nodes in a decision tree, that is, the number of rules. Thus, the number of rules will be considered as one objective to be optimized. While an evolutionary algorithm creates a varying size of decision trees, each decision tree will be tested on a given set of data for classification. The classification error rate will be the second objective. The continuous-valued attributes require partitioning into a discrete set of intervals. For simple control of VC dimension, we assume that the decision tree is a binary tree. Thus, the decision tree will have a single threshold for every internal node to partition the continuous-valued range into two intervals. The threshold is one of major components to form a pattern classifier.

We are interested in minimizing two objectives, classification error rate and tree size in a single evolutionary run. In the multi-objective optimization, the rank cannot be linearly ordered. The Pareto scoring in EMO approach has been popular and it is applied to maintain a diverse population over the two objectives. A dominance rank is thus defined in the Pareto distribution. A vector $X = (x_1, x_2, ..., x_m)$ for m objectives is said to *dominate* $Y = (y_1, y_2, ..., y_m)$ (written as $X \prec Y$) if and only if X is partially less than Y, that is,

$$(\forall i \in 1, ..., m, x_i \leq y_i) \wedge (\exists i \in 1, ..., m, x_i < y_i)$$

A *Pareto optimal set* is said to be the set of vectors that are not dominated by any other vector.

$$\{X = (x_1, ..., x_m) | \neg(\exists Y = (y_1, ..., y_m), Y \prec X)\}$$

To obtain a Pareto optimal set, a dominating rank method [11] is applied in this work. Individuals of rank 0 in a Pareto distribution are dominated by no other members and individuals of rank n are dominated only by individuals of rank k for $k < n$. The highest rank is zero, for an element which has no dominator. Fig. 11.2 shows an example of dominating rank method.

In the experiments, tournament selection of group size four is used for Pareto optimization. The tournament selection initially partitions the whole population into multiple groups for the fitness comparison; members in each group are randomly chosen among the population. Inside the tournament group, Pareto score of each member is compared each other and ranked. A higher rank of genomes in the group have more probability of reproducing themselves for the next generation. In our approach, a population is initialized with a random size of tree chromosomes. For each group of four members, the two best chromosomes using a dominating rank method are first selected in a group and then they reproduce themselves; more than one chromosome may have tie rank scores and in this case chromosomes will be randomly selected among multiple non-dominated individuals. A subtree crossover over a copy

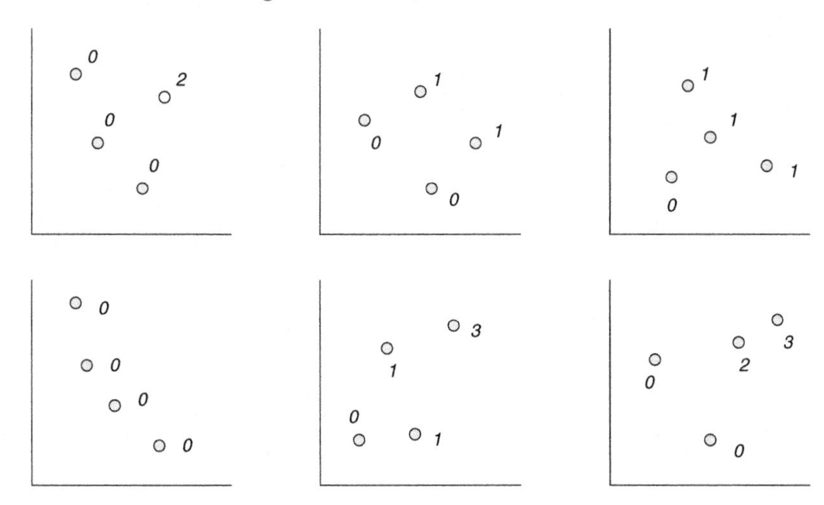

Fig. 11.2. Dominating rank method in a tournament selection of group size four (x-axis and y-axis represent the number of rules and classification error, respectively. each number represents the dominating rank, and a small rank is a better solution since two objectives should be minimized.)

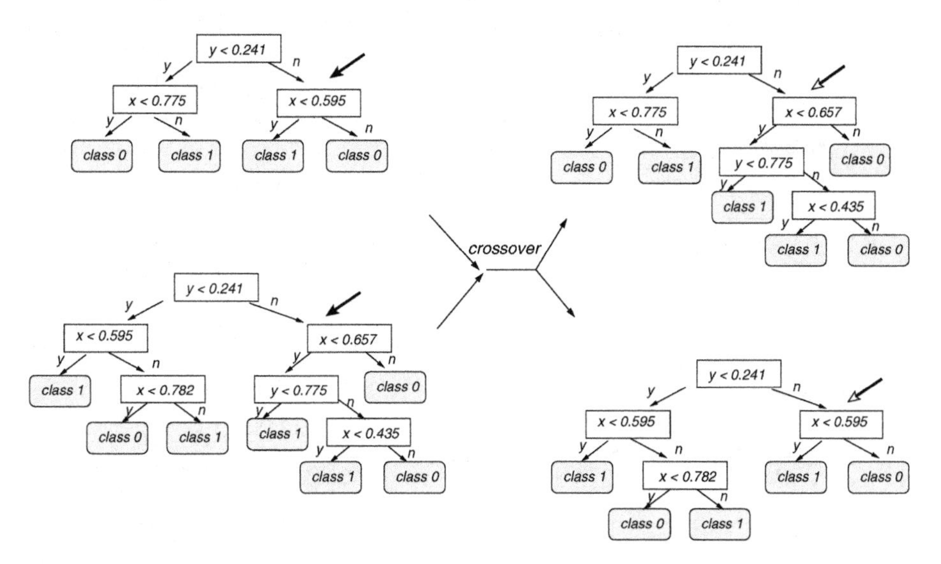

Fig. 11.3. Crossover operator (arrows: crossover point)

of two best chromosomes, followed by a mutation operator, will produce two new offspring. These new offspring replace the two worst chromosomes in the group. The crossover operator swaps subtrees of two parent chromosomes where the crossover point can be specified at an arbitrary branch – see Fig. 11.3.

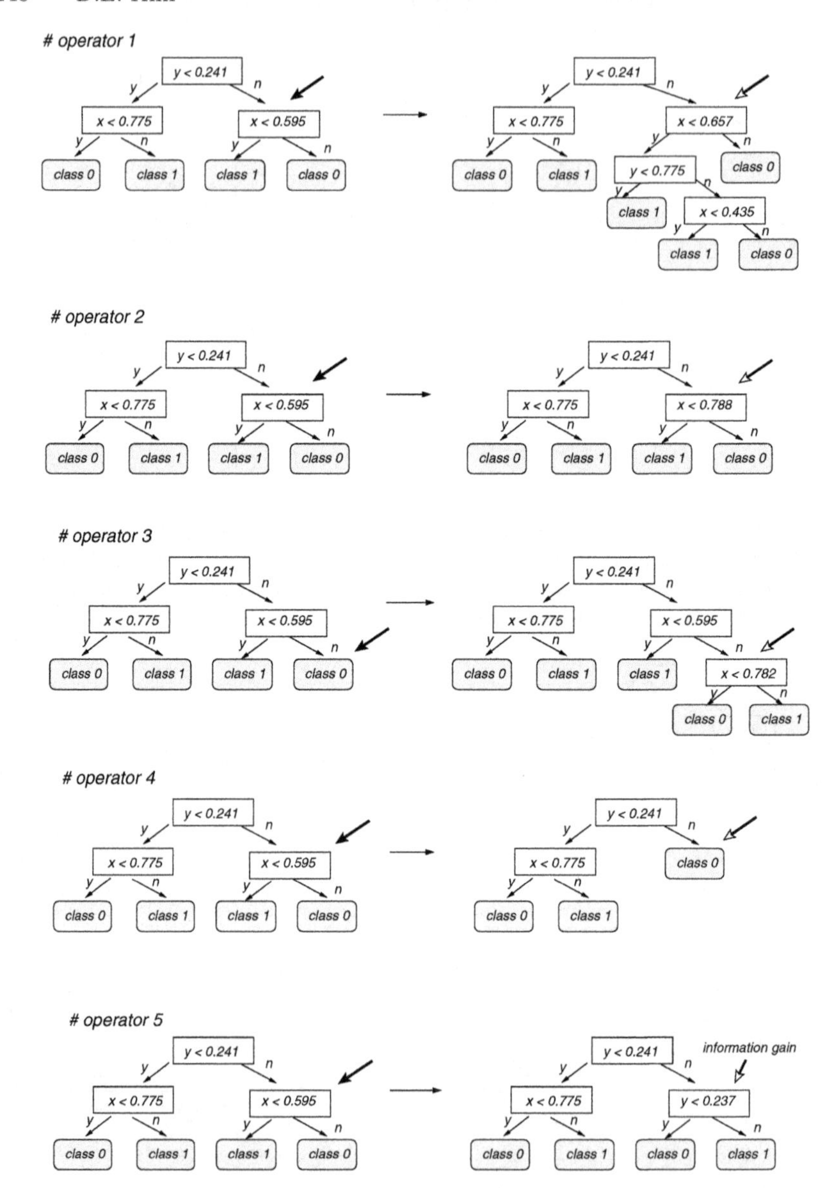

Fig. 11.4. Mutation operators (arrows: mutation point)

The mutation has five different operators as shown in Fig. 11.4. The first operator deletes a subtree and creates a new random subtree. The subtree to be replaced will be randomly chosen in a decision tree. The second operator first picks up a random internal node and then changes the attribute or its threshold. This keeps the parent tree and modifies only one node. The third operator chooses a leaf node and then splits it into two nodes. This will assist incremental evolution by adding one more decision boundary. The fourth operator selects a branch of a subtree and reduces it into a leaf node with random class. It will have the effect of removing redundant subtrees. The fifth operator sets a random attribute in a node and chooses one of the possible candidate thresholds randomly. The candidate thresholds can be obtained at boundary positions[2] by sorting the instances according to the selected variable (the threshold to maximize the information gain is also located at such a boundary position [9]). The last operator has an effect of choosing desirable boundaries based on information gain, but the random selection of the thresholds avoids local optimization only based on information gain. Thus, the last mutation operator[3] accelerates a fast speed of convergence in classification and the other four operators provide a variety of trees in a population. In this work, crossover rate 0.6 and mutation rate 0.2 were used.

In the initialization of the population or the recombination of trees, we have a limit for the tree size. The minimum size of leaf nodes is 2 and the maximum size of leaf nodes is set to 35 or 25; some data set does not need the exploration of as many nodes as 35, because a small number of leaf nodes are sufficient. If the number of leaf nodes in a new tree exceeds the limit, a new random subtree is generated until the limit condition is satisfied. A single run of the EMO method over training examples will lead to the Pareto optimal solutions over classification performance and the number of rules. Each non-dominated solution in a discrete space of tree size represents the minimized error fitness for each number of decision rules. The elitism strategy has been significantly effective for EMO methods [27]. In this work, an elitist pool is maintained, where each member is the best solution for every size of decision trees under progress. For each generation, every member in the elitist pool will be reproduced.

11.2.4 Variable-node C4.5

The well-known tree induction algorithm C4.5 efficiently generates decision trees using information gain. The algorithm can produce a varying size of decision trees by controlling the parameter of a minimum number of object in the branches. When the minimum number of objects in the two branches increases from two to one hundred or more sequentially, we can collect a set of pairs (classification performance, tree size) for each parameter value. That is, we have a collection of pattern classifiers each of which has a different size. Then we extract the best performance as well as the best pattern classifier

---

[2] We first try to find adjacent samples which generates different classification categories and then the middle point of the adjacent samples by the selected attribute is taken as a candidate threshold.

[3] There is a possibility of using only information gain splitting, but we use this method instead to allow more diverse trees in structure and avoid local optimization.

for each size of trees from the database. We will call this method as variable-node C4.5 to distinguish it from the conventional C4.5 method by default parameter setting. With the variable-node C4.5, the best pattern classifiers are obtained by heuristic measurement, while the EMO tries to find the best classifier for a given size of trees using evolutionary search mechanism.

When the cross validation process is applied, we will estimate the average performance over training set and test set with many trials. It is assumed that the average performance over a given size of trees by variable-node C4.5 will represent the estimation over all the best C4.5 decision trees obtained with a given size of trees. In case that a certain size of trees may be missing by the given splitting method, we calculate the average performance of classification for available tree sizes.

11.3 Experiments

11.3.1 EMO with Artificial Data

The EMO method was first tested on a set of artificial data with some noise as shown in Fig. 11.5; we will show an application of the EMO method to minimize the training error, not generalization error to show how it works, and the experiments minimizing the generalization error will be provided in section 3.2. The data set contains 150 samples with two classes. When the C4.5 tree induction program was applied, it generated 3 rules as shown in Fig. 11.5(a). It produced a 29.3 % training error rate (44 errors). For reference, a neural network (seven nodes in a hidden layer) trained with the back-propagation algorithm achieves a 14.7 % error rate (22 example misclassifications); however, the performance could be improved with better parameter setting. Evolving decision trees with 1000 generations and a population size of 200 by the EMO approach produced Pareto trees. Fig. 11.6 shows an example of the best tree chromosomes. With only two rules allowed, 43 examples were misclassified (28.7 % error rate) as shown in Fig. 11.5(f), and it was better than the C4.5 method. Moreover, six rules was sufficient to obtain the same performance as neural networks with seven hidden nodes. As the number of rules increases, decision boundaries are added and the training error performance improves.

In many cases, the best boundaries evolved for a small number of rules also belong to the classification boundaries for a large number of rules; new boundaries can provide better solutions in some cases. A small number of rules are relatively easily evolved, but a large number of rules needs a long time to find the best trees since more rules have more parameters to be evolved. Large trees tend to be evolved sequentially from the base of small trees or a small number of rules. Thus, incremental evolution from small to large structures of decision trees is operated with the EMO approach. We note that in Fig. 11.5, more rules improve the classification performance for the training data by adding decision boundaries, but some rules support only a small number

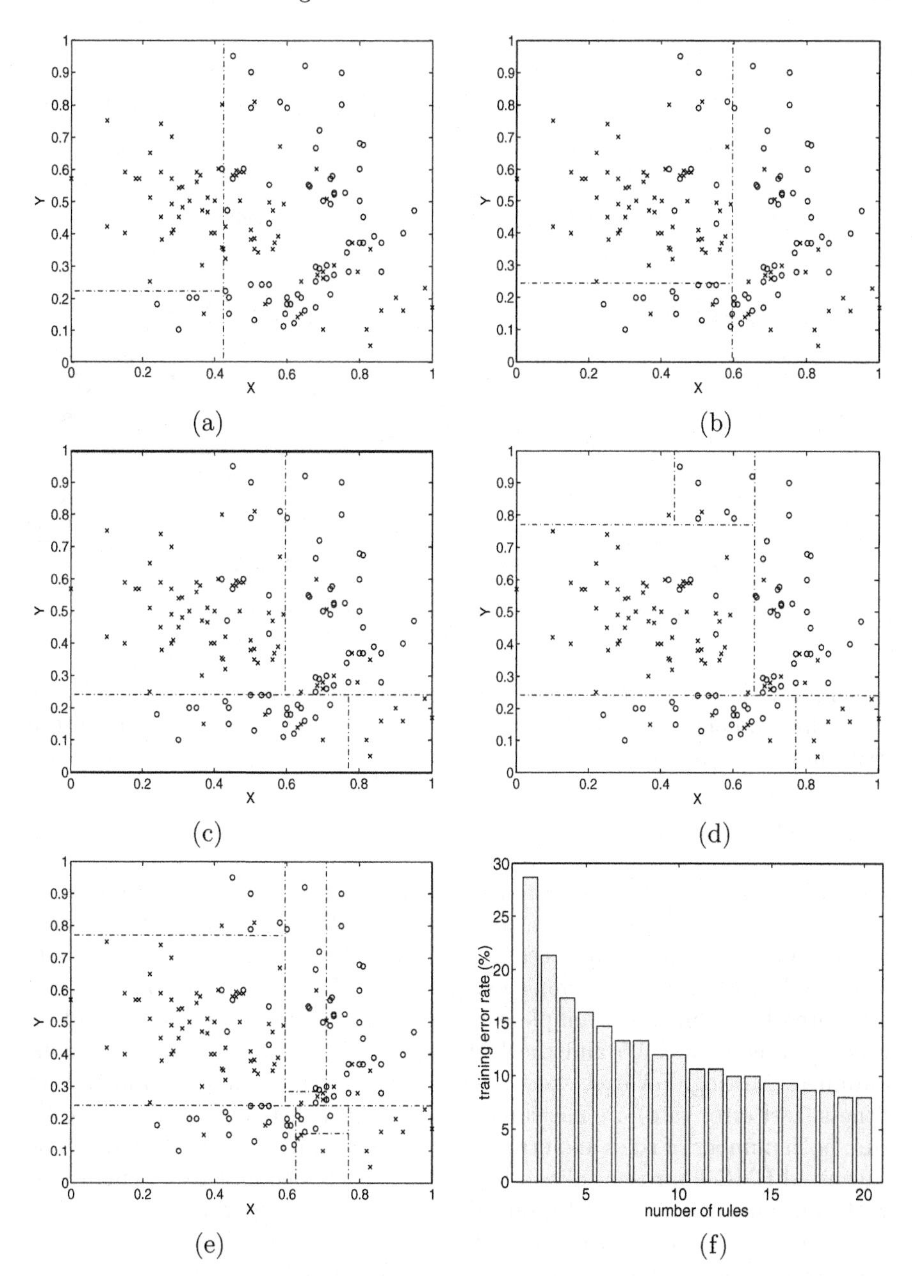

Fig. 11.5. Artificial data and EMO (a) data set and C4.5 decision boundaries ($\circ$: class 0, $\times$: class 1) (b) 3 rules from EMO (c) 4 rules from EMO (d) 6 rules from EMO (e) 9 rules from EMO (f) an example of EMO result

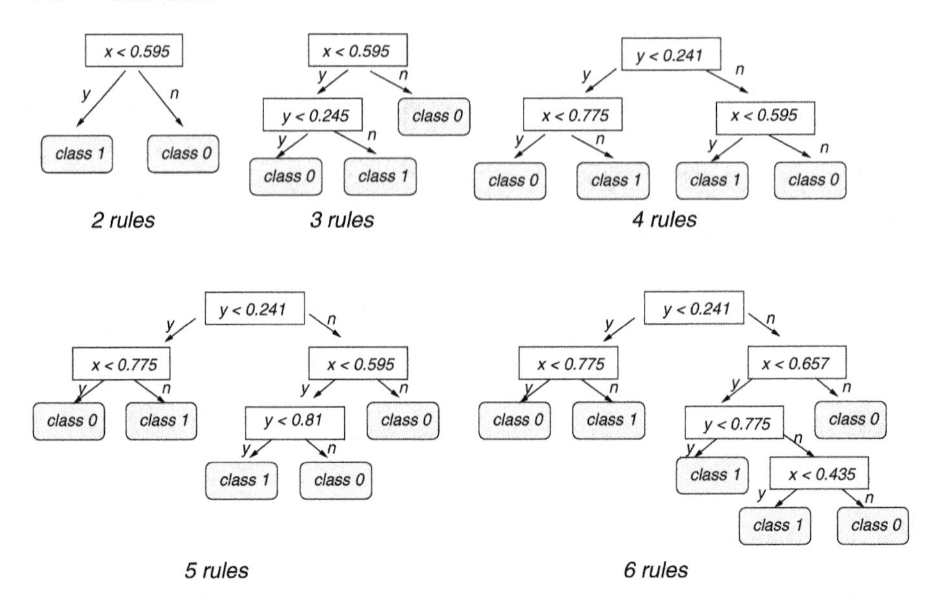

Fig. 11.6. An example of best chromosomes by EMO

of samples. The rules may cause over-specialization problem to degrade the generalization performance. This is the motivation of our work using structural risk minimization.

11.3.2 Machine Learning Data

For the general case, the suggested evolutionary approach has been tested on several sets of data[4] (*iris*, *wine*, *ionosphere*, *ecoli*, *pima*, *wpbc*, *glass*, *bupa*) in the UCI repository [1] and the artificial data in Fig. 11.5. These data sets are mostly for machine learning experiments. Classification error rates are estimated by running the complete 10-fold cross-validation ten times, and we used variable-node C4.5 and the EMO approach as well as C4.5 by default parameter setting. For each size of decision trees, 95% confidence intervals of fitness (test error rate) are measured by assuming t-distribution. For the C4.5 run, both number of rules and error rate will be examined with t statistic. The suggested EMO approach takes a population size of 500 and 1000 generations with tournament selection of group size four for each experiment.

Evolutionary computation was able to attain a hierarchy of structure for classification performance. There exists the best number of rules to show the minimum generalization error as expected from the structural risk minimization. Fig. 11.7(a)-(b) shows that the artificial data have four decision rules as the best structure of decision trees and that the *ionosphere* data have six

[4] Some sets of data include missing attribute values. In that case, the data sample is removed.

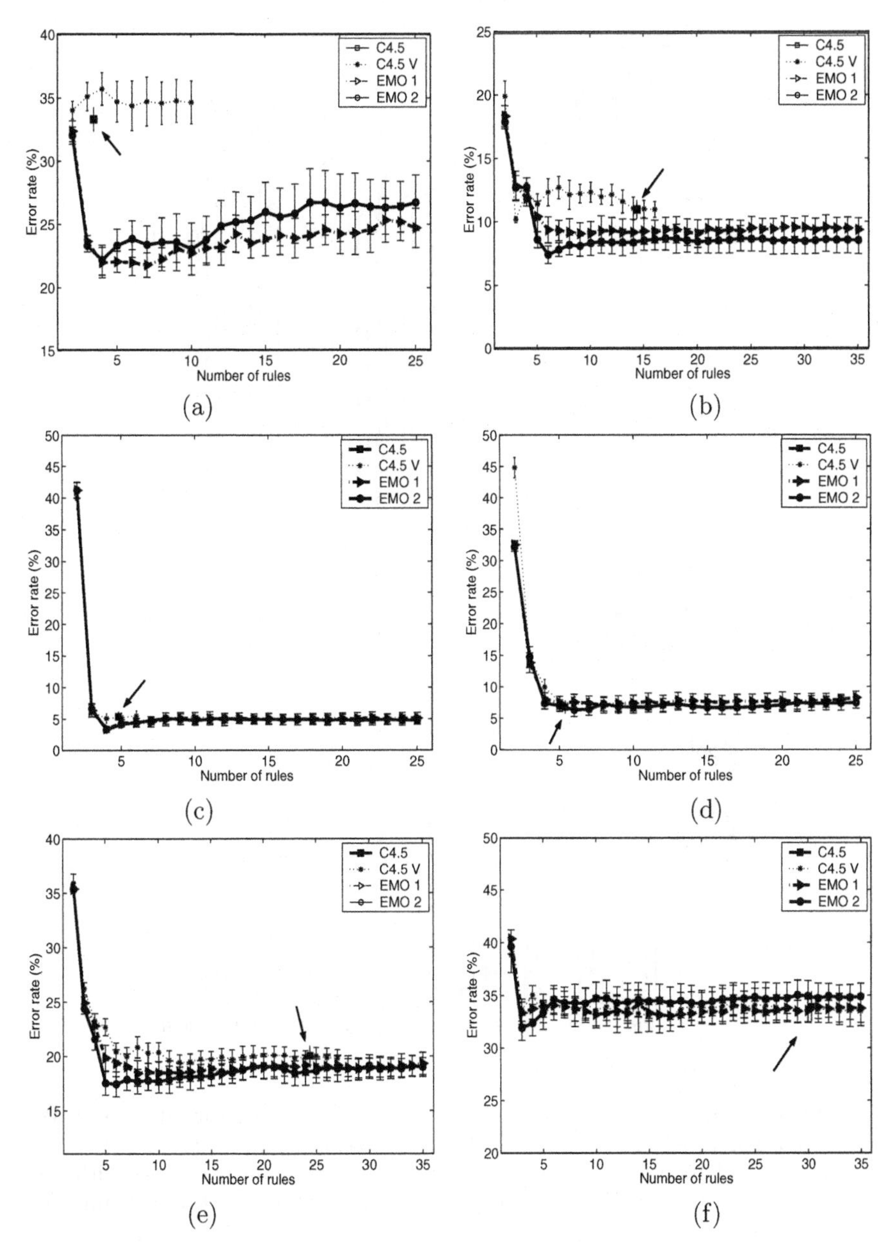

Fig. 11.7. Generalization performance with varying number of rules 1 (arrow: C4.5 with default parameters, *: C4.5 with varying number of nodes, ▷: EMO result with 100 generations, ○: EMO result with 1000 generations) (a) artificial data (b) *ionosphere* data (c) *iris* data (d) *wine* data (e) *ecoli* data (f) *bupa* data

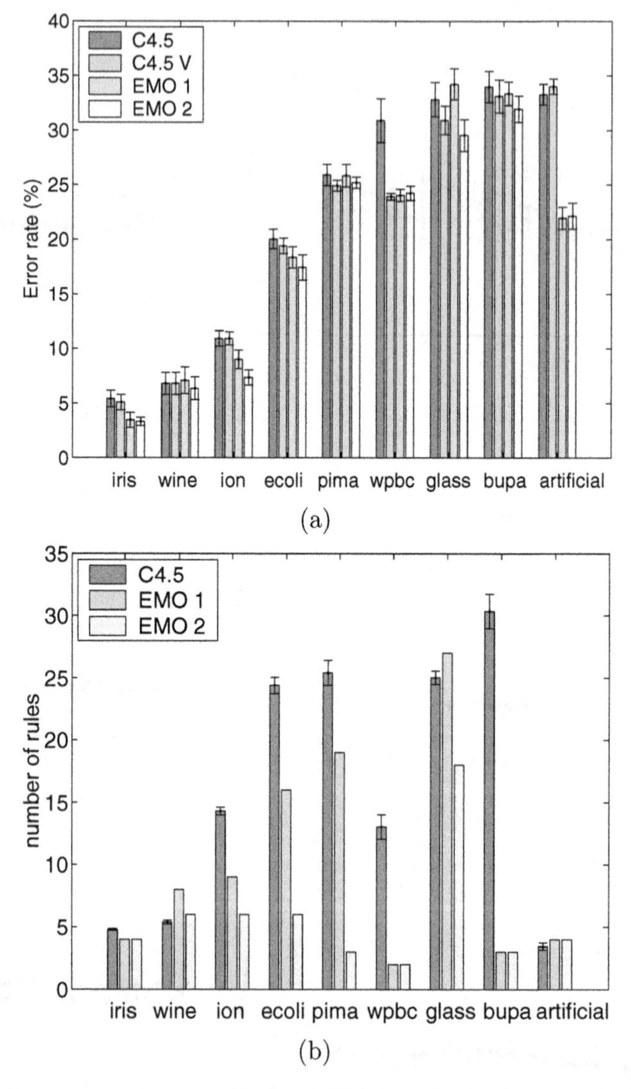

Fig. 11.8. Comparison between C4.5 and EMO method (a) error rate in test data with C4.5 and EMO (EMO 1 and EMO 2 represent the EMO running with 100 generations and 1000 generations, respectively) (b) the number of rules with C4.5 and EMO (the number of rules for EMO is determined by selecting the minimum error rate)

rules. If the tree size is larger than the best tree size, then the generalization performance degrades or has no improvement. More generations tend to show a better curve for the best structure of trees. This test validation process can easily determine a desirable number of rules. The EMO even with 100 gener-

Table 11.1. Data classification errors in C4.5 and variable-node C4.5

	C4.5				variable node C4.5	
data	pattern	attr.	error (%)	# rules	error (%)	# rules
artificial	150	2	33.3 ± 0.9	3.5 ± 0.3	34.0 ± 0.7	2
ionosphere	351	34	10.9 ± 0.7	14.3 ± 0.3	10.9 ± 0.6	16
iris	150	4	5.4 ± 0.8	4.8 ± 0.1	5.1 ± 0.7	4
wine	178	13	6.8 ± 1.0	5.4 ± 0.2	6.8 ± 1.0	9
ecoli	336	7	20.0 ± 0.9	24.4 ± 0.7	19.4 ± 0.7	12
pima	768	8	25.9 ± 1.0	25.4 ± 1.0	24.9 ± 0.5	13
wpbc	194	32	30.9 ± 2.0	13.1 ± 1.0	23.9 ± 0.3	2
glass	214	9	32.8 ± 1.6	25.0 ± 0.5	30.9 ± 1.3	9
bupa	345	6	34.0 ± 2.0	30.4 ± 1.9	33.1 ± 1.5	17

Table 11.2. Data classification errors in the EMO method

	EMO (100 gen.)		EMO (1000 gen.)	
data	error (%)	# rules	error (%)	# rules
artificial	21.9 ± 1.1	4	22.1 ± 1.2	4
ionosphere	9.0 ± 0.9	9	7.4 ± 0.7	6
iris	3.5 ± 0.7	4	3.3 ± 0.4	4
wine	7.1 ± 1.2	8	6.3 ± 1.0	6
ecoli	18.3 ± 1.0	16	17.4 ± 1.2	6
pima	25.8 ± 1.0	19	25.2 ± 0.5	3
wpbc	24.0 ± 0.6	2	24.2 ± 0.7	2
glass	34.2 ± 1.4	27	29.5 ± 1.5	18
bupa	33.4 ± 1.1	3	31.9 ± 1.2	3

ations is better than the C4.5 induction tree in the test error rates for these two sets of data, artificial and *ionosphere* data. Variable-node C4.5 does not produce a V-shape curve (in Fig. 11.1) for generalization performance with the VC-dimension, tree size, but instead irregular type of performance curve. Its performance is significantly worse than the EMO performance in most of cases. Fig. 11.7 shows that the variable-node C4.5 is mostly worse in performance than the EMO method for each number of rules, and if we choose the best structure to minimize structural risk, the EMO method outperforms the variable-node C4.5 in generalization performance for all cases except *pima* and *wpbc* data.

We collected the best model complexity and the corresponding performance into Table 11.1-11.2. For reference, we showed the performance of C4.5 by default parameters. Variable-node C4.5 often finds better performance than C4.5 by default parameter setting. Among the collection of varying sizes of trees, there exists some tree better in performance than C4.5 result for most of data sets.

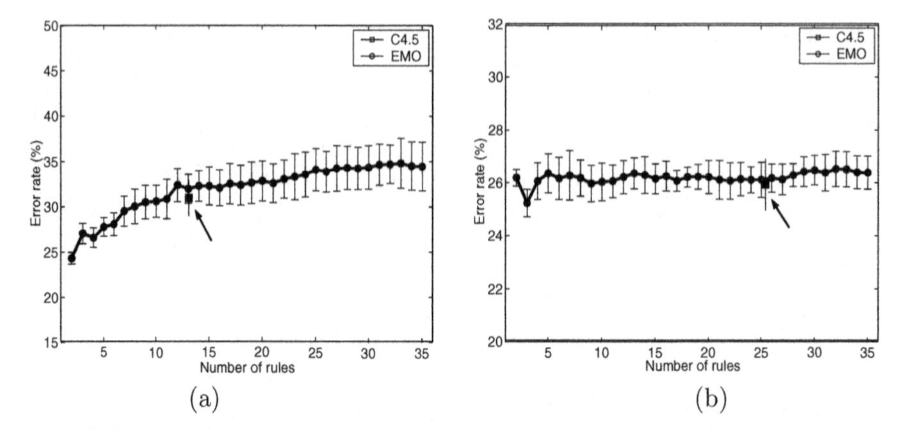

Fig. 11.9. Generalization performance with varying number of rules 2 (arrow: C4.5 with default parameter, ○: EMO result with 1000 generations) (a) *wpbc* data (b) *pima* data

In our experiments, the EMO with 100 generations outperforms C4.5 in the test error rate for all the data except *wine* and *glass*. The EMO with 1000 generations improves both the error rate and the number of rules, and it is better than C4.5 in the test error rate for all the data. Table 11.1-11.2 and Fig. 11.8 show that the EMO is significantly better than C4.5 in error rate with 95 % confidence levels for many experimental data. The EMO method with *wine* and artificial data have a little higher number of rules, but it is due to the fact that the EMO finds an integer number of rules in a discrete space. The other data experiments show that the best number of rules in decision trees by the EMO is significantly smaller than the number of rules by C4.5 induction trees, which is determined by information gain. It confirms that some rules from C4.5 are redundant and thus C4.5 may suffer from an over-specialization problem. In some cases variable-node C4.5 finds a small number of rules, but the rule set has much worse performance in classification than the best rule set by the EMO.

An interesting result is obtained for the *wpbc* and *pima* data (see Fig. 11.9). Two sets of data have a bad prediction performance, regardless of the number of rules. The performance of C4.5 is worse than or similar to that of two or three rules evolved. Investing longer training time on *wpbc* and *pima* data does not improve validation performance. It is presumed that more consistent data are required for the two sets of data.

11.4 Discussion

We evaluated generalization performance under various model complexities, with the pattern classifiers found by the EMO method. An algorithm of minimizing structural risk was applied to the Pareto sets of (performance, structure). For the comparison with C4.5, we tried to provide a variety of model complexities using variable-node C4.5, that is, by controlling the parameter of a minimum number of objects in the branches. The performance of C4.5 with varying number of nodes is shown in Fig. 11.7; it is mostly worse than our method in classification performance. With the parameter control, a specific number of decision rules can be missing. Generally it is hard to generate a consecutive number of leaf nodes or a large size of trees with C4.5. It would be a better comparison with the suggested approach if more sophisticated tree induction method with pruning or stopping growing can be tested. As an alternative, other tree induction algorithms that grow trees by one node can be tested and compared with the EMO method. We leave the comparison between the suggested approach and other methods under the same model complexities to future work.

Given any arbitrary data set for future prediction, we can apply the cross-validation process to find the best pattern classifier minimizing structural risk. The data set is divided into training and test set randomly and then the suggested EMO can be applied to the training data. This procedure can be repeated with many trials to obtain the average generalization error. The generalization performance under a variety of structure complexity can determine the best structure or best size of trees. The above method was applied to each training set in the 10-fold cross validation, and the best structure followed the result in Table 11.2 in most of cases. In our approach, the evolved decision trees have axis-parallel decision boundaries. The pattern classification may have a limitation of fitting nonlinear sample patterns, although we can find the best number of classifiers. Applying our EMO approach to linear regression trees or neural network classifiers would produce better classification performance.

If one of the objectives in EMO is discrete, elitism can be easily applied to evolutionary computation by keeping a pool of the best solutions for each discrete genotype. In the EMO approach, all members in the elitist pool were reproduced every generation, where each member corresponds to each size of trees. If chromosomes are linearly ranked by only error rate performance instead of Pareto dominating rank, it fails to produce uniform Pareto-optimal solutions, since the evolutionary run sticks to an one-objective solution. When we tested varying number of members in the elitist pool for reproduction in a new population, we found more members reproduced in the elite pool can significantly improve training performance.

We showed the performance of the EMO with different number of generations, 100 and 1000. In some case, 100 generations are sufficient to find the best model structure. Yet more generations often produce the result that the best model structure shifts to a smaller size of trees. It is believed that better training performance of the EMO can find better model structure to minimize structural risk, but it is still an open question how much training we need for desirable generalization performance.

The computing time of evolutionary computation requires much more time than C4.5. For example, a single EMO run with a population size of 500 and 100 generations over *pima* data takes about 22 seconds while a single run of C4.5 application takes only 0.1 second (Pentium computer). A single EMO run for *iris*, *wine*, *ionosphere*, *ecoli*, *wpbc*, *glass*, and *bupa* took roughly 3 seconds, 6 seconds, 28 seconds, 8 seconds, 12 seconds, 7 seconds, 8 seconds, respectively. Generally the EMO needs much more computing time to find the best performance for every size of trees, but it can improve the classification performance significantly in most of the data sets.

11.5 Conclusions

In this chapter, we introduced an evolutionary multiobjective optimization with two objectives, classification performance and tree size (number of rules), for decision tree classification. The proposed EMO approach searches for the best accuracy rate of classification for each different size of trees. By structural risk minimization, we can find a desirable number of rules for a given error bound. The performance of the best rule set is better than that of C4.5 or variable-node C4.5, although it takes more computing time. In particular, it can reduce the model complexity, the size of trees dramatically. It can also help to evaluate how difficult it is to classify a given set of data examples. Many researchers have used *pima* and *wpbc* in their experiments, but the distribution of error rates over the size of trees implies that these data cannot expect prediction. In addition, we can indirectly determine if a given set of data requires more consistent data or whether it includes many noisy samples.

For future study, the suggested method can be compared with the bagging method [5], which is one of the promising methods to obtain good accuracy rates. The bagging process may also be applied to the best rules obtained from the proposed method. The decision tree evolved in the present work has the simple form of a binary tree. The EMO approach can be extended to more complex trees such as trees with multiple thresholds or linear regression trees. The result can also be compared with that obtained from neural networks.

References

[1] C. Blake, E. Keogh, and C.J. Merz. *UCI* repository of machine learning databases. In *Proc. of the Fifth Int. Conf. on Machine Learning*, 1998.
[2] S. Bleuler, M. Brack, L. Thiele, and E. Zitzler. Multiobjective genetic programming: Reducing bloat using SPEA2. In *Congress on Evolutionary Computation*, pages 536–543. IEEE Press, 27-30 May 2001.
[3] M.C.J. Bot. Improving induction of linear classification trees with genetic programming. In *Genetic and Evolutionary Computation Conference*, pages 403–410. Morgan Kaufmann, 2000.

[4] M.C.J. Bot and W.B. Langdon. Application of genetic programming to induction of linear classification trees. In *Proceedings of the 3rd European Conference on Genetic Programming*, pages 247–258, 2000.
[5] L. Breiman. Bagging predictors. *Machine Learning*, 24(2):123–140, 1996.
[6] L. Breiman, J. Friedman, R. Olshen, and C. Stone. *Classification and Regression Trees.* Wadsworth International Group., 1984.
[7] C.A. Coello Coello, D.A. van Veldhuizen, and G.A. Lamont. *Evolutionary Algorithms for Solving Multi-Objective Problems.* Kluwer Academic, 2002.
[8] E.D. de Jong and J.B. Pollack. Multi-objective methods for tree size control. *Genetic Programming and Evolvable Machines*, 4(3):211–233, 2003.
[9] U.M. Fayyad. *On the induction of decision trees for multiple concept learning.* Ph.D. dissertation, EECS department, University of Michigan, 1991.
[10] U.M. Fayyad and K.B. Irani. Multi-interval discretization of continuous-valued attributes for classification learning. In *Proceedings of IJCAI'93*, pages 1022–1027. Morgan Kaufmann, 1993.
[11] C. M. Fonseca and P. J. Fleming. Genetic algorithms for multiobjective optimization: Formulation, discussion and generalization. In *Proc. of the Fifth Int. Conf. on Genetic Algorithms*, pages 416–423. Morgan Kaufmann, 1993.
[12] S. Geman. Neural networks and the bias/variance dilemma. *Neural computation*, pages 1–58, 1992.
[13] P. Geurts. *Contributions to decision tree induction: bias/variance tradeoff and time series classification.* Ph.D. thesis, University of Liege, Belgium, 2002.
[14] P. Geurts and L. Wehenkel. Investigation and reduction of discretization variance in decision tree induction. In *European Conference on Machine Learning, LNAI 1810*, pages 162–170. Springer Verlag, 2000.
[15] A.A. Freitas G.L. Pappa and C.A.A. Kaestner. Attribute selection with a multiobjective genetic algorithm. In *Proc. of the 16th Brazilian Symposium on Artificial Intelligence*, pages 280–290. Springer-Verlag, 2002.
[16] K.B. Irani and V.A. Khaminsani. Knowledge based automation of semiconductor manufacturing. In *SRC Project Annual Review Report*, The University of Michigan, Ann Arbor, 1991.
[17] D. Kim. Evolving internal memory for T-maze tasks in noisy environments. *Connection Science*, 16(3):183–210, 2004.
[18] D. Kim. Structural risk minimization on decision trees using an evolutionary multiobjective optimization. In *Proc. of European Conf. on Genetic Programming, LNCS 3003*, pages 338–348. Springer-Verlag, 2004.
[19] D. Kim and J. Hallam. An evolutionary approach to quantify internal states needed for the woods problem. In *From Animals to Animats 7*, pages 312–322. MIT Press, 2002.
[20] W. B. Langdon. Data structures and genetic programming. In P.J. Angeline and K.E. Kinnear, editors, *Advances in Genetic Programming 2*, pages 395–414. MIT press, Cambridge, MA, 1996.
[21] J. Mingers. An empirical comparison of selection measures for decision-tree induction. *Machine Learning*, 4(2):227–243, 1989.
[22] T. M. Mitchell. *Machine Learning.* McGraw Hill, 1997.
[23] J.R. Quinlan. Induction of decision trees. *Machine Learning*, 1(1):81–106, 1986.
[24] J.R. Quinlan. Improved use of continuous attributes in *C4.5*. *Journal of Artificial Intelligence Approach*, 4:77–90, 1996.

[25] J.R. Quinlan and R. Rivest. Inferring decision trees using the minimum description length principle. *Information and Computation*, 80(3):227–248, 1996.
[26] V.N. Vapnik. *The nature of statistical learning theory*. Springer Verlag, 1995.
[27] E. Zitzler. *Evolutionary Algorithms for Multiobjective Optimization: Methods and Applications*. Ph.D. thesis, Swiss Federal Institute of Technology, 1999.

12

Multi-objective Learning Classifier Systems

Ester Bernadó-Mansilla[1], Xavier Llorà[2], and Ivan Traus[3]

[1] Department of Computer Engineering, Enginyeria i Arquitectura La Salle, Universitat Ramon Llull. Quatre Camins, 2. 08022 Barcelona, Spain
esterb@salleurl.edu

[2] Illinois Genetic Algorithms Lab, University of Illinois at Urbana-Champaign, 104 S. Mathews Ave., Urbana, IL 61801, USA
xllora@illigal.ge.uiuc.edu

[3] Conducive Corp., 55 Broad Street, 3rd Floor, New York, NY 10004, USA
itraus@conducivecorp.com

Summary. Learning concept descriptions from data is a complex multiobjective task. The model induced by the learner should be *accurate* so that it can represent precisely the data instances, *complete*, which means it can be generalizable to new instances, and *minimum*, or easily readable. Learning Classifier Systems (LCSs) are a family of learners whose primary search mechanism is a genetic algorithm. Along the intense history of the field, the efforts of the community have been centered on the design of LCSs that solved these goals efficiently, resulting in the proposal of multiple systems. This paper revises the main LCS approaches and focuses on the analysis of the different mechanisms designed to fulfill the learning goals. Some of these mechanisms include implicit multiobjective learning mechanisms, while others use explicit multiobjective evolutionary algorithms. The paper analyses the advantages of using multiobjective evolutionary algorithms, especially in Pittsburgh LCSs, such as controlling the so-called *bloat* effect, and offering the human expert a set of concept description alternatives.

12.1 A Multiobjective Motivation

Classification is a central task in data mining and machine learning applications. It consists of inducing a model that describes the target concept represented in a dataset. The dataset is formed by a set of instances, where each instance is described by a set of features and an associated class. The model describes the relation between the features and the available classes, hence it can be used to explain the hidden structure of the dataset and to classify newly collected instances whose associated class is unknown.

Classification may be regarded as an inherently multiobjective task. Such a task requires inducing a knowledge representation that represents the target concept completely and consistently. In many domains, the induced model

E. Bernadó-Mansilla et al.: *Multi-objective Learning Classifier Systems*, Studies in Computational Intelligence (SCI) **16**, 261–288 (2006)
www.springerlink.com

should also be easily interpretable, which often means a compact representation, and take few computational resources especially in the exploitation stage but also in the training stage. All these objectives are usually opposed and classification schemes try to balance them heuristically.

Multiobjective learning is also present in learning classifier systems (LCSs). LCSs evolve a model of the target concept by the use of genetic algorithms (GAs). Particularly, genetic algorithms search for a hypothesis or a set of hypotheses representing the target concept. The hypotheses can be represented as rule sets, prototype sets, or decision trees [42, 43, 13]. Two main approaches are defined under the LCS framework. The so-called Michigan approach codifies a population of rules, where each individual of the population represents a single rule; thus, a partial solution. The whole population is needed to represent the whole target concept. The Pittsburgh approach codifies a complete knowledge representation (either a ruleset, instance set, or induction tree) in each individual. Thus, the best individual evolved is the solution to the classification problem. In both cases, genetic algorithms are used as search algorithms, seeking to optimize the multiobjective learning goals. In many cases [63, 23, 4], the multiobjective goals are dealt implicitly by the interaction of several components. In other cases [10, 45], the multiobjective goals are directly codified in the fitness function.

The later case benefits from the fact that the search is performed by a genetic algorithm and uses experience gained from the field of multiobjective evolutionary algorithms to get a hypothesis optimizing simultaneously the learning goals. One of the advantages associated with this approach is the improvement of the search in the space of possible hypotheses, minimizing the probabilities of falling into local minima. Moreover, the solution obtained is not only a single hypothesis, but a set of compromise solutions representing different hypotheses along the Pareto front. This offers the human expert the possibility of choosing among a set of alternatives. Besides, this allows to explore the possibilities of combining different hypotheses, which opens the area up to the field of ensemble classifiers [17].

This paper revises the main LCS approaches through the prism of multiobjective learning. First, we give a short history of LCS, summarizing early approaches of LCSs which were significant and inspired nowadays competent LCSs. Then, we focus our attention on the most successful systems of both the Michigan and Pittsburgh approaches. We focus on these types of systems separately, since they have different types of difficulties to achieve the learning goals, due to their different architecture. In Michigan LCSs, learning is centered on the evolution of a minimum set of rules where each rule should be accurate and maximally general. We revise how the XCS classifier system [63] achieves implicitly these goals, and compare it with MOLeCS [10], a multiobjective learning classifier. In Pittsburgh LCSs, learning implies a search through a space of rulesets, evolving an accurate, general, and minimum ruleset. The use of variable-size individuals causes *bloat* [60], i.e., an unnecessary growth of individuals without fitness improvement. Evolutionary multiobjective algorithms are useful to limit this effect while achieving the learning goals. Finally, we discuss open issues on multiobjective learning and directions for further research.

12.2 Learning Classifier Systems: A Short Overview

12.2.1 Cognition and Environment Interaction

The origins of the learning classifiers systems field can be placed within the work by Holland on adaptive systems theory and his early proposals on *schemata processors* [32]. This laid the basis for the first practical implementation of a classifier system which was called CS-1 (Cognitive System Level One) [34]. CS-1 was devised as a cognitive system capable of capturing the events from the environment and reacting to them appropriately. Its architecture was composed of a set of receptors collecting messages into a message list. These messages were matched against a set of rules to decide which actions should be sent to the effectors. When reward was received from the environment, an epochal algorithm distributed payoff into the list of active rules that triggered that action. A genetic algorithm was run to discover new rules by means of crossover and mutation.

The system inspired many approaches which were named under the framework of the Michigan approach. The early Michigan approach can be summarized as having the following properties. The system learns incrementally from the interaction with the environment. It usually starts from a random population of rules which is evaluated with a reinforcement learning scheme by means of a credit assignment algorithm. The *bucket brigade* algorithm [34] was a classical credit assignment algorithm, where the quality (strength) of rules is assigned according to the payoff prediction, i.e., the payoff that the rule would receive from the environment if its action was selected. The GA task is to discover new promising rules, while ensuring the co-evolution of a diverse set of rules. The GA uses the strength calculated by the credit assignment algorithm as the fitness of each rule, thus biasing the search towards highly rewarded rules. The maintenance of different rules along the evolution is addressed by a non-generational scheme and the use of niching techniques [29].

The major problems that arise with traditional Michigan LCSs are the achievement of accurate generalizations and the co-evolution and maintenance of different niches in the population. There is a delicate balance between competition of individuals, which compete to occupy its place in the population, and the co-operation needed to achieve jointly a set of rules. This balance is difficult to maintain for Michigan LCSs. The Boole classifier system [62] tried to balance this equilibrium by the use of fitness sharing and a GA applied to the niches defined by the action sets (sets of rules with similar conditions and the same action). COGIN [31] avoided competition among rules by restricting new classifiers to cover instances not covered by other classifiers of the population.

Despite the known difficulties, Michigan approaches succeeded in many applications. NEWBOOLE [16], which inherits from Boole, was tested in medical datasets and its performance found competitive with respect to CN2 [22] and backpropagation neural networks [57]. Additionally, NEWBOOLE obtained a compact and interpretable ruleset. Further work on NEWBOOLE derived into EpiCS [35], which was adapted to the requirements of epidemiologic databases with mechanisms for risk assessment, and unequal distribution of classes, among others. Results are also competitive when compared with other classifier schemes.

12.2.2 Genetic Algorithm based Concept Learners

Coexisting with the early developments of Michigan LCSs, a parallel line of LCSs was also under research named as the Pittsburgh approach. It emerged with LS-1 classifier system [58], which inspired a main classifier scheme called GABL [23].

The Pittsburgh approach can be characterized as a "classical" genetic algorithm, whose individuals codify a set of rules (in fact, later approaches also introduced other types of codifications). Each individual represents a whole ruleset; thus, a complete solution to the learning problem. The main learning cycle usually operates as in a generational genetic algorithm. The fitness of each individual is evaluated as the performance of its ruleset as a whole, usually under a supervised learning scheme. Fitness usually considers the classification accuracy of the ruleset against the training set of samples.

The first approaches such as GABL considered only classification accuracy on each individual's fitness. Individuals were codified as fixed length rulesets, where each rule was codified as a binary string. Posterior versions introduced incremental learning (GABIL) and the use of two specific operators, *adding alternatives* and *dropping condition*, to bias the search towards general rules and minimal rulesets respectively.

In Pittsburgh LCSs each individual codifies a complete ruleset, so there is no need for cooperation among individuals as in the Michigan approach. This way, the operation of the GA is simpler; the GA converges to a single solution and niching algorithms are not needed. However, since Pittsburgh LCSs search in the space of possible rulesets, the search space is large and usually takes more computational resources than the Michigan approach. Another difficulty is that few control can be applied at the rule level; e.g., rule's generalization, which may encourage the formation minimal rulesets, can hardly be tuned. The two additional GABIL's operators were designed with this purpose. GIL [36] was another proposal that included a large set of operators acting at different levels, whose purpose was also gaining control over the type of rules evolved, but at the expense of an increased parameterization.

The use of variable sized individuals allowed increased flexibility in Pittsburgh LCSs, but caused excessive growth of individuals, with the inclusion of useless rules inside the individuals or the evolution of overspecific rules. Some solutions added parsimony pressures in the fitness function, as in [27] while others addressed these issues from a multiobjective problem perspective [45]. Some recent developments are GALE [39], and GAssist[4]. The later will be analyzed as an example of implicit multiobjective learner. We will also study in detail some multiobjective evolutionary learners, such as MOLS-GA and MOLS-ES [45].

12.2.3 Hybrid Approaches

Due to the known different difficulties of the Michigan and the Pittsburgh approaches, several hybrid systems were proposed such as REGAL and GA-MINER. REGAL [28] uses a distributed architecture, composed of several local GAs evolving separately a single rule covering a partial set of training examples, and a global GA that forms the complete ruleset. The key point of the system relies on the distribution of examples and the combination of the partial solutions.

GA-MINER [25] is a system designed for pattern discovery in large databases. Each individual in the population has a single rule, but its evaluation is done independently from other rules in the population; so it does not need any credit assignment algorithm. The formation of rulesets is performed using an incremental update strategy based on heuristics. The main goal of GA-MINER is to find some interesting patterns in the dataset, instead of fully covering the database as expected in a classification task. Under this framework, the evolution of a complete ruleset is not necessary and in fact, this is not guaranteed by the system.

12.2.4 Accuracy-based Approaches

XCS [63, 64] represents a major development of Michigan style LCSs. Previous Michigan LCSs had identified difficulties to achieve accurate generalizations and maintain a balanced co-evolution of different niches in the population. This led to suboptimal rulesets that could hardly represent the target concept. XCS differs from traditional LCSs on the definition of rule fitness, which is based on the accuracy of the payoff prediction rather than on the prediction itself, and the use of a niche GA. These aspects have resulted in a strong tendency to evolve accurate and maximally general rules, favoring the achievement of knowledge representations that, besides being complete and accurate, tend to be minimal [37].

Other accuracy-based approaches have been studied recently [11]. Particularly, UCS (sUpervised Classifier System) shares many features with XCS, such as the generalization mechanisms. The main difference is that UCS is designed specifically for supervised learning problems. The experiments showed that UCS is more suitable to classification problems with large spaces, specially those with high number of classes or with highly unequal distribution of examples.

We will study XCS in more detail and compare it with MOLeCS, a multiobjective Michigan LCS.

12.3 Multiobjective Optimization

Prior to the study of the different multiobjective mechanisms of the Michigan and Pittsburgh LCSs, we briefly describe the notation we will use to refer to multiobjective issues.

In a multiobjective optimization problem (MOP) [61] a solution $\mathbf{x} \in \Omega$ is represented as a vector of n decision variables $\mathbf{x} = (x_1, \ldots, x_n)$, where Ω is the decision variable space. We want to optimize k objectives which are defined as $f_i(\mathbf{x})$, with $i = 1 \ldots k$. These objectives are grouped in a vector function denoted as $F(\mathbf{x}) = (f_1(\mathbf{x}), \ldots, f_k(\mathbf{x}))$, where $F(\mathbf{x}) \in \Lambda$. F is a function which maps points from the decision variable space Ω to the objective function space Λ:

$$\begin{aligned} F : \Omega &\longmapsto \Lambda \\ \mathbf{x} &\longmapsto \mathbf{y} = F(\mathbf{x}) \end{aligned} \tag{12.1}$$

Without loss of generality, we can define a MOP as the problem of minimizing a set of objectives $F(\mathbf{x}) = (f_1(\mathbf{x}), \ldots, f_k(\mathbf{x}))$, subject to some constraints $g_i(\mathbf{x}) \leq 0$, $i = 1, \ldots, m$. These constraints are necessary for problems where there are invalid solutions in Ω.

A solution that minimizes all the objectives and satisfies all constraints may not exist. Sometimes, the minimization of a certain objective implies a degradation in another objective. Then, there is not a global optimum that minimizes all the objectives simultaneously. In this context, the concept of optimality must be redefined. Vilfredo Pareto [53] introduced the concept of dominance and Pareto optimum to deal with this issue.

In general terms, a vector $\mathbf{u}$ *dominates* another vector $\mathbf{v}$, written as $\mathbf{u} \preceq \mathbf{v}$, if and only if every component u_i is less than or equal to v_i, and at least there is one component in $\mathbf{u}$ which is strictly less than the corresponding component in $\mathbf{v}$. This can be formulated as follows:

$$\mathbf{u} \preceq \mathbf{v} \iff \forall i \in 1, \ldots, k\,,\; u_i \leq v_i \wedge \exists i \in 1, \ldots, k : u_i < v_i \tag{12.2}$$

The concept of *Pareto optimality* is based on the dominance definition. Thus, a solution $\mathbf{x} \in \Omega$ is Pareto optimal if there is not any other solution $\mathbf{x}' \in \Omega$ whose objective vector $\mathbf{u}' = F(\mathbf{x}')$ dominates $\mathbf{u} = F(\mathbf{x})$. In other words, a solution whose objectives can not be improved simultaneously by any other solution is Pareto optimum.

The set of all solutions whose objective vectors are not dominated by any other objective vector is called the Pareto optimal set $\mathcal{P}^*$:

$$\mathcal{P}^* := \{\mathbf{x}_1 \mid \nexists\, \mathbf{x}_2 : \mathbf{F}(\mathbf{x}_2) \preceq \mathbf{F}(\mathbf{x}_1)\} \tag{12.3}$$

Analogously, the set of all vectors $\mathbf{u} = F(\mathbf{x})$ such that $\mathbf{x}$ belongs to the Pareto optimal set is called the Pareto Front $\mathcal{PF}^*$:

$$\mathcal{PF}^* := \{\mathbf{u} = \mathbf{F}(\mathbf{x}) = (f_1(\mathbf{x}), \ldots, f_k(\mathbf{x})) \mid \mathbf{x} \in \mathcal{P}^*\} \tag{12.4}$$

12.4 Multiobjective Learning in Michigan Style LCSs

Michigan style LCSs codify each individual as having a single rule. Thus, learning in Michigan LCSs implies codifying the learning goals at the rule level. The approach is to maximize accuracy and generality in each rule, which consequently could combine into a consistent, complete and minimal ruleset. This section revises two particular Michigan LCSs, representative of two different approaches. The first one, XCS, uses an accuracy-based fitness, while generalization is achieved mainly by a niche GA applied in a frequency basis. The later, MOLeCS, defines a multiobjective fitness which directly guides the search process to optimize these goals.

12.4.1 XCS: Implicit Multiobjective Learning

Description of XCS

XCS [63, 64] represents the target concept in a set of rules. This ruleset is incrementally evaluated by means of interacting with the environment, through a *reinforcement learning scheme*, and is improved by a search mechanism based on a *genetic algorithm.*

Each rule's quality is estimated by a set of parameters. The main parameters are: a) *prediction*, an estimate of the reward that the rule would receive from the environment, b) *accuracy*, the accuracy of the prediction and c) *fitness*, which is based only on accuracy.

The basic training cycle performs as follows. At each time step, an input x is presented to the system. Given x, the system builds a match set [M], which is formed by all the classifiers in the population whose conditions are satisfied by the input example. If no classifiers match, then the covering operator is triggered. It creates new classifiers matching the current input.

During training, XCS explores the consequences of all classes for each input. Therefore, given a match set [M], XCS selects randomly one of the classes proposed in [M] and sends it to the environment. All classifiers proposing that class are classified as belonging to the action set [A]. The environment returns a reward which is used to update the parameters of the classifiers in [A], and then the GA may be triggered. In test mode, XCS proposes the best class from those advocated in [M], and the update and search mechanisms are disabled.

If we run XCS in training, the GA is triggered when the average time since the last occurrence of the GA in the action set exceeds a threshold θ_{GA}. The

GA takes place in the action set, rather than in the whole population. It selects two parents from the current [A] with probability proportional to fitness, and applies crossover and mutation. The resulting offspring are introduced into the population. If the population is full, a classifier is selected for deletion. The deletion probability is proportional to the size of the action sets where the classifier has participated. If the classifier is experienced and poorly fit, its deletion probability is increased by an inverse proportion of its fitness.

XCS also uses subsumption as an additional mechanism to favor the generalization of rules. Whenever a classifier is created by the GA, it is checked for subsumption with its parents before being inserted into the population. If one of the classifier's parents is sufficiently experienced, accurate, and more general than the classifier, then the classifier is discarded, and the parent's numerosity is increased. This process is called *GA subsumption.*

Implicit Multiobjective Learning

XCS receives inputs from the instances available in the training set and receives feedback of its classifications in the form of rewards. The environment is designed to give a maximum reward if the system predicts the correct classification and a minimum reward (usually zero) otherwise. XCS's goal is to maximize the rewards received from the environment and in doing so it tries to get a *complete, consistent and minimal representation* of the target concept. We analyze the role of each component in achieving this compound goal.

The role of covering is to enforce a complete coverage of the input space. Whenever an input is not covered by any classifier, covering creates an initial classifier from where XCS launches its learning and search mechanisms in that part of the space.

The task of the reinforcement component is to evaluate the classifier's parameters from the reward received from the environment. This is done incrementally. For each example, it computes the prediction error of the classifier, gets a measure of accuracy, and then updates fitness based on this computed accuracy.

The search component is based on a genetic algorithm. Basing fitness on accuracy makes the genetic algorithm to search for accurate rules. Generalization is stressed by the application of the GA to the niches defined by the action sets. This is explained by Wilson's generalization hypothesis [63]: given two accurate classifiers with one of them matching a subset of the input states matched by the other, then the more general one will win because it participates in more action sets and thus has more reproductive opportunities. As a consequence, the more general classifier will tend to displace the specific classifier, resulting in compact representations. Subsumption is included to encourage generalization and compactness as well.

The niche GA is also designed to favor the maintenance of a diverse set of rules which jointly represent the target concept. This is achieved through different mechanisms: a) the GA's triggering mechanism, which tries to balance the application of the GA among all the action sets, b) selection, which is applied locally to the action sets, c) crossover, performing a kind of restricted mating, and d) the deletion algorithm, which tends to delete resources from the most numerous action sets. In XCS, the niches are defined by the action sets, which are formed by a set of classifiers matching a common set of input states and a common class.

12.4.2 MOLeCS: Explicit Multiobjective Learning

From Multiobjective Rulesets to Multiobjective Rules

MOLeCS (MultiObjective Learning Classifier System) [10] is a Michigan style LCS that evolves a ruleset describing the target concept by the use of multi-objective evolutionary algorithms.

As a Michigan-style LCS, MOLeCS evolves a population of rules which jointly represent the target concept. Thus, each individual is a rule and the whole ruleset is formed by all individuals in the population. The system's goal is to achieve a complete, consistent and compact ruleset by means of multi-objective evolutionary algorithms. However, these goals cannot be directly defined into the GA search, since the GA search is performed at the rule level, not at the ruleset level. Therefore, these goals are adapted to the mechanisms of a Michigan style LCS by defining two objectives at the rule level: generalization and accuracy. Each rule should maximize simultaneously generalization and accuracy. If fitness was only based on accuracy, the GA search would be biased towards accurate but too specific rules. This would result in an enhancement of the solution set (i.e., need of more rules) and also poor coverage of the feature space. On the contrary, basing fitness on generality would result in low performance, in terms of classification accuracy. The solution was to balance these two characteristics at the same time. The hypothesis was that by guiding the search towards general and accurate rules would result in minimum, complete and accurate set of rules.

Description of MOLeCS

Each individual in MOLeCS codifies a rule (classifier) of type: *condition* → *class*. Each rule r_i is evaluated against the available instances in the training dataset, and two measures are computed:

$$generalization(r_i) = \frac{\#\text{ covered examples }(r_i)}{\#\text{ examples in the training set}} \tag{12.5}$$

$$accuracy(r_i) = \frac{\#\text{ correctly classified examples }(r_i)}{\#\text{ covered examples }(r_i)} \tag{12.6}$$

MOLeCS defines the multiobjective problem as the simultaneous maximization of accuracy and generalization. Thus, the fitness of each rule is assigned according to a multiobjective evaluation strategy. Several MOEA techniques were explored in MOLeCS, being a method based on lexicographic ordering the most successful one. Particularly, the approach taken was that of sorting the population according to the accuracy objective and in case of rules equally accurate, sort them by the generalization objective. The interpretation was that of searching for accurate rules being as general as possible. Once sorted, fitness was assigned according to the ranking. Other strategies based on Pareto dominance and the aggregating approach were also considered but found less successful.

Once the fitness assignment phase is performed, the GA proceeds to the selection and recombination stages. In each iteration, G individuals are selected with stochastic universal sampling (SUS) [7]. Then, they undergo crossover with probability p_c and mutation with probability p_m per allele. The resulting offspring are introduced into the parental population. The replacement method was designed to achieve a complete coverage of the feature space.

In fact, it was argued that promoting general classifiers was not sufficient to reach a complete ruleset. The genetic algorithm could tend, due to the genetic drift [30], to a single general and accurate classifier and usually one classifier does not suffice to represent the whole target concept. Therefore, the system should enforce the *co-evolution* of a set of diverse fit rules by niching mechanisms. Niching in MOLeCS is performed in the replacement stage. Particularly, deterministic crowding is applied, where the child replaces the most similar parent only if it has greater fitness.

Once the system has learned, it is used under an exploit or test phase. It works as follows. An example coming from the test set is presented. Then, the system finds the matching rules and applies the fittest rule to predict the associated class. As explained before, in case of equally fit rules, the most accurate rule is chosen.

12.4.3 Results

XCS's generalization hypothesis [64] explains that the accuracy-based fitness coupled with the niche GA favor the evolution of compact rulesets consisting of the most general accurate rules. Additionally, the optimally hypothesis [37] argues that XCS tends to evolve minimum rulesets. These hypotheses are supported by theoretical studies on the pressures induced by the interaction of XCS's components [21, 19]. XCS has also demonstrated to be highly competitive with other machine learning schemes such as induction trees, and nearest neighbors, among others, in real world classification problems [65, 13]. Recent studies are investigating the domain of competence of XCS in real world domains, i.e., to what kind of problems XCS is suited and poorly suited to [12].

MOLeCS was tested in artificial classification problems, of type of multiplexer and parity problems, often used in the LCS community, as well as in real world datasets. The multiobjective evolutionary algorithms let the system evolve accurate and maximally general rules which together formed compact representations. A comparison with a single-objective approach maximizing only each rule's accuracy demonstrated that the multiobjective optimization was necessary to overcome the tendency of evolving too many specific rules which would lead the system towards suboptimal solutions partly representing the target concept. Optimizing simultaneously generality and accuracy was a better approach than a single optimization approach. However, giving the same importance to these objectives with techniques such as Pareto ranking caused the system evolve overgeneral rules, preventing other maximally general rules from being explored and maintained in the population. This consequently resulted in poor accurate rulesets. The best approach taken was that of introducing the decision preferences a priori, i.e., during the search process, so that the algorithm could find accurate rules as general as possible. This led to the evolution of nearly optimal rulesets, with results highly competitive with respect to other LCSs such as XCS in real world datasets.

The main difficulty of MOLeCS was identified within the niching mechanisms. MOLeCS presented difficulties to stabilize a niche population and obtain a diverse set of rules which jointly represented a complete ruleset. Niching mechanisms such as deterministic crowding were only useful for limited datasets which could be easily represented by small rulesets. Switching towards implicit niching mechanisms such as those of XCS, would include an extra generalization pressure, as it is explained by Wilson's generalization hypothesis. Having both an explicit pressure towards generalization by means of a multiobjective approach and an implicit generalization pressure produced by the niche GA could break the delicate equilibrium of evolutionary pressures inside LCSs and lead to overgeneral rules. However, this remains an unexplored area that would benefit from further research.

12.5 Multiobjective Learning in Pittsburgh Style LCSs

The previous section presented how Michigan classifier systems evolve a population of rules that classify a particular. However, the Pittsburgh style classifier systems evolve a population of individuals, each of them a variable-length ruleset that represents a complete solution to the problem. Such an approach greatly simplifies the evolutionary process. For instance, the fitness of each candidate individual can be computed using the accuracy of the ruleset classifying the dataset. Then, the evaluated individuals undergo the traditional selection and recombination mechanisms.

Such simplicity, however, comes with a price to pay, the tradeoff among accuracy and generalization. As in Michigan approaches, Pittsburgh classifier systems should provide compact and accurate individuals. The rest of this section revises the implications of such a goal on Pittsburgh classifier systems and presents some systems and results that exploit multiobjective optimization ideas to achieve it.

12.5.1 A Generalization Race?

Evolution of variable size individuals causes *bloat* [60]. The term, named within the field of genetic programming (GP), refers to the code growth of individuals without any fitness improvement.

Langdon [38] attributes bloat to two possible reasons, labeled as "fitness causes bloat" and "natural code is protective". The former refers to the fact that selection does not distinguish between individuals with the same fitness but different size. Thus, for each individual with a given fitness, there is a large set (possibly infinite) of individuals with the same fitness and larger codes. A search guided by fitness becomes a random walk among individuals with different sizes and equivalent fitness. The second cause of bloat considers that neutral code that does not influence fitness tends to be protective, i.e., reduces the probability that the genetic operators disrupt useful code. Thus, they do not improve fitness but their maintenance in the population is favored by the genetic operators.

In Pittsburgh LCSs, bloat may arise in two different forms: (1) the addition of useless rules, or (2) the evolution of over-specific rules. Without limitation on the number of rules in the ruleset, the search space becomes too large, and solutions are far from being optimal. Contributions to address this issue have been worked out from both the GP field and the LCSs field.

Some of the approaches taken in the GP field consist of imposing a *parsimony pressure* toward compact individuals (see for example, [59]) by varying fitness or through specially tailored operators. Recently, an approach proposed by Bleuler, Brack, Thiele, and Zitzler [15] uses a multiobjective evolutionary algorithm to optimize two objectives: maximize fitness and minimize size. In their approach they use the SPEA2 algorithm [68, 70, 69].

In Pittsburgh classifier systems, some approaches also used a parsimony pressure, while others directly codified a multiobjective evolutionary approach. The next sections detail them.

12.5.2 Implicit Multiobjective Learning

Parsimony Pressure

The classical approach to addressing bloat and multiobjective learning goals was to introduce a parsimony pressure in the fitness function, in such a way that the fitness of larger individuals was decreased [9, 3, 27]. For example, in [27] the *bloat* is controlled by a step fitness function: when the number of rules of an individual exceeds a certain maximum, its fitness is decreased abruptly. One problem of this approach is to set this threshold value appropriately.

Bacardit and Garrell [3] define a similar fitness function, as well as a set of operators for the deletion of introns[4] [49] (rules that are not used in the classification) and a tournament-based selection operator that considers the size of individuals. The authors argue that the *bloat* control has an influence over the generalization capability of the solutions. It has been observed that shorter rulesets tend to have more generalization capabilities [14, 3, 49].

Therefore, the use of a parsimony pressure has beneficial effects: it controls the unlimited growth of individuals, increases the efficiency in the search process and leads to solutions with better generalization. Nevertheless, the parsimony pressure must be balanced appropriately. An excessive pressure toward small individuals could result in premature convergence leading to compact solutions but with suboptimal fitness [49], or even in a total population failure (population collapses with individuals of minimal size). Soule and Foster [59] showed that the effect of the parsimony pressure can be measured by calculating explicitly the relationship between the size and the performance of individuals within the population.

Based on these results, it seems that a multiobjective approach may overcome some of these difficulties. The first step toward introducing multiobjective pressures in Pittsburgh classifier systems is to use a linear combination of the accuracy of a given individual and its generality –understood as inversely proportional to its size. The next section revises an approach of this type, which is based on the *minimum description length* principle [5].

GAssist

GAssist [5] is a Pittsburgh LCS descendant of *GABIL*. The system applies a near-standard generational *GA* that evolves individuals that represent complete problem solutions. An individual consists of an ordered, variable–length ruleset. Tournament selection is used.

A special fitness function based on the Minimum Description Length (*MDL*) principle [56] is used, balancing the accuracy and the complexity of an individual. The *MDL* formulation used is defined as follows:

$$MDL = W \cdot \mathrm{TL} + \mathrm{EL} \tag{12.7}$$

where TL is for the complexity of the ruleset, which considers the number of rules and the number of relevant attributes in each rule, and EL is a measure of the error of the individual on the training set. W is a constant that adjusts the relation between TL and EL. An adaptive mechanism is used to avoid domain-specific tuning of the constant.

The system also uses a rule deletion operator to further control the bloat effect. The algorithm deletes the rules that do not have been activated by any example, the so-called introns. The authors argue that the growth of

[4] Non-coding segments. In GP literature this concept has also been termed *non-effective code* [8].

introns in the individuals is protective, in the sense that they prevent the crossover operator from disrupting the useful parts of the individual. However, as the code growths, the chances of improving the fitness of the individual by recombination also decrease. Hence, removing part of this useless code through an appropriate balance is beneficial to the search process.

A comparison of GAssist with XCS [2] showed similar performance in terms of accuracy rate. However, the number of rules evolved by GAssist was much smaller that the number of rules of XCS. The comparison of GAssist with XCS also arose a certain difficulty of GAssist in handling multiple classes and huge search spaces. XCS was found to overfit in some datasets, where the generalization pressure due to the niche reproduction hardly applied because of data sparsity. Section 12.5.4 gives more details by means of a comparison of these systems in a selected set of classification problems.

12.5.3 Explicit Multiobjective Learning

By using a multiobjective evolutionary approach, we can explicitly search for a set of rules that minimizes the misclassification error and the size (number of rules). Besides controlling the number of rules (*bloat*) dynamically, this would allow the formation of compromise hypotheses. This explicit tradeoff formation let us explore the generalization capabilities of the hypotheses that form the Pareto front. In certain environments like data mining, where the extraction of explanatory models is desirable, high quality general solutions (in terms of accuracy out of sample, or compactness of hypotheses) are useful. For instance, the presence of noise in the dataset may lead to accurate but overfitted solutions. Maintaining a Pareto front of compromise solutions we can identify the overfitted perturbations of high quality general hypotheses. Therefore, evolving a set of different compromise solutions between accuracy and generalization, we can postpone the decision of picking the "best ruleset" to the final user (decision maker), or combine them all using some *bagging* technique [17, 41, 39].

Classification and Multiobjective Optimization

Multiobjective Evolutionary Algorithms can be applied to trade off between two objectives: the accuracy of an individual, and its size.

Let's define $\mathbf{x}$ as an individual that is a complete solution to the classification problem; $\mathcal{D}$ the training dataset for the given problem; $|\mathcal{D}|$ number of instances in $\mathcal{D}$; $miss(\mathbf{x}, \mathcal{D})$ the number of incorrectly classified instances of $\mathcal{D}$ performed by $\mathbf{x}$; and finally, $size(\mathbf{x})$ a measure of the current size of $\mathbf{x}$ (e.g. the number of rules it contains). Using this notation, a multiobjective approach can be defined as follows:

$$minimize\ F(\mathbf{x}) = (f_e(\mathbf{x}), f_s(\mathbf{x})) \tag{12.8}$$

$$f_e(\mathbf{x}) = \frac{miss(\mathbf{x}, \mathcal{D})}{|\mathcal{D}|} \tag{12.9}$$

$$f_s(\mathbf{x}) = size(\mathbf{x}) \tag{12.10}$$

The misclassification error corresponds to the number of instances incorrectly classified divided by the size of the training set. Searching for rulesets minimizing the misclassification error means to search for rulesets covering correctly as many instances as possible. Minimizing the number of rules of the ruleset seeks to search for general and compact rulesets, which moreover avoid the bloat effect. In fact, this is a simple multiobjective approach. Other types of proposals could include more objectives such as measures of generalization of the rulesets, or coverage (number of instances covered by a ruleset).

MOLS-GA

Some approaches have addressed learning rulesets in Pittsburgh LCS architectures from a multiobjective perspective. MOLS-GA and MOLS-ES [45] are two examples. They represent two similar approaches to the multiobjective problem, being different on whether they base the search mechanism on a GA (MOLS-GA) or an evolution stratgey (MOLS-ES). Since they do not offer significant differences on the multiobjective approach itself, we will center our study on MOLS-GA for being more representative of Pittsburgh LCSs.

MOLS-GA codifies a population of individuals, where each individual represents a complete representation of the target concept. The available knowledge representations are rulesets or instance sets [39, 42, 43]. If the problem's attributes are nominal, MOLS-GA uses rulesets, represented by the ternary alphabet (`0`, `1`, `#`) often used in other LCSs [33, 29, 63]. Otherwise, if the problem is defined by continuous-valued attributes, instance sets—based on a nearest neighbor classification—are used.

The GA learning cycle works as follows. First, the fitness of each individual in the population is computed. This is done on a multiobjective basis, taking into account the misclassification error and the size of each individual.

The individuals of the population are sorted in equivalent classes. These classes are determined by the Pareto fronts that can be defined among the population. That is, given a population of individuals $\mathcal{I}$, the first equivalence class $\mathcal{I}^0$ is the set of individuals which belongs to the evolved Pareto optimal set $\mathcal{I}^0 = \mathcal{P}^*(\mathcal{I})$. The next equivalence class $\mathcal{I}^1$ is computed without considering the individuals in $\mathcal{I}^0$, as $\mathcal{I}^1 = \mathcal{P}^*(\mathcal{I} \setminus \mathcal{I}^0)$, and so forth. Figure 12.1 shows an example of the different equivalence classes that appear in a population at a given iteration. This plot is obtained with the multiplexer (`mux`) problem. In this example, the population is classified into nine different fronts. The left front is $\mathcal{I}^0$, which corresponds to the non-dominated vectors of the population. The next front to the right represents $\mathcal{I}^1$ and so on.

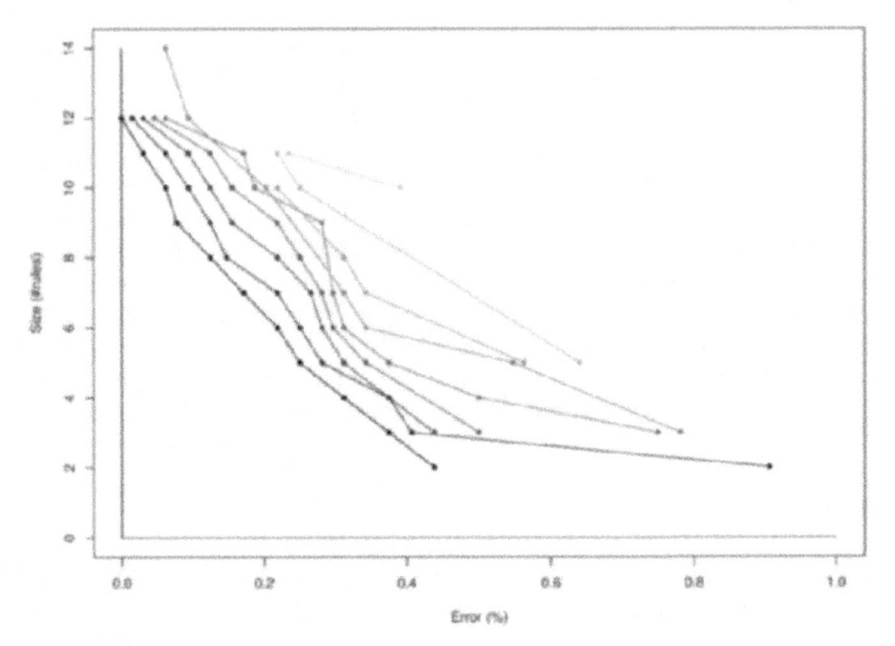

Fig. 12.1. Sorted population fronts at a given iteration of MOLS in the `mux` problem [45].

Once the population of individuals $\mathcal{I}$ is sorted, fitness values are assigned. Since the evolution must bias the population toward non-dominated solutions, we impose the constraint:

$$fitness(\mathcal{I}^i) > fitness(\mathcal{I}^{i+1}) \tag{12.11}$$

Thus, the evolution will try to guide the population toward the left part of the plot, i.e., the real Pareto front. The fitness of each individual depends on the front where the individual belongs. That is, all the individuals of the same equivalence class $\mathcal{I}^i$ receive the same constant value $(n-i)\delta$, where n is the number of equivalence classes and δ is a constant. Moreover, in order to spread the population along the Pareto front, a *sharing* function is applied. Thus, the final fitness of an individual j in a given equivalence class $\mathcal{I}^i$ is:

$$fitness(\mathcal{I}^i_j) = \frac{(n-i)\delta}{\sum_{k \in \mathcal{I}} \phi(d_{\mathcal{I}^i_j \mathcal{I}_k})} \tag{12.12}$$

where $\phi(d_{\mathcal{I}^i_j \mathcal{I}_k})$ is the sharing function [30]. The *sharing* function is computed using the Euclidean distance between the multiobjective vectors. The radius of the *sharing* function σ_{sh} was set to $\sigma_{sh} = 0.1$.

After individuals are evaluated, selection is applied using a tournament selection algorithm [50, 6] with elitism. Elitism is often applied in evolutionary multiobjective optimization algorithms and it usually consists of keeping the solutions of the Pareto front evolved in each generation [61]. MOLS-GA performs similarly: it keeps all the distinct solutions of the evolved Pareto front, and also a 30% of the individuals with the lowest error. This guarantees that the best compromise solutions evolved so far are not lost, as well as the solutions with the lowest error, which are important to drive the evolution toward accurate solutions.

After selection, crossover and mutation are applied. Crossover is adapted from two-point crossover to individuals of variable size; cut points can occur anywhere inside the ruleset, but they should be equivalent in both parents so that valid offspring can be obtained. The mutation consists in generating a random new gene value.

What is the Purpose of the Pareto Front?

The main purpose of the evolved Pareto front is to keep solutions with different tradeoffs between error and size. Coevolving these compromise solutions, we can delay the need of choosing a solution until the evolution is over and the evolved Pareto front is provided. However, this decision is critical for achieving a high quality generalization when tested with unseen instances of the target concept.

The *decision maker* has several hypotheses among which to choose, all provided by the classification front. The *decision maker* can be a human or an expert system with some knowledge about the problem. A typical approach is to select a solution from the evolved Pareto front. Two strategies may be considered, as shown in figure 12.2. The first one (*best-accuracy*) chooses the solution $\mathbf{x}$ of the front with the best accuracy, that is, the one that minimizes $f_e(\mathbf{x})$. On the other hand, the second one (*best-compromise*) picks the hypothesis of the front that minimizes the objective vector $\mathbf{u} = F(\mathbf{x})$. Thus, the selected solution is the one that balances equally both objectives. In other words, the solution $\mathbf{x}$ that minimizes $|F(\mathbf{x})| = \sqrt{f_e(\mathbf{x})^2 + f_s(\mathbf{x})^2}$.

We could instead benefit from the combination of the multiple hypotheses obtained in the evolved Pareto front. Such a technique is inspired in the *bagging* technique [17, 41], which consists of training several classifiers and combining them all into an ensemble classifier. The bagging technique tends to reduce the deviation among runs, and the combined solution often improves the generalization capability of the individual solutions [39]. An adaptation of such an strategy to the combination of solutions from the Pareto front could give similar benefits as shown elsewhere [40].

12.5.4 Experiments and Results

Through a Pareto Front Glass

Figure 12.1 shows an example of how the sorting of a population of candidate solutions produces a clear visualization of the tradeoff between accuracy and generality.

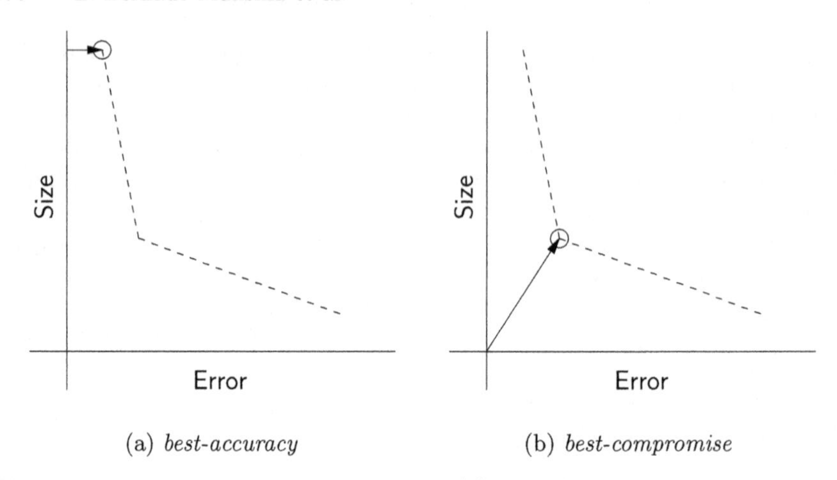

Fig. 12.2. Decision maker strategies using (a) the solution with the lowest error (*best accuracy*) and (b) the solution closer to the origin (*best compromise*).

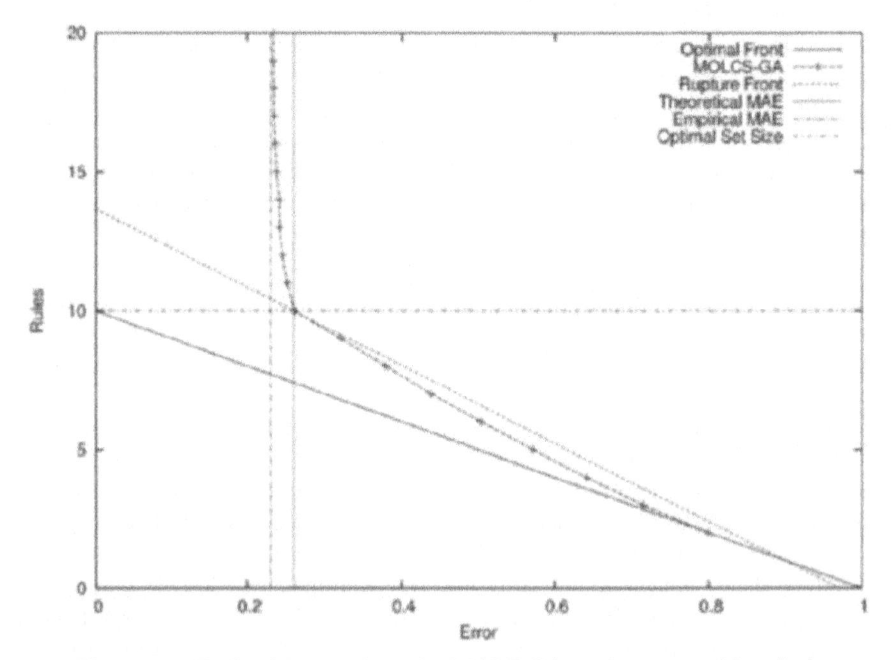

Fig. 12.3. Evolved Pareto front by MOLS-GA in the `LED` problem [44].

Such Pareto fronts arise an interesting property, as shown by Llorà and Goldberg [44]. In the presence of noise in the dataset, the Pareto front presents a clear rupture around the maximally general and accurate solutions. Such rupture point is achieved around the minimal achievable error (MAE) [44] for the dataset at hand. The *rupture point* indicates the place where the evolved

Pareto front abruptly changes its slope. The front that appears to the left of the *rupture point* is the result of the deviation of the empirical MAE from its theoretical value. This has an interesting interpretation. All the points that define this segment of the front are over-fitted solutions. This means that they are learning some misleading noisy pattern or a solution too specific to the current training dataset. If any of these solutions is tested using new instances not previously shown in the training phase, they would produce a significant drop in accuracy. Thus, this leads to a reduction of the generalization capabilities (in terms of classification accuracy) of the solutions kept in that part of the front. Moreover, these solutions are closer to the *bloat* phenomenon, because very small (misleading) improvements require a large individual growth. All these problems disappear when we force the theoretical and the empirical MAE to be the same.

Results

The work by Llorà, Goldberg, Traus, & Bernadó-Mansilla [45] evaluated the performance of different multiobjective LCSs on nine different datasets, which are described in Table 12.1. Two of them are *artificial datasets.* `Mux` is the eleven input multiplexer, widely used by the LCS community [63]. `Led` is the seven-segments problem [18], obtained from the UCI repository [47]. The remaining datasets also belong to the UCI repository: *bupa liver disorders* (`bpa`), *Wisconsin breast cancer* (`bre`), *glass* (`gls`), *ionosphere* (`ion`), *iris* (`irs`), *primary tumor* (`prt`), and *sonar* (`son`). These datasets contain categorical and numeric attributes, as well as binary and n-ary classification tasks.

We summarize some of the results published in [45], where two explicit multiobjective Pittsburgh approaches are compared with other schemes. Particularly, we compare MOLS-GA and MOLS-ES with GAssist and XCS. We also include a comparison with non-evolutionary schemes: IB1 [1], C4.5 [54, 55], and PART [26]. Their algorithms were obtained from the *Weka* package [67] and ran with their default configuration.

The results were obtained from *stratified ten-fold cross-validations runs* [48, 67] using the different learning algorithms on the selected datasets. MOLS-GA used a *best-accuracy* strategy for the test phase, whereas MOLS-ES used *best-compromise.* Table 12.2 shows the classification accuracy obtained by each method. Observe that LCSs in general are highly competitive with the non-evolutionary schemes. Moreover, MOLS-GA and MOLS-ES offer good classification accuracies when compared with GAssist and XCS. There is not a clear winner in all the datasets. Instead, all the methods seem to perform equivalently.

Table 12.3 shows the average and standard deviation of the number of rules obtained by the LCS approaches. Observe that the three Pittsburgh-based methods offer fairly simple rulesets, composed of 3 to 15 rules approximately. On the contrary, XCS gets much larger rulesets. In fact, the ruleset evolved by Pittsburgh LCSs is that obtained by the best individual of the population, while the ruleset obtained by XCS considers all the population. If we consider that GAssist evolves a population of 400 individuals, the number of explored rules is equivalent to that evolved by XCS. Moreover, the ruleset of a Pittsburgh LCSs operates as an activation list; i.e., the rules form a hierarchy going from the most general to the most specific ones. The first rules tend to cover many examples (the general case), while the last rules codify the exceptions to the general rule. XCS does not use such a hierarchy so that the number of necessary rules is higher. Despite these issues, the final XCS's ruleset should be still further processed. Most of the rules are only product of the recent exploration of the algorithm and are not relevant to performance. Other rules overlap partially with others; therefore, they could be compacted. Research on reduction algorithms and further processing of the evolved rules is being conducted by several authors to get a ruleset easily interpretable to human experts [66, 24, 51]. In [51] some reduction algorithms reported a reduction of 97% without a significant degradation in classification accuracy. For example, the `bpa` ruleset could be reduced to 90 rules, which is still higher that Pittsburgh LCSs, but more reasonable to human experts.

MOLS-GA is in average the method which gets smaller rulesets for similar values of classification accuracy. This probably means that MOLS-GA evolves better Pareto fronts than MOLS-ES. Figure 12.4 shows two examples for the `bre` and the `prt` datasets. See that the first one belongs to the case where MOLS-ES gets better performance, which corresponds to a better evolved Pareto front. The second plot corresponds to the more general case where MOLS-GA obtains a better front, resulting in higher performance. MOLS-GA also gets smaller rulesets than GAssist in average, which means that the explicit multiobjective approach is effective to further reduce the size of the ruleset.

Table 12.1. Summary of the datasets used in the experiments.

id	Data set	Size	Missing values(%)	Numeric Attributes	Nominal Attributes	Classes
`bpa`	*Bupa Liver Disorders*	345	0.0	6	-	2
`bre`	*Wisconsin Breast Cancer*	699	0.3	9	-	2
`gls`	*Glass*	214	0.0	9	-	6
`ion`	*Ionosphere*	351	0.0	34	-	2
`irs`	*Iris*	150	0.0	4	-	3
`led`	*Led (10% noise)*	2000	0.0	-	7	10
`mux`	*Multiplexer (11 inputs)*	2048	0.0	-	1	2
`prt`	*Primary Tumor*	339	3.9	-	17	22
`son`	*Sonar*	208	0.0	60	-	2

The Pareto fronts presented in figure 12.4 also suggest another interesting analysis of the results. For instance, the front presented in figure 4(b) shows an interesting resemblance to the fronts obtained in the noisy `led` problem

Table 12.2. Comparison of classification accuracy on selected datasets. The table shows the mean and standard deviation of each method on a *stratified ten-fold cross-validation* experiment [45, 2].

id	MOLS-GA	MOLS-ES	GAssist	XCS	C4.5	PART	IB1
bpa	76.5±13.4	68.7±6.7	61.5±8.3	65.4±6.9	65.8±6.9	65.8±10.0	64.2±9.1
bre	96.0±1.1	96.1±2.2	95.9±2.5	96.7±2.5	95.4±1.6	95.3±2.2	95.9±1.5
gls	67.1±9.3	63.4±7.3	68.2±9.3	70.5±8.5	68.5±10.4	69.0±10.0	66.4±10.9
ion	91.5±3.6	92.8±2.7	92.4±5.0	89.6±3.1	89.8±0.5	90.6±0.9	90.9±3.8
irs	99.3±1.9	95.3±3.1	95.3±5.7	94.7±5.3	95.3±3.2	95.3±3.2	95.3±3.3
led	74.9±13.7	74.4±3.4	n.a.	74.5±0.2	74.9±0.2	75.1±0.3	74.3±3.7
mux	100.0±0.0	100.0±0.0	100.0±0.0	100.0±0.0	99.9±0.2	100.0±0.0	99.8±0.2
prt	51.2±15.8	40.6±5.7	47.8±8.1	39.8±6.6	41.6±6.4	41.6±6.4	42.5±6.3
son	90.8±9.1	71.6±12.5	77.2±8.9	77.5±3.6	71.5±0.5	73.5±2.2	83.6±9.6

Table 12.3. Comparison of ruleset sizes obtained by different methods on selected datasets. The table shows the mean and standard deviation of each method on a *stratified ten-fold cross-validation* experiment [45, 2].

id	MOLS-GA	MOLS-ES	GAsssist	XCS
bpa	10.1±0.3	15.1±4.0	7.2±1.2	2377.5±125.7
bre	9.4±1.5	8.5±1.6	11.7±2.6	2369.5±114,5
gls	4.2±0.4	12.2±2.7	8.8±1.4	2709±98.9
ion	6.9±0.3	12.8±3.8	2.1±0.3	4360±134.9
irs	2.7±0.1	5.8±1.1	4.3±0.9	730±75.8
led	14.9±0.8	14.7±2.0	n.a.	102.5±10.2
mux	10.0±0.3	12.7±1.9	11.2±0.5	640±45.5
prt	9.3±0.5	12.0±5.0	16.0±3.8	1784±45.8
son	11.6±0.5	13.1±4.0	8.3±1.4	4745±123.2

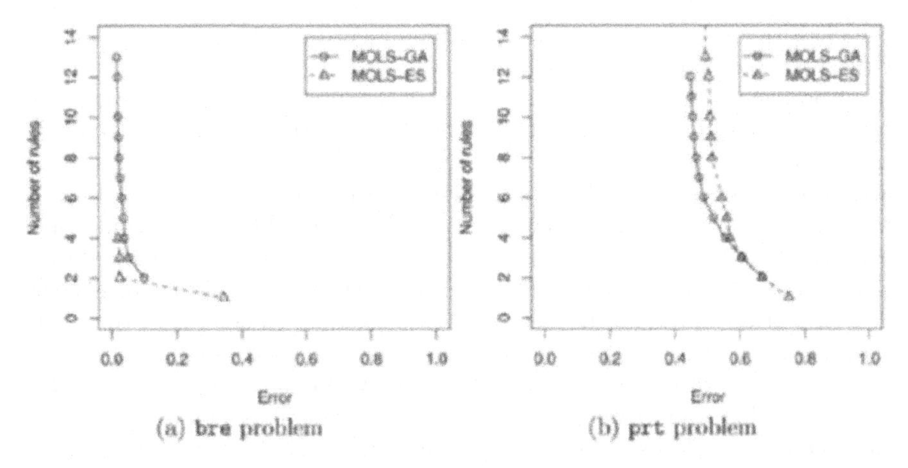

(a) bre problem (b) prt problem

Fig. 12.4. Pareto fronts achieved in real problems by MOLS-GA and MOLS-ES [45].

(see figure 12.3). This clearly points out to the presence of inconsistencies in the `prt` dataset that bounds the MAE. However, further work on the usage of the MAE measure is still needed, as well as exploring its connections to *probably approximately correct* models in the *computational learning theory* field [48].

12.6 Current Research Trends and Further Directions

Learning classifier systems address the complex multiobjective task either balancing the learning goals implicitly or by means of evolutionary multiobjective algorithms adapted to the particular architectures.

Within the field of Michigan LCSs, the best approach, XCS, does not use explicit multiobjective evolutionary algorithms. The combination of accuracy-based fitness and the niche GA, with the addition of other heuristic components, achieves the appropriate balance to favor the creation and maintenance of maximally general rules. In fact, the equilibrium is delicate and studies on pressures point out how to balance this equilibrium through appropriate parameter settings. Despite this difficult balance, it does not seem feasible to add explicit multiobjective algorithms to further emphasize the generalization of rules, because this probably would favor overgeneral rules. However, this still remains an unexplored research area which may be analyzed to strength generalization in cases where this is scarcely favored by the niche GA, such as in problems with unequal distribution of examples per class (see for example [52]). Explicit multiobjective algorithms are also able to achieve accurate and maximally general rules in other types of Michigan LCSs. The main difficulty is found on the architecture rather than on the multiobjective evaluation itself. The main problem is to stabilize different niches co-operating to form a complete ruleset. More research on niching algorithms is needed, combined with restricted mating methodologies.

A problem common in Michigan LCSs is the difficulty to control the number of rules forming the final ruleset. For binary attributes, favoring generalization of rules is enough to achieve a minimum ruleset. But for continuous attributes and real-valued representations, Michigan LCSs tend to produce high number of rules that overlap partially. Pittsburgh LCSs using some kind of pressure on the number of maximum rules produce rulesets much reduced than those of Michigan LCSs. Although some reduction algorithms have been designed in the Michigan framework to get compact rulesets [20], they still provide larger rulesets. This imposes a limitation to their interpretability by human experts. Stressing generalization and favoring removal of useless rules during training could help evolving smaller rulesets. This would also reduce the computational cost, since the system would maintain less rules during training.

Within the Pittsburgh approach, implicit and explicit multiobjective approaches are able to balance accuracy and generalization. Moreover, the explicit usage of multiobjective techniques provides an output with a clear tradeoff among both objectives. A Pareto front provides a way of sorting the final population of candidate solutions. Such a front may be explored for a particular solution, or just combined to form an ensemble of classifiers, as it is currently being analyzed in [40]. Such research also needs to face the readability of the final solution. Measures of rule interpretability must be defined. Once the proper measure is ready, it can simply be integrated in the evolutionary process as a third objective to be optimized. Many domains, such as medical domains, specify the cost of misclassifying each of the classes, instead of using a single measure of error. Multiobjective algorithms can also be appropriate to balance the different types of misclassification errors and provide a set of alternatives. However, we should analyze whether the addition of more objectives involves a harder search, and what kind of solutions can be evolved.

Another key research area for Pittsburgh classifier systems is the identification of overfitting situations from the evolved Pareto fronts, as pointed out in the chapter. However, such approach assumes that the global pressure toward accuracy and generality will lead the rulesets towards the desired learning goals. This is a clearly top-down approach. Recently, the work of Wilson in 1995 [63] and the major shift in the way in which fitness was computed in the Michigan approach have been revisited by researchers. Accuracy has become a central element in the process of computing the fitness of rules (or classifiers). Initial attempts to apply Wilson's ideas to Pittsburgh-style classifier systems are on their way [46]. When such ideas are combined with estimation of distribution algorithms, a bottom-up approach to Pittsburgh classifiers would start to emerge—as the compact classifier system (CCS) has preliminary shown.

Acknowledgments

The first author acknowledges the support of Enginyeria i Arquitectura La Salle, Ramon Llull University, as well as the support of Ministerio de Ciencia y Tecnología under Grant TIC2002-04036-C05-03 and Generalitat de Catalunya under Grant 2002SGR-00155.

This work was sponsored in part by the Air Force Office of Scientific Research, Air Force Materiel Command, USAF (F49620-03-1-0129), and by the Technology Research, Education, and Commercialization Center (TRECC), at University of Illinois at Urbana-Champaign, administered by the National Center for Supercomputing Applications (NCSA) and funded by the Office of Naval Research (N00014-01-1-0175). The US Government is authorized to reproduce and distribute reprints for Government purposes notwithstanding any copyright notation thereon.

The views and conclusions contained herein are those of the authors and should not be interpreted as necessarily representing the official policies or endorsements, either expressed or implied, of the Air Force Office of Scientific Research, the Technology Research, Education, and Commercialization Center, the Office of Naval Research, or the U.S. Government.

References

[1] D. Aha and D. Kibler. Instance-based learning algorithms. *Machine Learning*, 6:37–66, 1991.

[2] J. Bacardit and M. V. Butz. Data Mining in Learning Classifier Systems: Comparing XCS with GAssist. In *Seventh International Workshop on Learning Classifier Systems (IWLCS-2004)*. LNAI, Springer (in press), 2004.

[3] J. Bacardit and J. M. Garrell. Métodos de generalización para sistemas clasificadores de Pittsburgh. In *Primer Congreso Español de Algoritmos Evolutivos y Bioinspirados (AEB'02)*, pages 486–493, 2002.

[4] J. Bacardit and J. M. Garrell. Bloat Control and Generalization Pressure using the Minimum Description Length Principle for a Pittsburgh Approach Learning Classifier System. In *Proceedings of the 6th International Workshop on Learning Classifier Systems*. LNAI, Springer (in press), 2003.

[5] J. Bacardit and J. M. Garrell. Evolving Multiple Discretizations with Adaptive Intervals for a Pittsburgh Rule-Based Learning Classifier System. In *Proceedings of the Genetic and Evolutionary Computation Conference*, volume 2724 of *LNCS*, pages 1818–1831. Springer, 2003.

[6] T. Bäck. Generalized convergence models for tournament- and (μ, λ)-selection. *Proceedings of the Sixth International Conference on Genetic Algorithms*, pages 2–8, 1995.

[7] J. E. Baker. Reducing bias and inefficieny in the selection algorithm. In *Genetic Algorithms and their Applications: Proceedings of the Second International Conference on Genetic Algorithms*, pages 14–21, 1987.

[8] W. Banzhaf and W. B. Langdon. Some Considerations on the Reason for Bloat. *Genetic Programming and Evolvable Hardware*, 3(1):81–91, 2002.

[9] J. K. Bassett and K. A. De Jong. Evolving Behaviors for Cooperating Agents. In *Foundations of Intelligent Systems: 12th International Symposium*, volume 1932 of *LNAI*, pages 157–165. Springer-Verlag Berlin Heidelberg, 2000.

[10] E. Bernadó-Mansilla and J. M. Garrell. MOLeCS: Using Multiobjective Evolutionary Algorithms for Learning. In *Evolutionary Multi-Criterion Optimization, First International Conference, EMO 2001*, volume 1993 of *LNCS*, pages 696–710. Springer Verlag, 2001.

[11] E. Bernadó-Mansilla and J. M. Garrell. Accuracy-Based Learning Classifier Systems: Models, Analysis and Applications to Classification Tasks. *Evolutionary Computation*, 11(3):209–238, 2003.

[12] E. Bernadó-Mansilla and T. K. Ho. Domain of Competence of XCS Classifier System in Complexity Measurement Space. *IEEE Transactions on Evolutionary Computation*, 9(1):82–104, 2005.

[13] E. Bernadó-Mansilla, X. Llorà, and J. M. Garrell. XCS and GALE: a Comparative Study of Two Learning Classifier Systems on Data Mining. In *Advances in Learning Classifier Systems, 4th International Workshop*, volume 2321 of *LNAI*, pages 115–132. Springer, 2002.

[14] E. Bernadó-Mansilla, A. Mekaouche, and J. M. Garrell. A Study of a Genetic Classifier System Based on the Pittsburgh Approach on a Medical Domain. In *12th International Conference on Industrial and Engineering Applications of Artificial Intelligence and Expert Systems, IEA/AIE-99*, pages 175–184, 1999.
[15] S. Bleuler, M. Brack, L. Thiele, and E. Zitzler. Multiobjective genetic programming: Reducing bloat using SPEA2. In *Proceedings of the 2001 Congress on Evolutionary Computation CEC2001*, pages 536–543. IEEE Press, 2001.
[16] P. Bonelli, A. Parodi, S. Sen, and S. W. Wilson. NEWBOOLE: A fast GBML System. In *Seventh International Conference on Machine Learning*, pages 153–159. Morgan Kaufmann, 1990.
[17] L. Breiman. Bagging predictors. *Machine Learning*, 24(2):123–140, 1996.
[18] L. Breiman, J. Friedman, R. Olshen, and C. Stone. *Classification and Regression Trees.* Wadsworth International Group, 1984.
[19] M. V. Butz. *Rule-based Evolutionary Online Learning Systems: Learning Bounds, Classification, and Prediction.* PhD thesis, University of Illinois, 2004.
[20] M. V. Butz, P. L. Lanzi, X. Llorà, and D. E. Goldberg. Knowledge Extraction and Problem Structure Identification in XCS. In *Parallel Problem Solving from Nature (PPSN-2004)*, volume 3242 of *LNCS*, pages 1051–1060. Springer, 2004.
[21] M. V. Butz and M. Pelikan. Analyzing the Evolutionary Pressures in XCS. In *Proceedings of the Genetic and Evolutionary Computation Conference (GECCO'2001)*, pages 935–942. San Francisco, CA: Morgan Kaufmann, 2001.
[22] P Clark and T. Niblett. The CN2 induction algorithm. *Machine Learning*, 3(4):261–283, 1989.
[23] K. A. De Jong and W. M. Spears. Learning Concept Classification Rules Using Genetic Algorithms. In *Proceedings of the International Joint Conference on Artificial Intelligence*, pages 651–656. Sidney, Australia, 1991.
[24] P. W. Dixon, D. W. Corne, and M. J. Oates. A Preliminary Investigation of Modified XCS as a Generic Data Mining Tool. In *Advances in Learning Classifier Systems, 4th International Workshop*, volume 2321 of *LNCS*, pages 133–150. Springer, 2002.
[25] I. W. Flockhart. GA-MINER: Parallel Data Mining with Hierarchical Genetic Algorithms. Technical Report EPCC-AIKMS-GA-MINER-REPORT 1.0, University of Edinburgh, 1995.
[26] E. Frank and I. H. Witten. Generating Accurate Rule Sets Without Global Optimization. In *Machine Learning: Proceedings of the Fifteenth International Conference*, pages 144–151. Morgan Kaufmann, 1998.
[27] J. M Garrell, E. Golobardes, E. Bernadó-Mansilla, and X. Llorà. Automatic Diagnosis with Genetic Algorithms and Case-Based Reasoning. *Artificial Intelligence in Engineering*, 13:367–372, 1999.
[28] A. Giordana and F. Neri. Search-Intensive Concept Induction. *Evolutionary Computation*, 3(4):375–416, 1995.
[29] D. E. Goldberg. *Genetic Algorithms in Search, Optimization and Machine Learning.* Addison-Wesley Publishing Company, Inc., 1989.
[30] D. E. Goldberg and J. Richardson. Genetic algorithms with sharing for multimodal function optimization. In *Proceedings of the Second International Conference on Genetic Algorithms*, pages 41–49, 1987.
[31] D. P. Greene and S. F. Smith. Competition-based induction of decision models from examples. *Machine Learning*, 13:229–257, 1993.

[32] J. H. Holland. Processing and processors for schemata. In *Associative Information Processing*, pages 127–146, 1971.
[33] J. H. Holland. *Adaptation in Natural and Artificial Systems: An Introductory Analysis with Applications to Biology, Control and Artificial Intelligence.* MIT Press/ Bradford Books edition, 1975.
[34] J. H. Holland and J. S. Reitman. Cognitive systems based on adaptive algorithms. *Pattern Directed Inference Systems*, pages 313–329, 1978.
[35] J. H. Holmes. Discovering Risk of Disease with a Learning Classifier System. In *Proceedings of the Seventh International Conference of Genetic Algorithms (ICGA97)*, pages 426–433. Morgan Kaufmann, 1997.
[36] C. Janikow. *Inductive Learning of Decision Rules in Attribute-Based Examples: a Knowledge-Intensive Genetic Algorithm Approach.* PhD thesis, University of North Carolina at Chapel Hill, July 1991.
[37] T. Kovacs. XCS Classifier System Reliably Evolves Accurate, Complete and Minimal Representations for Boolean Functions. In *Soft Computing in Engineering Design and Manufacturing*, pages 59–68. Springer-Verlag, 1997.
[38] W. B. Langdon and R. Poli. Fitness causes bloat: Mutation. *Genetic Programming: First European Conference*, pages 37–48, 1998.
[39] X. Llorà. *Genetic Based Machine Learning using Fine-grained Parallelism for Data Mining.* PhD thesis, Enginyeria i Arquitectura La Salle. Ramon Llull University, Barcelona, Catalonia, European Union, February, 2002.
[40] X. Llorà, J. Bacardit, I. Traus, and E. Bernadó-Mansilla. Where to go once you evolved a bunch of promising hypotheses? In *Learning Classifier Systems, 6th International Workshop, IWLCS 2003*, LNAI. Springer (in press), 2005.
[41] X. Llorà and J. M. Garrell. Automatic Classification and Artificial Life Models. In *Proceedings of Learning00 Workshop.* IEEE and Univesidad Carlos III, 2000.
[42] X. Llorà and J. M. Garrell. Evolving Partially-Defined Instances with Evolutionary Algorithms. In *Proceedings of the 18th International Conference on Machine Learning (ICML'2001)*, pages 337–344. Morgan Kaufmann Publishers, 2001.
[43] X. Llorà and J. M. Garrell. Knowledge-Independent Data Mining with Fine-Grained Parallel Evolutionary Algorithms. In *Proceedings of the Genetic and Evolutionary Computation Conference (GECCO'2001)*, pages 461–468. Morgan Kaufmann Publishers, 2001.
[44] X. Llorà and D. E. Goldberg. Bounding the effect of noise in Multiobjective Learning Classifier Systems. *Evolutionary Computation*, 11(3):279–298, 2003.
[45] X. Llorà, D. E. Goldberg, I. Traus, and E. Bernadó-Mansilla. Accuracy, Parsimony, and Generality in Evolutionary Learning Systems via Multiobjective Selection. In *Learning Classifier Systems, 5th International Workshop, IWLCS 2002*, volume 2661 of *LNAI*, pages 118–142. Springer, 2003.
[46] X. Llorà, K. Sastry, and D. E. Goldberg. The Compact Classifier System: Scalability Analysis and First Results. In *Proceedings of the IEEE Conference on Evolutionary Computation.* IEEE press (in press), 2005.
[47] C. J. Merz and P. M. Murphy. UCI Repository for Machine Learning Data-Bases [http://www.ics.uci.edu/~mlearn/MLRepository.html]. *Irvine, CA: University of California, Department of Information and Computer Science*, 1998.
[48] T. M. Mitchell. *Machine Learning.* McGraw-Hill, 1997.

[49] P. Nordin and W. Banzhaf. Complexity Compression and Evolution. In *Proceedings of the Sixth International Conference on Genetic Algorithms*, 1995.

[50] C. K. Oei, D. E. Goldberg, and S. J. Chang. Tournament selection, niching, and the preservation of diversity. IlliGAL Report No. 91011, University of Illinois at Urbana-Champaign, Urbana, IL, 1991.

[51] A. Orriols Puig and E. Bernadó-Mansilla. Analysis of Reduction Algorithms in XCS Classifier System. In *Recent Advances in Artificial Intelligence Research and Development*, volume 113 of *Frontiers in Artificial Intelligence and Applications*, pages 383–390. IOS Press, 2004.

[52] A. Orriols Puig and E. Bernadó-Mansilla. The Class Imbalance Problem in Learning Classifier Systems: A Preliminary Study. In *Learning Classifier Systems, 7th International Workshop, IWLCS 2005*, LNAI. Springer (in press), 2005.

[53] V. Pareto. *Cours d'Economie Politique, volume I and II.* F. Rouge, Lausanne, 1896.

[54] R. Quinlan. Induction of decision trees. *Machine Learning*, 1(1):81–106, 1986.

[55] R. Quinlan. *C4.5: Programs for Machine Learning.* Morgan Kaufmann Publishers, 1993.

[56] J. Rissanen. Modeling by shortest data description. *Automatica*, vol. 14:465–471, 1978.

[57] D.E. Rumelhart, J.L. McClelland, and the PDP Research Group. *Parallel Distributed Processing, Vol. I, II.* The MIT Press, 1986.

[58] S. F. Smith. Flexible Learning of Problem Solving Heuristics through Adaptive Search. In *Proceedings of the 8th International Joint Conference on Artificial Intelligence*, pages 422–425, 1983.

[59] T. Soule and J. Foster. Effects of code growth and parsimony pressure on populations in genetic programming. *Evolutionary Computation*, 6(4):293–309, Winter 1998.

[60] W. A. Tackett. *Recombination, selection, and the genetic construction of computer programs.* Unpublished doctoral dissertation, University of Southern California, 1994.

[61] D. A. Van Veldhuizen and G. B. Lamont. Multiobjective evolutionary algorithms: Analyzing the state-of-the-art. *Evolutionary Computation*, 8(2):125–147, 2000.

[62] S. W. Wilson. Classifier System Learning of a Boolean Function. Technical Report RIS 27r, The Rowland Institute for Science, 1986.

[63] S. W. Wilson. Classifier Fitness Based on Accuracy. *Evolutionary Computation*, 3(2):149–175, 1995.

[64] S. W. Wilson. Generalization in the XCS Classifier System. In *Genetic Programming: Proceedings of the Third Annual Conference*, pages 665–674. San Francisco, CA: Morgan Kaufmann, 1998.

[65] S. W. Wilson. Mining Oblique Data with XCS. In *Advances in Learning Classifier Systems: Proceedings of the Third International Workshop*, volume 1996 of *LNAI*, pages 158–176. Springer-Verlag Berlin Heidelberg, 2001.

[66] S. W. Wilson. Compact Rulesets from XCSI. In *Advances in Learning Classifier Systems, 4th International Workshop*, volume 2321 of *LNAI*, pages 197–210. Springer, 2002.

[67] I. H. Witten and F. Eibe. *Data Mining. Practical Machine Learning Tools and Techniques with Java Implementations.* Morgan Kaufmann, 2000.

[68] E. Zitzler. *Evolutionary Algorithms for Multiobjective Optimization: Methods and Applications.* PhD thesis, Swiss Federal Institute of Technology (ETH) Zurich, 1999.

[69] E. Zitzler. SPEA2: Improving the Strength Pareto Evolutionary Algorithm. Technical report 103, Swiss Federal Institute of Technology (ETH) Zurich, Gloriastrasse 35, CH-8092 Zurich, May, 2001.

[70] E. Zitzler, K. Deb, and L. Thiele. Comparison of Multiobjective Evolutionary Algorithms: Empirical Results. *Evolutionary Computation*, 8(2):173–195, 2000.

Part III

Multi-Objective Learning for Interpretability Improvement

13

Simultaneous Generation of Accurate and Interpretable Neural Network Classifiers

Yaochu Jin, Bernhard Sendhoff and Edgar Körner

Honda Research Institute Europe
Carl-Legien-Str. 30
63073 Offenbach/Main, Germany
yaochu.jin@honda-ri.de

Summary. Generating machine learning models is inherently a multi-objective optimization problem. Two most common objectives are accuracy and interpretability, which are very likely conflicting with each other. While in most cases we are interested only in the model accuracy, interpretability of the model becomes the major concern if the model is used for data mining or if the model is applied to critical applications. In this chapter, we present a method for simultaneously generating accurate and interpretable neural network models for classification using an evolutionary multi-objective optimization algorithm. Lifetime learning is embedded to fine-tune the weights in the evolution that mutates the structure and weights of the neural networks. The efficiency of Baldwin effect and Lamarckian evolution are compared. It is found that the Lamarckian evolution outperforms the Baldwin effect in evolutionary multi-objective optimization of neural networks. Simulation results on two benchmark problems demonstrate that the evolutionary multi-objective approach is able to generate both accurate and understandable neural network models, which can be used for different purpose.

13.1 Introduction

Artificial neural networks are linear or nonlinear systems that encode information with connections and units in a distributed manner, which are more or less of biological plausibility. Over the past two decades, various artificial neural networks have successfully been applied to solve a wide range of tasks, such as regression and classification, among many others.

Researchers and practitioners are mainly concerned with the accuracy of neural networks. In other words, neural networks should exhibit high prediction or classification accuracy on both seen and unseen data. To improve the accuracy of neural networks, many efficient learning algorithms have been developed [8, 37].

Nevertheless, there are also many cases in which it becomes essential that human users are able to understand the knowledge that the neural network

Y. Jin et al.: *Simultaneous Generation of Accurate and Interpretable Neural Network Classifiers*, Studies in Computational Intelligence (SCI) **16**, 291–312 (2006)
www.springerlink.com

has learned from the training data. However, conventional learning algorithms do not take the interpretability of neural networks into account and thus the knowledge acquired by neural networks is often not transparent to human users. To address this problem, it is often necessary to extract symbolic or fuzzy rules from trained neural networks [7, 15, 24, 25]. A common drawback of most existing rule extraction method is that the rules are extracted after the neural network has been trained, which incurs additional computational costs.

This chapter presents a method for generating accurate and interpretable neural models simultaneously using the multi-objective optimization approach [28]. A number of neural networks that concentrate on accuracy and interpretability to a different degree will be generated using a multi-objective evolutionary algorithm combined with life-time learning, where accuracy and complexity serve as two conflicting objectives.

It has been shown that evolutionary multi-objective algorithms are well suited for and very powerful in obtaining a set of Pareto-optimal solutions in one single run of optimization [11, 12].

The idea of introducing Pareto-optimality to deal with multiple objectives in constructing neural networks can be traced back to the middle of 1990's [29]. In this work, two objectives, namely, the mean squared error and the number of hidden neurons are considered. In another work [39], the error on training data and the norm of the weights of neural networks are minimized.

The successful development in evolutionary multi-objective optimization [11, 12] has encouraged many researchers to address the conflicting, multiple objectives in machine learning using multi-objective evolutionary algorithms [2, 3, 4, 10, 16, 26, 27, 30, 32, 41]. Most of the work focuses on improving the accuracy of a single or an ensemble of neural networks. Multi-objective approach to support vector machines [21] and fuzzy systems [22, 40] has also been studied. Objectives in training neural networks include accuracy on training or test data, complexity (number of hidden neurons, and number of connections, and so on), diversity, and interpretability. It should be pointed out that a trade-off between the accuracy on the training data and the accuracy on test data does not necessarily result in a trade-off between accuracy and complexity.

Section 13.2 discusses different aspects in generating accurate and interpretable neural networks. General model selection criteria in machine learning are also briefly described. Section 13.3 shows that any formal neural network regularization methods can be treated as multi-objective optimization problems. The details of the evolutionary multi-objective algorithm, together with the life-time learning method will be provided in Section 13.4. The proposed method is verified on two classification examples in Section 13.5, where a population of Pareto-optimal neural networks are generated. By analyzing the Pareto front, both accurate neural network (in the sense of approximation accuracy on test data) and interpretable neural networks can be identified. A short summary of the chapter is given in Section 13.6.

13.2 Tradeoff between Accuracy and Interpretability

13.2.1 Accurate Neural Networks

Theoretically, feedforward neural networks with one hidden layer are universal approximators, see e.g., [18]. However, a large number of hidden neurons is required to approximate a given complex function to an arbitrary degree of accuracy.

A problem in creating accurate neural networks is that a very good approximation of the training data can result in a poor approximation of unseen data, which is known as overfitting. To avoid overfitting, it is necessary to control the complexity of the neural network, which in turn will limit the accuracy that can be achieved on the training data. Thus, when we talk about accurate neural network models, we mean that the neural network should have good accuracy on both training data and unseen test data.

13.2.2 Interpretability of Neural Networks

The knowledge acquired by neural networks is distributed on the weights and connections and is therefore not easily understandable to human users. To improve the interpretability of neural networks, the basic approach is to extract symbolic or fuzzy rules from trained neural networks [7, 15, 24]. According to [7], approaches to rule extraction from trained neural networks can be divided into two main categories. The first category is known as decompositional approach, in which rules are extracted from individual neurons in the hidden and output layer. In the second approach, the pedagogical approach, no analysis of the trained neural network is undertaken and the neural network is treated as a "black-box" that provides input-output data pairs for generating rules.

We believe the decompositional approach is more interesting, because the resulting neural network architecture reflects the problem structure in a way, particularly if structure optimization, structural regularization or pruning has been performed during learning [23, 25, 36], which often reduces the complexity of neural networks. Complexity reduction procedure prior to rule extraction is also termed skeletonization [14]. Decompositional approaches to rule extraction from trained neural network can largely be divided into three steps: neural network training, network skeletonization, and rule extraction [15].

13.2.3 Model Selection in Machine Learning

It is interesting to notice that we need to control the complexity of neural networks properly, no matter when we want to generate accurate or interpretable neural network models. However, the complexity for an accurate model and an interpretable model is often different. Usually, the complexity of an interpretable model is lower than an accurate model due to the limited cognitive capacity of human beings. Unfortunately, the complexity of interpretable neural networks may be insufficient for neural networks to achieve a good accuracy. Thus, a trade-off between accuracy and interpretability is often unavoidable.

Model selection is an important topic in machine learning. Traditionally, model selection criteria have been proposed for creating accurate models. The task of model selection is to choose the best model in the sense of accuracy for a set of given data, assuming that a number of models are available. Several criteria have been proposed based on the Kullback-Leibler Information Criterion [9]. The most popular criteria are Akaike's Information Criterion (AIC) and Bayesian Information Criterion (BIC). For example, model selection according to the AIC is to minimize the following criterion:

$$AIC = -2\ log(\mathcal{L}(\theta|y,\ g) + 2\ K, \tag{13.1}$$

where, $\mathcal{L}(\theta|y,\ g)$ is the maximized likelihood for data y given a model g with model parameter θ, K is the number of effective parameters of g.

The first term of Equation (13.1) reflects how good the model approximates the data, while the second term is the complexity of the model. Usually, the higher the model complexity is, the more accurate the approximation on the training data will be. Obviously, a trade-off between accuracy and model complexity has to be taken into account in model selection.

Model selection criteria have often been used to control the complexity of models to a desired degree in model generation. This approach is usually known as regularization in the neural network community [8]. The main purpose of neural network regularization is to improve the generalization capability of a neural network by control its complexity. By generalization, it is meant that a trained neural network should perform well not only on training data, but also on unseen data.

13.3 Multi-objective Optimization Approach to Complexity Reduction

13.3.1 Neural Network Regularization

Neural network regularization can be realized by including an additional term that reflects the model complexity in the cost function of the training algorithm:

$$J = E + \lambda \Omega, \tag{13.2}$$

where E is an error function, Ω is the regularization term representing the complexity of the network model, and λ is a hyperparameter that controls the strength of the regularization. The most common error function in training or evolving neural networks is the mean squared error (MSE):

$$E = \frac{1}{N} \sum_{i=1}^{N} (y^d(i) - y(i))^2, \tag{13.3}$$

where N is the number of training samples, $y^d(i)$ is the desired output of the i-th sample, and $y(i)$ is the network output for the i-th sample. For the sake of clarity, we assume that the neural network has only one output. Refer to [8] for other error functions, such as the Minkowski error or cross-entropy.

Several measures have also been suggested for denoting the model complexity Ω. A most popular regularization term is the squared sum of all weights of the network:

$$\Omega = \frac{1}{2} \sum_{k} w_k^2, \tag{13.4}$$

where k is an index summing up all weights. This regularization method has been termed *weight decay.*

One weakness of the weight decay method is that it is not able to drive small irrelevant weights to zero, when gradient-based learning algorithms are employed, which may result in many small weights [37]. An alternative is to replace the squared sum of the weights with the sum of absolute value of the weights:

$$\Omega = \sum_{i} |w_i|. \tag{13.5}$$

It has been shown that this regularization term it is able to drive irrelevant weights to zero [31].

Both regularization terms in equations (13.4) and (13.5) have also been studied from the Bayesian learning point of view, which are known as the Gaussian regularizer and the Laplace regularizer, respectively.

A more direct measure for model complexity of neural networks is the number of weights contained in the neural network:

$$\Omega = \sum_{i} \sum_{j} c_{ij}, \tag{13.6}$$

where c_{ij} equals 1 if there is connection from neuron j to neuron i, and 0 if not. It should be noticed that the above complexity measure is not generally applicable to gradient-based learning methods.

A comparison of the three regularization terms using multi-objective evolutionary algorithms has been implemented [27]. Different to the conclusions reported in [31] where the gradient-based learning method has been used, it has been shown that regularization using the sum of squared weights is able to change (reduce) the structure of neural networks as efficiently as using the sum of absolute weights. Thus, no significant difference has been observed from the Gaussian and Laplace regularizers when evolutionary algorithms are used for regularization. In other words, the difference observed in [31] is mainly due to the gradient learning algorithm, but not the regularizers themselves.

13.3.2 Multi-objective Optimization Approach to Regularization

It is quite straightforward to notice that neural network regularization in equation (13.2) can be reformulated as a bi-objective optimization problem:

$$\min \{f_1, f_2\} \tag{13.7}$$
$$f_1 = E, \tag{13.8}$$
$$f_2 = \Omega, \tag{13.9}$$

where E is defined in equation (13.3), and Ω is one of the regularization terms defined in equation (13.4), (13.5), or (13.6).

Note that regularization is traditionally formulated as a single objective optimization problem as in Equation (13.2) rather than a multi-objective optimization problem as in equation (13.7). In our opinion, this tradition can be mainly attributed to the fact that traditional gradient-based learning algorithms are not able to solve multi-objective optimization problems.

13.3.3 Simultaneous Generation of Accurate and Interpretable Models

If evolutionary algorithms are used to solve the multi-objective optimization problem in Equation (13.7), multiple solutions with a spectrum of model complexity can be obtained in a single optimization run. Neural network models with a sufficient but not overly high complexity are expected to perform a good approximation on both training and test data. Meanwhile, interpretable symbolic rules can be extracted from those of a lower complexity [28].

An important issue is how to choose models that are of good accuracy. One possibility is to employ model selection criteria we mentioned above, such as AIC, BIC, cross-validation, and bootstrap.

With the Pareto front that trades off between training error and complexity at hand, an interesting question arisen is that if it is possible to pick out the accurate models without resorting to conventional model selection criteria. In this chapter, a first step towards this direction is made by analyzing the performance gain with respect to complexity increase of the obtained Pareto-optimal solutions.

In order to generate interpretable neural networks, rules are extracted from the simplest Pareto-optimal neural networks based on the decompositional approach.

To illustrate the feasibility of the suggested idea, simulations are conducted on two widely used benchmark problems, the breast cancer diagnosis data and the iris data.

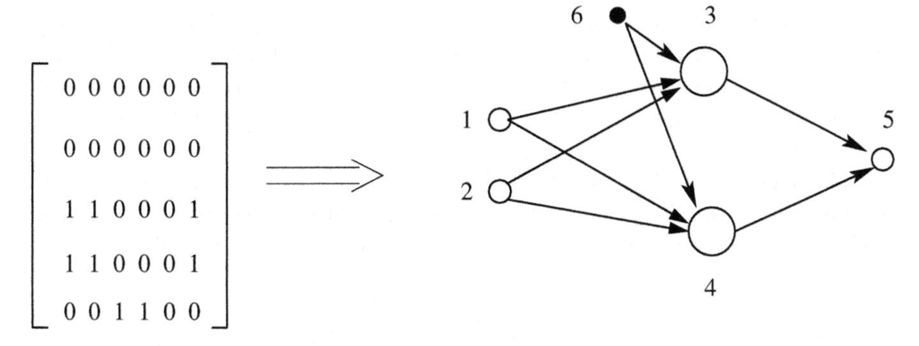

Fig. 13.1. A connection matrix and the corresponding network structure.

13.4 Evolutionary Multi-objective Model Generation

13.4.1 Parameter and Structure Representation of the Network

A connection matrix and a weight matrix are employed to describe the structure and the weights of the neural networks. The connection matrix specifies the structure of the network, whereas the weight matrix determines the strength of each connection. Assume that a neural network consists of M neurons in total, including the input and output neurons, then the size of the connection matrix is $M \times (M+1)$, where an element in the last column indicates whether a neuron is connected to a bias value. In the matrix, if element $c_{ij}, i = 1, ..., M, j = 1, ..., M$ equals 1, it means that there is a connection between the i-th and j-th neuron and the signal flows from neuron j to neuron i. If $j = M+1$, it indicates that there is a bias in the i-th neuron. Fig. 13.1 illustrates a connection matrix and the corresponding network structure. It can be seen from the figure that the network has two input neurons, two hidden neurons, and one output neuron. Besides, both hidden neurons have a bias.

The strength (weight) of the connections is defined in the weight matrix. Accordingly, if the c_{ij} in the connection matrix equals zero, the corresponding element in the weight matrix must be zero too.

13.4.2 Evolution and Life-time Learning

A genetic algorithm has been used for optimizing the structure and weights of the neural networks. Binary coding is adopted representing the neural network structure and real-valued coding for encoding the weights. Five genetic operations have been introduced in the evolution of the neural networks, four of which mutate the connection matrix (neural network structure) and one of which mutates the weights. The four mutation operators are the insertion of a hidden neuron, deletion of a hidden neuron, insertion of a connection and deletion of a connection [19]. A Gaussian-type mutation is applied to mutate the weight matrix. No crossover has been employed in this algorithm.

After mutation, an improved version of the Rprop algorithm [20] has been employed to train the weights. This can be seen as a kind of life-time learning (the first objective only) within a generation. Both Baldwin effect and Lamarckian evolution have been examined in this work, which will be discussed further in Section 13.4.3.

The Rprop learning algorithm [34] is believed to be a fast and robust learning algorithm. Let w_{ij} denotes the weight connecting neuron j and neuron i, then the change of the weight (Δw_{ij}) in each iteration is as follows:

$$\Delta w_{ij}^{(t)} = -\text{sign}\left(\frac{\partial E^{(t)}}{\partial w_{ij}}\right) \cdot \Delta_{ij}^{(t)}, \tag{13.10}$$

where $sign(\cdot)$ is the sign function, $\Delta_{ij}^{(t)} \geq 0$ is the step-size, which is initialized to Δ_0 for all weights. The step-size for each weight is adjusted as follows:

$$\Delta_{ij}^{(t)} = \begin{cases} \xi^+ \cdot \Delta_{ij}^{(t-1)} & , \text{ if } \frac{\partial E^{(t-1)}}{\partial w_{ij}} \cdot \frac{\partial E^{(t)}}{\partial w_{ij}} > 0 \\ \xi^- \cdot \Delta_{ij}^{(t-1)} & , \text{ if } \frac{\partial E^{(t-1)}}{\partial w_{ij}} \cdot \frac{\partial E^{(t)}}{\partial w_{ij}} < 0 \\ \Delta_{ij}^{(t-1)} & , \text{ otherwise} \end{cases}, \tag{13.11}$$

where $0 < \xi^- < 1 < \xi^+$. To prevent the step-sizes from becoming too large or too small, they are bounded by $\Delta_{\text{min}} \leq \Delta_{ij} \leq \Delta_{\text{max}}$.

One exception must be considered. After the weights are updated, it is necessary to check if the partial derivative changes sign, which indicates that the previous step might be too large and thus a minimum has been missed. In this case, the previous weight change should be retracted:

$$\Delta w^{(t)} = -\Delta_{ij}^{(t-1)}, \text{ if } \frac{\partial E^{(t-1)}}{\partial w_{ij}} \cdot \frac{\partial E^{(t)}}{\partial w_{ij}} < 0. \tag{13.12}$$

Recall that if the weight change is retracted in the t-th iteration, the $\partial E^{(t)}/\partial w_{ij}$ should be set to 0.

In reference [20], it is argued that the condition for weight retraction in equation (13.12) is not always reasonable. The weight change should be retracted only if the partial derivative changes sign and if the approximation error increases. Thus, the weight retraction condition in equation (13.12) is modified as follows:

$$\Delta w^{(t)} = -\Delta_{ij}^{(t-1)}, \text{ if } \frac{\partial E^{(t-1)}}{\partial w_{ij}} \cdot \frac{\partial E^{(t)}}{\partial w_{ij}} < 0 \textbf{ and } E^{(t)} > E^{(t-1)}. \tag{13.13}$$

It has been shown on several benchmark problems in [20] that the modified Rprop (termed as Rprop$^+$ in [20]) exhibits consistently better performance than the Rprop algorithm.

13.4.3 Baldwin Effect and Lamarckian Evolution

When life-time learning is embedded in evolution, two strategies can often be adopted. In the first approach, the fitness value of each individual will be replaced with the new fitness resulted from learning. However, the allele of the chromosome remains unchanged. In the second approach, both the fitness and the allele are replaced with the new values found in the life-time learning. These two strategies are known as the Baldwin effect and the Lamarkian evolution, which are two terms borrowed from biological evolution [5, 6].

13.4.4 Elitist Non-dominated Sorting

In this work, the elitist non-dominated sorting method proposed in the NSGA-II algorithm [13] has been adopted. Assume the population size is N. At first, the offspring and the parent populations are combined. Then, a non-domination rank and a local crowding distance are assigned to each individual in the combined population. In selection, all non-dominated individuals (say there are N_1 non-dominated solutions) in the combined population are passed to the offspring population, and are removed from the combined population. Now the combined population has $2N - N_1$ individuals. If $N_1 < N$, the non-dominated solutions in the current combined population (say there are N_2 non-dominated solutions) will be passed to the offspring population. This procedure is repeated until the offspring population is filled. It could happen that the number of non-dominated solutions in the current combined population (N_i) is larger than the left slots ($N - N_1 - N_2 - ... - N_{i-1}$) in the current offspring population. In this case, the $N - N_1 - N_2 - ... - N_{i-1}$ individuals with the largest crowding distance from the N_i non-dominated individuals will be passed to the offspring generation.

13.5 Simulation Results

13.5.1 Benchmark Problems

To illustrate the feasibility of the idea of generating accurate and interpretable models simultaneously using the evolutionary multi-objective optimization approach, the breast cancer data and the iris data in the UCI repository of machine learning database [33] are studied in this work. The breast cancer data contains 699 examples, each of which has 9 inputs and 2 outputs. The inputs are: clump thickness (x_1), uniformity of cell size (x_2), uniformity of cell shape (x_3), marginal adhesion (x_4), single epithelial cell size (x_5), bare nuclei (x_6), bland chromatin (x_7), normal nucleoli (x_8), and mitosis (x_9). All inputs are normalized, to be more exact, $x_1, ..., x_9 \in \{0.1,\ 0.2,\ ..., 0.8,\ 0.9,\ 1.0\}$. The two outputs are complementary binary value, i.e., if the first output is 1, which means "benign", then the second output is 0. Otherwise, the first output is

0, which means "malignant", and the second output is 1. Therefore, only the first output is considered in this work. The data samples are divided into two groups: one training data set containing 599 samples and one test data set containing 100 samples. The test data are unavailable to the algorithm during the evolution.

The iris data has 4 attributes, namely, Sepal-length, Sepal-width, Petal-length and Petal-width. The 150 data samples can be divided into 3 classes, Iris-Setosa (class 1), Iris-Versicolor (class 2), and Iris-Virginica (class 3), each class having 50 samples. In this work, 40 data samples for each class are used for training and the rest 10 data samples are for test.

13.5.2 Experimental Setup

The population size is 100 and the optimization is run for 200 generations. One of the five mutation operations is randomly selected and performed on each individual. The standard deviation of the Gaussian mutations applied on the weight matrix is set to 0.05. The weights of the network are initialized randomly in the interval of $[-0.2, 0.2]$. In the Rprop$^+$ algorithm, the step-sizes are initialized to 0.0125 and bounded between $[0, 50]$ during the adaptation, and $\xi^- = 0.2$, $\xi^+ = 1.2$, which are the default values recommended in [20] and 50 iterations are implemented in each local search.

Although a non-layered neural network can be generated using the coding scheme described in Section 13.3, only feedforward networks with one hidden layer are generated in this research. The maximum number of hidden nodes is set to 10. The hidden neurons are nonlinear and the output neurons are linear. The activation function used for the hidden neurons is as follows,

$$g(z) = \frac{x}{1 + |x|}, \tag{13.14}$$

which is illustrated in Fig. 13.2.

13.5.3 Comparison of Baldwin Effect and Lamarckian Evolution

To compare the performance of the Baldwin effect and Lamarkian evolution in the context of evolutionary multi-objective optimization of neural networks, the Pareto fronts achieved using the the two strategies are plotted in Fig. 13.3 for the breast cancer data and in Fig. 13.4 for the iris data.

From these results, it can be seen that the solution set from the Lamarkian evolution shows clearly better performance in all aspects including closeness, distribution and spread. The spread of the Pareto-optimal solutions is particularly poor for the breast cancer data, which might indicate that efficient life-time learning becomes more important for harder learning tasks, recalling that the breast cancer data has more input attributes and more data samples. Thus, only the results using the Lamarckian evolution will be considered in the following sections.

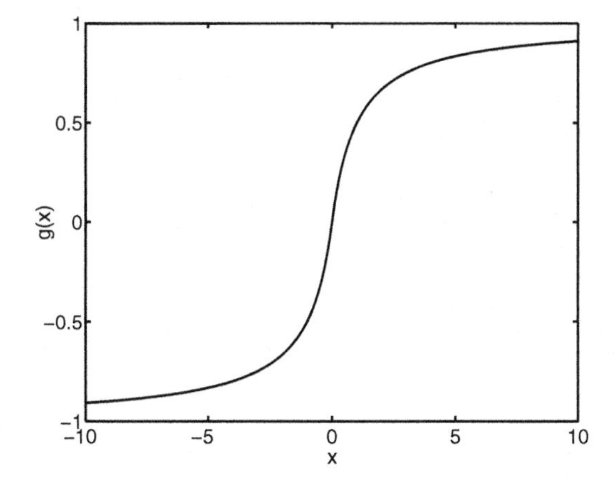

Fig. 13.2. The activation function of the hidden nodes.

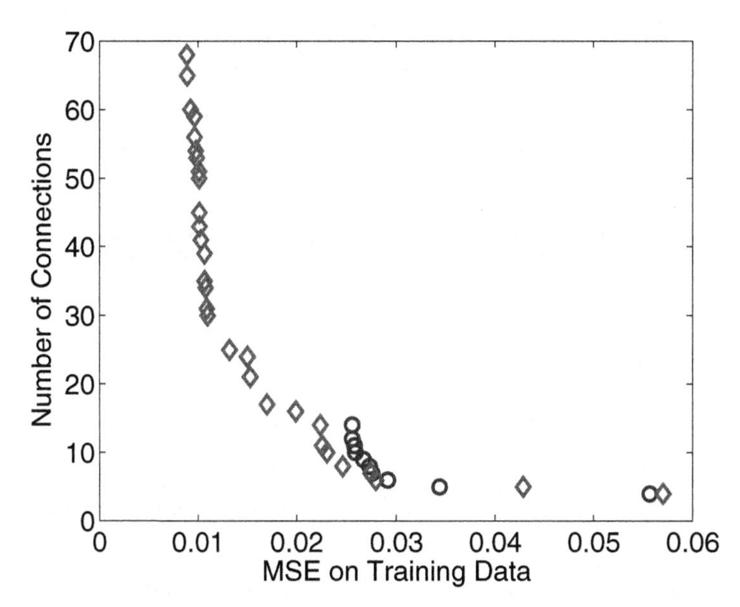

Fig. 13.3. Pareto fronts obtained using the Baldwin effect (circles) and Lamarckian evolution (diamonds) for the breast cancer data.

13.5.4 Rule Extraction from Simple Neural Networks

To obtain interpretable neural networks, we attempt to extract symbolic rules from simple neural networks. For the breast cancer data, the simplest Pareto-optimal neural network has only 4 connections: 1 input node, one hidden node and 2 biases, see Fig 13.5(a). The mean squared error (MSE) of the network on training and test data are 0.0546 and 0.0324, respectively.

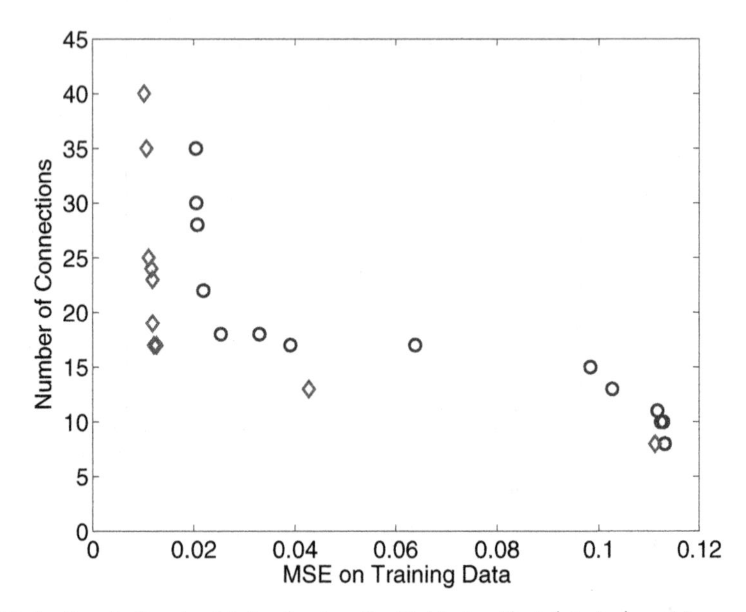

Fig. 13.4. Pareto fronts obtained using the Baldwin effect (circles) and Lamarckian evolution (diamonds) for the iris data.

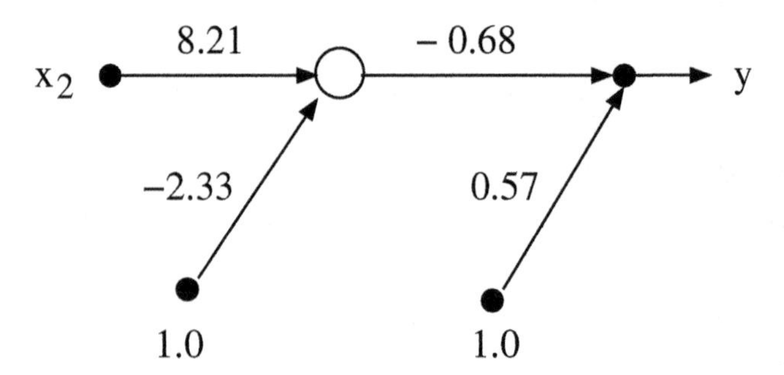

Fig. 13.5. The simplest neural network for the breast cancer data.

Assuming that a case can be decided to be "malignant" if $y < 0.25$, and "benign" if $y > 0.75$, we can then derive that if $x_2 > 0.4$, which means that $x_2 \geq 0.5$, then "malignant" and "benign" if $x_2 < 0.22$, i.e., then $x_2 \leq 0.2$.

From such a simple network, the following two symbol-type rules can be extracted (denoted as MOO_NN1):

$$\text{R1: If } x_2 \text{ (uniformity) } \geq 0.5, \text{ then malignant;} \tag{13.15}$$

$$\text{R2: If } x_2 \text{ (uniformity) } \leq 0.2, \text{ then benign.} \tag{13.16}$$

Based on these two simple rules, only 2 out of 100 test samples will be misclassified, and 4 of them cannot be decided with a predicted value of 0.49,

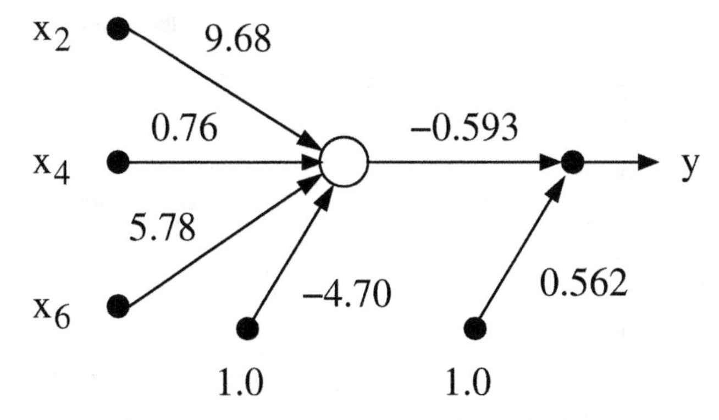

Fig. 13.6. The next simplest neural network on the breast cancer data.

which is very ambiguous. The prediction results on the test data are presented in Fig 13.5(b).

Now let us look at the second simplest network, which has 6 connections in total. The connection and weights of the network are given in Fig. 13.6(a), and the prediction results are provided in Fig. 13.6(b). The MSE of the network on training and test data are 0.0312 and 0.0203, respectively.

In this network, x_2, x_4 and x_6 are present. If the same assumptions are used in deciding whether a case is benign or malignant, then we could extract the following rules: (denoted as MOO_NN2)

$$
\begin{aligned}
\text{R1: If} \quad & x_2 \text{ (uniformity) } \geq 0.6 \quad \text{or} \\
& x_6 \text{ (bare nuclei) } \geq 0.9 \quad \text{or} \\
& x_2 \text{ (uniformity) } \geq 0.5 \wedge x_6 \text{ (bare nuclei) } \geq 0.2 \quad \text{or} \\
& x_2 \text{ (uniformity) } \geq 0.4 \wedge x_6 \text{ (bare nuclei) } \geq 0.4 \quad \text{or} \\
& x_2 \text{ (uniformity) } \geq 0.3 \wedge x_6 \text{ (bare nuclei) } \geq 0.5 \quad \text{or} \\
& x_2 \text{ (uniformity) } \geq 0.2 \wedge x_6 \text{ (bare nuclei) } \geq 0.7, \text{ then malignant;}
\end{aligned}
\tag{13.17}
$$

$$
\begin{aligned}
\text{R2: If} \quad & x_2 \text{ (uniformity) } \leq 0.1 \wedge x_6 \text{ (bare nuclei) } \leq 0.4 \quad \text{or} \\
& x_2 \text{ (uniformity) } \leq 0.2 \wedge x_6 \text{ (bare nuclei) } \leq 0.2, \text{ then benign;}
\end{aligned}
\tag{13.18}
$$

Compared to the simplest network, with the introduction of two additional features x_6 and x_4 (although the influence of x_4 is too small to be reflected in the rules), the number of cases that are misclassified has been reduced to 1, whereas the number of cases on which no decision can be made remains to be 4, although the ambiguity of the decision for the four cases did decrease.

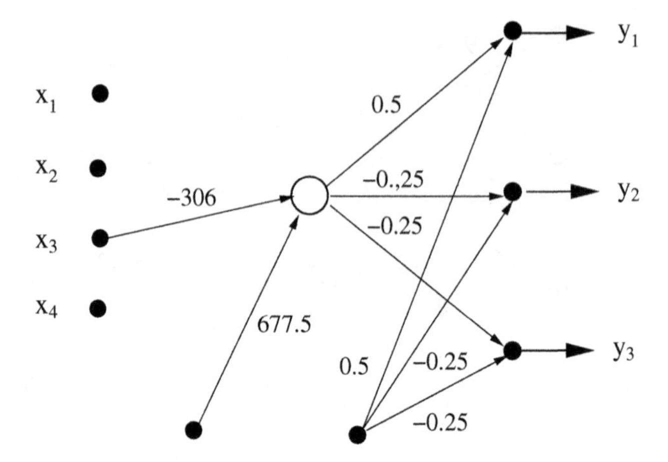

Fig. 13.7. The simplest neural network for the iris data.

Similar analysis has been done on the Iris data. The simplest neural network has 8 connections, where only attribute 1 has been selected, refer to Fig. 13.7.

From the neural network structure, it can be seen that the network can only distinguish the class Iris-Setosa (class 3) from others, but is still too simple to distinguish between Iris-Versicolor and Iris-Virginica. One rule can be extracted from the neural network:

$$\text{R1: If } x_3(\text{Petal-Length}) < 2.2, \text{ then Setosa;} \tag{13.19}$$

The second simple Pareto-optimal neural network has 13 connections with 2 inputs, see Fig. 13.8. Though the structure is still very simple, the network is able to separate three classes correctly. By analyzing the structure of the neural network, the following three rules can be extracted:

$$\text{R1:If } x_3(\text{Petal-Length}) \leq 2.2, x_4(\text{Petal-Width}) < 1.0, \text{ then Setosa;} \tag{13.20}$$

$$\text{R2:If } x_3(\text{Petal-Length}) > 2.2, x_4(\text{Petal-Width}) < 1.4, \text{ then Versicolor;} \tag{13.21}$$

$$\text{R3:If } \quad x_4(\text{Petal-Width}) > 1.8, \quad \text{ then Virginica;} \tag{13.22}$$

From the above analyses, we have been successful in extracting understandable symbolic rules from the simple neural networks generated using the evolutionary multi-objective algorithm. Thus, we demonstrate that the idea to generate interpretable neural networks by reducing the complexity of the neural networks is feasible.

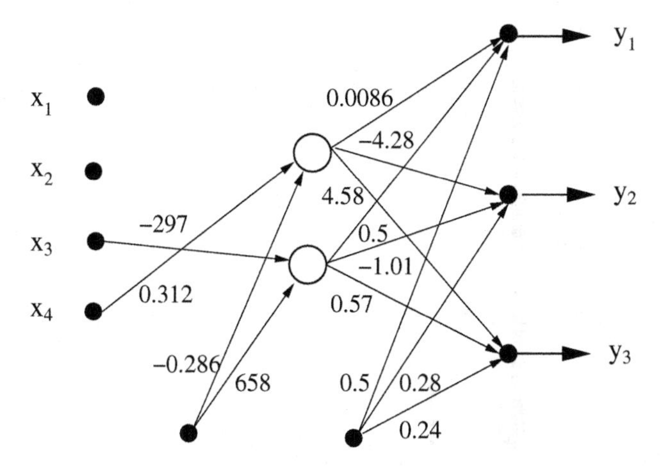

Fig. 13.8. The second simplest neural network for the iris data.

13.5.5 Identifying Accurate Neural Network Models

One advantage of evolutionary multi-objective optimization approach to neural network generation is that a number of neural networks with a spectrum of complexity can be obtained in one single run. With these neural networks that trade off between training error and model complexity, we hope that we can identify those networks that are most likely to have good generalization capability, i.e., perform well both on training data and unseen test data.

In multi-objective optimization, the "knee" of a Pareto front is often considered interesting. The reason is that for the solutions at the knee, a small improvement in one objective will cause large deterioration in the other objective. However, such a knee is not always obvious in some applications. Thus, it is important to define a quantitative measure for the knee according to the problem at hand. In [17], some interesting work has been presented to determine the number of clusters by exploit the Pareto front in multi-objective data clustering.

If we take a closer look at the Pareto fronts that we obtained for the two benchmark problems, we notice that when the model complexity increases, the accuracy on training data also increases. However, the amplitude in accuracy increase is not always the same when the number of connections increases. Thus, we introduce the concept of "normalized performance gain (NPG)", which is defined as follows:

$$NPG = \frac{MSE_i - MSE_j}{N_i - N_j}, \tag{13.23}$$

where MSE_i, MSE_j, N_i, N_j are the MSE on training data, and the number of connections of the i-th and j-th Pareto optimal solutions.

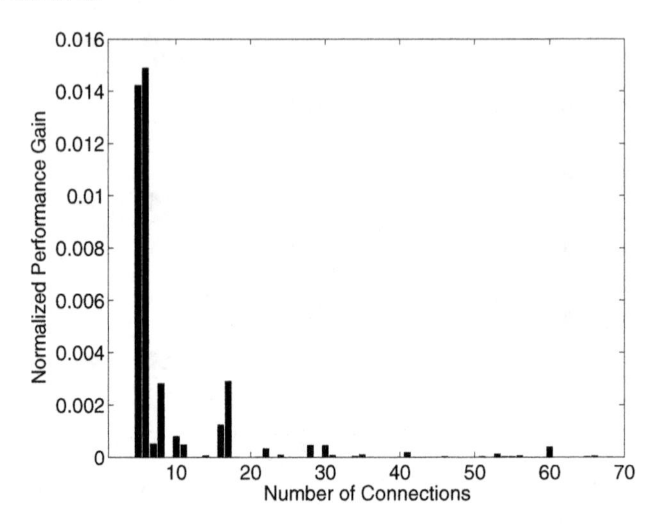

Fig. 13.9. Normalized performance gain for the breast cancer data.

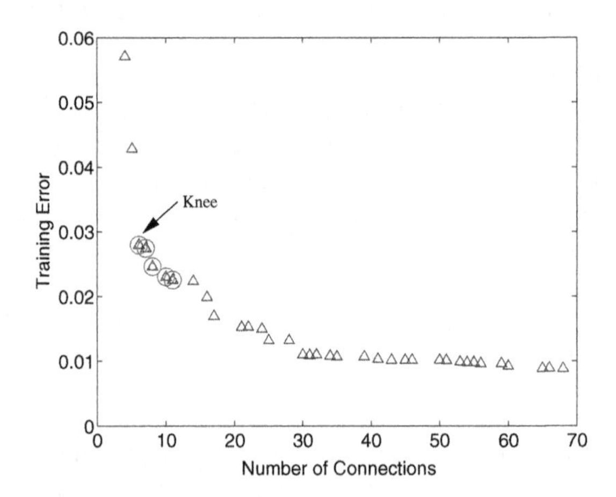

Fig. 13.10. Complexity versus training error for the breast cancer data. The five solutions right to the knee are circled.

The normalized performance gain on the breast cancer data is illustrated in Fig. 13.9. We notice that the largest NPG is achieved when the number of connections is increased from 4 to 6 and from 6 to 7. After that, the NPG decreases almost to zero when the number of connections is increased to 14, though there is some small fluctuations. Thus, we denote the Pareto optimal solution with 7 connections as the knee of the Pareto front, see Fig 13.10. To investigate if there is any correlation between the knee point and the approximation error on test data, we examine the test error of the neighboring five Pareto-optimal solutions right to the knee, which are circled in Fig.13.10.

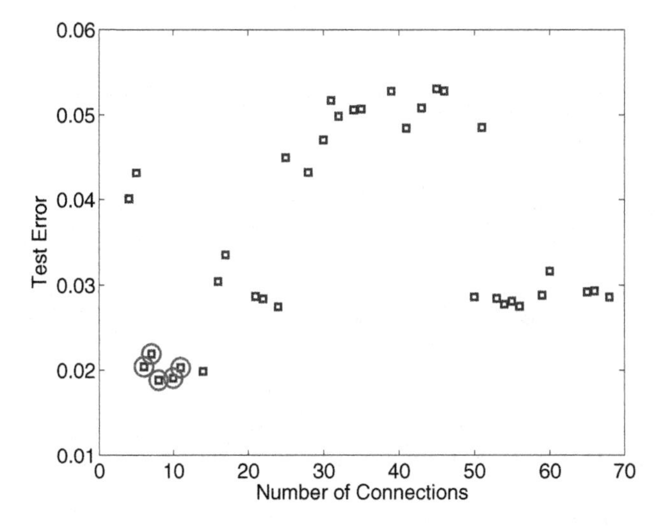

Fig. 13.11. Complexity versus test error for the breast cancer data. The circled solutions are those neighbor solutions right to the knee on the Pareto front.

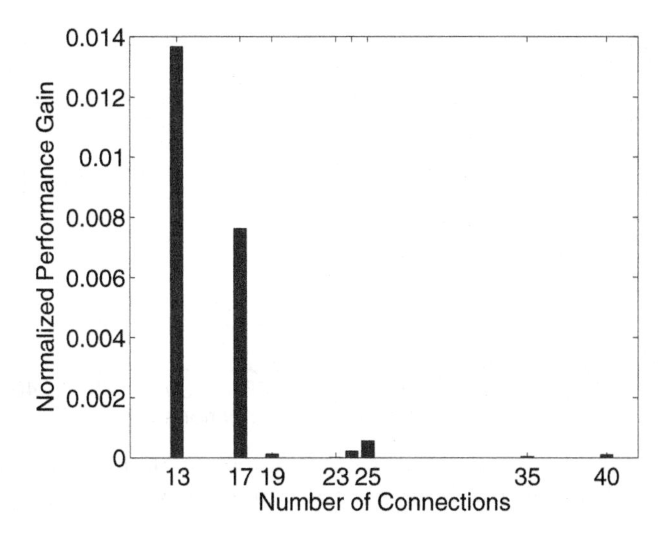

Fig. 13.12. Normalized performance gain for the iris data.

Very interestingly, we can see from Fig 13.11 that all the five solutions belong to the cluster also have the smallest error on test data.

We do the same analysis on the iris data. The NPG on the iris data is presented in Fig. 13.12. Similarly, the NPG is large when the number of connections increases from 8 to 13 and from 13 to 17. After that, the NPG decreases dramatically. Thus, the solution with 14 connections is considered as the knee of the Pareto front, see Fig. 13.13.

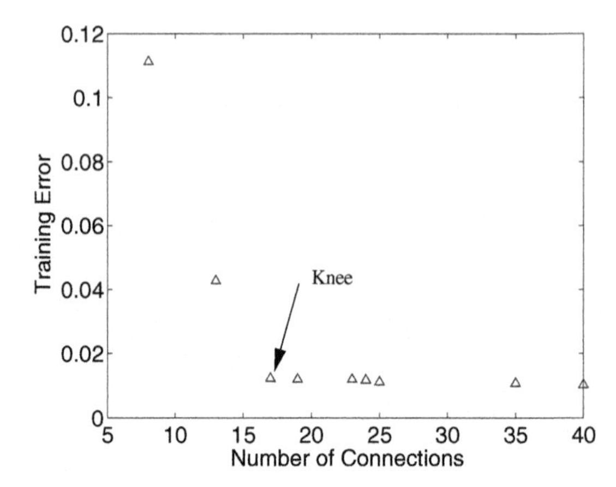

Fig. 13.13. Complexity versus training error for the iris data.

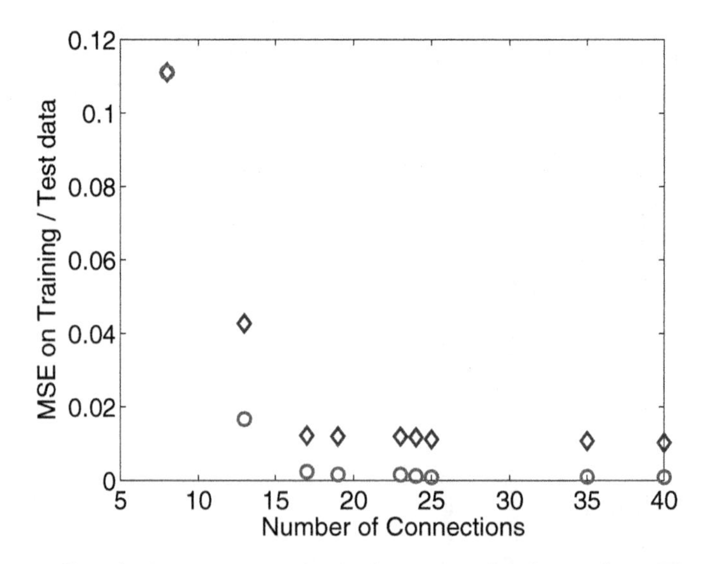

Fig. 13.14. Complexity versus training and test error for the iris data. The training errors are denoted by diamonds and the test error by circles.

However, when we look at the test error of all the solutions, we find that no observable overfitting has occurred with the increase in complexity. Actually, the test error shows very similar behavior to the training data with respect to the increase in complexity. In fact, the test errors are even smaller than the training error, refer to Fig. 13.14. If we take a look at the training and test data as shown in Fig. 13.15, this turns out to be natural, because none of the test samples (denoted by filled shapes) are located in the overlapping part of the different classes.

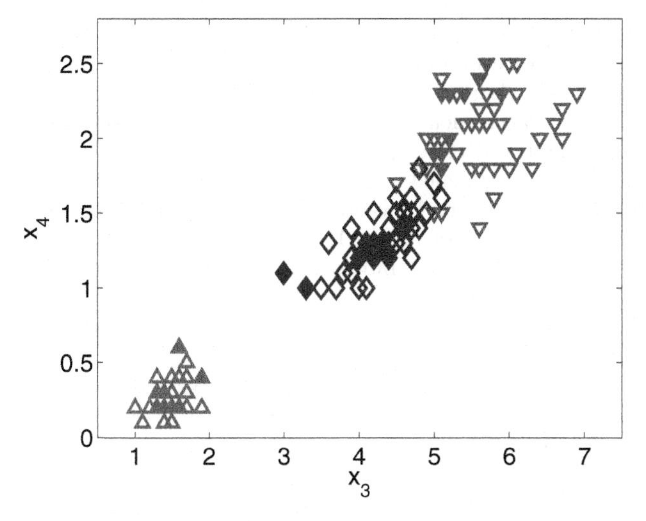

Fig. 13.15. Location of the training and test samples in the $x_3 - x_4$ space of the iris data. Training and test data of class 1 are denoted by triangles and filled triangles, class 2 by diamonds and filled, and class 3 by triangles down and filled triangles down, respectively.

13.6 Conclusions

Generating accurate and interpretable neural networks simultaneously is very attractive in many applications. This work presents a method for achieving this goal using an evolutionary multi-objective algorithm. The accuracy on the training data and the number of connections of the neural network are taken as the two objectives in optimization. We believe the complexity is a very important objective for learning models in that both generalization performance and interpretability are closely related to the complexity.

By analyzing the achieved Pareto-optimal solutions that trade off between training accuracy and complexity, we are able to choose neural networks with a simpler structure for extracting interpretable symbolic rules. In the meantime, neural networks that have good accuracy on unseen data can also be picked out by choosing those solutions in the neighborhood of the knee of the Pareto front, if overfitting occurs.

Much work remains to be done in the future. First it is interesting to compare our method with model selection criteria such as cross-validation or bootstrap methods. Second, this method should be verified on more benchmark problems.

References

[1] H.A. Abbass. A memetic Pareto evolutionary approach to artificial neural networks. *Proc. of the 14th Australian Joint Conf. on Artificial Intelligence,*

pages 1–12, 2001
[2] H.A. Abbass. An evolutionary artificial neural networks approach for breast cancer diagnosis. *Artificial Intelligence in Medicine*, 25(3):265–281, 2002
[3] H.A. Abbass. Pareto neuro-ensemble: Constructing ensembles of neural networks using multi-objective optimization. *Congress on Evolutionary Computation*, pages 2074–2080, 2003
[4] H.A. Abbass. Speeding up back-propagation using multi-objective evolutionary algorithms. *Neural Computation*, 15(11):2705–2726, 2003
[5] D.H. Ackley, M.L. Littman. Interactions between learning and evolution. *Artificial Life*, 2:487–509, 1992
[6] D.H. Ackley, M.L. Littman. A case for Lamarckian evolution. *Artificial Life*, 3:3–10, 1994
[7] R. Andrews, J. Diederich, and A. Tickle. A survey and critique of techniques for extracting rules from trained artificial neural networks. *Knowledge Based Systems*, 8(6):373–389, 1995
[8] C. M. Bishop. *Neural Networks for Pattern Recognition.* Oxford University Press, Oxford, UK, 1995
[9] K.P. Burnham and D.R. Anderson. *Model Selection and Multimodel Inference.* Springer, New York, second edition, 2002
[10] Evolutionary framework for the construction of diverse hybrid ensembles. In: *Proc. of 13th European Symposium on Artificial Neural Networks.* pages 253–258, 2005
[11] C. Coello Coello, D. Veldhuizen, and G. Lamont. *Evolutionary algorithms for solving multi-objective problems.* Kluwer Academic, New York, 2002
[12] K. Deb. *Multi-objective Optimization Using Evolutionary Algorithms.* Wiley, Chichester, 2001
[13] K. Deb, S. Agrawal, A. Pratap, and T. Meyarivan. A fast elitist non-dominated sorting genetic algorithm for multi-objective optimization: NSGA-II. In *Parallel Problem Solving from Nature*, volume VI, pages 849–858, 2000
[14] W. Duch, R. Adamczak, and K. Grabczewski. Extraction of logical rules from backpropagation networks. *Neural Processing Letters*, 7:1–9, 1998.
[15] W. Duch, R. Setiono, and J. Zurada. Computational intelligence methods for rule-based data understanding. *Proceedings of the IEEE*, 92(5):771–805, 2004.
[16] J.E. Fieldsend, S. Singh. Optimizing forecast model complexity using multi-objective evolutionary algorithms. In *Applications of Multi-objective Evolutionary Algorithms*, C. Coello Coello, G.B. Lamont (eds.), pages 675–700. World Scientific, 2004
[17] J. Handl and J. Knowles. Exploiting the trade-off - The benefits of multiple objectives in data clustering. *Evolutionary Multi-Criterion Optimization*, LNCS 3410, pages 547–560, 2005
[18] K. Hornik, M. Stinchcombe, H. White. Multilayer feedforward networks are universal approximators. *Neural Networks*, 2(5):359–366, 1989
[19] M. Hüsken, J. E. Gayko, and B. Sendhoff. Optimization for problem classes – Neural networks that learn to learn. In Xin Yao and David B. Fogel, editors, *IEEE Symposium on Combinations of Evolutionary Computation and Neural Networks (ECNN 2000)*, pages 98–109. IEEE Press, 2000
[20] C. Igel and M. Hüsken. Improving the Rprop learning algorithm. In *Proc. of the 2nd ICSC Int. Symposium on Neural Computation*, pages 115–121, 2000

[21] C. Igel. Multi-objective model selection for support vector machines. *Evolutionary Multi-Criterion Optimization*, LNCS 3410, pages 534–546, 2005
[22] H. Ishibuchi, T. Yamamoto. Evolutionary multiobjective optimization for generating an ensemble of fuzzy rule-based classifiers. In *Genetic and Evolutionary Computation Conference*, pages 1077–188, 2003
[23] M. Ishikawa. Rule extraction by successive regularization. *Neural Networks*, 13:1171–1183, 2000
[24] Y. Jin. *Advanced Fuzzy Systems Design and Applications*. Springer, Heidelberg, 2003
[25] Y. Jin, B. Sendhoff. Extracting interpretable fuzzy rules from RBF networks. *Neural Processing Letters*, 17(2):149–164, 2003
[26] Y. Jin, T. Okabe, B. Sendhoff. Evolutionary multi-objective optimization approach to constructing neural network ensembles for regression. In *Applications of Multi-objective Evolutionary Algorithms*, C. Coello Coello, G.B. Lamont (eds.), pages 653–673. World Scientific, 2004
[27] Y. Jin, T. Okabe, B. Sendhoff. Neural network regularization and ensembling using multi-objective evolutionary algorithms. In *Congress on Evolutionary Computation*, pages 1–8. IEEE, 2004
[28] Y. Jin, B. Sendhoff, E. Körner. Evolutionary multi-objective optimization for simultaneous generation of signal-type and symbol-type representations. *Evolutionary Multi-Criterion Optimization*, LNCS 3410, pages 752–766, 2005
[29] K. Kottathra, Y. Attikiouzel. A novel multicriteria optimization algorithm for the structure determination of multilayer feedforward neural networks. *Journal of Network and Computer Applications*, 19:135–147, 1996.
[30] M.A. Kupinski, M.A. Anastasio. Multiobjective genetic optimization of diagnostic classifiers with implementations for generating receiver operating characteristic curves. *IEEE Transactions on Medical Imaging*, 18(8):675–685, 1999
[31] D.A. Miller, J.M. Zurada. A dynamical system perspective of structural learning with forgetting. *IEEE Transactions on Neural Networks*, 9(3):508–515, 1998
[32] S. Park, D. Nam, C.H. Park. Design of a neural controller using multiobjective optimization for nonminimum phase. *Proc. of IEEE Int. Conf. on Fuzzy Systems* , I:533-537, 1999
[33] L. Prechelt. PROBEN1 - a set of neural network benchmark problems and benchmarking rules. Technical Report 21/94, Fakultät für Informatik, Universität Karlsruhe, 1994
[34] M. Riedmiller, H. Braun. A direct adaptive method for faster backpropgation learning: The RPROP algorithm. In *IEEE Int. Conf. on Neural Networks*, volume 1, pages 586–591, New York, 1993
[35] R. Setiono. Generating concise and accurate classification rules for breast cancer diagnosis. *Artificial Intelligence in Medicine*, 18:205–219, 2000
[36] R. Setiono, H. Liu. Symbolic representation of neural networks. *IEEE Computer*, 29(3):71–77, 1996
[37] R.D. Reed, R.J. Marks II. *Neural Smithing*. The MIT Press, 1999
[38] I. Taha, J. Ghosh. Symbolic interpretation of artificial neural networks. *IEEE Transactions on Knowledge and Data Engineering*, 11(3):448–463, 1999
[39] R. de A. Teixeira, A.P. Braga, R. H.C. Takahashi, R. R. Saldanha. Improving generalization of MLPs with multi-objective optimization. *Neurocomputing*, 35:189–194, 2000

[40] H. Wang, S. Kwong, Y. Jin, W. Wei, K.F. Man. Agent-based evolutionary approach for interpretable rule-based knowledge extraction. *IEEE Transactions on Systems, Man, and Cybernetics, Part C: Applications and Reviews*, 35(2):143–155, 2005

[41] S. Wiegand, C. Igel, U. Handmann. Evolutionary multi-objective optimization of neural networks for face detection. *Int. Journal of Computational Intelligence and Applications*, 4:237–253, 2004

14

GA-Based Pareto Optimization for Rule Extraction from Neural Networks

Urszula Markowska-Kaczmar, Krystyna Mularczyk

Wroclaw University of Technology,
Institute of Applied Informatics
Wyb. Wyspianskiego 27, 50-370 Wroclaw, Poland
urszula.markowska-kaczmar@pwr.wroc.pl

Summary. The chapter presents a new method of rule extraction from trained neural networks, based on a hierarchical multiobjective genetic algorithm. The problems associated with rule extraction, especially its multiobjective nature, are described in detail, and techniques used when approaching them with genetic algorithms are presented. The main part of the chapter contains a thorough description of the proposed method. It is followed by a discussion of the results of experimental study performed on popular benchmark datasets that confirm the method's effectiveness.

14.1 Introduction

For many years neural networks have been successfully used for solving various complicated problems. The areas of their applications include communication systems, signal processing, the theory of control, pattern and speech recognition, weather prediction and medicine. Neural networks are used especially in situations when algorithms for a given problem are unknown or too complex. In order to solve a problem, a network does not need an algorithm; instead, it must be provided with training examples that enable it to learn the correct solutions. The most commonly used neural networks have the input and output data presented in the form of vectors. Such a pair of vectors is an example of the relationship that a network should learn. The training consists of a certain iterative process of modifying the network's parameters, so that the answer given by the network for each of the input patterns is as close to the desired one as possible.

The unquestionable advantage of neural networks is their ability of generalization, which consists in giving correct answers for new input patterns that had not been used for training. Their resistance to errors is equally important. This means that networks can produce the right answers also in the case of noisy or missing data. Finally, it should be emphasized that they are fast

U. Markowska-Kaczmar and K. Mularczyk: *GA-Based Pareto Optimization for Rule Extraction from Neural Networks*, Studies in Computational Intelligence (SCI) **16**, 313–338 (2006)
www.springerlink.com

while processing data and relatively cheap to build. Nevertheless, a serious disadvantage is that neural networks produce answers in a way that is incomprehensible for humans. Because in certain areas of applications, for instance in medicine, the trustworthiness of a system that supports people's work is essential, the domain of extracting knowledge from neural networks has been developed since the 90s [1]. This knowledge describes in an understandable way the performance of a neural network that solves a given problem. Knowledge extraction may also be perceived as a tool of verifying what has actually been learned by a network.

Rapid development of data storage techniques has been observed in the last years. Data is presently perceived as a source of knowledge and exploited by a dynamically evolved area of computer science called *data mining*. Its methods aim at finding new and correct patterns in the processed data. The problems that data mining deals with include classification, regression, grouping and characteristics. The majority of them may be approached by means of neural networks. In this case knowledge extraction from a network could make its behavior more trustworthy.

The discovered knowledge, if presented in a comprehensible way, could be used for building expert systems as a source alternative or supplementary to the knowledge obtained from human experts. Rule extraction from neural networks might be used for this purpose as well as perceived as a method of automatic rule extraction for an expert system. The system itself could be built as a hybrid consisting of a neural network and a set of rules that describe its performance. Such a hybrid would combine the advantages of both components; it would work fast, which is typical of neural networks, and, on the other hand, it would offer a possibility of explaining the way of reaching a particular conclusion, which is the main quality of expert systems. Moreover, if the data changed, the system's reconstruction could be performed relatively quickly by repeating the process of training the network and extracting knowledge.

Skeptics could ask why knowledge should be extracted from a network and not directly from data. Networks' ability to deal with noisy data should be emphasized again. It tends to be much easier to extract knowledge after having the data processed by a network.

However, if we imagine a network with dozens of inputs with various types, the problem of finding restrictions that describe the conditions of belonging to a given class becomes NP–hard. Searching such a vast space of solutions may be efficiently performed by genetic algorithms – another nature–based technique, which became the basis for a method of rule extraction called MulGEx (**Mul**tiobjectve **G**enetic **Ex**tractor) that will be presented in this chapter.

14.2 The State-of-the-Art in Rule Extraction from Neural Networks

The research on effective methods of acquiring knowledge from neural networks has been carried out for 15 years, which testifies not only to the importance of the problem, but also to its complexity.

The majority of methods apply to networks that perform the classification task, although works concerning rule extraction for the regression task appear as well [17]. The developed approaches may be divided into three main categories: global, local and mixed. This taxonomy bases on the degree to which a method examines the structure of a network.

Global methods treat a network as a black box, observing only its inputs and responses produced at the outputs. In other words, a network provides the method with training patterns. The examples include VIA [21] that uses a procedure similar to classical sensibility analysis, BIO-RE [20] that applies truth tables to extract rules, Ruleneg [8], where an adaptation of PAC algorithm is applied, or [15], based on inversion. It is worth mentioning that in such approaches the architecture of a neural network is insignificant. Other algorithms in this group treat the rule extraction as a machine learning task where neural network delivers training patterns. In this case different machine learning algorithms are used, for example genetic algorithms or decision tree methods.

The opposite approach is the local one, where the first stage consists of describing the conditions of a single neuron's activation, i.e., in determining the values of its inputs that produce an output equal to 1. Such rules are created for all neurons and concatenated on the basis of mutual dependencies. Thus we obtain rules that describe the relations between the inputs and outputs for the entire network. As examples one may mention Partial RE [20], M of N [7], Full RE [20], RULEX [2]. The problem of rule extraction by means of the local approach is simple if the network is relatively small. Otherwise different approaches are developed in order to reduce the architecture of a neural network. Some methods group the hidden neurons' activations, substituting a cluster of neurons by one neuron, other optimize the structure of a neural network introducing a special training procedure or using genetic algorithms [16], [10].

The last group encompasses mixed methods that combine the two approaches described above.

Most of the methods concern multilayer perceptrons (MLP networks). However, methods dedicated to other networks are developed as well, e.g., [18], [5], [9].

Knowledge acquired from neural networks is represented as crisp prepositional rules. Fuzzy rules [14], first order rules and decision trees are used, too.

There are many criteria of the evaluation of the extracted rules' quality. The most frequently used include:

- fidelity,
- accuracy,
- consistency,
- comprehensibility.

In [1] these four requirements are abbreviated to FACC. *Accuracy* is determined on the basis of the number of previously unseen patterns that have been correctly classified. *Fidelity* stands for the degree to which the rules reflect the behavior of the network they have been extracted from. *Consistency* occurs if, during different training sessions, the network produces sets of rules that classify unseen patterns in the same way. *Comprehensibility* is defined as the ratio of the number of rules to the number of premises in a single rule. These criteria are discussed in detail by Ghosh and Taha in [20].

In real applications not all of them may be taken into account and their weight may be different. The main problem in rule extraction is that these criteria, especially fidelity and comprehensibility, tend to be contradictory. The least complex sets, consisting of few rules, cover usually only the most typical cases that are represented by large numbers of training patterns. If we want to improve a given set's fidelity, we must add new rules that would deal with the remaining patterns and exceptions, thus making the set more complicated and less understandable. Therefore rule extraction requires finding a compromise between different criteria, since their simultaneous optimization is practically infeasible. A good algorithm of rule extraction should have the following properties [21]:

- independence from the architecture of the network,
- no restrictions on the process of a network's training,
- guaranteed correctness of obtained results,
- a mechanism of accurate description of a network's performance.

A method that would meet all these requirements (or at least the vast majority) has not been developed yet. Some of the methods are applicable to enumerable or real data only, some require repeating the process of training or changing the network's architecture, some require providing a default rule that is used if no other rule can be applied in a given case. The methods differ in computational complexity (that is not specified in most cases). Hence the necessity of developing new methods. This work is an attempt to fill this gap for a network that solves the problem of classification.

14.3 Problem Formulation

The presented method of extracting rules from a neural network belongs to the group of global methods. It treats the network as a black box (Fig. 14.1)that provides it with training patterns. For this reason the architecture of a network, i.e., the way of connecting individual neurons, is insignificant.

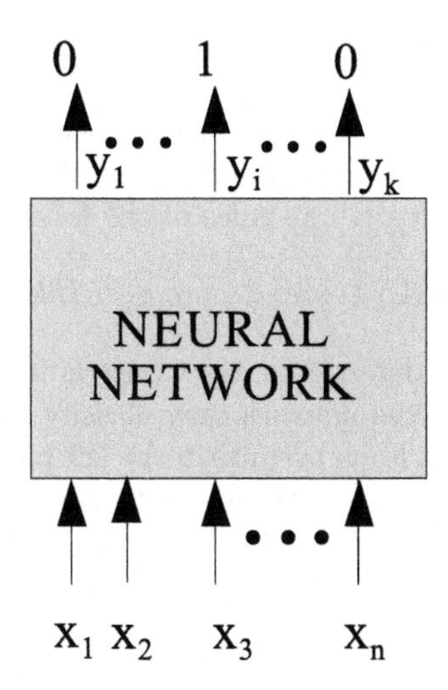

Fig. 14.1. Neural network as a black box.

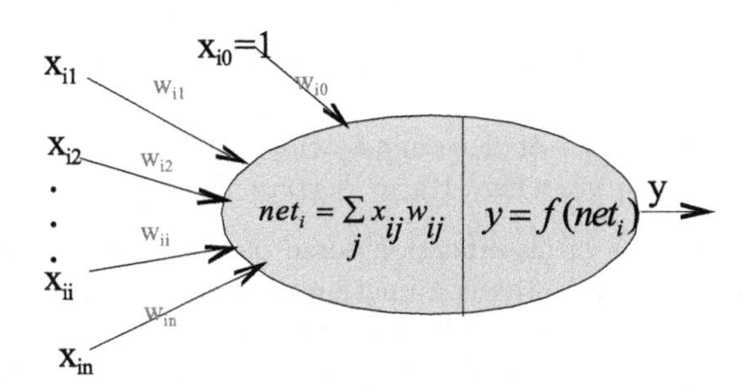

Fig. 14.2. Model of neuron.

Every neuron (Fig. 14.2) in a neural network performs simple operations of addition and multiplication by calculating its total input (for the i-th neuron $net_i = \sum x_{ij} w_{ij}$, where x_{ij} is the signal on the j-th input and w_{ij} is the weight of this connection), which is followed by applying the activation function $f(net)$ and producing the neuron's output.

However, because of a large number of neurons in a network and the parallelism of processing, it is difficult to describe clearly how a network solves a problem. Broadly speaking, the knowledge of this problem is encoded in the network's architecture, weights and activation functions of individual neurons.

Let's assume that a network solves a classification problem. This resolves itself into dividing objects (patterns) into k mutually separable classes C_1, C_2, ..., C_k.

Every p-th pattern will be described as an input vector $x_p = [x_{p1}, x_{p2}, ..., x_{pm}]$ presented to the network. After the training the network's output indicates the class that the pattern belongs to. The class is encoded using the

"1 of k" rule, i.e., only one of k outputs may produce the value 1 and the remaining ones must be equal to 0.

Our goal is to describe the performance of such a network in a comprehensible way. Undoubtedly, the most popular way of representing the extracted knowledge is the 0-order logic, i.e., rules of the following form:

$$IF\, prem_1\, AND\, prem_2 ... prem_n\; THEN\, class_v, \tag{14.1}$$

A single premise $prem_i$ determines the constraints that must be imposed on the i-th input so that the network may classify the pattern into the class included in the conclusion. Each premise in the left part of the rule corresponds to a constraint imposed on the i-th attribute X_i. Depending on the type of the attribute, a constraint is defined in one of the following ways:

- *For a real type of attribute* (discrete and continuous) it introduces the bounds of the range of acceptable values, i.e., $X_i \in [Value_{min}, Value_{max}]$.
- *For enumerable attributes* – a constraint represents a subset of permissible values, i.e., $X_i = Value_1\, OR\, X_i = Value_2\, OR\, ...X_i = Value_k$, where $Value_j$ belongs to the set of all possible values defined within a given type.
- *For logical attributes* – it determines which of the two values they must take, i.e., $X_i = Value$, where $Value \in \{true, false\}$.

During the process of classification some of the attributes may prove to be insignificant. Therefore the conjunction of premises does not necessarily have to contain constraints for all attributes. Some of the premises may be eliminated, which indicates that all values of corresponding attributes are accepted by the rule.

The *IF...THEN...* notation is very natural and intuitive. One cannot expect every doctor, for instance, to be willing to acquaint themselves with the notation used in the first order logic (the calculus of predicates). This simple reason explains the popularity of the above-mentioned solution. The majority of classification problems, where there are no dependencies between the attributes, can be expressed by means of the 0-order logic. Therefore the purpose of a rule extraction algorithm is to produce one or more sets of rules, presented in an understandable way, e.g. in the above-mentioned notation, that would meet the criteria defined in the previous section. The most important features are that these sets reflect the performance of the network both for training patterns and the previously unseen ones, and that the number and complexity of the rules they consist of are kept at a relatively low level.

The problem of classification, even in the presence of one class only, is connected with multimodal optimization, since in a general case the patterns belonging to one class cannot be covered by one rule. On the other side, if we recall the criteria that an extracted set of rules should fulfill, the problem of rule extraction turns out to be multiobjective.

14.4 Pareto Optimization

The most common way of dealing with multiobjetive problems is weighted aggregation, but its main disadvantage consists in the necessity for choosing weights that determine the relative importance of the criteria. This decision rests entirely on the user and must be made before the algorithm is run. The weights are used for fitness calculation and influence the degree to which particular criteria are optimized, forcing the algorithm to respect certain priorities. For example, in rule extraction one must decide whether rules should be very accurate or rather more concise and comprehensible. The problem is that choosing the correct weights is not necessarily easy and can be done onlyapproximately. A significant role is played by intuition and therefore the results produced by the algorithm may prove not to be satisfactory. Another drawback is that depending on the purpose of rule extraction, different criteria may be important and different sets of weights may be required to meet all the needs. In both cases the algorithm requires rerunning with modified parameters, which, especially for complex problems, tends to be very time-consuming.

Pareto optimization is an alternative to scalarization that enables avoiding the operation of converting the values of criteria into a single value. All criteria are equally taken into account and therefore the purpose of the algorithm is to produce a whole set of solutions with different merits and disadvantages. Such an algorithm would create complex and accurate rules as well as more general and comprehensible ones, leaving the final choice of the best solution to the user. The method in question bases on the concept of Pareto domination – a relation of quasi order, denoted by the symbol $\prec$, that makes it possible to compare two solutions represented as vectors consisting of the values of individual criteria. Let's assume that there are two solutions – x and y, and a function f that is used for measuring the solutions' quality, such that:

$$f(x) = [f_1(x), f_2(x), ..., f_m(x)], \tag{14.2}$$

$$f(y) = [f_1(y), f_2(y), ..., f_m(y)]. \tag{14.3}$$

On the assumption that all criteria are minimized, the relation of dominance is defined in the following way:

$$f(x) \prec f(y) \quad \Leftrightarrow \quad (\forall k = 1, ..., m) \quad f_k(x) \leq f_k(y) \quad \wedge \quad (\exists k) \quad f_k(x) < f_k(y). \tag{14.4}$$

This means that $f(x)$ dominates $f(y)$ if two conditions are met, namely: all the criteria values (vector elements) in x are at least as good as the corresponding values in y, and at least one of them is better than in y. Its worth noticing that two solutions may not be bound by this relation at all, for example:

$$f_1(x) < f_1(y) \wedge f_2(x) > f_2(y) \quad \Rightarrow \quad \neg(f(x) \prec f(y)) \wedge \neg(f(y) \prec f(x)). \tag{14.5}$$

In such a case both solutions are considered equally valuable. A multiobjective algorithm, unlike a single objective one, aims at finding an entire set of non-dominated solutions that are located as close to the real Pareto front as possible. Moreover, an additional requirement is that these solutions should be distributed evenly in the space to ensure that all criteria are equally taken into account. Genetic algorithms are particularly suitable for performing multiobjective optimization, because they operate on a large number of individuals simultaneously, which facilitates optimizing a whole set of solutions. Besides, due to niching techniques the second condition is met automatically.

14.5 Multimodality and Multiobjectiveness in Genetic Algorithms

Genetic algorithms aim at finding the global extreme, which is a great advantage, but in some cases, when finding all the optima is the basis of a correct solution, may prove to be undesirable. That's why the mechanism of niching was developed, which enables individuals to remain at the local optima. Several methods are used for approaching such problems [13], among others:

- *Iterations* - a genetic algorithm is rerun several times; if all optima may be found with equal probability, one gets a chance of finding them due to independent computations.
- *Parallel computations* - where a set of populations is evolved independently.
- *Sharing-based methods* - the sharing function determines by how much an individual's fitness is reduced, depending on the number of other individuals in its neighborhood. The original fitness is divided by the value of the sharing function, which is proportional to the number of individuals surrounding a given one and inversely proportional to the distance from them.

The existence of several objectives in the problem to be solved is reflected in the method of evaluating individuals in a genetic algorithm. The objective function takes the form of a vector, therefore it can not be directly used as the fitness function. Such problems in GA-based methods are solved by applying scalarization, which is the most common approach. In this case the fitness function is a weighted sum of elements, where each represents one of the objectives. The problem becomes how to define the weights. Choosing the right weights may be a problem, because the objectives are usually mutually exclusive and several solutions representing different trade-offs between them may exist.

Pareto optimization may also be used in this case. It consists in comparing individuals by means of domination, which allows either rank assignment or performing the tournament selection. The first Pareto-based fitness assignment method, proposed by Goldberg in [6], belongs to the most commonly used ones. It consists in assigning the highest rank to all nondominated solutions. Afterwards, these solutions are temporarily removed from the population and the procedure is repeated for the remaining individuals that receive a lower rank value. The process of rank assignment is carried out as follows:

$Temp := Population$
$n = 0$
$while\,(Temp \neq 0) \quad do$
$\{$
$Nondominated = \{x \in Temp / (\neg \exists y \in Temp) f(y) \prec f(x)\}$
$(\forall x \in Nondominated) \quad rank(x) := n$
$Temp := Temp \backslash Nondominated$
$n := n + 1$
$\}$

Niching may be used at every stage of this algorithm to help preserve diversity in the population. This optimisation has been introduced into the NSGA algorithm [19].

Another method, proposed by Fonseca and Fleming in 1993, is based on the idea that the rank of an individual depends on the number of other individuals in the population that dominate it. Zitzler and Thiele modified it to develop the Strength Pareto Approach, which solves the problem of niching when elitism is introduced to a multiobjective genetic algorithm [23], [4]. In the case of multiobjective optimisation, elitism requires storing all nondominated individuals found during the course of evolution in an external set. These individuals participate in reproduction along with the ones from the standard population, but they are not mutated, which prevents them from losing their quality. Fitness assignment in the Strength Pareto Evolutionary Algorithm (SPEA) is performed in the following way (N denotes the size of the population):

- For each individual i in the external set do:

$$f_i = \frac{card\{j \in Population | i \prec j\}}{N+1}$$

- For each individual j in the population, do:

$$f_j = \sum_{i, i \prec j} f_i + 1.$$

Because in this method the lowest fitness values are assigned to the best individuals, i.e., $f_i \in [0,1)$ and $f_j \in [1,N)$, the fitness function must be modified so that selection may be carried out.

14.6 MulGEx as A New Method of Rule Extraction

MulGEx belongs to black box methods. It does not require any special network architecture nor use the information encoded in the network itself. Moreover, it does not impose any constraints on the types of attributes in the input patterns. Its main idea consists in introducing genetic algorithms working on two levels as in [12]. MulGEx is based on the concept of Pareto optimization, but scalarization has been implemented on both levels to allow comparison. In the case of the lower-level algorithm this is less important, for it always produces one solution at a time, but on the upper level the choice influences the algorithm's performance significantly. The difference has been shown by experimental study.

The following criteria have been used for evaluating the produced solutions:

- *coverage* - the number of patterns classified correctly by a rule or set,
- *error* - calculated on the basis of the number of misclassified patterns,
- *complexity* - depending on the number of premises in a single rule and rules in a set.

The first two correspond to *fidelity* and determine to what extent the answers given on the basis of the extracted rules are identical to those produced by the network. *Complexity* has been introduced to evaluate the solutions' *comprehensibility*. The remaining FACC requirements cannot be used as criteria during the process of rule extraction. *Accuracy* may be measured only after the algorithm has stopped and the final set of rules is applied to previously unseen patterns. *Consistency* requires running the algorithm several times and comparing the obtained results.

One of the most important decisions that needs to be made before implementing a genetic algorithm concerns the form of an individual, i.e., the way of encoding data in the chromosome. This influences not only the genetic operators, but also the entire algorithm's performance. Two alternative approaches are used in rule extraction, namely Michigan and Pitt [13]. The first one consists in encoding a single rule in each individual. This allows rules to be optimized efficiently, but gives no possibility of evaluating how the extracted rules work as a set. Moreover, one must implement one of the niching techniques as well as solve the problem of contradictory rules that might be evolved. In the Pitt approach an individual contains an entire set of rules, which eliminates the drawbacks of the previous method. However, because of the variable length of chromosomes, i.e., the changing number of rules in the evolved sets, more complicated genetic operators have to be implemented. The proposed method combines both these approaches. The main idea is shown in Fig. 14.3.

The first step consists in evolving single rules by a low-level genetic algorithm. Focusing on individual rules allows adjusting their parameters precisely.

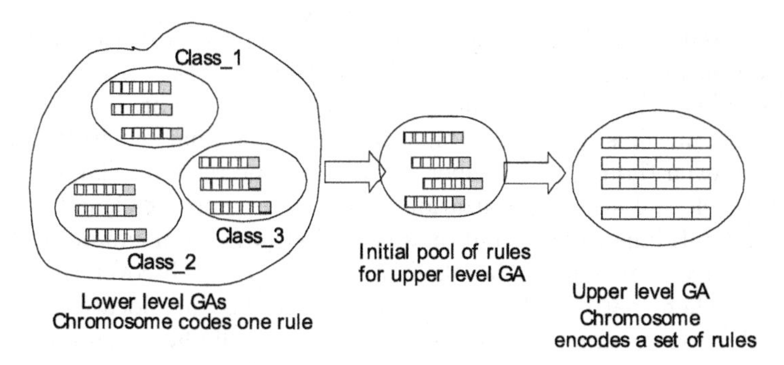

Fig. 14.3. The main idea of MulGEx.

The results are passed to the upper-level algorithm that performs further optimization by working on sets built of the obtained rules and evaluating how well these rules cooperate.

14.7 Details of the Method

The algorithms on both levels will be described in detail in this section. The most significant elements in the design of a GA-based application will be presented, namely the way of encoding individuals, genetic operators and fitness function.

14.7.1 The Lower-level Algorithm

The lower level algorithm delivers the initial pool of rules for the upper level algorithm. It operates on chromosomes representing single rules (the Michigan approach). In order to reduce the complexity of rule extraction, several independently evolved populations are introduced, one for each of the classes defined in the problem. This is a solution to the problem of multimodality that guarantees that the algorithm finds relatively precise rules for all classes, including those represented by a small number of training patterns. Each of the genetic algorithms on the lower level is run several times and produces one rule at a time. This sequential approach to rule extraction solves the problem of multimodality within the classes. The best rule evolved is stored in an external set (the upper level genetic algorithm's initial pool). Afterwards, all training patterns recognized by this rule are removed from the input set used by the algorithm for fitness calculation (sequential covering approach [22]). Then a new population is created in a random way and the algorithm is rerun.

At this stage the choice of the method of dealing with multiobjectiveness is of little significance, since only one rule is extracted at a time. Both approaches, i.e., scalarization and Pareto optimization, have been implemented and compared. Choosing the most appropriate weights is relatively easy, because complexity is rarely taken into account at this stage. In most cases it is reasonable to select rules with a very small error value or, if possible, with no error at all. This guarantees high fidelity of the final set evolved by the algorithm, even though the number of rules may be relatively large.

Nevertheless, Pareto optimization tends to be more efficient here, especially for multidimensional solution spaces and large numbers of classes. At an early stage of evolution the algorithm attempts to discover the areas in the space where patterns belonging to a given class are located. The first rules with coverage greater than 0 evolved have usually large error values and should be subsequently refined in order to reduce misclassification. A single objective algorithm combines *coverage* and *error* into one value and therefore does not take into account the potential usefulness of such rules. These rules are assigned low fitness values because of high error and may be excluded from reproduction, thus being eliminated from the population. This hinders the process of evolution. A Pareto-based genetic algorithm considers all objectives separately, therefore such rules are very valuable (as non-dominated) and will be optimized by means of genetic operators with a strong probability in the next generations. As a result, the initial error will be reduced. Therefore the number of generations needed to evolve satisfying rules tends to be lower if Pareto optimization is used. In this case, however, the choice of the best rule is not straightforward because of a limited possibility of comparing individuals. Because of this, having stopped the algorithm, one must temporarily apply certain weights to scalarize the fitness function.

The number of generations created within such an iteration is chosen by the user who may decide to evolve a constant number of generations every time or to remove the best rule when no improvement has been detected for a given period of time. Improvement occurs when an individual with the highest fitness so far appears in a single objective algorithm or when a new non-dominated solution is found in a multiobjective one.

The process of extraction for a given class completes when there are no more patterns corresponding to this class, or when the user decides to stop the algorithm after noticing that newly found rules cover too few patterns (which means that the algorithm may have started taking exceptions and noisy data into account). As soon as evolution in all the populations has stopped, all rules stored in the external set are passed to the upper-level algorithm for further processing.

The Form of the Chromosome

The chromosome on this level is presented in Fig. 14.4. An individual consists of a set of premises and a conclusion that is identical for all individuals within a given population. The conclusion contains a single integer - the number of the appropriate class. The form of a single premise depends on the type of the corresponding input in the neural network and may represent a constraint imposed on a binary, enumerable or real value.

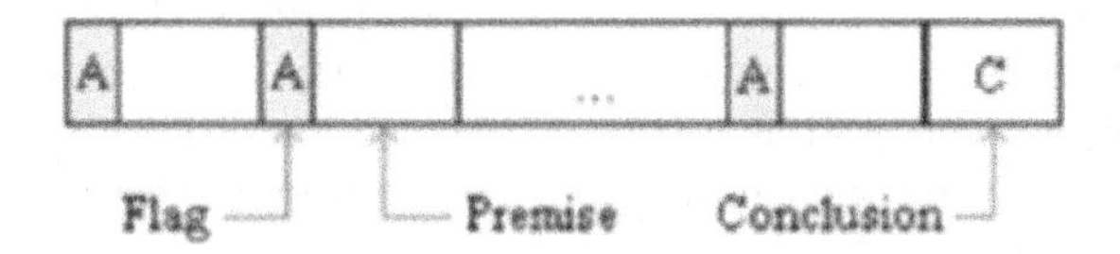

Fig. 14.4. A schema of chromosomes on the lower level.

The appropriate genes designed for all above-mentioned types of attributes are presented in Fig. 14.5a – 14.5c. Every premise is accompanied by a logical flag A that indicates whether the constraint is active. If the flag is set to *false*, all values of a given attribute are accepted by the rule.

Fig. 14.5. Genes representing different types of premises, depending on the type of attribute.

A constraint defined for a *binary attribute* has been implemented as a single logical variable (Fig. 14.5a). In order to match a given rule, a pattern must have the same value of this attribute as the premise, unless the flag indicates that this particular constraint is inactive. An *enumerable attribute* requires an array of logical variables, where every element corresponds to one value within the type (Fig. 14.5b). All elements set to true represent together the subset of accepted values. A given rule can be applied to a pattern if the value of the attribute in the pattern belongs to the subset specified within the premise.

A constraint imposed on a *real value* consists of two real variables representing the minimal and maximal value of the attribute that can be accepted by the rule (Fig. 14.5c). The number of premises in a rule is constant and equal to the number of the network's input attributes. Premises are eliminated by deactivating their flags, which indicates that certain constraints are not taken into account when applying the rule to the patterns.

Genetic Operators

Having defined the form of the chromosome, one must introduce appropriate genetic operators to allow reproduction and mutation.

The process of exchanging genetic information between individuals is based on uniform crossover. Every logical or real variable in the chromosome is copied from one of the offspring's two parents. The choice of the parent is performed randomly, with equal probability. A logical premise is therefore copied entirely from one parent. The subset of values in an enumerable one is a random combination of those encoded in the parents' chromosomes. In the case of real premises, the newly created range may be either copied entirely from one parent or created as a sum or intersection of both parents' ranges. The conclusion of the rule may be copied from either of the individuals, since its value is the same within a given population.

Mutation consists of modifying individual premises in a way that depends on their type. Some of the logical variables in premises corresponding to binary or enumerable attributes, chosen randomly usually with a small probability, may be assigned the opposite value, which changes the set of patterns accepted by the rule. Constraints imposed on real attributes may be altered by modifying one of the bounds of the range of accepted values. This is done by adding a random value to a given real variable within the premise. The algorithm becomes more effective if small changes are more probable than larger ones, which prevents individuals from being modified too rapidly and losing their qualities. Mutation may also influence the flags attached to premises. This results in individual constraints being activated or deactivated and makes rules either more specific and complex or more general. For obvious reasons mutation cannot affect the conclusion that remains constant in the course of evolution.

The Fitness Function and the Method of Selection

The quality of an individual is measured in the following way: The first step consists in calculating the number of patterns classified correctly by the rule (*coverage*) as well as the number of misclassified patterns (*error*). This requires checking whether the neural network's response, given for every pattern that the rule can be applied to, matches the conclusion. Additionally, the number of active premises is determined in order to evaluate the *complexity* of the rule. These criteria of quality assessment have to be gathered into a single fitness value.

In single objective optimization, all of them are multiplied by certain weights and the results are summed up to produce a scalar value. Since fitness cannot be negative, the obtained values may need to be scaled. This is the simplest solution, however, the information concerning the values of individual criteria is lost. The weights used for fitness calculation are chosen by the user and the function takes the form presented by Eq. 14.6 (the objectives that are minimized are negated).

$$Fitness_{scalar} = W_c \cdot coverage - W_e \cdot error - W_x \cdot complexity \qquad (14.6)$$

The multiobjective approach requires creating a vector containing all values of the criteria (Eq. 14.7). Comparing two individuals in this case must be based on the relation of domination. Fitness is calculated on the basis of Goldberg's method, unless the option of retaining the best individuals (elitism) has been chosen. In the latter case, the procedure of fitness assignment proposed in the Strength Pareto Approach is used.

$$Fitness_{Pareto} = [coverage, error, complexity] \tag{14.7}$$

The last element that needs to be defined when designing a genetic algorithm is the method of selection. Various possibilities have been proposed here. The roulette wheel technique has been implemented in MulGEx, which implies that all individuals have a chance to be chosen for reproduction, but the probability of them being selected is proportional to their fitness value.

14.7.2 The Upper-level Algorithm

The upper-level algorithm in MulGEx is multiobjective in the Pareto sense, although scalarization has also been implemented to allow comparison. It operates on entire sets of rules that are initially created on the basis of the results obtained from the lower-level one.

Every individual in the initial population consists of all the rules produced by the lower-level genetic algorithms (Fig. 14.3). This implies that the maximal fidelity achieved at the previous stage is retained and further optimization does not require adding new rules. The purpose of the upper-level algorithm consists mostly in improving comprehensibility by either excluding certain rules from the sets or eliminating individual premises. Naturally, this process should be accompanied by possibly low deterioration of *fidelity*, i.e., the algorithm must decide which rules are the least significant. Gradual simplification of the sets results in producing various solutions with different qualities, which is the main advantage of a multiobjective algorithm. Other modifications of the rules are also possible at this stage, but their influence on the quality of evolved solutions is usually of little importance.

Evolution stops after a given number of generations has been created since the beginning or since the last improvement. The result produced by the algorithm is a set of non-dominated solutions that correspond to sets of rules with different qualities and drawbacks. The final choice of the most appropriate one is left to the user and is usually done by applying weights to the criteria to enable comparing non-dominated solutions. Due to the multiobjective approach the user may examine several solutions at a time without having to rerun the algorithm with different parameters. This allows identifying noisy data. Normally, increasing the complexity of a set is accompanied by improved fidelity. However, if the observed improvement is very small after a new rule has been added, this may indicate that this rule covers exceptions that haven't been eliminated by a neural network. The multiobjective approach helps to determine the optimal size of a rule set, which is not possible in the case of scalarization, where weights have to be carefully adjusted for the algorithm to produce satisfying results.

As previously, designing the algorithm requires defining the form of an individual, followed by genetic operators, fitness evaluation and the method of selection.

The Form of the Chromosome

An individual takes the form of a set of rules that describes the performance of the network, which implies that rules corresponding to all classes are included in every set.

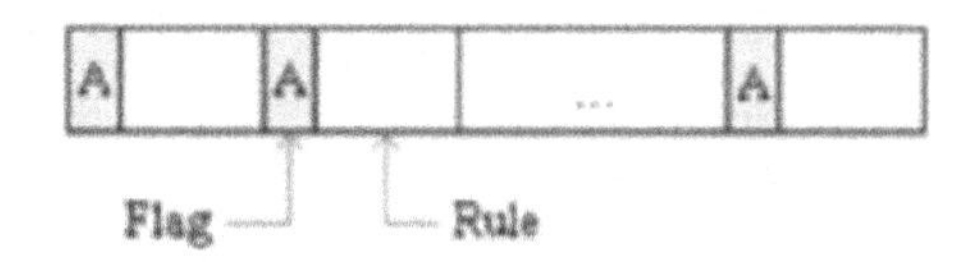

Fig. 14.6. A schema of chromosomes on the upper level.

Rules are accompanied by binary flags whose purpose is the same as in the chromosome on the lower level. Namely, setting a flag to *false* results in the temporary exclusion of a given rule from the set. This allows adjusting the size of the set without having to introduce variable-length chromosomes. The form of chromosome on this level is presented in Fig. 14.6.

Genetic Operators

The hierarchical structure of the chromosome requires defining specialized genetic operators.

Crossover exchanges corresponding rules between the parents in a uniform way. Rules are copied into the offspring entirely, including the flag.

The operator of mutation is more complicated, for it should enable modification on two levels - in relation to entire sets and individual rules. Therefore mutation is performed in two steps. First, every flag attached to a rule may change its value with a small probability, which results in adding or removing the rule from the set. Afterwards, premises inside the rule may be modified by applying the mutation operator used on the lower level. This helps to adjust the ranges of accepted values inside the rule or reduce the rule's complexity at the cost of fidelity.

It is worth noticing that in the lower-level algorithm some rules were evolved on the basis of a reduced set of patterns. At this stage the whole set is used for evaluation. Moreover, individual rules cooperate to perform the classification task and are assessed together. Because of these new conditions further optimization of rules might be possible and that is why the operator of mutation in the upper-level algorithm may alter the internal structure of rules that was developed before.

Evaluation of Individuals

The criteria used for the evaluation of individuals are calculated on the basis of the criteria introduced for rules on the lower level.

Coverage depends on the number of patterns classified correctly by the set as a whole, whereas *error* is the sum of errors made by each of the active rules. *Complexity* is defined as the number of active rules increased by the overall number of active premises within them.

The process of creating new generations resembles the one introduced on the lower level. Fitness is assigned to individuals depending on the type of the algorithm, according to one of the methods described in the previous subsection. Again, Pareto optimization requires creating a vector of the values of objectives (Eq. 14.7). Scalarization implemented for the purpose of comparison consists in applying weights to the objectives so that a single value is obtained.

The difference between genetic algorithms on both levels is that at this stage the purpose of the algorithm is to produce a whole set of solutions simultaneously, therefore niching proves to be very useful. To this end the technique of sharing function has been implemented to reduce the fitness of those individuals that are surrounded by many others. However, this is not necessary in the case of the SPEA algorithm, since niching is guaranteed by the method of fitness assignment.

14.8 Experimental Study

The purpose of experiments is to verify the algorithm's effectiveness and versatility. The method should be tested on various data, which aims at checking whether it has the following properties:

- independence on the types of attributes,
- effectiveness in the presence of superfluous attributes,
- resistance to noise,
- scalability.

A good practice is to perform experiments on well-known benchmark data sets, so that the results may be compared with those obtained by means of other methods. MulGEx has been tested on four sets from [3] and presented in Table 14.1. The data contain different types of attributes, which allows verifying the method's versatility. Resistance to noise was tested on the *LED-24* data set, where 2% noise had been added to make the classification task more difficult. Moreover, this data allows to observe whether the algorithm can detect superfluous attributes, since not all of them participate in determining the class that a given pattern belongs to (17 attributes are insignificant and have been chosen randomly). The evaluation of scalability requires patterns with a large number of attributes, hence the *Quadrupeds* set.

Table 14.1. Data sets used for the tests

Name (examples)	Type of attributes	Class	No of examples
Iris (150 instances)	3 continuous	Setosa	50
		Versicolour	50
		Virginica	50
LED-24 (1000 instances)	24 binary (17 superfluous), 2% noise	10 classes	about 100 per one class
Monk-1 (432 instances)	6 enumerable	0	216
		1	216
Quadrupeds (100 instances)	72 continuous	Dog	29
		Cat	21
		Horse	27
		Giraffe	23

All sets were processed by a neural network in order to reduce noise and eliminate unusual data. A multi-layer feed-forward network with one hidden layer was provided with the data and trained using the back-propagation algorithm. The results of the training are gathered in Table 14.2. The subsequent experiments performed to evaluate MulGex were conducted both on raw as well as processed sets, unless the network achieved maximal accuracy.

Table 14.2. Results of the network training

data set	Number of neurons in the hidden layer	Accuracy %
Iris	2	96,7
LED-24	3	95
Monk-1	3	100
Quadrupeds	2	100

The optimal values of the algorithm's parameters depend on the properties of the data set, especially on the number of attributes. During the experiments the best effectiveness of rule extraction was achieved when the probability of crossover was relatively low, for example 0.5. The reason is that even a small modification of a rule may change its *coverage* and *error* significantly. Therefore if the fittest individuals exchange genetic information during reproduction, the offspring may lose the most desirable properties. Low probability of crossover guarantees that some of the individuals are copied directly into the new population. Introducing elitism may also be helpful in this case, for it ensures that the best individuals are always present in the next generation. The probability of mutation should be inversely proportional to the number of attributes and during the tests was usually set not to exceed 0.01. The size of the population influences the efficiency of the algorithm as well. Theoretically, increasing the number of individuals reduces the number of generations needed to find satisfactory solutions. The more individuals, the more potential solutions may be examined and compared at a time. On the other hand, because the process of calculating *coverage* and *error* for each individual is very time-consuming, large populations cause deterioration in the algorithm's performance. Therefore a reasonable compromise must be found. Rule extraction for the above-mentioned data sets was successfully carried out when populations consisted of 20 - 50 individuals.

The results of the experiments performed both on raw data and sets processed by the network are gathered in Table 14.3 and Table 14.4, respectively. In each case multiobjective algorithms on both levels were stopped after having evolved 100 generations without improvement. MulGEx produced sets of non-dominated solutions with various properties. Then, solutions containing different numbers of rules were selected and their fidelity, i.e., the difference between *coverage* and *error*, was determined. Every test was carried out 5 times and the average result as well as the standard deviation were calculated.

The results of the experiments indicate that MulGEx is suitable for solving problems with various numbers and types of attributes, as well as with different sets of classes defined for a given problem. In the case of *Iris*, *Monk-1* and *Quadrupeds* the algorithm succeeded in finding a relatively small rule set that covered all training patterns without misclassification. This was not possible for *LED-24*, where the presence of noise resulted in increasing *complexity* or *error*, depending on the weights chosen by the user.

The experiments confirm the fact that the criteria used for evaluating solutions are contradictory. The simplest sets, consisting of few rules, cover only the most typical examples. When more rules are added to the set, its fidelity is improved at the cost of complexity.

At this point it should be mentioned that MulGEx satisfies the requirement that the produced solutions should be distributed evenly in the space.

Table 14.3. The results of experiments on raw data

Iris		LED-24		Monk-1		Quadrupeds	
Rules	*Fidelity*	*Rules*	*Fidelity*	*Rules*	*Fidelity*	*Rules*	*Fidelity*
1	50.0± 0.0	3	326.0± 1.0	1	108.0± 0.0	1	27.4± 2.5
2	96.8± 0.4	6	608.2± 6.4	2	180.0± 0.0	2	53.0± 5.0
4	143.6± 0.5	9	843.2± 12.6	4	324.0± 0.0	3	76.8± 2.8
6	147.4± 0.5	12	918.4± 5.0	6	396.0± 0.0	4	97.8± 2.9
8	149.4± 0.5	15	924.8± 4.7	7	432.0± 0.0	5	99.3± 1.2

Table 14.4. The results of experiments on data processed by a network

Iris		LED-24	
Rules	*Fidelity*	*Rules*	*Fidelity*
1	50.0± 0.0	3	336.2± 2.0
2	98.2± 0.4	6	642.0± 3.6
4	143.0± 0.7	9	897.6± 4.0
6	147.6± 1.1	12	981.8± 4.3
7	149.5± 0.9	15	987.4± 4.7

The evolved rule sets represent different trade-offs between the objectives and contain sets with maximal fidelity as well as sets consisting of one rule only and various intermediate solutions. The results obtained for *LED-24* for raw and processed data differ significantly. The *fidelity* of sets containing the same number of rules is higher for data passed through a neural network, which means that in this case a single rule covers statistically more patterns and misclassifies less. This is possible due to neural networks' ability of reducing noise, which facilitates rule extraction.

The Pareto approach in the upper-level algorithm has been compared with scalarization. To this end separate tests for the *Iris* data set have been performed. The lower-level algorithm was run 5 times and produced initial rule sets. In all cases Pareto optimization was used at this stage and rules with very high error weight were selected, so that the initial set could achieve maximal fidelity. Afterwards the upper-level one was executed several times with different settings. When the Pareto approach was used, it was run only once during each of the tests and a whole set of solutions was created. Then the user could apply weights to allow the algorithm to select the most appropriate solution. In the case of scalarization weights for the objectives had to be determined beforehand, therefore separate runs were necessary to produce results for all indicated weights' combinations.

The results (Table 14.5) show that for the *Iris* data set there is hardly any difference in the quality of solutions between the two approaches. However, it must be emphasized that scalarization requires rerunning the algorithm every time when the user decides to change the weights, which is very time-consuming. If we take into account the fact that adjusting weights is not easy, it proves to be much more convenient to find a set of solutions in a single run and then apply different weights in search of the most satisfying one.

Another advantage of Pareto optimization is that one may analyze all solutions simultaneously. Fig. 14.7 shows the values of *fidelity* and *complexity* for rule sets produced during one of the experiments. Before choosing a solution the user may evaluate which one would be the most satisfactory, based on the shape of the Pareto front. In the presented example very high *fidelity* may be achieved by a set with *complexity* equal to 6. If *complexity* is increased by adding new rules or premises, the improvement of *fidelity* is relatively low.

Table 14.5. The comparison of Pareto approach and scalarization (*Cov* stands for coverage, *Comp* – for complexity)

Weights			Pareto			Scalarization		
Cov.	*Error*	*Comp.*	*Cov.*	*Error*	*Comp.*	*Cov.*	*Error*	*Comp.*
1	1	1	143.6± 1.5	1.0± 0.0	8.6± 1.5	143.2± 1.5	1.0± 0.0	8.2± 1.6
1	1	10	143.0± 0.0	5.4± 0.9	6.0± 0.0	142.0± 1.7	6.0± 1.0	6.0± 0.0
1	10	1	141.2± 1.6	0.0± 0.0	9.2± 1.6	141.2± 1.6	0.0± 0.0	9.2± 1.6
10	1	1	150.0± 0.0	4.0± 2.1	18.4± 3.9	150.0± 0.0	2.0± 2.0	22.0± 2.9

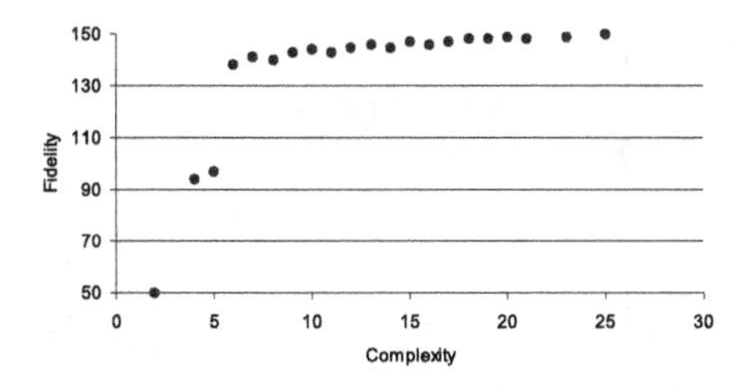

Fig. 14.7. The values of the criteria for rules sets obtained for *Iris*.

This may indicate that more complicated sets cover noisy data and are not valuable for the user. On the other hand, if *complexity* is reduced below 6, *fidelity* deteriorates rapidly, which suggests that an important rule has been eliminated. Therefore, having this additional information due to the Pareto approach, the user may choose the most appropriate weights when selecting one particular solution.

14.8.1 Evaluation of Rules by Visualization

The quality of rules produced by MulGEx for *LED-24* may be easily evaluated due to the possibility of visualizing the data. The first attributes in *LED-24* represent seven segments of a *LED* display presented in Fig. 14.8. The remaining ones are superfluous and are successfully excluded by the algorithm from participating in classification. The experiment was carried out for data that had been processed by a network.

The presented results were produced by the multiobjective lower-level algorithm and the following weights: *coverage* = 10, *error* = 10, *complexity* = 1 were used for selecting the most satisfactory solutions after the algorithm had stopped. The size of population was set to 20 and the probabilities of crossover and mutation to 0.9 and 0.01, respectively. The option of saving the best individuals (elitism) was chosen. Evolution was stopped after 100 generations without improvement and the best rule in each of the populations was selected. Table 14.6 and Fig. 14.9 show an example of rules produced for *LED-24*. In Fig. 14.9 thick black lines denote segments that are lit, missing lines stand for segments that are off and thin grey lines correspond to seg-

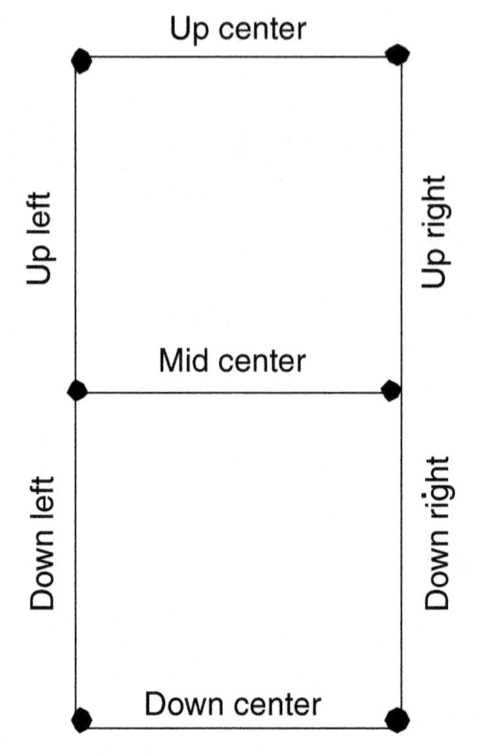

Fig. 14.8. The segments of *LED*.

ments whose state is insignificant when assigning a pattern to a given class (these segments are represented by inactive premises).

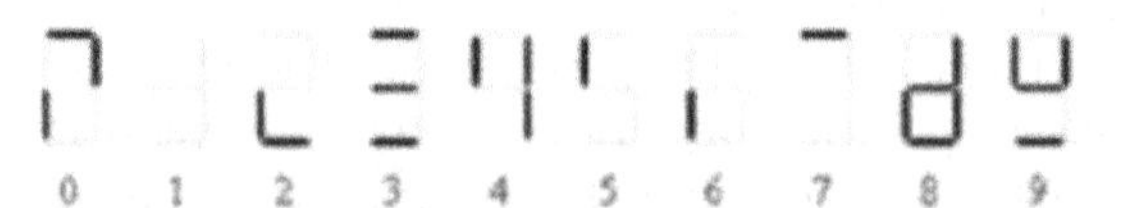

Fig. 14.9. Visualization of rules for *LED-24*.

The obtained rules presented in the form of a schema in Fig. 14.9 prove to be very comprehensible for humans. One may easily notice that the results resemble digits that appear on a LED display, which confirms the effectiveness of MulGEx. At the same time the algorithm succeeded in reducing the rules' complexity. The state of some of the segments proves to be insignificant when a certain pattern is classified as one of the digits.

Table 14.6. Rules produced by the lower-level algorithm for *LED-24*

Rule	Coverage	Error
IF Up center AND Up right AND NOT Mid center AND Down left THEN 0	114	0
IF NOT Up center AND NOT Up left AND NOT Down left THEN 1	83	0
IF NOT Up left AND Down left AND NOT Down right AND Down center THEN 2	118	0
IF Up center AND NOT Up left AND Mid center AND NOT Down left AND Down center THEN 3	110	0
IF Up left AND Up right AND NOT Down left AND Down right AND NOT Down center THEN 4	82	0
IF Up left AND NOT Up right AND NOT Down left THEN 5	92	3
IF NOT Up right AND Down left THEN 6	103	0
IF Up center AND NOT Up left AND NOT Mid center AND NOT Down left THEN 7	102	0
IF Up right AND Mid center AND Down left AND Down right AND Down center THEN 8	104	0
IF Up left AND Up right AND Mid center AND NOT Down left AND Down center THEN 9	81	0

14.9 Comparison to Other Methods

The vast majority of rule extraction algorithms are based on the scalar approach. Therefore comparing their effectiveness may not be appropriate because of different goals pursued by their authors. Nevertheless two different approaches have been implemented in MulGEx to allow evaluating their performance.

The results of experiments performed for MulGEx may be confronted with those obtained for a different multiobjective rule extraction algorithm, namely GenPar [11]. This is facilitated by the fact that the same data sets were used in both cases. Nevertheless the relative effectiveness of these methods may be evaluated only approximately because of certain differences in the way of performing experiments and the scope of collected information.

GenPar was tested on data processed by a neural network only. Moreover, during the course of evolution coverage and error were combined into a single objective, which reduced the number of objectives by one, and only coverage was measured during the tests. The results for GenPar, presented in [11] and shown in Table 14.7, contain information on the quality of evolved sets expressed in terms of coverage and the number of rules.

The results for GenPar indicate that its effectiveness is lower than that of MulGEx. Although fidelity achieved by solutions with the same numbers of rules are similar, in all cases GenPar failed to produce a solution with the

Table 14.7. The results of experiments for GenPar (*size* – number of rules, *fid* – fidelity equivalent to coverage) on the basis of [11]

Iris		LED-24		Quadrupeds	
size	*fid*	*size*	*fid*	*size*	*fid*
1	50	1	87	2	182
2	94	4	324	4	471
4	137	5	473	–	–
5	144	13	928	–	–

maximal fidelity possible. The reason may be that GenPar is based exclusively on the Pitt approach (entire rule sets are encoded in individuals). This implies that sets are evaluated as a whole and the performance of single rules is not taken into account, which prevents rules from being intensively optimized by adjusting values within the premises. The superiority of MulGEx lies in introducing a lower-level algorithm that evolves single rules before sets are formed and allows constant chromosome length on the upper-level.

14.10 Conclusion

Neural networks, in spite of their efficiency in solving various problems, have no ability of explaining their answers and presenting acquired knowledge in a comprehensible way. This serious disadvantage leads to developing numerous methods of knowledge extraction. Their purpose is to produce a set of rules that would describe a network's performance with the highest fidelity possible, taking into account its ability to generalize, in a comprehensible way. These objectives are contradictory, therefore finding a satisfactory solution requires a compromise.

Because the problem of rule extraction is very complex, genetic algorithms are commonly used for this purpose, especially for networks that solve the problem of classification. Since the fitness function is scalar, standard algorithms can deal with one objective only. Multiobjective problems are approached in many ways; the most common one is to aggregate the objectives into a single value, but this approach has many drawbacks that can be eliminated by introducing proper multiobjective optimization based on domination in the Pareto sense.

The proposed method, MulGEx, combines multiobjective optimization with a hierarchical structure to achieve high efficiency. It is network architecture independent and can be used regardless of the number and types of attributes in the input vectors. Although the algorithm has been designed to extract rules from neural networks, it can be also used with raw data. In this case, however, its efficiency may be deteriorated because the presence of noise in the data requires more complicated rule sets to describe the process of classification.

MulGEx consists of genetic algorithms working on two levels. The lower-level one produces single rules (independently evolved populations correspond to the classes defined for a given classification problem) and passes them to the upper-level one for further optimization. The latter algorithm evolves entire sets of rules that describe the performance of a network.

Due to Pareto optimization MulGEx produces an entire set of solutions with different properties simultaneously, varying from complex ones with very high fidelity to the simplest ones that cover only the most typical data. The usefulness of such solutions depends on the purpose of rule extraction. In most cases relatively general rules are considered valuable for a classification problem, for they provide information about the most important properties that objects belonging to a given class should have. Nevertheless, if we attempt to find hidden knowledge by means of *data mining*, we might be interested rather in more specific rules that cover a smaller number of patterns, but achieve high fidelity. The final choice of the most appropriate solution is left to the user who may select several solutions without the need for rerunning the algorithm. The effectiveness and versatility of MulGEx has been confirmed by experiments and the superiority of the Pareto approach over scalarization has been pointed out.

References

[1] Andrews, R., J. Diedrich, and A. Tickle: 1995, 'Survey and critique of techniques for extracting rules from trained artificial neural networks'. *Knowledge-Based Systems* **8**(6), 373–389.

[2] Andrews, R. and S. Geva: 1994, 'Rule extraction from constrained error back propagation MLP'. In: *Proceedings of 5th Australian Conference on Neural Networks, Brisbane, Quinsland.* pp. 9–12.

[3] Blake, C. C. and C. Merz: 1998, 'UCI Repository of Machine Learning Databases'. *University of California, Irvine, Dept. of Information and Computer Sciences.*

[4] Coello, Carlos A. Van Veldhuizen, D. A. and G. B. Lamont: 2002, *Evolutionary Algorithms for Solving Multi-Objective Problems.* Kluwer Academic Pblisher. New York.

[5] Fu, X. and L. Wang: 2001, 'Rule extraction by genetic algorithms based on a simplified RBF neural network'. In: *Proceedings Congress on Evolutionary Computation.* pp. 753–758.

[6] Goldberg, D. E.: 1989, *Genetic Algorithms in Search, Optimization and machine Learning.* Addison-Wesley Publishing Company, Reading, Massachusetts.

[7] Hayashi, Y., R. Setiono, and K. Yoshida: 2000, 'Learning M of N concepts for medical diagnosis using neural networks'. *Journal of Advanced Computational Intelligence* **4**(4), 294–301.

[8] Hayward, R., C. Ho-Stuart, J. Diedrich, and P. E.: 1996, 'RULENEG: Extracting rules from trained ANN by stepwise negation'. Technical report, Neurocomputing Research Centre Queensland University Technology Brisbane, Old.

[9] Jin, Y. and B. Sendhoff: 2003, 'Extracting interpretable fuzzy rules from RBF networks'. *Neural Processing Letters* **17**(2), 149–164.

[10] Jin, Y., B. Sendhoff, and E. Koerner: 2005, 'Evolutionary multi-objective optimization for simultaneous generation of signal-type and symbol-type representations'. In: *The Third International Conference on Evolutionary Multi-Criterion Optimization.* pp. 752–766.

[11] Markowska-Kaczmar, U. and P. Wnuk-Lipinski: 2004, 'Rule Extraction from Neural Network by Genetic Algorithm with Pareto Optimization'. In: L. Rutkowski (ed.): *Artificial Intelligence and Soft Computing.* pp. 450–455.

[12] Markowska-Kaczmar, U. and R. Zagorski: 2004, 'Hierarchical Evolutionary Algorithms in the Rule Extraction from neural Networks'. *Fourth International Symposium on Engineering of Intelligent Systems, EIS.* Funchal, Portugal.

[13] Michalewicz, Z.: 1996, *Genetic Algorithms + Data Structures = Evolution Programs.* Springer Verlag Berlin Heidelberg.

[14] Mitra, S. and H. Yoichi: 2000, 'Neuro-Fuzzy Rule Generation: Survey in Soft Computing Framework'. *IEEE Transactions on Neural Networks* **11**(3), 748–768.

[15] Palade, V., D. C. Neagu, and R. J. Patton: 2001, 'Interpretation of Trained Neural Networks by Rule Extraction'. *Fuzzy Days 2001, LNC 2206* pp. 152–161.

[16] Santos, R. Nievola, J. and A. Freitas: 2000, 'Extracting comprehensible rules from neural networks via genetic algorithms'. *Symp. on Combinations of Evolutionary Computation and Neural Network* **1**, 130–139.

[17] Setiono, R., W. K. Leow, and J. Zurada: 2002, 'Extraction of Rules from Artificial Neural Networks for Nonlinear Regression'. *IEEE Transactions on Neural Networks* **13**(3), 564–577.

[18] Siponen, M., J. Vesanto, O. Simula, and P. Vasara: 2001, 'An Approach to Automated Interpretation of SOM'. In: *In Proceedings of Workshop on Self-Organizing Map 2001(WSOM2001).* pp. 89–94.

[19] Srinivas, N. and K. Deb: 1993, 'Multiobjective optimizatio using nondominated sorting in genetic algorithms'. Technical report, Department of Mechanical Engineering, Indian Institute of Technology, Kanpur, India.

[20] Taha, I. and J. Ghosh: 1999, 'Symbolic Interpretation of Artificial Neural Networks'. *IEEE Transactions on Knowledge and Data Engineering* **11**(3), 448–463.

[21] Thrun, S. B.: 1995, 'Extracting rules from artificial neural networks with distributed representation'. In: G. Tesauro, D. Touretzky, and T. Leen (eds.): *Advances in Neural Information Processing Systems.*

[22] Weijters, A. and J. Paredis: 2000, 'Rule Induction with Genetic Sequential Covering Algorithm (GeSeCo)'. In: *Proceedings of the second ICSC Symposium on Engineering of Intelligent Systems, (EIS 2000).* pp. 245–251.

[23] Zitzler, E. and L. Thiele: 1999, 'Multiobjective Evolutionary Algorithms: A Comparative Case Study and Strength Pareto Approach'. *IEEE Transaction on Evolutionary Computation* **3**(4), 257–271.

15

Agent Based Multi-Objective Approach to Generating Interpretable Fuzzy Systems*

Hanli Wang[1], Sam Kwong[1], Yaochu Jin[2], and Chi-Ho Tsang[1]

[1] Department of Computer Science, City University of Hong Kong, 83 Tat Chee Avenue, Kowloon, Hong Kong, China
{wanghl@cs,cssamk@,wilson@cs}.cityu.edu.hk

[2] Honda Research Institute Europe, 63073, Offenbach/Main, Germany
yaochu.jin@honda-ri.de

Summary. Interpretable fuzzy systems are very desirable for human users to study complex systems. To meet this end, an agent based multi-objective approach is proposed to generate interpretable fuzzy systems from experimental data. The proposed approach can not only generate interpretable fuzzy rule bases, but also optimize the number and distribution of fuzzy sets. The trade-off between accuracy and interpretability of fuzzy systems derived from our agent based approach is studied on some benchmark classification problems in the literature.

15.1 Introduction

The fundamental concept of fuzzy reasoning was first introduced by Zadeh [1] in 1973, and over the last thirty years thousands of commercial and industrial fuzzy systems have been successfully developed. One of the most important motivations to build up a fuzzy system is to let users gain a deep insight into an unknown system through the human-understandable fuzzy rules. Another main attraction lies in the characteristics of fuzzy systems: the ability to handle complex, nonlinear, and sometimes mathematically intangible dynamic systems. However, when the fuzzy rules are extracted by the traditional methods, there is often a lack of interpretability in the resulting fuzzy rules due to two main factors: 1) it is difficult to determine the fuzzy sets information such as the number and membership functions, and 2) rule bases are usually of high redundancy. Therefore, fuzzy systems considering both the interpretability and accuracy are very desirable, especially when they are generated from the training data. Hence, increasing attention has been paid to improve the interpretability of fuzzy systems [2]-[18].

* This work is partly supported by the City University of Hong Kong Strategic Grant 7001615.

H. Wang et al.: *Agent Based Multi-Objective Approach to Generating Interpretable Fuzzy Systems*, Studies in Computational Intelligence (SCI) **16**, 339–364 (2006)
www.springerlink.com

In order to study the trade-off between the accuracy and interpretability of fuzzy systems, the multi-objective evolutionary algorithm (MOEA) is very suitable to be applied. Indeed, the genetic fuzzy system is one of the most successful approaches to hybridize fuzzy systems with the learning and adaptation abilities [19]. A main advantage of using MOEA is that many solutions, each of which represents an individual fuzzy system, can be obtained in a single run. And the accuracy and interpretability issues can be incorporated into multiple objectives to evaluate candidate fuzzy systems. Thus, the improvement of interpretability and the trade-off between the accuracy and interpretability can be easily studied. Recently, more efforts have been made in the MOEA domain to extract interpretable fuzzy systems [4], [17]-[18], [20]-[30]. In [4], a single MOEA is proposed for fuzzy modeling to solve the function approximation problems. This algorithm has a variable-length, real-coded representation, where each individual of the population contains a variable number of rules between 1 and *max* which is defined by the decision maker. In our former studies [17]-[18], we have a detailed discussion about some important issues of interpretability and propose an agent based approach to generate interpretable fuzzy systems for function approximation problems, experimental results have demonstrated that our methods are more powerful than that in [4]. Compared with other methods [20]-[30], where the number of fuzzy sets is predetermined and the distributions of fuzzy sets are not much considered, our agent based approach can not only extract interpretable fuzzy rule bases but also optimize the number and distribution of fuzzy sets through the hierarchical chromosome formulation [31]-[33] and interpretability-based regulation method. In this chapter, we extend our research scope from the function approximation problems to the classification problems based on the proposed agent-based multi-objective approach. In addition, more innovative issues about interpretability of fuzzy systems are presented and the agent-based approach is updated accordingly.

The chapter is organized as follows. Section 15.2 discusses the interpretability issues of fuzzy systems. The proposed agent-based multi-objective approach is discussed in Section 15.3. In Section 15.4, the experimental results are presented on some benchmark classification problems. Finally, we conclude this work and some potential future works are discussed in Section 15.5.

15.2 Interpretability of Fuzzy Systems

The most important motivation to apply fuzzy systems is that they use linguistic rules to infer knowledge similar to the way humans think. Methods for constructing fuzzy systems from training data should not be limited to finding the best approximation of data only. It is also preferable to extract knowledge from training data in the form of fuzzy rules that can be easily understood and interpreted. Interpretability (also called transparency) of fuzzy systems is the criterion that indicates whether a fuzzy system is easily interpreted or not. In our previous work [17] and [18], we have discussed some important interpretability issues in detail. In the following, we will have a quick review about these issues and further propose two more innovative concepts about interpretability: overlapping and relativity.

15.2.1 Completeness and Distinguishability

The discussion of completeness and distinguishability is necessary if fuzzy systems are generated automatically from data. The partitioning of fuzzy sets for each fuzzy variable should be complete and well distinguishable. The completeness means that for each input variable, at least one fuzzy set is fired. The concepts of completeness and distinguishability of fuzzy systems are usually expressed through a fuzzy similarity measure [2], [3], [7], and [34]. In [34], similarity between fuzzy sets is defined as the degree to which the fuzzy sets are equal. In fact, if the similarity of two fuzzy sets is too big, then it indicates that the two fuzzy sets overlap too much and the distinguishability between them is poor.

Let A and B be two fuzzy sets of the fuzzy variable x with the membership functions $u_A(\mathrm{x})$ and $u_B(\mathrm{x})$, respectively. We use the following similarity measure between fuzzy sets [34]:

$$S(A,B) = \frac{M(A \cap B)}{M(A \cup B)} = \frac{M(A \cap B)}{M(A) + M(B) - M(A \cap B)} \tag{15.1}$$

where $M(A)$ denotes the cardinality of A, and the operators $\cap$ and $\cup$ represent the intersection and union. There are several methods to calculate the similarity. One computationally effective form in [11] and [12] is described in (15.2) on a discrete universe $U = \{x_j | j = 1, 2, \cdots, m\}$. $\wedge$ and $\vee$ are the minimum and maximum operators, respectively.

$$S(A,B) = \frac{\sum_{j=1}^{m} [u_A(x_j) \wedge u_B(x_j)]}{\sum_{j=1}^{m} [u_A(x_j) \vee u_B(x_j)]} \tag{15.2}$$

15.2.2 Consistency

Another important issue about interpretability is the consistency among fuzzy rules and the consistency with *a priori* knowledge. Consistency among fuzzy rules means that if two or more rules are simultaneously fired, then their conclusions should be coherent [7, 5]. The consistency with *a priori* knowledge means that the fuzzy rules generated from data should not be in conflict with the expert knowledge or heuristics. A definition of consistency and its calculation method is given in [7]. Another important factor about consistency is the inclusion relation [27]. If the antecedents of two fuzzy rules are compatible with an input vector and the antecedents of one rule is included in those of the other rule, the former rule should have a larger weight than the latter to calculate the output value. In [18], we proposed a method to calculate the updated rule weight to represent the inclusion relation.

15.2.3 Utility

Even if the partitioning of fuzzy variables is complete and distinguishable, it is not guaranteed that each of the fuzzy sets be used by at least one rule. We use the term utility to describe such cases. If a fuzzy system is of sufficient utility, then all of the fuzzy sets are utilized as antecedents or consequents by fuzzy rules. Whereas, a fuzzy system of insufficient utility indicates that there exists at least one fuzzy set that is not utilized by any of the rules. Then the unused fuzzy sets should be removed from the rule base resulting in a more compact fuzzy system.

15.2.4 Compactness

A compact fuzzy system means that it has the minimal number of fuzzy sets and fuzzy rules. In addition, the number of fuzzy variables is also worth being considered.

15.2.5 Overlapping

From human heuristics, for any input value, the number of maximum membership values (i.e., 1) should not be greater than one. In other words, the region of maximum membership value of fuzzy sets should not overlap. We use the following two figures to illustrate the overlapping concept. Figure 15.1

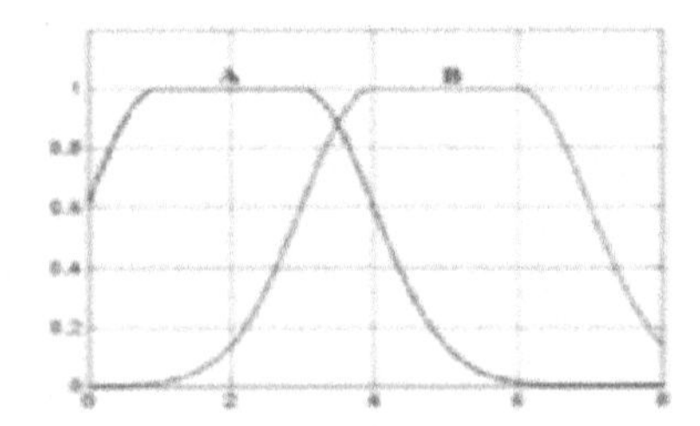

Fig. 15.1. Appropriate distribution about overlapping

shows an appropriate distribution between fuzzy sets A and B because the regions of the maximum membership value of A and B do not overlap, whereas in Fig. 15.2 such regions overlap each other. So regarding the input value 3.5 in Fig. 15.2, the membership of A and B are the same and equal to 1, i.e., 3.5 completely belongs to both of these two fuzzy sets, it will cause a bad

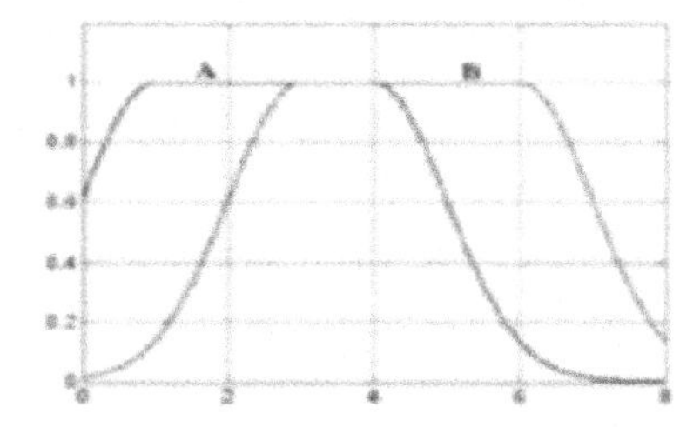

Fig. 15.2. Inappropriate distribution about overlapping

interpretability of fuzzy systems, e.g., 3.5 is both completely big and completely small. Therefore, the overlapping should be avoided when we build fuzzy systems, especially the fuzzy system is constructed from sampling data.

15.2.6 Relativity

Relativity is associated with the relative bounds of definition domain of two fuzzy sets. We will discuss the relativity issue with the aid of trapezoidal membership functions because other types of membership functions can also be easily transformed to the trapezoidal type such as the triangular and Gaussian membership functions, etc. Let the parameter vector $[a_1, a_2, a_3, a_4]$ represent the membership function parameters of fuzzy set A and $[b_1, b_2, b_3, b_4]$ of B, where a_1 is the lower bound of the definition domain of A, a_2 is the left center, a_3 is the right center and a_4 is the upper bound. The relative position of A and B is shown in Fig. 15.3. In this case, it is very hard to interpret the input values in the ranges $[b_1, c_1]$ and $[c_2, a_4]$, where c_1 and c_2 are the abscissas at which A and B intersect. For instance, A and B are assigned the linguistic

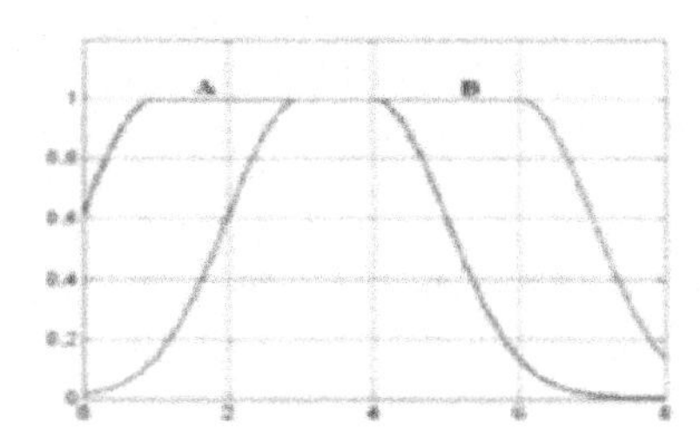

Fig. 15.3. Inappropriate distribution about relativity

term *small* and *big*, and considering the input value 2, the membership degree of *small* is lower than that of *big*. Obviously, this interpreted result is in contradiction with the common sense. Thus, we propose the following condition to guarantee a better interpretability for relativity (Fig. 15.4).

$$If\ a_3 < b_2,\ then\ a_1 \leq b_1\ and\ a_4 \leq b_4 \tag{15.3}$$

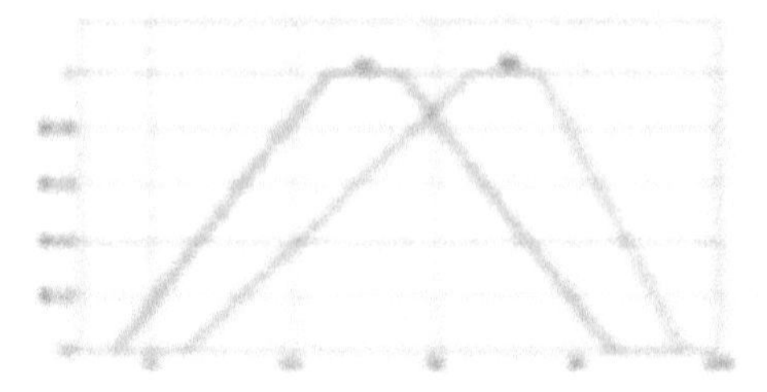

Fig. 15.4. Appropriate distribution about relativity

15.3 Agent-Based Multi-Objective Approach

As we mentioned earlier, when fuzzy systems are extracted using the traditional methods, the interpretability will very likely be lost. First, the numbers of fuzzy sets and rules are usually larger than necessary. Second, the distribution of fuzzy sets is inappropriate. Both of these two factors make it hard to interpret the resulted fuzzy systems. The generation of fuzzy systems from training data can be described as finding a small number of combination of fuzzy sets, each of which is used as the antecedent or consequent part of a fuzzy rule. While this task seems to be simple at a first glance, it is in fact very difficult especially in the case of high-dimensional problems, because the number of possible combinations of linguistic values exponentially increases. Worse, it is almost impossible to have a good understanding about an unknown complex system, thus giving the linguistic terms for each fuzzy variable in advance becomes very hard, even not to mention extracting interpretable fuzzy rule bases from the combination of fuzzy sets.

To address this problem, we propose an agent-based multi-objective approach to construct fuzzy systems from training data. Agents are computational entities capable of autonomously achieving goals by executing needed actions. Although there seems to be little general agreement on a precise definition of what constitutes an intelligent agent, four common agent characteristics have been proposed in [35]: situatedness, autonomy, adaptability, and sociability. It turns out to be particularly difficult to build up a multi-agent system where the agents do what they should do in the light of the afore mentioned characteristics. Several challenging and intertwined issues have been proposed in [36]-[39].

In our agent based framework (Fig. 15.5), there are two kinds of agents in the multi-agent system: the Arbitrator Agent (AA) and the Fuzzy Set Agent (FSA). These FSAs obtain information from the AA in which the information is expressed in terms of training data. Thus, the situatedness of agents is realized. According to the available training data, the FSA can autonomously determine its own fuzzy sets information such as the number and distribution with the aid of hierarchical chromosome formulation and interpretability-based regulation method, then extract fuzzy rules according to the available fuzzy sets using the Pittsburgh-style approach [40]. So the autonomy and

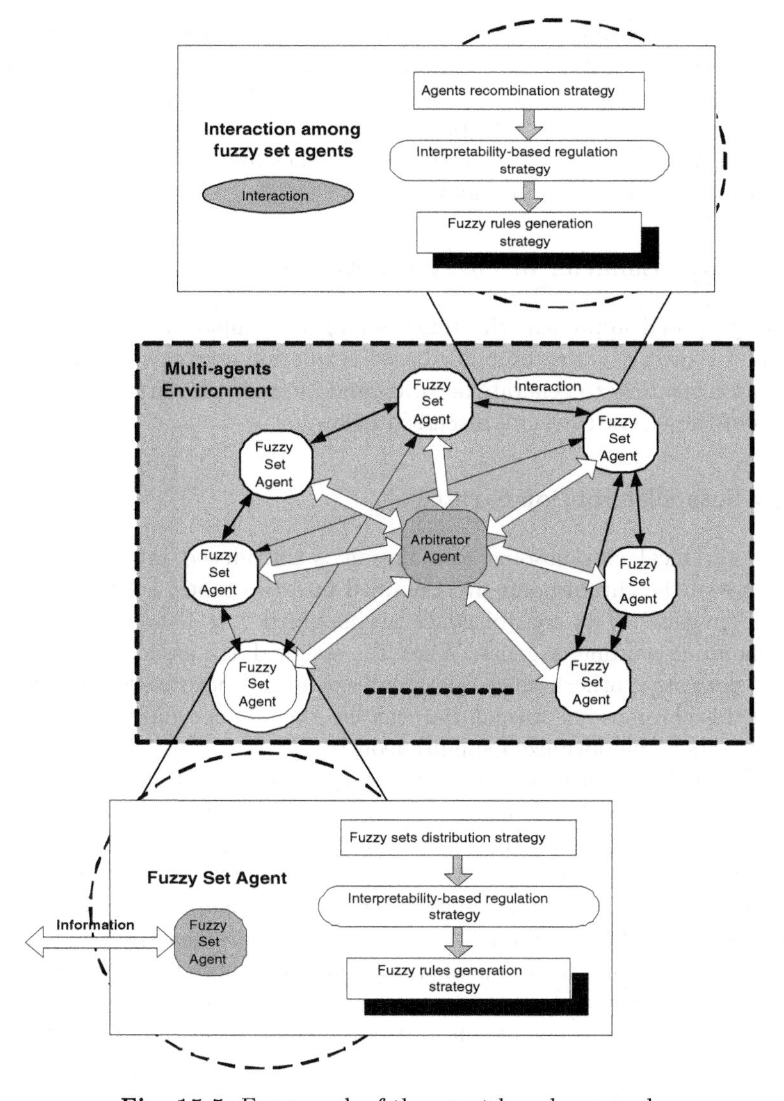

Fig. 15.5. Framework of the agent-based approach

adaptability of agents are achieved. As far as the social behavior is concerned, the FSA is able to cooperate and compete with other FSAs. In the proposed approach, the FSAs cooperatively exchange their fuzzy sets information by ways of crossover and mutation of the hierarchical chromosome, and generate offspring FSAs. After the self-evolving of FSAs, the AA uses the NSGA-II [41] algorithm to evaluate the FSAs, and judge which FSAs should survive and be kept to the next population, whereas the obsolete agents are die. In [18], the agent based approach is mainly applied to generate interpretable fuzzy systems for function approximation problems. And the simulation results have demonstrated that the proposed approach is very effective to study the trade-off between the accuracy and interpretability of fuzzy systems. In this work, we update the interpretability-based regulation strategy according to the newly proposed interpretability issues and apply the agent-based approach to classification problems.

15.3.1 Intra Behavior of Fuzzy Set Agents

In the proposed approach, the fuzzy set agents employ the fuzzy sets distribution strategy, interpretability-based regulation strategy, and fuzzy rules generation strategy to generate accurate and interpretable fuzzy systems. The details of these strategies are discussed below.

Fuzzy Sets Distribution Strategy

The hierarchical chromosome formulation is introduced in [31]-[33], where the genes of the chromosome are classified into two types: control genes and parameter genes. The control genes are assigned 1 or 0 to manipulate the corresponding parameter genes. When 1 is assigned, the associated parameter gene is activated, otherwise the parameter gene is deactivated. The effectiveness of this chromosome formulation enables the number and the distribution of fuzzy sets to be optimized simultaneously.

For each fuzzy variable x_i, we determine the possible maximal number of fuzzy sets M_i so that it can sufficiently represent this fuzzy variable. For N dimensional problems, there are totally $M_1 + M_2 + \cdots + M_N$ possible fuzzy sets. So there are $M_1 + M_2 + \cdots + M_N$ control genes coded as 0 or 1, where 1 is assigned to represent that the corresponding parameter gene is activated to join the evolutionary process, and 0 means deactivated. We apply the Gaussian combinational membership function (abbreviated as Gauss2mf) to describe antecedent fuzzy sets, i.e., a combination of two Gaussian functions. The Gauss2mf depends on four parameters σ_1, c_1, σ_2 and c_2 as given by:

$$\mu(x;\sigma_1,c_1,\sigma_2,c_2) = \begin{cases} \exp\left[\frac{-(x-c_1)^2}{2\sigma_1^2}\right] & : \; x < c_1 \\ 1 & : \; c_1 \le x \le c_2 \\ \exp\left[\frac{-(x-c_2)^2}{2\sigma_2^2}\right] & : \; c_2 < x \end{cases} \tag{15.4}$$

where σ_1 and c_1 determine the shape of the leftmost curve. The shape of the rightmost curve is specified by σ_2 and c_2. Hence the parameter vector $[\sigma_1, c_1, \sigma_2, c_2]$ is used to represent one parameter gene. The Gauss2mf is a kind of smooth membership functions, so the resulting model will in general have a high accuracy in fitting the training data. Another characteristic of Gauss2mf is that the completeness of fuzzy system is guaranteed because the Gauss2mf covers the universe sufficiently. An example of the relationship between control

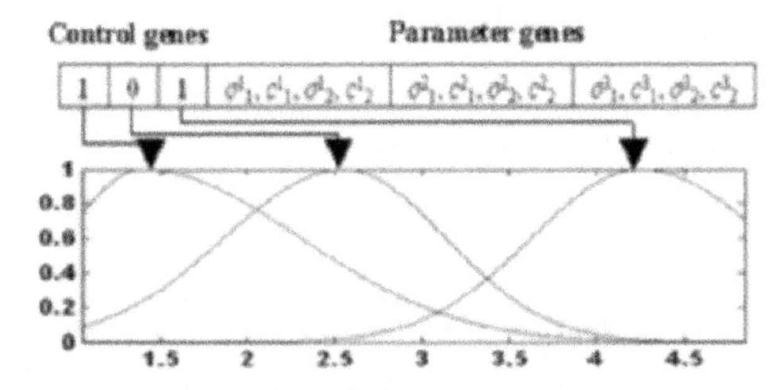

Fig. 15.6. Example of control genes and parameter genes

genes and parameter genes is given in Fig. 15.6. The initializations of control genes and parameter genes are given as follows.

A. Initialization of Control Genes

The fuzzy set agents initialize control genes using a uniformly random number in the interval [0, 1]. For example, *rand* is a uniformly random number in [0, 1]. If $rand > 0.5$, then the target control gene is assigned to 1, otherwise, 0 is assigned.

B. Initialization of Parameter Genes

As far as the parameter genes are concerned, the initialization step is a little bit complicated. In this work, we define a fuzzy set A using the membership function $u_A(x; a_1, a_2, a_3, a_4)$, where a_1, a_2, a_3, and a_4 are the lower bound, left center, right center and upper bound of the definition domain, respectively ($a_1 \leq a_2 \leq a_3 \leq a_4$). However, we use the Gauss2mf in (15.4) as the membership function, so it is not easy to obtain a_1 and a_4 like the trapezoidal type of membership functions. We need to calculate a_1 and a_4 using a very small number ε (e.g. 0.001) which is regarded as equal to zero: $u_A(a_1; a_1, a_2, a_3, a_4) = u_A(a_4; a_1, a_2, a_3, a_4) = \varepsilon$. The relationship between the parameter list $[\sigma_1, c_1, \sigma_2, c_2]$ and $[a_1, a_2, a_3, a_4]$ is described as:

$$\sigma_1 = \sqrt{-\frac{(a_1 - c_1)^2}{2 \ln \varepsilon}} \tag{15.5}$$

$$c_1 = a_2 \tag{15.6}$$

$$\sigma_2 = \sqrt{-\frac{(a_4 - c_2)^2}{2 \ln \varepsilon}} \tag{15.7}$$

$$c_2 = a_3 \tag{15.8}$$

Suppose A belongs to the variable x with the lower bound lb and upper bound ub, respectively. First we initialize the parameter c_1 and c_2 as:

$$c_1 = min\{candi_1, candi_2\} \quad (15.9)$$

$$c_2 = max\{candi_1, candi_2\} \quad (15.10)$$

in order to guarantee the relation:

$$c_1 \leq c_2 \quad (15.11)$$

where

$$candi_1 = (ub - lb) \times rand_1 + lb \quad (15.12)$$

$$candi_2 = (ub - lb) \times rand_2 + lb \quad (15.13)$$

$rand_1$ and $rand_2$ are two random numbers in [0,1]. In addition, we also guarantee c_1 is not greater than ub and c_2 is not less than lb:

$$c_1 = min(c_1, ub) \quad (15.14)$$

$$c_2 = max(c_2, lb) \quad (15.15)$$

In the next step, the parameters σ_1 and σ_2 are derived as:

$$\sigma_1 = \sqrt{-\frac{[(ub - lb) \times rand_3]^2}{2 \ln \varepsilon}} \quad (15.16)$$

$$\sigma_2 = \sqrt{-\frac{[(ub - lb) \times rand_4]^2}{2 \ln \varepsilon}} \quad (15.17)$$

where $rand_3$ and $rand_4$ are two random numbers in the interval [0, 1].

Interpretability-Based Regulation Strategy

After the fuzzy set distribution strategy, the interpretability-based regulation strategy is applied to the initial fuzzy sets to obtain a better distribution about interpretability. This strategy includes the following actions.

A. Merging Similar Fuzzy Sets

A method to calculate the similarity measure between two fuzzy sets is given in (15.2). If the similarity value is greater than a given threshold, then we merge these two fuzzy sets to generate a new one. Considering two fuzzy sets A and B with the membership functions $u_A(x;\ a_1,\ a_2,\ a_3,\ a_4)$ and $u_B(x;\ b_1,\ b_2,\ b_3,\ b_4)$, the resulting fuzzy set C with the membership function $u_C(x;\ c_1,\ c_2,\ c_3,\ c_4)$ is derived from merging A and B by:

$$c_1 = min(a_1, b_1) \quad (15.18)$$

$$c_2 = \lambda_2 a_2 + (1 - \lambda_2) b_2 \quad (15.19)$$

$$c_3 = \lambda_3 a_3 + (1 - \lambda_3) b_3 \quad (15.20)$$

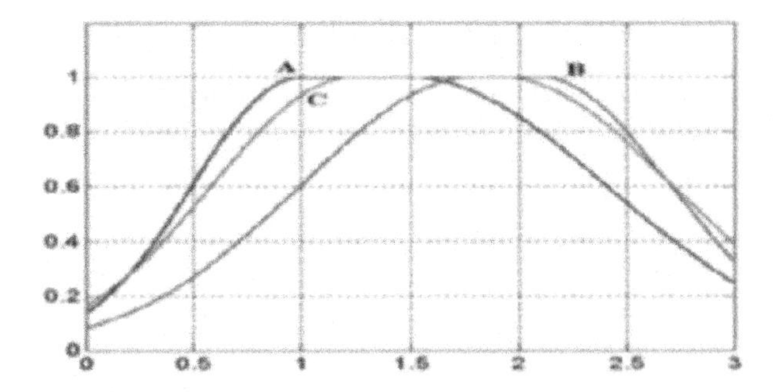

Fig. 15.7. Merging A and B to create C

$$c_4 = max(a_4, b_4) \tag{15.21}$$

The parameters λ_2, $\lambda_3 \in [0, 1]$ determines the relative importance of A and B to generate C. The threshold for merging similar fuzzy sets plays an important role in the improvement of interpretability. According to our experience, values in the range [0.4, 0.7] may be a good choice. In our approach, we set the threshold equal to 0.6. Figure 15.7 illustrates the case of merging A and B to create C.

B. Removing Fuzzy Sets Similar to Universal Set or Singleton Set

If the similarity value between a fuzzy set and the universal set $U(u_U(x) = 1, \forall x \in X)$ is greater than the upper bound (θ_U) or smaller than the lower bound (θ_S), then we can remove it. In the first case, the fuzzy set is very similar to the universal set and in the latter case similar to a singleton set. Neither of these cases is desirable for interpretable rule system generation. We set $\theta_U = 0.9$, $\theta_S = 0.05$ in this work.

C. Overlapping Regulation

If two fuzzy sets overlap, i.e., the maximum membership regions of these two fuzzy sets overlap, we should regulate their relative positions to break down the overlapping. Assume there are two fuzzy sets A and B represented by Gauss2mf in the form of $[\sigma_1^A, c_1^A, \sigma_2^A, c_2^A]$ and $[\sigma_1^B, c_1^B, \sigma_2^B, c_2^B]$, the following two cases illustrate the regulation action.

Case 1: $c_1^A \leq c_1^B \leq c_2^B \leq c_2^A$
This means the maximum membership region of A includes that of B, and we use the following equations to regulate these two sets and generate two new fuzzy sets A' and B' in the form of $[\sigma_1^{A'}, c_1^{A'}, \sigma_2^{A'}, c_2^{A'}]$ and $[\sigma_1^{B'}, c_1^{B'}, \sigma_2^{B'}, c_2^{B'}]$.

$$\sigma_1^{A'} = \sigma_1^A, c_1^{A'} = c_1^A, \sigma_2^{A'} = \sigma_2^B, c_2^{A'} = c_1^B \tag{15.22}$$

$$\sigma_1^{B'} = \sigma_1^{B}, c_1^{B'} = c_2^{B}, \sigma_2^{B'} = \sigma_2^{A}, c_2^{B'} = c_2^{A} \tag{15.23}$$

An example of the fuzzy sets distribution before and after overlapping regulation is shown in Figs. 15.8 and 15.9, respectively.

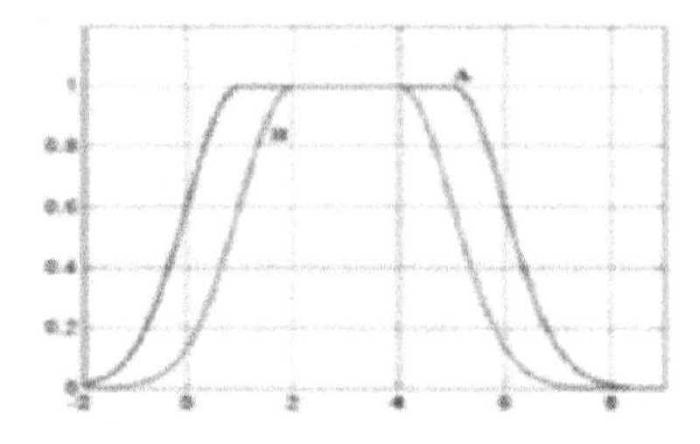

Fig. 15.8. Before regulation in case 1

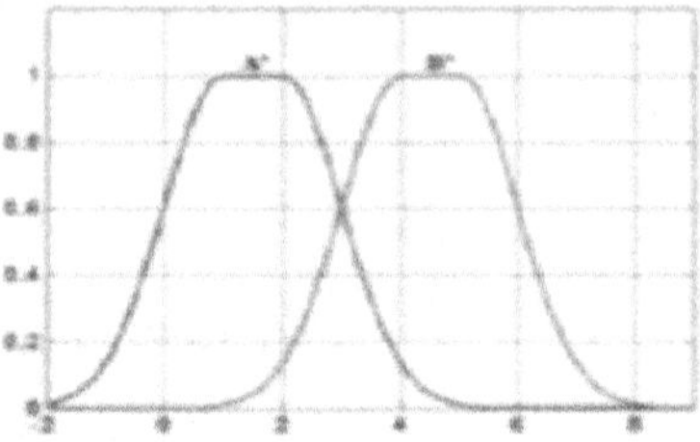

Fig. 15.9. After regulation in case 1

Case 2: $c_1^A \leq c_1^B \leq c_2^A \leq c_2^B$
This indicates the maximum membership regions of A and B overlap in part. And the following equations are applied to generate two fuzzy sets A' and B' (Figs. 15.10 and 15.11).

$$\sigma_1^{A'} = \sigma_1^{A}, c_1^{A'} = c_1^{A}, \sigma_2^{A'} = \sigma_2^{A}, c_2^{A'} = c_1^{B} \tag{15.24}$$

$$\sigma_1^{B'} = \sigma_1^{B}, c_1^{B'} = c_2^{A}, \sigma_2^{B'} = \sigma_2^{B}, c_2^{B'} = c_2^{B} \tag{15.25}$$

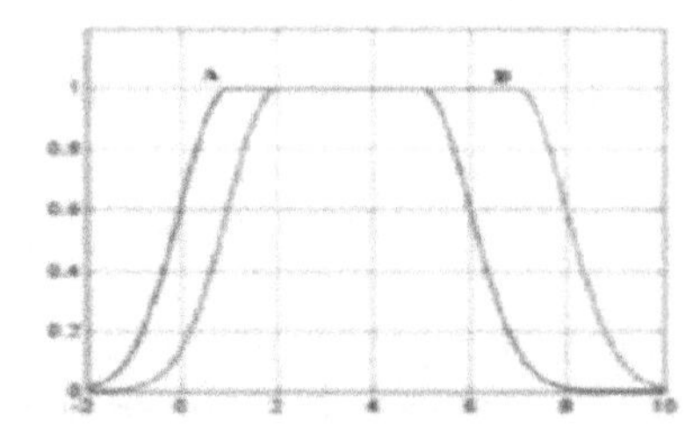

Fig. 15.10. Before regulation in case 2

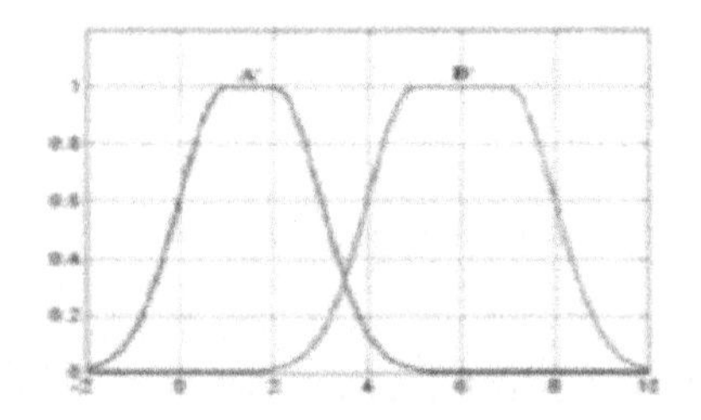

Fig. 15.11. After regulation in case 2

D. Relativity Regulation

Assume there are two fuzzy sets A ($u_A(x;\ a_1,\ a_2,\ a_3,\ a_4)$) and B ($u_B(x;\ b_1,\ b_2,\ b_3,\ b_4)$), and $a_3 < b_2$. If (15.3) is not satisfied, we should regulate their parameters to generate two fuzzy sets A' and B' in the following action (Figs. 15.12 and 15.13).

$$If\ a_1 > b_1,\ then\ a_1' = b_1,\ b_1' = a_1 \tag{15.26}$$

$$If\ a_4 > b_4,\ then\ a'_4 = b_4,\ b'_4 = a_4 \tag{15.27}$$

After implementing the interpretability-based regulation strategy, we suppose that the fuzzy set agent obtains a fuzzy system with $M_1^a + M_2^a + \cdots + M_N^a$ sets, where $0 \leq M_i^a \leq M_i$ and the case that M_i^a is equal to zero indicates the variable x_i does not participate in the generation of fuzzy systems.

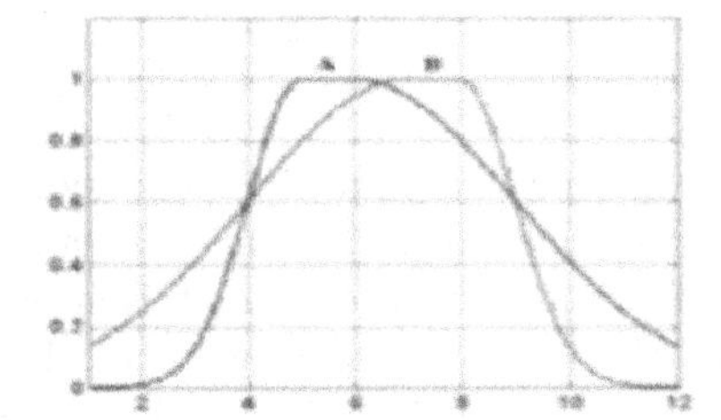

Fig. 15.12. Before relativity regulation

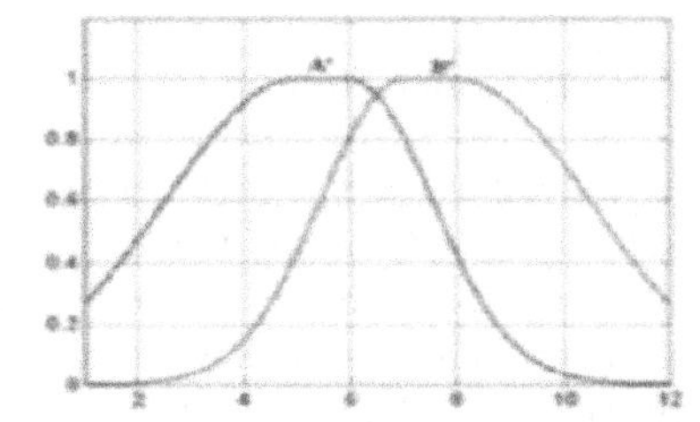

Fig. 15.13. After relativity regulation

Fuzzy Rules Generation Strategy

During the generation of rule base, the fuzzy set agents use the Pittsburgh-style approach to extract rules. We use the Pittsburgh-style approach [40] because each fuzzy rule base can be considered as one solution and many solutions can be obtained in a single run. So the multi-objective optimization algorithms can be applied easily to select better solutions. Suppose there are N fuzzy variables, M_i^a is the number of active fuzzy sets for variable x_i. The *don't care* conditions (i.e., incomplete rules) are also considered, so the total maximum number of possible fuzzy rules is $(M_1^a+1)\times(M_2^a+1)\times\cdots\times(M_N^a+1)$ for N-dimensional problems. The task of fuzzy set agents in this stage is to find a small number of rules considering both the accuracy and interpretability. In the following, we will discuss how the fuzzy set agents achieve these goals.

A. Initialization of Rule Base Population

In the Pittsburgh-style genetic based machine learning approach, the search for a compact rule base with high performance ability corresponds to the evolution of a population of fuzzy rule bases. In this work, each fuzzy rule is coded as an array of length N, where the *ith* element of the array indicates which fuzzy set of the *ith* fuzzy variable is fired. The *ith* element is denoted as c_i and initially set to an integer between 0 and M_i^a with the same probability $1/(M_i^a+1)$. If c_i is greater than zero, it indicates that the $c_i th$ fuzzy set of the *ith* variable is fired, whereas if c_i is equal to zero, then the *ith* variable does not play a role in rule generation. Then the fuzzy set agent sets the population size N_{pop}, i.e., the number of fuzzy rule bases. For each fuzzy rule base, it is represented as a concatenated string of the length $N \times N_{rule}$, where N_{rule} is a predefined integer to describe the size of the initial rule base. In this concatenated string, each substring of length N represents a single fuzzy rule. Because the heuristic procedure [21] is applied to calculate rule consequents for classification problems, the rule consequents are not coded in the concatenated string. The fuzzy rule bases are randomly initialized so that the cell value of the concatenated string represents one of the fuzzy sets of the corresponding variable or is equal to zero indicating *don't care* conditions.

B. Crossover and Mutation

Offspring rule bases are generated by crossover and mutation. One-point crossover is used in this work. The crossover operation randomly selects a cutoff point for each parent to generate offspring rule bases. A mutation operation randomly replaces each element of the rule base string with another linguistic value if a probability test is satisfied. Elimination of existing rules and addition of new rules are also used as mutation operations to change the number of rules. Note that the crossover and mutation operations maybe introduce the same rules, the fuzzy set agent will check the offspring fuzzy rule base to delete the same rules after the crossover and mutation operations.

C. Calculation of Rule Consequents

For the classification problems, the heuristic procedure [21] is applied to generate rule consequents from the training pattern data. Suppose the pattern classification problem is a c class problem in the n-dimensional pattern space with continuous attributes. The rule for classification problems is expressed in the following form:

R_i: If x_1 is A_{i1} and, $\cdots$, and x_n is A_{in}, then Class C_i with $CF = CF_i$

where R_i is the label of the ith fuzzy rule, $A_{i1}, \cdots, A_{in}$ are the antecedent fuzzy sets, C_i is the consequent class (i.e., one of the given c classes), and CF_i is the grade of certainty of R_i. For each class, calculate the sum of the compatibility grades of the training patterns $X^i = [x_1^i, x_2^i, \cdots, x_n^i]$ for the rule R_j as:

$$\beta_{Class\ h}(R_j) = \sum_{X^i \in Class\ h} u_j(X^i), \ h = 1, 2, \cdots, c \tag{15.28}$$

where $\beta_{Class\ h}(R_j)$ is the sum of the compatibility grades of the training patterns in Class h for R_j. Find Class $\hat{h}_j$ that has the maximum value of $\beta_{Class\ h}(R_j)$ as the consequent class for R_j:

$$\beta_{Class\ \hat{h}_j}(R_j) = max\{\beta_{Class\ 1}(R_j), \cdots, \beta_{Class\ c}(R_j)\} \tag{15.29}$$

If the maximum value of $\hat{h}_j$ can not be uniquely specified, that is, there are some classes that have the same maximum value, the fuzzy rule R_j is removed from the rule base. Then we calculate the certainty factor CF_j as follows:

$$CF_j = \frac{\left[\beta_{Class\ \hat{h}_j}(R_j) - \bar{\beta}\right]}{\sum_{h=1}^{c} \beta_{Class\ h}(R_j)} \tag{15.30}$$

where

$$\bar{\beta} = \frac{\sum_{h \neq \hat{h}_j} \beta_{Class\ h(R_j)}}{c-1} \tag{15.31}$$

D. Evaluation Criteria and Selection Mechanism

For each fuzzy set agent, three criteria are used to evaluate the rule base candidates: 1) the number of fuzzy rules, 2) the total length of fuzzy rules, i.e., the total number of rule antecedents in the rule base, and 3) the accuracy in terms of the classification rate. In this work, the single winning rule method [21] is used to calculate the classification rate. For each training pattern data X^i, the winner rule R_j is determined as:

$$u_j(X^i) \cdot CF_j = max\{u_k(X^i) \cdot CF_k \mid k = 1, 2, \cdots, R\} \tag{15.32}$$

where R is the number of fuzzy rules. If the predicted class is not the actual one or more than one fuzzy rules have the same maximum result, the classification error increases one.

Based on the above three criteria, the fuzzy set agent uses the NSGA-II [41] algorithm to generate the rule base candidates. In order to compare the rule base candidates, we predefine the preference for the three criteria. The accuracy is assigned the first priority and the other two criteria about the interpretability are assigned the same second priority. In other words, we firstly compare two candidates according to the accuracy only. If these two candidates have the same accuracy level based on our preference, then the other two criteria are compared to determine which candidate is better. If one candidate is better than the other based on the accuracy preference, then it is no need to compare the other two criteria. In the current work, we use the difference of the accuracy value of rule base candidates to design the preference. If the difference is less than or equal to a predefined value, then the candidates have the same accuracy level, otherwise we can determine which candidate is better without comparing the other two criteria. We take such measures because a fuzzy system constructed from training data should have a certain degree of accuracy. If we do not pay more attention to accuracy, the fuzzy sets agents in the long run maybe generate solutions of high accuracy, however, it would take a lot of computational time to achieve this goal. Another advantage is that we can adjust the preference to meet different trade-off requirements. Suppose there are $N_{par} + N_{offs}$ candidates, where N_{par} is the parent population size and N_{offs} is the number of offspring resulting from the crossover and mutation operations. The fuzzy set agent adopts the elitism strategy to select N_{par} best candidates from the mixed populations.

In order to guarantee the sufficient utility mentioned in Sect. 15.2.3, the fuzzy set agent identifies the unused active fuzzy sets and flips their corresponding control genes from 1 to 0 at the end of the self evolution.

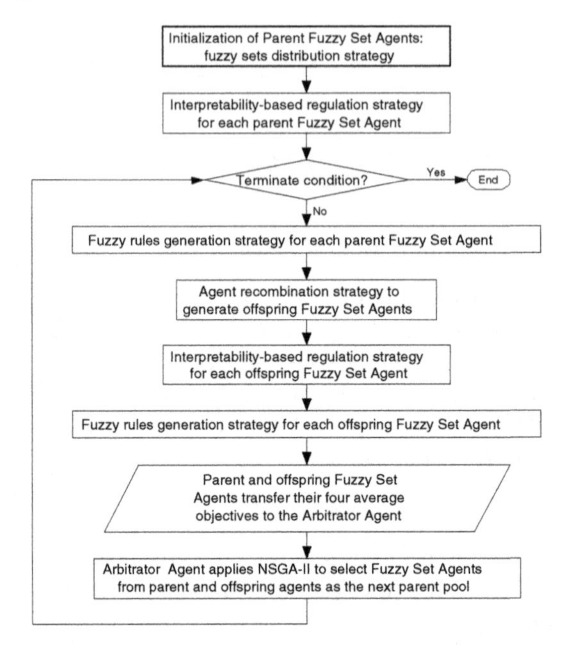

Fig. 15.14. Flowchart of agent based approach

15.3.2 Interaction among Agents

The fuzzy set agents can interact with each other. Assume the number of offspring agents (N_{aoff}) is less than or equal to the number of parent agents (N_{acur}) in the current population: $N_{aoff} \leq N_{acur}$. We select N_{aoff} fuzzy set agents from the current agent population and use the crossover and mutation operations to generate N_{aoff} offspring agents, i.e., two parent agents generate two offspring agents. The N_{aoff} parent agents are different with one another and selected randomly with the same probability. It is because such a selection mechanism is simple and easy to implement, and the mating restriction is not incorporated in the current research. Then crossover and mutation operations are implemented on both the control genes and parameter genes of two paired parent agents and two offspring agents are generated. The offspring agents use the interpretability-based regulation strategy and fuzzy rules generation strategy to obtain rule bases. Therefore, the cooperation among fuzzy sets agents are achieved by exchanging fuzzy sets information and generating offspring agents. Then four criteria including the three aforementioned criteria and the number of fuzzy sets are transferred to the arbitrator agent. As mentioned previously, the accuracy is predefined as the first priority and the other three criteria are predefined the same second priority. The arbitrator agent applies the NSGA-II algorithm to evaluate the parent and offspring fuzzy set agents and select N_{acur} best agents to become the next agent population. The elitist fuzzy set agents considering both the accuracy and interpretability can survive from the competition, whereas worse ones are discarded. The flowchart of the proposed approach is shown in Fig. 15.14.

15.4 Experimental Results

In order to examine the performance of the agent based multi-objective approach, we use two benchmark classification problems reported in the literature: Iris data and Wine data. Matlab 6.1 is applied to implement the experiments.

15.4.1 Iris Data

Iris data [42] contains 150 pattern instances with 4 attributes from 3 classes. The four attributes are sepal length, sepal width, petal length, and petal width. Iris Setosa, Iris Versicolour, and Iris Virginica are the three classes with 50 instances each. In the current work, we use all the 150 instances to train the fuzzy set agents.

In this example, we use 20 fuzzy set agents, each of which has 5 fuzzy rule base solutions, one hundred fuzzy systems are obtained in total. We run our agent based multi-objective approach ten times. Similar experimental results have been observed in all the runs. To simplify our discussion, we only present the simulation results in one of the ten runs. The *Pareto front* [43] after 100 iterations of the evolution is given in Fig. 15.15, which shows the trade-off among multi-objectives of the available non-dominated solutions. The upper left figure illustrates the trade-off between the accuracy and fuzzy sets number, the upper right figure shows the trade-off between the accuracy and fuzzy rules number, the lower left figure is for the trade-off between the accuracy and fuzzy rules total length, and the lower right figure represents the trade-off among the three objectives: accuracy, fuzzy sets number, and fuzzy rules number. There are seven different forms of non-dominated solutions out of one hundred. In Table 15.1, we compare our approach with other methods in the research arena. From the simulation results, we find that the necessary number of fuzzy variables is less than the input number (4 for Iris data). To express clearly, the second column is shown in Table 15.1, where the number before the brace represents the number of input variables, and the numbers in the brace denote the number of fuzzy sets for each input variable in order. Due to space limit, we only use the first solution in Table 15.1 as an example to show the fuzzy sets distribution (Figs. 15.16-15.18) and rule base (Table 15.2).

15.4.2 Wine Data

Wine data [42] contains 178 pattern instances with 13 attributes from 3 classes. The instances are distributed as follows: 59 instances for class 1, 71

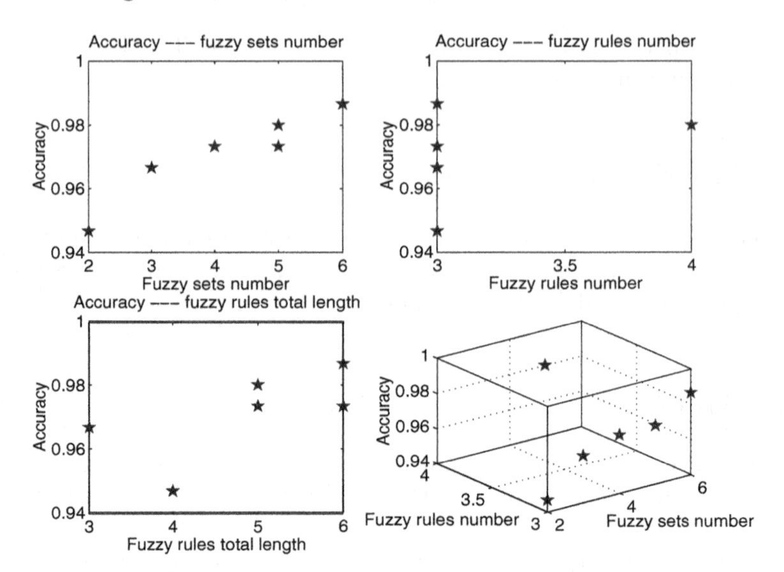

Fig. 15.15. Pareto front about fuzzy systems of Iris data

Table 15.1. Comparison results for Iris data

Ref.	Accuracy	Input	No. Sets	No. Rules	Rules leng.
[12]	0.973	2{0,0,3,2}	5	3	6
[2]	0.993	4{3,3,3,3}	12	3	12
[24]	0.973	4{5,5,5,5}	20	4.6(Avg.)	N/A
[44]	1.000	4{4,5,4,5}	18	5	18
[45]	0.993	4{3,3,3,3}	12	6	24
1	0.987	3{0,2,3,1}	6	3	6
2	0.980	3{1,0,2,2}	5	4	5
3	0.973	4{1,1,1,2}	5	3	5
4	0.973	3{0,1,2,1}	4	3	6
5	0.973	3{1,0,2,1}	4	3	6
6	0.967	2{0,0,2,1}	3	3	3
7	0.947	2{0,0,1,1}	2	3	4

Table 15.2. Rule base for solution 1 in Table 15.1

R_1: If x_2 is big and x_3 is small, then class 1 with $CF = 0.998$;
R_2: If x_2 is small and x_3 is big, then class 3 with $CF = 0.395$;
R_3: If x_3 is medium and x_4 is small, then class 2 with $CF = 0.894$.

Antecedent parameters:

x_2: small=[0.419,2.032,0.193,2.750] big=[0.476,3.178,0.513,3.738]
x_3: small=[0.104,1.043,0.479,2.003] medium=[0.425,2.606,0.33,4.589]
big=[0.784,5.180,1.285,5.712]
x_4: small=[0.098,0.200,0.178,1.420]

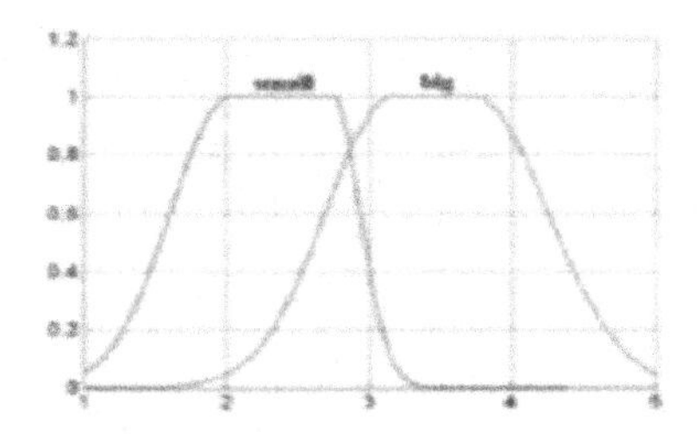

Fig. 15.16. Attribute 2: Sepal width

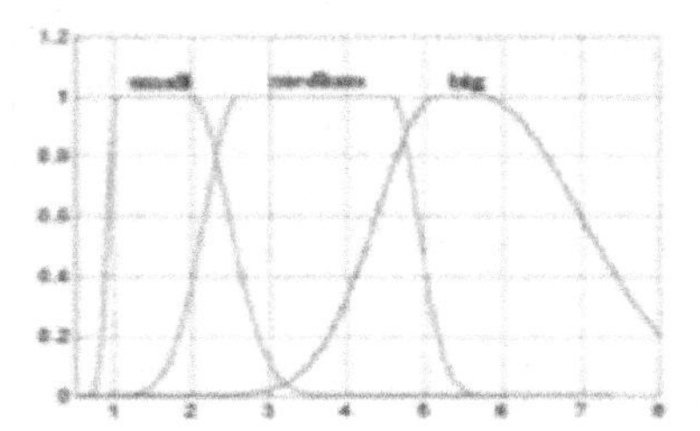

Fig. 15.17. Attribute 3: Petal length

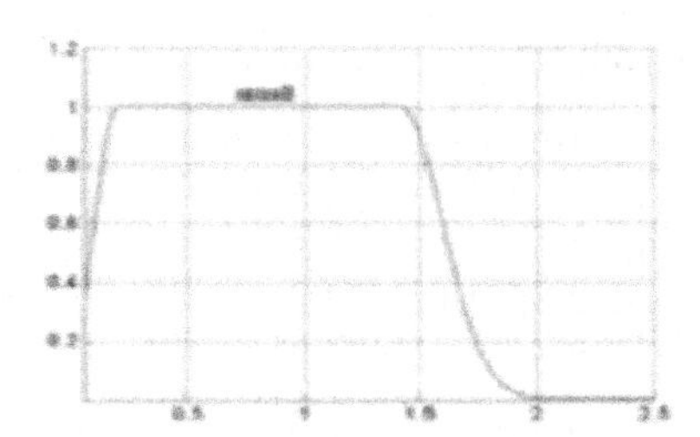

Fig. 15.18. Attribute 4: Petal width

instances for class 2, and 48 instances for class 3. In the current work, all the 178 instances are used for training. There are 20 fuzzy set agents in the multi-agent system, each of which has 5 rule bases. Ten runs have been conducted to test our agent based approach. Similar to Iris data, the results of one trial from the ten are presented in this work. The trade-off among the multiple objectives about the non-dominated fuzzy system solutions for Wine data is shown in Fig. 15.19. Also we compare our results with other methods in Table 15.3. Due to the space limit, only one simple fuzzy system is illustrated using the sixteenth solution in Table 15.3 as an example. The distributions of fuzzy sets are shown in Figs. 15.20-15.29, and the rule base is given in Table 15.4.

From the experimental results of the studied problems, the advantage of our approach can be easily seen: diversified fuzzy systems emphasizing both the accuracy and interpretability are obtained. This is obviously illustrated in Figs. 15.15 and 15.19, and quantified in Tables 15.1 and 15.3. From these tables, it is also noticed that the accuracy of our results is compatible to or

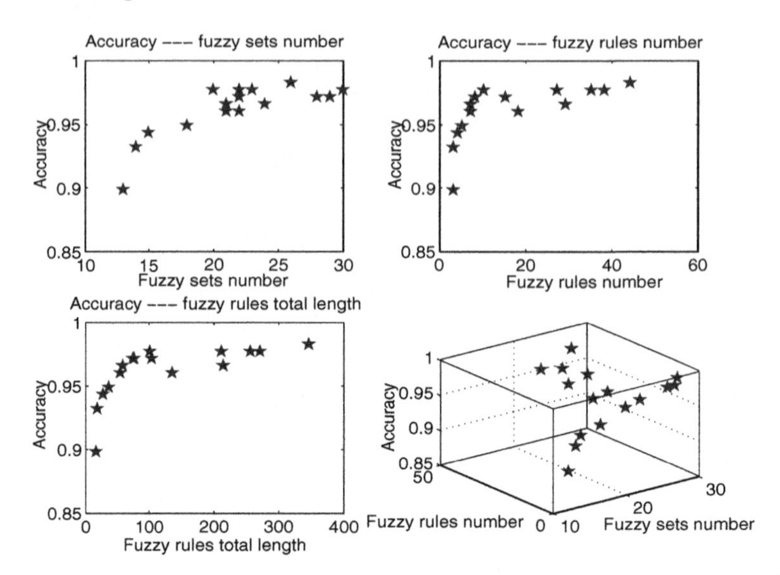

Fig. 15.19. Pareto front about fuzzy systems of Wine data

Table 15.3. Comparison results for Wine data

Ref.	Accuracy	Input	No. Sets	No. Rules	Rules leng.
[23]	0.985	13{5,5,5,5,5,5,5,5,5,5,5,5,5}	65	60	N/A
[24]	0.998	13{5,5,5,5,5,5,5,5,5,5,5,5,5}	65	21.2(Avg.)	N/A
1	0.983	12{3,1,1,2,2,3,3,1,0,3,2,3,2}	26	44	346
2	0.978	13{1,4,3,3,3,2,1,2,1,3,2,2,3}	30	10	101
3	0.978	13{1,1,2,1,2,2,1,2,1,2,2,1,2}	20	38	271
4	0.978	12{1,1,2,3,2,2,1,2,0,3,2,1,2}	22	35	256
5	0.978	12{2,1,2,3,2,2,1,2,0,3,2,1,2}	23	27	211
6	0.972	13{1,3,3,4,2,2,1,2,1,3,2,2,3}	29	8	74
7	0.972	13{1,4,3,4,2,2,1,1,1,2,2,2,3}	28	8	76
8	0.972	10{2,2,2,1,1,0,0,2,3,0,2,3,4}	22	15	104
9	0.966	11{1,2,2,2,0,0,3,3,2,3,2,3,1}	24	7	58
10	0.966	11{1,2,2,2,0,0,1,3,2,3,2,2,1}	21	29	214
11	0.961	11{1,2,2,2,0,0,1,3,2,3,2,2,1}	21	18	136
12	0.961	11{1,2,2,2,0,0,3,3,1,3,2,2,1}	22	7	54
13	0.949	10{2,1,0,0,0,1,2,2,3,2,2,2,1}	18	5	36
14	0.944	9{2,1,0,0,0,0,2,2,2,2,1,2,1}	15	4	27
15	0.933	10{2,1,2,0,0,1,1,0,1,1,2,2,1}	14	3	18
16	0.899	10{2,1,1,1,0,1,0,1,1,3,1,0,1}	13	3	16

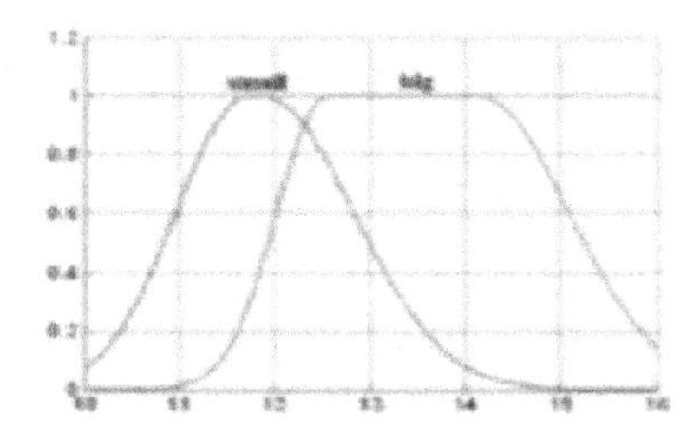

Fig. 15.20. Fuzzy sets of attribute 1

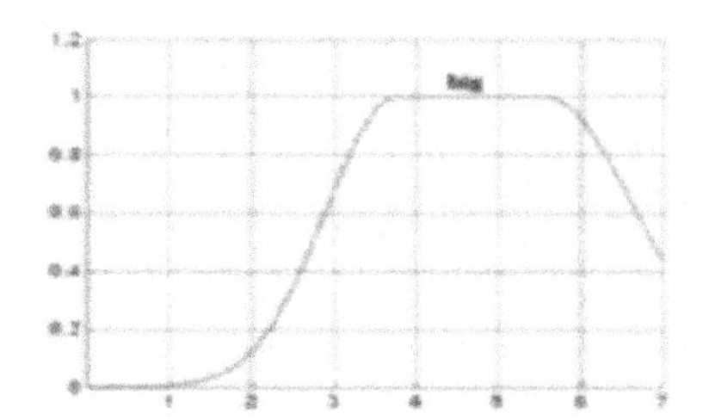

Fig. 15.21. Fuzzy sets of attribute 2

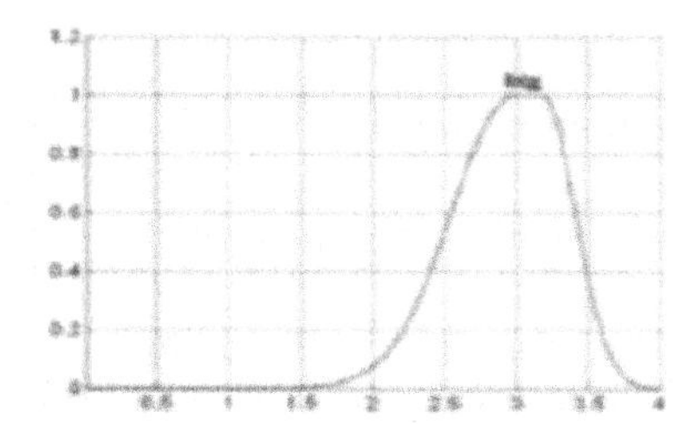

Fig. 15.22. Fuzzy sets of attribute 3

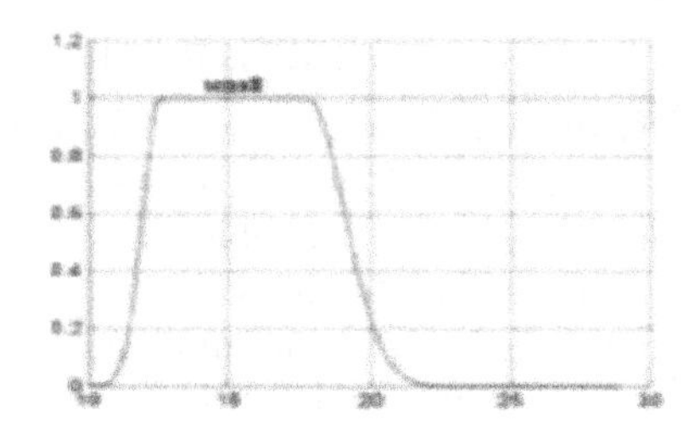

Fig. 15.23. Fuzzy sets of attribute 4

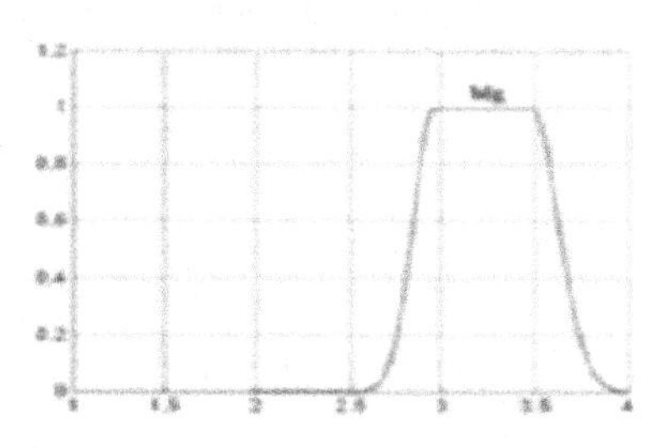

Fig. 15.24. Fuzzy sets of attribute 6

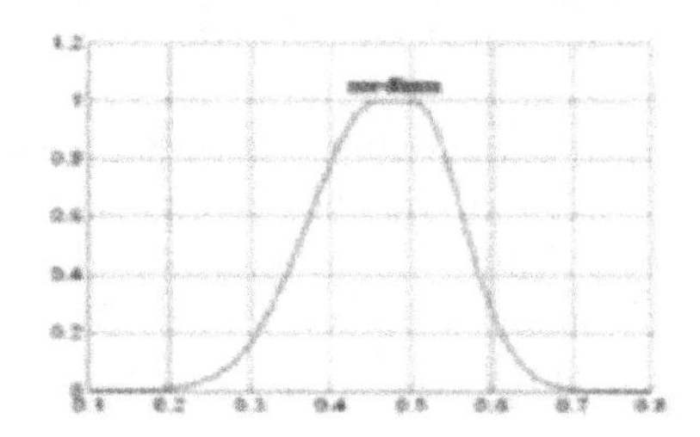

Fig. 15.25. Fuzzy sets of attribute 8

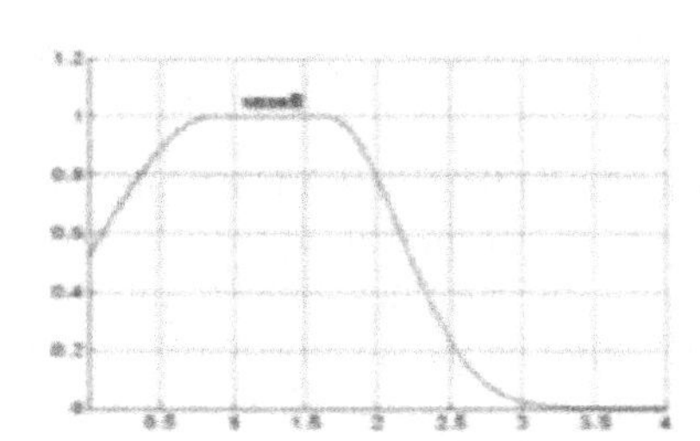

Fig. 15.26. Fuzzy sets of attribute 9

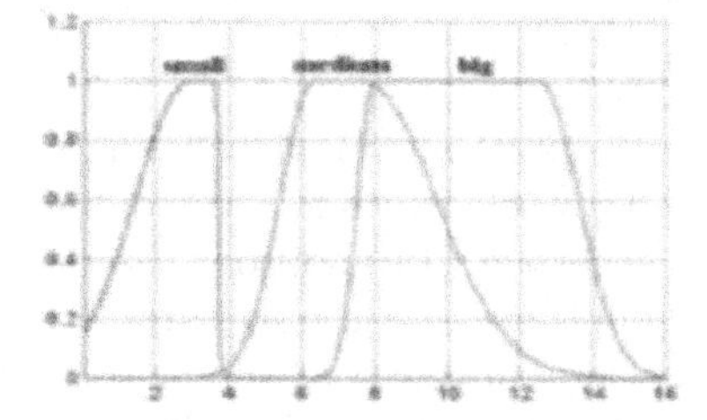

Fig. 15.27. Fuzzy sets of attribute 10

a little worse than other methods known in the literature. However, our approach can get more compact fuzzy rule bases, and more importantly, the distribution of fuzzy sets can be autonomously generated with fewer number of rules needed to represent fuzzy variables when compared with other methods. The trade-off between the accuracy and interpretability of fuzzy systems could be easily understood according to our solutions, most of which have bet-

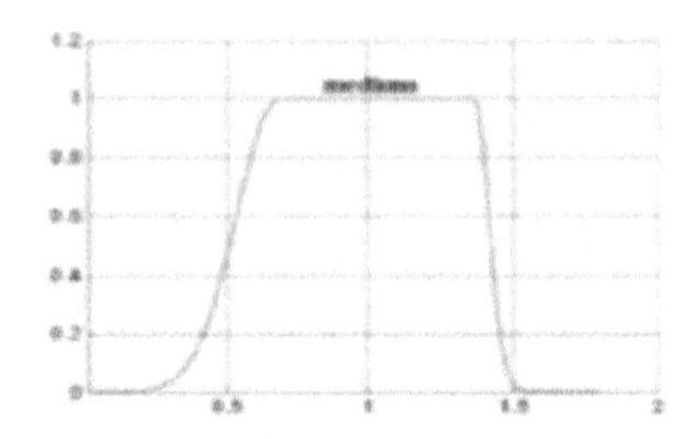

Fig. 15.28. Fuzzy sets of attribute 11

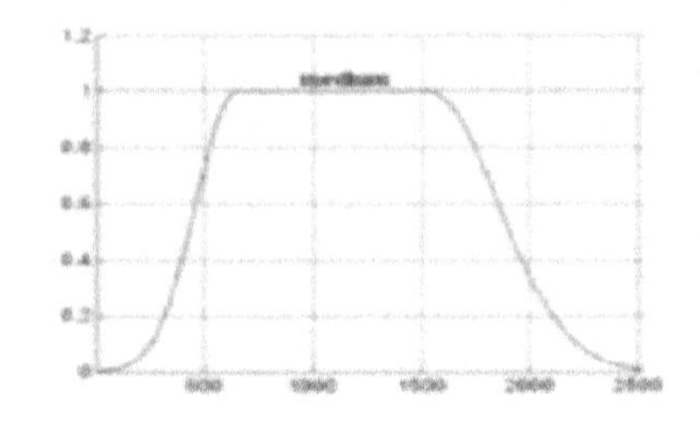

Fig. 15.29. Fuzzy sets of attribute 13

Table 15.4. Rule base for solution 16 in Table 15.3

Rule base
R_1: If x_1 is small, x_2 is big, x_8 is medium, and x_{10} is big, then class 3 with $CF = 0.9987$;
R_2: If x_2 is big, x_3 is big, x_4 is small, x_9 is small, and x_{10} is small, then class 2 with $CF = 0.9494$;
R_3: If x_1 is big, x_4 is small, x_6 is big, x_8 is medium, x_{10} is medium, x_{11} is medium, and x_{13} is medium, then class 1 with $CF = 0.9876$.
Antecedent parameters:
x_1: small=[0.749,11.724,0.942,11.887] big=[0.522,12.554,0.945,14.143]
x_2: big=[0.857,3.759,1.154,5.527]
x_3: big=[0.436,2.991,0.225,3.181]
x_4: small=[0.612,12.585,1.300,17.805]
x_6: big=[0.116,2.962,0.143,3.483]
x_8: medium=[0.083,0.460,0.061,0.504]
x_9: small=[0.758,0.868,0.496,1.658]
x_{10}: small=[1.493,2.936,0.061,3.629] medium=[0.927,6.326,2.001,7.676] big=[0.445,7.969,1.086,12.500]
x_{11}: medium=[0.148,0.683,0.051,1.365]
x_{13}: medium=[185.4,649.8,320.7,1531.3]

ter interpretability issues about fuzzy systems. Different forms of rule bases and fuzzy sets which focus on different aspects of interpretability and accuracy are generated. In addition, the number of fuzzy variables can be automatically optimized, e.g. only three out of four input variables participate in the fuzzy system construction for Iris data (solution 1 in Table 15.1) and similar cases are also observed for Wine data (Table 15.3). This indicates that more compact fuzzy systems can be generated through our agent based approach, which considers not only the number of fuzzy sets, but also the number of fuzzy variables.

15.5 Conclusion and Future Works

In this chapter, we propose an agent based multi-objective approach to generate interpretable fuzzy systems to solve the classification problems. Compared with other methods reported in the literature, one of the contributions of our agent based approach is that more compact fuzzy systems can be generated, which not only consider the fuzzy rule base, but also the number of fuzzy sets for each fuzzy variable. And the distributions of fuzzy sets are continuously optimized during the agent based approach based on the interpretability issues we have proposed. Simulation results have demonstrated that the agent based multi-objective approach can generate diversified fuzzy systems with a good trade-off between the accuracy and interpretability.

In the future work, some data mining techniques related to dimension reduction such as SUD [46] and SCM [47], etc. can be applied ahead to determine more important variables. It will result in using fewer variables to train the agents to generate more compact fuzzy systems. Additionally, some clustering techniques such as the fuzzy C-Means technique can be used to obtain initial fuzzy sets distributions for the agents. Therefore, agents in our architecture can evolve from a good starting point. In addition, the agent based approach should be improved, especially the interaction mechanism. In the current work, the agents communicate with each other through the crossover and mutation of the hierarchical chromosome. It is a very easy way to implement. However, more systematic and effective mechanism should be further explored to improve the agents' capability for cooperation and competition. It is also valuable to incorporate the preference of human beings. Users may provide some preferences that reflect their attitudes towards the relative importance of interpretability and accuracy. Currently, we predefine the preference and the first priority is set to the accuracy in the NSGA-II algorithm. Similarly, different priorities can be assigned to different objectives from different users. In the future research, the preference would be incorporated dynamically during the process of fuzzy system generation. This is an interactive manner between users and computer algorithms.

References

[1] L.A. Zadeh. Outline of a new approach to the analysis of complex systems and decision processes, *IEEE Trans. Syst., Man, Cybern.*, Vol. SMC-3, pp. 28-44, Jan. 1973

[2] G. Castellano, A.M. Fanelli, E. Gentile and T. Roselli. A GA-based approach to optimize fuzzy models learned from data, *Proc. Workshop on Approx. Learning in Evol. Comput., GECCO 2002*, pp. 5-8, Jul. 2002

[3] S. Guillaume. Designing Fuzzy Inference Systems from Data: An interpretability Oriented Review, *IEEE Trans. Fuzzy Syst.*, Vol.9, No.3, pp. 426-443, Jun. 2001

[4] F. Jiménez, A. F. Gómez-Skarmeta, H. Roubos, and R. Babuška. Accurate, Transparent, and Compact Fuzzy Models for Function Approximation and Dynamic Modeling through Multi-objective Evolutionary Optimization, *EMO 2001*, pp. 653-667, 2001
[5] Y. Jin. Advanced Fuzzy Systems Design and Applications. Physica/Springer, Heidelberg, 2003
[6] Y. Jin, W. von Seelen and B. Sendhoff. An approach to Rule-Based Knowledge Extraction, *IEEE Conf. Fuzzy Syst.*, Vol. 2, pp. 1188-1193, May 1998
[7] Y. Jin, W. von Seelen, and B. Sendhoff. On Generating FC3 Fuzzy Rule Systems from Data Using Evolution Strategies, *IEEE Trans. Syst., Man, Cybern. B*, Vol. 29, No. 6, pp.829-845, Dec. 1999
[8] Y. Jin. Fuzzy Modeling of high-dimensional systems: Complexity reduction and interpretability improvement, *IEEE Trans. Fuzzy Syst.*, Vol. 8, No. 2, pp. 212-221, 2000
[9] Y. Jin and B. Sendhoff. Extracting Interpretable Fuzzy Rules from RBF Networks, *Neural Processing Letters*, 17(2), pp. 149-164, 2003
[10] I. Rojas, H. Pomares, J. Ortega, and A. Prieto. Self-Organized Fuzzy System Generation from training Examples, *IEEE Trans. Fuzzy Syst.*, Vol. 8, No. 1, pp. 23-36, Feb. 2000
[11] H. Roubos and M. Setnes. GA-Fuzzy Modeling and Classification: Complexity and Performance, *IEEE Trans. Fuzzy Syst.*, Vol. 8, No. 5, pp. 509-522, Oct. 2000
[12] H. Roubos and M. Setnes. Compact and Transparent Fuzzy Models and Classifiers Through Iterative Complexity Reduction, *IEEE Trans. Fuzzy Syst.*, Vol. 9, No. 4, pp. 516-524, Aug. 2001
[13] L. Wang and J. Yen. Exacting Fuzzy Rules for System Modeling Using a Hybrid of Genetic Algorithms and Kalman Filter, *Fuzzy Sets Syst.*, Vol. 101, pp. 353-362, 1999
[14] J. Yen and L. Wang. Application of Statistical Information Criteria for Optimal Fuzzy Model Construction, *IEEE Trans. Fuzzy Syst.*, Vol. 6, No. 3, pp. 362-372, 1998
[15] J. Yen and L. Wang. Simplifying Fuzzy Rule-Based Models Using Orthogonal Transformation Methods, *IEEE Trans. Syst., Man, Cybern. B*, Vol. 29, No. 1, pp. 13-24, 1999
[16] J. Casillas, O. Cordón, F. Herrera, and L. Magdalena (Eds.). *Interpretability Issues in Fuzzy Modeling*, SPRINGER, BERLIN 2003
[17] H. Wang, S. Kwong, Y. Jin, W. Wei, and K. F. Man. Multi-objective hierarchical genetic algorithm for interpretable fuzzy rule-based knowledge extraction, *Fuzzy Sets Syst.*, Vol. 149, No. 1, pp. 149-186, Jan. 2005
[18] H. Wang, S. Kwong, Y. Jin, W. Wei, and K. F. Man. Agent-Based Evolutionary Approach for Interpretable Rule-Based Knowledge Extraction, *IEEE Trans. Syst., Man, Cybern. C*, Vol. 35, No. 2, pp. 143-155, May 2005
[19] O. Cordón, F. Gomide, F. Herrera, F. Hoffmann, and L. Magdalena. Ten years of genetic fuzzy systems: current framework and new trends, *Fuzzy Sets Syst.*, Vol. 141, No. 1, pp. 5-31, Jan. 2004
[20] N. Xiong. Evolutionary learning of rule premises for fuzzy modeling, *Int. Journal of Syst. Sci.*, Vol. 32, No. 9, pp. 1109-1118, 2001

[21] H. Ishibuchi, K. Nozaki, N. Yamamoto, and H. Tanaka. Selecting Fuzzy If-Then Rules for Classificaiton Problems Using Genetic Algorithms, *IEEE Trans. Fuzzy Syst.*, Vol. 3, No. 3, pp. 260-270, Aug. 1995
[22] H. Ishibuchi, T. Nakashima, and T. Kuroda. A Hybrid Fuzzy Genetics-based Machine Learning Algorithm: Hybridization of Michigan Approach and Pittsburgh Approach, *IEEE Int. Conf. Syst., Man, Cybern.*, pp. 296-301, Oct. 1999
[23] H. Ishibuchi, T. Nakashima, and T. Murata. Performance Evaluation of Fuzzy Classifier Systems for Multidimensional Pattern Classification Problems, *IEEE Trans. Syst., Man, Cybern. B*, Vol. 29, No. 5, pp. 601-618, Oct. 1999
[24] H. Ishibuchi, T. Nakashima, and T. Kuroda. A Hybrid Fuzzy GBML Algorithm for Designing Compact Fuzzy Rule-Based Classification Systems, *9th IEEE Int. Conf. Fuzzy Syst.*, pp.706-711, May 2000
[25] H. Ishibuchi, T. Nakashima and T. Murata. Multi-objective optimization in linguistic rule extraction from numerical data, *EMO 2001*, pp. 588-602, 2001
[26] H. Ishibuchi and T. Nakashima. Effect of Rule Weight in Fuzzy Rule-based Classification Systems, *IEEE Trans. Fuzzy Syst.*, Vol. 9, No. 4, pp. 506-515, 2001
[27] H. Ishibuchi and T. Nakashima. Three-Objective Optimization in Linguistic Function Approximation, *in Proc. Evol. Comput.*, pp. 340-347, May 2001
[28] H. Ishibuchi and T. Yamamoto. Fuzzy rule selection by multi-objective genetic local search algorithms and rule evaluation measures in data mining, *Fuzzy Sets Syst.*, Vol. 141, No. 1, pp. 59-88, Jan. 2004
[29] H. Ishibuchi and S. Namba. Evolutionary Multiobjective Knowledge Extraction for High-Dimensional Pattern Classification Problems, *PPSN VIII, LNCS 3242*, pp. 1123-1132, 2004
[30] T. Murata, S. Kawakami, H. Nozawa, M. Gen, and H. Ishibuchi. Three-Objective Genetic Algorithms for Designing Compact Fuzzy Rule-Based Systems for Pattern Classification Problems, *GECCO 2001*, pp. 485-492, Jul. 2001
[31] K. F. Man, K. S. Tang and S. Kwong. Genetic Algorithms: Concepts and Applications, *IEEE Trans. Ind. Electron.*, Vol. 43, No. 5, pp. 519-534, Oct. 1996
[32] K. S. Tang, K. F. Man, S. Kwong and Q. He. Genetic Algorithms and Their Applications, *IEEE Signal Processing Mag.*, Vol. 13, No. 6, pp. 22-37, Nov. 1996
[33] K. S. Tang, K. F. Man, Z. F. Liu and S. Kwong. Minimal Fuzzy Memberships and Rules Using Hierarchical Genetic Algorithms, *IEEE Trans. Ind. Electron.*, Vol. 45, No. 1, pp. 162-169, Feb. 1998
[34] M. Setnes, R. Babuška, U. Kaymak, and H. R. van Nauta Lemke, Similarity Measures in Fuzzy Rule Base Simplification, *IEEE Trans. Syst., Man, Cybern. B*, Vol. 28, No. 3, pp. 376-386, Jun. 1998
[35] K. P. Sycara. The Many Faces of Agents, *AI Mag.*, Vol. 19, No. 2, pp. 11-12, 1998
[36] A. H. Bond and L. Gasser, An analysis of problems and research in DAI, *in Readings in Distributed Artificial Intelligence, A. H. Bond et al., Eds.*, 1988, pp. 3-35, 1988
[37] B. Moulin and B. Chaib-Draa. An overview of distributed artificial intelligence, *in Foundations of Distributed Artificial Intelligence, G. M. P. O'Hare et al., Eds.*, John Wiley & Sons Inc. New York, 1996, pp. 3-55
[38] N. R. Jennings, K. Sycara, and M. Wooldridge. A roadmap of agent research and development, *Autonomous Agents and Multi-Agent Systems*, pp. 7-38, 1998

[39] G. Weiss (Eds.). *Multiagent Systems: A Modern Approach to Distributed Artificial Intelligence*, MIT Press, Cambridge, Massachusetts, 1999

[40] S. F. Smith. *A learning system based on genetic adaptive algorithms*, Doctoral Disseration, Department of Computer Science, University of Pittsburgh, 1980

[41] K. Deb, A. Pratap, S. Agarwal, and T. Meyarivan. A fast and elitist multiobjective genetic algorithm: NSGA-II, *IEEE Trans. Evol. Comput.*, Vol. 6, No. 2, pp. 182-197, Apr. 2002

[42] UCI Machine Learning Repository [Online]. Available: http:// www.ics.uci.edu / ~mlearn /MLRepository.html

[43] K. Deb. *Multi-Objective Optimization using Evolutionary Algorithms*, John Willy & Sons, Chichester, UK, 2001, pp. 28-46

[44] M. Russo. Genetic Fuzzy Learning, *IEEE Trans. Evol. Comput.*, Vol. 4, No. 3, pp. 259-273, Sept. 2000

[45] N. Xiong and L. Litz. Identifying Flexible Structured Premises for Mining Concise Fuzzy Knowledge, *in Interpretability Issues in Fuzzy Modeling, J. Casillas et al., Eds.*, 2003, pp. 54-76

[46] M. Dash, H. Liu, and J. Yao. Dimensionality Reduction for Unsupervised Data, *Proc. 9th Int. Conf. Tools Artif. Intell.*, pp. 532-539, Nov. 1997

[47] X. Fu and L. Wang. Data Dimensionality Reduction With Application to Simplifying RBF Network Structure and Improving Classification Performance, *IEEE Trans. Syst., Man, Cybern. B*, Vol. 33, No. 3, pp. 399-409, Jun. 2003

16

Multi-objective Evolutionary Algorithm for Temporal Linguistic Rule Extraction

Gary G. Yen

School of Electrical and Computer Engineering
Oklahoma State University
Stillwater, OK 74078-5032, USA
gyen@okstate.edu

Summary. Autonomous temporal linguistic rule extraction is an application of growing interest due to its relevance to both decision support systems and fuzzy controllers. In the presented work, rules are evaluated using three qualitative metrics based on their representation on the truth space diagram. Performance metrics are then treated as competing objectives and Multiple Objective Evolutionary Algorithm is used to search for an optimal set of non-dominated rules. Novel techniques for data pre-processing and rule set post-processing are developed that deal directly with the delays involved in dynamic systems. Data collected from a simulated hot and cold water mixer and a two-phase vertical column is used to validate the proposed procedure.

16.1 Introduction

When performing process management decisions, a control algorithm makes use of the knowledge of the plant in the form of a model. In most cases, this knowledge is commonly derived from the first principles and/or laboratory and pilot plant experiments; and often such "ideal" knowledge is of little practical use under real world complications due to unaccounted factors and modeling uncertainties.

Human operators, on the other hand, make use of another type of model when in charge of process management decisions. After a long time in contact with the plant, process operators are capable of attaining some understanding of what factors govern the process and derive relationships between process variables based on intuition and past experience. This process was best described in [23] as "a cognitive skill of experienced process operators that fits the current facts about the process and enables the operators to assess process behavior and predict the effects of possible control actions." However, the knowledge attained in this fashion also presents critical deficiencies since wrong impressions on what is going on with the process will lead to operator misjudgment as described in [17]. Furthermore, incoherencies inside

G.G. Yen: *Multi-objective Evolutionary Algorithm for Temporal Linguistic Rule Extraction*, Studies in Computational Intelligence (SCI) **16**, 365–383 (2006)
www.springerlink.com

such knowledge propagate itself as "mis-knowledge" or "technical folklore" are passed down from one generation of process operators to the next.

By making use of linguistic information in the form of IF/THEN logical statements or rules, Expert Systems and Fuzzy Logic Controllers (FLCs) are technologies capable of enabling better process monitoring and control. FLCs have found applications in a variety of fields such as robotics [7], automated vehicles [5], and process control [19], to name a few. Expert Systems have been used in the Chemical Process Industry (CPI) to control, monitor [6], and understand process behaviors. Other applications of such knowledge based systems have been in operator training and for planning and scheduling of operations in control and maintenance [11], especially for getting a plant back online after a failure or abnormal operating condition.

Expert Systems can be built from knowledge inserted by human experts or acquired from historic data from the system. Knowledge bases made by polling information from experienced personnel not only incorporate the before mentioned "technical folklore," but also is intrinsically incomplete. Such rules pertain only to information that is critical or obvious to the operators; it is related to information just necessary for them to maintain desired plant conditions. Such information does not incorporate the knowledge of events that are lesser in significance or rarer in occurrence, but which affect the operation of the plant nonetheless. A complete rule base should possess information on almost all plant events that have an effect on the desired output or may change the variable under control. Finally, knowledge collected from experts is usually in the form of static rules loosely related to the real numerical world. Due to its lack of a mechanism to deal with the temporal behavior of the process, the rigid, non-adaptive knowledge devised in this fashion becomes inadequate for complete supervisory control of dynamic systems. Therefore, the solution lies on the development of an algorithm capable of autonomously extracting and improving a dynamic rule set for an expert system directly from process data.

As pointed out in an extensive survey of Knowledge Discovery [8], many technical fields are exploring rule extraction from data. Data Mining views the rule extraction problem as one of finding patterns that describe the information in a database [14]. Machine Learning on the other hand approaches the same problem by generating deterministic finite-state automata with generalizing abilities capable of describing the relationship between antecedents and consequents [25]. Yet, the most appreciable contributions made in process applications are all based on computational intelligence paradigms, such as Neural Networks, Fuzzy Logic, or Evolutionary Computation [9, 16, 12, 27]. Still, temporal reasoning is missing in most studies.

It is fundamental for the modeling of a dynamic system that the model used incorporates the concept of time. In [18], a temporal restrictor is first proposed as an entity that restricts a fuzzy proposition, expressed as 'A be B T,' which translates into 'A is satisfying the fuzzy predicate B, taking into account the temporal restrictor T.' The temporal restrictor is expressed as a membership function. The conjunction *be* is a generic verb of being which can take different temporal connotations according to the rule. In this way, it is possible to add to the extracted rules information in the form of 'temperature was hot one_hour_ago.' However, this approach is rudimentary and incomplete; it does not account for persistence of excitation, and treating time as a restricting factor greatly limits the generalization of the rules.

Bakshi and Stephanopoulos exploited temporal knowledge and proposed reducing the dimensionality in [2] by describing process behavior with *triangles.* The design is specifically crafted for fault detection applications. However, our goal is to explore relations between general variables, not just discrete faults. Data mining in a time series database was approached in [15] by using three different temporal attributes: *length*, *slope*, and *fluctuation* (signal-to-noise ratio over time). Nevertheless, the application is not targeted on rule extraction, but on time-series prediction.

Based on the widely applied Autoregressive Moving Average (ARMA) [1] models, Tsai and Wu [24] proposed to incorporate temporal relationships into fuzzy rules by matching antecedents with consequents a fixed number of time steps in the future. In [21] the architecture was extended to allow different discrete time delays to be used for each antecedent in single consequent rules. Due to the usage of discrete time delays however, the representation capability of the rule set was largely affected by the particular choice of time delays and it displayed great sensitivity to noise, especially related to datasets composed of data sampled from continuous systems. Displaying applications related to the stock market and the weather, Last et al. [15] applied stochastic preprocessing techniques to improve the meaningfulness of the data provided to the rule extraction mechanism. Based on the concept of internal clocks that biological organisms use for the learning of period and interval timing, Carse and Fogarty [3] proposed the usage of a temporal membership function for the averaging of sampled data in order to generate crisp values related to fuzzy time periods. Applications of such an approach have been documented in distributed adaptive routing control in packet switched communication networks [3]. Therefore, in this study, a particular fuzzy delay is assigned to the temporally averaged consequent of each rule, generating a structure of the form:

IF *condition 1* ***AND*** *condition 2* ***AND*** *condition 3...* ,
THEN *after a certain fuzzy delay, a control variable will be such.*

The statement between the IF and the THEN conjunction is the *antecedent* while the statement after the THEN conjunction is the *consequent.*

A crucial step in the autonomous extraction of rules is the method used to validate and compare those that are created. An optimal rule should be accurate, properly describes the dynamic relationship between its antecedent and consequent, and possesses enough data to support it. In a methodology introduced in [20], three metrics based on a Truth Space Diagram (TSD) capable of encapsulating and measuring each of these three goals were introduced and tested. However, it was also shown that in the general scenario such metrics cannot be independently optimized due to inherent confliction among them.

Multiple Objective Evolutionary Algorithm (MOEA) is a tool capable of performing efficient searches on high dimensional spaces to locate the Pareto front, a set of solutions that contain the best rule for each possible trade-off between competing goals. A growing research field, MOEA has already demonstrated successful applications in solving challenging benchmark problems [26] and real world applications [4].

In this chapter, three metrics developed in [20] are used under a novel dynamic treatment of the data to evaluate linguistic rules against process data. MOEA is then introduced to locate inside the high dimensional rule space the Pareto front of the antecedents that best describe (in the sense of different combination of metrics) a given consequent.

The remainder of the chapter is organized as follows. Section 16.2 describes the three metrics generation process, detailing the novel temporal data pre-processing, the development of the truth space diagrams, and the choice of relevant metrics. Section 16.3 introduces the use of MOEA for the automated extraction of rules and the necessary rule set post-processing. The simulation results of applying the proposed algorithm to the Hot and Cold water mixer and a two-phase vertical column are explored in Section 16.4, following by some final remarks and conclusions in Section 16.5.

16.2 Rule Evaluation

In order to guide and automate process management decisions, a rule must display three basic characteristics: high accuracy, precise antecedent/consequent relationship, and sufficient data support. However, it is seldom possible to maximize these three characteristics at the same time. For example, if an event is observed only a single time, it is trivial to develop a rule with 100% accuracy, however it will lack support from historical data and the probability that it will describe a whole family of similar events with comparative accuracy is small. For this reason there is a need to develop three qualitative metrics, each focusing on one of such competing characteristics.

In [20] a series of metrics were suggested, each capable of specifically representing a different quality of a rule. All metrics were designed with a qualitative measure manifested through the Truth Space Diagram (TSD). In the first part of this section, the efficiency of the TSD is further enhanced with the introduction of a novel pre-processing strategy for the representation of the temporal behavior of the plant into the linguistic rules. The concept of the TSD is then introduced taking in consideration the enhanced temporal representation and finally the three metrics of interest are introduced.

16.2.1 Data Pre-processing

In previous applications of the TSD methodology, data pre-processing took place in the following manner: 1) The consequent data was shifted backwards in time by fixed intervals so that each set of antecedents matched three consequent values corresponding to short, medium, and long delays; 2) The crisp input-output data was fuzzified. Fuzzification proceeded through the application of Equation (16.1) by using triangular membership functions to classify each variable into three fuzzy categories – low, medium, and high:

$$\mu_x^{i,j} = \begin{cases} \frac{a_j - x_i}{a_j - b_j}, & x_i \in [a_j, b_j] \\ 0, & \text{otherwise,} \end{cases} , \tag{16.1}$$

where j=1 to 3, i=1 to n, x_i is the crisp numerical value of the ith input or output variable, $\mu_x^{i,j}$represents the fuzzy membership value of x_i in the jth fuzzy category, a_j and b_j are the fuzzy set break points for category j, and n is the maximum number of datasets in the input-output data. Figure 16.1 illustrates the fuzzy classification of one variable into three fuzzy categories.

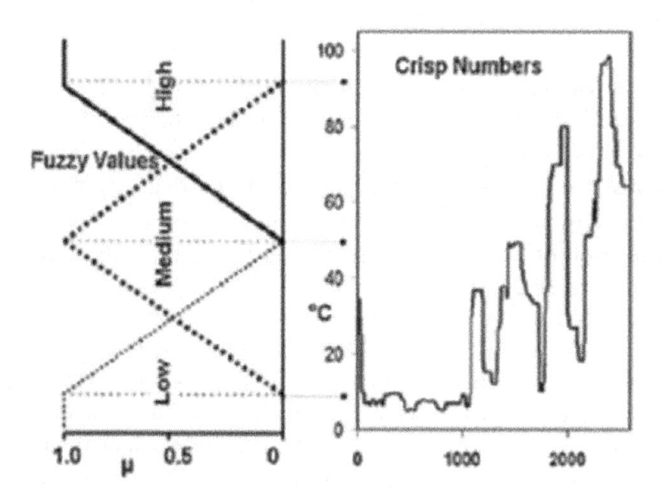

Fig. 16.1. Fuzzy classification procedure for the antecedents. In this case, temperature with centers at 10°, 50°, and 90°C.

The fuzzification of the physical crisp data in such a manner leads to great generalization capabilities, inherent noise rejection, and direct rule interpretation by human operators. Although ideal for the treatment of the antecedents, the manner through which the dynamic temporal element is incorporated into the consequent lacks such benefits. In essence, consequences for plant characteristics at any given time are only observed at discrete instants. Since a consequent's time delay contains variability as much as physical characteristics of a system, although rule extraction is possible, it requires extensive data in order to determine the true correlations through time. Moreover, the previous approach relies on *accurate* knowledge on the inherent major delays of the plant in order to set up the number of iterations that correspond exactly to the operator's understanding of small, medium, and long delays.

In the present work, such deficiencies are addressed by dealing with time uncertainties in a novel manner that is different in essence from the fuzzification of physical variables. The application of such approach leads to a meaningful linguistic description that maintains all the previously stated benefits while better capturing the temporal characteristics of dynamic plants. Instead of simply shifting the data to obtain a single measurement to represent a delay, averaged values of a consequent are obtained for each fuzzy delay region through Equation (16.2) and it is those averaged values that are then classified into the membership function of the consequent by Equation (16.3).

$$y_\delta^i(t) = \frac{\sum_{k=t_\delta^1}^{t_\delta^3} M_\delta(k) \cdot y^y(t+k)}{\sum_{k=t_\delta^1}^{t_\delta^3} M_\delta(k)}, \tag{16.2}$$

$$\mu_{y,\delta}^{i,j}(t) = \frac{a_j - y_\delta^i(t)}{a_j - b_j}, \tag{16.3}$$

where $y^i(t)$ is the crisp measurement of the ith consequent at time t, $y_\delta^i(t)$ corresponds to its arithmetical average for a given fuzzy delay δ, $\delta \in$[short , medium , long], t_δ^i are the fuzzy set break points ($i \in$[1, 2, 3]) for category δ, $\mathrm{M}_\delta(t)$ denotes the membership function of a fuzzy delay δ, and $\mu_{y,\delta}^{i,j}(t)$ is the fuzzy membership value of the ith consequent for a fuzzy delay δ. An example of the application of the procedure involving Equations (16.2) and (16.3) can be seen in Figure 16.2, where the fuzzy membership value of the consequent $y^1(t)$ is calculated at time 20s for *medium* delay and *low* temperature, i.e. $\mu_{y,m}^{1,l}(20)$.

By processing the process data from the antecedents using (1) and that from the consequents with (2-3), crisp data is translated into linguistic variables. It is important to note that although each antecedent relates to a single linguistic variable, due to the introduction of the fuzzy delay, each consequent is represented by three fuzzy variables, each related to a different delay membership function.

16.2.2 Truth Space Diagram

The TSD is a two-dimensional space in which a series of metrics capable of quantifying the quality of a particular cause-and-effect rule can be obtained. Each TSD relates to a single rule. For every data point extracted either from mathematical simulations, pilot plant experiments, or real-word sensory data, a point is plotted in the TSD according to its truth of the antecedent *Ta* and the truth of the consequent *Tc*. Both parameters are calculated as geometrical

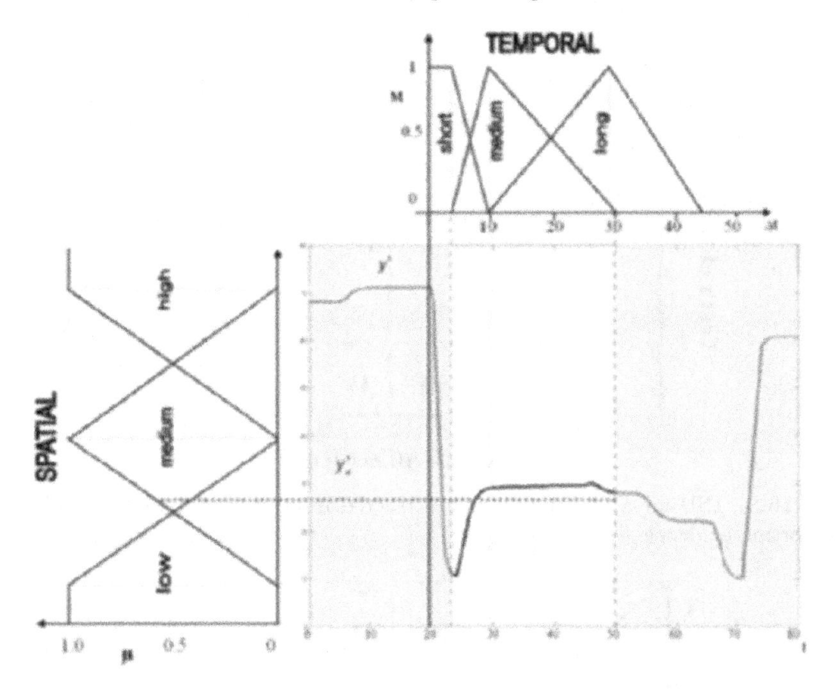

Fig. 16.2. Proposed physical and temporal two-step fuzzification procedure.

means of the fuzzy membership function of each variable of the antecedents and consequents. Hence, the truth space delimited by Ta and Tc is bounded between 0 and 1 in which a value equal to 0 means absolute false while a value of 1 means absolute truth.

Designed in this fashion, the TSD represents a one-to-one mapping from the dataset from the real (numerical) space to a new (truth) space defined by the linguistic statements of a specific rule. The TSD can be divided into four quadrants and each quadrant provides different information about the linguistic rule. For example, consider point A in Figure 16.3. The values for Ta and Tc are high for this data point, i.e. the predicted consequent follows the appointed antecedent or the cause and effect match according to the relevant rule statement. This reveals that the information expressed in the linguistic rule is contained within the numerical data. Hence, many points in Quadrant II (denoted as Quad **II**) of the TSD reflect the validity of the rule in question. Consequentially, points in Quadrant IV (denoted as Quad **IV**) show that the rule statement is false, i.e. what the antecedent of the rule express does not lead, in most cases, to the predicted consequent. An example of this can be seen from data point B in Figure 16.4. Similarly, points in Quadrant I demonstrate the incompleteness of the rule, since the predicted consequent was due to an event(s) other than the one expressed in the antecedents of the rule. Finally, the presence of a cluster of points in Quadrant III show the possibility

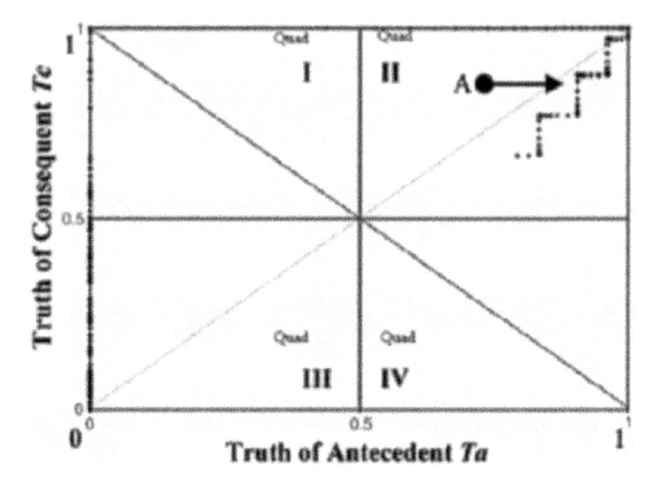

Fig. 16.3. TSD for a meaningful rule extracted from process data with sufficient supporting evidence.

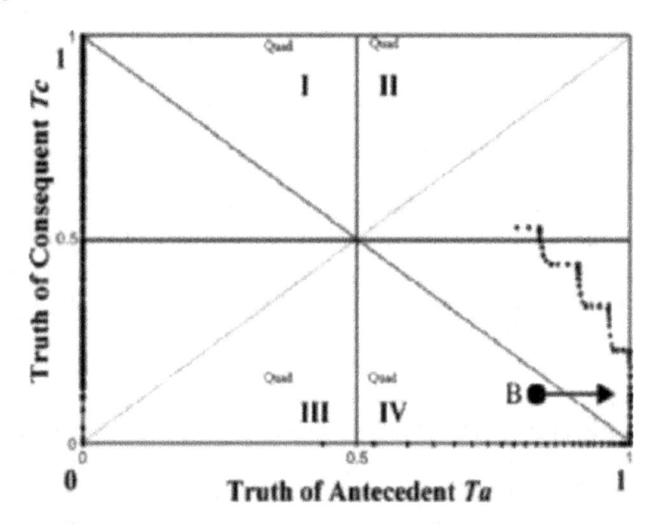

Fig. 16.4. TSD for a rule that was proven inaccurate in a significant number of points in the process data.

of that a rule is valid, however the amount of data currently available does not allow yet for a conclusion to be drawn with enough confidence. The points that lie on the vertical and horizontal axis show that either the antecedent or the consequent of a particular rule were not expressed in the data.

16.2.3 Numerical Metrics

As mentioned previously, the goal of the presented work is to extract rules that present high accuracy, precise antecedent/consequent relationship, and that are supported by sufficient data. By using the TSD, it is possible to obtain metrics for each of these conflicting goals. In order to transform the problem into one of minimization however, the actual metrics of interest are converted into: rule inaccuracy, antecedent/consequent mismatch, and lack of supporting evidence in the dataset. To improve the performance of the rule extraction algorithm, all metrics presented here are normalized to the interval [0,1].

Metric 1: *rule inaccuracy* – A rule is deemed inaccurate when its antecedent is observed but the consequence that follows after the prescribed delay does not match the predicted behavior. As mentioned previously, in the TSD this concept relates to the points in the Quadrant IV (i.e. set Q4), which relate to high truth of the antecedent but low truth of the consequent. The number of data points in Q4 (i.e. n_4) can therefore be used as a relative measure of inaccuracy, however it is necessary to normalize this number by dividing n_4 by the total number of data points in which the rule antecedents were observed with sufficient confidence, i.e. the sum of the points in Quadrant II (n_2) and in Quadrant IV (n_4). Equation (16.4) summarizes m_1, the rule inaccuracy metric.

$$m_1 = \begin{cases} \frac{n_4}{n_2+n_4}, & n_2 + n_4 > 0 \\ 1, & \text{otherwise} \end{cases} . \tag{16.4}$$

Metric 2: *antecedent/consequent mismatch* – from a good rule it is expected that the value of Tc should match the value of the Ta. In other words, the intensity in which the antecedents are observed should be equal to the intensity of the resulted consequent. In real world scenarios however, Tc is affected by both the quality of the rule and the quality of the available data (e.g. noise corruption). By analyzing the data points in Quadrant II (i.e. set Q2), it is possible to measure antecedent/consequent mismatch directly by summing all distances from each data point in it to the diagonal of the TSD. Defining Ta_i and Tc_i respectively as the truth of the antecedent and the truth consequent for rule i, $\|.\|$ as the Euclidian norm, and since 0.3536 is the maximum distance to the diagonal, this second metric is stated as shown in Equation (16.5):

$$m_2 = \begin{cases} \frac{\sum_{i \in Q2} \|Ta_i - Tc_i\|}{0.3536 \cdot n_2}, & n_2 > 0 \\ 1, & \text{otherwise} \end{cases} . \tag{16.5}$$

Metric 3: *lack of supporting evidence in the dataset* – Since there is a need for sufficient information inside the available data for any conclusion to be drawn, this metric is crucial for the success of any data driven rule extraction method. Using the TSD representation, Equation (16.6) is built for this purpose.

$$m_3 = 1 - \frac{n_{TSD}}{n_{data}}, \tag{16.6}$$

where n_{TSD} is the total number of data points mapped in the TSD excluding points on the abscissa ($Ta = 0$); and n_{data} is the total number of data points available in the dataset.

16.3 Rule Extraction

Evolutionary Algorithms (EA) is commonly regarded as a family of stochastic search procedures that is inspired by computational models of natural evolutionary processes to develop computer based optimization problem solving systems [22]. Being a population based algorithm, in EA each candidate solution is represented as an individual. When evolving towards better solutions, the individuals that better meet the optimization goal (individuals with greater *fitness*) have a greater probability of being selected to take part in the creation of the individuals of the new generation.

For problems that have multiple conflicting goals that cannot be directly combined into a single scalar measure of fitness, Multiple Objective Evolutionary Algorithm (MOEA) provides a method through which a population of solutions can evolve towards a set of solutions within which no solution is better than another in all optimization goals. By defining that an individual *dominates* another when all of its optimization goals are closer to the ideal values than those of the other individual, such a set can be referred to as the non-dominated set. The non-dominated set among all possible solutions is called the Pareto front and its determination is then the ultimate goal of MOEA. Therefore, as shown in (7), in MOEA the fitness F is a vector of the optimization goals, in this case represented by the three goodness metrics.

$$F = [m_1, m_2, m_3]. \tag{16.7}$$

In the present work, MOEA in the form presented in [26] is used once for every consequent to evolve an initial random population of related rules towards the Pareto front of the tri-dimensional space defined by F. Therefore, for each consequent, a set of equally good (in the sense of the minimization of the three previously defined metrics) antecedents is extracted based on their relative success.

As laid down in the pseudocode in Table 16.1, the first step of implementing an MOEA algorithm is the generation of the initial population of candidate solutions. In order to guarantee an unbiased and diverse population while maintaining a low computational demand, 20 initial individuals are generated with random antecedent values, clearly 20 is an ad hoc choice that needs to be quantified in future research. In the following step, the three metrics are calculated for the rules formed by the antecedents of each individual and the consequent related to the current MOEA run. It is also in this step that each individual is assigned a rank value according to their relative success in minimizing the elements of the fitness vector F (i.e. the concept of Pareto optimality). In particular, the ranking scheme discussed in [10] is implemented, in which an individual is assigned a rank value equal to one plus the number of individuals it is dominated by.

Table 16.1. MOEA pseudocode

1. Generate initial population;
2. Evaluate the metrics of all individuals and rank them;
3. for (i=1: *maximum_generation*)
4. Choose parents with probability inversely proportional to their ranks;
5. Perform crossover operation on parents to generate new individuals;
6. With probability equal to the mutation rate, perform mutation procedure;
7. Evaluate all three metrics on the new individuals;
8. Update population ranking.

The third step in the presented pseudocode is the first in its main loop and it relates to the selection of two individuals that will be involved in the generation of new individuals to the population. The selection is performed stochastically by assigning a greater selection probability to individuals with smaller rank values, and therefore individuals with greater fitness. The individuals in this way chosen are denominated parents and, in step 4, part of their individual solutions is exchanged in the operation termed crossover. Through the crossover operation, two new individuals (solutions) are formed, combining elements of both parents. In the following step, mutation, another biologically inspired process, may affect with a specific probability (defined as the mutation rate) the newly generated individuals. In MOEA, mutation takes place by randomly modifying an arbitrary portion of the solution related to a given individual. Independent of the occurrence of mutation in step 5, on step 6 the fitness vector F is evaluated for the two new individuals, followed by the updating of the ranks of all individuals in the population. A generation is then concluded and the algorithm returns to step 3 until a maximum number of generations is reached.

After MOEA generates a set of non-dominated rules for each consequent, thresholds are used over each metric to eliminate outliers and establish minimum acceptable performances (e.g. minimum degree of accuracy required of a rule). Another post-MOEA data processing involves removal of time-redundant information from the rule set. If, for instance, a consequent should develop quickly and remain unchanged for a long time throughout the dataset, the antecedents would be credited with both short and long-term effects even though the long-term effect is only a matter of persistence. Time redundancy then refers to rules with equal antecedents and physical consequents, but with different consequent delays. In such cases, the rule related to the longer consequent delay is removed.

16.4 Simulation Results and Discussion

The results of the application of the proposed rule extraction algorithm to two examples are presented in this section. The first is a proof-of-the-concept computer simulation. The simulation pertains to a hot and cold water mixer that is sufficiently challenging to demonstrate the impact of each of the algorithm sub-systems while at the same time it remains simple enough to be intuitively understandable. The second pertains to the results of rule extraction over process data collected from the operation of a laboratory scale two phase column. Different from the computer simulation, the two phase column is an actual plant, subject to real world noise levels, sensor calibrations and other implementation imperfections.

16.4.1 Hot and Cold Water Mixer

To demonstrate the feasibility and clarify implementation details of the proposed process for extracting temporal cause and effect relationships, data was acquired from a Hot and Cold water mixer simulator shown in Figure 16.5. The simulator incorporates real world dynamics such as transport and measurement delays and is capable of adding deviations such as measurement bias and process drifts that have an ARMA stochastic behavior, noise and valve "sticktion." This is a simple example, but incorporates behaviors which are representative of a majority of unit operations within the CPI. The simulation was non-linear, had multiple inputs and its dynamics (such as hydrodynamic delay) depended upon operation conditions. A detailed description containing the mathematical model of the Hot and Cold water mixer simulator is available in [27].

For the purpose of generating data, the four input variables were manipulated, the flow of each input tube (F_1 and F_2) changing randomly at every 20 seconds and the input temperatures (T_1 and T_2) changing randomly at every 40 seconds. The periods of manipulation of the variables were shifted so as not to lead into two changes occurring at the same time. Their effect on the temperature at the output of the mixer stream (T_3) was measured over time. All flow variables were restricted to the interval [0,30] kg/min and the temperature variables to [0,100] ^{o}C.

According to the proposed data pre-processing procedure, physical variables were fuzzified with centers at 10, 50 and 90 ^{o}C for the temperatures and at 2.5, 15 and 25 kg/min for low, medium and high flow rates respectively. For the fuzzy delay, centers were placed at 3, 7 and 20 seconds for short, medium and long delays respectively. Note that long delay rules will be harder to calculate since the input variables will change randomly at faster rates. The dataset is intentionally devised in this form to challenge the rule extraction procedure with data of different degrees of quality.

The proposed MOEA based on the three selected metrics was implemented over the pre-processed data generating a non-dominated set of rule candidates

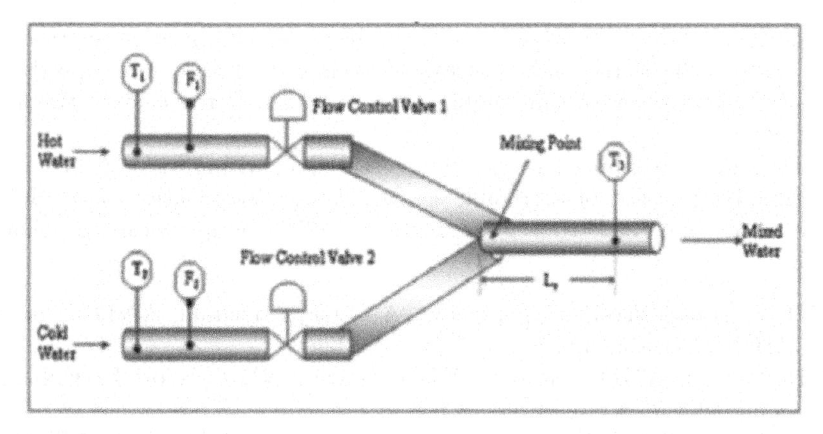

Fig. 16.5. The hot and cold water simulator used for validation of the rule extraction algorithm.

for each consequent. An initial population of 20 randomly generated individuals was allowed to evolve through the course of 200 maximum generations. The individuals received a rank equal to one plus the number of individuals it was dominated by. At each generation, parents were chosen according to a probability inversely proportional to their rank. For the generation of new individuals, crossover was implemented with a single crossover point and a mutation rate of 0.01 was used. An elitism scheme was implemented to guarantee that all non-dominated solutions were preserved during the evolution process. Figure 16.6 displays the obtained non-dominated set containing 13 antecedent combinations relating to high temperature at T_3 after a long delay.

For post-processing, the minimum acceptable accuracy of the extracted rules was set at 90%, a minimum of 1% of the information inside the observed dataset was necessary to validate a rule, and a maximum spread of 0.6 around

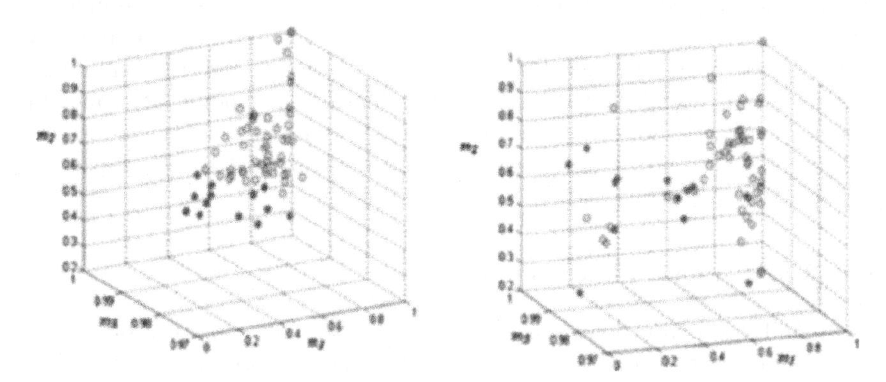

Fig. 16.6. Distributions of individuals related to two different consequents in the metric space at generation 200. Filled circles form the non-dominated set.

the diagonal of the TSD was allowed. In terms of the minimization metrics m_1, m_2, and m_3, the corresponding thresholds were 0.1, 0.99 and 0.6 respectively. Finally, rules that were time-redundant were removed to generate the final rule set.

Through the outlined process, the presented algorithm was capable of extracting 49 rules out of a possible set of 729 rule combinations (containing both "good" and "bad" rules). Some examples of the obtained rules are shown below:

- **IF** T_1 is high **AND** F_1 is medium **AND** T_2 is medium **AND** F_2 is low **THEN** after a medium delay T_3 will be high.
- **IF** T_1 is low **AND** F_1 is low **AND** T_2 is low **AND** F_2 is low **THEN** after a long delay T_3 will be low.
- **IF** T_1 is medium **AND** F_1 is low **AND** T_2 is high **AND** F_2 is high **THEN** after a short delay T_3 will be high.

Good rules are those that express the phenomenological-based, cause-and-effect mechanism as a logical relation between their antecedent and consequent parts. Consequentially, bad rules are defined as those that are inconsistent with the process phenomena. Therefore, in order to evaluate the quality of the 49 extracted rules, each one of them had its antecedents implemented in the simulator and those that demonstrated matching consequents were deemed good rules. As a result, 5 of those rules were rejected demonstrating a success ratio of 89.8% of the proposed rule extraction algorithm. Moreover, most rejected rules pointed to borderline consequents (e.g. the measured T_3 would be 78°C, when the maximum value acceptable for a medium fuzzy range was 70°C). Such scenarios reflect the choice of fuzzy membership function centers, left at the discretion of the operator.

As mentioned previously, any rule extraction procedure can only produce results as good as the data provided. This simulation was intentionally designed to provide much sparser and more noise corrupted data for the extraction of rules related to long delays. Among the 44 good rules in the final set, only 6 of those portrayed long delays, while the expected from a fully representative dataset would be one third of the total. Since the algorithm minimizes inaccuracy (m_1) while at the same time evaluating the amount of supporting evidence (m_3), the lower number of long delay rules extracted demonstrates the success of the procedure in avoiding unsupported rules to be presented to the operator in the final set.

16.4.2 Two-Phase Flow Column

Gas-liquid two-phase flows are defined as the flow of a mixture of the two phases, flowing together, through a system. The description of a two-phase flow in pipes is highly intricate due to the various existence of the interface between the two phases [27]. For gas-liquid two-phase flows, the variety of interface forms depends on the flow rates, phase properties of the fluid and on the inclination and the geometry of the tube. Generally, for vertical gas-liquid two-phase flows, the flow regimes are mainly determined by the phase flow rates. In this case, Bubbly, Slug, Churn and Annular are four significant regimes that can be recognized as standard patterns in the chemical industry. The characteristics of these four patterns are shown in Figure 16.7. Each of these four patterns has a distinguished air/water density and flow speed ratio. These characteristics have a strong influence on pressure drop and heat and mass transfer mechanisms in a system, and are very important in the chemical process industries.

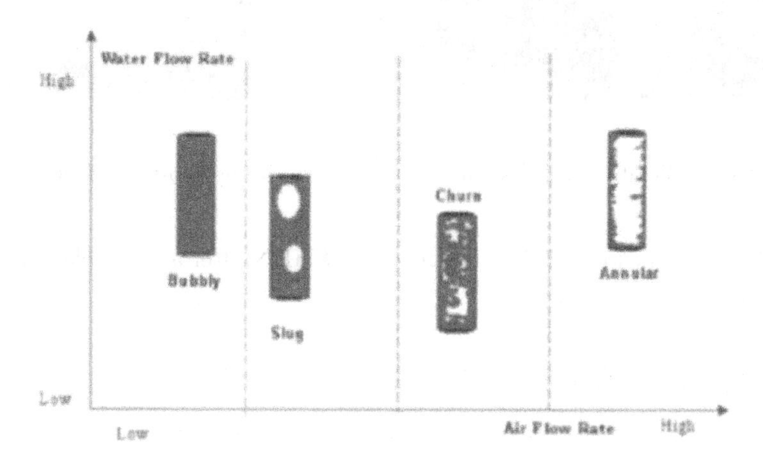

Fig. 16.7. Water/air flow ratio of four major two-phase vertical flow patterns.

Using the laboratory-scale vertical two-phase column shown in Figure 16.8, real process data (subjected to ambient noise) was acquired from the pressure drop in the column (ΔP) while independently varying the flow rates of air (F_a) and water (F_w). As illustrated in Figure 16.9, the two inputs were alternately modified at every 30 seconds (150 samples).

Due to the second order nature of the response of the two-phase column, higher fuzzyfication definition was required. Therefore, five membership functions were used to generate the levels *very low*, *low*, *medium*, *high* and *very high*; leading to 375 possible rules. Using the same procedure discussed for the hot and cold water mixer simulator, 22 satisfactory rules were extracted from the two-phase column data. Some examples of the extracted rule set are shown below.

- **IF** F_a is high **AND** F_w is very low **THEN** after a short delay ΔP will be very low.
- **IF** F_a is medium **AND** F_w is low **THEN** after a medium delay ΔP will be low.

(a)

(b)

Fig. 16.8. Details of the base (a) and the top (b) of the two-phase flow column.

- **IF** F_a is very low **AND** F_w is medium **THEN** after a long delay ΔP will be medium.

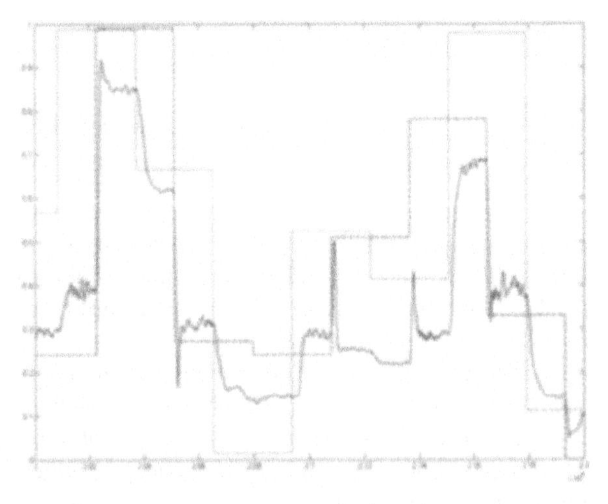

Fig. 16.9. Two-phase column pressure drop (solid) in response to variations in water flow (dashed) and air flow (dot-dashed).

16.5 Conclusions

As demonstrated though the application of the procedure on the data collected from the simulated hot and cold water mixer, the proposed rule extraction procedure succeeded in autonomously generating a viable rule set from a less than completely representative data set. The use of MOEA as an optimization algorithm allowed for three conflicting metrics to be evaluated simultaneously leading to the final extraction of optimal non-dominated rule sets. Both pre-processing, involving the representation of each rule inside a TSD, and post-processing, which allowed for the removal of time-redundant rules, were applied successfully and with beneficial outcomes.

The presented work assumes that all antecedents remain unaltered until the consequent is observed, however its ultimate goal is to achieve rules that specifically account for antecedent persistence such as: "**IF** (the reactor temperature has been high for a short period of time) **AND** (the speed is in manual for an extended period of time) **THEN** (in a short while the product will be slightly yellow)." To achieve such a goal, persistence must also be added to the dynamic persistence of the rule, which will lead to different data pre-processing requirements. To further improve the dynamic representation of complex plants, there may also be need to explore novel operators as opposed to the logical ones currently in use, such as "shortly succeeded by." Future work also focuses on the autonomous definition of fuzzy membership centers and on procedures capable of improving the quality of the extracted rule set as new batches of data becomes available over time.

References

[1] K. Astrom, T. Soderstrom. Uniqueness of the maximum likelihood estimates of the parameters of an ARMA model. *IEEE Transactions on Automated Control*, 19:769 -773, 1974

[2] B. Bakshi, G. Stephanopoulos. Representation of process trends - IV: Introduction of real-time patterns from operating data for diagnosis and supervisory control. *Computers & Chemical Engineering*, 18:303–332, 1994

[3] B. Carse, T. Fogarty. Evolutionary learning of temporal behavior using discrete and fuzzy classifier systems. In: *Proceedings of the International Symposium on Intelligent Control*, pp.183–188, 1995

[4] F. Chen, Z. Chen, Z. Jiao. A novel processing for multiple gases detection. In: *Proceedings of the World Congress on Intelligent Control and Automation*, pp. 2186–2189, 2002

[5] K. Cheng, J. Chen. A fuzzy-nets training scheme for controlling nonlinear systems. *Computers & Industrial Engineering*, 31:425–428, 1996

[6] C. Cimander, T. Bachinger, C. Mandenius. Integration of distributed multi-analyzer monitoring and control in bioprocessing based on a real-time expert system. *Journal of Biotechnology*, 103:327–348, 2003

[7] C. Collewet, G. Rault, S. Quellec, P. Marchal. Fuzzy adaptive controller design for the joint space control of an agricultural robot. *Fuzzy Sets and Systems*, 99:1–25, 1998
[8] M. Craven, J. Shavlik. Rule extraction: where do we go from here? *Technical Report 99-1*, University of Wisconsin Machine Learning Research Group, Madison, WI, 1999
[9] M. Delgado, A. Gomez-Skarmeta, F. Martin. A fuzzy clustering-based rapid prototyping for fuzzy rule-based modeling. *IEEE Transactions on Fuzzy Systems*, 5:223–233, 1997
[10] C. Fonseca, P. Fleming. Genetic algorithms for multiobjective optimization: formulation, discussion and generalization. In: *Proceedings of the International Conference on Genetic Algorithms*, pp.416–423, 1993
[11] D. Fonseca, G. Knapp. An expert system for reliability centered maintenance in the chemical industry. *Expert Systems with Applications*, 19:45–57, 2000
[12] X. Fu, L. Wang. Rule extraction by genetic algorithms based on a simplified RBF neural network. In: *Proceedings of the Congress on Evolutionary Computation*, pp.753–758, 2001
[13] F. Hewitt. Measurement of Two Phase Flow Parameters. Academic: London, 1978
[14] I. Jagielska. Linguistic rule extraction from neural networks for descriptive data mining. In: *Proceedings of the Knowledge-Based Intelligent Electronic Systems Conference*, pp.89–92, 1998
[15] M. Last, Y. Klein, A. Kandel. Knowledge discovery in time series databases. *IEEE Transactions on Systems, Man and Cybernetics, Part B: Cybernetics*, 31:160–169, 2001
[16] S. Mitra, Y. Hayashi. Neuro-fuzzy rule generation: survey in soft computing framework. *IEEE Transactions on Neural Networks*, 11:748–768, 2000
[17] E. Oshima. Computer aided plant operation. *Computers & Chemical Engineering*, 7:311–329, 1983
[18] A. Pesonen, A. Wolski. Quantified and temporal fuzzy reasoning for active monitoring in rapidbase. In: *Proceedings of Symposium on Tool Environments and Development Methods for Intelligent Systems*, pp.227–242, 2000
[19] R. Rhinehart, P. Murugan. Improve process control using fuzzy logic. *Chemical Engineering Process*, 91:60–65, 1996
[20] N. Sharma. Metrics for evaluation of the goodness of linguistic rules. MS Thesis, Oklahoma State University, School of Chemical Engineering, 2003
[21] N. Sisman, F. Alpaslan. Temporal neurofuzzy MAR algorithm for time series data in rule-based systems. In: *Proceedings of the International Conference on Knowledge-Based Intelligent Electronic Systems*, pp.316–320, 1998
[22] M. South, C. Bancroft, M. Willis, M. Tham. System identification via genetic programming. In: *Proceedings of the UKACC International Conference on Control*, 912–917, 1996
[23] G. Stephanopoulos, C. Han. Intelligent systems in process engineering: A review. *Computers & Chemical Engineering*, 20:743–791, 1996
[24] C. Tsai, S. Wu. A study for second-order modeling of fuzzy time series. In: *Proceedings of the IEEE International Fuzzy Systems Conference*, pp.719–725, 1999

[25] A. Vahed C. Omlin. Rule extraction from recurrent neural networks using a symbolic machine learning algorithm. In: *Proceedings of the International Conference on Neural Information Processing*, pp.712–717, 1999
[26] G. Yen, H. Lu. Dynamic multiobjective evolutionary algorithm: adaptive cell-based rank and density estimation. *IEEE Transactions on Evolutionary Computation*, 7:253–274, 2003
[27] G. Yen, H. Lu. Acoustic emission data assisted process monitoring. *ISA Transactions*, 41:273–282, 2002

17

Multiple Objective Learning for Constructing Interpretable Takagi-Sugeno Fuzzy Model

Shang-Ming Zhou[1] and John Q. Gan[2]

[1] Department of Computer Science, University of Essex, CO4 3SQ, UK
shmzhou@sohu.com
[2] Department of Computer Science, University of Essex, CO4 3SQ, UK
jqgan@essex.ac.uk

Summary. This chapter discusses the interpretability of Takagi-Sugeno (TS) fuzzy systems. A new TS fuzzy model, whose membership functions are characterized by linguistic modifiers, is presented. The tradeoff between global approximation and local model interpretation has been achieved by minimizing a multiple objective performance measure. In the proposed model, the local models match the global model well and the erratic behaviors of local models are remedied effectively. Furthermore, the transparency of partitioning of input space has been improved during parameter adaptation.

17.1 Introduction

In data-driven fuzzy modeling, interpretation preservation during adaptation is one of the most important issues [1]-[24]. Traditional fuzzy modeling schemes, such as neuro-fuzzy method [27], often generate fuzzy models with poor interpretable partitioning of input space due to their accuracy-oriented properties. However, if model interpretability criterion is considered, the model's global performance will be typically inferior to what can be achieved with a global performance criterion [21], that is to say, the global system accuracy and the model interpretability are conflicting objectives. The central task of interpretable fuzzy modeling is to construct a fuzzy model with good trade-off between global system accuracy and model interpretability. Moreover, in interpretable fuzzy modeling, the conciseness of fuzzy rules is also an important factor. Hence, interpretable fuzzy modeling is a typical multiple objective learning problem in nature [10]-[24].

Currently, most efforts for interpretable fuzzy modeling are put into generating transparent input space partitioning i.e., interpretable fuzzy sets. Although there exists no unified standard for selecting membership functions (MFs) during adaptation, in the interests of preserving or enhancing interpretability, some researchers have suggested semantic criteria to guide the

S.-M. Zhou et al.: *Multiple Objective Learning for Constructing Interpretable Takagi-Sugeno Fuzzy Model*, Studies in Computational Intelligence (SCI) **16**, 385–403 (2006)
www.springerlink.com

adaptation of MFs. de Oliveira proposed several semantic criteria for designing MFs, such as distinguishability of MFs, normalization of MFs, a moderate number of linguistic terms per variable, natural zero positioning, and coverage of the universe of discourse, etc. [9], which have been proved to be reasonable [10] [11].

However, in the Takagi-Sugeno (TS) fuzzy model [28], a most widely investigated paradigm in fuzzy modeling, there is another type of model interpretation that needs to be further decrypted, i.e., the interaction between the global model and local models. These local models in the TS fuzzy model often exhibit eccentric behaviors that are hard to be interpreted, as shown in Fig.17.1, thus such a model could not be interpreted in terms of individual rules, which imposes limits to the model application.

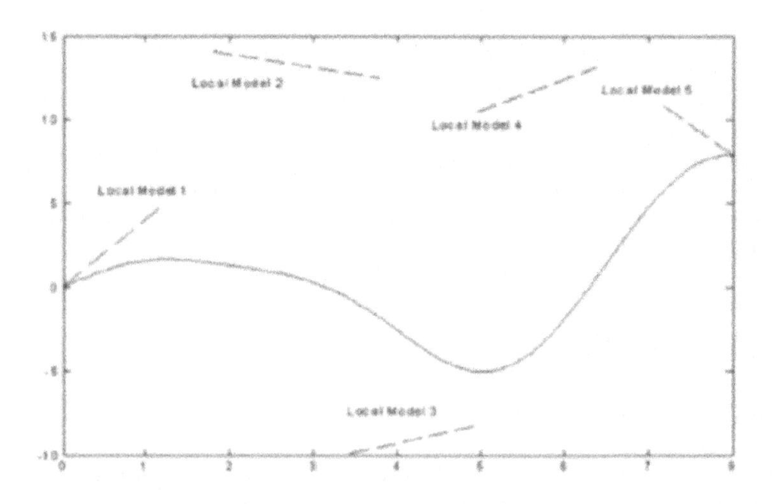

Fig. 17.1. TS model with uninterpretable local models

In case there is no *a priori* knowledge available for local models in a TS system, one may accept the basic heuristic that if the local models match the global model well in the local regions, then the local models are considered to possess good interpretability. In such a way, these local linear models possess the abilities to reflect the local properties of the system to be modeled, which is crucial for the success of local linear models' applications to nonlinear state estimation, fusion and predictive control [29] [30] [31]. In [19], Gan and Harris analyzed the relationship between fuzzy local linearization and local basis function expansion and the role of the local models. Yen et al proposed a new scheme in which multiple objectives focusing on global learning and local learning are aggregated into a single objective function [20]. Johansen and

Babuska treated the combination of global learning and local learning as a standard multi-objective identification problem and used the Pareto-optimal solutions to determine the trade-off between the possibly conflicting objectives of global model accuracy and local model interpretation [21].

As a matter of fact, the MFs' shapes, particularly the core regions and overlapping of adjacent MFs, impact the TS local models greatly. While, in fuzzy cluster analysis, "good" clusters are actually not very fuzzy [25]. Although fuzzy algorithms are used in data clustering, the aim of the clustering is to generate a "harder" partitioning of the data set [26], by which a better interpretation of input space partitioning can be achieved. The requirements directly related to this criterion in fuzzy modeling are that under the condition of preserving the global accuracy at a satisfactory level, the fuzzy sets should have large core regions, and that adjacent fuzzy sets should be less overlapped. However, traditional data-driven algorithms for rule generation, such as neuro-fuzzy algorithms, usually generate fuzzy sets with "too much" overlap due to their accuracy-oriented properties. In order to obtain MFs with large bounded core, Hoppner and Klawonn [24] proposed a novel clustering algorithm by using the distance to the Voronoi cell of a cluster rather than to the cluster prototype, and furthermore they assigned a "reward" to membership degrees that are near to 0 and 1.

This chapter proposes to use linguistic modifiers to characterize the MFs and multiple objectives learning for interpretability improvement and accuracy preservation. In order to measure the degree of linguistic modification, as an extension of the fuzziness measure proposed by Yager [39], we propose an index of fuzziness to evaluate the performance of linguistic modification for MFs with adjustable crossover points. A tradeoff between global approximation and local model interpretation can be achieved by minimizing a performance measure that combines the global error measure and the proposed index of fuzziness. By using the linguistic modifiers, as the MFs become less overlapped and possess larger core regions along with updating, the desired situation would emerge more possibly: there is only one rule applicable or dominating at a time, and the consequents are forced to represent the local behaviors of the system, which leads to good local model interpretability.

17.2 Definition of a Multiple Objective Optimisation Problem

Interpretable fuzzy modelling is a typical multiple objective optimisation problem in nature. The general multiple objective optimisation problem is posed as follows,

$$\begin{array}{ll} \min & J(x) = (J_1(x), J_2(x), \cdots, J_k(x)) \\ s.t. & x \in \Omega \end{array} \tag{17.1}$$

where $J(x)$ is the vector of objectives, $J_i(\cdot) : \Omega \rightarrow \Re$, and $\Omega$is the space of feasible solutions. These objectives are usually incommensurate and in conflict with one another with respect to their minimum points. In most cases, it is unlikely for the different objectives to be optimized by the same alternative parameter choices. Hence, some trade-off between the criteria is needed to ensure a satisfactory design.

In multiple objective optimization, one important concept is the *Pareto optimality* [32] [33], as the concept of "optimality" in single objective optimization does not apply directly in the multiple objective setting. Essentially, a vector $x^* \in \Omega$ is said to be Pareto optimal for multiple objective optimization if there is no other solution that improves at least one of the objective functions$J_i(\cdot)$ without detriment to another objective function. Pareto optimal solutions are also known as *efficient*, *non-dominated*, or *non-inferior* solutions. An alternative to the concept of *Pareto optimality*, which yields a single solution point, is the idea of *compromise solution* [34]. It entails minimizing the difference between the potential optimal point and a utopia point.

In practice, problems with multiple objectives can be solved by different techniques. A good survey of these techniques can be found in [33]. The most common approach to multiple objective optimization is the weighted sum method which aggregates the multiple objectives into one scalar objective:

$$J(x) = \sum_{i=1}^{k} w_i J_i(x) \tag{17.2}$$

where w_i are the weights typically set by the decision-maker such that $\sum_{i=1}^{k} w_i = 1$ and $w_i > 0$. The problem can then be optimized using a standard unconstrained optimization algorithm. Actually, (17.2) is a special case of the weighted exponential sum, a widely investigated utility function for multiple objective optimization:

$$J(x) = \sum_{i=1}^{k} w_i \left[J_i(x)\right]^p, \; J_i(x) > 0 \; \forall i \tag{17.3}$$

or

$$J(x) = \sum_{i=1}^{k} \left[w_i J_i(x)\right]^p, \; J_i(x) > 0 \; \forall i \tag{17.4}$$

One of the concerns in interpretable fuzzy modeling is under what conditions solutions of (17.2), (17.3) or (17.4) are Pareto optimal[35]. Zadeh proved that if all $w_i > 0$, the minimum of (17.2) is Pareto optimal [36], i.e. minimizing (17.2) is sufficient for Pareto optimality. The same property is also preserved in (17.3) [37] and (17.4) [38].

The weighted sum method for multiple objective optimization has been widely adopted in fuzzy modeling for improving model interpretability [10]

[20] [22] [23], in which various criteria were developed as the objectives for model interpretability with the consideration of global system accuracy. In this chapter, the weighted sum strategy is used to solve the multiple objective optimization problem in constructing a TS model with good local model interpretability and good global system performance. The key issue is how to define an objective function leading to TS local models with good interpretability.

17.3 A TS Model Using Linguistic Modifiers as MFs

17.3.1 The Takagi-Sugeno Fuzzy Model

In this chapter, the TS fuzzy model in the following form will be addressed:

$$\begin{aligned} R_i:\ & if\, x_1\, is\, A_{i_1}^{(1)}\, and \cdots and\, x_n\, is\, A_{i_n}^{(n)}\, then \\ & y_i = a_{i0} + a_{i1}x_1 + \cdots + a_{in}x_n \end{aligned} \tag{17.5}$$

where x_j are input variables, y_i is the output variable of the ith local model, $A_{i_j}^{(j)}$ are fuzzy sets about x_j, $a_{ij}$are the consequent parameters that have to be identified in terms of given data sets, R_i is the ith rule of the TS system, and $1 \le i_1 \le L_1, \ldots, 1 \le i_n \le L_n,\ 1 \le i \le L = \prod_{j=1}^{n} L_j$, with $L_j$being the number of fuzzy sets about x_j. The global output of the system is calculated by

$$y = \sum_{i=1}^{L} w_i y_i \tag{17.6}$$

where w_i is the normalized firing strength of rule R_i:

$$w_i = \tau_i / \sum_{i=1}^{L} \tau_i \tag{17.7}$$

and τ_i is called the firing strength of rule R_i, which is computed as $\tau_i = \prod_{j=1}^{n} A_{i_j}^{(j)}(x_j)$.

It can be seen that in this TS model, given the fuzzy sets about every variable on its domain of discourse, the rule base includes all the possible combinations of these fuzzy sets to cover the whole input space. For the sake of representing the rules clearly, we sort the rules as follows: corresponding to a combination of premise fuzzy sets$A_{i_1}^{(1)}, \ldots, A_{i_n}^{(n)}$, the rule is indexed as i in the rule base, where $i = \sum_{j=1}^{n-1} [(i_j - 1) \cdot \prod_{q=j+1}^{n} L_q] + i_n$.

In this chapter, the linguistic modifiers is proposed to use as the MFs for the above TS model, so the MFs of fuzzy sets $A_{i_j}^{(j)}$ are obtained as follows:

$$A_{i_j}^{(j)}(x_j; \beta_{i_j-1,j}, C_{i_j,j}^{(1)}, \beta_{i_j,j}, C_{i_j,j}^{(2)}, \beta_{i_j+1,j}, p_j)$$
$$= \begin{cases} \dfrac{1}{\mu_{C_{i_j,j}^{(1)}}^{p_j-1}} \left(\dfrac{x_j - \beta_{i_j-1,j}}{\beta_{i_j,j} - \beta_{i_j-1,j}} \right)^{p_j}, & \beta_{i_j-1,j} \le x_j < C_{i_j,j}^{(1)} \\ 1 - \dfrac{1}{\left(1 - \mu_{C_{i_j,j}^{(1)}}\right)^{p_j-1}} \left(\dfrac{\beta_{i_j,j} - x_j}{\beta_{i_j,j} - \beta_{i_j-1,j}} \right)^{p_j}, & C_{i_j,j}^{(1)} \le x_j < \beta_{i_j,j} \\ 1 - \dfrac{1}{\left(1 - \mu_{C_{i_j,j}^{(2)}}\right)^{p_j-1}} \left(\dfrac{x_j - \beta_{i_j,j}}{\beta_{i_j+1,j} - \beta_{i_j,j}} \right)^{p_j}, & \beta_{i_j,j} \le x_j < C_{i_j,j}^{(2)} \\ \dfrac{1}{\mu_{C_{i_j,j}^{(2)}}^{p_j-1}} \left(\dfrac{\beta_{i_j+1,j} - x_j}{\beta_{i_j+1,j} - \beta_{i_j,j}} \right)^{p_j}, & C_{i_j,j}^{(2)} \le x_j < \beta_{i_j+1,j} \end{cases} \tag{17.8}$$

where$C_{i_j,j}^{(1)}$ and $C_{i_j,j}^{(2)}$ are the left and right crossover points of $A_{i_j}^{(j)}$ respectively, $\beta_{i_j-1,j}$, $\beta_{i_j,j}$and $\beta_{i_j+1,j}$are the cores of fuzzy sets $A_{i_j-1}^{(j)}$, $A_{i_j}^{(j)}$ and $A_{i_j+1}^{(j)}$ individually and $\beta_{i_j-1,j} < \beta_{i_j,j} < \beta_{i_j+1,j}$, while,$\mu_{C_{i_j,j}^{(1)}}$ and $\mu_{C_{i_j,j}^{(2)}}$ are evaluated as follows:

$$\mu_{C_{i_j,j}^{(1)}} = \frac{C_{i_j,j}^{(1)} - \beta_{i_j-1,j}}{\beta_{i_j,j} - \beta_{i_j-1,j}}; \quad \mu_{C_{i_j,j}^{(2)}} = \frac{\beta_{i_j+1,j} - C_{i_j,j}^{(2)}}{\beta_{i_j+1,j} - \beta_{i_j,j}} \tag{17.9}$$

17.3.2 Convergence Analysis of the Linguistic Modifier

For the sake of simplicity in analysis, a fuzzy set is denoted as A_k whose MF $A_k(x; \beta_{k-1}, C_k^{(1)}, \beta_k, C_k^{(2)}, \beta_{k+1}, p)$is defined correspondingly in terms of (17.8). It can be seen from (17.8) that the first two rows of expressions are used to adjust the left part of the MF with center β_k, while the last two rows the right part of the MF. Because the MF defined by the modifier is symmetrical, without loss of generality, it is only needed to analyze the convergence of the first two rows of expressions.

If $\beta_{k-1} \le x < C_k^{(1)}$, then

$$A_k(x; \beta_{k-1}, C_k^{(1)}, \beta_k, C_k^{(2)}, \beta_{k+1}, p) =$$
$$A_k(x; \beta_{k-1}, C_k^{(1)}, \beta_k, C_k^{(2)}, \beta_{k+1}, p-1) \left(\frac{A_k^0(x)}{\mu_{C_k^{(1)}}} \right)$$

where $A_k^0(x) = \max\left\{0, \min\left(\frac{x-\beta_{k-1}}{\beta_k - \beta_{k-1}}, \frac{\beta_{k+1} - x}{\beta_{k+1} - \beta_k} \right)\right\}$, i.e., the triangular function. Since $A_k^0(x) \le \mu_{C_k^{(1)}}$, hence

$$A_k(x;\beta_{k-1},C_k^{(1)},\beta_k,C_k^{(2)},\beta_{k+1},p) \leq A_k(x;\beta_{k-1},C_k^{(1)},\beta_k,C_k^{(2)},\beta_{k+1},p-1)$$

that is to say, as $p(>1)$ increases the membership degrees of the points in this area will decrease, and

$$\lim_{p\to\infty} A_k(x;\beta_{k-1},C_k^{(1)},\beta_k,C_k^{(2)},\beta_{k+1},p) = \lim_{p\to\infty} A_k^0(x)\left(\frac{A_k^0(x)}{\mu_{C_k^{(1)}}}\right)^{p-1} = 0$$

If $C_k^{(1)} \leq x < \beta_k$, then

$$1 - A_k(x;\beta_{k-1},C_k^{(1)},\beta_k,C_k^{(2)},\beta_{k+1},p) =$$
$$\left(1 - A_k(x;\beta_{k-1},C_k^{(2)},\beta_k,C_k^{(2)},\beta_{k+1},p-1)\right)\left(\frac{1-A_k^0(x)}{1-\mu_{C_k^{(1)}}}\right)$$

since $A_k^0(x) \geq \mu_{C_k^{(1)}}$, hence,

$$A_k(x;\beta_{k-1},C_k^{(1)},\beta_k,C_k^{(2)},\beta_{k+1},p) \geq A_k(x;\beta_{k-1},C_k^{(1)},\beta_k,C_k^{(2)},\beta_{k+1},p-1)$$

that is to say, as $p(>1)$ increases the membership degrees of the points in this area will increase, and

$$\lim_{p\to\infty} A_k(x;\beta_{k-1},C_k^{(1)},\beta_k,C_k^{(2)},\beta_{k+1},p) =$$
$$\lim_{p\to\infty}\left[1-\left(1-A_k^0(x)\right)\left(\frac{1-A_k^0(x)}{1-\mu_{C_k^{(1)}}}\right)^{p-1}\right] = 1$$

This proves the desired convergence property of the linguistic modifier, that is to say, as the linguistic modifier parameters increase, the ε-insensitive cores of fuzzy sets will become bigger and bigger, and at the same time the overlapping among adjacent fuzzy sets will become smaller and smaller. In such a way, the local models will be forced to dominate the local behaviors of the system, and tend to become the tangents of the global model. As a result, improvements could be made in not only the interpretability of local models but also the transparency of partitioning of input space. However, based on MFs with less overlapping and larger core regions, the global approximation ability of the TS model could be degraded. In the following section we propose a scheme to make the linguistic modifiers optimally adjusted so that the accuracy and interpretability of the model can be balanced in terms of a multiple objective performance measure.

17.4 Fuzzy Model Interpretability Improvement Based on a Multiple Objective Learning Scheme

17.4.1 Fuzziness Measure of A Fuzzy Set

The proposed fuzziness measure is based on the distance between a fuzzy set A and an ordinary (crisp) set $\underline{A}$ near to A. $\underline{A}$ is defined as follows:

$$\underline{A}(x) = \begin{cases} 1 & if\ C^{(1)} \leq x \leq C^{(2)} \\ 0 & otherwise \end{cases} \tag{17.10}$$

where $C^{(1)}$ and $C^{(2)}$ are the left and right crossover points of fuzzy set A respectively. Given a data set $\{x(k)\}_{k=1}^{N}$ on domain x, the index of fuzziness, $F(A)$, is defined based on the distance between A and $\underline{A}$ as follows:

$$F(A) = \frac{2}{N^{1/r}} d_r(A, \underline{A}) \tag{17.11}$$

where N is the length of the data set, and r is the order of the distance d_r between A and $\underline{A}$. Obviously, in case $C^{(1)} = C^{(2)} = C$ and $A(C)$=0.5, $F(A)$ becomes the classic fuzziness measure proposed by Yager [39]. Particularly, in case $r = 2$ (Euclidean distance is used), (17.12) defines a *quadratic index of fuzziness*,

$$F_q(A) = \frac{2}{\sqrt{N}} \sqrt{\sum_{k=1}^{N} (A(x(k)) - \underline{A}(x(k)))^2} \tag{17.12}$$

In this chapter, the *quadratic index of fuzziness* is used in the proposed learning algorithm.

17.4.2 A Multiple Objective Function and the Learning Algorithm

For a given data set $\{(x(k), d(k))\}_{k=1}^{N}$, a hybrid learning scheme is employed to update the consequent parameters a_{ij} and the premise parameters p_j. In the first part of the algorithm, the premise parameters p_j are fixed (starting with fixed value 1), and the consequent parameters a_{ij} are identified by least squares estimates in terms of the global accuracy measure. In the second pass, the newly obtained consequent parameters a_{ij} are fixed, and the premise parameters p_j are updated by a gradient descent algorithm in terms of a multiple objective performance measure as defined below:

$$J = \varphi E + \theta F \tag{17.13}$$

whereφ and θ are two positive constants satisfying the condition:$\varphi + \theta = 1$. φ and θ are usually determined according to user preferences, objectives priorities etc.. E is the global accuracy measure:

$$E = \frac{1}{2}\sum_{k=1}^{N} \|d(k) - y(k)\|^2 \tag{17.14}$$

and F is the index of fuzziness of the TS model defined as

$$F = \sum_{i_1=1}^{L_1} \cdots \sum_{i_n=1}^{L_n} \sum_{j=1}^{n} F(A_{i_j}^{(j)}) \tag{17.15}$$

$F(A_{i_j}^{(j)})$is the *quadratic index of fuzziness* defined by (17.12). It can be seen that the minimum of (17.13) is a Pareto optimal solution.

Least-squares Estimates for Consequent Parameters

In order to identify the consequent parameters in the TS model, we reformulate some expressions in (17.5)-(17.7). Defining a base matrix M as follows:

$$M = \begin{bmatrix} M_1^T(1) & \cdots & M_L^T(1) \\ \\ M_1^T(N) & & M_L^T(N) \end{bmatrix}_{N \times L(n+1)} \tag{17.16}$$

where $M_i^T = (w_i \; w_i x_1 \; \cdots \; w_i x_n)$, and the consequent parameters are represented by a column vector $a = (a_{10} \; a_{11} \cdots \; a_{1n} \; a_{20} \; a_{21} \cdots \; a_{2n} \; \cdots \; \cdots \; a_{L0} \; a_{L1} \; \cdots \; a_{Ln})^T$, we reformulate the TS model as follows:

$$M \cdot a = d \tag{17.17}$$

where $d = (d(1) \; \cdots \; d(N))^T$. Because the consequent parameters in a do not make any contribution to the index of fuzziness of the TS model, they can be identified practically based on the global approximation accuracy objective E defined in (17.14). Since the number of training data pairs is usually greater than $L \times (n+1)$, this is a typical ill-posed problem and generally there does not exist exact solution for vector a. The least-squares estimate of a can be obtained by $a^* = M^+ d$, where M^+ is the Moore-Penrose inverse of matrix M.

Updating the Premise Parameters

The premise parameters are updated in terms of the multiple objective function defined in (17.13), which aims at striking a good trade-off between the global approximation ability and the interpretability of local models. The equation for updating the linguistic modifier parameters is as follows:

$$p_j(t+1) = p_j(t) - \rho \frac{\partial J}{\partial p_j} \tag{17.18}$$

where t is the iteration step, ρ is the learning rate, and

$$\frac{\partial J}{\partial p_j} = \varphi \frac{\partial E}{\partial p_j} + \theta \frac{\partial F}{\partial p_j} \tag{17.19}$$

$$\frac{\partial E}{\partial p_j} = -\sum_{k=1}^{N} \sum_{i=1}^{L} (d(k) - y(k)) y_i(k) \frac{\partial w_i(k)}{\partial p_j} \tag{17.20}$$

$$\frac{\partial w_i(k)}{\partial p_j} = \frac{\sum_{l=1}^{L} \tau_l(k) \frac{\partial \tau_i(k)}{\partial p_j} - \tau_i(k) \sum_{l=1}^{L} \frac{\partial \tau_l(k)}{\partial p_j}}{\left(\sum_{l=1}^{L} \tau_l(k) \right)^2} \tag{17.21}$$

$$\frac{\partial \tau_i(k)}{\partial p_j} = \prod_{q \neq j} A_{i_q}^{(q)}(x_q(k)) \frac{\partial A_{i_j}^{(j)}(x_j(k))}{\partial p_j} \tag{17.22}$$

$$\frac{\partial F}{\partial p_j} = \sum_{1 \le i_j \le L_j} \frac{\partial F(A_{i_j}^{(j)})}{\partial p_j} \tag{17.23}$$

$$\frac{\partial F(A_{i_j}^{(j)})}{\partial p_j} = \frac{2}{\sqrt{N}} \frac{\sum_{k=1}^{N} (A_{i_j}^{(j)}(x_j(k)) - \underline{A}_{i_j}^{(j)}(x_j(k))) \frac{\partial A_{i_j}^{(j)}(x_j(k))}{\partial p_j}}{\sqrt{\sum_{k=1}^{N} (A_{i_j}^{(j)}(x_j(k)) - \underline{A}_{i_j}^{(j)}(x_j(k)))^2}} \tag{17.24}$$

Furthermore,
For $\beta_{i_j-1,j} < x_j < C_{i_j,j}^{(1)}$

$$\frac{\partial A_{i_j}^{(j)}(x_j)}{\partial p_j} = A_{i_j}^{(j)}(x_j(k)) \ln \left[\frac{x_j - \beta_{i_j-1,j}}{\mu_{C_{i_j,j}^{(1)}} (\beta_{i_j,j} - \beta_{i_j-1,j})} \right] \tag{17.25}$$

For $C_{i_j,j}^{(1)} \le x_j < \beta_{i_j,j}$

$$\frac{\partial A_{i_j}^{(j)}(x_j)}{\partial p_j} = \left[A_{i_j}^{(j)}(x_j(k)) - 1 \right] \ln \left[\frac{\beta_{i_j,j} - x_j}{(1 - \mu_{C_{i_j,j}^{(1)}})(\beta_{i_j,j} - \beta_{i_j-1,j})} \right] \tag{17.26}$$

For $\beta_{i_j,j} < x_j < C_{i_j,j}^{(2)}$

$$\frac{\partial A_{i_j}^{(j)}(x_j)}{\partial p_j} = \left[A_{i_j}^{(j)}(x_j) - 1 \right] \ln \left[\frac{x_j - \beta_{i_j,j}}{(1 - \mu_{C_{i_j,j}^{(17.2)}})(\beta_{i_j+1,j} - \beta_{i_j,j})} \right] \tag{17.27}$$

For $C_{i_j,j}^{(2)} \le x_j < \beta_{i_j+1,j}$

$$\frac{\partial A_{i_j}^{(j)}(x_j)}{\partial p_j} = A_{i_j}^{(j)}(x_j(k)) \ln \left[\frac{\beta_{i_j+1,j} - x_j}{\mu_{C_{i_j,j}^{(2)}}(\beta_{i_j+1,j} - \beta_{i_j,j})} \right] \tag{17.28}$$

For others

$$\frac{\partial A_{i_j}^{(j)}(x_j)}{\partial p_j} = 0 \tag{17.29}$$

It should be noted that in (17.25)~(17.28), in order to avoid zero values of the variables in the logarithmic function, (17.29) is used for the centers of fuzzy sets, which is reasonable due to the property $\lim_{x \to 0} x \ln x = 0$ and the definition of MFs in (17.8).

Rule Base Refinement

Because all possible combinations of fuzzy sets about one-dimensional input variables are considered in the rule base for this TS model, a common problem in fuzzy modeling, *i.e.*, the curse of dimensionality, would emerge. In this subsection we give a simple but efficient method to refine the rule base during adaptation.

For a given training data set $\{(x(k), d(k))\}_{k=1}^{N}$, the total firing strength of the ith rule R_i, received from all input samples, is obtained by

$$IR_i = \sum_{k=1}^{N} \tau_i(x(k)) \tag{17.30}$$

If

$$IR_i < firingLimit \tag{17.31}$$

then rule $R_i$will be removed, where *firingLimit* is a given lower limit of firing strength.

17.5 Experimental Results

In this section two examples are used to evaluate the proposed method in terms of local model interpretability and global accuracy. For the sake of visualizing experimental results on the transparency of local models well, the first example considers a system with one input variable and one output variable, characterized as follows:

$$y = 50(1 - \cos(\pi x/50)) \frac{\sin(2x)}{e^{x/5}} \tag{17.32}$$

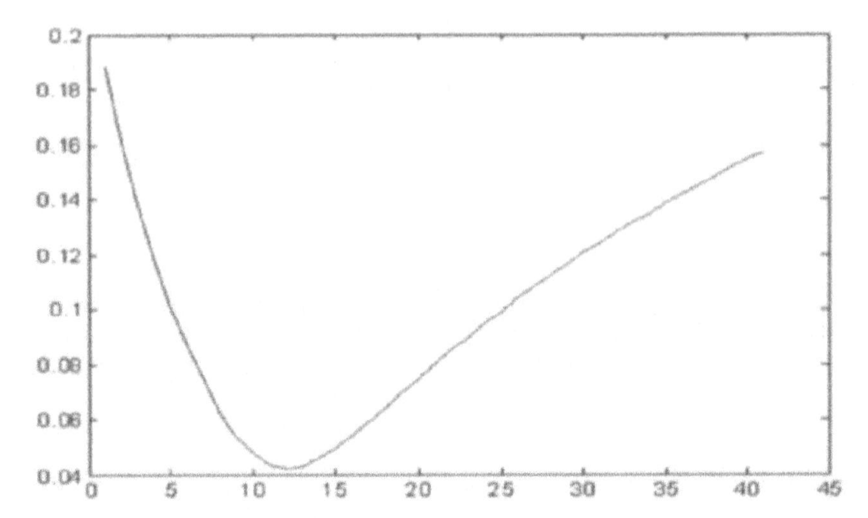

Fig. 17.2. Model approximation error vs linguistic modifier parameter updating

200 input-output data pairs were collected for the purpose of parameter identification for the TS model. The number of local models can be determined by examining the local properties of the data automatically or manually. In this example, there are 14 local areas with distinctive linearizations, so we set the number of fuzzy partitions on input variable x as 14. By clustering the 200 input-output data pairs using a Mercer kernel fuzzy c-means (MKFCM) clustering algorithm [40], 13 data centers on input variable x are obtained, which are then used as the crossover points of the linguistic modifiers. The minimum and maximum values of input variable x are used as the cores of 2 linguistic modifiers, and the cores of the remaining 12 linguistic modifiers are set by the midpoints of the corresponding crossover points generated above. $\varphi = 0.6$ and $\theta = 0.4$ for multiple objective learning. During adaptation, the firing strength lower limit is set to be 0.0001. After 12 iterations of updating the linguistic modifier parameter, the global approximation error of the proposed TS model arrives at a minimum, as shown in Fig.17.2, which indicates that if the modifier parameter is updated furthermore, the global approximation performance could be degraded. However, Fig.17.3 shows that a trade-off between the global accuracy and the fuzziness measure is found at the 27th iteration by the multiple objective function. It is noted that the training data fire all the 14 rules very well, and there is no rule being cancelled. The finally generated MFs are depicted in Fig.17.4. By using these 14 MFs in the TS model, it can be seen from Fig.17.5 that the corresponding 14 local models exhibit the desired good interpretability: they match the system well at the local areas, at the same time the global prediction accuracy is still preserved.

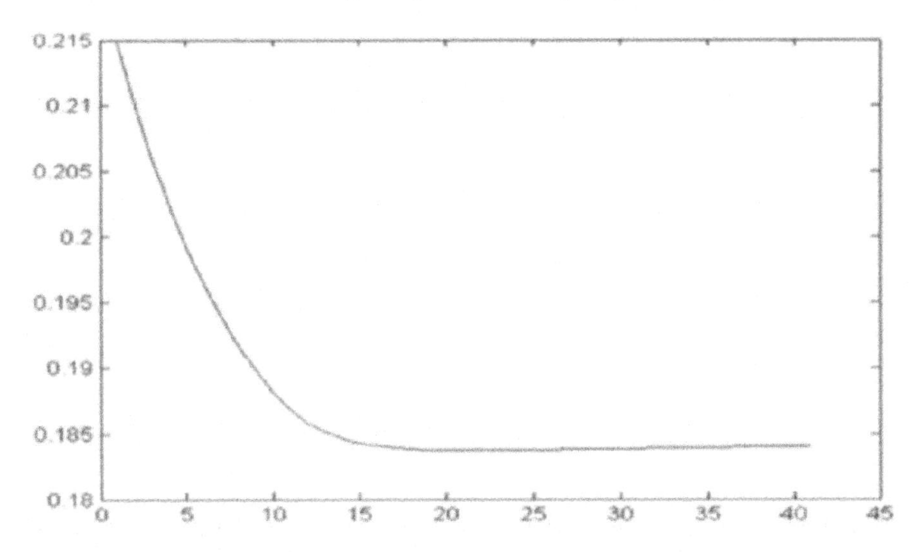

Fig. 17.3. Multiple objective performance measure

In comparison with the proposed method, the well-known ANFIS model [27] is used to model the same system (17.32), the 14 MFs generated by ANFIS method are shown in Fig.17.6. Although the ANFIS method can approximate the system very well, as depicted in Fig.17.7, obviously the interpretability of its 14 local models is poor: they exhibit erratic behaviours, and can not characterize the local activities of the modelled system well.

The second example is to model a dynamic system with two feedback input variables, one external input variable and one output variable, which is described as follows [41]:

$$y(k) = 0.3y(k-1) + 0.6y(k-2) + 0.6\sin(\pi u) + 0.3\sin(0.3\pi u) + 0.1\sin(5\pi u) \tag{17.33}$$

where $u = \sin(2\pi(k-1)/250)$ is the external input. We select $(y(k-1), y(k-2), u(k-1))$ as the three input variables of the TS model, and $y(k)$ the output variable. Initially, 2000 input-output data pairs were generated to build the proposed TS model, where the first 1500 data pairs were used as training data and the latter 500 pairs as test data. Four initial fuzzy sets were set up for each input variable, whose cores can be obtained by cluster analysis algorithms. In this chapter, they were obtained by the MKFCM algorithm [40]. $\varphi = 0.6$, $\theta = 0.4$. The firing strength lower limit is set as 0.0001. The finally generated fuzzy sets for the three input variables are depicted in Fig.17.8 with large core areas and small overlapping, which shows good distinguishability of partitioning of input space in terms of the criteria introduced in [9] [25] [26]. Fuzzy sets generated in this way would make the local models dominate the system behaviours in local regions, which leads to good local model interpretability.

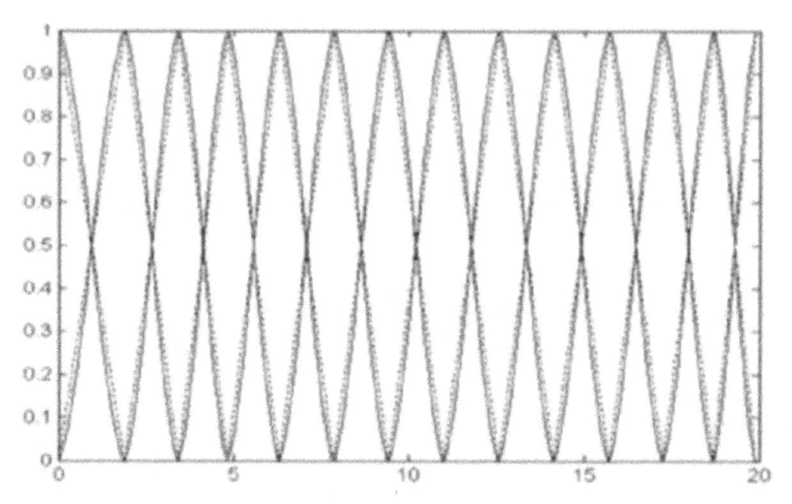

Fig. 17.4. MFs with modifier parameters equal to 1 in dotted line (DOL) and 1.428 in solid line (SL)

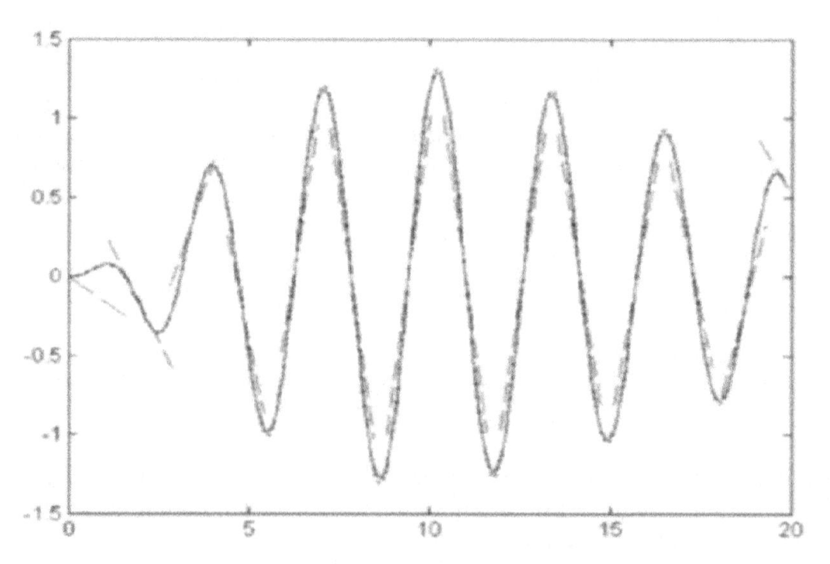

Fig. 17.5. Good interpretability of local models obtained by the proposed method : SL-desired output, DOL-model output, dashed lines (DALs)-local models

As shown in Fig.17.9, on both training data and test data, this dynamical system is well approximated by the built TS model. Although 64 rules are constructed in the initial rule base, 31 rules are preserved in the final rule base after the refinement by the proposed method.

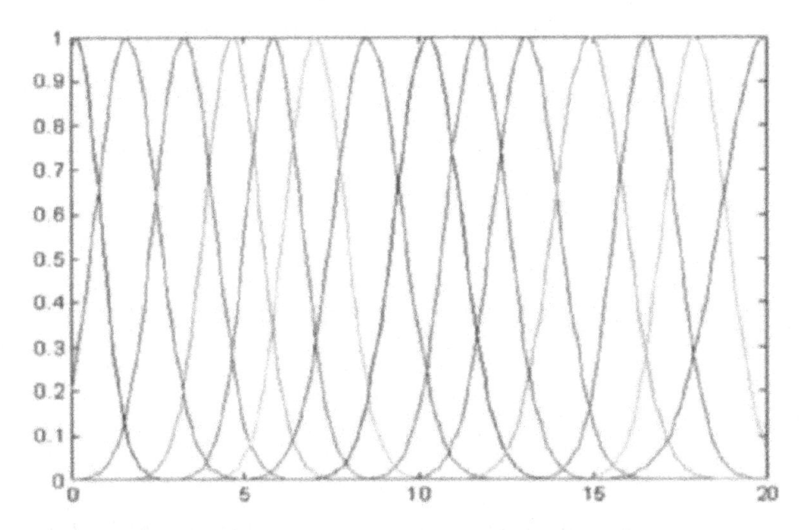

Fig. 17.6. Good interpretability of local models obtained by the proposed method : SL-desired output, DOL-model output, dashed lines (DALs)-local models

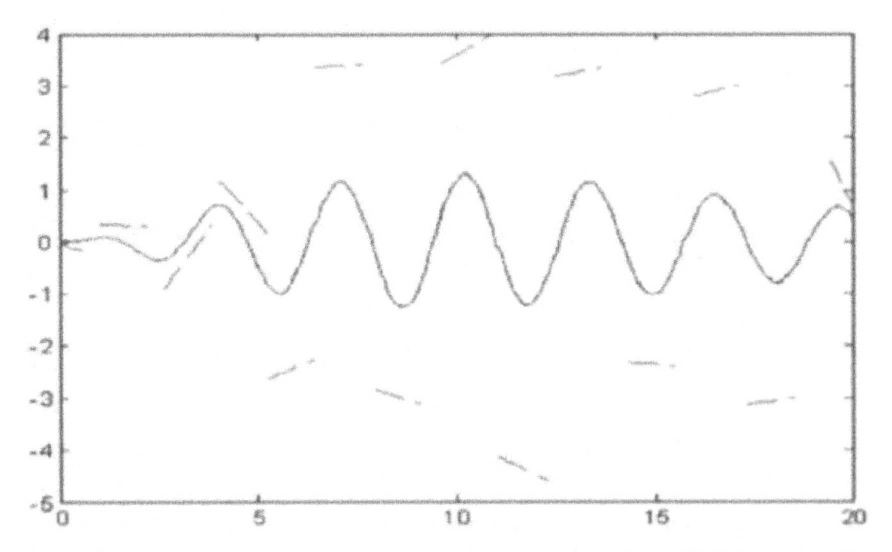

Fig. 17.7. Poor interpretability of local models produced by ANFIS method : SL-desired output, DOL-model output, DAL-local models

17.6 Conclusion

In this chapter, a new TS type fuzzy model, whose MFs are characterized by linguistic modifiers, is proposed. A tradeoff between global approximation and local model interpretation has been achieved by minimizing a multiple objec-

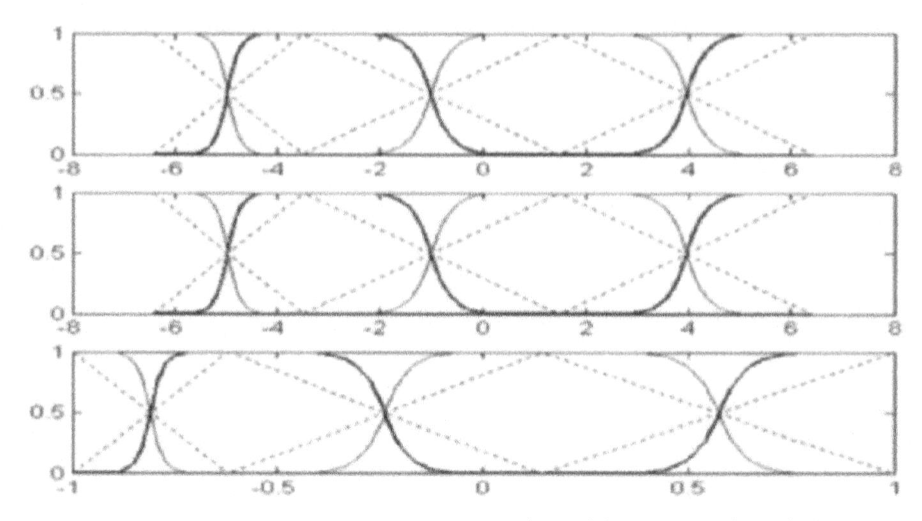

Fig. 17.8. MFs generated for input variables $y(k-1)$(top row), $y(k-2)$ (middle row), and $u(k-1)$(lowest row): DOL-initial triangular MFs, SL-final MFs

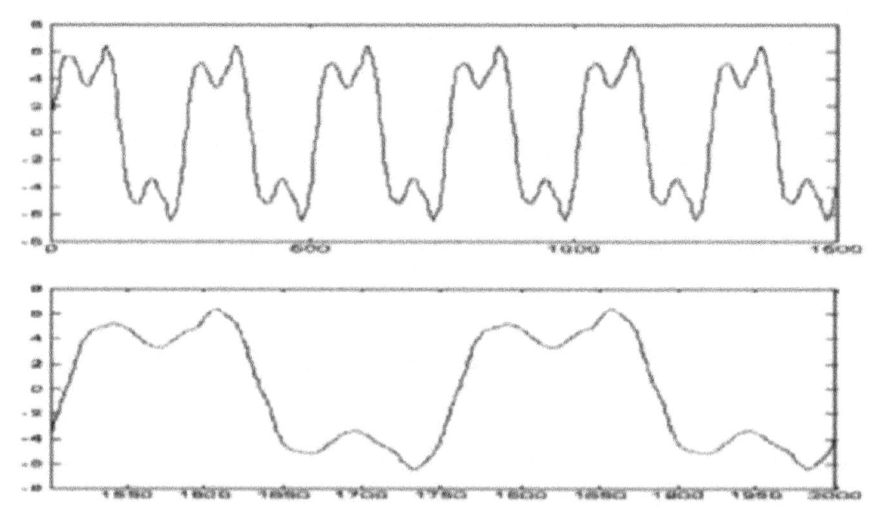

Fig. 17.9. Approximation results of the proposed model for training data (top) and test data (bottom): SL-desired output and DOL-model output.

tive performance measure. In the proposed model, the local models match the global model well and the erratic behaviors of local models are remedied greatly. Furthermore, the transparency of partitioning of input space has been improved during parameter adaptation. Due to the promising performance exhibited, the proposed method would have potential applications to fuzzy system modeling, particularly, to nonlinear state estimation and control problems. However, a problem is how to select the optimal weighting coefficients in a combined learning objective function. We will pay attention to the problem in the future research.

References

[1] Jang JSR, Sun CT, Mizutani E (1997) Neuro Fuzzy and Soft Computing. Englewood Cliffs, NJ: Prentice-Hall.

[2] Guillaume S (2001) Designing fuzzy inference systems from data: an interpretability-oriented review. IEEE Trans. on Fuzzy Systems, 9:426-443.

[3] Setnes M, Babuska R and Verbruggen HB (1998) Rule-based modeling: precision and transparency. IEEE Trans. on Systems, Man and Cybernetics-Part C, 28:165-169.

[4] Shen Q and Marin-Blazquez JG (2002) Microtuning of membership functions: accuracy vs.interpretability. Proc. of IEEE Int. Conf. on Fuzzy Systems (FUZZ-IEEE'02):168-173, Honolulu, Hawai, USA.

[5] Guillaume S and Charnomordic B (2004) Generating an interpretable family of fuzzy partitions from data. IEEE Trans. on Fuzzy Systems,12:324-335.

[6] Mucientes M and Casillas J (2004) Obtaining a fuzzy controller with high interpretability in mobile robots navigation, Proc. of IEEE Int. Conf. on Fuzzy Systems (FUZZ-IEEE'04):1637-1642, Budapest, Hungary, 2004.

[7] Drobics M, Bodenhofer U and Klement EP (2003) FS-FOIL: an inductive learning method for extracting interpretable fuzzy descriptions, Int. Journal of Approximate Reasoning,32:131-152.

[8] Babuska R and Verbruggen H (2003) Neuro-fuzzy methods for nonlinear system identification, Annual Reviews in Control, 27:73-85.

[9] de Oliveira JV (1999) Semantic constraints for membership function optimization, IEEE Trans.on Systems, Man and Cybernetics-Part A,29:128-138.

[10] Jin Y (2000) Fuzzy modeling of high-dimensional systems: complexity reduction and interpretability improvement, IEEE Trans. on Fuzzy Systems, 8:212-221.

[11] Penna-Reyes CA and Sipper M (2003) Fuzzy CoCo: balancing accuracy and interpretability of fuzzy models by means of coevolution. in J. Casillas, O. Cordon, F. Herrera, L Magdalena, (eds.) Interpretability issues in Fuzzy Modelling, Studies in Fuzziness and Soft Computing, Physica-Verlag.

[12] Mikut R, Jäkel L and Gröll L (2005) Interpretability issues in data-based learning of fuzzy systems, Fuzzy Sets and Systems, 150:179-197.

[13] Paiva RP and Dourado A (2004) Interpretability and learning in neuro-fuzzy systems, Fuzzy Sets and Systems, 147:17-38.

[14] Jimenez F, Sanchez G, Gomez-Skarmeta AF, Roubos H and Babuska R (2002) Fuzzy modeling with multi-objective neuro-evolutionary algorithms, Proc. of IEEE Int. Conf. on Systems, Man and Cybernetics (IEEE SMC'02), Yasmine Hammamet, Tunisia.

[15] Jin Y, von Seelen W and Sendhoff B (1999) On generating FC3(flexible, complete, consistent and compact) fuzzy rule systems from data using evolution strategies, IEEE Trans.on Systems, Man and Cybernetics, Part B: Cybernetics, 29:829-845.

[16] Castellano G, Fanelli AM and Mencar C (2002) A neuro-fuzzy network to generate human-understandable knowledge from data, Cognitive Systems Research, 3:125-144.
[17] Wang H, Kwong S, Jin Y, Wei W and Man KF (2005) Multi-objective hierarchical genetic algorithm for interpretable fuzzy rule-based knowledge extraction, Fuzzy Sets and Systems, 149:149-186.
[18] Delgado MR, Von Zuben F and Gomide F (2002) Multi-objective decision making: towards improvement of accuracy, interpretability and design autonomy in hierarchical genetic fuzzy systems, Proc. of IEEE Int. Conf. on Fuzzy Systems (FUZZ-IEEE'02):1222-1227, Honolulu, Hawai, USA.
[19] Gan Q and Harris CJ (1999) Fuzzy local linearization and local basis function expansion in nonlinear system modeling, IEEE Trans. on Systems, Man and Cybernetics-Part B., 29:559-565.
[20] Yen J, Wang L and Gillespie CW (1998) Improving the interpretability of TSK fuzzy models by combining global learning and local learning, IEEE Trans. on Fuzzy Systems, 6:530-537.
[21] Johansen TA and Babuska R (2003) Multi-objective identification of Takagi-Sugeno fuzzy models, IEEE Trans. on Fuzzy Systems, 11:847-860.
[22] Roubos H and Setnes M (2001) Compact and transparent fuzzy models and classifiers through iterative complexity reduction, IEEE Trans. on Fuzzy Systems, 9:516-524.
[23] Gomez-Skarmeta F and Jimenez F (1999) Fuzzy modeling with hybrid systems, Fuzzy Sets and Systems, 104:199-208.
[24] Hoppner F and Klawonn F (2001) A new approach to fuzzy partitioning, Proc. of the Joint 9 th IFSA Congress and 20 th NAFIPS Int. Conf.:1419-1424, Vancouver, Canada.
[25] Bezdek JC (1981) Pattern Recognition with Fuzzy Objective Function Algorithms. New York: Plenum.
[26] Gath I and Geva AB (1989) Unsupervised optimal fuzzy clustering, IEEE Trans. on Pattern Analysis and Machine Intelligence, 11:773-781.
[27] Jang JSR (1993) ANFIS: adaptive-network-based fuzzy inference system, IEEE Trans. on Systems, Man and Cybernetics, 23:665-685.
[28] Takagi T and Sugeno M (1985) Fuzzy identification of systems and its applications to modelling and control, IEEE Trans. on Systems, Man and Cybernetics, 15:116-132.
[29] MaGinnity S and Irwin GW (1996) Nonlinear state estimation using fuzzy local linear models, Int. J. of System Science, 28:643-656.
[30] Gan Q and Harris CJ (1999) Neurofuzzy local linearisation for model identification and state estimation of unknown nonlinear processes Int. J. of Intelligent Control and Systems, 3:389-407.
[31] Fink A, nelles Fink O and Isermann R (2002) Nonlinear internal model control for MISO systems based on local linear neuro-fuzzy models, Proc. of the 15 th IFAC World Congress, Barcelona, Spain.
[32] Statnikov RB and Matusov JB (1995) Multicriteria Optimization and Engineering, New York, Chapman and Hall.
[33] Marler RT and Arora JS (2004) Survey of multi-objective optimization methods for engineering, Structural Multidisciplinary Optimization, 26:369-395.

[34] Salukvadze ME (1971) On the optimization of vector functionals, 1. Programming of optimal trajectories, 2. Analytic construction of optimal controls, Automation and Remote Control, 31 (in Russian).
[35] Gomez-Skameta A, Jimenez F and Ibanez J (1998) Pareto-optimality in fuzzy modeling, Proc. of the 6 th European Congress on Intelligent Techniques and soft Computing, EUFIT'98:694-700, Aachen, Gemany, Sept..
[36] Zadeh LA(1963) Optimality and non-scalar-valued performance criteria, IEEE Trans. on Automatic Control, 8:59-60.
[37] Chankong V and Haimes YY (1983) Multiobjective Decision Making Theory and Methodology, New York, Elsevier Science Publishing.
[38] Zeleny M (1982) Multiple Criteria Decision Making,New York, McGraw Hill.
[39] Yager RR (1979) On the measure of fuzziness and negation part I: Membership in the unit interval, Int. J. of Gen. Syst.,5:221-229.
[40] Zhou SM and Gan JQ (2004) An unsupervised kernel based fuzzy c-means clustering algorithm with kernel normalization, Int. J. of Computational Intelligence and Applications, 4:355-373.
[41] Narendra KS and Parthsarathy K (1990) Identification and control of dynamical systems using neural networks, IEEE Trans. on Neural Networks, 1:4-27.

Part IV

Multi-Objective Ensemble Generation

18

Pareto-Optimal Approaches to Neuro-Ensemble Learning

Hussein Abbass

The ARC Centre for Complex Systems
Artificial Life and Adaptive Robotics (A.L.A.R.) Lab
School of Information Technology and Electrical Engineering
Australian Defence Force Academy
University of New South Wales, Canberra, ACT 2600, Australia
abbass@itee.adfa.edu.au

Summary. The whole is greater than the sum of the parts; this is the essence of using a mixture of classifiers instead of a single classifier. In particular, an ensemble of neural networks (we call neuro-ensemble) has attracted special attention in the machine learning literature. A set of trained neural networks are combined using a post-gate to form a single super-network. The three main challenges facing researchers in neuro-ensemble are:(1) which network to include in, or exclude from the ensemble; (2) how to define the size of the ensemble; (3) how to define diversity within the ensemble.

In this chapter, we will review our previous work on the use and advantages of evolutionary multi-objective optimization for evolving neural networks and, in specific, neuro-ensemble. We will highlight key research questions in this area and potential future directions.

18.1 Introduction

The paper of Kottatha and Attikiouzel [31] was the first to propose formulating the neural network learning problem as a multi-objective problem: one objective to minimize the mean square error (MSE) and the second objective to minimize the number of hidden units. The authors used a branch and bound technique and nonlinear goal programming to find a solution with the minimum MSE and number of hidden units. Unfortunately, goal programming requires prior decisions on the achievement function and the approach does not aim at generating the Pareto-optimal set. Instead, the method will potentially terminate with a single optimal solution.

In 1997, Wang and Wahl [59] published their paper on image reconstruction using a multi-objective neural network approach. The authors used the weighted sum method to combine the objectives. The two objectives were representing the smoothness of the image and the cross entropy between the

H. Abbass: *Pareto-Optimal Approaches to Neuro-Ensemble Learning*, Studies in Computational Intelligence (SCI) **16**, 407–427 (2006)
www.springerlink.com

original projection data and re-projection data due to the reconstructed image.

Park et. al. [43] used a standard multi-objective evolutionary programming method to evolve networks on two objectives in a control problem. Kupinski and Anastasio [34] used a Niche Pareto genetic algorithm to evolve neural networks using two objective functions: specificity and sensitivity of the classification.

Interestingly, it seems that none of the previous papers published after Kottatha's paper were aware of each other's work. In 2001 [1], I published a paper proposing the idea of explicitly modelling the neural network learning problem as a multi-objective optimization problem (MOP). At the time, I was not aware of the previous 4 papers, either because they were not focusing directly on the learning problem or because they were published in journals outside the journals I would normally look at for a literature review on neural networks.

The idea started with my interest in neural network and evolutionary multi-objective optimization (EMO); and was merely a matter of why not apply EMO to artificial neural networks (ANN). As the work became more mature, there was an urge to look deeply into this problem. Many in machine learning would argue that the learning problem has always been seen as a multi-objective problem; where we trade-off training error and generalization. As a matter of fact, the literature of machine learning avoided the explicit modelling of the learning problem as a MOP. First, we did not have a clear mechanism to create multiple networks simultaneously until methods such as bagging and boosting became popular in the machine learning literature. Second, even with these methods, it was easier to analyze the problem using a single objective than a multi-objective approach. Third, with a new wave of research focusing on support vector machines, researchers were more focused on finding different type of kernels and better ways to speed-up the algorithms developed in this area than the multi-objective approach.

By the time the more mature version of my paper was published in the Neural Computation Journal in 2003 [6], I believed more and more in the value of explicitly representing the neural network learning problem as a MOP. First, multi-objective optimization emerged during that time as an ideal way to promote diversity [7]. Second, there are different ways for formulating the problem, each way results in an interesting set of Pareto-optimal solutions with different features. Third, the applications of the EMO approach to ANN was found very successful on different domains including robotics [54, 52, 53, 55, 56] and breast cancer diagnosis [2].

One issue that remained unanswered during the initial stage of this work is which network to select. Ultimately, when a Pareto-optimal set of solutions is generated, there is only one decision (solution) that the decision maker needs. None of the work that was published on that topic attempted to answer this question. Kottatha and Attikiouzel [31] relied on the subjective decision of determining an appropriate achievement level. Kupinski and Anastasio [34] used the receiver operating characteristic curve to choose a network with a good balance, but when they discussed the complexity of the network, they left the choice to the user to decide what is appropriate. Even in my own papers at the time, the decision was somehow arbitrary by choosing the network with the minimum training error or evaluating on an independent validation set.

However, the neural network domain had the answer and offered a new use of the Pareto-optimal set; that is, to combine all, or some of the, solutions on the Pareto-optimal set into a super-individual or an ensemble. Many research studies in traditional optimization theory focused on how to choose a single solution from the Pareto-optimal set. The use of an ensemble to combine all solutions in the set was not only an efficient and theoretically sound (as we will see later in this chapter) way of doing it, it was also a novel use of the Pareto-optimal set.

By looking at one of the formulations I had for representing the learning problem as a multi-objective problem, I found a very good reason for combining the individuals on the Pareto-optimal set for this formulation into an ensemble. The reason being, individuals on the Pareto-optimal front represent an optimal bias-variance trade-off. This motivated me to extend the work on evolving neural networks using EMO to neuro-ensemble [5, 4].

Since my 2001 paper [1], a number of researchers extended this work and came up with more robust ways for evolving Pareto-optimal neural networks. I will only review my own work in this chapter, but the reader should consult the following references for better versions of my algorithm [12, 14, 13], other perspectives on this problem [27, 28], or merely getting inspired by the approach [60, 61, 40, 25].

The formation of a mixture of predictors (ensemble) can improve the prediction accuracy [33]. The literature of ensemble learning is very rich and there is no space to review this literature here. However, interested readers can refer to [9, 62, 18, 16, 15, 20, 26, 32, 35, 36, 37, 39, 44, 48, 57, 58, 64, 65, 66, 63, 67].

There have been many studies on ensembles of neural networks, see for example [33, 37, 38, 39] and [49] for a good review. In some studies, the networks are trained separately then combined to form an ensemble. In some other studies a group of networks is trained simultaneously while an explicit term is added to the error function to make them different. For example, in *negative correlation learning* (NCL), proposed by Liu and Yao [38], a mathematical penalty term describing the negative correlation between the networks is added to the conventional mean square error of each network and traditional *backpropagation* (BP) [45] is used for network training. In [41], an analysis of NCL revealed that the penalty function in NCL acts to maximize the average difference between each network and the mean of the population; while the intended aim of anti-correlation mechanisms is to maximize the average difference between pairs of population members, which is not necessarily the same thing. In [39], an *evolutionary algorithm* (EA) is used in conjunction with NCL to evolve the ensemble. The k-means algorithm was used to cluster the individuals in the population. The fittest individual in each cluster is

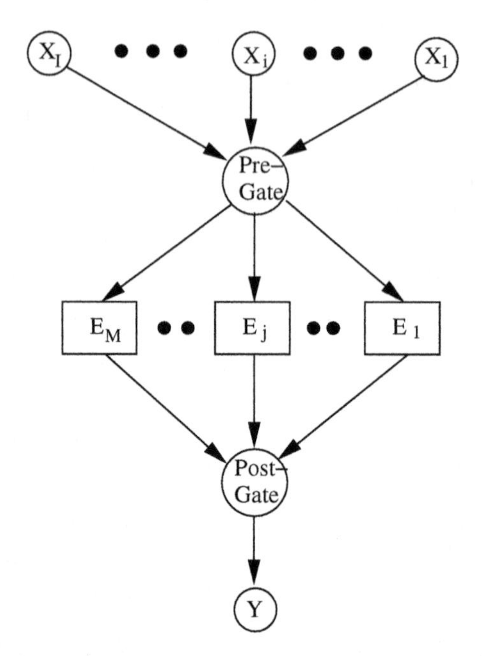

Fig. 18.1. A generic description to ensemble research.

used as a member in the ensemble. The authors found no statistical difference between the use of the population as a whole in the ensemble and the use of a subset. It is not clear however how to choose the value of k.

Figure 18.1 presents a generic representation of research in ensembles of learning machines. Two types of gates can be identified: pre-gate and post-gate. A pre-gate is used as a mechanism to decide on which input instance affects which member of the ensemble. Research in this area is mainly concerned with getting members of the ensemble to specialize on different parts of the input space. For example, we can see a pre-gate as a result of clustering the input space into k clusters, with the pre-gate deciding on which member of the k clusters being assigned to which members of the M learning machines in the ensemble. The study of this pre-gate falls under the umbrella of competitive-learning, where the main objective of the learning machine is to compete against other learning machines on the inputs.

A post-gate is used to decide on how to combine the different outputs made by the members in the ensemble. The majority of research in ensemble-learning is concerned with the post-gate, while attempting to implicitly construct a pre-gate. The motivation for this learning approach comes from the area of expert systems, where a committee of experts is used to make decisions. If one and only one committee member responds to each situation, there will be no conflict within the committee and the problem will become similar to the use of pre-gates with one-to-one or many-to-one mapping between the clusters in the input space and the ensemble members. However, if a group of committee members responds differently to the same input instance, the problem would then be how to combine the different opinions of the committee members. In this last case, the research would fall into the study of the post-gate. Obviously, we can have a pre-gate as well as a post-gate, where different group of experts get specialized on different parts of the input space, while a post gate is used within each group. Up to our knowledge, little research has been done on this last situation. This chapter is concerned with post-gates.

Three key open research questions remain in the literature:

1. On what basis should a network be included in, or excluded from the ensemble?
2. How should the ensemble size be determined?
3. How to ensure that the networks in an ensemble are different?

The objective of this work is to attempt to answer these three questions. This is a challenging task and this work should be seen as only an attempt to provide a theoretically-sound answer to these questions. The rest of the chapter is organized as follows: In Section 18.2, background materials are covered followed by an explanation of the methods in Section 18.3. Results are discussed in Section 18.4 and conclusions are drawn in Section 18.5.

18.2 Multiobjective Optimization and Neural Networks

Consider a *multiobjective optimization problem* (MOP) as presented below:-

$$\textbf{Optimize } F(\mathbf{x} \in \Upsilon) \tag{18.1}$$

$$\textbf{Subject to: } \Upsilon = \{\mathbf{x} \in R^n | G(\mathbf{x}) \leq 0\} \tag{18.2}$$

Where $\mathbf{x}$ is a vector of decision variables $(x_1, \ldots, x_n)$ and $F(\mathbf{x} \in \Upsilon)$ is a vector of objective functions $(f_1(\mathbf{x} \in \Upsilon), \ldots, f_K(\mathbf{x} \in \Upsilon))$. Here $f_1(\mathbf{x} \in \Upsilon), \ldots, f_K(\mathbf{x} \in \Upsilon)$, are functions on R^n and Υ is a nonempty set in R^n. The vector $G(\mathbf{x})$ represents a set of constraints.

The aim is to find the vector $\mathbf{x}^* \in \Upsilon$ which optimizes $F(\mathbf{x} \in \Upsilon)$. Without any loss of generality, we assume that all objectives are to be minimized. We note that any maximization problem can be transformed to a minimization one by multiplying the former by -1.

The core aspect in the discussion of a *multiobjective optimization problem* (MOP) is the possible conflict that arises in the case of optimizing the objectives simultaneously. At a certain point, one objective can just be improved at the expense of at least another objective. The principle of dominance (Figure 18.2) in MOP allows a partial order relation that works as follows: a solution does not have an advantage to be included in the set of optimal solutions unless there is no solution that is better than the former when measured on all objectives. A non–dominated solution is called Pareto-optimal.

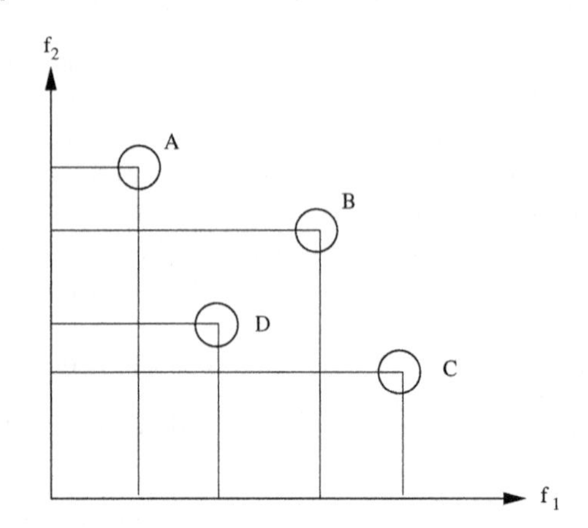

Fig. 18.2. The concept of dominance in multiobjective optimization. Assuming that both f_1 and f_2 are to be maximized, D is dominated by B since B is better than D when measured on all objectives. However, A, B and C are non–dominated since none of them is better than the other two when measured on all objectives.

A MOP can be solved in different ways. One method is by taking a weighted sum of the objectives. However, this is not an efficient way to solve the problem [46]. First, the weighted sum method could only generate a single non-dominated solution at a time. Second, it assumes convexity of the Pareto-optimal front. Third, the question of determining the appropriate values for the weights remains unsolved; that is, we will need to run the algorithm a number of times with different weights. Another method for solving MOPs is through the use of goal programming, where an aspiration level is set for each objective, transforming each objective to an equality after adding underachievement and overachievement deviations, then a single objective is constructed through the minimization of a prioritized and/or weighted sum of the relevant deviations. This method is biased to the choice of the aspiration levels.

A third method for solving MOP, ϵ constraint, is to sequentially optimize each objective. The way it works is as follows: select one of the objective and construct a single objective optimization problem ignoring the other objectives. The optimum of the single objective problem in conjunction with the corresponding objective function are used to form a constraint while ϵ is the right hand side of this constraint representing the amount the decision maker is willing to sacrifice in the corresponding optimal objective value. After adding this constraint to the constraint set of the problem, another objective is selected and the process is repeated until either all objectives are selected or the solution does not change. An obvious drawback to this approach is that it is sensitive to the order of optimizing the objectives and the value of ϵ.

A fourth group of methods for solving MOPs generate a single specific Pareto-optimal solution. Compromise programming and Benson's method are representatives of this class. A fifth group of methods are called interactive methods, where an interactive session is established with the decision maker to find satisfactory solutions.

The last group of methods is evolutionary algorithms (EAs) [17, 19]. EAs offered something different from all other methods. Being population based, they are able to generate a set of near-Pareto-optimal solutions in a single run. In addition, they do not require assumptions of convexity, differentiability, and/or continuity as traditional optimization problems do. EAs with local search are usually used to improve the performance of EAs to get closer to the actual optimal or, the Pareto-optimal set in case of MOPs.

In evolutionary artificial neural network (EANN), an evolutionary algorithm is used for training the ANN. A major advantage to the evolutionary approach over BP alone is the ability to escape a local optimum. The major disadvantage of the EANN approach is that it is computationally expensive, as the evolutionary approach is normally slow.

Recently, the problem of simultaneous optimization of the network architecture and the corresponding training error has been casted as a multiobjective optimization problem [6]. It was found that by combining BP with an EMO algorithm, a considerable reduction in the computational cost can be achieved [6]. The multiobjective optimization outputs a set of solutions, where no solution in the set is better than all others when compared on all objectives. The Pareto-optimal set provided a new insight into the learning problem, where an obvious question that emerged from the study is how to utilize the information embedded in the set as a whole. One answer to this question is to form an ensemble, which is the focus of this chapter.

18.2.1 Evolutionary Artificial Neural Networks

The following notations will be used for a single hidden layer ANN:

- I and H are the number of input and hidden units respectively.
- $\mathbf{X}^p \in \mathbf{X} = (x_1^p, x_2^p, \ldots, x_I^p), p = 1, \ldots P$, is the pth pattern in the input feature space $\mathbf{X}$ of dimension I, and P is the total number of patterns.
- Without any loss of generality, $\mathbf{Y}_o^p \in \mathbf{Y}_o$ is the corresponding scalar of pattern $\mathbf{X}^p$ in the hypothesis space $\mathbf{Y}_o$.
- w_{ih} and w_{ho}, are the weights connecting input unit i, $i = 1 \ldots I$, to hidden unit h, $h = 1 \ldots H$, and hidden unit h to the output unit o respectively. The number of outputs is assumed to be 1 in this chapter.
- $\Theta_h(\mathbf{X}^p) = \sigma(a_h)$; $a_h = \sum_{i=1}^{I} w_{ih} x_i^p$, $h = 1 \ldots H$, is the hth hidden unit's output corresponding to the input pattern $\mathbf{X}^p$, where a_h is the activation of hidden unit h, and $\sigma(.)$ is the activation function taken in this chapter to be the logistic function $\sigma(z) = \frac{1}{1+e^{-Dz}}$, with D the function's sharpness or steepness and is taken to be 1.
- $\hat{Y}_o^p = \sigma(a_o)$; $a_o = \sum_{h=1}^{H} w_{ho} \Theta_h(\mathbf{X}^p)$ is the network's output and a_o is the activation of output unit o corresponding to the input pattern $\mathbf{X}^p$.

In this chapter, we use the quadratic error function by squaring the difference between the predicted and actual output. We will also make use of the traditional BP algorithm as a local search method. For a complete description of BP, see for example [23].

18.2.2 Negative Correlation Learning

We have chosen to compare our work against the *negative correlation learning* (NCL) method of [39] because the latter can be seen as one of the recent work which uses an evolutionary approach to evolve an ensemble of neural networks and has also been used successfully for feature selection tasks [11].

Negative Correlation Learning was proposed by Liu and Yao [38] for training an ensemble of ANNs using Backpropagation [37] then it was combined with an evolutionary approach [39]. The ensemble is trained using the training set, where the output of the ensemble F^p is given by the following equation:

$$F^p = \frac{1}{M} \sum_{m=1}^{M} \hat{Y}^p(m) \tag{18.3}$$

The expected error of pattern p is given by the average error of the ensemble, $Error^p$, as defined by the following two Equations:

$$Error^p = \frac{1}{P} \sum_{p=1}^{P} Error^p(m) \tag{18.4}$$

$$Error^p(m) = \frac{1}{P} \sum_{p=1}^{P} \frac{1}{2} (\hat{Y}^p(m) - Y^p)^2 + \frac{1}{P} \sum_{p=1}^{P} \lambda \Phi^p(m) \tag{18.5}$$

Here, $\Phi^p(m)$ is the penalty function of network m and pattern p. This function represents the correlation term that we need to minimize. In NCL, the following is used as the anticorrelation measure:

$$\Phi^p(m) = (\hat{Y}^p(m) - F^p) \sum_{l \neq m} (\hat{Y}^p(l) - F^p) \tag{18.6}$$

The previous function has some desirable characteristics, as when it is combined with the mean square error, the result will be a nice tradeoff between the bias, variance and co-variance. Moreover, it does not change the BP algorithm much as it only adds a simple term to the derivative of the error as follows:

$$\frac{\partial Error^p(m)}{\partial \hat{Y}^p(m)} = (\hat{Y}^p(m) - Y^p) - \lambda(\hat{Y}^p(m) - F^p) \tag{18.7}$$

18.3 Formation of Neuro-ensembles

18.3.1 The Multiobjective Learning Problem

In casting the learning problem as a multiobjective problem, a number of possibilities appear to be useful. A successfully-tested possibility that was presented in [6] is to formulate the problem as follows:

Prob1

$$\textbf{Minimize } f_1 = \Psi \tag{18.8}$$

$$\textbf{Minimize } f_2 = \sum_o \sum_p (\hat{Y}_o^p - Y_o^p)^2 \tag{18.9}$$

$$\textbf{Subject to } \Theta_h(\mathbf{X}^p) = \sigma(a_h); a_h = \sum_{i=1}^{I} w_{ih} x_i^p, \quad h = 1 \dots H \tag{18.10}$$

$$\hat{Y}_o^p = \sigma(a_o); a_o = \sum_{h=1}^{H} w_{ho} \Theta_h(\mathbf{X}^p) \tag{18.11}$$

where Ψ is a measure for the architecture's complexity. This measure can be a counting of the number of synapses in the network, the number of hidden units, or a combination. We may note that the VC-dimension, which is a measure for the network capacity, for a feedforward neural network with sigmoid activation function is $O(\Gamma^2)$; Γ is the number of free parameters in the network [23]. By minimizing the number of connections or nodes, we control the VC-dimension.

Another alternative that we will use in this chapter is to divide the training set into two subsets using stratified sampling methods. Let $p1$ and $p2$ be the number of patterns/instances in the two subsets $Sub1$ and $Sub2$ respectively. The learning problem can be formulated as

Prob2

$$\textbf{Minimize } f_3 = \sum_o \sum_{p1} (\hat{Y}_o^{p1} - Y_o^{p1})^2 \tag{18.12}$$

$$\textbf{Minimize } f_4 = \sum_o \sum_{p2} (\hat{Y}_o^{p2} - Y_o^{p2})^2 \tag{18.13}$$

$$\textbf{Subject to } \Theta_h(\mathbf{X}^p) = \sigma(a_h); a_h = \sum_{i=1}^{I} w_{ih} x_i^p, \quad h = 1 \dots H, \quad p \in p1 \cup p2 \tag{18.14}$$

$$\hat{Y}_o^p = \sigma(a_o); a_o = \sum_{h=1}^{H} w_{ho} \Theta_h(\mathbf{X}^p), \quad p \in p1 \cup p2 \tag{18.15}$$

There are advantages of the formulation presented by **Prob2** over the one presented by **Prob1** [6]. To demonstrate these advantages, let us assume that we are able to generate the actual Pareto-optimal frontier for both Prob1 and Prob2. The Pareto-optimal frontier for **Prob1** will contain by definition very inferior networks in terms of training error. That is because a very small network will have an advantage in terms of f_1; therefore the frontier will contain networks which have VC-dimension less than that required to capture the underlying function in the training set. It becomes a problem in defining which network should be included in the ensemble since some of these networks are very inferior and will introduce a lot of noise in the ensemble. It is also difficult to draw a line between good and bad networks on the frontier. Apparently, one can leave it to another algorithm to decide on which network to include in the ensemble but this turns to be not simpler than training a single network that generalizes well. The formulation presented by **Prob2**, however, does not suffer from this problem. The worst training error that a network will have on $Sub1$, for example, is paired with the best training error the network will have on $Sub2$ by virtue of the Pareto-optimal definition. This entails, as we will see in the next subsection, that the range of networks on the actual Pareto-optimal frontier is bounded from right and left with the bias-variance trade-off of $Sub1$ and $Sub2$ respectively.

The third alternative is to add random noise to the training set to form a second objective

Prob3

$$\textbf{Minimize } f_3 = \sum_o \sum_p (\hat{Y}_o^p - Y_o^p)^2 \tag{18.16}$$

$$\textbf{Minimize } f_4 = \sum_o \sum_p (\hat{Y}_o^p - Y_o^p)^2 + N(0, 0.2) \tag{18.17}$$

$$\textbf{Subject to } \Theta_h(\mathbf{X}^p) = \sigma(a_h); a_h = \sum_{i=1}^{I} w_{ih} x_i^p, \tag{18.18}$$

$$h = 1 \ldots H, \quad p \in p1 \cup p2$$

$$\hat{Y}_o^p = \sigma(a_o); a_o = \sum_{h=1}^{H} w_{ho} \Theta_h(\mathbf{X}^p) \tag{18.19}$$

18.3.2 Bias-Variance Tradeoff

To understand the significance of the Pareto-optimal concept, let us assume that the actual Pareto-optimal frontier $Pareto - optimal_{actual}$ in a problem is composed of the following networks $\Omega_1, \ldots, \Omega_i, \ldots, \Omega_m$, where Ω_i is the ith network in a Pareto-optimal set of m networks. Let us assume that these networks are in an increasing order from left to right in terms of their performance on the first objective; that is, Ω_{i+1} has higher prediction accuracy on $Sub1$ than Ω_i, therefore the former has lower prediction accuracy on $Sub2$ than the latter by virtue of the Pareto-optimal definition.

Proposition 1. *If* $Pareto\!-\!optimal_{actual}$ *is the actual Pareto-optimal frontier for Prob2, then all networks on the frontier are a subset of the bias/variance trade-off set for Prob2.*

Let us define $error(\Omega_i, Sub1)$, $bias(\Omega_i, Sub1)$ and $variance(\Omega_i, Sub1)$ to be the training error, bias and variance of Ω_i corresponding to $Sub1$ respectively. The same can be done for $Sub2$. If $Pareto-optimal_{actual}$ is the actual Pareto-optimal frontier for this problem, then $error(\Omega_m, Sub1) < error(\Omega_{m-1}, Sub1)$ and $error(\Omega_m, Sub2) > error(\Omega_{m-1}, Sub2)$. Let us assume that the network's architecture is chosen large enough to have a zero bias for both $Sub1$ and $Sub2$. Then $variance(\Omega_m, Sub1) < variance(\Omega_{m-1}, Sub1)$ and $variance(\Omega_{m-1}, Sub2) < variance(\Omega_m, Sub2)$. Note that the variance represents the inadequacy of the information contained in $Sub1$ and $Sub2$ about the actual function from which $Sub1$ and $Sub2$ were sampled. Therefore, the Pareto-optimal frontier for $Prob2$ is also a bias/variance trade-off.

18.3.3 Assumptions

In $Prob2$, there is no guarantee that the ensemble size would be more than 1. Take the case where a network is large enough to over-fit on both $Sub1$ and $Sub2$, where the training error on each of them is zero. In this case, the Pareto-optimal set would have a single solution. It is therefore our first assumption that we choose a network size small enough not to over-fit the union of $Sub1$ and $Sub2$, while it is large enough to learn each of them independently. In $Prob3$, it is unlikely that the ensemble size will be 1 since it is impossible to find a network which satisfies both objectives and achieve error on both equal to 0.

Our second assumption relates to the algorithm. Getting to the actual Pareto-optimal frontier is always a challenge for any algorithm. We do not assume that we have to get to the actual Pareto-optimal frontier for this method to work. What we assume instead is that there is no guarantee that the Pareto-optimal set returned by our method is the optimal ensemble as being suggested by this chapter.

18.3.4 Formation of the Ensemble

The ensemble is formed from all networks on the Pareto-optimal frontier found by the evolutionary method. We have used three methods for forming the ensemble's gate. In voting, the predicted class is the one where the majority of networks agreed. In winner-take-all, the network with the largest activation in the ensemble is used for prediction. In simple averaging, the activations for all networks in the ensemble are averaged and the class is determined using this average.

18.3.5 The MPANN Algorithm

Recently, we developed the *Pareto-optimal differential evolution* (PDE) method [8]. The algorithm is an adaptation of the original *differential evolution* (DE) introduced by Storn and Price [51] for optimization problems over continuous domains. There is a large number of *evolutionary multiobjective optimization* (EMO) algorithms in the literature [17, 19], but we selected PDE in our experiments because it improved on twelve evolutionary algorithms for solving MOPs. These algorithms were FFGA [21], HLGA [22], NPGA [24], NSGA [50], RAND [68], SOEA [68], SPEA [68], VEGA [47], PAES, PAESgray, PAES98gray, PAES98 and PAES98mut3p [29, 30]. The reader may wish to consult [3, 8]. PDE generally improves over traditional differential evolution algorithms because of the use of Gaussian distribution, which spreads the children around the main parent in both directions of the other two supporting parents. Since the main parent is a non-dominated solution in the current population, it is found [3] that the Gaussian distribution helps to generate more solutions on the Pareto-optimal-front; which is a desirable characteristic in EMO. PDE was also tested successfully for evolving neural networks, where it is called *Memetic Pareto-optimal Artificial Neural Network* (MPANN) algorithm [1, 2, 6].

1. Create a random initial population of potential solutions. The elements of the weight matrix Ω are assigned uniformally distributed random values $U(0,1)$.
2. Apply BP to all individuals in the population.
3. Repeat
   a) Evaluate the individuals in the population and label the non-dominated ones.
   b) If the number of non-dominated individuals is less than 3 repeat the following until the number of non-dominated individuals is greater than or equal to 3 so that we have the minimum number of parents needed for crossover:
      i. Find a non-dominated solution among those who are not marked.
      ii. Mark the solution as non-dominated.
   c) Delete all dominated solutions from the population.
   d) Repeat
      i. Select at random an individual as the main parent α_1, and two individuals, α_2, α_3 as supporting parents.
      ii. **Crossover:** For all weights do
      $$\omega_{ih}^{child} \leftarrow \omega_{ih}^{\alpha_1} + N(0,1)(\omega_{ih}^{\alpha_2} - \omega_{ih}^{\alpha_3}) \tag{18.20}$$
      $$\omega_{ho}^{child} \leftarrow \omega_{ho}^{\alpha_1} + N(0,1)(\omega_{ho}^{\alpha_2} - \omega_{ho}^{\alpha_3}) \tag{18.21}$$
      iii. Apply BP to the child then add the child to the population.
   e) Until the population size is M
4. Until termination conditions are satisfied.

18.4 Experiments

18.4.1 Experimental Setup

Similar to [39], we have tested MPANN on two benchmark problems; the Australian credit card assessment problem and the diabetes problem, available by anonymous ftp from ice.uci.edu [10].

The Australian credit card assessment dataset contains 690 patterns with 14 attributes; 6 numeric and 8 discrete (with 2 to 14 possible values). The predicted class is binary - 1 for awarding the credit and 0 for not. To be consistent with the literature [42], the dataset is divided into 10 folds where class distribution is maintained in each fold. Cross–validation is used where we run the algorithm with 9 out of the 10 folds for each data set, then we test with the remaining one. Similar to [39], the number of generations is 200, the population size 25, the learning rate for BP 0.003, the number of hidden units is set to 5, and the number of epochs BP was applied to an individual is set to 5.

The diabetes dataset has 768 patterns; 500 belonging to the first class and 268 to the second. It contains 8 attributes. The classification problem is difficult as the class value is a binarized form of another attribute that is highly indicative of a certain type of diabetes without having a one-to-one correspondence with the medical condition of being diabetic [42]. To be consistent with the literature [42], the dataset was divided into 12 folds where class distribution is maintained in each fold. Cross–validation is used where we run the algorithm with 11 out of the 12 folds for each dataset and then we test with the remaining one. Similar to [39], the number of generations is 200, the population size 25, the learning rate for BP 0.003, the number of hidden units is set to 5, and the number of epochs BP was applied to an individual is set to 5.

The results of this experiment are presented in Table 18.1. It is interesting to see that the training error is slightly worse than the test error for *Prob*2. In a normal situation, this might be interpreted as under-training. The situation here is very different indeed. Let us first explain how this training error is calculated. Recall that we have divided the training set into two subsets using stratified samples. By virtue of the Pareto-optimal definition, the Pareto-optimal frontier would have networks which are over-fitting on the first subset but not on the second and other networks which are over-fitting on the second subset but not on the first. However, the sum of the bias of a machine trained on the first subset and another machine trained on the second subset would be less than a machine trained on the union of the two subsets. Therefore, in a conventional situation, where the whole training set is used as a single training set, a biased machine would usually perform better on the training and worse on the test. In *Prob*2, this is unlikely to occur since the bias and the variance are averaged over the entire Pareto-optimal-set. For *Prob*3, however, it is clear that the performance on the training is better than on testing as would be expected since the whole training set is used for training the networks.

Table 18.1. Accuracy rates of MPANN2 for the Australian Credit Card and the diabetes datasets.

Australian credit card dataset				
	Prob2		Prob3	
Simple averaging				
	Training	Testing	Training	Testing
Mean	0.849	0.865	0.850	0.844
SD	0.019	0.043	0.017	0.058
Min	0.803	0.797	0.809	0.724
Max	0.877	0.928	0.869	0.913
Majority voting				
	Training	Testing	Training	Testing
Mean	0.854	0.862	0.852	0.844
SD	0.018	0.049	0.015	0.056
Min	0.812	0.783	0.821	0.724
Max	0.879	0.928	0.871	0.913
Winner-rakes-all				
	Training	Testing	Training	Testing
Mean	0.847	0.858	0.834	0.824
SD	0.018	0.044	0.027	0.053
Min	0.805	0.797	0.768	0.724
Max	0.876	0.913	0.866	0.9
Diabetes dataset				
	Prob2		Prob3	
Simple averaging				
	Training	Testing	Training	Testing
Mean	0.769	0.777	0.753	0.744
SD	0.023	0.032	0.019	0.034
Min	0.737	0.723	0.738	0.690
Max	0.808	0.84	0.795	0.806
Majority voting				
	Training	Testing	Training	Testing
Mean	0.771	0.779	0.755	0.744
SD	0.023	0.033	0.019	0.034
Min	0.737	0.723	0.738	0.690
Max	0.81	0.84	0.795	0.806
Winner-rakes-all				
	Training	Testing	Training	Testing
Mean	0.767	0.777	0.753	0.746
SD	0.022	0.037	0.020	0.032
Min	0.737	0.723	0.738	0.690
Max	0.807	0.84	0.795	0.802

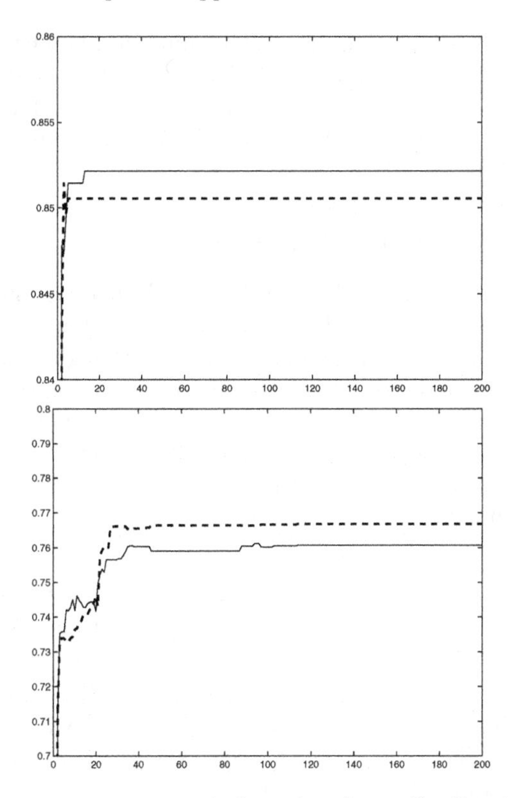

Fig. 18.3. The average accuracy rate for networks on the Pareto-optimal frontier found in each evolutionary generation. Upper figure represents the Australian credit card dataset while the bottom figure represents the diabetes dataset. The x-axis is the generation number and the y-axis is the accuracy rate. The dotted line represents the first training subset while the dashed line represents the second subset.

We can also observe in Table 18.1 that the standard deviation on the training set is always smaller than on the test. This should be expected given that the networks are trained on two subsets which constitute the entire training set for *Prob*2 or on the entire training set as in *Prob*3.

Figure 18.3 shows the accuracy achieved over each of the two training subsets averaged over all networks on the Pareto-optimal frontier found in each evolutionary generation for one of the ten runs. For the Australian credit card dataset, it is clear that convergence occurred early from generation 20 while for the diabetes dataset, convergence occurred from generation 90 and onwards. It is also interesting to see that the accuracy on both subsets for both datasets does not intersect, which may imply that one subset is usually difficult to learn than the other. This implies that the Pareto-optimal set is not empty. On the average, there was 4 networks on the Pareto-optimal front for the Australian credit card dataset and 6 networks on the Pareto-optimal front for the diabetes dataset. The minimum number of networks on the Pareto-optimal front was 3 for both datasets and the maximum was 5 for the Australian credit cards and 11 for the diabetes. The final ensemble was formed from 4 and 6 networks for the Australian credit card and the diabetes datasets respectively.

18.4.2 Comparisons with Other Work

We compare our results against Backprob[42] and Liu et.al. [39]. In Table 18.2, we find that the Pareto-optimal-based ensemble is better than BP and equivalent to Liu et.al. with smaller ensemble size. Although it is desirable in a new method to perform better than other methods on some dataset, the main contribution from our perspective here is that the decision on which network to include in, or exclude from the ensemble is determined automatically by the definition of the Pareto-optimal frontier. Also, the ensemble's size is automatically determined.

Table 18.2. Comparing MPANN against other work for the Australian credit card and diabetes data sets. The three rates for MPANN and Liu et. al. represent the performance using simple average, vote, winner-takes-all strategies, respectively.

	Accuracy Rate on Test Set	
Algorithm	Australian	Diabetes
This chapter (Prob2)	0.865,0.862,0.858	0.777,0.779,0.777
Backprob[42]	0.846	0.749
Liu et.al. [39]	0.855,0.857,0.865	0.766,0.764,0.779

18.5 Conclusion

In this chapter, we cast the problem of training artificial neural networks as a multiobjective optimization problem and use the resultant Pareto-optimal frontier to form an ensemble. The method provides a theoretically-sound approach for the formation of neuro-ensembles while answering three main questions in the neural network ensemble regarding the criteria for including a network in, or excluding it from, the ensemble, the Pareto-optimal set defines the size of the ensemble, and the Pareto-optimal concept ensures that the networks in the ensemble are different.

References

[1] H.A. Abbass. A memetic pareto evolutionary approach to artificial neural networks. In M. Stumptner, D. Corbett, and M. Brooks, editors, *Proceedings*

of the 14th Australian Joint Conference on Artificial Intelligence (AI'01), pages 1–12, Berlin, 2001. Springer-Verlag.

[2] H.A. Abbass. An evolutionary artificial neural network approach for breast cancer diagnosis. *Artificial Intelligence in Medicine*, 25(3):265–281, 2002.

[3] H.A. Abbass. The self-adaptive pareto differential evolution algorithm. In *Proceedings of the IEEE Congress on Evolutionary Computation (CEC2002)*, volume 1, pages 831–836, Piscataway, NJ, 2002. IEEE Press.

[4] H.A. Abbass. Pareto neuro-ensemble. In *Proceedings of the 16th Australian Joint Conference on Artificial Intelligence (AI'03)*, pages 554–566, Berlin, 2003. Springer-Verlag.

[5] H.A. Abbass. Pareto neuro-evolution: Constructing ensemble of neural networks using multi-objective optimization. In *Proceedings of the The IEEE Congress on Evolutionary Computation (CEC2003)*, volume 3, pages 2074–2080. IEEE-Press, 2003.

[6] H.A. Abbass. Speeding up back-propagation using multiobjective evolutionary algorithms. *Neural Computation*, 15(11):2705–2726, 2003.

[7] H.A. Abbass and K. Deb. Searching under multi-evolutionary pressures. In C.M. Fonseca, P.J. Fleming, E. Zitzler, K. Deb, and L. Thiele, editors, *Proceedings of the Second International Conference on Evolutionary Multi-Criterion Optimization (EMO 2003)*, number 2632 in Lecture Notes in Computer Science (LNCS), pages 391–404. Springer-Verlag, Berlin, 2003.

[8] H.A. Abbass, R.A. Sarker, and C.S. Newton. PDE: A pareto-frontier differential evolution approach for multi-objective optimization problems. In *Proceedings of the IEEE Congress on Evolutionary Computation (CEC2001)*, volume 2, pages 971–978, Piscataway, NJ, 2001. IEEE Press.

[9] R.E. Banfield, L.O. Hall, K.W. Bowyer, and W.P. Kegelmeyer. A new ensemble diversity measure applied to thinning ensembles. In T. Windeatt and F. Roli, editors, *Proceedings of the 4th International Workshop on Multiple Classifier Systems*, Lecture Notes in Computer Science, pages 306–316, Guilford, UK, June 2003. Springer-Verlag.

[10] C.L. Blake and C.J. Merz. UCI repository of machine learning databases, http://www.ics.uci.edu/~mlearn/mlrepository.html. *University of California, Irvine, Dept. of Information and Computer Sciences*, 1998.

[11] G. Brown, X. Yao, J. Wyatt, H. Wersing, and B. Sendhoff. Exploiting ensemble diversity for automatic feature extraction. In L. Wang, J.C. Rajapakse, K. Fukushima, S.Y. Lee, and X. Yao, editors, *Proceedings of the 9th International Conference on Neural Information Processing (ICONIP'02)*, pages 1786–1790, 2002.

[12] A. Chandra and X. Yao. DIVACE: Diverse and accurate ensemble learning algorithm. In *Proc. of the Fifth International Conference on Intelligent Data Engineering and Automated Learning (IDEAL'04)*, pages 619–625. Lecture Notes in Computer Science 3177, Springer, 2004.

[13] A. Chandra and X. Yao. Ensemble learning using multi-objective evolutionary algorithms. *Mathematical Modelling and Algorithms*, to appear, 2005.

[14] A. Chandra and X. Yao. Evolutionary framework for the construction of diverse hybrid ensembles. In *Proc. of the 13th European Symposium on Artificial Neural Networks (ESANN'05)*, pages 253–258, 2005.

[15] S.-B. Cho and J.-H. Ahn. Speciated neural networks evolved with fitness sharing technique. In *Proceedings of the 2001 Congress on Evolutionary Computation*, volume 1, pages 390–396, 2001.
[16] S.-B. Cho, J.-H. Ahn, and S.-I. Lee. Exploiting diversity of neural ensembles with speciated evolution. In *Proceedings. IJCNN '01. International Joint Conference on Neural Networks*, volume 2, pages 808–813, 2001.
[17] C.A. Coello, D.A. Van Veldhuizen, and G.B. Lamont. *Evolutionary Algorithms for Solving Multi-Objective Problems.* Kluwer Academic, New York, 2002.
[18] Padraig Cunningham and John Carney. Diversity versus quality in classification ensembles based on feature selection. In *11th European Conference on Machine Learning*, volume 1810, pages 109–116, Barcelona, Catalonia, Spain, May 31 - June 2 2000. Springer, Berlin.
[19] K. Deb. *Multi-objective optimization using evolutionary algorithms.* John Wiley & Sons,, New York, 2001.
[20] T.G. Dietterich. Ensemble methods in machine learning. *Lecture Notes in Computer Science*, 1857, 2000.
[21] C.M. Fonseca and P.J. Fleming. Genetic algorithms for multiobjective optimization: Formulation, discussion and generalization. *Proceedings of the Fifth International Conference on Genetic Algorithms, San Mateo, California*, pages 416–423, 1993.
[22] P. Hajela and C.Y. Lin. Genetic search strategies in multicriterion optimal design. *Structural Optimization*, 4:99–107, 1992.
[23] S. Haykin. *Neural networks - a comprehensive foundation.* Printice Hall, USA, 2 edition, 1999.
[24] J. Horn, N. Nafpliotis, and D.E. Goldberg. A niched pareto genetic algorithm for multiobjective optimization. *Proceedings of the First IEEE Conference on Evolutionary Computation*, 1:82–87, 1994.
[25] C. Igel. Multi-objective model selection for support vector machines. In *Proc. of the International conference on evolutionary multi-objective optimization*, pages 534–546. LNCS 3410, Springer-Verlag, 2005.
[26] D. Jimenez and N. Walsh. Dynamically weighted ensemble neural networks for classification. In *Proceedings of the 1998 International Joint Conference on Neural Networks*, pages 753–756, 1998.
[27] Y. Jin, T. Okabe, and B. Sendhoff. Neural network regularization and ensembling using multi-objective evolutionary algorithms. In *Proc. of the 2004 Congress on Evolutionary Computation (CEC'2004)*, volume 1, pages 1–8. IEEE-Press, 200.
[28] Y. Jin, B. Sendhoff, and E. Korner. Evolutionary multi-objective optimization for simultaneous generation of signal-type and symbol-type representations. In *Proc. of the International Conference on Evolutionary Multi-Criterion Optimization*, pages 752–766. LNCS 3410, Springer-Verlag, 2005.
[29] J. Knowles and D. Corne. The pareto archived evolution strategy: a new baseline algorithm for multiobjective optimization. *In 1999 Congress on Evolutionary Computation, Washington D.C., IEEE Service Centre*, pages 98–105, 1999.
[30] J. Knowles and D. Corne. Approximating the nondominated front using the pareto archived evolution strategy. *Evolutionary Computation*, 8(2):149–172, 2000.

[31] K. Kottathra and Y. Attikiouzel. A novel multicriteria optimization algorithm for the structure determination of multilayer feedforward neural networks. *Journal of Network and Computer Applications*, 19:135–147, 1996.
[32] A. Krogh and J. Vedelsby. Neural network ensembles, cross validation and active learning. In G. Tesauro, D.S. Touretzky, and T.K. Len, editors, *Advances in Neural Information Processing System 7*, volume 7, pages 231–238. MIT Press, 1995.
[33] Anders Krogh and Jesper Vedelsby. Neural network ensembles, cross validation, and active learning. In G. Tesauro, D.S. Touretzky, and T.K. Leen, editors, *Advances in Neural Information Processing Systems*, volume 7, pages 231–238. MIT Press, 1995.
[34] M.A. Kupinski and M.A. Anastasio. Multiobjective genetic optimization of diagnostic classifiers with implications for generating receiver operating characteristic curves. *IEEE Transactions on Medical Imaging*, 18(8):675–685, 1999.
[35] Kuncheva L.I. That elusive diversity in classifier ensembles. In *Proc IbPRIA 2003, Mallorca, Spain, 2003, Lecture Notes in Computer Science*, pages 1126–1138. Springer-Verlag, 2003.
[36] Y. Liu and X. Yao. Evolving modular neural networks which generalise well. In *Proc. of 1997 IEEE International Conference on Evolutionary Computation*, pages 605–610, Indianapolis, USA, 1997.
[37] Y. Liu and X. Yao. Ensemble learning via negative correlation. *Neural Networks*, 12:1399–1404, 1999.
[38] Y. Liu and X. Yao. Simultaneous training of negatively correlated neural networks in an ensemble. *IEEE Transactions on Systems, Man, and Cybernetics, Part B: Cybernetics*, 29(6):716–725, 1999.
[39] Y. Liu, X. Yao, and T. Higuchi. Evolutionary ensembles with negative correlation learning. *IEEE Trans. Evolutionary Computation*, 4(4):380–387, 2000.
[40] M.H. Luerssen. Graph grammar encoding and evolution of automata networks. In *Proc. Twenty-Eighth Australasian Computer Science Conference*, pages 229–238. Australian Computer Society, 2005.
[41] R. McKay and H.A. Abbass. Anti-correlation: A diversity promoting mechanisms in ensemble learning. *The Australian Journal of Intelligent Information Processing Systems (AJIIPS)*, 7(3/4):139–149, 2001.
[42] D. Michie, D.J. Spiegelhalter, and C.C. Taylor. *Machine learning, neural and statistical classification.* Ellis Horwood, 1994.
[43] S. Park, D. Nam, and C.H. Park. Design of a neural controller using multiobjective optimization for nonminimum phase systems. In *IEEE International on Fuzzy Systems Conference Proceedings*, pages 533–537. IEEE, 1999.
[44] B.E. Rosen. Ensemble learning using decorrelated neural networks. *Connection Science. Special Issue on Combining Artificial Neural: Ensemble Approaches*, 8(3/4):373–384, 1996.
[45] D.E. Rumelhart, G.E. Hinton, and R.J. Williams. Learning internal representations by error propagation. In J.L. McClelland D.E. Rumelhart and the PDP Research Group Eds, editors, *Parallel Distributed Processing: Explorations in the Microstructure of Cognition., Foundations, 1, 318,*. MIT Press Cambridge, 1986.
[46] Y. Sawaragi, H. Nakayama, and T. Tanino. *Theory of multiobjective optimization*, volume 176 of *Mathematics in science and engineering.* Academic Press Inc, Harcourt Brace Jovanovich, 1985.

[47] J.D. Schaffer. Multiple objective optimization with vector evaluated genetic algorithms. *Genetic Algorithms and their Applications: Proceedings of the First International Conference on Genetic Algorithms*, pages 93–100, 1985.
[48] A. Sharkey. *Combining Artificial Neural Nets. Ensemble and Modular Multi-Net Systems.* Springer-Verlag New York, Inc., 1998.
[49] A.J.C. Sharkey. On combining artificial neural nets. *Connection Science*, 8:299–313, 1996.
[50] N. Srinivas and K. Dev. Multiobjective optimization using nondominated sorting in genetic algorithms. *Evolutionary Computation*, 2(3):221–248, 1994.
[51] R. Storn and K. Price. Differential evolution: a simple and efficient adaptive scheme for global optimization over continuous spaces. Technical Report TR-95-012, International Computer Science Institute, Berkeley, 1995.
[52] J. Teo and H.A. Abbass. Coordination and synchronization of locomotion in a virtual robot. In L. Wang, J.C. Rajapakse, K. Fukushima, S.Y. Lee, and X. Yao, editors, *Proceedings of the 9th International Conference on Neural Information Processing (ICONIP'02)*, volume 4, pages 1931–1935, Singapore, 2002. Nanyang Technological University, ISBN 981-04-7525-X.
[53] J. Teo and H.A. Abbass. Multi-objectivity for brain-behavior evolution of a physically-embodied organism. In R. Standish, M.A. Bedau, and H.A. Abbass, editors, *Artificial Life VIII: The 8th International Conference on Artificial Life*, pages 312–318. MIT Press, Cambridge, MA, 2002.
[54] J. Teo and H.A. Abbass. Trading-off mind complexity and locomotion in a physically simulated quadruped. In L. Wang, K.C. Tan, T. Furuhashi, J.H. Kim, and X. Yao, editors, *Proceedings of the 4th Asia-Pacific Conference on Simulated Evolution And Learning (SEAL'02)*, volume 2, pages 776–780, Singapore, 2002. Nanyang Technological University, ISBN 981-04-7523-3.
[55] J. Teo and H.A. Abbass. Elucidating the benefits of a self-adaptive Pareto EMO approach for evolving legged locomotion in artificial creatures. In *Proceedings of the 2003 Congress on Evolutionary Computation (CEC2003)*, volume 2, pages 755–762. IEEE Press, Piscataway, NJ, 2003.
[56] J. Teo and H.A. Abbass. Automatic generation of controllers for embodied legged organisms: A Pareto evolutionary multi-objective approach. *Evolutionary Computation*, 12(3):355–394, 2004.
[57] K. Tumer and J. Ghosh. Error correlation and error reduction in ensemble classifiers. *Connection Science*, 8(3-4):385–403, 1996.
[58] N. Ueda and R. Nakano. Generalization error of ensemble estimators. In *IEEE International Conference on Neural Networks*, volume 1, pages 90–95, 1996.
[59] Y. Wang and F. Wahl. Multi-objective neural network for image reconstruction. *IEE Proceedings - Visions, Image and Signal Processing*, 144(4):233–236, 1997.
[60] Z. Wang, X. Yao, and Y. Xu. An improved constructive neural network ensemble approach to medical diagnoses. In *Proc. of the Fifth International Conference on Intelligent Data Engineering and Automated Learning (IDEAL'04)*, pages 572–577. Lecture Notes in Computer Science 3177, Springer, 2004.
[61] S. WIEGAND, C. IGEL, and U. HANDMANN. Evolutionary multi-objective optimization of neural networks for face detection. *International Journal of Computational Intelligence and Applications*, 4(3):237–253, 2005.
[62] R.B. Xtal. Accuracy bounds for ensembles under 0 - 1 loss. Technical report, Mountain Information Technology Computer Science Department, University of Waikato, New Zealand, June June 24, 2002.

[63] X. Yao. Evolving artificial neural networks. *Proceedings of the IEEE*, 87:1423–1447, 1999.

[64] X. Yao and Y. Liu. Ensemble structure of evolutionary artificial neural networks. In *IEEE International Conference on Evolutionary Computation (ICEC'96)*, pages 659–664, Nagoya, Japan, 1996.

[65] X. Yao and Y. Liu. A new evolutionary system for evolving artificial neural networks. *IEEE Transactions on Neural Networks*, 8:694–713, 1997.

[66] X. Yao and Y. Liu. Making use of population information in evolutionary artificial neural networks. *IEEE Transactions on Systems, Man, and Cybernetics, Part B: Cybernetics*, 28:417–425, 1998.

[67] G. Zenobi and P. Cunningham. Using diversity in preparing ensemble of classifiers based on different subsets to minimize generalization error. In *Lecture Notes in Computer Science*, volume 2167, pages 576–587, 2001.

[68] E. Zitzler and L. Thiele. Multiobjective evolutionary algorithms: A comparative case study and the strength pareto approach. *IEEE Transactions on Evolutionary Computation*, 3(4):257–271, 1999.

19

Trade-Off Between Diversity and Accuracy in Ensemble Generation

Arjun Chandra, Huanhuan Chen and Xin Yao

The Centre of Excellence for Research in Computational Intelligence and Applications (CERCIA)
School of Computer Science, The University of Birmingham
Edgbaston, Birmingham B15 2TT, UK
{A.Chandra,H.Chen,X.Yao}@cs.bham.ac.uk

Summary. Ensembles of learning machines have been formally and empirically shown to outperform (generalise better than) single learners in many cases. Evidence suggests that ensembles generalise better when they constitute members which form a diverse and accurate set. Diversity and accuracy are hence two factors that should be taken care of while designing ensembles in order for them to generalise better. There exists a trade-off between diversity and accuracy. Multi-objective evolutionary algorithms can be employed to tackle this issue to good effect. This chapter includes a brief overview of ensemble learning in general and presents a critique on the utility of multi-objective evolutionary algorithms for their design. Theoretical aspects of a committee of learners viz. the bias-variance-covariance decomposition and ambiguity decomposition are further discussed in order to support the importance of having both diversity and accuracy in ensembles. Some recent work and experimental results, considering classification tasks in particular, based on multi-objective learning of ensembles are then presented as we examine ensemble formation using neural networks and kernel machines.

19.1 Introduction

One of the main issues in machine learning research is that of generalisation. Generalisation refers to the prediction ability of a base learner (or learning machine). The better a predictor performs on unseen input-output mappings, the better it is said to possess the ability to generalise. The 'bias-variance dilemma' is a theoretical result which illustrates the importance of generalisation in neural computing research. This dilemma was explained very well in [31].

An ensemble or a committee of learning machines has been shown to outperform (generalise better than) single learners both theoretically and empirically [13]. Tumer and Ghosh [57] present the formal proof of this. According to Dietterich [24], ensembles form one of the main research directions as far as

A. Chandra et al.: *Trade-Off Between Diversity and Accuracy in Ensemble Generation*, Studies in Computational Intelligence (SCI) **16**, 429–464 (2006)
www.springerlink.com

machine learning research is concerned. Brown [13] gives an extensive survey of ensemble methods. A theoretical account of why ensembles perform better than single learners is also presented here [13].

Although ensembles perform better than their members, constructing them is not an easy task. As Dietterich [24] points out, the key to successful ensemble methods is to construct base models (individual predictors) which perform better than random guessing individually and which are at least somewhat uncorrelated as far as making errors on the training set is concerned. The statement essentially tells that in order for an ensemble to work properly it should have a diverse and accurate set of predictors (also mentioned in one of the seminal works on diversity in classifier ensembles by Hansen and Salamon [33]). Krogh and Vedelsby [41] formally show that an ideal ensemble is one that consists of highly correct (accurate) predictors which at the same time disagree as much as possible (i.e. substantial diversity amongst members is exhibited). This has also been tested and empirically verified as referred to in [51] and shown in [52]. Thus, diversity and accuracy are two key issues that should be taken care of when constructing ensembles. There exists a trade-off between these two terms as will be shown later on and is also mentioned in [60].

Given that ensembles generalise better as compared to a single predictor, ensemble research has become an active research area and has seen an influx of researchers coming up with myriad of algorithms trying to improve the prediction ability of such aggregate systems in recent years. Brown [13] gives a taxonomy of methods for creating diversity. Yates and Partridge [66] show that there exists a hierarchical relationship between the ways in which diversity has and can be enforced while creating ensembles with each method having its own diversity generating potential. Additionally, Dietterich [24] states that one of the promising and open research areas is that of combining ensemble learning methods to give rise to new ensemble learning algorithms. Moreover, incorporating evolution in segments of machine learning has been a widely studied area (e.g. evolving neural networks [9, 56, 64], evolving neural network ensembles [47]) with the main thrust here being that evolution has been used as another fundamental form of adaptation in addition to learning [64] making neural systems adapt to a dynamic environment effectively and efficiently [64].

A more recent approach to ensemble learning views learning as a multi-objective optimisation problem [3]. Recently, we proposed an algorithm called DIVACE (DIVerse and Accurate Ensemble Learning Algorithm) [17, 20] which uses good ideas from Negative Correlation Learning (NCL)[45] and the Memetic Pareto Artificial Neural Network (MPANN)[1, 3] algorithm, and formulates the ensemble learning problem as a multi-objective problem explicitly within an evolutionary setup aiming at finding a good trade-off between diversity and accuracy. One very strong motivation for the use of evolutionary multi-criterion optimisation in the creation of an ensemble in both DIVACE[17, 20] and MPANN [3] is that due to the presence of multiple conflicting objectives, the evolutionary approach engenders a *set* of near optimal solutions. The presence of more than one optimal solution indicates that if one uses multi-objectivity while creating ensembles, one can actually generate an ensemble automatically where the member networks would inadvertently be near optimal [15, 17].

We recently attempted to integrate the aforementioned ideas into a co-evolutionary framework [18] with a view to synthesising new evolutionary ensemble learning algorithms stressing on the fact that multi-objective evolutionary optimisation is a formidable ensemble construction technique. DIVACE, as an idea, gives us a perfect base on which to develop a framework wherein various ensemble methods can be combined and diversity enforcement can be tackled at multiple levels of abstraction. The evolutionary framework gives a simple yet effective means for the development of new ensemble learning algorithms.

This chapter considers an overview of ensemble methods as a means to achieve good generalisation. Theoretical aspects of a committee of learners, in particular, the bias-variance-covariance decomposition of the generalisation error and ambiguity decomposition are discussed in order to support the importance of having both diversity and accuracy and explaining the trade-off between them in the construction of such aggregate learners. A theoretical extension of NCL[45] to accommodate classification tasks is also presented. Multi-objective evolutionary ensemble learning as a solution to achieving a good trade-off and its application to such tasks, describing recent work alluded to above, is further considered.

19.2 Ensembles Tackling the Bias-Variance Dilemma

19.2.1 Bias-Variance Dilemma and Generalisation

We first present the all pervasive notion of the 'bias-variance' dilemma that is one of the most intriguing and much researched on issues in neural computation literature. In order for a learned machine/predictor to exhibit good input-output mappings for unseen data, the predictor should 'know' the exact mapping function it is trying to model. In practise, however, this is not possible, i.e. a predictor cannot learn the exact function it wants to learn mainly due to the dearth in the amount of data present for it to train on and due to noise in this data.

Given that a learning algorithm is presented with only a subset of the training patterns from the infinite set to generate a base model (if we are to model a curve then it is fairly obvious that a curve has infinite number of points and what is presented to a learning algorithm is just a subset of these points), a key point that must always be taken into account is that this algorithm should be kept from memorising the exact representation of this data or else it would simply act like an archive of this data and would not be able to store the information present in it. The idea behind training is simply to inculcate a statistical model (a statistical oracle) within a base model manifested in the settings of its parameters and its complexity. In most training algorithms, there exists a bound on the amount of training necessary beyond which the base model generated would tend towards becoming more and more inflexible in that it would try to 'memorise' the exact representation of the data, thus making it more susceptible to giving a wrong input-output mapping for unseen inputs. There also exists a lower bound on the amount of training necessary as below this the system (base model or predictor) would be more or less similar to what would happen if we consider random guessing. As stated by Bishop in [7] "the goal of network training is not to learn an exact representation of the training data itself, but rather to build a statistical model of the process which generates the data".

This statistical model can be called the network function. *The goal is to make the network function as fitting as possible to the training data and at the same time let it remain smooth enough such that if the predictor is trained using another related data set and fitted on it, the network function does not become significantly different.* This statement exposes what is expressed by two much touted terms in neural computation research: *bias* and *variance*. Bias and variance are conflicting terms as far as achieving a good fit to the training data is concerned and there exists a trade-off between them [7], in that, if we try to reduce the bias, the variance increases and vice-versa. Geman et al. [31] explain this trade-off very well and show that the expected error function can be decomposed into *bias* and *variance* components. Let the actual mapping function be $f(x)$ and the desired function be $\Phi(x)$ which actually is the conditional average $\langle\phi|x\rangle$, ϕ being the mean of the distribution of the output for a given x. Consider the mean square error function,

$$(f(x) - \langle\phi|x\rangle)^2 \text{ or } (f(x) - \Phi(x))^2 , \qquad (19.1)$$

the value of which depends on the data set used for training. The expectation operator $E\{.\}$ is used to get rid of this dependence. The use of this operator gives us an error function (Equation 19.2) which is independent of the initial conditions for a predictor as well as the choice of the training set:

$$E\{(f(x) - \Phi(x))^2\} \qquad (19.2)$$

The bias-variance decomposition just alluded to (and according to [31]), can be done as follows:

$$\begin{aligned}E\{(f(x)-\Phi(x))^2\} &= E\{(f(x)-E\{f(x)\}+E\{f(x)\}-\Phi(x))^2\} \\ &= E\{[(f(x)-E\{f(x)\})+(E\{f(x)\}-\Phi(x))]^2\} \\ &= E\{(f(x)-E\{f(x)\})^2+(E\{f(x)\}-\Phi(x))^2+ \\ &\qquad 2\,(f(x)-E\{f(x)\})\,(E\{f(x)\}-\Phi(x))\} \\ &= E\{(f(x)-E\{f(x)\})^2+(E\{f(x)\}-\Phi(x))^2\} \\ &= E\{(f(x)-E\{f(x)\})^2\}+(E\{f(x)\}-\Phi(x))^2 \\ &= Variance + Bias^2 \qquad (19.3)\end{aligned}$$

Let us say we have a training set D and there are two curves (polynomial functions) expressing the functional mapping this data set represents. If we consider two extrema for these polynomial functions viz. a straight line/constant linear function and a function that fits the points perfectly, we can come up with very lucid definitions for these two terms. Alternatively, we can call the constant function as our simple model, i.e. a simple network function/mapping, and the perfectly fitting function can be called our complex model. The more complex a model, the more configurable parameters it has [12].

For a clear description of what bias and variance really are, let us consider Figure 19.1. The data set D can be seen as dots in the figure. The figure also shows 3 curves: desired function, simple model and complex model.

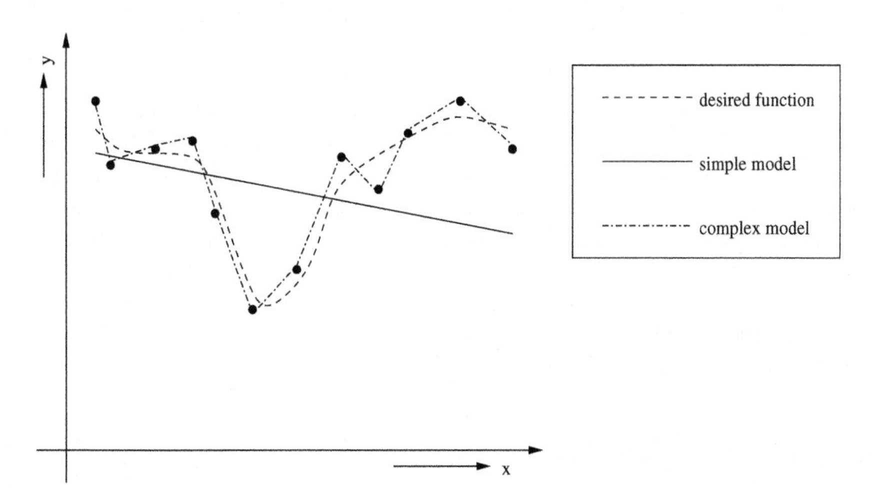

Fig. 19.1. The role played by bias and variance with respect to the network function complexity.

Bias

In simple terms, bias signifies the network function being different (on an average) from the desired mapping. It is the extent to which the average of the network function (over all data sets) differs from the desired function [7]. From equation 19.3,

$$Bias^2 = (E\{f(x)\} - \Phi(x))^2 \tag{19.4}$$

With the simple model, the network function $f(x)$ is independent of the data set D and differs from the desired function $\Phi(x)$ mainly because D was not considered at all. Consequently, the simple model can be thought of as a random function and it has a high bias.

With the complex model, the bias term becomes very small. This happens because from equation 19.4,

$$E\{f(x)\} = complex\ overfitted\ function \approx \langle \phi | x \rangle \text{ or } \Phi(x) \tag{19.5}$$

Hence obviously,

$$(\Phi(x) - \Phi(x))^2 = 0 \tag{19.6}$$

Variance

Variance can also be termed as the *sensitivity* of a network to the training set. It expresses the notion of the extent to which the network function is sensitive to a particular choice of the training set [7]. Here again, from equation 19.3 we have

$$Variance = E\{(f(x) - E\{f(x)\})^2\}. \tag{19.7}$$

Equation 19.7 describes how different the network function is from its expected value over all possible weight initialisations and choice of training sets. A point to emphasise here is that, $E\{f(x)\}$ depends on the choice of the training set which means that if our network function $f(x)$ is independent of the training set, the variance term for such a model is zero. However, for a network function that fits perfectly on the training data, this term plays the main role as far as the mean square error is concerned (in which case the bias term is zero).

The simple model, being independent of the training set D, will exhibit *zero* variance. Here,

$$E\{f(x)\} = constant\ linear\ function = f(x).$$

So, putting this value in equation 19.7 we have

$$E\{f(x) - f(x)\} = 0. \tag{19.8}$$

For the complex model, from equations 19.5 and 19.7,

$$E\{(f(x) - E\{f(x)\})^2\} = E\{(f(x) - \Phi(x))^2\}. \tag{19.9}$$

Since $f(x)$ and $\Phi(x)$ are not equal, due mainly to the fact that $f(x)$ is largely dependent on the choice of the training set D and always overfits it, we could imagine the variance as being quite large.

The Dilemma

The dilemma here is that, if we try to use a simple model, we get a large bias and small variance. If we use a complex model, we get a small bias, but a large variance. All in all, the average generalisation error remains large. We need to find an optimal balance between bias and variance in order to reduce this error and make it as small as possible. Obviously, the solution here is to choose a model which is neither too simple, nor too complex, i.e. a model that exhibits an optimum trade-off between bias and variance. Figure 19.2 shows this situation very clearly. The simplest strategy to achieve a good trade-off

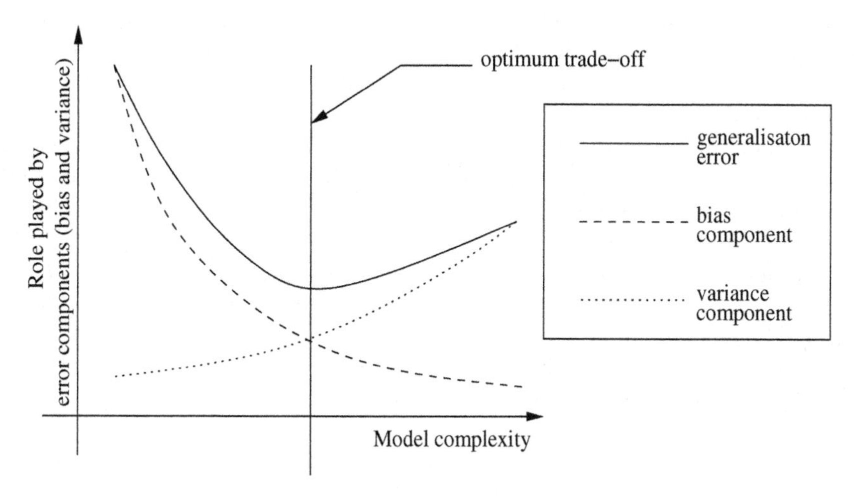

Fig. 19.2. Bias and variance components of the generalisation error Vs. the model complexity.

between bias and variance is to use more data (well distributed over the input space) for training [7]. As the data points increase, a more complex model can be utilised [7]. A simple way to look at this is that having an infinitely large number of data points to train on would actually hard-code the desired function into the network making the network function identical to the desired function which essentially means that there would be zero bias (both desired and network function being exactly the same) and zero variance (due to the use of all data that exist). This results in the network function being smoother as compared to what would happen if we use a small training set. Bishop [7] gives a comprehensive account of techniques in tackling this dilemma. Some of these are regularisation, early stopping, training with noise, soft weight sharing, growing and pruning algorithms etc.

Bias-variance Decomposition for Classification

Historically, the bias-variance decomposition was initially deduced for regression using the squared loss function. We use the squared loss function for classification (considering outputs of a learner as posterior probabilities of a class, enabling the usage of regression results for the bias-variance decomposition for classification tasks) as a general theme throughout the chapter, but for the sake of completeness, we consider elaborating on the bias-variance decomposition characteristics for the 0/1 loss function as well. As machine learning researchers paid more attention to classification problems, using 0/1 loss function as the evaluation criterion, several decompositions for this function were proposed [39, 30, 34]. Domingos [26] proposed a unified bias-variance decomposition framework in which bias and variance are defined for an arbitrary loss function. We introduce bias-variance decomposition for the 0/1 loss function [59], based on this framework.

In order to do the bias-variance decomposition in the unified framework, we need to define (definitions can be found in [59, 26]) some notions: *optimal prediction* $y_*(x)$, *main prediction* y_m, *bias* and *net–variance* (which includes *biased variance* and *unbiased variance*).

Consider a set D of data sets $D_i = \{x_j, t_j\}_{j=1}^m$, $D = \{D_i\}_{i=1}^n$, $t_j \in \{-1, 1\}$, $x_j \in X$. Define $f_{D_i} = \mathcal{L}(D_i)$ as the model f_{D_i} produced by a learner $\mathcal{L}$ using a training set D_i. The output for a data point x is $f_{D_i}(x) = y$. The goal of learning is to minimize $E_t[L(t, y)]$.

The optimal prediction y_* at data point x is defined as

$$y_*(x) = \arg\min_y E_t[L(t, y)]$$

where the subscript t denotes that the expectation is taken with respect to all possible values of t.

The main prediction y_m for a loss function at data point x is defined as

$$y_m = \arg\min_{y'} E_D[L(y, y')]$$

For 0/1 loss function, $y_m = \arg\max(p_1, p_{-1})$, where $p_1 = P_D(y = 1|X = x)$ and $p_{-1} = P_D(y = -1|X = x)$.

After defining the optimal prediction and main prediction, we can define the bias.

The *bias* $B(x)$ is:

$$B(x) = L(y_*, y_m)$$

Considering the 0/1 loss when the data set is noise free, $y_* = t$ and $y_m = \arg\max(p_1, p_{-1})$, and the bias is:

$$B(x) = \begin{cases} 1 \text{ if } y_m \neq t \\ 0 \text{ if } y_m = t \end{cases} = \left| \frac{y_m - t}{2} \right| = \begin{cases} 1 \text{ if } p_{corr}(x) \leq 0.5 \\ 0 \text{ otherwise} \end{cases}$$

Intuitively, variance should be defined as the expectation over the difference between the main prediction and the actual prediction. So the *net-variance* $V(x)$ is defined as:

$$V(x) = E_D[L(y_m, y)]$$

For 0/1 loss function:

$$V(x) = \frac{1}{n}\sum_{i=1}^{n} L(y_m, y_{D_i}) = \frac{1}{n}\sum_{i=1}^{n} \begin{cases} 1 \text{ if } y_m \neq y_{D_i} \\ 0 \text{ if } y_m = y_{D_i} \end{cases}$$

where $y_{D_i} = f_{D_i}(x)$.

This variance can be decomposed into two terms: the *biased variance* V_b, which increases the error and the *unbiased variance* V_u, which reduces the error. So, the net variance is the difference of the two terms: $V = V_u - V_b$.

Figure 19.3 shows the different effects of biased and unbiased variance on the error

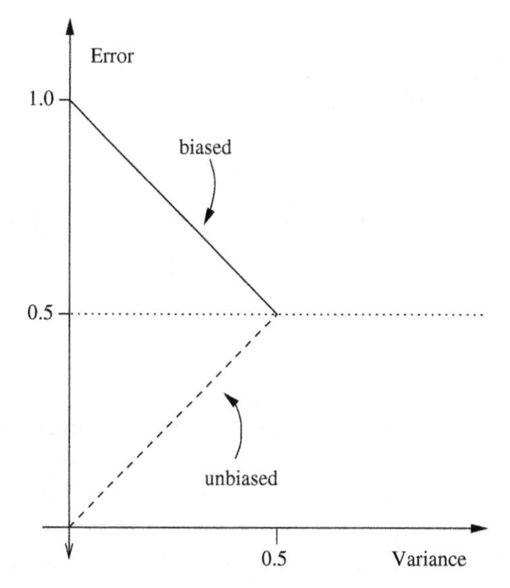

Fig. 19.3. Effects of biased and unbiased variance on the error [59].

The *average loss on the example* x (with no noise) $E_D(x)$, is given by

$$E_D(x) = B(x) + V_u(x) - V_b(x).$$

For a set of test set $\mathcal{T} = \{x_j, t_j\}_{j=1}^{r}$, *average bias* can be defined as

$$E_x[B(x)] = \frac{1}{r}\sum_{j=1}^{r} B(x_j) = \frac{1}{r}\sum_{j=1}^{r} \left|\frac{y_m(x_j) - t_j}{2}\right|.$$

The *average net-variance*, $E_x[V(x)]$ is

$$\begin{aligned} E_x[V(x)] &= \frac{1}{r}\sum_{j=1}^{r} V(x_j) \\ &= \frac{1}{nr}\sum_{j=1}^{r}\sum_{i=1}^{n} L(y_m(x_j), y_{D_i}(x_j)) \\ &= \frac{1}{nr}\sum_{j=1}^{r}\sum_{i=1}^{n} \begin{cases} 1 \text{ if } y_m(x_j) \neq y_{D_i}(x_j) \\ 0 \text{ if } y_m(x_j) = y_{D_i}(x_j) \end{cases} \end{aligned}$$

The *average unbiased variance*, $E_x[V_u(x)]$ is

$$E_x[V_u(x)] = \frac{1}{nr}\sum_{j=1}^{r} V_u(x_j) = \frac{1}{r}\sum_{j=1}^{r}\sum_{i=1}^{n} \begin{cases} 1 \text{ if } ((B(x_j) = 0) \wedge (y_m(x_j) \neq y_{D_i}(x_j))) \\ 0 \text{ otherwise} \end{cases}$$

The *average biased variance*, $E_x[V_b(x)]$ is

$$E_x[V_b(x)] = \frac{1}{nr}\sum_{j=1}^{r} V_b(x_j) = \frac{1}{r}\sum_{j=1}^{r}\sum_{i=1}^{n} \begin{cases} 1 \text{ if } ((B(x_j) = 1) \wedge (y_m(x_j) \neq y_{D_i}(x_j))) \\ 0 \text{ otherwise} \end{cases}$$

The average loss on all the examples (with no noise) is

$$\begin{aligned} E_{D,x}[L(t,y)] &= E_x[B(x)] + E_x[V_u(x)] - E_x[V_b(x)] \\ &= E_x[B(x)] + E_x[(1 - 2B(x))V(x)] \end{aligned}$$

The above formula represents a unified framework for the bias-variance decomposition for any loss function. It serves as a powerful tool for analysing learning processes and learning machines [59]. From the formula, we can see the antipodal effects of variance on the error. In the unbiased case, variance increase the error, while in the biased case, variance decrease the error. Additionally, the effects of noise on the error are the same as variance. This property is interesting in that, we can improve the performance of classifiers by adding some small appropriate noise or enlarging the variance in the biased case. Experiments on benchmark datasets illustrating the utility of the framework can be found in [59].

19.2.2 Ensembles Tackling the Bias-variance Dilemma

Need for Ensembles

There are formal and empirical proofs [13, 45, 47] that ensembles generalise better than single predictors. This section will mainly elaborate on some key reasons which make an aggregate system such as this work. An ensemble, as the name suggests, is a collection of learning machines essentially used to replace a single learner/predictor in order to achieve a co-operative (or competitive) decision when mapping an input space onto an output space.

One of the major advantages of using an integrated system such as this is that ensembles tend to be more robust/reliable as compared to single learners. Graceful degradation is impossible to achieve with a single learner because of the fact that if a network does not perform well, it will not perform well. With an ensemble of networks, there is always a chance that some subset of the networks will work even if others are not performing well at all. Some researchers have addressed the issue of designing ensembles from a software engineering perspective [55, 66, 61] where the idea of N-version programming/multi-version programming has been considered. Multi-version programming deals with producing multiple versions of a program such that each fails independently (or in an uncorrelated manner) [44] and where these versions can be combined by some means such as a majority vote to give a more reliable system [55]. Sharkey et al. [55] refer to this as being a reliability through redundancy approach which is a standard way of designing conventional software (as shown by Littlewood and Miller in [44]) and hardware [53] for safety critical systems.

Perhaps a more defining viewpoint is that of Liu and Yao [47] where they stress the fact that ensembles adopt a divide and conquer strategy where the individuals learn to subdivide the problem (decompose the problem) at hand and solve the respective subproblems more efficiently. There are various ways this decomposition can take place and diversity plays an important role in order for the decomposition to be successful, as will be seen shortly.

The need for ensembles has its reasons in the nature of machine learning algorithms and the properties of the problem formulated. The problem of designing a learner is often posed as a search problem. Machine learning algorithms usually work by searching for the best possible hypothesis (learner) in the space of possible hypotheses **H** [24].

Firstly, the space **H** is very large and a large amount of training data is required to restrict the search to the most feasible regions. By eliminating the infeasible regions in accordance with the training data, we might end up with good approximations of the optimal hypothesis. Not combining these may lead to loss of useful information as all the remaining hypotheses are equally attractive (equally accurate).

Secondly, the nature of learning algorithms in them, more often than not, being greedy search algorithms applying local optimisation techniques (e.g. gradient decent for multi-layer perceptrons) gets them stuck in a local optima. In turn, the use of search heuristics can be attributed to the fact that the problems posed are difficult. Finding the best hypothesis which is optimal both in terms of complexity and accuracy is considered as an NP-complete problem [8]. In theory, even if the best hypothesis can be found, the use of heuristics prevents it from finding it in practise. The best solution here is to train base models starting from different points in the hypothesis space and combine these hypotheses found.

Additionally, the deficiencies of learning algorithms and the nature of search spaces together let ensembles provide a complementary solution. According to [24], there may be problems for which there is no hypothesis in the possible space of hypotheses due to limitations in the representation capabilities of learning algorithms [60], i.e. the actual solution is not part of the infinitum of solutions that can be obtained from the algorithms. In such cases, there may be good approximations that can be combined in some way to actually shift the resulting model towards the true hypothesis. As Valentini et al. [60] state "...combination of hypotheses... can expand the space of representable functions, embracing also the true one". Sections 19.2.2 and 19.2.2 give a more formal view on why ensembles work.

Ambiguity Decomposition

Brown et al. [13] give a very clear description on the utility of ensembles (ensembles of neural networks) in the regression context although the whole idea presented can be applied to a classification task (if we consider the outputs of the predictors as the posterior class probabilities). However, no formal proof has been proposed for base models that output non-ordinal values, although there is empirical evidence which models the unknown formalism [13].

Suppose that we have trained a neural network using backpropagation (hereafter referred to as BP) on some data set D. For any test pattern $(\mathbf{x}, \mathbf{y})$ with input $\mathbf{x}$, we will get some prediction for $\mathbf{y}$, which might not be the same as $\mathbf{y}$. Training another network on the same data set may result in another hypothesis where the output might again not be the same as what is desired but it might also *differ* from the output of the first network.

According to [13], if many such networks are considered, each differing on their initial weight settings (each network, to begin with, has its weights initialised randomly), the outputs from each network for the testing data point $(\mathbf{x}, \mathbf{y})$ will follow a distribution dependent on the data set D used for training them and the weight initialisations. Consequently, this would also be true for the errors that they make. The mean of the distribution is the expected value $E\{f(\mathbf{x})\}$ for any given input $\mathbf{x}$.

Since we have many networks, each will have a different estimate/prediction of the output given an input $\mathbf{x}$ but *if we take the average of these predictions we might get a better estimate of the output as compared to estimates from individual networks.* This, in essence, is what an ensemble does. Consequently, if we have an infinite number of networks, we would certainly come very close to if not the same as the exact value for the output $\mathbf{y}$ from $E\{f(\mathbf{x})\}$ given an input $\mathbf{x}$, provided the individual networks *differ* in their predictions and hence *differ* in the errors that they make. This point emphasises the notion of *diversity* which is a very important ingredient in constructing good ensembles as will be seen later.

The main idea is that if we combine estimates from a number of predictors which have been trained on the same data set (yet have different error estimates for any given test input mainly because of the different random weight initialisations) a combined estimate of the outputs is likely to be better than the output from individual predictors.

Following is a more formal result as to the effectiveness of ensemble methods, which was presented by Krogh and Vedelsby [41] and is also considered in [13].

Let f_{ens} be the ensemble output given by

$$f_{ens} = \sum_i w_i f_i,$$

where f_i is the output of network i. f_{ens} is a convex combination, i.e. $\sum_i w_i = 1$ and $\forall i$, $w_i \geq 0$. Let Φ be the desired output. The average quadratic error of the member networks, i.e. $\sum_i w_i(f_i - \Phi)^2$, can be expanded as follows:

$$\begin{aligned}\sum_i w_i(f_i - \Phi)^2 &= \sum_i w_i(f_i - f_{ens} + f_{ens} - \Phi)^2 \\ &= \sum_i w_i[(f_i - f_{ens}) + (f_{ens} - \Phi)]^2 \\ &= \sum_i w_i[(f_i - f_{ens})^2 + (f_{ens} - \Phi)^2 + 2(f_i - f_{ens})(f_{ens} - \Phi)]. \end{aligned} \quad (19.10)$$

In equation 19.10,

$$\begin{aligned}\sum_i w_i 2(f_i - f_{ens})(f_{ens} - \Phi) &= 2(f_{ens} - \Phi)\sum_i w_i(f_i - f_{ens}) \\ &= 2(f_{ens} - \Phi)(f_{ens} - f_{ens}) \\ &= 0\end{aligned}$$

and

$$\sum_i w_i(f_{ens} - \Phi)^2 = (f_{ens} - \Phi)^2.$$

Hence,

$$\begin{aligned}\sum_i w_i(f_i - \Phi)^2 &= \sum_i w_i(f_i - f_{ens})^2 + (f_{ens} - \Phi)^2 \\ \Rightarrow (f_{ens} - \Phi)^2 &= \sum_i w_i(f_i - \Phi)^2 - \sum_i w_i(f_i - f_{ens})^2. \end{aligned} \quad (19.11)$$

Equation 19.11 says that the mean square error/quadratic error of the ensemble (the left hand side of the equation) is guaranteed to be less than or equal to the average quadratic error of the member networks (the first term in the right hand side of the equation).

Mathematically,

$$(f_{ens} - \Phi)^2 \leq \sum_i w_i (f_i - \Phi)^2, \tag{19.12}$$

In Equation 19.11, the second term on the right hand side (i.e. $\sum_i w_i (f_i - f_{ens})^2$) will always be greater than or equal to zero. This is often called the ambiguity term and it tells how different individual members within the ensemble are for some test pattern (data point), i.e. emphasises diversity. It means that if this term is somehow made large, the left hand side will become small, i.e. the error estimate of the ensemble for a given data point will decrease. In other words, if we have more diversity in the ensemble, the accuracy of the combined predictor will increase. It can also be seen that if we try to maximise this term i.e. cross some limit (let us call it a trade-off line), the first term (i.e. $\sum_i w_i (f_i - \Phi)^2$) may also increase, which then would cancel out the whole effect caused due to the amplification of the ambiguity term. Here, the first term signifies the accuracy of an individual for some test pattern. *The second term signifies that we should have diverse members in the ensemble but there is a limit/trade-off line crossing which would make the individual predictors less accurate.* This is what is often referred to as the *trade-off between diversity and accuracy* in ensembles. In the classification context, according to Hansen and Salamon [33] and as also mentioned in [13], a necessary and sufficient condition for a majority voting ensemble of classifiers to be more accurate than any of its component classifiers is that the components be both accurate and diverse. So the trade-off applies to both regression and classification problems.

A more general decomposition of the quadratic error function given in [58, 13] for the ensemble is presented next in order to help fully understand the way in which individual predictors affect the ensemble as a whole. This is achieved by decomposing the quadratic error into *bias*, *variance* and *covariance*. Since we are concerned with the error expectation on future/unseen data points and need to be consistent with the bias-variance decomposition literature, we need to view the expected value of the error over all possible choices of data sets on which the predictors could be trained on and over all possible parameter initialisations.

The bias-variance-covariance decomposition will also be used to explain the *sensitivity* of the quadratic error of an ensemble towards the *covariance* (in the output estimates) between the individual networks. We will see that the less correlated (having low covariance) the networks are, the better is the ensemble formed.

Bias-Variance-Covariance Decomposition

Specifically, from equation 19.3,

$$E\{(f(x) - \Phi(x))^2\} = E\{(f(x) - E\{f(x)\})^2\} + (E\{f(x)\} - \Phi(x))^2 \tag{19.13}$$

In case of an ensemble, let $f(x)$ be the ensemble output. Considering simple averaging as our combination rule,

$$f(x) = \frac{1}{M} \sum_{i=1}^{M} f_i(x),$$

where M is the number of networks in the ensemble. For simplicity, we will write

$$f(x) = \frac{1}{M} \sum_{i} f_i. \tag{19.14}$$

The variance component of Equation 19.3 can be further subdivided into 2 parts: variance and covariance. Substituting Equation 19.14 into the variance component of Equation 19.3 we have

$$\begin{aligned}
&Variance\,(ensemble) \qquad (19.15)\\
&= E\{(\frac{1}{M}\sum_i f_i - E\{\frac{1}{M}\sum_i f_i\})^2\}\\
&= E\{(\frac{1}{M}\sum_i f_i - E\{\frac{1}{M}\sum_i f_i\})(\frac{1}{M}\sum_j f_j - E\{\frac{1}{M}\sum_j f_j\})\}\\
&= E\{(\frac{1}{M}\sum_i f_i - \frac{1}{M}\sum_i E\{f_i\})(\frac{1}{M}\sum_j f_j - \frac{1}{M}\sum_j E\{f_j\})\}\\
&= \frac{1}{M^2}E\{(\sum_i f_i - \sum_i E\{f_i\})(\sum_j f_j - \sum_j E\{f_j\})\}\\
&= \frac{1}{M^2}E\{[\sum_i (f_i - E\{f_i\})][\sum_j (f_j - E\{f_j\})]\}\\
&= \frac{1}{M^2}E\{[\sum_i (f_i - E\{f_i\})][\sum_{j=1,j\neq i}^{M} (f_j - E\{f_j\}) + (f_i - E\{f_i\})]\}\\
&= \frac{1}{M^2}E\{[\sum_i (f_i - E\{f_i\})][\sum_{j=1,j\neq i}^{M} (f_j - E\{f_j\})]\} +\\
&\quad \frac{1}{M^2}E\{\sum_i (f_i - E\{f_i\})^2\}
\end{aligned}$$

$$= \frac{1}{M^2}E\{\sum_i \sum_{j=1,j\neq i}^{M} (f_i - E\{f_i\})(f_j - E\{f_j\})\} + \frac{1}{M^2}E\{\sum_i (f_i - E\{f_i\})^2\} \tag{19.16}$$

$$= \frac{1}{M^2}Covariance + \frac{1}{M^2}Variance. \tag{19.17}$$

We also use a variant of the *Covariance* term in DIVACE [17, 20]. This variant was initially used in one of the established ensemble learning algorithms called Negative Correlation Learning (NCL) proposed by Liu and Yao [45] as a regularisation term.

The bias-variance-covariance decomposition can now be expressed as:

$$E\{(\frac{1}{M}\sum_i f_i - \Phi(x))^2\} = (E\{f(x)\} - \Phi(x))^2 +$$

$$\frac{1}{M^2}E\{\sum_i \sum_{j=1, j\neq i}^{M} (f_i - E\{f_i\})(f_j - E\{f_j\})\} +$$

$$\frac{1}{M^2}E\{\sum_i (f_i - E\{f_i\})^2\}. \quad (19.18)$$

Equation 19.18 expresses the fact that the quadratic error of the ensemble depends on bias, variance and also on the relationships between individual members of the ensemble. The *covariance* term here can be said to indicate the *diversity* or *disparity* between the member networks as far as their error estimates are concerned.

Hence, it is believed that the more diverse the individual members an ensemble has, the less correlated they would be, which seems obvious. This suggests that the covariance term should be as low as possible. The lower the covariance term, the less the error correlation amongst the networks, which implies reduced error and better performance at the ensemble level. This is the main reason why *diversity* in neural network ensembles is extremely important. This is also true for any kind of ensemble, be it one having support vector machines as its members, decision trees, or for that matter, a mixture of various types of learning machines. A thorough discussion of diversity with regard to neural network ensembles is covered in [12, 13].

Ensembles Reducing Bias

As is evident from the previous discussion, ensembles reduce the expected error (Equation 19.2) by reducing the variance term in Equation 19.3. However, Valentini et al. [60] refer to the fact that ensembles reduce either one or both terms of Equation 19.3, i.e. they may reduce bias too.

Some ensemble methods have been shown to have a greater effect on the reduction in the expected error. This is due to them acting on both the bias and variance terms of Equation 19.3. These include two of the most widely studied ensemble approaches, namely, Bagging [10] (more precisely, a variant thereof [11]) and Boosting [28, 48]. Apart from Bagging and Boosting, another technique called Error-Correcting Output Coding proposed by Dietterich and Bakiri [25] also tends to reduce both terms.

Freund and Schapire [29] presented an argument on the capability of Boosting to reduce the bias component of the error function given by Equation 19.3. Bauer and Kohavi [6] further confirmed the results. They showed that bagging can reduce the bias term of the expected error.

Given that these methods reduce bias too, a further reduction in the error is obvious as, being ensemble methods, these more often than not reduce the variance part of the expected error function. Of course, boosting sometimes increases the variance resulting in an increase in the expected error but that depends on the data set being worked upon. In general however, boosting and bagging do reduce the variance portion of the error and at times the bias as well.

Relationship Between NCL and the Diversity-Accuracy Trade-off

Here we explain the reason why negative correlation learning has been successful in many applications, focusing our discussion on diversity and accuracy by exploring its relationship with both ambiguity and bias-variance-covariance decompositions. We also extend negative correlation learning to classification problems and show that it has somewhat different forms from regression problems.

Negative Correlation Learning(NCL):
Assume that there is a training set $D = \{(x_1, d_1), ..., (x_N, d_N)\}$, where $x \in R^P$, $d \in \{-1, 1\}$, N is the size of the training set. For the simple averaging case,

$$f(n) = \frac{1}{M} \sum_{n=1}^{M} f_i(n), \tag{19.19}$$

where M is the number of the individual neural networks in the ensemble, $f_i(n)$ is the output of network on the nth training pattern, and $f(n)$ is the output of the ensemble on the nth training pattern. Negative correlation learning introduces a correlation penalty term into the error function of each individual network in the ensemble so that all the networks can be trained simultaneously and interactively on the same training set. The error function for a network in NCL is defined by

$$E_i = \frac{1}{N} \sum_{n=1}^{N} E_i(n) = \frac{1}{N} \sum_{n=1}^{N} \frac{1}{2}(f_i(n) - d(n))^2 + \frac{1}{N} \sum_{n=1}^{N} \lambda p_i(n), \tag{19.20}$$

where $E_i(n)$ is the value of the error function of network i at the presentation of the nth training pattern. The first term on the right side of Equation 19.20 is the empirical risk function of network i. In the second term, p_i, is a correlation penalty function. The purpose of minimising p_i is to negatively correlate each network's error with errors for the rest of the ensemble. The parameter $0 \leq \lambda \leq 1$ is used to adjust the strength of the penalty. The penalty function, p_i, has the form:

$$p_i(n) = (f_i(n) - f(n)) \sum_{j \neq i} (f_j(n) - f(n)). \tag{19.21}$$

The partial derivative of $E_i(n)$, with respect to the output of network i on the nth training pattern, is

$$\begin{aligned}\frac{\partial E_i(n)}{\partial f_i(n)} &= f_i(n) - d(n) + \lambda \frac{\partial p_i(n)}{\partial f_i(n)} \\ &= f_i(n) - d(n) + \lambda \sum_{j \neq i} (f_j(n) - f(n)) \\ &= f_i(n) - d(n) + \lambda (f_i(n) - f(n)) \\ &= (1-\lambda)(f_i(n) - d(n)) + \lambda (f(n) - d(n)),\end{aligned} \tag{19.22}$$

where we have made use of the assumption that $f(n)$ has constant value with respect to $f_i(n)$. The standard back-propagation algorithm is used for weight adjustments in the mode of pattern-by-pattern updating. Weight update of all the individual networks is performed simultaneously using Equation 19.22 after the presentation of each training pattern. Negative correlation learning from Equation 19.22 is a simple extension to the standard BP algorithm. In fact, the only modification that is needed is to calculate an extra term of the form $\lambda(f_i(n) - f(n))$for the ith network.

During the training process, all the individual networks interact with each other through their penalty terms in the error functions. Each network f_i minimises not only the difference between $f_i(n)$ and $d(n)$, but also the difference between $f(n)$ and $d(n)$. That is, negative correlation learning considers errors that all other networks have learned while training a network. For $\lambda = 0$, there are no correlation penalty terms in the error functions of the individual networks, and the individual networks are just trained independently. For $\lambda = 1$, from (4) we get

$$\frac{\partial E_i(n)}{\partial f_i(n)} = f(n) - d(n). \tag{19.23}$$

Note that the empirical risk function of the ensemble for the nth training pattern is defined by

$$E_{ens}(n) = \frac{1}{2}(\frac{1}{M}\sum_{n=1}^{M} f_i(n) - d(n))^2. \tag{19.24}$$

The partial derivative of $E_{ens}(n)$ with respect to f_i on the nth training pattern is

$$\frac{\partial E_{ens}(n)}{\partial f_i(n)} = \frac{1}{M}(\frac{1}{M}\sum_{n=1}^{M} f_i(n) - d(n)) = \frac{1}{M}(f(n) - d(n)). \tag{19.25}$$

In this case, we get

$$\frac{\partial E_i(n)}{\partial f_i(n)} \propto \frac{\partial E_{ens}(n)}{\partial f_i(n)}. \tag{19.26}$$

The minimisation of the empirical risk function of the ensemble is achieved by minimising the error functions of the individual networks. From this point of view, negative correlation learning provides a novel way to decompose the learning task of the ensemble into a number of subtasks for different individual networks.

A Theoretical Basis for NCL:

We will now show that NCL can be clearly related to both the ambiguity and bias-variance-covariance decompositions. We first note that, due to the property that the sum of deviations around a mean is equal to zero, we have

$$(f_i - f) = -\sum_{j \neq i}(f_j - f). \tag{19.27}$$

Now, we can rearrange the penalty term as follows

$$p_i = -(f_i - f)^2. \tag{19.28}$$

As can be seen, the penalty function is now necessarily negative and each network minimises its penalty by moving its output further away from the ensemble output, the mean response of the networks.

So the error of ensemble is

$$\begin{aligned} E &= \frac{1}{M}\sum_{i=1}^{M} E_i = \frac{1}{M}\sum_{i=1}^{M}\{\frac{1}{N}\sum_{n=1}^{N}\frac{1}{2}(f_i(n) - d(n))^2 + \frac{1}{N}\sum_{n=1}^{N}\lambda p_i(n)\} \\ &= \frac{1}{N}\sum_{n=1}^{N}\{\frac{1}{M}\sum_{i=1}^{M}\frac{1}{2}(f_i(n) - d(n))^2 - \frac{1}{M}\sum_{i=1}^{M}\lambda((f_i(n) - f(n))^2)\} \end{aligned} \tag{19.29}$$

When $\lambda = 1$, the formula is the average form of ambiguity decomposition. We know that the ambiguity term is related to the bias-variance-covariance decomposition. The expected value of the average individual squared error gives the bias component, while the expected value of the ambiguity term gives us the variance and covariance components. When training a simple ensemble, we only minimise the errors of the individual networks, and therefore only explicitly influence the bias component in the ensemble. However when training with NCL, we use the individual error (accuracy) plus the ambiguity (or diversity) as a penalty term. The expected value of the ambiguity term provides the missing second half of the ensemble error that is not included in simple ensemble learning (i.e. when $\lambda = 0$) [1]. It therefore can be seen that

[1] Note that λ describes the emphasis one wants to put on ambiguity or diversity and hence can be tuned to manage the trade-off between accuracy and diversity [14].

NCL succeeds because it trains the individual networks with error functions which more closely approximate the individual's contribution to ensemble error, than that used by simple ensemble learning.

Extending NCL for Classification Problems:
Ambiguity decomposition of the quadratic error by Krogh and Vedelsby [41] was a very encouraging result for regression ensemble research. The next question is whether this kind of decomposition can be applied to classification problem. In this section we will present the ambiguity decomposition for classification with 0/1 error function.

Suppose the classification task is to use an ensemble comprising N component neural networks to approximate a function $f : \mathbb{R}^m \rightarrow L$ where L is the set of class labels, and the predictions of the component networks are combined through majority voting where each component network votes for a class and the class label receiving the most number of votes is regarded as the output of the ensemble. Without loss of generality, here we assume that L contains only two class labels, i.e., the function to be approximated is $f : \mathbb{R}^m \rightarrow \{-1, +1\}$.

Now assume there are N instances, for a single arbitrary data point, d denotes the target on this instance, f_i is the actual output of the ith component neural network, $d \in \{-1, +1\}$ and $f_i \in \{-1, +1\}(i = 1, 2, ..., N)$ respectively. It is obvious that if the actual output of the ith component network is correct according to the expected output, then $f_i d = +1$, otherwise $f_i d = -1$.

The output of ensemble is defined by

$$f_{ens} = Sgn(\sum_{i=1}^{N} w_i f_i), \tag{19.30}$$

where $Sgn(x)$ is a function defined as:

$$Sgn(x) = \begin{cases} 1 & \text{if } x > 0 \\ 0 & \text{if } x = 0 \\ -1 & \text{if } x < 0 \end{cases} \tag{19.31}$$

For classification problems, the error of the ensemble at an arbitrary data point is defined by

$$Error(f_{ens} \cdot d) = Error(Sgn(\sum_{i=1}^{N} w_i f_i) \cdot d), \tag{19.32}$$

where $Error(x)$ is a function defined as

$$Error(x) = \begin{cases} 1 & \text{if } x = -1 \\ 0.5 & \text{if } x = 0 \\ 0 & \text{if } x = 1 \end{cases} \tag{19.33}$$

The average error of N component neural networks is $\sum_{i=1}^{N} w_i \cdot Error(f_i \cdot d)$.

Now, let us see the difference between the error of the ensemble and the average error of the N component neural networks

$$
\begin{aligned}
&= Error(f_{ens} \cdot d) - \sum_{i=1}^{N} w_i \cdot Error(f_i \cdot d) \\
&= Error(Sgn(\sum_{j=1}^{N} w_i f_j) \cdot d) - \sum_{i=1}^{N} w_i \cdot Error(f_i \cdot d) \\
&= \frac{1}{N} \sum_{i=1}^{N} (Error(Sgn(\sum_{j=1}^{N} w_i f_j) \cdot d) - N w_i \cdot Error(f_i \cdot d)).
\end{aligned}
$$

From the definition of $Error(x)$, (Equation 19.33), the function $Error(x)$ can been seen as a linear function $f(x) = -\frac{1}{2}(x-1)$, $x \in \{-1, 0, 1\}$. So, the expression $Error(x) - a \cdot Error(y)$ can be seen as a combination of two functions. Hence, $Error(x) - a \cdot Error(y) = -\frac{1}{2}((x-1) - (y-1)a) = \frac{1}{2}((y-1)a - (x-1))$, $x, y \in \{-1, 0, 1\}$.

We now get the following form for the difference,

$$
\begin{aligned}
&= \frac{1}{N} \sum_{i=1}^{N} (Error(Sgn(\sum_{j=1}^{N} w_i f_j) \cdot d) - N w_i \cdot Error(f_i \cdot d)) \\
&= \frac{1}{2N} \sum_{i=1}^{N} ((f_i \cdot d - 1) N w_i - Sgn(\sum_{j=1}^{N} w_i f_j) \cdot d + 1) \\
&= \frac{1}{2N} \sum_{i=1}^{N} (f_i d N w_i - f_{ens} \cdot d + 1 - N w_i) \\
&= \frac{d}{2} \sum_{i=1}^{N} (w_i f_i - \frac{1}{N} f_{ens}).
\end{aligned}
$$

So,

$$
Error(f_{ens} \cdot d) = \sum_{i=1}^{N} w_i \cdot Error(f_i \cdot d) - \frac{d}{2} \sum_{i=1}^{N} (\frac{1}{N} f_{ens} - w_i f_i). \tag{19.34}
$$

The second term on the right hand side i.e., $\frac{d}{2} \sum_{i=1}^{N} (\frac{1}{N} f_{ens} - w_i f_i)$, is the ambiguity term; we want it to be positive and as large as possible. Since $f_{ens} = Sgn(\sum_{i=1}^{N} w_i f_i)$, so $\sum_{i=1}^{N} w_i f_i = \mid \sum_{i=1}^{N} w_i f_i \mid \cdot Sgn(\sum_{i=1}^{N} w_i f_i) = s \cdot Sgn(\sum_{i=1}^{N} w_i f_i)$, where $s \in [0, 1]$ is the absolute value of $\sum_{i=1}^{N} w_i f_i$. So we can obtain the following formula for the ambiguity term:

$$\frac{d}{2}\sum_{i=1}^{N}(\frac{1}{N}f_{ens} - w_i f_i) = \frac{d}{2}(f_{ens} - \sum_{i=1}^{N} w_i f_i)$$
$$= \frac{d}{2}(f_{ens} - s \cdot f_{ens}) = \frac{1-s}{2} \cdot d \cdot f_{ens}. \quad (19.35)$$

From equation 19.35, $d \cdot f_{ens} > 0$ requires that the output of the ensemble be correct.

As can be seen, the decomposition (Equation 19.34) is made up of two terms. The first, $\sum_i w_i \cdot Error(f_i \cdot d)$, is the weighted average error of the individuals. The second (ambiguity) term is positive when $d \cdot f_{ens} > 0$, i.e., the output of ensemble is correct. The result is different from the ambiguity decomposition for regression in that, only in the situation when the output of the ensemble is correct, its performance improves. Otherwise, the correct rate of the ensemble is lower than the weighed average of the individual errors.

So, for classification problems, the formula of negative correlation learning is

$$E_i = \frac{1}{N}\sum_{n=1}^{N} E_i(n) = \frac{1}{N}\sum_{n=1}^{N} Error(f_i(n) \cdot d(n)) + \frac{1}{N}\sum_{n=1}^{N} \lambda p_i(n), \quad (19.36)$$

where $p_i(n) = d(n)(f_i(n) - f_{ens}(n))$. This tells us that the main difference in NCL for regression from NCL for classification problems is that they have different $p_i(n)$. Use of this form of NCL will be demonstrated as part of our future work.

19.2.3 Diversity in Ensembles

From the bias-variance-covariance decomposition (Equation 19.18) we know that ensembles usually lower the expected error by having members which are uncorrelated (encouraging diversity) in producing errors (lowering the covariance term). This makes the variance term of the expected error function (Equation 19.3) low, hence reducing error or enhancing performance of the predictor formed.

Having uncorrelated failures suggests that the ensemble should have members which individually work well on different parts of the training set. Making errors on different, rather disjoint parts of the training set would result in a situation where it is possible to get the correct output for each training pattern when the outputs of individuals are combined.

Sharkey et al.[55] suggest that for an ensemble to have good diversity, the component predictors should show different patterns of generalisation (i.e. have uncorrelated failures with respect to the test set). Uncorrelated failures with respect to the test set is achieved by estimating uncorrelated failures on the training set and this is what most ensemble methods try to do.

There is a large body of research discussing holistic classification schemes for ensemble learning methods. Valentini et al. [60] categorise ensemble methods as being either generative or non-generative depending on whether they actively or passively boost the accuracy and diversity of the base learners. The usual stance taken in some schemes is of using diversity as the categorising criterion. Sharkey [54] presents one such scheme. However, Brown et al. [13] recently proposed a new classification scheme which they feel covers a majority of the ensemble methods developed to date. This scheme includes, to our knowledge, more ensemble methods than any other scheme and so we took the ideas expressed in it while constructing our framework as will be seen shortly (Section 19.3.2).

Seeking a predictor from within a hypothesis space and using a set of predictors thus found results in the formation of an ensemble. Brown et al. [13] use this idea as the basis for classifying ensemble methods and state that in order to construct an ensemble, predictors from different points in the hypothesis space are desirable. This is determined by the way in which this space is explored. Accordingly, diverse ensembles can be constructed by making the components learn in three main ways as given in Table 19.1.

Table 19.1. Ensemble classification scheme from Brown et. al. [13].

Class	Description
A	by making the component learners start the learning process with different initial conditions. They call this the “Starting point in hypothesis space” [13] class.
B	by manipulating the training set used, using different architecture or using different learning algorithms to train the components. This is called “Set of accessible hypotheses” [13].
C	by modifying the trajectory used by the components in the hypothesis space using penalty function (as a regularisation term) methods and global search methods such as evolutionary algorithms. More formally referred to as “Traversal of hypothesis space” class as per [13].

Apart from the above discussed classification scheme, Yates and Partridge [66] came up with a scheme which puts diversity generating methods into various levels depending on the efficacy of each in enforcing diversity within an ensemble. According to them, methodological diversity is the most effective diversity generating method. This is followed by training set structure, architecture of the learning machine and initial conditions for the base learners in this order. Figure 19.4 shows this ordering.

Building on the efficacy of methodological diversity, Wang et al. [61, 62] developed a method of constructing ensembles using both neural networks and decision trees as base models. Also, Langdon et al.[43] have tried combining neural networks and decision trees. Woods et al. [63] go one step further

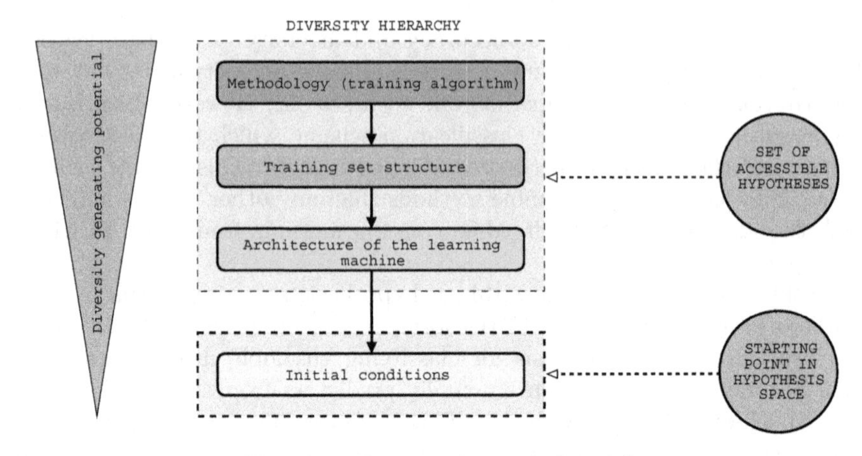

Fig. 19.4. Diversity hierarchy[16, 66].

by constructing an ensemble using four methodologically different types of learners. These heterogeneous ensembles are commonly known as hybrid ensembles.

As can be seen in Figure 19.4, this hierarchy considers two out of the three main diverse ensemble constructing methods from the scheme shown in Table 19.1. We recently proposed an evolutionary framework for constructing ensembles which considers enforcement of diversity at all these levels incorporating class **C** from Table 19.1 as well, thus taking care of both, the classification scheme and the diversity hierarchy. The framework is discussed later on in the chapter.

19.3 Multi-objective Evolution for Ensembles

Multi-objectivity in ensembles, as an area of research, has not been explored extensively yet. According to our knowledge, the idea of designing neural networks within a multi-objective setup was first considered by Kottathra and Attikiouzel [40] where they used a branch and bound method to determine the number of hidden neurons (the second objective being the mean square error) in feed forward neural networks. Kupinski and Anastasio [42] used the Niched Pareto GA [36] to learn classifiers, taking sensitivity (true negative rate) and specificity (true positive rate) as two objectives to optimise. ROC [49] analysis of the resulting solutions (in the Pareto set) is said to have resulted in better ROC curves when compared with generating curves using the conventional approach. Recently, Abbass [4] proposed an evolutionary multi-objective neural network learning approach where the multi-objective problem formulation essentially involved setting up of two objectives viz. complexity of the network and the training error (quadratic error/mean square error). The network complexity here could mean the number of synaptic weights, number of hidden units or a combination of both. An algorithm called MPANN was proposed which uses Pareto differential evolution [5]. MPANN was later on considered [3, 2] for learning and formation of neural network ensembles, albeit, with a different multi-objective formulation (as opposed to that in [4]). Based on multi-objective evolutionary optimisation, taking inspiration from MPANN [3, 2], and exploiting the main idea of NCL [45] (which belongs to class **C** of the scheme discussed in the previous section) we recently proposed DIVACE [17, 20]. Reformulating the regularisation process as a multi-objective optimisation problem, as is done in DIVACE by reformulating NCL, was also studied in [22, 38]. Apart from classification, multi-objective evolutionary optimisation for constructing ensembles for regression problems, finding trade-offs between regression accuracy and complexity of networks, was exhaustively studied in [37]. In cases where individual networks overfit the training data, regression accuracy was shown to improve.

The main reason for using a multi-objective evolutionary approach to designing ensembles is that multi-objectivity enforces the search/ optimisation process (the search process here being finding neural networks with good overall performance) to yield a set of near optimal solutions instead of just a single solution. Getting a set of solutions essentially means that we get a set of near optimal neural networks. These near optimal neural networks could in turn be used as members of an ensemble. With a population based approach we will, in essence, be generating a set of networks and the underlying multi-objective framework would take care of the selection of a set of near optimal solutions/networks from the population. The whole process of generating a Pareto set of neural networks which could be used as members of an ensemble will be automatic as the whole population would, with the passage of time, move towards the Pareto front. Thus, the idea of using multi-objective evolutionary algorithms for the purpose of designing neural network ensembles is very promising.

The main problem in actually using such an approach is the formulation of the multi-objective optimisation problem such that not only the final ensemble created have accurate members but the members also be uniformly distributed on the Pareto optimal front, i.e. diversity is catered for. Remarkably, for multi-objective optimisation to be effective, the optimisation process should lead to convergence to the Pareto optimal front while at the same time maintaining as diverse a distribution of solutions as possible on the Pareto front [23]. A striking parallel between multi-criterion optimisation and the necessity of having diverse enough members within an ensemble is evident. Hence, formulating a problem properly would surely do justice to both (multi-criterion optimisation and ensembles as disparate yet related computational paradigms).

Thus, incorporating the idea of multi-objective evolutionary optimisation into constructing ensembles is an attractive proposition. Next, we present some recent work on integrating ensemble learning with multi-objective evolutionary optimisation.

19.3.1 Diverse and Accurate Ensemble Learning Algorithm (DIVACE)

DIVACE [17, 20] takes in ideas from MPANN and NCL algorithms. For the evolutionary process, MPANN was used. Diversity was treated as a separate objective and the negative correlation penalty function from NCL was used to quantify it. The two objectives on which to optimise the performance of the ensemble were accuracy and diversity.

Objective 1 – Accuracy. Given a training set T with N patterns. For each network k in the ensemble,

$$\textbf{(Minimise)}\ \text{Accuracy}_k = \frac{1}{N}\sum_{i=1}^{N}\left(f_k^i - o^i\right)^2, \tag{19.37}$$

where o^i is the desired output and f_k^i the posterior probability of the class (classification task) or the observed output (regression task) for one training sample i.

Objective 2 – Diversity. From NCL, the correlation penalty function is used as the second objective on which to optimise the ensemble performance. Let N be the number of training patterns and let there be M members in the ensemble, so for each member k, the following term gives an indication of how different it is from other members.

$$\textbf{(Minimise)}\ \text{Diversity}_k = \sum_{i=1}^{N}\left(f_k^i - f^i\right)\left[\sum_{j\neq k, j=1}^{M}\left(f_j^i - f^i\right)\right], \tag{19.38}$$

where f^i is the ensemble output for a training sample i. In the information theoretic sense, mutual information is a measure of the correlation between two random variables. A link between the diversity term used here (equation 19.38) and mutual information was shown in [46]. Minimisation of mutual information between variables extracted (outputs) by two neural networks can be regarded as a condition to ensure that they are different. It has been shown that negative correlation learning, due to the use of the penalty function, can minimise mutual information amongst ensemble members [46, 65]. Hence the use of this penalty function as the diversity term in DIVACE.

It was mentioned in [17] that DIVACE is in no way limited to the use of one particular term for diversity. Also, different accuracy measures and evolutionary processes could well be used. The idea was to address the diversity-accuracy trade-off in a multi-objective evolutionary setup and DIVACE was shown to achieve a good trade-off. More details on the algorithm can be found in [17, 20].

19.3.2 A Framework for the Evolution of Hybrid Ensembles

Designing hybrid ensembles as a field of research is still in its infancy. There are many issues which one should consider while tackling this relatively new paradigm in ensemble research.

One reason for the lack of substantial literature in this area or the relative inactivity could be attributed to the fact that heterogeneous ensembles as the name implies, deal with the combination of base learners that are trained using different training methodologies (algorithms) which entails finding a suitable mixing mechanism. Brute force or even a search through a predefined search space (as some researchers have tried [61, 62, 63, 32]) would obviously result in a hybrid ensemble but would have a big disadvantage.

The hypothesis space for even a single predictor is supposedly very large and machine learning algorithms usually end up in some local optima while searching for a good hypothesis. An ensemble is a set of hypotheses and finding a good ensemble means searching for a solution in an extended hypothesis space or a combined hypothesis space. Searching for an optimum in this space can be rather cumbersome. The combined hypothesis space will be enormous in most cases and finding the best performing mixture of learners in that case would be an NP-complete problem for which an optimal solution can hardly be found without resorting to stochastic search mechanisms.

Some researchers have suggested using evolutionary algorithms [43] to evolve the mixture of individual predictors (mix) which intuitively looks very plausible but then again the types of individuals apropos, learning algorithms one wants to use, has to be determined beforehand.

Even when evolving the mix, the fact that individual predictors may in themselves not be optimum for the mix should be taken into account while designing hybrid ensembles. This basically reflects the fact that the search space is indeed very large as not only one wants to have good enough individuals (which should be found from their respective hypothesis spaces), the combined space also need to be searched through.

Another point here to make is that a good mix of learners can only be good with respect to the decision combination strategy utilised. Decision combination is another active area of research when it comes to multiple classifier systems or ensembles in general. Empirical studies on previously proposed combination strategies are presented in [35].

A possible evolutionary framework for the construction of diverse hybrid ensembles can be described by Figure 19.5. As can be seen, there are three levels of evolution present. First is the evolution of the mix i.e. evolving the mixture of the various types of predictors. Second, we consider evolution of the ensemble based on the structure of the training set (given the mix). A process similar to the original DIVACE forms the third and final evolutionary level.

The framework shows that the hybrid ensemble at Level 2 will have subsets of different types of predictors. These subsets can, in themselves, be consid-

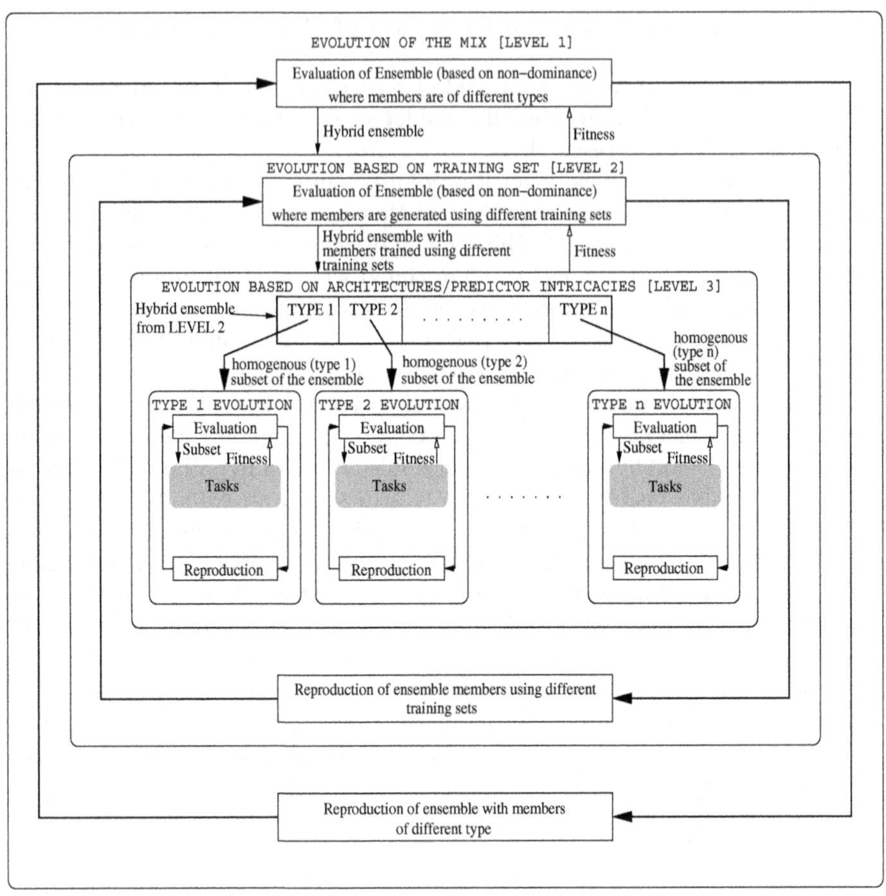

Fig. 19.5. The proposed framework [16, 18, 19].

ered as homogeneous ensembles and evolved in accordance with DIVACE, keeping the other subsets fixed. Level 3 can be called as the DIVACE stage where reproduction depends on the evolutionary factor(s) chosen for a given predictor type. The factors could be architectures or weights in case of NNs, kernel function in case of SVMs, architectures of RBFNs etc.

Level 3, due to its subset evolution process, *enforces competition* between the various subsets. This competition makes the framework as a whole model a co-evolutionary learning strategy where subspecies (subsets) compete with each other to stay in the ensemble. Additionally, these very species cooperate at the immediate higher level (Level 2) and compete again at Level 1. *The framework therefore embodies both competitive and cooperative co-evolution within a multi-objective and multi-level evolutionary setup.* The choice on the placement of these levels essentially depends on the prior knowledge available. However, the above mentioned co-evolutionary theme would be most effective if we keep this very ordering in the levels or at least let the innermost level stay where it is as it makes more sense to have a hybrid set of predictors and then let the subsets compete with each other than to enforce competition without having a mix. Details on the range of problems the framework tries to solve can be found in [19].

A majority of the ideas expressed by researchers in order to construct ensembles having both diverse and accurate members so as to have good generalisation capabilities have been incorporated into the framework. We can hence say that the framework represents a generic model from which new hybrid ensemble construction algorithms can be developed. It represents a class of ensemble learning methodologies aimed at enforcing diversity and accuracy explicitly within an evolutionary framework.

DIVACE-II (as a successor of DIVACE) can be thought of as being one instance of the framework and tries to find an ensemble composed of feed forward neural networks, radial basis function neural networks and support vector machines. We model all three levels in our algorithm, however, due to the computationally intensive nature of evolutionary methods, one should limit the ensemble construction approach to as fewer levels as possible depending on domain knowledge. More on DIVACE-II can be found in [18, 19]. A brief discussion of results from it, together with those from DIVACE is taken up in the next section.

19.4 Application of the Algorithms on Benchmark Datasets

Both DIVACE and DIVACE-II were tested on 2 benchmark data sets (Australian credit card assessment dataset and Diabetes dataset), available by anonymous ftp from ice.uci.edu in /pub/machine-learning-databases. We also experimented with the multi-objective formulation for both. Pairwise Failure Crediting (PFC) [20] was recently proposed as a diversity measure which credits individuals in the ensemble with differences in the failure patterns[2], taking each pair of individuals and accruing credits in a manner similar to implicit fitness sharing [21, 27]. We consider results obtained using this and the NCL penalty function term diversity measures with both algorithms.

We compare[3] DIVACE-II with MPANN (both variants from [3] - we refer to these as MPANN1 and MPANN2 here), DIVACE and EENCL [47] due to

[2] A failure pattern is a string of 0s and 1s indicating success or failure of the learning machine on the training instances in the original training set.

[3] The comparison is mainly with respect to the majority voting combination rule as DIVACE-II only employs this decision fusion strategy

the experimental setup [4] in all these being similar and a direct comparison with other evolutionary ensemble construction methods being difficult [47]. We also compare the algorithms with 21 other learning algorithms from the literature.

Table 19.2 shows interesting properties of the DIVACE-II (both versions) in that, the mean test accuracy is higher than the mean training accuracy for both datasets, which mainly suggests that (on an average) the generalisation ability of DIVACE-II is good i.e. it does not seem to overfit. MPANN2, DIVACE and EENCL on the other hand have higher mean training accuracies and so it can be said that, although these methods hold promise and do show good signs of generalisation, DIVACE-II performs even better due to its test accuracy being much higher. MPANN1 is the other algorithm having a mean test accuracy greater than its mean training accuracy but here again, the mean test accuracy (and mean training accuracy) is not better than DIVACE-II. DIVACE-II does compare well with previously studied approaches.

Table 19.2. Confidence intervals with a confidence level of 95% for training and testing of DIVACE-II and other algorithms on both datasets. Results computed using accuracy/correct classification rates obtained from 10 and 12 folds for the Australian and Diabetes datasets respectively.

	Training		Testing	
Algorithm	Australian	Diabetes	Australian	Diabetes
DIVACE-II	$.877 \pm .005$	$.771 \pm .004$	$.895 \pm .0223$	$.789 \pm .0146$
DIVACE-II with PFC	$.876 \pm .003$	$.776 \pm .003$	$.889 \pm .0159$	$.785 \pm .0222$
MPANN1	$.854 \pm .011$	$.771 \pm .013$	$.862 \pm .0303$	$.779 \pm .0186$
MPANN2	$.852 \pm .009$	$.755 \pm .011$	$.844 \pm .0347$	$.744 \pm .0192$
DIVACE	$.867 \pm .004$	$.783 \pm .003$	$.857 \pm .0303$	$.766 \pm .0322$
EENCL	$.891 \pm .006$	$.802 \pm .004$	$.857 \pm .0241$	$.764 \pm .0237$

Tables 19.3 and 19.4 put both DIVACE and DIVACE-II against 24 previously studied learning algorithms. All approaches are not evolutionary and this has been done mainly due to the lack of results in the field. The tables shows the average test error rates [5] (lower means better) for many learning algorithms.

The algorithms used for comparison can be categorised into five classes: evolutionary ensemble methods (MPANN [3], EENCL [47]), statistical meth-

[4] n-fold cross validation used here. $n = 10$ for Australian and $n = 12$ for Diabetes dataset. Learning rate for NNs is not the same as that used in [3, 47] for DIVACE-II as the evolutionary process is inherently very different and we use methodologically different learners. Moreover, we evolve the population for 50 generations as opposed to 200 in [3, 47].

[5] The two rates for both DIVACE and DIVACE-II are from the use of different diversity measures (original formulation and PFC respectively).

Table 19.3. Comparison of DIVACE-II with other learning algorithms from [50] in terms of the average test error rates for the Australian credit card assessment dataset. Results are averaged on 10-fold cross validation.

Algorithm	Error Rate	Algorithm	Error Rate
DIVACE-II	0.105, 0.111	IndCART	0.152
DIVACE	0.138, 0.125	NewID	0.181
MPANN1	0.135	AC^2	0.181
MPANN2	0.156	Baytree	0.171
EENCL	0.135	NaiveBay	0.151
Discrim	0.141	CN2	0.204
Quadisc	0.207	C4.5	0.155
Logdisc	0.141	ITrule	0.137
SMART	0.158	Cal5	0.131
ALLOC80	0.201	DIPOL92	0.141
k-NN	0.181	BP	0.154
CASTLE	0.148	RBF	0.145
CART	0.145	LVQ	0.197

ods (Discrim, Quadisc, Logdisc, SMART, ALLOC80, k-NN, CASTLE, NaiveBay), decision trees based methods (CART, IndCART, NewID, AC^2, Baytree, Cal5, C4.5), rule based methods (CN2, ITrule) and neural network based methods (BP, LVQ, RBF, DIPOL92). Details of the algorithms in the latter four classes can be found in [50]. The error rates refer to the percentage of wrong classifications on the test set.

As is evident from the tables, DIVACE-II has been able to achieve a good generalisation performance (due to it having lower average test error rates on both datasets). It is generally better than most of the algorithms shown for both datasets which shows the promise of not only DIVACE-II but also the evolutionary hybrid ensemble construction framework from which it is derived. Also, DIVACE compares well with most of the algorithms and is better than a majority of them.

In general, it can be said that both DIVACE and DIVACE-II certainly do find a good trade-off between diversity and accuracy. Moreover, since DIVACE-II is based on DIVACE and has outperformed it on both datasets considered here, the idea of enforcing diversity at multiple levels under a multi-objective evolutionary setup to find a good trade-off between diversity and accuracy is very much valid.

19.5 Conclusion

This chapter considers an overview of ensemble methods as a means to achieve good generalisation. Theoretical aspects of a committee of learners are also discussed in order to support the importance of having both diversity and

Table 19.4. Comparison of DIVACE-II with other learning algorithms from [50] in terms of the average test error rates for the Diabetes dataset. Results are averaged on 12-fold cross validation.

Algorithm	Error Rate	Algorithm	Error Rate
DIVACE-II	0.211, 0.215	IndCART	0.271
DIVACE	0.227, 0.226	NewID	0.289
MPANN1	0.221	AC^2	0.276
MPANN2	0.254	Baytree	0.271
EENCL	0.221	NaiveBay	0.262
Discrim	0.225	CN2	0.289
Quadisc	0.262	C4.5	0.27
Logdisc	0.223	ITrule	0.245
SMART	0.232	Cal5	0.25
ALLOC80	0.301	DIPOL92	0.224
k-NN	0.324	BP	0.248
CASTLE	0.258	RBF	0.243
CART	0.255	LVQ	0.272

accuracy and explaining the trade-off between them in the construction of such aggregate learners.

A brief overview of a popular ensemble learning algorithm viz. Negative Correlation Learning (NCL) is carried out. NCL aims at achieving a good trade-off by directly making use of the ambiguity decomposition for weight updates of member networks. The algorithm is extended for classification tasks. We find that NCL has somewhat different forms for such tasks in having a different interpretation for the penalty function term.

Multi-objective evolutionary ensemble learning is also presented as a solution to achieving a good trade-off between diversity and accuracy. Some recent work (learning algorithms viz. DIVACE and DIVACE-II together with the framework for the construction of diverse hybrid ensembles) along these lines are reviewed.

DIVACE tackles the ensemble learning problem within a multi-objective evolutionary setup. It uses diversity and accuracy as objectives and which are two factors that are in conflict and much required within an ensemble in order that it generalises well or at least generalises better than the individual members that it is made of. The explicit use of diversity and accuracy in the formulation of the problem tries to make the evolutionary process search for a good trade-off by continuously moving towards the Pareto front.

Hybrid ensembles are also discussed as promoting diversity. Bringing together diversity enforcement mechanisms with DIVACE at the backdrop results in this multi-level ensemble learning framework (and the algorithm DIVACE-II) where individual predictors trained using disparate learning methodologies are generated automatically by successively competing and cooperating with each other. The trade-off between diversity and accuracy is addressed at various levels and results in better generalisation as demonstrated by recent results on the application of the approach on benchmark datasets.

References

[1] H. A. Abbass. A memetic pareto evolutionary approach to artificial neural networks. In *Proceedings of the 14th Australian Joint Conference on Artificial Intelligence*, pages 1–12, Berlin, 2000. Springer-Verlag.

[2] H. A. Abbass. Pareto neuro-ensemble. In *16th Australian Joint Conference on Artificial Intelligence*, pages 554–566, Perth, Australia, 2003. Springer.

[3] H. A. Abbass. Pareto neuro-evolution: Constructing ensemble of neural networks using multi-objective optimization. In *The IEEE 2003 Conference on Evolutionary Computation*, volume 3, pages 2074–2080. IEEE Press, 2003.

[4] H. A. Abbass. Speeding up backpropagation using multiobjective evolutionary algorithms. *Neural Computation*, 15(11):2705–2726, November 2003.

[5] H. A. Abbass, R. Sarker, and C. Newton. Pde: A pareto-frontier differential evolution approach for multi-objective optimization problems. In *Proceedings of the IEEE Congress on Evolutionary Computation (CEC2001)*, volume 2, pages 971–978. IEEE Press, 2001.

[6] E. Bauer and R. Kohavi. An empirical comparison of voting classification algorithms: Bagging, boosting, and variants. *Machine Learning*, 36(1-2):105–139, 1999.

[7] C. M. Bishop. *Neural Networks for Pattern Recognition.* Oxford University Press, 1995.

[8] A. Blum and R. L. Rivest. Training a 3-node neural network is NP-complete. In *Machine Learning: From Theory to Applications*, pages 9–28, 1993.

[9] E. Boers, M. Borst, and I. Sprinkhuizen-Kuyper. Evolving artificial neural networks using the "baldwin effect". Technical Report 95-14, Leiden Unversity, Deptartment of Computer Science, The Netherlands, 1995.

[10] L. Breiman. Bagging predictors. *Machine Learning*, 24(2):123–140, 1996.

[11] L. Breiman. Using adaptive bagging to debias regressions. Technical Report 547, University of California, Berkeley, 1999.

[12] G. Brown. *Diversity in Neural Network Ensembles.* PhD thesis, School of Computer Science, University of Birmingham, 2004.

[13] G. Brown, J. Wyatt, R. Harris, and X. Yao. Diversity creation methods: A survey and categorisation. *Journal of Information Fusion (Special issue on Diversity in Multiple Classifier Systems)*, 6:5–20, March 2005.

[14] G. Brown and J. L. Wyatt. The use of the ambiguity decomposition in neural network ensemble learning methods. In T. Fawcett and N. Mishra, editors, *20th International Conference on Machine Learning (ICML'03)*, Washington DC, USA, August 2003.

[15] A. Chandra. Evolutionary approach to tackling the trade-off between diversity and accuracy in neural network ensembles. Technical report, School of Computer Science, The University of Birmingham, UK, April 2004.

[16] A. Chandra. Evolutionary framework for the creation of diverse hybrid ensembles for better generalisation. Master's thesis, School of Computer Science, The University of Birmingham, Birmingham, UK, September 2004.

[17] A. Chandra and X. Yao. DIVACE: Diverse and Accurate Ensemble Learning Algorithm. In *Proc. 5th Intl. Conference on Intelligent Data Engineering and Automated Learning (LNCS 3177)*, pages 619–625, Exeter, UK, August 2004. Springer-Verlag.
[18] A. Chandra and X. Yao. Evolutionary framework for the construction of diverse hybrid ensembles. In M. Verleysen, editor, *Proc. 13th European Symposium on Artificial Neural Networks*, pages 253–258, Brugge, Belgium, April 2005. d-side.
[19] A. Chandra and X. Yao. Evolving hybrid ensembles of learning machines for better generalisation. *Neurocomputing (submitted)*, 2005.
[20] A. Chandra and X. Yao. Ensemble learning using multi-objective evolutionary algorithms. *Journal of Mathematical Modelling and Algorithms (to appear)*, 2006.
[21] P. J. Darwen and X. Yao. Every niching method has its niche: Fitness sharing and implicit sharing compared. In *Proc. of the 4th International Conference on Parallel Problem Solving from Nature (PPSN-IV), (LNCS-1141)*, pages 398–407, Berlin, September 1996. Springer-Verlag.
[22] R. de Albuquerque Teixeira, A. P. Braga, R. H. Takahashi, and R. R. Saldanha. Improving generalization of mlps with multi-objective optimization. *Neurocomputing*, 35:189–194, 2000.
[23] K. Deb. *Multi-Objective Optimization Using Evolutionary Algorithms*. Chichester, UK : Wiley, 2001.
[24] T. G. Dietterich. Machine-learning research: Four current directions. *The AI Magazine*, 18(4):97–136, 1998.
[25] T. G. Dietterich and G. Bakiri. Error-correcting output codes: a general method for improving multiclass inductive learning programs. In T. L. Dean and K. McKeown, editors, *Proceedings of the Ninth AAAI National Conference on Artificial Intelligence*, pages 572–577, Menlo Park, CA, 1991. AAAI Press.
[26] P. Domingos. A unified bias-variance decomposition and its applications. In *Proceedings of the Seventeenth International Conference on Machine Learning*, pages 231–238, Stanford, CA, USA, 2000.
[27] S. Forrest, R. E. Smith, B. Javornik, and A. S. Perelson. Using genetic algorithms to explore pattern recognition in the immune system. *Evolutionary Computation*, 1(3):191–211, 1993.
[28] Y. Freund and R. Schapire. A short introduction to boosting. *Journal of Japanese Society for Artificial Intelligence*, 14(5):771–780, 1999.
[29] Y. Freund and R. E. Schapire. Experiments with a new boosting algorithm. In *Proceedings of the 13th International Conference on Machine Learning*, pages 148–156. Morgan Kaufmann, 1996.
[30] J. Friedman. On bias, variance, 0/1 loss and the curse of dimensionality. *Data Mining and Knowledge Discovery*, 1:55–77, 1997.
[31] S. Geman, E. Bienenstock, and R. Doursat. Neural networks and the bias/variance dilemma. *Neural Computation*, 4(1):1–5, 1992.
[32] S. Gutta, J. Huang, I. F. Imam, and H. Wechsler. Face and hand gesture recognition using hybrid classifiers. In *Proceedings of the 2nd International Conference on Automatic Face and Gesture Recognition (FG '96)*, pages 164–170. IEEE Computer Society, 1996.
[33] L. K. Hansen and P. Salamon. Neural network ensembles. *IEEE Transactions on Pattern Analysis and Machine Intelligence*, 12(10):993–1001, 1990.

[34] T. Heskes. Bias/variance decomposition for likelihood-based estimators. *Neural Computation*, 10:1425–1433, 19998.

[35] T. K. Ho, J. J. Hull, and S. N. Srihari. Decision combination in multiple classifier systems. *IEEE Transactions on Pattern Analysis and Machine Intelligence*, 16(1):66–75, January 1994.

[36] J. Horn and N. Nafpliotis. Multiobjective Optimization using the Niched Pareto Genetic Algorithm. Technical Report IlliGAl Report 93005, University of Illinois, Urbana-Champaign, July 1993.

[37] Y. Jin, T. Okabe, and B. Sendhoff. *Applications of Evolutionary Multi-objective Optimization (Advances in Natural Computation)*, volume 1, chapter Evolutionary multi-objective approach to constructing neural network ensembles for regression, pages 653–672. World Scientific, 2004.

[38] Y. Jin, T. Okabe, and B. Sendhoff. Neural Network Regularization and Ensembling Using Multi-objective Evolutionary Algorithms. In *2004 Congress on Evolutionary Computation (CEC'2004)*, volume 1, pages 1–8, Portland, Oregon, USA, June 2004. IEEE Service Center.

[39] E. Kong and T. Dietterich. Error - correcting output coding correct bias and variance. In *Proceedings of The XII International Conference on Machine Learning*, pages 313–321, San Francisco, CA, USA, 1995.

[40] K. Kottathra and Y. Attikiouzel. A novel multicriteria optimization algorithm for the structure determination of multilayer feedforward neural networks. *Journal of Network and Computer Applications*, 19:135–147, 1996.

[41] A. Krogh and J. Vedelsby. Neural network ensembles, cross validation, and active learning. *Neural Information Processing Systems*, 7:231–238, 1995.

[42] M. A. Kupinski and M. A. Anastasio. Multiobjective genetic optimization of diagnostic classifiers with implications for generating receiver operating characteristic curves. *IEEE Transactions on Medical Imaging*, 18(8):675–685, August 1999.

[43] W. B. Langdon, S. J. Barrett, and B. F. Buxton. Combining decision trees and neural networks for drug discovery. In *Genetic Programming, Proceedings of the 5th European Conference, EuroGP 2002*, pages 60–70, Kinsale, Ireland, 3-5 April 2002.

[44] B. Littlewood and D. R. Miller. Conceptual modeling of coincident failures in multiversion software. *IEEE Transactions on Software Engineering*, 15(12):1596–1614, December 1989.

[45] Y. Liu and X. Yao. Ensemble learning via negative correlation. *Neural Networks*, 12(10):1399–1404, 1999.

[46] Y. Liu and X. Yao. Learning and evolution by minimization of mutual information. In J. J. M. Guervós, P. Adamidis, H.-G. Beyer, J.-L. Fernández-Villacañas, and H.-P. Schwefel, editors, *Parallel Problem Solving from Nature VII (PPSN-2002)*, volume 2439 of *LNCS*, pages 495–504, Granada, Spain, 2002. Springer Verlag.

[47] Y. Liu, X. Yao, and T. Higuchi. Evolutionary ensembles with negative correlation learning. *IEEE Transactions on Evolutionary Computation*, 4(4):380, November 2000.

[48] R. Meir and G. Raetsch. An introduction to boosting and leveraging. *Advanced lectures on machine learning*, pages 118–183, 2003.

[49] C. E. Metz. Basic principles of roc analysis. *Seminars in Neuclear Medicine*, 8(4):283–298, 1978.

[50] D. Michie, D. Spiegelhalter, and C. Taylor. *Machine Learning, Neural and Statistical Classification.* Ellis Horwood Limited, 1994.
[51] D. Opitz and R. Maclin. Popular ensemble methods: An empirical study. *Journal of Artificial Intelligence Research*, 11:169–198, 1999.
[52] D. W. Opitz and J. W. Shavlik. Generating accurate and diverse members of a neural-network ensemble. *Neural Information Processing Systems*, 8:535–541, 1996.
[53] T. Schnier and X. Yao. Using negative correlation to evolve fault-tolerant circuits. In *Proceedings of the 5th International Conference on Evolvable Systems: From Biology to Hardware (ICES'2003)*, pages 35–46. Springer-Verlag. Lecture Notes in Computer Science, Vol. 2606, March 2003.
[54] A. Sharkey. *Multi-Net Systems*, chapter Combining Artificial Neural Nets: Ensemble and Modular Multi-Net Systems, pages 1–30. Springer-Verlag, 1999.
[55] A. Sharkey and N. Sharkey. Combining diverse neural networks. *The Knowledge Engineering Review*, 12(3):231–247, 1997.
[56] K. O. Stanley and R. Miikkulainen. Evolving neural networks through augmenting topologies. *Evolutionary Computation*, 10(2):99–127, 2002.
[57] K. Tumer and J. Ghosh. Analysis of decision boundaries in linearly combined neural classifiers. *Pattern Recognition*, 29(2):341–348, February 1996.
[58] N. Ueda and R. Nakano. Generalization error of ensemble estimators. In *Proceedings of International Conference on Neural Networks*, pages 90–95, 1996.
[59] G. Valentini and T. G. Dietterich. Bias-variance analysis of support vector machines for the development of svm-based ensemble methods. *Journal of Machine Learning Research*, 5(1):725–775, 2004.
[60] G. Valentini and F. Masulli. Ensembles of learning machines. In R. Tagliaferri and M. Marinaro, editors, *Neural Nets WIRN Vietri-2002 (LNCS 2486)*, pages 3–19. Springer-Verlag, June 2002.
[61] W. Wang, P. Jones, and D. Partridge. Diversity between neural networks and decision trees for building multiple classifier systems. In *Proc. Int. Workshop on Multiple Classifier Systems (LNCS 1857)*, pages 240–249, Calgiari, Italy, June 2000. Springer.
[62] W. Wang, D. Partridge, and J. Etherington. Hybrid ensembles and coincident-failure diversity. In *Proceedings of the International Joint Conference on Neural Networks, 2001)*, volume 4, pages 2376–2381, Washington, USA, July 2001. IEEE Press.
[63] K. Woods, W. Kegelmeyer, and K. Bowyer. Combination of multiple classiers using local accuracy estimates. *IEEE Transactions on Pattern Analysis and Machine Intelligence*, 19:405–410, 1997.
[64] X. Yao. Evolving artificial neural networks. In *Proceedings of the IEEE*, volume 87, pages 1423–1447. IEEE, September 1999.
[65] X. Yao and Y. Liu. Evolving neural network ensembles by minimization of mutual information. *International Journal of Hybrid Intelligent Systems*, 1(1), January 2004.
[66] W. Yates and D. Partridge. Use of methodological diversity to improve neural network generalization. *Neural Computing and Applications*, 4(2):114–128, 1996.

20

Cooperative Coevolution of Neural Networks and Ensembles of Neural Networks

Nicolás García-Pedrajas

University of Córdoba
Campus Universitario de Rabanales
Córdoba (Spain)
npedrajas@uco.es

Summary. Cooperative coevolution is a recent paradigm in the area of evolutionary computation focused on the evolution of coadapted subcomponents without external interaction. In cooperative coevolution a number of species are evolved together. The cooperation among the individuals is encouraged by rewarding the individuals according to their degree of cooperation in solving a target problem. The work on this paradigm has shown that cooperative coevolutionary models present many interesting features, such as specialization through genetic isolation, generalization and efficiency. Cooperative coevolution approaches the design of modular systems in a natural way, as the modularity is part of the model. Other models need some *a priori* knowledge to decompose the problem *by hand*. In most cases, either this knowledge is not available or it is not clear how to decompose the problem.

This chapter describes how cooperative coevolution can be applied to the evolution of neural networks and ensembles of neural networks. Firstly, we present a model that develops subnetworks (modules) instead of whole networks. These modules are combined making up a network, by means of a cooperative coevolutionary algorithm. Secondly, we present a general framework for designing neural network ensembles by means of cooperative coevolution. The proposed model has two main objectives: first, the improvement of the combination of the trained individual networks; second, the cooperative evolution of such networks, encouraging collaboration among them, instead of a separate training of each network. In addition, a population of ensembles is evolved, improving the combination of networks and obtaining subsets of networks to form ensembles that perform better than the combination of all the evolved networks.

We also show how the multiobjective evaluation of the fitness of modules, networks, and ensembles can improve the performance of the model. For each element (module, network, and ensemble), different objectives are defined, considering not only its performance in the given problem, but also its cooperation with the rest of the networks. The results show the usefulness of the multiobjective approach.

N. García-Pedrajas: *Cooperative Coevolution of Neural Networks and Ensembles of Neural Networks*, Studies in Computational Intelligence (SCI) **16**, 465–490 (2006)
www.springerlink.com

20.1 Introduction

In the area of neural network design one of the main problems is finding suitable architectures for solving specific problems. The choice of such architecture is very important, as a network smaller than needed would be unable to learn and a network larger than needed would end in over-training.

The problem of finding a suitable architecture and the corresponding weights of the network is a very complex task (for a review see [52]). Modular systems are often used in machine learning as an approach for solving these complex problems. Moreover, in spite of the fact that small networks are preferred because they usually lead to a better performance, the error surfaces of such networks are more rugged and have few good solutions [44]. In addition, there is much neuropsychological evidence showing that the brain of humans and other animals consists of modules, which are subdivisions in identifiable parts, each one with its own purpose and function [6].

Evolutionary computation [18] is a set of global optimization techniques that have been widely used in the last few years for training and automatically designing neural networks. Some efforts have been made in designing modular [5] neural networks with these techniques(e.g. [53]), but in almost all of them the design of the networks is helped by methods outside evolutionary computation, or the application area for those models is limited to very specific architectures.

Cooperative coevolution [39] is a recent paradigm in the area of evolutionary computation focused on the evolution of coadapted subcomponents without external interaction. In cooperative coevolution a number of species are evolved together. The cooperation among the individuals is encouraged by rewarding the individuals based on how well they cooperate to solve a target problem. The work on this paradigm has shown that cooperative coevolutionary models present many interesting features, such as specialization through genetic isolation, generalization and efficiency [39]. Cooperative coevolution approaches the design of modular systems in a natural way, as the modularity is part of the model. Other models need some *a priori* knowledge to decompose the problem *by hand*. In most cases, either this knowledge is not available or it is not clear how to decompose the problem.

The automatic design of artificial neural networks has two basic sides: parametric learning and structural learning. In structural learning, both architecture and parametric information must be learned through the process of training. Basically, we can consider three models of structural learning: Constructive algorithms, destructive algorithms, and evolutionary computation.

Constructive algorithms [14] [21] [37] start with a small network (usually a single neuron). This network is trained until it is unable to continue learning, then new components are added to the network. This process is repeated until a satisfactory solution is found. Destructive methods, also known as pruning algorithms [40], start with a big network, which is able to learn but usually ends in over-fitting, and try to remove the connections and nodes that are not useful.

Evolutionary computation has been widely used in the last few years to evolve neural network architectures and weights. There have been many applications for parametric learning [47] and for both parametric and structural learning [53] [2] [34] [45] [29] [4] [49] [3]. These works fall in two broad categories of evolutionary computation: genetic algorithms and evolutionary programming.

Genetic algorithms are based on a representation independent of the problem, usually the representation is a string of binary, integer or real numbers. This representation (the genotype) encodes a network (the phenotype). This is a dual representation scheme. The ability to create better solutions in a genetic algorithm relies mainly on the operation of *crossover*. This operator forms offspring by recombining representational components from two members of the population.

Evolutionary programming [11] is, according to many authors, the most suitable paradigm of evolutionary computation for evolving artificial neural networks [2]. Evolutionary programming uses a natural representation for the problem. Once the representation scheme has been chosen, mutation operators specific to the representation scheme are defined. Evolutionary programming offers a major advantage over genetic programming when evolving artificial neural networks, the representation scheme allows to manipulate networks directly.

The use of evolutionary learning for designing neural networks dates from no more than two decades (see [52] or [43] for reviews). However, a lot of work has been made in these two decades, leaving many different approaches and working models, for instance, [34], [3], or [53]. Evolutionary computation has been used for learning connection weights and for learning both architecture and connection weights. The main advantage of evolutionary computation is that it performs a global exploration of the search space to avoid becoming trapped in local minima, as it usually happens with local search procedures.

G. F. Miller *et al.* [30] proposed that evolutionary computation is a very good candidate to be used to search the space of topologies, because the fitness function associated with that space is complex, noisy, non-differentiable, multi-modal and deceptive.

Almost all the current models try to develop a global architecture, which is a very complex problem. Although, some attempts have been made in developing modular networks [25] [41], in most cases the modules are combined only after the evolutionary process has finished and not following a cooperative coevolutionary model.

Few authors have devoted their attention to the cooperative coevolution of subnetworks. Some authors have termed this kind of cooperative evolution (where the individuals must cooperate to achieve a good performance) *symbiotic evolution* [31]. More formally, we should speak of *mutualism*, that is, the cooperation of two individuals from different species that benefits both organisms.

R. Smalz and M. Conrad [45] developed a cooperative model with two populations: a population of nodes, divided into clusters, and a population of networks that are combinations of neurons, one from each cluster. Both populations are evolved separately.

B. A. Whitehead and T. D. Choate [49] developed a cooperative-competitive genetic model for Radial-Basis Function (RBF) neural networks. In this work there is a population of genetically encoded neurons that evolves both the centers and the widths of the radial basis functions. There is just one network that is formed by the whole population of RBF's.

D. W. Opitz and J. W. Shavlik [36] developed a model called ADDEMUP (*Accurate and Diverse Ensemble Maker giving United Predictions*). They evolved a population of networks by means of a genetic algorithm and combined the networks in an ensemble with a linear combination. The competition among the networks is encouraged with a diversity term added to the fitness of each network.

D. E. Moriarty and R. Miikkulainen [31] developed an genuine cooperative model, called SANE, which had some common points with the work of R. Smalz and M. Conrad [45]. In this work they propose two populations: one of nodes and another of networks that are combinations of the individuals from the population of nodes. Zhao *et al.* [56] proposed a framework for cooperative coevolution, and applied that framework to the evolution of RBF networks. Nevertheless, their work, rather than a finished model, is an open proposal that aims at the definition of the problems to be solved in a cooperative environment.

S-B. Cho and K. Shimohara [6] developed a modular neural network evolved by means of genetic programming. Each network is a complex structure formed by different modules which are codified by a tree structure.

This chapter is organized as follows: Section 20.2 describes the application of cooperative coevolution to network design; Section 20.3 shows how the use of multiobjective optimization can improve that application; Section 20.4 uses the same principles to the effective evolution of ensembles of neural networks; finally Section 20.5 states the conclusions we can extract from the three described models.

20.2 COVNET: A Cooperative Coevolutionary Model

COVNET [16] is a cooperative coevolutionary model, that is, several species are coevolved together. Each species is a subnetwork that constitutes a partial solution of a problem; the combination of several individuals from different species makes up the network that must be applied to the specific problem. The population of subnetworks, *modules*, is made up by several subpopulations[1] that evolve independently. Each one of these subpopulations constitutes a species. The combination of individuals from these different subpopulations that coevolve together is the key factor of the model.

The evolution of coadapted subcomponents must address four major issues: problem decomposition, interdependence among subcomponents, credit assignment and maintenance of diversity. Cooperative coevolution gives a framework where these issues could be faced in a natural way. The problem decomposition is intrinsic in the model. Each population will evolve different species that must cooperate in order to be rewarded with high fitness values. There is no need to any *a priori* knowledge to decompose the problem *by hand.* The interdependence among the subcomponents comes from the fact that the fitness of each individual depends on how well the individual works together with the members of other species. Credit assignment is made using a method developed together with the coevolutionary model (see Section 20.2.3). The diversity is maintained along the evolution due to the fact that each species is evolved without exchanging genetic material among them. This is an important aspect of the cooperative coevolutionary model. Exchanging genetic material between two different species (that it, subpopulations) will usually produce non-viable offspring. Moreover, the mixing of genetic material might reduce the diversity of the populations.

A module is made up of a variable number of nodes with free interconnection among them, that is, each node could have connections from input nodes, from other nodes of the module, and to output nodes. More formally a module could be defined as follows:

Definition 1. (Module) *A module is a subnetwork formed by: a set of nodes with free interconnection among them, the connection of these nodes from the input and the connections of the nodes to the output. It cannot have connections with a node belonging to another module.*

The input and output layers of the modules are common, they are the input and output layers of the network. It is important to note that the genotype of the module has a one-to-one mapping to the phenotype, as the many-to-one mapping between them is one of the main sources of deception [2].

In the same way we define a network as a combination of modules, and can be defined:

Definition 2. (Network) *A network is the combination of a finite number of modules. The output of the network is the sum of the outputs of all the modules that constitute the network.*

In practice all the networks of the population must have the same number of modules, and this number, N, is fixed along the evolution.

---

[1] Each subpopulation evolves independently, so we can talk of subpopulations or species indistinctly, as each subpopulation will constitute a different species.

As there is no restriction in the connectivity of the module, the transmission of the impulse along the connections must be defined in a way that avoids recurrence. The transmission has been defined in three steps:

1. Each node generates its output as a function of the inputs of the module (that is, the inputs of the whole network): $p_i = f^i\left(\sum_{j=0}^{n} w_{i,j}x_j\right)$, this value is called *partial output.*
2. These partial outputs are propagated along the connections. Then, each node generates its output: $y_i = f^i\left(\sum_{j=0}^{n} w_{i,j}x_j + \sum_{j=1}^{h} w_{i,n+j}p_j\right)$.
3. Finally, each node of the output layer of the module generates its output: $o_j = f^{output}\left(\sum_{i=1}^{h} w_{i,n+h+j}y_i\right)$.

These three steps are repeated over all the modules. The actual output vector of the network is the sum of the output vectors generated by each module.

As the modules must coevolve to develop different behaviors we have N_P independent subpopulations of modules that evolve separately. The network will always have N_P modules, each one from one different subpopulation of modules. Nevertheless, the task is not only developing cooperative modules but also obtaining the best combinations. For that reason we have also a population of networks. This population keeps track of the best combinations of modules and evolves as the population of modules evolves.

Niche creation is implicit, as the subpopulations must coevolve complementary behaviors in order to get useful networks, as the combination of several modules with the same behavior when they receive the same inputs would not produce networks with a good fitness value. So, there is no need of calculating a *fitness sharing* measure that can bias the evolutionary process.

20.2.1 Module Population

The module population is formed by N_P subpopulations. Each subpopulation consists of a fixed number of modules. The population is subject to the operations of replication and mutation. Crossover is not used due to its disadvantages in evolving artificial neural networks [2]. With these features the algorithm falls in the class of evolutionary programming [11].

There is no limitation in the structure of the module or in the connections among the nodes. There are only one restriction to avoid unnecessary complexity in the resulting modules, there can be no connections to an input node or from an output node.

The algorithm for the generation of a new module subpopulation is similar to other models proposed in the literature, such as GNARL [2], EPNet [53], or the genetic algorithm developed by G. Bebis *et al.* [3] The steps for generating the subpopulations are the following:

- The modules of the initial subpopulation are created randomly. The number of nodes of the module, h, is obtained from a uniform distribution: $0 \leq h \leq h_{max}$. Each node is created with a number of connections, c, taken from a uniform distribution: $0 \leq c \leq c_{max}$. The initial value of the weights is uniformly distributed in the interval $[w_{min}, w_{max}]$.
- The new subpopulation is generated replicating the best $P\%$ of the former population. The remaining $(1 - P)\%$ is removed and replaced by mutated copies of the best $P\%$. An individual of the best $P\%$ is selected by roulette selection and mutated. This mutated copy substitutes one of the worst $(1 - P)\%$ individuals.
- There are two types of mutation: parametric and structural. The severity of the mutation is determined by the relative fitness, F_r, of the module. Given a module ν its relative fitness is defined as:

$$F_r = e^{-\alpha F(\nu)}. \tag{20.1}$$

where $F(\nu)$ is the fitness value of module ν.

Parametric mutation consists of a local search algorithm in the space of weights, a simulated annealing algorithm. This algorithm performs random steps in the space of weights. Each random step affects all the weights of the module.

Structural mutation is more complex because it implies a modification of the structure of the module. The behavioral link between parents and their offspring must be enforced to avoid generational gaps that produce inconsistency in the evolution. There are four different structural mutations: Addition of a node with no connections to enforce the behavioral link with its parent [53] [2], deletion of a node, addition of a connection with 0 weight, and deletion of a connection.

All the above mutations are made in the mutation operation on the module. For each mutation there is a minimum value, Δ_m, and a maximum value, Δ_M. The number of elements (nodes or connections) involved in the mutation is calculated as follows:

$$\Delta = \Delta_m + F_r(\nu)(\Delta_M - \Delta_m). \tag{20.2}$$

So, before making a mutation the number of elements, Δ, is calculated, if $\Delta = 0$ the mutation is not actually carried out.

There is no migration among the subpopulations. So, each subpopulation must develop different behaviors of their modules, that is, different species of modules, in order to compete with the other subpopulations for conquering its own niche and to cooperate to form networks with high fitness values.

20.2.2 Network Population

The network population is formed by a fixed number of networks. Each network is the combination of one module of each subpopulation of modules. So the networks are strings of integer numbers of fixed length. The value of the numbers is not significant as they are just labels of the modules. The network population is evolved using the *steady-state* genetic algorithm [50].

Crossover is made at module level, using a standard two-point crossover. So the parents exchange their modules to generate their offspring. Mutation is also carried out at module level. When a network is mutated one of its modules is selected and is substituted by another module of the same subpopulation selected by means of a roulette algorithm.

During the generation of the new module population some modules of every population are removed and substituted. The removed modules are also substituted in the networks. This substitution has two advantages: first, poor performing modules are removed from the networks and substituted by potentially better ones; second, the new modules have the opportunity to participate in the networks immediately after their creation.

The general structure of the two populations is shown in Figure 20.1. In this case each member of the individuals population is a network and each member of each subpopulation of subcomponents is a module.

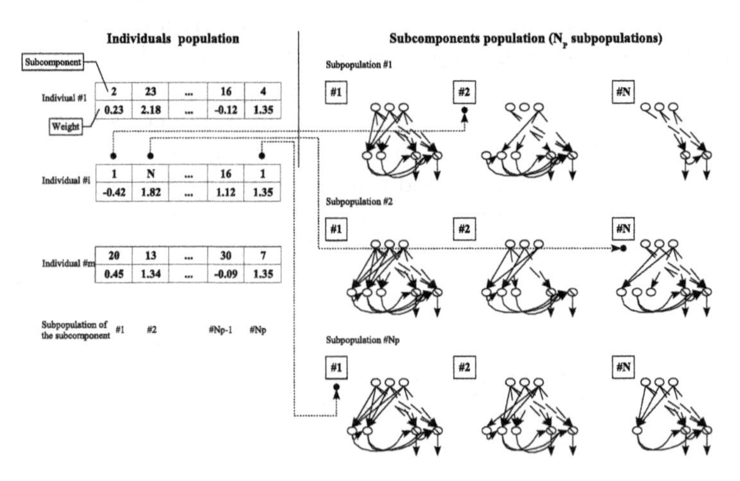

Fig. 20.1. Populations of individuals and subcomponents. This general structure is common to the three different models explained in this chapter.

20.2.3 Fitness Assignment

The assignment of fitness to networks is straightforward. Each network is assigned a fitness in function of its performance in solving a given problem. If the model is applied to classification, the fitness of each network is the number of patterns of the training set that are correctly classified; if it is applied to regression, the fitness is the sum of squared errors, and so on. In the test problems below the classification is made following the criterion of the maximum: the pattern is assigned to the class whose corresponding output is the highest one. Ties are resolved arbitrarily assigning the pattern to a *default class*.

Assigning fitness to the modules is a much more complex problem. In fact, the assignment of fitness to the individuals that form a solution in cooperative evolution is one of its key issues. The performance of the model highly depends on that assignment. A discussion of the matter can be found in the Introduction of [39].

A credit assignment must fulfill the following requirements to be useful:

- It must enforce competition among the subpopulations to avoid two subpopulations developing similar responses to the same features of the data.
- It must enforce cooperation. The different subpopulations must develop complementary features that together could solve the problem.
- It must measure the contribution of a module to the fitness of the network, and not only the performance of the networks where the module is present. A module in a good network must not get a high fitness if its contribution to the performance of the network is not significant. Likewise, a module in a poor performing network must not be penalized if its contribution to the fitness of the network is positive.

Some methods for calculating the fitness of the modules have been tried. The best one consists of the weighted sum of three different criteria. These criteria, for obtaining the fitness of a module ν in a subpopulation π, are:

- *Substitution* (σ). k networks are selected using an elitist method, that is, the best k networks of the population. In these networks the module of subpopulation π is substituted by the module ν. The fitness of the network with the module of the population π substituted by ν is measured. The fitness assigned to the module is the averaged difference in the fitness of the networks with the original module and with the module substituted by ν. This criterion has two important features:
  - It encourages the modules to compete within the subpopulations, rewarding the modules most *compatible* with the modules of the rest of the subpopulation. This is true even for a distributed representation, because it has been shown that such representation is also modular. Moreover, as the modules have no connection among them, they are more independent than in a standard network.
  - As many of the modules are competing with their parents, this criterion allows to measure if an offspring is able to improve the performance of its parent.
- *Difference*(δ). The module is removed from all the networks where it is present. The fitness is measured as the difference in performance of these networks. This criterion enforces competition among subpopulations of

modules preventing more than one subpopulation from developing the same behavior.

- *Best k (β_k).* The fitness is the mean of the fitness values of the best k networks where the module ν is present. Only the best k are selected because the importance of the worst networks of the population must not be significant. This criterion rewards the modules in the best networks, and does not penalize a good module if it is in some poor performing networks.

Considered independently none of these criteria is able to fulfill the three desired features above mentioned. Nevertheless, when the weighted sum of all of them is used they have proved to give a good performance in the problems used as tests. Typical values of the weights of the components of the fitness used in the experiment are $\lambda_\delta \simeq 2\lambda_\sigma \simeq 60\lambda_{\beta_n}$.

In order to encourage small modules, we have included a regularization term in the fitness of the module. Being n_n the number of nodes of the module and n_c the number of connections, the *effective fitness*, f_i', of the module is calculated following:

$$f_i' = f_i - \rho_n n_n - \rho_c n_c. \tag{20.3}$$

The values of the coefficients must be in the interval $0 < \rho_n, \rho_c << 1$ in order to avoid the regularization term introducing a high bias in the learning process.

So, the equation of the effective fitness of the module ν of subpopulation π is the following:

$$f_\nu^\pi = \lambda_\sigma \sigma + \lambda_\delta \delta + \lambda_{\beta_k} \beta_k - \rho_n n_n - \rho_c n_c, \tag{20.4}$$

if the expression above is negative for any of the modules of a subpopulation, then the fitness values of all the modules of that subpopulation are shifted, as we have mentioned above, as following:

$$f_\nu^\pi = f_\nu^\pi - \min\{f_i^\pi\}_{i=1}^N, \tag{20.5}$$

where N is the number of modules of the module subpopulation. In the experiments reported below the values used for these parameters are $\lambda_\sigma = 3.50$, $\lambda_\delta = 1.45$, $\lambda_{\beta_k} = 0.05$, $\rho_n = 0.25$, and $\rho_c = 0.025$.

20.2.4 Experiments

In order to test the efficiency of this approach we have compared COVNET with a modular network formed by 4 experts, each one a standard multi-layered perceptron. We have used four problems from the UCI repository [20]. Table 20.1 shows the results of the models for the four problems. The table shows how COVNET is able to outperform the modular network in all problems. The comparison between the two methods is made by means of a Student's t-test [1]. This test tells us if the difference between the means of two repeated experiments are significantly different from a statistical point of view. At a confidence level of 95% we can say that the two experiments have different means if the p-value of the test, that is, the probability that a variate would assume a value greater than or equal to the observed value strictly by chance, is less than 0.05. So, the differences in mean are statistically significant as it is shown by the t-test in Table 20.1.

Table 20.1. Test error results of COVNET. The mean and standard deviation of the test error averaged over the 30 runs of the algorithm are shown.

Problem	COVNET		Modular		t-test
	Mean	σ	Mean	σ	
Card	0.1157	0.0104	0.1374	0.0106	0.0000
Glass	0.3723	0.0239	0.3904	0.0612	0.0136
Heart	0.1426	0.0279	0.1941	0.0310	0.0000
Pima	0.1990	0.0220	0.2299	0.0274	0.0000

20.3 MOBNET

The results of the previous section show how combining different criteria the assignment of credit to subcomponents could be improved. As we have mentioned, improving the fitness assignment is one of the key issues in cooperative coevolution. However, the combination of several criteria to obtain the fitness of each subcomponent has two major problems:

1. The relevance of each criterion must be pondered. This is a complex and time consuming task, involving a long process of trial and error.
2. The ranges of the objectives are different. In most cases if the objectives are not weighted, some of the objectives could be overshadowed by objectives with higher values. This problem becomes even harder if we take into account that the ranges of the different objectives are different in every run of the algorithm.

These two problems become more important as more objectives are introduced in the evaluation of the fitness of an individual, preventing the addition of interesting criteria.

On the other hand, if we approach this problem as a multi-objective optimization task, we will benefit from many advantages. First, there is no need to weight the objectives as their relevance and range do not matter any more. Secondly, the solutions based on Pareto optimality guarantee the diversity of the final population. And third, there is an underlying theory applicable to the problem.

20.3.1 Multiobjective Optimization

The principles of multi-objective optimization are quite different from those of single objective optimization. The main goal in a single-objective optimization task is to find the global optimal solution. However, in a multi-objective optimization problem, there are more than one objective function, each of which may have a different individual optimal solution. If there is sufficient difference in the optimal solutions corresponding to different objectives, the objective functions are often known as conflicting to each other.

Multi-objective optimization with such conflicting objective functions gives rise to a set of optimal solutions, instead of one optimal solution, as none of the solutions could be considered to be *the best one*, because none of them is better than any other with respect to all objective functions. These optimal solutions have a special name – Pareto-optimal solutions.

The basic concept in multi-objective optimization is the concept of domination. Being $x^{(i)}$ a solution and $\mathbf{f}(x^{(i)}) = (f_1(x^{(i)}), f_2(x^{(i)}), \ldots, f_M(x^{(i)}))$ the evaluation of that solution for every objective, this concept is defined as follows:

Definition 3. *(Domination) A solution $x^{(1)}$ is said to dominate another solution $x^{(2)}$ if both the following conditions are true:*

1. *The solution $x^{(1)}$ is no worse than $x^{(2)}$ in all objectives, that is, $f_j(x^{(1)}) \not\prec f_j(x^{(2)})$ for all $j = 1, 2, \ldots, M$ objectives, where $\prec$ denotes worse.*
2. *The solution $x^{(1)}$ is strictly better than the solution $x^{(2)}$ in at least one objective, that is, $f_{\bar{j}}(x^{(1)}) \succ f_{\bar{j}}(x^{(2)})$ for at least one $\bar{j} \in \{1, 2, \ldots, M\}$, where $\succ$ denotes better.*

Similar to the definitions of local and global optimal solutions in single-objective optimization, the following definitions are their equivalent in multi-objective optimization:

Definition 4. *(Local Pareto-optimal Set) If for every member x in a set $\underline{P}$, there exists no solution y satisfying $\|y - x\|_\infty \leq \epsilon$, where ϵ is a small positive number, which dominates any member in the set $\underline{P}$, then the solutions belonging to the set $\underline{P}$ constitute a local Pareto-optimal set.*

Definition 5. *(Global Pareto-optimal set) If there exists no solution in the search space which dominates any member in the set $\underline{P}$, then the solutions belonging to the set $\underline{P}$ constitute a global Pareto-optimal set.*

There are two primary goals that a multi-objective optimization algorithm must achieve:

1. Guide the search towards the global Pareto-optimal region.
2. Maintain the population diversity in the Pareto-optimal front.

The multi-objective evolutionary algorithms are based on the concept of non-domination, and almost all of them follow the principles sketched by D. Goldberg [18][2]. Among the most successful algorithms are the multi-objective genetic algorithm of Fonseca and Fleming [12], the niched Pareto genetic algorithm of Horn, Nafploitis and Goldberg [22], the strength Pareto evolutionary algorithm of Zitzler and Theile [57] and the non-dominated sorting genetic algorithm of Srinivas and Deb [46]. The algorithm we have applied is basically an adaptation of the latter to evolutionary programming.

20.3.2 Multi-objective Evaluation of Fitness

MOBNET [15] (after *Multi OBjective evolutionary NETwork*) improves COVNET model introducing the multi-objective evaluation of the fitness of the modules and networks. As in COVNET the network population is formed by a fixed number of networks. Each network is the combination of one module of each subpopulation of modules.

The multi-objective algorithm is common to both populations, modules and networks. We will consider a population of individuals where individual i has a vector of objectives values $x^{(i)}$. The population has N individuals and M objectives are considered.

The proposed algorithm is based on the concept of Pareto optimality explained above. It has common points with other multi-objective evolutionary algorithms. This algorithm is basically an adaptation of NSGA [46] to evolutionary programming. The idea underlying NSGA is the use of a ranking selection method to emphasize current non-dominated individuals and a niching method to maintain diversity in the population.

The algorithm consists of two stages. First, the successive non-dominated fronts[3] are obtained and every individual of these fronts is assigned an equal dummy fitness, $\mathrm{F}_{\mathrm{dummy}}$. Second, the members of every front share their fitness. The sharing procedure must guarantee that none of the member of a front gets a higher fitness than any of the member of the previous front.

Once the individuals of a non-dominated front are assigned their fitness, they are not considered any more for obtaining the new non-dominated front. That is the reason why we talk about *successive non-dominated fronts.*

The problem of fitness sharing among the members of the same Pareto front is crucial, as the performance of the algorithm depends highly on it. The standard method of explicit fitness sharing [19] cannot be applied as the individual are not in a Euclidean space. So we must define a measure

[2] Although the first practical implementation was suggested by J. D. Schaffer [42].

[3] A non-dominated front is a subset of individuals that are not dominated by any member of the population.

of diversity for modules and networks in order to apply the algorithm. Our interest is focused on keeping the behavioral diversity among the individuals, so the measures are based on the functional diversity of the modules and networks. These measures are the following:

- *Module's Functional diversity.* This measure is based on comparing the discrepancy between the outputs of the two modules. In order to obtain this measure for two modules n_1 and n_2, $fd(n_1, n_2)$, we set in turn every input connection to 1 and the rest of the connections to 0, and get the output of the modules.
  The functional diversity is the average distance of the output of the two modules over all the inputs. This measure is used for applying a standard fitness sharing algorithm.
  Given a set of n_k modules of the k-th non-dominated front each having a dummy fitness of f_{dummy}, the sharing procedure is performed for each solution $i = 1, 2, \ldots, n_k$.
- *Network's functional diversity.* The same idea of the above method is applied to the network population. The measure of network functional diversity is based on the functional diversity of modules. Given two networks $n^i = (n_1^i, n_2^i, \ldots, n_m^i)$ and $n^j = (n_1^j, n_2^j, \ldots, n_m^j)$ their functional diversity, ϕ, is defined as:

$$\phi(n^i, n^j) = \sum_{k=1}^{m} fd(n_k^i, n_k^j). \tag{20.6}$$

  With this distance measure the above sharing algorithm is applied, substituting the measure $fd(n_i, n_j)$ by $\phi(n^i, n^j)$.

Module's Objectives

As we have stated, the introduction of multi-objective optimization gives us the freedom to define many objectives to be fulfilled by the modules. The diversity that guarantees the Pareto-optimal solutions allows the combination of modules that are good in different objectives.

Many different objectives could be defined, specially when the available *a priori* knowledge of a problem allows the definition of some of the features of the networks. We have used the three objectives of COVNET, difference, substitution, and best, and added the following additional objectives:

- *Complexity* (3 objectives). Three objectives play the role of a regularization term. These objectives are the number of nodes, the number of connections and the sum of the weights of the modules. All of them are taken in negative value, as we want to minimize them. In most problems selecting just the number of nodes is enough to control the size of the networks. If smaller networks are needed, the other two objective could be used.

- *Count of networks.* Number of networks where the module participates. It encourages the module to be part of the networks. Moreover, when a module participates in more networks, its fitness is more accurately estimated and must be rewarded.
  This objective also avoids, in part, the negative effect after the elimination of a module that participates in many networks during the evolutionary process.

All the members of all the subpopulations of modules evolve with the same subset of objectives.

Network's Objectives

Each individual of the population of networks has also several objectives. Initially only the performance of the network was considered, as this performance can be measured for networks. But, as the module population evolves better considering more objectives that can guide the evolution towards the most interesting subcomponents, this same principle could be applied to the evolution of networks.

So we also applied to networks the multi-objective evolutionary algorithm developed for modules. The objectives defined are the following:

- *Performance.* Performance of the network in solving the given task. As we have applied the model only to classification tasks, the performance of the network is just the number of patterns from the training set correctly classified.
- *Fitness of each module.* As many objectives as module subpopulations. With this objective we intend to encourage the combination of the best individuals of the module subpopulations.
  This objective has also a very interesting *side effect*, as the modules with the highest fitness values will survive during the evolution, the combination of modules with high fitness values are more durable and so there is more time for evolving to useful networks.

20.3.3 Experiments

In order to test the efficiency of this approach we have compared MOBNET with a standard multi-layered perceptron trained using the `quickprop` algorithm. We have used four problems from the UCI repository. Table 20.2 shows the results of the models for the three problems. The table shows how MOBNET is able to outperform the modular network in all problems. The differences in mean are statistically significant as it is shown by the t-test.

These results also show that MOBNET is able to obtain better results than COVNET. In this way, the approach using multi-objective optimization outperforms its counterpart without multi-objective optimization.

Table 20.2. Test error results of MOBNET. The mean and standard deviation of the test error averaged over the 30 runs of the algorithm are shown.

Problem	MOBNET		MLP		t-test
	Mean	σ	Mean	σ	
Card	0.1138	0.0115	0.1273	0.0176	0.0009
Glass	0.3516	0.0559	0.4025	0.0481	0.0000
Heart	0.1363	0.0278	0.1559	0.0221	0.0037
Pima	0.1984	0.0176	0.2156	0.0212	0.0012

20.4 Cooperative Ensemble of Neural Networks

The cooperative model, based on two separate populations that evolve cooperatively, can also be applied to the evolution of ensembles of neural networks [17]. If we adapt the model to the evolution of ensembles of neural networks, we have two populations:

- *Population of networks.* This population consists of a number of independent subpopulations of networks. The independent evolution of subpopulations is an effective way of keeping the networks of different populations diverse. The absence of genetic material exchange among subpopulations also tends to produce more diverse networks whose combination is more effective. Every subpopulation is evolved using evolutionary programming.
- *Population of ensembles.* Each member of the population of ensembles is an ensemble formed by a network from every network subpopulation. Each network has an associated weight.
  The population of ensembles keeps track of the best combinations of networks, selecting the subsets of networks that are promising for the final ensemble.

Additionally, every network is subject to back-propagation training throughout its evolution with a certain probability. In this way, the network is allowed to learn from the training set, but it is also prevented from being too similar to the rest of the networks by means of its evaluation using different objectives. The back-propagation algorithm is implemented as a mutation operator.

Our basic network is a generalized multi-layer perceptron (GMLP), as defined in [48]. It consists of an input layer, an output layer and a number of hidden nodes interconnected among them.

Given a GMLP with m inputs, N hidden nodes, and n outputs, and $\mathbf{X}$ and $\mathbf{Y}$ being the input and output vectors respectively, it is defined by the equations [48]:

$$\begin{aligned} x_i &= X_i, & 1 \leq i \leq m \\ h_i &= \textstyle\sum_{j=1}^{i-1} w_{ij} x_j, & m < i \leq m+N+n \\ x_j &= f(h_j), & m < j \leq m+N+n \\ Y_i &= x_{i+m+N}, & 1 \leq i \leq n. \end{aligned} \tag{20.7}$$

where w_{ij} is the weight of the connection from node j to node i.

The network population is formed by N_P subpopulations. These networks are not fully connected. When a network is initialized, each connection is created with a given probability. The population is subject to operations of replication and mutation. Crossover is not used due to its potential disadvantages [2] that it has for evolving neural networks. With these features the algorithm falls in the class of evolutionary programming [11].

The algorithm for the evolution of the subpopulations of networks is basically the same used in COVNET and MOBNET for the evolution of modules. The ensemble population is formed by a fixed number of ensembles. Each ensemble is the combination of one network from each subpopulation of networks with an associated weight.

The objective of any method for developing network ensembles must be obtaining accurate networks as diverse as possible. Diversity is assured by means of three different mechanisms:

1. Coevolution of genetically isolated subpopulation of networks. As there is no exchange of genetic material among the members of the different subpopulations, diversity among the subpopulations is preserved.
2. Fitness-sharing in the evaluation of the networks. When a network is evaluated, we use fitness-sharing for decreasing the fitness of the networks functionally close to each other.
3. Objectives of diversity. Additionally, each network is evaluated using one or more objectives of diversity. Again, diverse networks are rewarded. The evaluation of the ensembles also includes an objective rewarding the ensembles formed by diverse networks.

20.4.1 Multi-objective Evaluation of Network and Ensemble Fitness

If we approach the problem of evaluation of the fitness of networks and ensembles as a multi-objective optimization task, we will benefit from the same advantages that we have mentioned for the evolution of modular networks: (i) there is no need to weight the different objectives, (ii) the solutions based on Pareto optimality guarantee the diversity of the final population, and (iii) there is an underlying theory applicable to the problem.

The next two sections describe the objectives that have been defined for networks and ensembles. These are only a subset of all the objectives that may be interesting for a given task. One of the advantages of the model is that it allows the introduction of any useful objective without modifying the general structure of the system.

Some of the following objectives as the counterparts of module objectives adapted to the evolution of ensembles instead of the evolution of modular networks.

Individual Networks Objectives

The objectives of the networks could be grouped into four sets: objectives of performance, objectives of regularization, objectives of cooperation and objectives of diversity. These objectives are the following:

Objectives of performance

These objectives measure the performance of the network from three different complementary points of view. The objectives are the following:

- *Performance.* As we use bagging, this measure of performance is the number of patterns correctly classified by the network pondered by the weight of each pattern.
- *Shared performance.* This objective enforces the networks to classify different patterns [27]. In this way, the networks that are able to accurately classify patterns that are incorrectly classified by many ensembles are rewarded. Each pattern receives a value, p_i, that measures the number of ensembles that correctly classify the pattern, namely:

$$p_i = m_{\mathrm{ok}}/M, \tag{20.8}$$

  where m_{ok} is the number of ensembles that classify the pattern correctly, and M is the number of ensembles.
  With this objective the networks that classify *difficult* patterns are rewarded, even if the total number of patterns correctly classified by the network is not high.
- *Ensembles.* Average performance of the ensembles where the network is present. In order to reward the best collaborating networks, this objective measures the average performance of the ensembles where the network participates. When a network does not participate in any ensemble, the objective cannot be calculated. In such a case the objective receives a value of 0.

Objectives of regularization

In order to reward small networks, many measures may be included as regularization terms. These measures can be taken from network pruning [40], or regularization theory [51]. Most authors use the weight decay term proposed in many papers [23], $\sum_{i,j} |w_{ij}|$, other authors propose a cost function of the form [40], $\sum_{i,j}(w_{ij})^2$. Both measures have a strong impact on the evolution of the network due to the heavy constraint that is imposed on the weights. For that reason, we have used the following less restrictive term:

- *Regularization objective.* This term is taken from [33]. The idea is to model the weights of the network using a mixture of two Gaussians, a narrow (n) one, and a broad (b) one:

$$p(w) = \pi_n \frac{1}{\sqrt{2\pi}\sigma_n} e^{-(x-\mu_n)^2/2\sigma_n^2} + \pi_b \frac{1}{\sqrt{2\pi}\sigma_b} e^{-(x-\mu_b)^2/2\sigma_b^2} \tag{20.9}$$

where the parameters of the distributions, π, σ, and μ, are obtained by means of an Expectation-Maximization (EM) algorithm [10]. The effect of this regularization term is a kind of *soft* version of weight-sharing in which the learning process decides itself which weights should be tied together.

Objectives of cooperation

These two objectives explicitly promote cooperation among the networks. Instead of evaluating the performance of networks or ensembles, these objectives evaluates the relevance of the network within the ensembles where it participates and how well it cooperates with the rest of the members of those ensembles. These two objectives are the adaptation of difference and substitution criteria to networks.

Objectives of diversity

In the development of ensembles of neural networks, we must take into account the source of the error on an ensemble. The ensemble generalization error can be expressed:

$$E = \bar{E} - \bar{A}, \tag{20.10}$$

where $\bar{E} = \sum_i \alpha_i E_i$ is the weighted average of the individual networks' generalization errors and $\bar{A} = \sum_i \alpha_i A_i$ is the weighted average of the diversity among these networks. From this point of view, the objective of any ensemble is to obtain highly correct networks that disagree as much as possible.

Maintaining diversity needs some kind of speciation mechanism [22]. The most common techniques are: fitness sharing [19] [26], crowding [9], implicit fitness sharing [22] [32], local mating [8], and negative correlation [28]. Nevertheless most of these methods may bias the evolutionary process.

The importance of the diversity of combined networks in an ensemble have been stated by many authors [38] [25] [28] [24]. Nevertheless, some works raise some doubts about the usefulness of diversity measures in building classifier ensembles in real-world classification problems [24]. Moreover, it is very difficult to determine what measures of diversity are the most suitable for a given task. Our approach takes advantage of the cooperative and multiobjective environment we are using. We define 4 objectives regarding diversity, and the evolutionary process will combine networks that are good at different objectives. The number of diversity measures is enormous, so we have selected four of the most widely used, each one centered on a different idea. These objectives are the following:

- *Correlation.* Following Liu and Yao [25] we introduce an objective that measures the correlation of the error of each individual network with the ensembles where it participates. The error correlation of network i in an ensemble of networks, P_i, is measured using:

$$P_i = \frac{1}{N} \sum_{n=1}^{P} p_i(\mathbf{x}),$$
$$p_i(\mathbf{x}) = (f_i(\mathbf{x}) - F(\mathbf{x})) \sum_{j \neq i} (f_j(\mathbf{x}) - F(\mathbf{x})) \tag{20.11}$$

  where P is the number of training patterns, $f_i(\mathbf{x})$ is the output of network i for pattern $\mathbf{x}$, and $F(\mathbf{x})$ is the output of the ensemble for pattern $\mathbf{x}$. With this measure a network must learn what all other networks have not yet learned.
  The value used as objective is $-\bar{p}_i$, that is, the average error correlation of the network over all the ensembles where it is present.
  From both theoretical and experimental results [7] it has been shown that, if the individual networks in an ensemble are unbiased, the most effective combination of them occurs when the errors of the individual networks are negatively correlated. As a consequence, the mutual information between each individual and the rest of the population should be minimized [28] to improve the estimation of the ensemble.
- *Functional diversity.* The measure we have chosen is the average Euclidean distance among the outputs of the two networks.
  This measure is used to test the discrepancy among the outputs of the networks. For two networks, f_i and f_j, the functional diversity, $\phi(f_i, f_j)$, is defined as:

$$\phi(f_i, f_j) = \frac{1}{P} \sum_{k=1}^{P} \| f_i(\mathbf{x}_k) - f_j(\mathbf{x}_k) \|. \tag{20.12}$$

- *Mutual information.* The mutual information between two networks, f_i and f_j, is given by [28]:

$$I(f_i, f_j) = -\frac{1}{2} \log(1 - \rho_{ij}^2), \tag{20.13}$$

  where ρ_{ij} is the correlation coefficient between f_i and f_j.
- *Yule's Q statistics.* This objective [24] measures whether the mistakes of the classifiers are uncorrelated, and it is one of the various statistics to assess the similarity of two classifiers outputs. We use the standard Yule's Q statistics [54] for two classifiers.

Ensemble Objectives

The same principles of multi-objective evolution can also be applied to the evolution of ensembles. So, we have defined two objectives to be considered in the evaluation of ensemble fitness.

These two objectives for the ensembles are:

- *Performance.* This objective is just the performance of the network measured as the number of patterns classified correctly.
- *Ambiguity.* If the in-correlation among the networks that form the ensemble is increased, *without increasing the individual errors*, the global error of the ensemble is reduced. The ambiguity is defined over the training set.

As in MOBNET the multi-objective algorithm is common to both populations, networks and ensembles, and is the same algorithm adapted to MOBNET from NSGA.

20.4.2 Experiments

The use of the ten objectives is not advisable, because some of the objectives share the same principles. So, we did an initial experiment in order to test whether all the objectives were useful in the evolution. From these experiments we obtain a subset of 6 objectives to be used for all the problems. This subset is made up of: {*Difference, Substitution, Ensembles, Shared performance, Regularization, Yule's Q*}. These objectives were obtained with the criterion of selecting at least one objective from each group and within each group selecting the best performing one in the initial experiment.

In order to test the efficiency of this approach we have compared the cooperative ensemble with a standard ensemble. The standard ensemble is made up of 25 networks. It is known [55] that the diversity and the accuracy of the ensemble usually plateau at some size between 10 and 50 members. Moreover, Opitz and Maclin [35] have found after some exhaustive experiments that the error of the ensemble does not decrease after adding 25 networks. We use Ada boosting [13] to train the ensemble.

We have used ten problems from the UCI repository. Table 20.3 shows the results of the models for the ten problems. For Soybean and Vowel problems the Ada method could not be applied, as it requires that each individual classifier has an accuracy above 50%.

The table shows how the cooperative ensemble is able to outperform the standard ensemble in all problems. The differences in mean are statistically significant as it is shown by the t-test.

20.5 Conclusions

In this chapter we have shown how the use of cooperative coevolution is useful in the evolution of neural networks and ensembles of neural networks. We

Table 20.3. Test error results of the cooperative ensemble. The mean and standard deviation of the test error averaged over the 30 runs of the algorithm are shown.

Problem	C oop. ensemble		Ada		t-test
	Mean	σ	Mean	σ	
Cancer	0.0123	0.0047	0.0249	0.0046	0.0000
Card	0.1217	0.0103	0.1390	0.0144	0.0000
Gene	0.1238	0.0105	0.1429	0.0092	0.0000
Glass	0.2289	0.0482	0.3145	0.0259	0.0000
Heart	0.1196	0.0207	0.1569	0.0260	0.0000
Horse	0.2674	0.0343	0.2985	0.0233	0.0001
Pima	0.1969	0.0170	0.2076	0.0213	0.0347
Sonar	0.1436	0.0153	0.2253	0.0173	0.0000
Soybean	0.0761	0.0083	–	–	–
Vowel	0.4587	0.0269	–	–	–

have presented a cooperative coevolutionary model for the design of artificial neural networks. This model is based on the coevolution of several species of subnetworks that must cooperate to form networks for solving a given problem. Instead of trying to evolve whole networks, a task that is not feasible in many problems or ends up with poorly performing neural networks, we evolve these subnetworks that must cooperate in solving the given task.

We have presented a new method for assigning credit to the individuals of the different species that cooperate to form a network. This method is based on the combination of three criteria. The criteria enforce competition within species and cooperation among species. The same idea underlying this method could be applied to other models of cooperative coevolution. The extension of this model using multi-objective optimization obtains very good results in real-world problems.

This model has proved to perform better than standard algorithms in two real problems of classification. Moreover, it has shown better results than the methods of training modular neural networks by means of gradient descent, e.g. the backpropagation learning rule, and it has achieved better results for two of the three tested problems than the results reported in the literature and comparable results in the other problem, and with less complexity than most models.

We have also shown how cooperative coevolution offers a framework for all the steps of the design and training of network ensembles. The simultaneous evolution of all the networks that form the ensemble allows us to obtain more cooperative networks with a performance significantly above the performance of classic ensemble methods.

The multi-objective evaluation of the fitness of the subcomponents introduces the possibility of enforcing several aspects of the networks that are interesting for a better performance of the ensembles. In this work we have proposed a set of general objectives that can be applied for any problem. However, the definition of other sets of objectives that may be adequate for a given problem might improve the performance of the model.

Acknowledgement

The author would like to acknowledge Rosa Moya-Sánchez for her helping in the final version of this paper.

References

[1] T. W. Anderson. *An introduction to multivariate statistical analysis.* Wiley Series in Probability and Mathematical Statistics. John Wiley & Sons, New York, 2nd edition, 1984.
[2] P. J. Angeline, G. M. Saunders, and J. B. Pollack. An evolutionary algorithm that constructs recurrent neural networks. *IEEE Transactions on Neural Networks*, 5(1):54 – 65, January 1994.
[3] G. Bebis, M. Georgiopoulos, and T. Kasparis. Coupling weight elimination with genetic algorithms to reduce network size and preserve generalization. *Neurocomputing*, 17:167–194, 1997.
[4] M. V. Borst. *Local structure optimization in evolutionary generated neural networks architectures.* PhD thesis, Leiden University, The Netherlands, August 1994.
[5] T. Caelli, L. Guan, and W. Wen. Modularity in neural computing. *Proceedings of the IEEE*, 87(9):1497–1518, September 1999.
[6] S-B. Cho and K. Shimohara. Evolutionary learning of modular neural networks with genetic programming. *Applied Intelligence*, 9:191–200, 1998.
[7] R. T. Clemen and R. L. Winkler. Limits for the precision and value of information from dependent sources. *Operations Research*, 33:427–442, 1985.
[8] R. J. Collins and D. R. Jefferson. Selection in massively parallel genetic algorithms. In *Proceedings of the Fourth International Conference on Genetic Algorithms*, pages 249–256, San Mateo. CA, 1991. Morgan Kaufmann.
[9] K. A. de Jong. *An analysis of the behavior of a class of genetic adaptive systems.* PhD thesis, University of Michigan, Ann Arbor, MI, 1975.
[10] A. Dempster, N. Laird, and D. Rubin. Maximum-likelihood from incomplete data via the EM algorithm. *J. Royal Statist. Soc. Ser. B*, 39(1–38), 1977.
[11] D. B. Fogel. *Evolving artificial intelligence.* PhD thesis, University of California, San Diego, 1992.
[12] C. M. Fonseca and P. J. Flemming. Genetic algorithms for multi-objective optimization: Formulation, discussion, and generalization. In S. Forrest, editor, *Proceedings of the Fifth International Conference on Genetic Algorithms*, pages 416–423, Urbana, IL, 1993. Morgan Kauffman Publishers.
[13] Y. Freund and R. Schapire. Experiments with a new boosting algorithm. In *Proc. of the Thirteenth International Conference on Machine Learning*, pages 148–156, Bari, Italy, 1996.

[14] S. Gallant. *Neural-Network Learning and Expert Systems.* MIT Press, Cambridge, MA, 1993.
[15] N. García-Pedrajas, C. Hervás-Martínez, and J. Muñoz-Pérez. Multiobjective cooperative coevolution of artificial neural networks. *Neural Networks*, 15(10):1255–1274, November 2002.
[16] N. García-Pedrajas, C. Hervás-Martínez, and J. Muñoz-Pérez. COVNET: A cooperative coevolutionary model for evolving artificial neural networks. *IEEE Transactions on Neural Networks*, 14(3):575–596, May 2003.
[17] N. García-Pedrajas, C. Hervás-Martínez, and D. Ortiz-Boyer. Cooperative coevolution of artificial neural network ensembles for pattern classification. *IEEE Transactions on Evolutionary Computation*, 9(3):271–302, June 2005.
[18] D. E. Goldberg. *Genetic Algorithms in Search, Optimization and Machine Learning.* Addison–Wesley, Reading, MA, 1989.
[19] D. E. Goldberg and J. Richardson. Genetic algorithms with sharing for multimodal function optimization. In *Proceedings of the Second International Conference on Genetic Algorithms*, pages 148 – 154, San Mateo, CA, 1987. Morgan Kaufmann.
[20] S. Hettich, C.L. Blake, and C.J. Merz. UCI repository of machine learning databases, 1998. http://www.ics.uci.edu/~mlearn/MLRepository.html.
[21] V. Honavar and V. L. Uhr. Generative learning structures for generalized connectionist networks. *Information Science*, 70(1/2):75–108, 1993.
[22] J. Horn, D. E. Goldberg, and K. Deb. Implicit niching in a learning classifier system: Natures's way. *Evolutionary Computation*, 2(1):37 – 66, 1994.
[23] M. Ishikawa. A structural learning algorithm with forgetting of link weights. Technical Report TR-90-7, Electrotechnical Laboratory, Tsukuba-City, Japan, 1990.
[24] L. Kuncheva and C. J. Whitaker. Measures of diversity in classifier ensembles and their relationship with the ensemble accuracy. *Machine Learning*, 51(2):181–207, May 2003.
[25] Y. Liu and X. Yao. Ensemble learning via negative correlation. *Neural Networks*, 12(10):1399–1404, December 1999.
[26] Y. Liu and X. Yao. Simultaneous training of negatively correlated neural networks in an ensemble. *IEEE Trans. on Systems, Man, and Cybernetics, Part B: Cybernetics*, 26(6):716–726, 1999.
[27] Y. Liu, X. Yao, and T. Higuchi. Evolutionary ensembles with negative correlation learning. *IEEE Transactions on Evolutionary Computation*, 4(4):380–387, November 2000.
[28] Y. Liu, X. Yao, Q. Zhao, and T. Higuchi. Evolving a cooperative population of neural networks by minimizing mutual information. In *Proc. of the 2001 IEEE Congress on Evolutionary Computation*, pages 384–389, Seoul, Korea, May 2001.
[29] V. Maniezzo. Genetic evolution of the topology and weight distribution of neural networks. *IEEE Transactions on Neural Networks*, 5(1):39–53, January 1994.
[30] G. F. Miller, P. M. Todd, and S. U. Hedge. Designing neural networks. *Neural Networks*, 4:53–60, 1991.
[31] D. E. Moriarty and R. Miikkulainen. Efficient reinforcement learning through symbiotic evolution. *Machine Learning*, 22:11 – 32, 1996.

[32] D. E. Moriarty and R. Miikkulainen. Forming neural networks through efficient and adaptive coevolution. *Evolutionary Computation*, 4(5):373–399, 1997.
[33] S. J. Nowlan and G. E. Hinton. Simplifying neural networks by soft weight-sharing. *Neural Computation*, 4(4):473–493, 1992.
[34] S. V. Odri, D. P. Petrovacki, and G. A. Krstonosic. Evolutional development of a multilevel neural network. *Neural Networks*, 6:583–595, 1993.
[35] D. Opitz and R. Maclin. Popular ensemble methods: An empirical study. *Journal of Artificial Intelligence Research*, 11:169–198, 1999.
[36] D. W. Opitz and J. W. Shavlik. Actively searching for an effective neural network ensemble. *Connection Science*, 8(3):337–353, 1996.
[37] R. Parekh, J. Yang, and V. Honavar. Constructive neural-network learning algorithms for pattern classification. *IEEE Transactions on Neural Networks*, 11(2):436–450, March 2000.
[38] M. P. Perrone and L. N. Cooper. When networks disagree: Ensemble methods for hybrid neural networks. In R. J. Mammone, editor, *Neural Networks for Speech and Image Processing*, pages 126–142. Chapman – Hall, 1993.
[39] M. A. Potter and K. A. de Jong. Cooperative coevolution: An architecture for evolving coadapted subcomponents. *Evolutionary Computation*, 8(1):1–29, 2000.
[40] R. Reed. Pruning algorithms – A survey. *IEEE Transactions on Neural Networks*, 4:740 – 747, 1993.
[41] B. E. Rosen. Ensemble learning using decorrelated neural networks. *Connection Science*, 8(3):373–384, december 1996.
[42] J. D. Schaffer. *Some experiments in machine learning using vector evaluated genetic algorithm.* PhD thesis, Vanderbilt University, Nashville, TN, 1984.
[43] J. D. Schaffer, L. D. Whitley, and L. J. Eshelman. Combinations of genetic algorithms and neural networks: A survey of the state of the art. In L. D. Whitley and J. D. Schaffer, editors, *Proceedings of COGANN-92 International Workshop on Combinations of Genetic Algorithms and Neural Networks*, pages 1–37, Los Alamitos, CA, 1992. IEEE Computer Society Press.
[44] Y. Shang and B. W. Wah. Global optimization for neural networks training. *IEEE Computer*, 29(3):45–54, 1996.
[45] R. Smalz and M. Conrad. Combining evolution with credit apportionment: A new learning algorithm for neural nets. *Neural Networks*, 7(2):341 – 351, 1994.
[46] N. Srinivas and K. Deb. Multi-objective function optimization using non-dominated sorting genetic algorithms. *Evolutionary Computation*, 2(3):221–248, 1994.
[47] A. J. F. van Rooij, L. C. Jain, and R. P. Johnson. *Neural Networks Training Using Genetic Algorithms*, volume 26 of *Series in Machine Perception and Artificial Intelligence*. World Scientific, Singapore, 1996.
[48] P. J. Werbos. *The Roots of Backpropagation: From Ordered Derivatives to Neural Networks and Political Forecasting.* Wiley, New York, NY, 1994.
[49] B. A. Whitehead and T. D. Choate. Cooperative–competitive genetic evolution of radial basis function centers and widths for time series prediction. *IEEE Transactions on Neural Networks*, 7(4):869–880, July 1996.
[50] D. Whitley. The GENITOR algorithm and selective pressure. In Morgan Kaufmann Publishers, editor, *Proc 3rd International Conf. on Genetic Algorithms*, pages 116–121, Los Altos, CA, 1989.

[51] P. M. Williams. Bayesian regularization and pruning using a Laplace prior. *Neural Computation*, 7:117–143, 1995.

[52] X. Yao. Evolving artificial neural networks. *Proceedings of the IEEE*, 9(87):1423–1447, 1999.

[53] X. Yao and Y. Liu. A new evolutionary system for evolving artificial neural networks. *IEEE Transactions on Neural Networks*, 8(3):694–713, May 1997.

[54] G. Yule. On the association of attributes in statistics. *Phil. Trans.*, 194:257–319, 1900.

[55] G. Zenobi and P. Cunningham. Using diversity in preparing ensembles of classifiers based on different feature subsets to minimize generalization error. In L. de Raedt and P. Flach, editors, *12th European Conference on Machine Learning (ECML 2001)*, LNAI 2167, pages 576–587. Springer–Verlag, 2001.

[56] Q. F. Zhao, O. Hammami, Kuroda K, and K. Saito. Cooperative coevolutionary algorithm - How to evaluate a module? In *Proc. 1st IEEE Symposium of Evolutionary Computation and Neural Networks*, pages 150–157, San Antonio, TX, May 2000.

[57] E. Zitzler and L. Thiele. Multi-objective optimization using evolutionary algorithms – A comparative case study. In *Problem Solving from Nature – PPSN V*, pages 292–301, Amsterdam, The Netherlands, September 1998. Springer – Verlag.

21

Multi-Objective Structure Selection for RBF Networks and Its Application to Nonlinear System Identification

Toshiharu Hatanaka, Nobuhiko Kondo and Katsuji Uosaki

Department of Information and Physical Sciences, Osaka University,
Suita, 565-0871, Japan
hatanaka@ist.osaka-u.ac.jp

Summary. Evolutionary multiobjective optimization approach to RBF networks structure determination is discussed in this chapter. The candidates of RBF network structure are encoded into the chromosomes in GAs and they evolve toward the Pareto optimal front defined by the several objective functions with regard to model accuracy and model complexity. Then, an ensemble of networks is constructed by using the Pareto optimal networks. We discuss its application to nonlinear system identification. Numerical simulation results indicate that the ensemble network is much more robust for the case of existence of outliers or lack of data, than the one selected based on information criteria.

21.1 Introduction

Mathematical models of the actual systems play important roles in many engineering problems, such as control system design, fault detection, fault diagnosis, signal processing and time series prediction. Hence, building mathematical model of unknown systems based on the observed input and output data is a fundamental issue of engineering and it is called system identification [1]. System identification algorithms have mainly been developed for linear systems, however, in many practical cases, actual existing systems have some kind of nonlinear properties, so the system models have to represent such properties appropriately. From this point of view nonlinear system identification has been studied and a plenty of identification algorithms have been developed in the last two decades [2, 3]. Most of these approaches give a good model under some criterion based on the prior knowledge. However, there are usually several demands to system model, the system model optimized under the specific criterion is not always an optimal model in means of the other criterion. For example, it is required that the model should be easy to handle and well explainable for the identification data set contaminated by observation

T. Hatanaka et al.: *Multi-Objective Structure Selection for RBF Networks and Its Application to Nonlinear System Identification*, Studies in Computational Intelligence (SCI) **16**, 491–505 (2006)
www.springerlink.com

noise, though these properties are mutually competitive. Hence, system identification method based on multiobjective optimization will be an useful tool, but there are a few studies have been conducted from this viewpoint [4, 5].

On the other hand, since multiobjective optimization is receiving much attentions in the field of system optimization, the evolutionary computation is much being studied as an efficient technique to providing the Pareto optimal solutions simultaneously [6, 7]. From this point of view, an application of multiobjective evolutionary computation to nonlinear system identification is proposed [8, 9]. These approaches deal with polynomial dynamic system model and give the optimal model set concerning model accuracy and complexity.

While, the use of artificial neural networks as a promising approach to nonlinear system identification has been studied due to powerful nonlinear mapping ability. [10, 11]. The primary importance in applying neural network to nonlinear system identification is to select its structure suitably. Then, several approaches to determine the neural network structure have been proposed. However, a general method of the structure determination has not established, because the optimal structure depends on a class of the objective system, application area, learning algorithm and so on. So the network structure is generally determined by trial and error or a heuristic method. In recent years, evolutionary neural network based on multiobjective optimization is receiving much attentions and is applied to time series forecasting and pattern recognition [12, 14, 13, 15]. In the field of nonlinear system identification, there generally exists a tradeoff between the model accuracy and the model complexity [2], so it makes the structure determination of the neural networks more difficult.
From this viewpoint, we have been considering multiobjective optimization based nonlinear system identification using the evolutionary algorithms. In this chapter, we deal with the static nonlinear system identification using RBF (Radial Basis Function) network, which is a kind of artificial neural network. RBF network has in their hidden layer a number of basis function which respond locally in input space. The network output is the linear sum of the basis function values. If the parameters of RBF networks, i.e. the number of basis functions and the widths and centers of each basis function, are determined, output layer weights can be calculated with the training data [16]. This parameter setting affects the quality of function approximation. Therefore we consider the structure determination problem of RBF networks as a multiobjective optimization problem that concerns with the model accuracy, the model complexity and the output layers' weights. Then a method of obtaining the candidates of model as a Pareto optimal set based on evolutionary algorithms is proposed. The designers will be able to select one model from the Pareto optimal set obtained by the proposed method according to their use or some criteria. In addition, by introducing the concept of the ensemble learning [17, 18], one model can be obtained by constructing the ensemble network constructed by the Pareto set. It is expected that the ensemble network is much robust than one of the Pareto models. Then by numerical simulations,

we compare the ensemble network with the models which are selected from the Pareto set by AIC(Akaike information criterion) [19] and BIC(Bayesian information criterion) [20]. In the Section 21.2, the concept of multiobjective optimization by GA is introduced. In the Section 21.3, an outline of RBF networks is described and the proposed method is introduced including ensemble of Pareto optimal RBF networks. Some numerical study results are shown in the Section 21.4 and concluding remarks are given at the last section.

21.2 Multiobjective Optimization by GA

21.2.1 Genetic Algorithms

Genetic algorithms (GAs) are a kind of stochastic search or optimization algorithms, originally proposed by Holland [21]. They have been invented based on natural genetics and evolution. The outline of a simple GA procedure is following way. Initially, the initial population of individuals which have a binary digit string as the "chromosome" is generated at random. Each bit of chromosome is called "gene" [22]. The "fitness", which is a measure of adaptation to environment, is calculated for each individual. Then, "selection" operation passing individuals to next generation is performed based on fitness value, and then "crossover" and "mutation" are performed on the selected individuals to generate new population by transforming chromosomes into offspring's ones. This procedure is continued until the end condition is satisfied. This algorithm is conforming to the mechanism of evolution, in which the genetic information changes for every generation and the individuals which adapt to environment better survive preferentially.

The properties of GAs are that GA is a stochastic parallel search based on the multi search points and GA requires only fitness value based on the objective functions. Particularly GA attracts attentions as a solver of multi-objective optimization problems due to parallel search.

21.2.2 Multiobjective Optimization

In the multiobjective optimization problems, there generally exists tradeoff among the objective functions. And so two concepts, "domination" and the "Pareto optimality", are considered.
Now, we consider the multiobjective optimization problem such as

$$\begin{array}{rl} \min & f_1(\boldsymbol{x}), \cdots, f_n(\boldsymbol{x}) \\ \text{subject to} & g_j(\boldsymbol{x}) \geq 0, j = 1, 2, \cdots, k \end{array}$$

where, $\boldsymbol{x}$ represents m dimensional decision variable $\boldsymbol{x} = (x_1, x_2, \cdots, x_m)^T$ and $f_i(\boldsymbol{x}),\ \ i = 1, 2, \cdots, n$ denote n objective functions. $g_i(\boldsymbol{x}),\ \ i = 1, 2, \cdots, k$ are the constraints.

$\boldsymbol{x}_1$ is said to "dominate" $\boldsymbol{x}_2$, if and only if

$$\forall i = 1, 2, \ldots, n \quad f_i(\boldsymbol{x}_1) \leq f_i(\boldsymbol{x}_2)$$

and

$$\exists j = 1, 2, \ldots, n \quad f_j(\boldsymbol{x}_1) < f_j(\boldsymbol{x}_2).$$

A solution $\boldsymbol{x}_0$ that is not dominated by any other $\boldsymbol{x}$ is called the "Pareto optimal solution" [7, 6]. Pareto optimal solution is considered to be the best solution comprehensively. And generally many Pareto optimal solutions exist simultaneously. Considering tradeoff among the objective functions, on multiobjective optimization problems it is important to obtain a Pareto optimal solution set.

21.2.3 Multiobjective GA Based on Ranking

A parameter *rank* is introduced in order to apply the concepts of domination and Pareto optimal to GA. Though there are some ranking methods, this study adopts ranking method proposed by Fonseca [6]. According to this ranking method, a *rank* of an individual $\boldsymbol{x}_i$ on a generation t is:

$$rank(\boldsymbol{x}_i, t) = 1 + p_i^{(t)}$$

where p_i is the total number of individuals which dominate $\boldsymbol{x}_i$. By calculating this *rank* for each individual and selecting based on it, a population can evolve toward a Pareto optimal solution set. Since GA is a multi-point search algorithm, GA is expected to find a Pareto optimal set in a single simulation run.

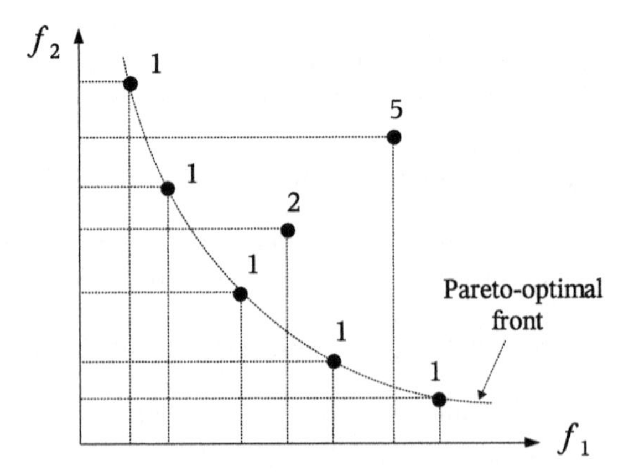

Fig. 21.1. Ranking method (two objective functions)

21.3 Construction of Pareto RBF Networks

21.3.1 RBF Network

RBF (Radial Basis Function) network is composed of three layers as shown in Fig. 21.2 and has basis functions which respond locally in input space. The basis function $\phi_j(\boldsymbol{x})$ in this study is defined by the Gaussian function,

$$\phi_j(\boldsymbol{x}) = \exp\left(-\frac{(\boldsymbol{x}-\boldsymbol{c}_j)^T(\boldsymbol{x}-\boldsymbol{c}_j)}{2\sigma_j^2}\right). \tag{21.1}$$

Here, $\boldsymbol{x}$ is input variable, $\boldsymbol{c}_j$ is center vector, and σ_j^2 is a parameter which decides function width. Using this $\phi_j(\boldsymbol{x})$, RBF network is constructed as follows:

$$u(\boldsymbol{x}) = w_0 + \sum_{j=1}^{m} w_j\phi_j(\boldsymbol{x}) \tag{21.2}$$

Here, m is the number of hidden units, i.e., the basis functions, and w_j are the output layer weights. RBF network will be determined if the parameters $m, \boldsymbol{c}_j, \sigma_j$, and w_j are estimated based on the data observed from the system. In this study, these parameters are estimated by two GAs. The parameters σ_j are assumed to be constant value for simplicity.

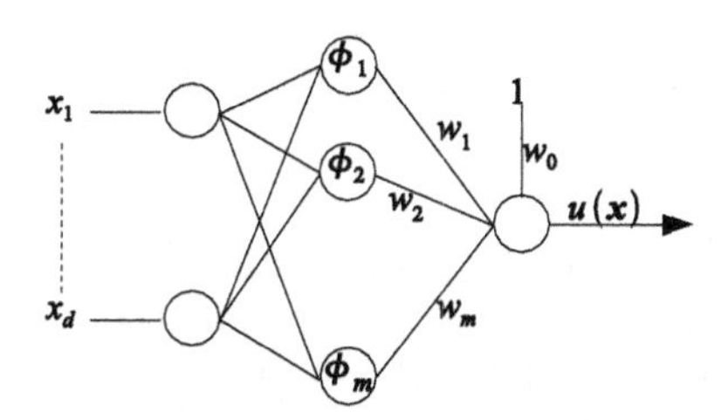

Fig. 21.2. An RBF network

21.3.2 Genetic Representation

In this study, we apply MOGA (Multiobjective Genetic Algorithm), based on NSGA–II proposed by Deb [7] to determine both the number of basis functions and the centers of them. The candidate of the center of basis function is assumed to be the position of the training data points. The chromosomes of MOGA population indicate that the data points are employed as centers of basis functions i.e. "1" represents that a basis function is located at the

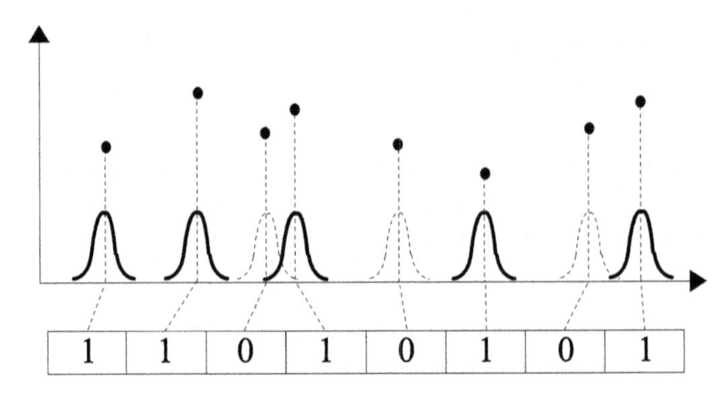

Fig. 21.3. Chromosome representation

corresponding training data point, as shown in Fig. 21.3. By this setting, the length of the chromosome becomes equal to the number of training data, the number of "1" gene in the chromosome indicates the number of basis functions and the locus of the "1" shows the center position of the basis functions. Then, the connection weights are estimated by real-coded GA, in which each individual represents directly the weight values. The first stage is multiobjective binary coded GA to examine the structure candidate of the RBF network and the second stage is real coded GA that estimates the connection weights for the candidate of RBF network structure represented by each individual in the first stage.

The overall flow diagram is shown in Fig. 21.4, in which two–stage GAs are used to RBF network determination.

21.3.3 Evaluation

It is generally demanded that the mathematical models not only can explain the relationship between input and output enough but also is simple in order to have the generalization ability. In this study three evaluation criteria are set for the evaluation of MOGA which determines the network architecture. The first objective function is the number of basis function. This objective indicates the complexity of the model. The second objective function is logMSE. This objective indicates the extent of a fit of the model to the training data. MSE(Mean Squared Error) is defined as :

$$MSE = \frac{1}{n} \sum_{i=1}^{n} \{y_i - \hat{y}_i\}^2 \tag{21.3}$$

Here, y_i is the observed output, $\hat{y}_i$ is the model output. The third objective function is the sum of the absolute value of weights. These three evaluation criteria are to be minimized. MSE is used for evaluation in real-coded GA which estimates the weight parameters.

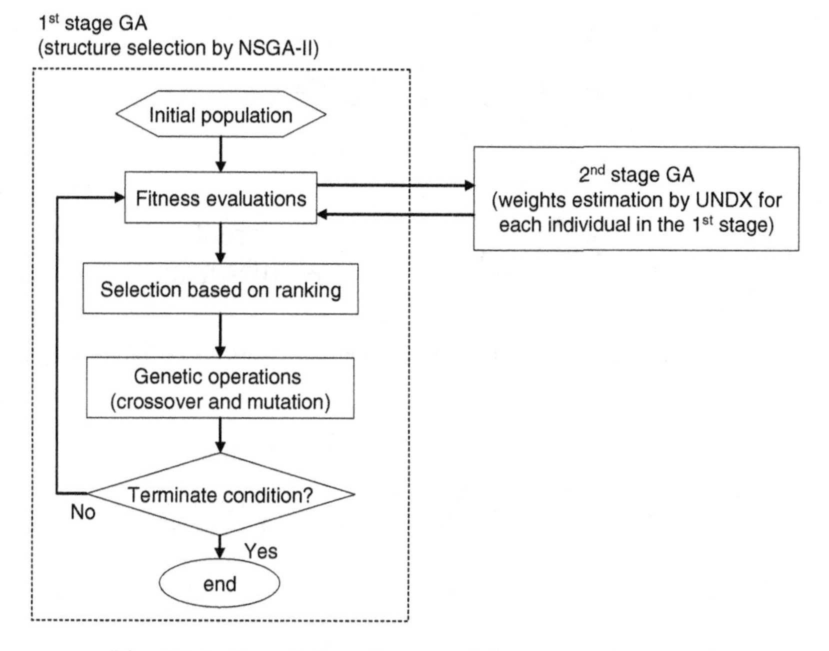

Fig. 21.4. Overall flow diagram of the proposed approach

21.3.4 Construction of Pareto RBF Networks

The Pareto optimal RBF network construction algorithm is consist of multiobjective genetic algorithm including real-coded GA. As multiobjective genetic algorithm we adopt NSGA-II which is one of multiobjective genetic algorithms and is known to have the capability to maintain diversity[7]. After estimating all the parameters of the network, *rank* is assigned for each individual by the concept of multiobjective optimization problem, in which three fitness functions are to be minimized. Then Pareto optimal individuals will be obtained in accordance with NSGA-II algorithm. In NSGA-II, we apply genetic operations which are the uniform crossover and the bit reversal mutation.
In real-coded GA, UNDX (Unimodal Normal Distribution Crossover) [23] is applied in the proposed method. UNDX generates two offsprings by normal random numbers which is determined by three parents, as shown in Figure 21.5. Basically offsprings are generated by normal distribution around segment connecting two parents. The third parent is used to determine the standard deviation of normal distribution.
MGG(Minimal Generation Gap) [24] is adopted as the generation alternation model of real-coded GA used in the proposed method to preserve the diversity of population. MGG procedure is described as follows, also refer to Fig. 21.6.

1. Plurality of real number vector is generated at random as the initial population.

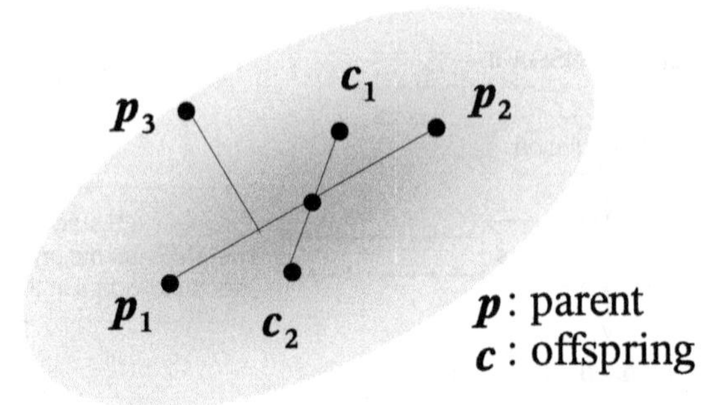

Fig. 21.5. Unimodal normal distribution crossover (UNDX)

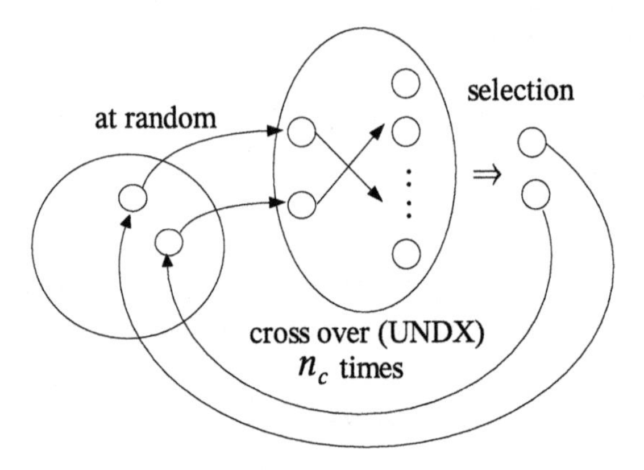

Fig. 21.6. Minimal generation gap (MGG) GA

2. Two parents are selected at random from population.
3. $2n_c$ offsprings are generated by applying UNDX to two parents n_c times. Here the third parent which determines the standard deviation of normal distribution is selected from population.
4. Fitness values of each offspring are calculated, then two individuals are selected from the set which is composed of two parents and all offsprings, then two parents are replaced by the selected two individual. The individuals selected here are elite and the individual selected by roulette selection in which the elite was pruned.
5. Continue 2 $\sim$ 4 until the end condition is met.

21.3.5 Pareto RBF Network Ensemble

Various models based on three criterion can be obtained by proposed method, so the designers will be able to select one model flexibly. On the other hand, there are the demand to obtain one model with good generalization ability. For instance, model selection by information criteria has been studied.

Recently the ensemble learning is receiving much attentions in the field of machine learning. In the ensemble learning, a monolithic model is constructed by combining several models. While some learning methods to make models constructing ensemble have been proposed, in this study ensemble is constructed of Pareto optimal models obtained by the proposed method.

Suppose that the number of Pareto models is m and the output of j-th network is $y_j(\boldsymbol{x})$, then the output of ensemble network $y^{EN}(\boldsymbol{x})$ is :

$$y^{EN}(\boldsymbol{x}) = \sum_{j=1}^{m} w_j y_j(\boldsymbol{x}) \tag{21.4}$$

Here, w_j is the weight on the output of j-th network, m indicates the number of obtained Pareto networks. In this study, w_j is assumed to be $1/m$ about every j, for simplicity.

In this study, difference of performance between the ensemble network and networks selected from Pareto optimal set based on information criteria is considered from numerical simulation results. In this work, the AIC (Akaike's Information Criterion) [19] and BIC (Bayesian Information Criterion) [20] are used as information criteria.

$$AIC = n \log MSE + 2(v+1), \tag{21.5}$$

$$BIC = n \log MSE + (v+1) \log n, \tag{21.6}$$

where, n is the sample size and v represents the number of parameters.

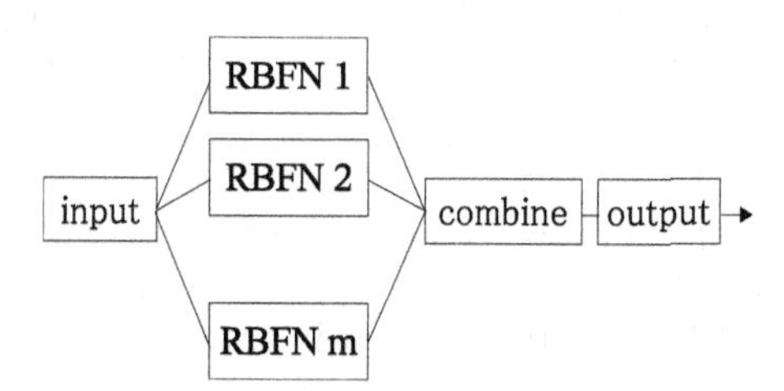

Fig. 21.7. An RBF network ensemble

21.4 Numerical Simulation

21.4.1 Function approximation problem

In the numerical simulation, the nonlinear function approximation problem is considered. Let the true function be:

$$v(x) = 2x + 3\sin(5\pi x) + \sin(10\pi x). \tag{21.7}$$

Training data set is sampled by

$$\begin{aligned} y_i &= v(\boldsymbol{x}_i) + \varepsilon_i, \quad i = 1, 2, \ldots, n \\ \varepsilon_i &\sim N(0, \sigma^2), \end{aligned} \tag{21.8}$$

where $\boldsymbol{x}_i$, $\boldsymbol{y}_i$ are input-output data and n is the number of training data. x_i are sampled from uniform distribution over $[0, 1]$. The observation data are disturbed by normal white noise ε_i with mean 0, variance σ^2.

In order to investigate difference of performance between the ensemble network and networks selected from Pareto optimal set based on information criteria, following numerical simulations have been implemented.

At first Pareto optimal RBF networks are constructed using training data. About the parameters of NSGA-II, population size is 50, crossover rate is 0.7, mutation rate is 0.1 and generation size is 10. The RBF width parameters σ_j^2, $j = 1, 2, \cdots$ are set to 0.01. MGG is iterated 10,000 times with population size 30 and n_c is 30. Next, MSE for test data not used in training is calculated to investigate the generalization ability.

21.4.2 Simulation 1

Pareto optimal RBF networks were constructed using training data with the observation noise $\varepsilon_i \sim N(0, 0.04)$. Then MSE for 50 test data not used in training was calculated for both the ensemble network and networks selected from Pareto optimal set based on AIC and BIC. The number of training data was changed as 30, 40, 50 and 60. For each number of training data, data set was changed 5 times respectively. Results are shown in Table 1.

21.4.3 Simulation 2

Next, the variance of the observation noise was assumed to become large with a certain probability. 90% of noise variance was 0.16 and 10% of it was 16. Then MSE for 50 test data not used in training was calculated for both

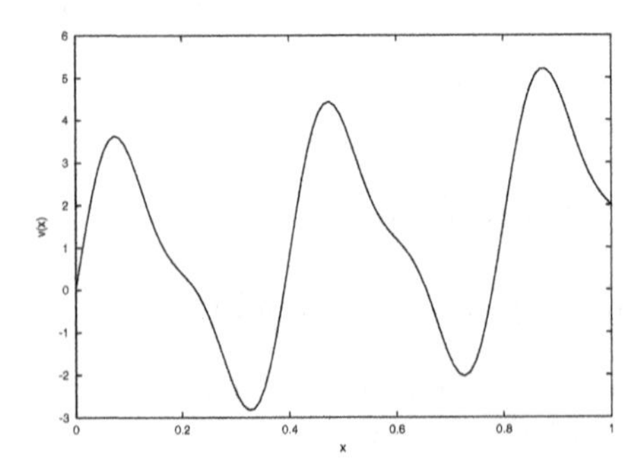

Fig. 21.8. True function of simulation 1

Table 21.1. Simulation 1 : The column of "data" indicates the number of training data, the column of "RBFN" indicates the number of obtained Pareto RBF networks. The column of "ensemble" indicates MSE of the ensemble network. The columns of "AIC" and "BIC" indicate MSE of the network which are selected by AIC and BIC, respectively.

data	RBFN	ensemble	AIC	BIC
30	50	0.191255	0.099932	0.088485
	48	0.171739	0.147592	0.147592
	50	0.435584	0.255772	0.255772
	36	0.630114	0.978955	0.652990
	50	0.174871	0.075501	0.075501
40	50	0.169304	0.102441	0.102441
	50	0.174922	0.060889	0.060889
	43	0.176029	0.049408	0.069194
	44	0.396571	0.045771	0.045771
	50	0.193280	0.047974	0.047974
50	43	0.162108	0.062830	0.062830
	46	0.217515	0.064232	0.114111
	42	0.186255	0.063561	0.074141
	41	0.202725	0.089901	0.089901
	50	0.090469	0.037580	0.037580
60	34	0.222999	0.096943	0.096943
	50	0.126545	0.056253	0.056253
	46	0.196389	0.101005	0.101005
	36	0.338770	0.108320	0.108320
	50	0.115663	0.077814	0.077814

the ensemble network and networks selected from Pareto optimal set based on AIC and BIC. The number of training data was changed as 30, 40, 50 and 60. For each number of training data, data set was changed 5 times respectively. Results are shown in Table 21.2.

21.4.4 Simulation 3

Next, 80% of noise variance was 0.16 and 20% of it was 16. Then MSE for 50 test data not used in training was calculated for both the ensemble network and networks selected from Pareto optimal set based on AIC and BIC. The number of training data was changed as 30, 40, 50 and 60. For each number of training data, data set was changed 5 times respectively. Results are shown in Table 21.3.

Table 21.2. Simulation 2

data	RBFN	ensemble	AIC	BIC
30	38	1.051167	0.957824	0.957824
	47	0.728419	0.348573	0.348573
	32	0.927173	2.525247	1.654921
	29	1.752871	1.738257	1.933985
	32	2.513687	2.191173	2.191173
40	17	0.778018	0.784632	1.036531
	46	0.507676	0.364317	0.610114
	16	0.978053	1.238106	1.238106
	22	2.217106	2.252523	2.427382
	19	0.758372	0.982137	0.982137
50	28	0.680729	0.821663	0.821663
	37	0.820927	0.602269	0.602269
	34	0.569077	0.771010	0.728347
	38	0.785553	0.874953	0.874953
	17	0.431460	0.830321	0.830321
60	16	0.957909	0.659941	0.659941
	15	0.814199	0.785704	0.785704
	24	0.894919	0.903971	0.903971
	34	0.320639	0.243167	0.243167
	8	0.734543	0.804033	0.804033

21.4.5 Simulation 4

Next, 70% of noise variance was 0.16 and 30% of it was 16. Then MSE for 50 test data not used in training was calculated for both the ensemble network and networks selected from Pareto optimal set based on AIC and BIC. The number of training data was changed as 30, 40, 50 and 60. For each number of training data, data set was changed 5 times respectively. Results are shown in Table 21.4.

21.4.6 Discussion of the results

In the first simulation result, the case of stationary noise with small variance is considered. There networks selected from Pareto optimal set based on information criteria have smaller test data MSE than the ensemble network. However, at the fourth row of Table 1, MSEs for test data are significantly large. In this simulation run the ensemble network has the smallest MSE for test data. Training data set lacks within the range from 0.7 to 0.85 in this simulation run. The ensemble network is expected to have good generalization ability in such cases.

In the simulation 2, 3 and 4, we consider the case where there are outliers in training data set. There the ensemble network often have smaller test data

Table 21.3. Simulation 3

data	RBFN	ensemble	*AIC*	*BIC*
30	35	1.022114	1.768824	1.768824
	32	3.090617	3.723074	3.723074
	30	1.372639	1.898516	1.898516
	36	4.202276	5.199522	5.199522
	28	16.542650	94.693893	94.693893
40	25	1.031609	1.284169	1.284169
	38	0.924128	1.045010	1.045010
	27	3.924914	2.424119	1.425429
	49	1.006428	1.434848	1.434848
	37	0.866881	0.956438	0.956438
50	13	1.228426	1.445913	1.445913
	21	0.838093	1.347125	1.347125
	32	1.411863	1.186526	1.186526
	33	1.168921	1.339410	1.733162
	11	0.709311	0.808769	0.808769
60	20	0.814219	1.081037	1.081037
	32	0.412720	0.445307	0.445307
	12	0.787953	0.778415	0.778415
	26	1.250772	0.974598	1.398607
	29	0.800697	0.943744	0.943744

MSE than networks selected from Pareto optimal set based on information criteria. The ensemble network may ease the effect of bad networks such as networks over fitting the outliers.

Summing up the numerical simulation results, the ensemble network has higher generalization ability than the single network that is selected based on the information criteria such as AIC or BIC for the case of existence of outliers or lack of data. This results indicate that the ensemble network is robust, though more simulation and discussion are needed.

21.5 Conclusions

In this study, we have proposed a method of obtaining a Pareto optimal RBF network set based on multiobjective evolutionary algorithms. Then, an application of the ensemble network constructed by such Pareto optimal models is also considered. Numerical simulation results indicate that the ensemble network is much more robust than networks selected based on information criteria. Reduction of the computational costs, improvement of the ensemble method, comparison to the conventional approaches, application to dynamic system identification and so on are the future works.

Table 21.4. Simulation 2-4

data	RBFN	ensemble	*AIC*	*BIC*
30	21	3.612076	4.014570	4.014570
	36	7.883370	8.813035	8.813035
	30	2.545823	5.239426	5.239426
	32	0.718388	0.729001	0.729001
	47	1.321716	4.653079	4.653079
40	35	5.670076	3.722443	3.722443
	19	3.039101	3.793278	1.407294
	12	1.278787	1.710274	1.710274
	30	3.257849	2.846025	2.846025
	26	0.617000	0.725918	0.658233
50	13	0.953733	0.990628	0.990628
	9	2.517913	3.027015	3.027015
	22	1.781070	2.862607	2.862607
	13	1.577241	1.458167	0.942308
	26	3.334088	4.163925	4.163925
60	31	1.097832	1.201612	1.424725
	7	2.583948	2.422122	2.422122
	25	1.172458	1.095895	2.169411
	27	0.799774	0.723667	0.723667
	10	0.930999	0.810712	0.810712

References

[1] Ljung L (1999) System Identification: Theory for the User, Prentice-Hall, NJ.

[2] Nelles O (2001) Nonlinear System Identification, Springer, Berlin.

[3] Yang Z J (1998) Identification of Nonlinear Systems, Journal of The Society of Instrument and Control Engineers, Vol.37, No.4, pp.249–255 (in Japanese).

[4] Johansen T A (2000) Multi-objective identification of FIR models, Proceedings of IFAC Symposium on System Identification SYSID2000, Vol. 3, pp.917–922.

[5] Johansen T A and Babuška R (2003) Multi-Objective Identification of Takagi-Sugeno Fuzzy Models, IEEE Transactions on Fuzzy Systems, Vol.11, No.6, pp. 847–860.

[6] Fonseca C M and Fleming P J (1993) Genetic algorithms for multiobjective optimization : Formulation, discussion and generalization, Proceedings of the Fifth International Conference on Genetic Algorithms, pp.416–423.

[7] Deb K (2001) Multi-objective Optimization using Evolutionary Algorithms, John Wiley & Sons, New York.

[8] Katya R-V , Fonseca C M and Flemming P J (1997) Multiobjective Genetic Programming : A Non-Linear System Identification Application, Genetic Programming 97 Conference, pp. 207–212.

[9] Hatanaka T, Uosaki K and Hossaka Y (2002) Application of Evolutionary Multi-Objective Optimization to Nonlinear System Identification, Proceedings of the SICE Kansai Branch Symposium 2002, pp.54–56 (in Japanese).

[10] Chen S, Billings S A (1992) Neural networks for nonlinear dynamic system modelling and identification, International Journal of Control. Vol. 56, no. 2, pp. 319–346.
[11] Sjoberg J, Zhang Q,inghua, Ljung L, et.al.(1995) Nonlinear black-box modeling in system identification: a unified overview, Automatica. Vol. 31, no. 12, pp. 1691–1724.
[12] Abbass H A (2003) Pareto Neuro-Evolution: Constructing Ensemble of Neural Networks Using Multi-objective Optimization, Proceedings of the 2003 IEEE Congress on Evolutionary Computation, Vol. 3, pp.2074–2080.
[13] Abbass H A (2003) Pareto Neuro-Ensemble, Proceedings of the 16th Australian Joint Conference on Artificial Intelligence, pp.554–566.
[14] Kondo N, Hatanaka T and Uosaki K (2004) Pareto RBF Networks Based on Multiobjective Evolutionary Computation, Proceedings of 2004 SICE Annual Conference, pp.2177–2182.
[15] Fieldsend, J E, Singh S (2005) Pareto evolutionary neural networks, IEEE Transactions on Neural Networks, Vol. 16, No.2, pp.338–354.
[16] Hatanaka T, Uosaki K and Kuroda T (2001) Structure Selection of RBF Neural Network Using Information Criteria, Proceedings of Fifth International Conference on Knowledge-Based Intelligent Information Engineering Systems and Allied Technologies, pp.166–170.
[17] Jin Y, Okabe T and Sendhoff B (2004) Neural network regularization and ensembling using multi-objective evolutionary algorithms, Proceedings of the 2004 IEEE Congress on Evolutionary Computation, pp.1–8.
[18] Islam M M, Yao X and Murase K (2003) A constructive algorithm for training cooperative neural network ensembles, IEEE Transactions on Neural Networks, Vol.14, No.4, pp.820–834.
[19] Akaike H (1974) A New Look at the Statistical Model Identification, IEEE Transactions on Automatic Control, Vol.19, pp. 716–723.
[20] Schwarz G(1978) Estimating the Dimension of a Model, Annals of Statistics, Vol.7, No.2, pp.461–464.
[21] Holland J (1975) Adaption in Natural and Artificial Systems, The University of Michigan Press, MI.
[22] Bäck T (1996) Evolutionary Algorithms in Theory and Practice, Oxford Press, New York.
[23] Ono I and Kobayashi S (1997) A Real-coded Genetic Algorithm for Function Optimization Using Unimodal Normal Distribution Crossover, Proceedings of 7th International Conference.on Genetic Algorithms, pp.246–253.
[24] Sato H, Ono I and Kobayashi S (1997) A New Generation Alternation Model of Genetic Algorithms and Its Assessment, Journal of Japanese Society for Artificial Intelligence, Vol.12, No.5, pp.734–744.

22

Fuzzy Ensemble Design through Multi-Objective Fuzzy Rule Selection

Hisao Ishibuchi and Yusuke Nojima

1-1 Gakuen-cho, Sakai, Osaka 599-8531, Japan
Department of Computer Science and Intelligent Systems
Osaka Prefecture University
{hisaoi,nojima}@cs.osakafu-u.ac.jp

Summary. The main advantage of evolutionary multi-objective optimization (EMO) over classical approaches is that a variety of non-dominated solutions with a wide range of objective values can be simultaneously obtained by a single run of an EMO algorithm. In this chapter, we show how this advantage can be utilized in the design of fuzzy ensemble classifiers. First we explain three objectives in multi-objective formulations of fuzzy rule selection. One is accuracy maximization and the others are complexity minimization. Next we demonstrate that a number of non-dominated rule sets (i.e., fuzzy classifiers) are obtained along the accuracy-complexity tradeoff surface from multi-objective fuzzy rule selection problems. Then we examine the effect of combining multiple non-dominated fuzzy classifiers into a single ensemble classifier. Experimental results clearly show that the combination into ensemble classifiers improves the classification ability of individual fuzzy classifiers for some data sets.

22.1 Introduction

A promising approach to the design of reliable classifiers is to combine multiple classifiers into a single one [2, 10]. Several methods have been proposed for generating multiple classifiers such as bagging [3] and boosting [12]. The point in classifier aggregation is to generate an ensemble of classifiers with high diversity. Ideally the classification errors by each classifier should be uncorrelated. In this chapter, we examine the use of evolutionary multiobjective optimization (EMO) algorithms for generating an ensemble of fuzzy rule-based classifiers with high diversity.

The main advantage of evolutionary multiobjective optimization (EMO) over classical approaches is that many non-dominated solutions can be simultaneously obtained by a single run of an EMO algorithm. When an EMO algorithm is used in the design of an ensemble classifier, a number of non-dominated classifiers with high diversity can be obtained by its single run. This advantage has already been utilized in some studies on the design of neural

H. Ishibuchi and Y. Nojima: *Fuzzy Ensemble Design through Multi-Objective Fuzzy Rule Selection*, Studies in Computational Intelligence (SCI) **16**, 507–530 (2006)
www.springerlink.com

network ensembles. For example, reference [1] formulated a two-objective optimization problem by dividing the given training patterns into two subsets. Mean squared errors (MSEs) of a neural network on the two subsets were used as two objectives to be minimized by an EMO algorithm. An ensemble classifier was constructed by combining non-dominated neural networks with respect to the two objectives. On the other hand, Chandra and Yao [4, 5] used a different formulation in order to increase the diversity of neural networks in a more direct manner. They formulated a two-objective optimization problem using an accuracy measure and a diversity measure. The MSE on all the given training patterns was used as the accuracy measure while the diversity measure was calculated for each neural network as the difference from the other individuals in each population during the execution of an EMO algorithm. An ensemble classifier was constructed by combining non-dominated neural networks on the accuracy-diversity tradeoff surface. Jin et al. [29, 30] used an EMO algorithm to minimize the MSE and a complexity measure. They defined the complexity of a neural network by the number of connections. Non-dominated neural networks on the accuracy-complexity tradeoff surface were used to construct an ensemble classifier. Oliveira et al. [32, 31] defined the complexity of a neural network by the number of input nodes. They used an EMO algorithm to find non-dominated feature subsets on the accuracy-complexity tradeoff surface. An ensemble classifier was constructed using the obtained non-dominated feature subsets.

In the field of fuzzy rule-based systems, the accuracy-complexity tradeoff is often referred to as the interpretability-accuracy tradeoff. This is because high interpretability is the main advantage of fuzzy rule-based systems over other nonlinear systems such as neural networks. EMO algorithms have been used to design fuzzy rule-based systems with high interpretability and high accuracy in some studies (e.g., [7], [27, 28], and [35, 36]). In these studies, a number of non-dominated fuzzy rule-based systems were obtained on the interpretability-accuracy tradeoff surface. The two-objective fuzzy rule selection method [14] was one of the first EMO-based approaches to the interpretability-accuracy tradeoff analysis of fuzzy rule-based systems. A number of non-dominated rule sets were found with respect to the classification accuracy and the number of fuzzy rules. This method was extended to the case of three objectives in [16] by considering the total number of antecedent conditions as an additional complexity measure. The three-objective fuzzy rule selection method was improved by using a state-of-the-art EMO algorithm [21] and a memetic EMO algorithm [18]. The same idea as the EMO-based multiobjective fuzzy rule selection was also used for the design of non-fuzzy rule-based classification systems [18]. In these studies on fuzzy and non-fuzzy rule selection, a data mining technique was used to find promising candidate rules. An idea of constructing fuzzy rule-based ensemble classifiers using EMO algorithms was proposed in [22]. Almost the same three-objective formulation was also used in [35].

In this chapter, we examine the following three formulations of multiobjective fuzzy rule selection for the design of fuzzy rule-based ensemble classifiers:

Problem 1 (P1): Maximize $f_1(S)$ and minimize $f_2(S)$,
Problem 2 (P2): Maximize $f_1(S)$ and minimize $f_3(S)$,
Problem 3 (P3): Maximize $f_1(S)$, minimize $f_2(S)$, and minimize $f_3(S)$,

where

S : A subset of fuzzy rules (i.e., an individual fuzzy rule-based classifier),
$f_1(S)$: The number of correctly classified training patterns by S,
$f_2(S)$: The number of fuzzy rules in S,
$f_3(S)$: The total number of antecedent conditions of fuzzy rules in S.

Since the number of antecedent conditions is often referred to as the rule length, $f_3(S)$ can be viewed as the total rule length of fuzzy rules in S.

An EMO algorithm is applied to the above-mentioned three fuzzy rule selection problems to find non-dominated rule sets with different interpretability-accuracy tradeoffs. We use the NSGA-II algorithm [9], which is one of the most well-known and frequently used EMO algorithms. An ensemble classifier is constructed from the obtained non-dominated rule sets for each rule selection problem. The three rule selection problems are compared with each other in terms of the generalization ability of the constructed ensemble classifiers. When we choose the members of each ensemble classifier, we examine three strategies for member selection: all the obtained non-dominated rule sets, a prespecified number of the best rule sets with respect to the classification accuracy on the training patterns, and non-dominated rule sets satisfying a prespecified minimum requirement for the classification accuracy.

This chapter is organized as follows. First we briefly describe some basic concepts in multiobjective optimization in Section 22.2 where we also describe the NSGA-II algorithm [9]. Next we explain our two-stage fuzzy rule selection method to find a number of non-dominated rule sets with different interpretability-accuracy tradeoffs in Section 22.3. The first stage is heuristic extraction of promising candidate fuzzy rules while the second stage is evolutionary multiobjective rule selection. Then we show experimental results in Section 22.4 where our two-stage fuzzy rule selection method is applied to the above-mentioned three multiobjective fuzzy rule selection problems. The three strategies for ensemble member selection are also examined in Section 22.4. Finally we conclude this chapter in Section 22.5.

22.2 Evolutionary Multiobjective Optimization

22.2.1 Some Basic Concepts in Multiobjective Optimization

In this subsection, we briefly describe some basic concepts in multiobjective optimization. Let us consider the following k-objective maximization problem:

$$\text{Maximize } \mathbf{f}(\mathbf{x}) = (f_1(\mathbf{x}),\ f_2(\mathbf{x}),\ \ldots,\ f_k(\mathbf{x})), \tag{22.1}$$

$$\text{subject to } \mathbf{x} \in \mathbf{X}, \tag{22.2}$$

where $\mathbf{f}(\mathbf{x})$ is the objective vector, $f_i(\mathbf{x})$ is the i-th objective to be maximized, $\mathbf{x}$ is the decision vector, and $\mathbf{X}$ is the feasible region in the decision space. When the following conditions are satisfied, a feasible solution $\mathbf{x} \in \mathbf{X}$ is said to be dominated by another feasible solution $\mathbf{y} \in \mathbf{X}$ (i.e., $\mathbf{y}$ dominates $\mathbf{x}$: $\mathbf{y}$ is better than $\mathbf{x}$):

$$\forall i, \quad f_i(\mathbf{x}) \leq f_i(\mathbf{y}) \text{ and } \exists j, f_j(\mathbf{x}) < f_j(\mathbf{y}). \tag{22.3}$$

If there is no feasible solution $\mathbf{y}$ that dominates $\mathbf{x}$, $\mathbf{x}$ is said to be a Pareto-optimal solution of the multiobjective optimization problem. The set of all the Pareto-optimal solutions is referred to as the Pareto-optimal solution set. The image of the Pareto-optimal solution set onto the objective space is the Pareto front. The dominance relation in 22.3 is also used to define non-dominated solutions in a population of solutions. When there is no solution $\mathbf{y}$ in a population that dominates $\mathbf{x}$, we refer to $\mathbf{x}$ as a non-dominated solution in that population. The concept of non-dominated solutions is used to evaluate solutions in EMO algorithms.

The task of EMO algorithms is to find all the Pareto-optimal solutions. It is, however, impractical to try to find all the Pareto-optimal solutions of a large-scale multiobjective optimization problem. In this case, EMO algorithms try to find a number of well-distributed near Pareto-optimal solutions.

The dominance relation in 22.3 is modified when it is applied to each of the three fuzzy rule selection problems in Section 22.1. For example, the dominance relation is modified for **P3** as follows: A rule set $S_{\mathbf{x}}$ is said to be dominated by another rule set $S_{\mathbf{y}}$ (i.e., $S_{\mathbf{y}}$ dominates $S_{\mathbf{x}}$: $S_{\mathbf{y}}$ is better than $S_{\mathbf{x}}$) when all the following inequalities hold:

$$f_1(S_{\mathbf{x}}) \leq f_1(S_{\mathbf{y}}), \quad f_2(S_{\mathbf{x}}) \geq f_2(S_{\mathbf{y}}), \quad f_3(S_{\mathbf{x}}) \geq f_3(S_{\mathbf{y}}), \tag{22.4}$$

and at least one of the following inequalities holds:

$$f_1(S_{\mathbf{x}}) < f_1(S_{\mathbf{y}}), \quad f_2(S_{\mathbf{x}}) > f_2(S_{\mathbf{y}}), \quad f_3(S_{\mathbf{x}}) > f_3(S_{\mathbf{y}}). \tag{22.5}$$

Roughly speaking, when a rule set $S_{\mathbf{x}}$ has lower classification accuracy and higher complexity than another rule set $S_{\mathbf{y}}$, $S_{\mathbf{x}}$ is said to be dominated by $S_{\mathbf{y}}$ in all the three fuzzy rule selection problems in Section 22.1. The task of EMO algorithms is to find a number of non-dominated rule sets with different interpretability-accuracy tradeoffs, which are Pareto-optimal or near Pareto-optimal solutions of each fuzzy rule selection problem.

22.2.2 NSGA-II Algorithm

The NSGA-II algorithm [9] is one of the most well-known and frequently used EMO algorithms in the literature. We use this algorithm in our two-stage fuzzy rule selection method since it has a number of advantages such as high performance, algorithmic simplicity, and high popularity. An outline of the NSGA-II algorithm is written as follows (for details, see [8, 9]):

Step 1 (Initialization): Generate an initial population with N_{pop} solutions where N_{pop} is the population size.

Step 2 (Creation of Offspring Population): Generate an offspring population by iterating the following procedures N_{pop} times:

(1) Choose a pair of parent solutions from the current population using binary tournament selection. Each solution is evaluated by Pareto ranking and a crowding measure.

(2) Generate an offspring from the selected parent solutions by crossover and mutation.

Step 3 (Generation Update): Combine the current population and the offspring population into a merged one. Then choose the best N_{pop} solutions from the merged population to construct the next population. Each solution is evaluated by Pareto ranking and a crowding measure in the same manner as the selection phase of parent solutions in Step 2.

Step 4 (Termination Test): If a prespecified stopping condition is not satisfied, return to Step 2. Otherwise terminate the execution of the algorithm. In the latter case, we choose all the non-dominated solutions in the merged population in Step 4 as the final solutions.

In Step 2 of the NSGA-II algorithm, each solution in the current population is evaluated in the following manner. First, Rank 1 is assigned to all the non-dominated solutions in the current population. All solutions with Rank 1 are tentatively removed from the current population. Next, Rank 2 is assigned to all the non-dominated solutions in the reduced current population. All solutions with Rank 2 are tentatively removed from the reduced current population. This procedure is iterated until all solutions are tentatively removed from the current population. In this manner, a different rank is assigned to each solution. Solutions with smaller ranks are viewed as being better than those with larger ranks. Among solutions with the same rank, an additional criterion called a crowding measure is taken into account. The crowding measure for a solution calculates the distance between its adjacent solutions with the same rank in the objective space (for details, see [8, 9]). Less crowded solutions with larger values of the crowding measure are viewed as being better than more crowded solutions with smaller values of the crowding measure. Solutions in the merged population in Step 4 of the NSGA-II algorithm are evaluated in the same manner based on Pareto ranking and the crowding measure.

22.3 Heuristic Rule Extraction and Evolutionary Multiobjective Rule Selection

In this section, we explain how a number of non-dominated fuzzy rule-based classifiers with different interpretability-accuracy tradeoffs can be obtained by our two-stage rule selection method.

22.3.1 Fuzzy Rule-Based Classifiers

Let us assume that we have$m$ training patterns $\mathbf{x}_p = (x_{p1}, \ldots, x_{pn})$, $p = 1, 2, \ldots, m$ from M classes where x_{pi} is the attribute value of the p–th training pattern for the i–th attribute ($i = 1, 2, \ldots, n$). For our n-dimensional M-class pattern classification problem, we use fuzzy rules of the following form:

$$\text{Rule } R_q : \text{ If } x_1 \text{ is } A_{q1} \text{ and } \ldots \text{ and } x_n \text{ is } A_{qn} \text{ then Class } C_q \text{ with } CF_q, \tag{22.6}$$

where R_q is the label of the q-th rule, $\mathbf{x} = (x_1, \ldots, x_n)$ is an n-dimensional pattern vector, A_{qi} is an antecedent fuzzy set, C_q is a class label, and CF_q is a rule weight. We define the compatibility grade of each training pattern $\mathbf{x}_p$ with the antecedent part $\mathbf{A}_q = (A_{q1}, \ldots, A_{qn})$ of the fuzzy rule R_q in 22.6 using the product operator as

$$\mu_{\mathbf{A}_q}(\mathbf{x}_p) = \mu_{A_{q1}}(x_{p1}) \cdot \mu_{A_{q2}}(x_{p2}) \cdot \ \ldots \ \cdot \mu_{A_{qn}}(x_{pn}), \quad p = 1, 2, \ldots, m, \tag{22.7}$$

where $\mu_{A_{qi}}(\,\cdot\,)$ is the membership function of A_{qi}.

For determining the consequent class C_q and the rule weight CF_q, we first calculate the confidence of the fuzzy rule "$\mathbf{A}_q \Rightarrow$ Class h" for each class h as follows (see the textbook on fuzzy data mining [17] for fuzzy versions of some basic concepts in data mining such as confidence and support):

$$c(\mathbf{A}_q \Rightarrow \text{Class } h) = \sum_{\mathbf{x}_p \in \text{Class } h} \mu_{\mathbf{A}_q}(\mathbf{x}_p) \Big/ \sum_{p=1}^{m} \mu_{\mathbf{A}_q}(\mathbf{x}_p), \quad h = 1, 2, \ldots, M. \tag{22.8}$$

The consequent class C_q is specified as the class with the maximum confidence:

$$c(\mathbf{A}_q \Rightarrow \text{Class } C_q) = \max\{c(\mathbf{A}_q \Rightarrow \text{Class } h) \mid h = 1, 2, \ldots, M\}. \tag{22.9}$$

Rule weights have a significant effect on the classification accuracy of a fuzzy rule-based classifier. Several methods have been examined to determine the rule weight of each fuzzy rule in the literature [25] where good results are obtained from the following specification:

$$CF_q = c(\mathbf{A}_q \Rightarrow \text{Class } C_q) - \sum_{\substack{h = 1 \\ h \neq C_q}}^{M} c(\mathbf{A}_q \Rightarrow \text{Class } h). \tag{22.10}$$

We use this definition in this chapter.

Let S be a fuzzy rule-based classifier (i.e., a set of fuzzy rules). When an input pattern $\mathbf{x}_p$ is to be classified by the fuzzy rule-based classifier S, a single winner rule R_w is chosen from S as follows:

$$\mu_{\mathbf{A}_w}(\mathbf{x}_p) \cdot CF_w = \max\{\mu_{\mathbf{A}_q}(\mathbf{x}_p) \cdot CF_q \mid R_q \in S\}. \tag{22.11}$$

The input pattern $\mathbf{x}_p$ is assigned to the consequent class C_w of the winner rule R_w. When multiple rules with different consequent classes have the same maximum value in (17.7), the classification of the input pattern $\mathbf{x}_p$ is rejected. The classification of $\mathbf{x}_p$ is also rejected when there is no compatible fuzzy rules with positive compatibility grades for $\mathbf{x}_p$. In this case, all fuzzy rules have the same maximum value of zero in the right-hand side in (17.7).

In this chapter, we use an ensemble of multiple fuzzy rule-based classifiers to classify input patterns. First an input pattern is classified by each individual fuzzy rule-based classifier using the single winner-based method in (17.7). Then the final classification is performed through the simple majority vote scheme based on the classification result by each individual classifier (see [15]) for various voting methods for fuzzy rule-based classifiers). When multiple classes have the same maximum number of votes, one class is randomly chosen among those classes with the maximum vote.

22.3.2 Heuristic Rule Extraction

Genetic rule selection was proposed for designing fuzzy rule-based classifiers with high accuracy and high comprehensibility in [14, 15] where a scalar fitness function was defined as the weighted sum of the first two objectives of our fuzzy rule selection: to maximize the number of correctly classified training patterns (i.e., to maximize $f_1(S)$) and to minimize the number of fuzzy rules (i.e., to minimize $f_2(S)$). That is, fuzzy rule selection was handled in the framework of single-objective optimization with the following scalar fitness function:

$$\text{fitness}(S) = w_1 \cdot f_1(S) - w_2 \cdot f_2(S), \tag{22.12}$$

where w_1 and w_2 are prespecified positive constants. As we have already mentioned in Section 22.1, the single-objective formulation in (17.8) has been extended to two-objective and three-objective formulations.

One difficulty in the design of fuzzy rule-based classifiers through rule selection is that the number of possible candidate fuzzy rules exponentially increases with the number of input variables. When we use K linguistic values and "*don't care*" as antecedent fuzzy sets for each of n attributes, the total number of possible combinations of those $(K+1)$ antecedent fuzzy sets is $(K+1)^n$. In the first stage of our two-stage fuzzy rule selection method, a prespecified number of promising candidate fuzzy rules are generated from those combinations in a heuristic manner using a data mining criterion. That is, the first stage is the heuristic rule extraction phase. In the field of data mining, rules are often evaluated by two rule evaluation criteria: support and confidence. In the same manner as the fuzzy version of the confidence in (17.4), the support of the fuzzy rule "$\mathbf{A}_q \Rightarrow$ Class h" for each class h is defined as follows:

$$s(\mathbf{A}_q \Rightarrow \text{Class } h) = \frac{1}{m} \sum_{\mathbf{x}_p \in \text{Class } h} \mu_{\mathbf{A}_q}(\mathbf{x}_p). \tag{22.13}$$

Various rule evaluation criteria were examined in [23] where good results were obtained from the following criterion:

$$f_{\text{SLAVE}}(R_q) = s(\mathbf{A}_q \Rightarrow \text{Class } C_q) - \sum_{\substack{h = 1 \\ h \neq C_q}}^{M} s(\mathbf{A}_q \Rightarrow \text{Class } h). \tag{22.14}$$

This is a modified version of a rule evaluation criterion used in an iterative fuzzy GBML (Genetics-Based Machine Learning) algorithm called SLAVE [13].

In the heuristic rule extraction phase (i.e., the first stage) of our two-stage fuzzy rule selection method, a prespecified number of candidate fuzzy rules with the largest values of the SLAVE criterion in (17.10) are found for each class. For designing fuzzy rule-based classifiers with high comprehensibility, only short fuzzy rules are examined as candidate fuzzy rules. This restriction on the rule length is consistent with the third objective (i.e., $f_3(S)$: the total rule length) of the three-objective rule selection problem (i.e., **P3** in Section 22.1).

22.3.3 Evolutionary Multiobjective Fuzzy Rule Selection

The second stage of our two-stage method is evolutionary multiobjective fuzzy rule selection where the NSGA-II algorithm is used to find a number of non-dominated rule sets from candidate fuzzy rules extracted in the first stage. Let us assume that N fuzzy rules (i.e., N/M fuzzy rules for each class where M is the number of classes) have already been extracted as candidate fuzzy rules using the SLAVE criterion in the first stage of our two-stage method. A subset S of the N candidate fuzzy rules is handled as an individual in the NSGA-II algorithm in the second stage. Each individual is represented by a binary string of the length N as

$$S = s_1 s_2 \cdots s_N, \tag{22.15}$$

where $s_j = 1$ and $s_j = 0$ mean that the j-th candidate rule is included in S and excluded from S, respectively. From this coding, we can see that the size of the search space in the second stage is 2^N, which depends on the number of candidate fuzzy rules (i.e., N) extracted in the first stage.

Since each individual is represented by a binary string, we can use standard genetic operations in the second stage (i.e., in the NSGA-II algorithm) for multiobjective fuzzy rule selection. We use uniform crossover and bit-flip mutation in our computational experiments.

In order to efficiently find non-dominated rule sets, we use two problem-specific heuristic tricks in the NSGA-II algorithm. One trick is the use of biased mutation where a larger probability is assigned to the mutation from 1 to 0 than that from 0 to 1. This is for efficiently decreasing the number of fuzzy rules in each rule set. The other trick is the removal of unnecessary rules, which is a kind of local search. Since each training pattern is classified by the single winner-based method in the rule set S, some fuzzy rules in S may be chosen as winner rules for no training patterns. We can remove those fuzzy rules without degrading the first objective (i.e., $f_1(S)$: the number of correctly classified training patterns). At the same time, the second objective (i.e., $f_2(S)$: the number of fuzzy rules) and the third objective (i.e., $f_3(S)$: the total rule length) are improved by removing unnecessary rules. Thus we remove all fuzzy rules that are not selected as winner rules for any training patterns from the rule set S. The removal of unnecessary rules is performed after the first objective is calculated for each rule set and before the second and third objectives are calculated.

The NSGA-II algorithm is applied to one of the three rule selection problems in Section 22.1. A number of non-dominated rule sets (i.e., non-dominated fuzzy rule-based classifiers) are obtained by its single run. Some (or all) of the obtained non-dominated fuzzy rule-based classifiers are used to construct an ensemble classifier. The performance of the constructed ensemble classifier is examined by the classification accuracy on test patterns. In this manner, the three formulations of multiobjective fuzzy rule selection are compared with each other in terms of the performance of ensemble classifiers.

22.4 Computational Experiments

22.4.1 Data Sets

We used six data sets with many numerical attributes: Wisconsin breast cancer, Diabetes, Glass, Cleveland heart disease, Sonar, and Wine. These data sets are available from the UCI ML repository (http://www.ics.uci.edu/∼mlearn/). Table 22.1 shows the number of attributes, the number of patterns, and the number of classes in each data set. Some data sets include incomplete patterns with missing values. Those patterns were not used in our computational experiments.

In the last two columns of Table 22.1, we show benchmark results on these data sets. They are error rates reported in [11] where six variants of the C4.5 algorithm [33, 34] were examined. The six variants were different from

Table 22.1. Data sets used in our computer simulations.

Data set	No. of attributes	No. of patterns	No. of classes	C4.5	
				Best	Worst
Breast W	9	683*	2	5.1	6.0
Diabetes	8	768	2	25.0	27.2
Glass	9	214	6	27.3	32.2
Heart C	13	297*	5	46.3	47.9
Sonar	60	208	2	24.6	35.8
Wine	13	178	3	5.6	8.8

$*$ Incomplete patterns with missing values are not included.

each other in their discretization methods of continuous attributes. The performance of each variant was evaluated by ten independent iterations (with different data partitions) of the whole ten-fold cross-validation (10-CV) procedure (i.e., 10 × 10-CV) in [11]. We used the same performance evaluation procedure in our computational experiments.

22.4.2 Experimental Conditions

All attribute values of each data set were normalized into real numbers in the unit interval [0, 1]. As antecedent fuzzy sets, we used "*don't care*" and 14 triangular fuzzy sets generated from four fuzzy partitions with different granularities in Fig. 22.1. We generated 300 fuzzy rules for each class as candidate rules in a greedy manner using the SLAVE criterion in (17.10). Thus the total number of candidate rules is $300M$ where M is the number of classes. The upper bound on the length of candidate fuzzy rules is two for the sonar data set with 60 attributes and three for the other data sets with 13 or less attributes.

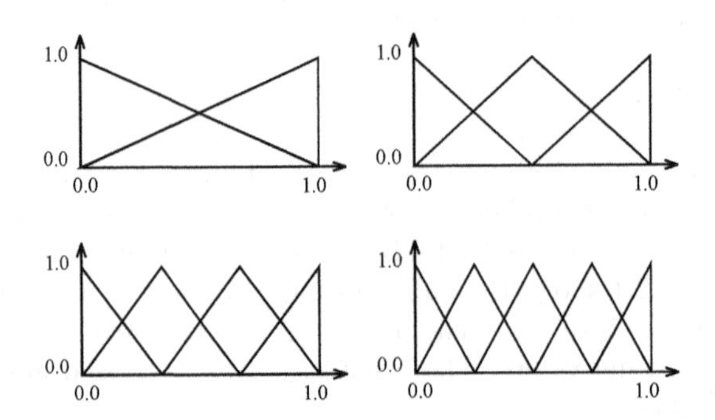

Fig. 22.1. Four fuzzy partitions used in our computer simulations.

The NSGA-II algorithm was employed to find non-dominated rule sets from $300M$ candidate fuzzy rules. We used the following parameter values in the NSGA-II algorithm:

- Population size: 200 strings
- Crossover probability: 0.8
- Biased mutation probabilities: $p_{\mathrm{m}}(0 \rightarrow 1) = 1/300M$ and $p_{\mathrm{m}}(1 \rightarrow 0) = 0.1$
- Stopping condition: 5000 generations

22.4.3 Illustrative Computational Experiments

Before comparing the three formulations of multiobjective fuzzy rule selection by examining the generalization ability of ensemble classifiers on the six data sets by the 10-CV procedure, we demonstrate how our two-stage fuzzy rule selection method works through illustrative computational experiments on the diabetes data set. For illustration purpose, we randomly divided the 768 patterns of this data set into 384 training patterns and 384 test patterns. Using the SLAVE criterion, we generated 600 candidate fuzzy rules (300 rules for each class) from the training patterns. In the heuristic extraction phase of candidate fuzzy rules, we examined fuzzy rules of length 3 or less using "*don't care*" and 14 antecedent fuzzy sets in Fig. 22.1.

The NSGA-II algorithm was employed for multiobjective fuzzy rule selection from the generated 600 candidate fuzzy rules. We applied the NSGA-II algorithm to **P1** with the two objectives: maximize $f_1(S)$ and minimize $f_2(S)$. After 5000 generations, we obtain 9 non-dominated rule sets (i.e., non-dominated fuzzy rule-based classifiers). It should be noted that these 9 non-dominated rule sets were obtained by a single run of the NSGA-II algorithm. This result clearly demonstrates that EMO algorithms can find a number of non-dominated solutions by their single run. One of the obtained 9 non-dominated rule sets was an empty rule set (i.e., $S = \phi$, $f_1(S) = 0$, $f_2(S) = 0$). Since no patterns are classifiable by an empty rule set, it is always excluded from ensemble classifiers. An ensemble classifier was constructed by combining the other 8 rule sets.

Each pattern was classified by each rule set (i.e., each individual fuzzy rule-based classifier) using the single winner-based fuzzy reasoning method to calculate the error rates on the training patterns and the test patterns. The performance of each rule set is shown in Fig. 22.2 where the rejection of classification is counted as an error (i.e., strictly speaking the vertical axis is the sum of the error rate and the rejection rate). It should be noted that some non-dominated rule sets are overlapping in Fig. 22.2. The second smallest rule set with only a single fuzzy rule is not shown in Fig. 22.2 because its error rates are out of the range of the vertical axis (i.e., 34.0% on the training patterns and 36.8% on the test patterns). Experimental results by the ensemble classifier are shown by the solid line in Fig. 22.2. For comparison, the reported results by the C4.5 algorithm in [11] are also shown in Fig. 22.2 (see Table 22.1).

We can observe a clear tradeoff structure between the error rate on the training patterns and the number of fuzzy rules in Fig. 22.2(a). That is, the interpretability-accuracy tradeoff structure is clearly shown in Fig. 22.2 (a) with respect to the classification accuracy of fuzzy rule-based classifiers on the training patterns. Such a tradeoff structure is not clear in Fig. 22.2(b) with respect to the generalization ability for the test patterns. The generalization ability is somewhat degraded by the increase in the number of fuzzy rules due to the overfitting to the training patterns in Fig. 22.2(b). We can see from Fig. 22.2(b) that the performance of the ensemble classifier is better than almost all the individual fuzzy rule-based classifiers. It should be noted that the computational experiments in this subsection were performed using 50% training patterns and 50% test patterns for illustration purpose. Thus we can not directly compare between the reported results by the C4.5 algorithm based on the 10-CV procedure and our experimental results in Fig. 22.2(b). In the next subsection, we use the 10-CV procedure for the performance evaluation of fuzzy rule-based classifiers and their ensemble classifiers.

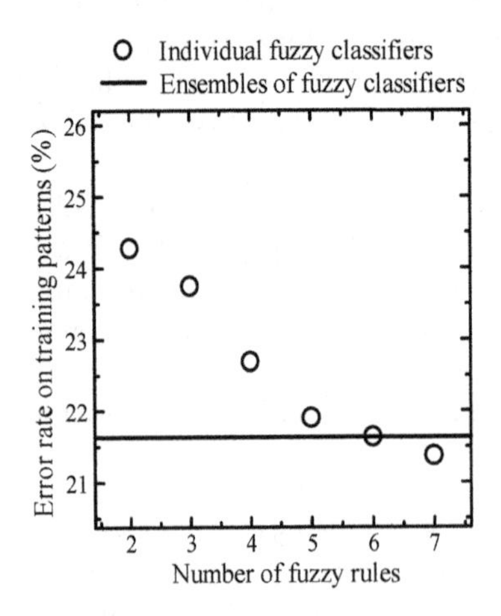

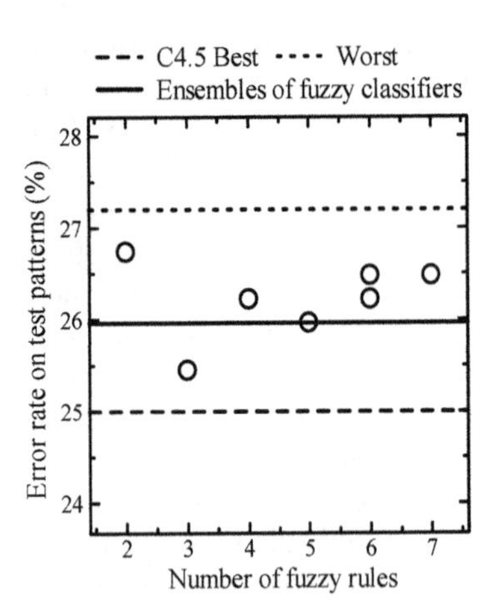

Fig. 22.2. Experimental results of a single run of the NSGA-II algorithm on the diabetes data set. The NSGA-II algorithm was applied to **P**1 with 50% training patterns and 50% test patterns. (a) Error rates on training patterns. (b) Error rates on test patterns.

In the same manner as Fig. 22.2, we also applied the NSGA-II algorithm to **P**2 with the two objectives: $f_1(S)$ and $f_3(S)$. Experimental results are shown in Fig. 22.3. The horizontal axis in Fig. 22.3 is the total number of antecedent conditions (i.e., $f_3(S)$: the total rule length) while it was the number of fuzzy rules (i.e., $f_2(S)$) in Fig. 22.2. As in Fig. 22.2 (a), we can observe a clear interpretability-accuracy tradeoff structure in Fig. 22.3(a). Such a tradeoff structure is not clear in Fig. 22.3(b) with respect to the accuracy on the test patterns. From the comparison between Fig. 22.2 and Fig. 22.3, we can see that more non-dominated rule sets were obtained in Fig. 22.3 from **P2** than Fig. 22.2 from **P1**. More specifically, 9 rule sets and 10 rule sets were obtained from **P1** and **P2**, respectively (7 rule sets and 8 rule sets are shown in Fig. 22.2 and Fig. 22.3, respectively). It should be noted again that some rule sets are overlapping in Fig. 22.3. An ensemble classifier was constructed using the obtained 9 non-dominated rule sets (excluding an empty rule set). Error rates by the ensemble classifier are also shown in Fig. 22.3 by the solid line. We can see from Fig. 22.2(b) that the ensemble classifier outperformed all the individual fuzzy rule-based classifiers and the reported results in [11].

The NSGA-II algorithm was also applied to the three-objective fuzzy rule selection problem (i.e., **P3**) in the same manner as Fig. 22.2 and Fig. 22.3. From a single run of the NSGA-II algorithm, 20 non-dominated rule sets were obtained. The classification performance of each rule set is shown in Fig. 22.4 together with that of their ensemble classifier. It should be noted that multiple non-dominated rule sets with the same number of fuzzy rules were obtained in Fig. 22.4. These non-dominated rule sets are different from each other in the total number of antecedent conditions (i.e., the total rule length). More non-dominated rule sets were obtained in Fig. 22.4 from the three-objective formulation than Fig. 22.2 and Fig. 22.3 from the two-objective formulations. In general, the number of obtained non-dominated solutions usually increases with the number of objectives. We can observe a clear interpretability-accuracy tradeoff structure in Fig. 22.4(a) with respect to the classification accuracy of fuzzy rule-based classifiers on the training patterns. This tradeoff structure is not clear in Fig. 22.4(b) with respect to the classification accuracy of fuzzy rule-based classifiers for the test patterns.

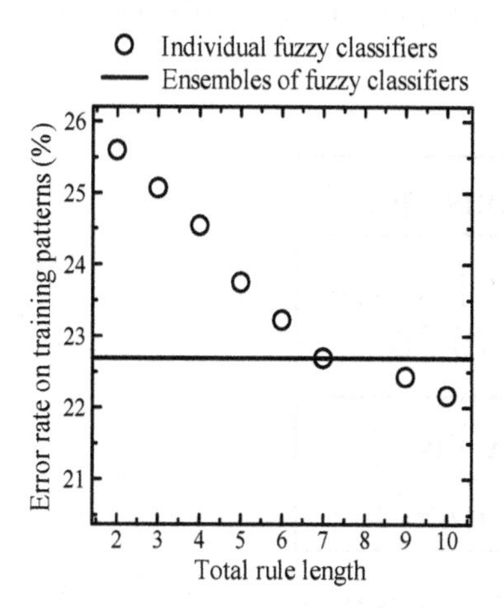

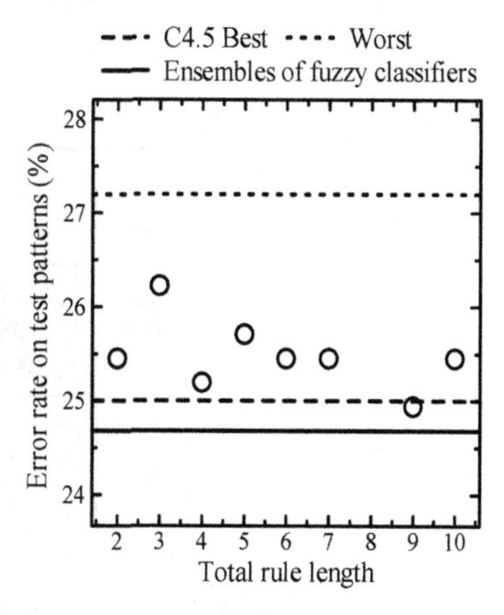

Fig. 22.3. Experimental results on the diabetes data set using **P2**. (a) Error rates on training patterns. (b) Error rates on test patterns.

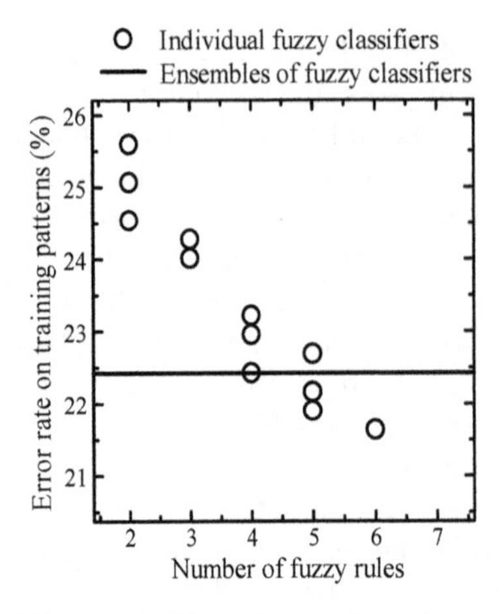

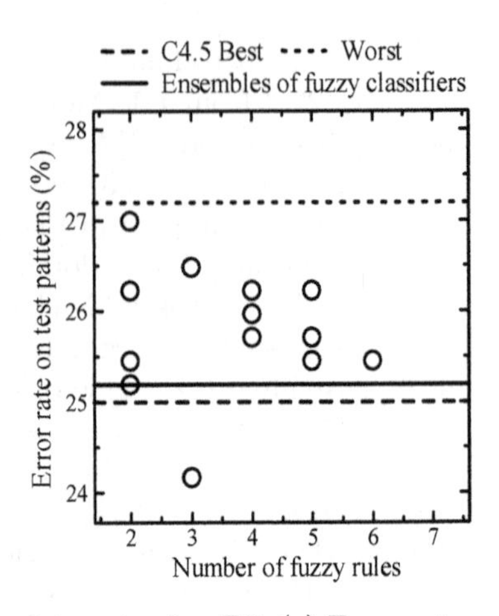

Fig. 22.4. Experimental results on the diabetes data set using **P**3. (a) Error rates on training patterns. (b) Error rates on test patterns.

Examples of non-dominated rule sets with the same number of fuzzy rules are shown in Fig. 22.5 and Fig. 22.6. Fig. 22.5 shows the simplest rule set in Fig. 22.4 with two fuzzy rules, a 25.6% error rate on the training patterns, and a 25.4% error rate on the test patterns. The total number of antecedent conditions (i.e., the total rule length) is 2 in Fig. 22.5. On the other hand, Fig. 22.6 shows the most complicated rule set among those with two fuzzy rules. This rule set corresponds to the left-bottom circle in Fig. 22.4(a) and Fig. 22.4(b), which has a 24.5% error rate on the training patterns and a 25.2% error rate on the test patterns. The total rule length of this rule set is 4 excluding "*don't care*" conditions.

	x_2	Consequent
R_1		Class 1 (0.36)
R_2		Class 2 (0.45)

Fig. 22.5. The simplest rule set with two fuzzy rules in Fig. 22.4.

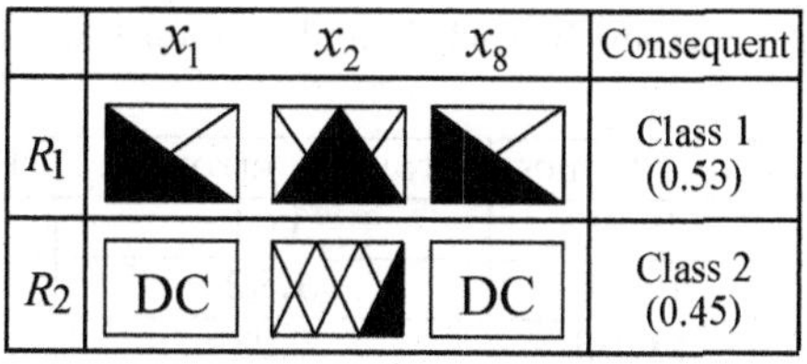

	x_1	x_2	x_8	Consequent
R_1				Class 1 (0.53)
R_2	DC		DC	Class 2 (0.45)

Fig. 22.6. The most complicated rule set with two fuzzy rules in Fig. 22.4.

22.4.4 Comparison of Three Formulations

In this subsection, we compare the three formulations in terms of the classification ability of ensemble classifiers designed by each formulation. The generalization ability of ensemble classifiers was evaluated by ten independent iterations of the whole 10-CV procedure (i.e., 10 × 10-CV) in our computational experiments in this subsection. When we constructed ensemble classifiers from non-dominated rule sets, we examined the following three strategies:

(i) The use of all the obtained non-dominated rule sets (excluding an empty rule set).

(ii) The use of a prespecified number of the best non-dominated rule sets in terms of their classification accuracy on training patterns. We examined 10 specifications of the number of classifiers in a single ensemble classifier: 1, 2, ..., 10. When the total number of obtained rule sets was less than the prespecified number, all of them (excluding an empty rule set) were used.

(iii) The use of non-dominated rule sets that have lower error rates on training patterns than a prespecified upper bound. We examined ten specifications of the upper bound on error rates: 5%, 10%, 15%, ..., 50%. In some cases, there were no qualified rule sets that had lower error rates than the prespecified upper bound. In those cases, we could not construct ensemble classifiers. When we could not construct ensemble classifiers more than 50 runs among 100 runs in 10 × 10-CV, we do not report the corresponding experimental results.

In the following, we report experimental results using each combination of the three formulations and the three ensemble strategies on each data set.

Wisconsin Breast Cancer Data Set: In Table 22.2, we show the average number of obtained non-dominated rule sets (excluding an empty rule set), the average error rates of those rule sets on training patterns and test patterns. Error rates of ensemble classifiers on test patterns (i.e., generalization ability of ensemble classifiers) are shown in Fig. 22.7. The horizontal axis of Fig. 22.7(a) is the upper bound on the number of rule sets in each ensemble classifier. The right-most label ∞ of the horizontal axis means that there is no upper bound on the number of rule sets (i.e., all the obtained non-dominated rule sets were used in ensemble classifiers). On the other hand, the horizontal axis of Fig. 22.7(b) is the upper bound on error rates on training patterns of rule sets used in ensemble classifiers. The upper bound of 99% (i.e., the right-

Table 22.2. Obtained rule sets by each formulation for the Wisconsin breast cancer data set.

Formulation	No. of rules	Training error (%)	Test error (%)
P1	9.50	6.04	7.84
P2	10.51	**5.82**	**7.40**
P3	**11.96**	6.23	7.84

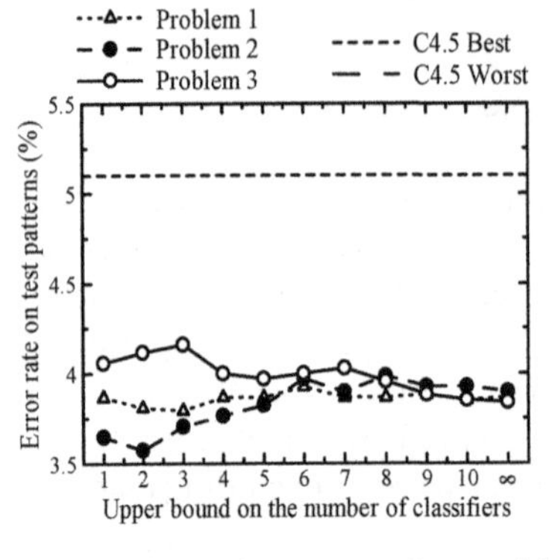

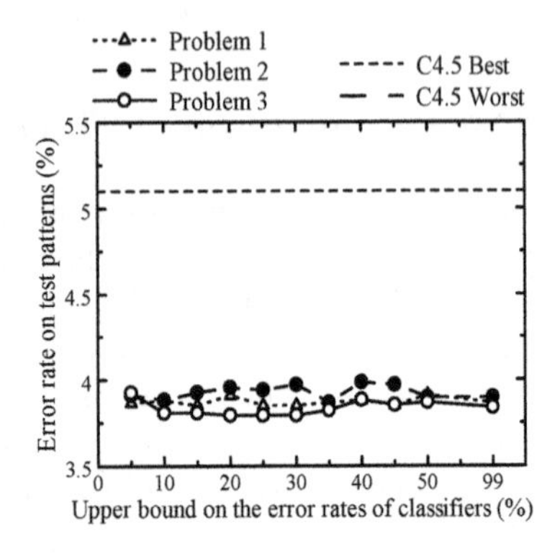

Fig. 22.7. Performance of ensemble classifiers on the Wisconsin breast cancer data set. (a) Second strategy for constructing ensembles. (b) Third strategy for constructing ensembles.

most label of the horizontal axis of Fig. 22.7(b)) means that all the obtained non-dominated rule sets (except for an empty rule set) were used in ensemble classifiers.

We can observe an improvement in error rates by combining more than three fuzzy rule-based classifiers in the case of **P3** in Fig. 22.7(a). We can also see that the performance of individual fuzzy rule-based classifiers and ensemble classifiers in Fig. 22.7 is much better than the best result of the C4.5 algorithm (i.e., the best average error rate: 5.1%) reported in [11].

Diabetes Data Set: In the same manner as the case of the Wisconsin breast cancer data set, we show experimental results on the diabetes data set in Table 22.3 and Fig. 22.8. In Fig. 22.8, the effect of combining multiple fuzzy rule-based classifiers into a single ensemble classifier is not clear. The best results were obtained from **P1** among the three formulations in Fig. 22.8. The performance of individual fuzzy rule-based classifiers and ensemble classifiers in Fig. 22.8 is slightly inferior to the best result of the C4.5 algorithm.

Glass Data Set: Experimental results on the glass data set are shown in Table 22.4 and Fig. 22.9. In Fig. 22.9, the best results were obtained from **P3** among the three formulations. We can observe a slight improvement in error rates by combining multiple fuzzy rule-based classifiers in the case of **P3** in Fig. 22.9(a). The performance of fuzzy rule-based ensemble classifiers in Fig. 22.9 is inferior to even the worst result of the C4.5 algorithm. It may

Table 22.3. Obtained rule sets by each formulation for the diabetes data set.

Formulation	No. of rules	Training error (%)	Test error (%)
P1	8.94	24.03	26.90
P2	13.73	23.93	26.74
P3	**16.46**	**23.90**	**26.39**

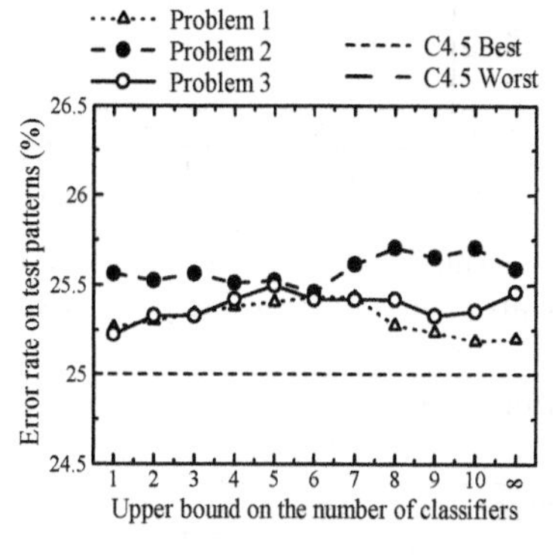

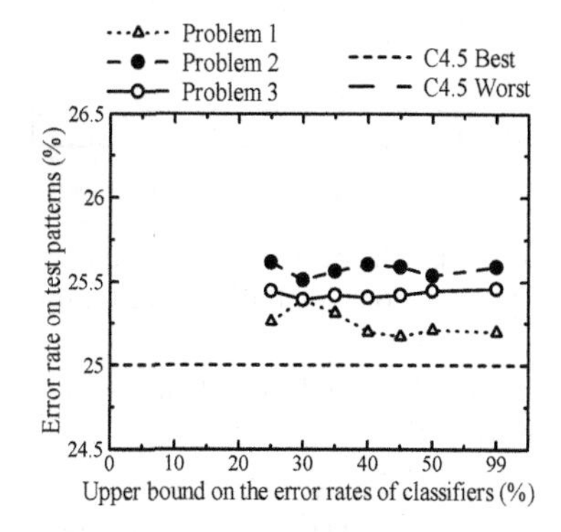

Fig. 22.8. Performance of ensemble classifiers on the diabetes data set. (a) Second strategy for constructing ensembles. (b) Third strategy for constructing ensembles.

Table 22.4. Obtained rule sets by each formulation for the glass data set.

Formulation	No. of rules	Training error (%)	Test error (%)
P1	25.08	**26.43**	**42.19**
P2	28.63	32.40	44.89
P3	**36.40**	33.43	44.50

suggest that the uniform fuzzy partitions of the input space in Fig. 22.1 did not work well on the glass data set. That is, adjustment of antecedent fuzzy sets seems to be necessary to obtain fuzzy rule-based classifiers with high performance on the glass data set.

Cleveland Heart Disease Data Set: Experimental results on the Cleveland heart disease data set are shown in Table 22.5 and Fig. 22.10. In Fig. 22.10, the best results were obtained from **P3** among the three formulations. We can observe a clear positive effect of combining multiple fuzzy rule-based classifiers in Fig. 22.10 for all the three formulations. When an ensemble classifier was constructed from all the non-dominated rule sets obtained from **P3**, its performance was comparable to the best result of the C4.5 algorithm in Fig. 22.10.

Sonar Data Set: Experimental results on the sonar data set are shown in Table 22.6 and Fig. 22.11. The best results were obtained from **P3** in Fig. 22.11. The effect of combining multiple fuzzy rule-based classifiers into a single ensemble classifier is not clear in Fig. 22.11 (a) while it is clear in

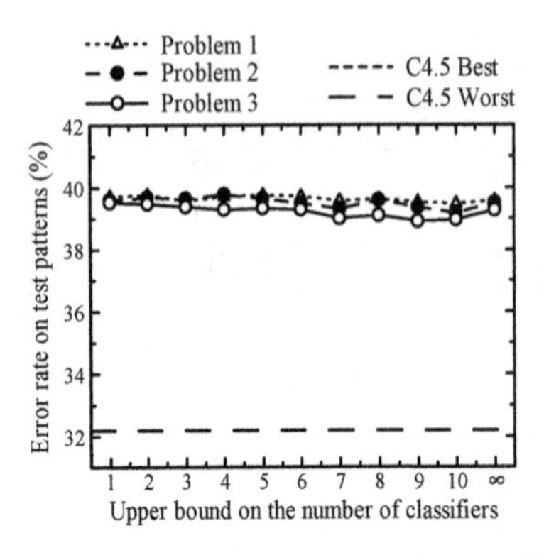

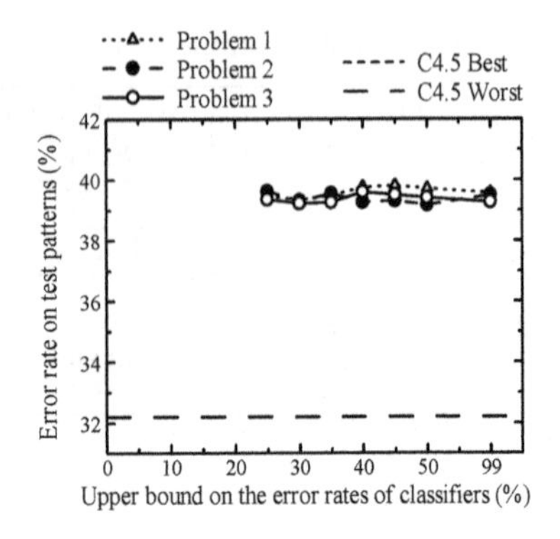

Fig. 22.9. Performance of ensemble classifiers on the glass data set. (a) Second strategy for constructing ensembles. (b) Third strategy for constructing ensembles.

Table 22.5. Obtained rule sets by each formulation for the Cleveland heart disease data set.

Formulation	No. of rules	Training error (%)	Test error (%)
P1	83.33	**30.15**	47.86
P2	86.50	32.77	47.92
P3	**87.04**	33.34	**47.60**

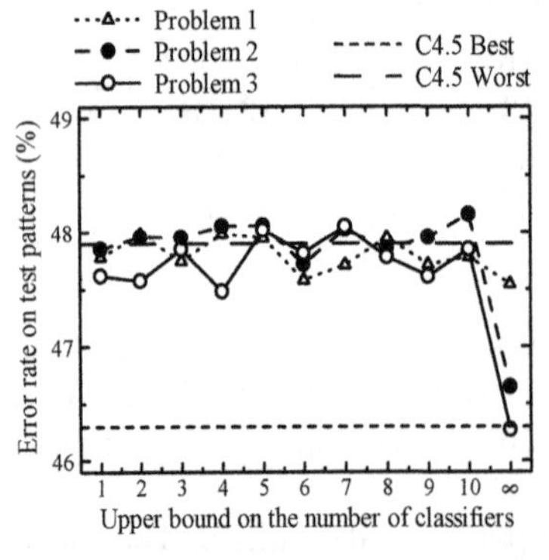

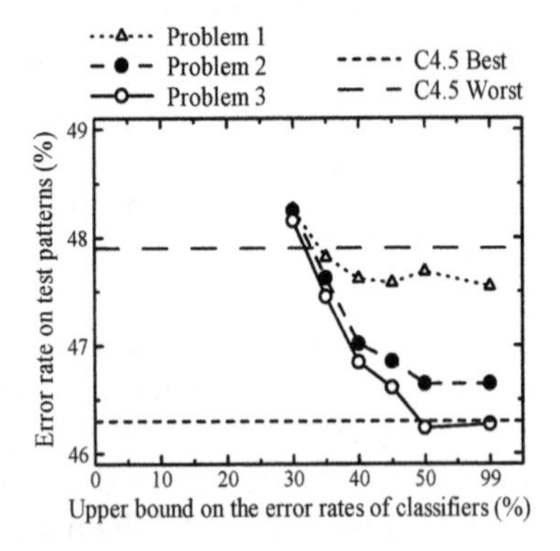

Fig. 22.10. Performance of ensemble classifiers on the Cleveland heart disease data set. (a) Second strategy for constructing ensembles. (b) Third strategy for constructing ensembles.

Fig. 22.11(b). We can also see that the experimental results in Fig. 22.11 by fuzzy rule-based classifiers and ensemble classifiers are better than the best result of the C4.5 algorithm.

Wine Data Set: Experimental results on the wine data set are shown in Table 22.7 and Fig. 22.12. The best results were obtained from **P3** in Fig. 22.12. The improvement in error rates by combining multiple fuzzy rule-based classifiers is clear in Fig. 22.12 in the case of **P3**. Experimental results from **P3** are better than the best result of the C4.5 algorithm (i.e., the best average error rate: 5.6%) in Fig. 22.12.

Table 22.6. Obtained rule sets by each formulation for the sonar data set.

Formulation	No. of rules	Training error (%)	Test error (%)
P1	9.69	**16.43**	27.66
P2	13.02	16.54	25.96
P3	**14.15**	17.03	**25.83**

Table 22.7. Obtained rule sets by each formulation for the wine data set.

Formulation	No. of rules	Training error (%)	Test error (%)
P1	6.23	14.87	19.10
P2	8.49	**12.92**	**17.98**
P3	**10.94**	15.59	18.95

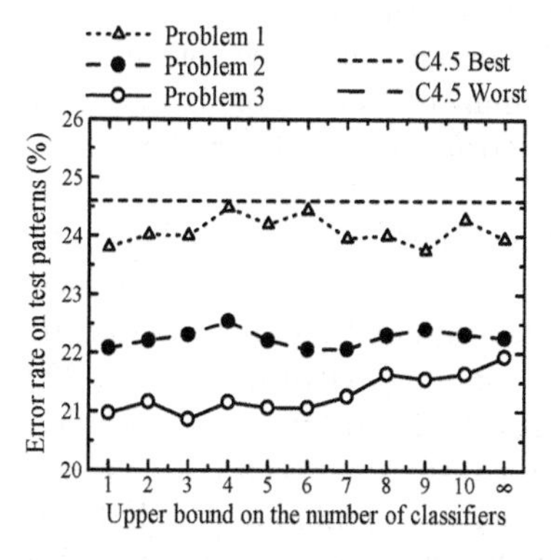

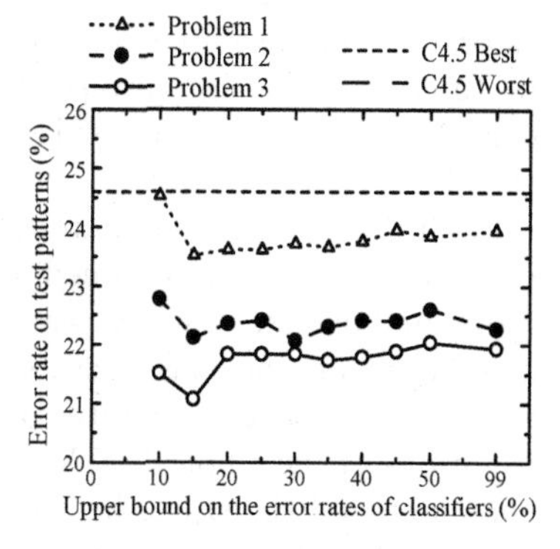

Fig. 22.11. Performance of ensemble classifiers on the sonar data sets. (a) Second strategy for constructing ensembles. (b) Third strategy for constructing ensembles.

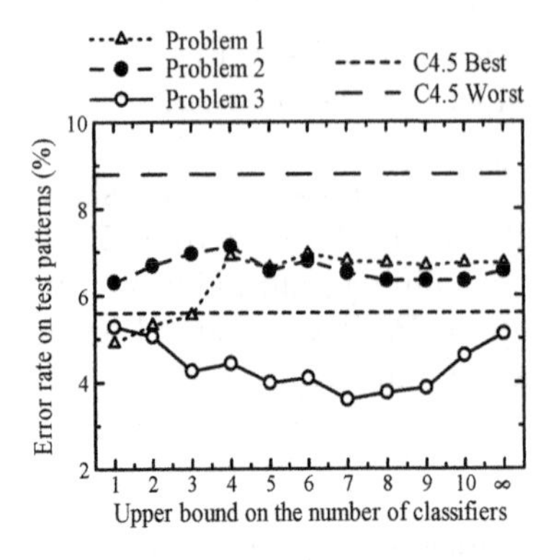

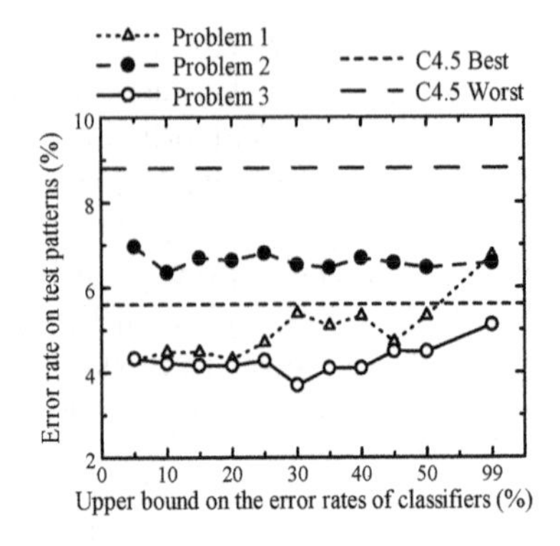

Fig. 22.12. Performance of ensemble classifiers on the wine data set. (a) Second strategy for constructing ensembles. (b) Third strategy for constructing ensembles.

22.4.5 Discussion on Experimental Results

In Table 22.8, we summarized the average number of obtained non-dominated rule sets and the average error rate of their ensemble classifiers with all the obtained non-dominated rule sets on test patterns for each data set. For comparison, we also cite the reported results by the C4.5 algorithm in [11] in the same manner as Table 22.1. The largest average number of obtained non-dominated rule sets and the smallest average error rate for each data set are highlighted by boldface in Table 22.8. In Table 22.8, more non-dominated rule sets were obtained from **P3** with the three objectives than **P1** and **P2** with the two objectives. The use of the total rule length (i.e., $f_3(S)$) in **P2** seems to lead to more non-dominated rule sets than the use of the total number of fuzzy rules (i.e., $f_2(S)$) in **P1**. The best average error rates were obtained from **P3** among the three formulations on average while the difference in average error rates of ensemble classifiers is not large among the three formulations in Table 22.8. We can also see that the performance of ensemble classifiers is comparable to the best results of the C4.5 algorithm except for the case of the glass data set.

In Table 22.9, we compare the generalization ability between the ensemble classifiers of all the obtained non-dominated rule sets and the single best individual classifier (which was chosen from the obtained non-dominated rule sets based on the classification accuracy on training patterns). Table 22.9 shows the average error rates on test patterns calculated by ten independent iterations of the whole 10-CV procedure. Better results between the ensemble classifiers and the single best individual classifier are highlighted by boldface in Table 22.9. From this table, we can see that the effect of combining multiple fuzzy rule-based classifiers depends on data sets and formulations. It improved the generalization ability of fuzzy rule-based classifiers for some data sets while it degraded the generalization ability for other data sets.

Table 22.8. Summary of experimental results.

Data set	Number of non-dominated rules			Error rates of ensembles			C4.5	
	P1	P2	P3	P1	P2	P3	Best	Worst
Breast W	9.50	10.51	**11.96**	3.87	3.90	**3.84**	5.1	6.0
Diabetes	8.94	13.73	**16.46**	25.20	25.59	25.46	**25.0**	27.2
Glass	25.08	28.63	**36.40**	39.57	39.50	39.27	**27.3**	32.2
Heart C	83.33	86.50	**87.04**	47.54	46.65	**46.27**	46.3	47.9
Sonar	9.69	13.02	**14.15**	23.95	22.26	**21.93**	24.6	35.8
Wine	6.23	8.49	**10.94**	6.74	6.57	**5.11**	5.6	8.8
Average	23.80	26.81	**29.49**	24.48	24.08	23.65	**22.32**	26.32

Table 22.9. Comparison in the generalization ability between individual and ensemble classifiers. Average error rates were calculated by ten independent iterations of the whole 10-CV procedure.

Data set	P1		P2		P3	
	Single Best	Ensemble	Single Best	Ensemble	Single Best	Ensemble
Breast W	**3.87**	**3.87**	**3.65**	3.90	4.06	**3.84**
Diabetes	25.26	**25.20**	**25.56**	25.59	**25.22**	25.46
Glass	39.71	**39.57**	39.59	**39.50**	39.52	**39.27**
Heart C	47.78	**47.54**	47.85	**46.65**	47.62	**46.27**
Sonar	**23.81**	23.95	**22.07**	22.26	**20.97**	21.93
Wine	**4.91**	6.74	**6.30**	6.57	5.28	**5.11**
Average	**24.22**	24.48	24.17	**24.08**	23.78	**23.65**

22.5 Concluding Remarks

In this chapter, we first demonstrated that a number of non-dominated rule sets (i.e., non-dominated fuzzy rule-based classifiers) were obtained by a single run of an EMO algorithm for each of the three fuzzy rule selection problems. Then we examined the effect of combining multiple non-dominated fuzzy rule-based classifiers into a single ensemble classifier. Our experimental results showed that the effect of combining multiple non-dominated fuzzy rule-based classifiers depended on data sets and formulations. For some data sets, we observed a clear improvement in average error rates on test patterns by the use of ensemble classifiers. For other data sets, ensemble classifiers did not outperform the single best individual classifier (which was chosen from the obtained non-dominated rule sets based on the classification accuracy on training patterns) independent of the number of classifiers to be combined.

As we have demonstrated in this chapter, non-dominated rule sets obtained by multiobjective fuzzy rule selection had a large diversity with respect to their classification rates on training patterns, the number of fuzzy rules, and the total rule length. This diversity, however, does not always mean the diversity of fuzzy rules in each rule set. Many non-dominated rule sets may share the same fuzzy rules or the same subset of fuzzy rules. There may not be a large diversity in classification results by non-dominated rule sets obtained by multiobjective fuzzy rule selection. A promising research direction is to combine a diversity-maintenance mechanism of rule sets (or their classification results) into evolutionary multiobjective search as in [4, 5] used for the design of ensemble neural network classifiers. The use of fuzzy genetics-based machine learning (e.g., [26, 35]) instead of fuzzy rule selection is also a promising research direction because no prescreening stage of candidate fuzzy rules is needed.

References

[1] H.A. Abbass. Pareto neuro-evolution: Constructing ensemble of neural networks using multi-objective optimization. *Proc. of Congress on Evolutionary Computation*, pages 2074–2080, 2003

[2] E. Bauer, R. Kohavi. An empirical comparison of voting classification algorithms: Bagging, boosting, and variants. *Machine Learning*, 36:105-139, 1999

[3] L. Breiman. Bagging predictors. *Machine Learning*, 24:123-140, 1996

[4] A. Chandra, X. Yao. DIVACE: Diverse and accurate ensemble learning algorithm. *Lecture Notes in Computer Science 3177: Intelligent Data Engineering and Automated Learning - IDEAL 2004*. Springer, Berlin, pp 619-625, 2005

[5] A. Chandra, X. Yao. Evolutionary framework for the construction of diverse hybrid ensemble. *Proc. of the 13th European Symposium on Artificial Neural Networks - ESANN 2005*, pp 253–258, 2005

[6] O. Cordon, F. Herrera, F. Hoffman, L. Magdalena. Genetic Fuzzy Systems, World Scientific, Singapore, 2001

[7] O. Cordon, M.J.D. Jesus, F. Herrera, L. Magdalena, P. Villar. A multiobjective genetic learning process for joint feature selection and granularity and contexts learning in fuzzy rule-based classification systems. In: J. Casillas, O. Cordon, F. Herrera, L. Magdalena(eds). Interpretability Issues in Fuzzy Modeling. Springer, Berlin, pp 79-99, 2003

[8] K. Deb. Multi-Objective Optimization Using Evolutionary Algorithms, John Wiley & Sons, Chichester, 2001

[9] K. Deb, A. Pratap, S. Agarwal, T. Meyarivan. A fast and elitist multiobjective genetic algorithm: NSGA-II. *IEEE Trans. on Evolutionary Computation*, 6:182-197, 2002

[10] T. G. Dietterich. An experimental comparison of three methods for constructing ensembles of decision trees: Bagging, boosting, and randomization. *Machine Learning*, 40:139-157, 2000

[11] T. Elomaa, J. Rousu. General and efficient multisplitting of numerical attributes. *Machine Learning*, 36:201-244, 1999

[12] Y. Freund, R.E. Schapire. A decision-theoretic generalization of on-line learning and an application to boosting. *Journal of Computer and System Sciences*, 55:119-139, 1997

[13] A. Gonzalez, R. Perez. SLAVE: A genetic learning system based on an iterative approach. *IEEE Trans. on Fuzzy Systems*, 7:176-191, 199
[14] H. Ishibuchi, T. Murata, I.B. Turksen. Single-objective and two-objective genetic algorithms for selecting linguistic rules for pattern classification problems. *Fuzzy Sets and Systems*, 89: 135-150, 1997
[15] H. Ishibuchi, T. Nakashima, T. Morisawa. Voting in fuzzy rule-based systems for pattern classification problems. *Fuzzy Sets and Systems*, 103: 223-238, 1999
[16] H. Ishibuchi, T. Nakashima, T. Murata. Three-objective genetics-based machine learning for linguistic rule extraction. *Information Sciences*, 136: 109-133, 2001
[17] H. Ishibuchi, T. Nakashima, M. Nii. Classification and modeling with linguistic information granules: Advanced approaches to linguistic data mining. Springer, Berlin, 2004
[18] H. Ishibuchi, S. Namba. Evolutionary multiobjective knowledge extraction for high-dimensional pattern classification problems. *Lecture Notes in Computer Science 3242: Parallel Problem Solving from Nature - PPSN VIII*. Springer, Berlin, pp 1123-1132, 2004
[19] H. Ishibuchi, K. Nozaki, N. Yamamoto, H. Tanaka. Construction of fuzzy classification systems with rectangular fuzzy rules using genetic algorithms. *Fuzzy Sets and Systems*, 65: 237-253, 1994
[20] H. Ishibuchi, K. Nozaki, N. Yamamoto, H. Tanaka. Selecting fuzzy if-then rules for classification problems using genetic algorithms. *IEEE Trans. on Fuzzy Systems*, 3: 260-270, 1995
[21] H. Ishibuchi, T. Yamamoto. Effects of three-objective genetic rule selection on the generalization ability of fuzzy rule-based systems. *Lecture Notes in Computer Science 2632: Evolutionary Multi-Criterion Optimization - EMO 2003*. Springer, Berlin, pp 608-622, 2003
[22] H. Ishibuchi, T. Yamamoto. Evolutionary multiobjective optimization for generating an ensemble of fuzzy rule-based classifiers. *Lecture Notes in Computer Science 2723: Genetic and Evolutionary Computation - GECCO 2003*. Springer, Berlin, pp 1077-1088, 2003
[23] H. Ishibuchi, T. Yamamoto. Comparison of heuristic criteria for fuzzy rule selection in classification problems. *Fuzzy Optimization and Decision Making*, 3: 119-139, 2004
[24] H. Ishibuch, T. Yamamoto. Fuzzy rule selection by multi-objective genetic local search algorithms and rule evaluation measures in data mining. *Fuzzy Sets and Systems*, 141: 59-88, 2004
[25] H. Ishibuchi, T. Yamamoto. Rule weight specification in fuzzy rule-based classification systems. *IEEE Trans. on Fuzzy Systems*, 2005 (in press).
[26] H. Ishibuchi, T. Yamamoto, T. Nakashima. Hybridization of fuzzy GBML approaches for pattern classification problems. *IEEE Trans. on Systems, Man, and Cybernetics - Part B: Cybernetics* 35: 359-365, 2005
[27] F. Jimenez, A.F. Gomez-Skarmeta, G. Sanchez, H. Roubos, R. Babuska. Accurate, transparent and compact fuzzy models for function approximation and dynamic modeling through multi-objective evolutionary optimization. *Lecture Notes in Computer Science 1993: Evolutionary Multi-Criterion Optimization - EMO 2001*. Springer, Berlin, pp 653-667.
[28] F. Jimenez, A.F. Gomez-Skarmeta, G. Sanchez, H. Roubos, R. Babuska. Accurate, transparent and compact fuzzy models by multi-objective evolutionary

algorithms. In: Casillas J, Cordon O, Herrera F, Magdalena L (eds) Interpretability Issues in Fuzzy Modeling. Springer, Berlin, pp 431-451, 2003
[29] Y. Jin, T. Okabe, B. Sendhoff. Neural network regularization and ensembling using multi-objective evolutionary algorithms. *Proc. of Congress on Evolutionary Computation - CEC 2004*, pp 1-8, 2004
[30] Y. Jin, T. Okabe, B. Sendhoff. Evolutionary multi-objective optimization approach to constructing neural network ensembles for regression. In: Coello CAC, Lamont GB (eds) *Applications of Multi-Objective Evolutionary Algorithms.* World Scientific, Singapore, pp 653-673, 2004
[31] L.S. Oliveira, M. Morita, R. Sabourin, F. Bortolozzi. Multi-objective genetic algorithms to create ensemble of classifiers. *Lecture Notes in Computer Science 3410: Evolutionary Multi-Criterion Optimization - EMO 2005.* Springer, Berlin, pp 592-606, 2005
[32] L.S. Oliveira, R. Sabourin, F. Bortolozzi, C.Y. Suen. Feature selection for ensembles: A hierarchical multi-objective genetic algorithm approach. *Proc. of 7th International Conference on Document Analysis and Recognition - ICDAR* 2003, pp 676-680, 2003
[33] J.R. Quinlan. C4.5: Programs for Machine Learning. Morgan Kaufmann, San Mateo.
[34] J.R. Quinlan. Improved use of continuous attributes in C4.5. *Journal of Artificial Intelligence Research*, 4: 77-90, 1996
[35] H. Wang, S. Kwong, Y. Jin, W. Wei, K.F. Man. Agent-based evolutionary approach for interpretable rule-based knowledge extraction. *IEEE Trans. on Systems, Man, and Cybernetics - Part C: Applications and Reviews*, 35: 143-155, 2005
[36] H. Wang, S. Kwong, Y. Jin, W. Wei, K.F. Man. Multi-objective hierarchical genetic algorithm for interpretable fuzzy rule-based knowledge extraction. *Fuzzy Sets and Systems* 149: 149-186, 2005

Part V

Applications of Multi-Objective Machine Learning

23

Multi-Objective Optimisation for Receiver Operating Characteristic Analysis

Richard M. Everson and Jonathan E. Fieldsend

School of Engineering, Computer Science and Mathematics,
University of Exeter, Exeter, UK, EX4 4QF.
R.M.Everson@exeter.ac.uk, J.E.Fieldsend@exeter.ac.uk

Summary. Receiver operating characteristic (ROC) analysis is now a standard tool for the comparison of binary classifiers and the selection operating parameters when the costs of misclassification are unknown.

This chapter outlines the use of evolutionary multi-objective optimisation techniques for ROC analysis, in both its traditional binary classification setting, and in the novel multi-class ROC situation.

Methods for comparing classifier performance in the multi-class case, based on an analogue of the Gini coefficient, are described, which leads to a natural method of selecting the classifier operating point. Illustrations are given concerning synthetic data and an application to Short Term Conflict Alert.

23.1 Introduction

One of the fundamental problems of machine learning is deciding to which class an unknown example belongs on the basis of a number of examples whose correct class is known. Applications abound, for example: automatically distinguishing harvested potatoes from clods of earth; detecting fraudulent financial transactions; clinical screening; and deciding whether aircraft are likely to pass dangerously close to each other. The cost of making the wrong classification ranges from almost negligible or slightly embarrassing to – in the case of safety critical systems – life threatening. *False positives*, for example the incorrect identification of clods as potatoes, are inevitable in most practical situations and attempting to limit their number leads to a reduction in the number of *true positives*. Selecting a classifier and its operating parameters to simultaneously maximise the true positive rate while minimising the false positive rate is thus a multi-objective optimisation problem, which we address in this chapter.

Given a classifier that yields estimates of the exemplar's probability of belonging to each of the classes and when the relative costs of misclassification are known, it is straightforward to determine the decision rule that minimises

R.M. Everson and J.E. Fieldsend: *Multi-Objective Optimisation for Receiver Operating Characteristic Analysis*, Studies in Computational Intelligence (SCI) **16**, 533–556 (2006)
www.springerlink.com

the average cost of misclassification. However, the true costs of misclassification are frequently unknown and difficult to determine precisely (e.g. [4, 1]). In such cases the practitioner must either guess the misclassification costs or explore the trade-off in classification rates as the decision rule is varied.

Receiver operating characteristic (ROC) analysis provides a convenient graphical display of the trade-off between true and false positive classification rates. Since its introduction in the medical and signal processing literatures [19, 38] ROC analysis has become a prominent method for selecting an operating point. The recent work of Provost and Fawcett [31, 30] reintroduced ROC analysis to the machine learning community; see [16, 20, 34] for a recent snapshot of methodologies and applications. The fundamental trade-off between true and false positive rates permits ROC analysis to be cast as a multi-objective optimisation problem. In this chapter we review the foundations of ROC analysis and the application of evolutionary algorithms to finding classifiers with optimal ROC curves. The methodology is illustrated on a synthetic problem and on a safety related system employed to raise a warning if two aircraft are likely to become dangerously close. The evolutionary optimisation point of view allows a straightforward generalisation of the two class classification methodology to multiple classes, which we describe in Section 23.5.

ROC analysis is frequently used for evaluating and comparing classifiers, the area under the ROC curve (AUC) or, equivalently, the Gini coefficient. Although the straightforward analogue of the AUC is unsuitable for more than two classes, in Section 23.6 we describe a straightforward generalisation of the Gini coefficient which quantifies the superiority of a classifier's performance to random allocation and permits the comparison of classifiers on a particular problem.

23.2 Risk and Cost

In general a classifier seeks to allocate an exemplar or measurement $\mathbf{x}$ to one of a number of classes, $\mathcal{A}_k$. For the time being we permit the number of classes Q to be greater than 2; we specialise to binary classification in Section 23.3 and return to multi-class ROC analysis in Sections 23.5 and 23.6.

Allocation of $\mathbf{x}$ to the incorrect class, say $\mathcal{A}_j$, usually incurs some, often unknown, cost denoted by λ_{kj}. We count the cost of a correct classification as zero: $\lambda_{kk} = 0$, but see Elkan [9] for a treatment of the general case. Denoting the probability under some decision rule or classifier of assigning an exemplar to $\mathcal{A}_j$ when its true class is in fact $\mathcal{A}_k$ as $p(\mathcal{A}_j \mid \mathcal{A}_k)$ the overall risk or expected cost is

$$R = \sum_{j,k} \lambda_{kj} p(\mathcal{A}_j \mid \mathcal{A}_k) \pi_k \tag{23.1}$$

where π_k is the prior probability of $\mathcal{A}_k$. The performance of some particular classifier may be conveniently be summarised by a *confusion matrix* or contingency table, $\hat{C}$, which summarises the results of classifying a set of examples. Each entry $\hat{C}_{kj}$ of the confusion matrix gives the number of examples, whose true class was $\mathcal{A}_k$, that were actually assigned to $\mathcal{A}_j$. Normalising the confusion matrix so that each row sums to unity gives the confusion rate matrix, which we denote by C, whose entries are estimates of the misclassification probabilities: $p(\mathcal{A}_j \,|\, \mathcal{A}_k) \approx C_{kj}$. Thus the expected risk is estimated as

$$R = \sum_{j,k} \lambda_{kj} C_{kj} \pi_k. \tag{23.2}$$

The expected risk can be written in terms of the posterior probabilities of classification to each class. The conditional risk or average cost of assigning $\mathbf{x}$ to $\mathcal{A}_j$ is

$$R(\mathcal{A}_j \,|\, \mathbf{x}) = \sum_k \lambda_{kj} p(\mathcal{A}_k \,|\, \mathbf{x}) \tag{23.3}$$

where $p(\mathcal{A}_k \,|\, \mathbf{x})$ is the posterior probability that $\mathbf{x}$ belongs to $\mathcal{A}_k$. If $\alpha(\mathbf{x}_n)$ is a decision rule or classifier that assigns $\mathbf{x}$ to one of the classes $\mathcal{A}_k$, then the expected overall risk is

$$R = \int R(\alpha(\mathbf{x}) \,|\, \mathbf{x}) p(\mathbf{x}) \, d\mathbf{x}. \tag{23.4}$$

The expected risk is then minimised, being equal to the Bayes risk, when $\mathbf{x}$ is assigned to the class with the minimum conditional risk (e.g., [8]). Choosing 'zero-one costs', $\lambda_{jk} = 1 - \delta_{jk}$, means that all misclassifications are equally costly and the conditional risk is equal to the class posterior probability. The optimum assignment is therefore to the class with the greatest posterior probability, which minimises the overall error rate.

When the costs of misclassification are known it is therefore straightforward make assignments to achieve the Bayes risk (provided, of course, that the classifier yields accurate assessments of the posterior probabilities $p(\mathcal{A}_k \,|\, \mathbf{x})$). However, costs are frequently unknown and difficult to estimate, particularly when there are many classes; in this case it is useful to be able to compare the classification rates as the costs vary.

23.3 Binary ROC Analysis

For binary classification, in which $\mathbf{x}$ is assigned either to $\mathcal{A}_1$ or $\mathcal{A}_2$, the conditional risk may be simply rewritten in terms of the posterior probability of assigning to $\mathcal{A}_1$, resulting in the rule: assign $\mathbf{x}$ to $\mathcal{A}_1$ if $p(\mathcal{A}_1 \,|\, \mathbf{x}) > t = \lambda_{12}/(\lambda_{12} + \lambda_{22})$. This decision rule reveals that there is, in fact, only one

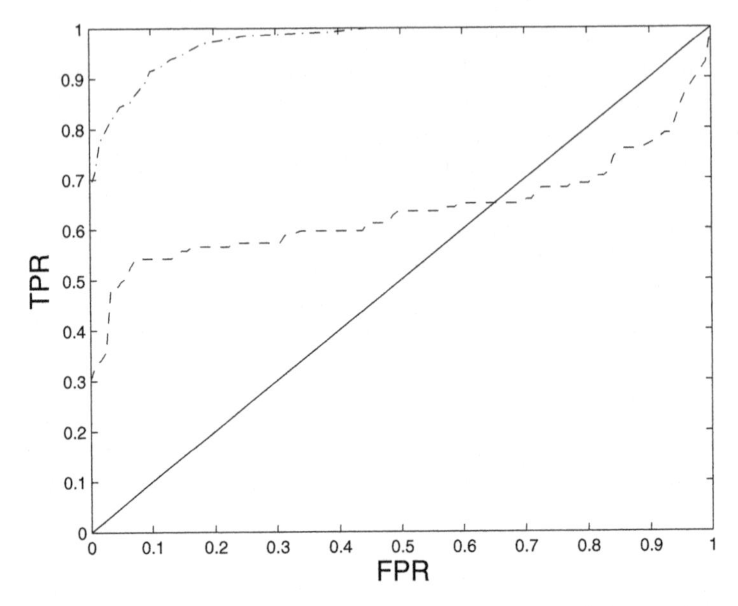

Fig. 23.1. ROC curves of maximum likelihood MLP (dashed-dotted line) and logistic regression (dashed line) classifiers. Curves traced out by varying costs. The ROC corresponding to random allocation is shown as the diagonal line.

degree of freedom in the binary cost matrix and, as might be expected, the entire range of classification rates for each class is swept out as the classification threshold t varies from 0 to 1. It is this variation of rates that the ROC curve exposes for binary classifiers. ROC analysis focuses on the classification of one particular class, say $\mathcal{A}_1$,[1] and plots the true positive classification rate for $\mathcal{A}_1$ versus the false positive rate as the threshold t or, equivalently, the ratio of misclassification costs is varied.

As an illustrative example we consider a two-class, two-feature synthetic data set based on a Gaussian mixture model data set introduced by Ripley [32]. Weights for the five components were $(0.16, 0.17, 0.17, 0.25, 0.25)$: the first 3 components, with component means at $(1,1)^T$, $(-0.7, 0.3)^T$ and $(0.3, 0.3)^T$, generate $\mathcal{A}_1$; while the remaining 2 components, with means at $(-0.3, 0.7)^T$ and $(0.4, 0.7)^T$, generate $\mathcal{A}_2$. The covariances of all components are isotropic: $\boldsymbol{\Sigma}_j = 0.03\mathbf{I}$. In the work described here 250 observations were used for training.

Figure 23.1 shows the ROC curve for a multi-layer perceptron (MLP) with 5 units in a single hidden layer, trained by minimising the cross entropy using quasi-Newton minimisation, which is tantamount to finding the maximum likelihood model, see for example, [3]. As illustrated by the figure, a range of true and false positive classification rates is available as the decision threshold

[1] Note that all the information about the other class is easily recovered.

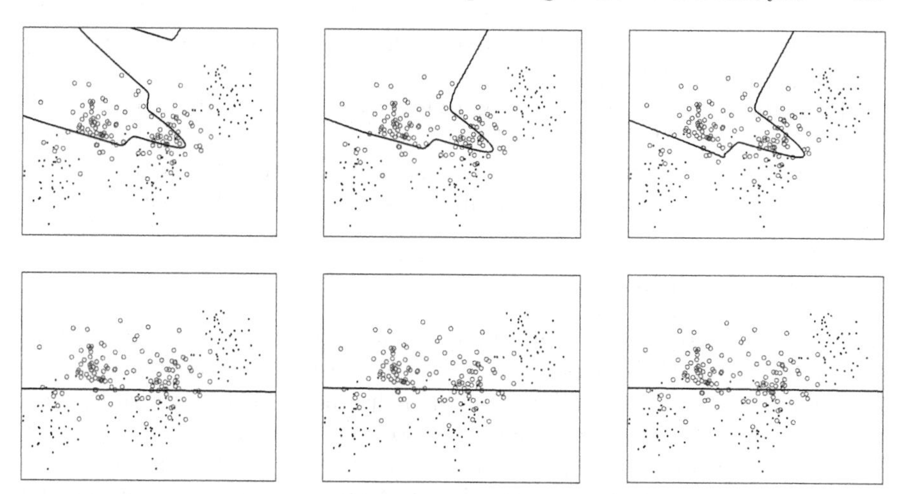

Fig. 23.2. Data points for an augmented version of Ripley's synthetic data and decision boundaries of maximum likelihood classifiers at different false positive rates for the 'circles' class found by varying the decision threshold t. *First row:* MLPs, *Second row:* Logistic regression classifiers. *First column:* FPR = 0.04, *Second column:* FPR = 0.072, *Third column:* FPR = 0.144.

t is varied. The figure also shows the ROC curve for a logistic regressor, whose ROC curve is clearly inferior to the MLP's because every point on the logistic regressor's curve is dominated by (at least) one point on the MLP's curve. This may be expected as the logistic regressor's decision boundaries are constrained to be hyper-planes (straight lines in this 2D situation) and it therefore has much less flexibility than the MLP.

Figure 23.2 shows the decision regions for the MLP and logistic regressor when the false positive rate for the 'circles' class is 0.040, 0.072 and 0.144. As the figure shows the decision boundaries for the logistic regressor at different thresholds are parallel to each other because contours of posterior probability $p(\mathcal{A}_k \mid \mathbf{x})$ are parallel straight lines, and there is little variation in the location of the decision boundary as the false positive rate varies from 0.04 to 0.144. By contrast, the MLP decision boundaries are curved, better fitting the data, and the decision boundaries for different thresholds are not parallel because contours of posterior probability are not parallel. Nonetheless both sets of decision boundaries show the same general trend: a higher true positive rate is achieved by moving the decision boundary so as to encompass more $\mathcal{A}_1$ (circles) observations, which means that more $\mathcal{A}_2$ observations are erroneously assigned as $\mathcal{A}_1$ resulting in an increased false positive rate.

The diagonal of the ROC plot (Figure 23.1) shows the performance of the classifier that allocates examples to $\mathcal{A}_1$ with constant probability, without regard for the features $\mathbf{x}$. If a classifier performs worse than random for some thresholds, such as the logistic regressor for FPR $\gtrsim 0.63$, then performance equivalent to reflecting the ROC curve in the diagonal is obtained merely by swapping the class labels.

Portions of the logistic regressor ROC curve are markedly concave, and Scott et al [33] and Provost annd Fawcett [30, 31] have shown that classifiers with operating characteristics on the convex hull of the ROC curve can be constructed by stochastically combining classifiers on the ROC curve itself.

23.3.1 Pareto Optimality

So far we have considered ROC curves for a single classifier as the decision threshold t is varied. It is useful, however, to consider the classifiers that are obtained by varying, not only the decision threshold, but the classifier parameters, such as the weights in a neural network or the thresholds in a decision tree. In general we denote these classifier parameters by $\mathbf{w}$. This leads naturally to a multi-objective optimisation problem which may be solved using current evolutionary methods. In order to permit straightforward generalisation to problems with more than two classes, rather than attempting to maximise the true positive rate and minimise the false positive rate, we consider the equivalent problem of minimising the false positive rates for both classes, which in terms of the confusion rate matrix are C_{12} and C_{21}. We therefore seek solutions to the multi-objective minimisation problem:

$$\text{minimise} \quad C_{jk}(\mathbf{w}, \boldsymbol{\lambda}) \quad \text{for all } j \neq k. \tag{23.5}$$

Here we have made explicit the dependence of the false positive rates on both the parameters $\mathbf{w}$ and the misclassification costs $\boldsymbol{\lambda}$. For notational convenience and because they will be treated as a single entity, we write the costs and classifier parameters as a single vector of generalised parameters, $\boldsymbol{\theta} = \{\boldsymbol{\lambda}, \mathbf{w}\}$; to distinguish $\boldsymbol{\theta}$ from the classifier parameters $\mathbf{w}$ we use the optimisation terminology *decision vector* to refer to $\boldsymbol{\theta}$.

If all the misclassification rates for one classifier with decision vector $\boldsymbol{\theta}$ are no worse than the classification rates for another classifier $\boldsymbol{\phi}$ and at least one rate is better, then the classifier and costs determined by $\boldsymbol{\theta}$ is said to *strictly dominate* that with decision vector $\boldsymbol{\phi}$. Thus $\boldsymbol{\theta}$ strictly dominates $\boldsymbol{\phi}$ (denoted $\boldsymbol{\theta} \prec \boldsymbol{\phi}$) iff:

$$\begin{aligned} C_{jk}(\boldsymbol{\theta}) &\leq C_{jk}(\boldsymbol{\phi}) \quad \forall j, k \quad \text{and} \\ C_{jk}(\boldsymbol{\theta}) &< C_{jk}(\boldsymbol{\phi}) \quad \text{for some } j, k. \end{aligned} \tag{23.6}$$

Less stringently, $\boldsymbol{\theta}$ *weakly dominates* $\boldsymbol{\phi}$ (denoted $\boldsymbol{\theta} \preceq \boldsymbol{\phi}$) iff:

$$C_{jk}(\boldsymbol{\theta}) \leq C_{jk}(\boldsymbol{\phi}) \quad \forall j, k. \tag{23.7}$$

A set E of decision vectors is said to be a *non-dominated* set if no member of the set is dominated by any other member:

$$\boldsymbol{\theta} \not\prec \boldsymbol{\phi} \quad \forall \boldsymbol{\theta}, \boldsymbol{\phi} \in E. \tag{23.8}$$

A solution to the minimisation problem (23.5) is thus *Pareto optimal* if it is not dominated by any other feasible solution, and the non-dominated set of all Pareto optimal solutions is known as the Pareto front. The Pareto optimal ROC curve may be thought of as the non-dominated set formed from the union of the ROC curves for each fixed parameter set $\mathbf{w}$; however, multi-objective evolutionary techniques permit more efficient location of the Pareto front when the classifier has many parameters.

Recent years have seen the development of a number of evolutionary techniques based on dominance measures for locating the Pareto front; see [5, 6] and [35] for recent reviews. Kupinski and Anastasio [25] and Anastasio et al [2] introduced the use of multi-objective evolutionary algorithms (MOEAs) to optimise ROC curves for binary problems, illustrating the method on a synthetic data set and for medical imaging problems; and we have used a similar methodology for locating optimal ROC curves for safety-related systems [13, 11]. In the following section we describe a straightforward evolutionary algorithm for locating the Pareto front for binary and multi-class problems. We illustrate the method on a synthetic problem for two different classifiers in Section 23.4.1.

23.4 Evolving Classifiers

The algorithm we describe is based on a simple analogue of mutation-based evolution (such as [15, 13, 22, 23, 27, 26]), but any recent elitist MOEA could equally well be used [5, 6, 7, 17, 35, 37].

The algorithm, an evolution strategy (ES), maintains a set or archive E of decision vectors, whose members are mutually non-dominating, which forms the current approximation to the Pareto front and is a source of elite solutions for evolution. As the computation progresses members of E are selected, copied and their decision vectors perturbed, and the objectives corresponding to the perturbed decision vector evaluated; if the perturbed solution is not dominated by any element of E, it is inserted into E and any members of E which are dominated by the new entrant are removed. Therefore the archive can only move towards the Pareto front: it is in essence a greedy search where the archive E is the current point of the search and perturbations to E that are not dominated by the current E are always accepted.

Algorithm 4 describes the procedure in more detail. The archive E is initialised by evaluating the misclassification rates for a number (here 100) of randomly chosen parameter values and costs, and discarding those which are dominated by another element of the initial set. Then at each generation a single element, $\boldsymbol{\theta}$ is selected from E (line 3 of Algorithm 4); selection may be uniformly random, but partitioned quasi-random selection (PQRS) [15] was used here to promote exploration of the front. PQRS prevents clustering of solutions in a particular region of the front biasing the search because they

Algorithm 4 Multi-objective evolution scheme for ROC surfaces.

1:	$E :=$ initialise()	
2:	`for` $t := 1 : T$	*Loop for T generations*
3:	$\{\mathbf{w}, \boldsymbol{\lambda}\} = \boldsymbol{\theta} :=$ select(E)	*PQRS*
4:	$\mathbf{w}' :=$ perturb$(\mathbf{w})$	*Perturb parameters*
5:	`for` $i := 1 : L$	*Loop over weight samples*
6:	$\boldsymbol{\lambda}' :=$ sample()	*Sample costs*
7:	$C :=$ classify$(\mathbf{w}', \boldsymbol{\lambda}')$	*Evaluate classification rates*
8:	$\boldsymbol{\theta}' := \{\mathbf{w}', \boldsymbol{\lambda}'\}$	
9:	`if` $\boldsymbol{\theta}' \not\preceq \boldsymbol{\phi} \; \forall \boldsymbol{\phi} \in E$	
10:	$E := \{\boldsymbol{\phi} \in E \mid \boldsymbol{\phi} \not\prec \boldsymbol{\theta}'\}$	*Remove dominated elements*
11:	$E := E \cup \boldsymbol{\theta}'$	*Insert* $\boldsymbol{\theta}'$
12:	`end`	
13:	`end`	
14:	`end`	

are selected more frequently, thus increasing the efficiency and range of the search.

The selected *parent* decision vector is copied, after which the costs $\boldsymbol{\lambda}$ and classifier parameters $\mathbf{w}$ are treated separately. The parameters $\mathbf{w}$ of the classifier are perturbed or, in the nomenclature of evolutionary algorithms, mutated, to form a *child*, $\mathbf{w}'$ (line 4). Here we seek to encourage wide exploration of parameter space by additively perturbing each of the parameters with a random number δ drawn from a heavy tailed distribution (such as the Laplacian density, $p(\delta) \propto e^{-|\delta|}$). The Laplacian distribution has tails that decay relatively slowly, thus ensuring that there is a high probability of exploring regions distant from the current solutions, facilitating escape from local minima [36].

With a proposed parameter set $\mathbf{w}'$ on hand the procedure then investigates the misclassification rates as the costs are varied with fixed parameters. In order to do this we generate L sample costs $\boldsymbol{\lambda}'$ and evaluate the misclassification rates for each of them. Since the misclassification costs are non-negative and sum to unity, a straightforward way of producing samples is to make draws from a Dirichlet distribution:

$$p(\boldsymbol{\lambda}) = Dir(\boldsymbol{\lambda} \mid \alpha_1, \dots, \alpha_D) \tag{23.9}$$

$$= \frac{\Gamma(\sum_{i=1}^{D} \alpha_i)}{\prod_{i=1}^{D} \Gamma(\alpha_i)} \left(1 - \sum_{i=1}^{D-1} \lambda_i\right)^{\alpha_D - 1} \prod_{i=1}^{D-1} \lambda_i^{\alpha_i - 1} \tag{23.10}$$

where the index i labels the $D \equiv Q(Q-1)$ off-diagonal entries in the cost matrix. Samples from a Dirichlet density lie on the simplex $\sum_{kj} \lambda_{kj} = 1$. The $\alpha_{jk} \geq 0$ determine the density of the samples; since we have no preference for particular costs here, we set all the $\alpha_{kj} = 1$ so that the simplex (that is, cost space) is sampled uniformly with respect to Lebesgue measure.

The misclassification rates for each cost sample $\boldsymbol{\lambda}'$ and classifier parameters $\mathbf{w}$ are used to make class assignments for each example in the given dataset (line 7). Usually this step consists of merely modifying the posterior probabilities $p(\mathcal{A}_k \mid \mathbf{x})$ to find the assignment with the minimum expected cost and is therefore computationally inexpensive as the probabilities need only be computed once for each $\mathbf{w}'$. The misclassification rates $C_{kj}(\boldsymbol{\theta}')$ $(j \neq k)$ comprise the objective values for the decision vector $\boldsymbol{\theta}' = \{\mathbf{w}', \boldsymbol{\lambda}\}$ and decision vectors that are not dominated by any member of the archive E are inserted into E (line 11) and any decision vectors in E that are dominated by the new entrant are removed (line 10). Since artificially limiting the archive size may inhibit convergence, the archive is permitted to grow without limit. Although managing the number of solutions in the archive has not proved a computational bottleneck, data structures to efficiently maintain and query large archives may be used for very large archives [15, 21].

A $(\mu + \lambda)$ evolution strategy (ES) is defined as one in which μ decision vectors are selected as parents at each generation and perturbed to generate λ offspring.[2] The set of offspring and parents are then truncated or replicated to provide the μ parents for the following generation. Although Algorithm 4 is based on a $(1 + 1)$-ES, it is interesting to note that each parent $\boldsymbol{\theta}$ is perturbed to yield L offspring, all of whom have the classifier parameters $\mathbf{w}'$ in common. With linear costs, evaluation of the objectives for many $\boldsymbol{\lambda}'$ samples is inexpensive. Nonlinear costs could be incorporated in a straightforward manner, although it would necessitate complete reclassification for each $\boldsymbol{\lambda}'$ sample and it would therefore be more efficient to resample $\mathbf{w}$ with each $\boldsymbol{\lambda}'$.

Although we have assumed complete ignorance as to the misclassification costs, some imprecise information may often be available; for example the approximate bounds on the ratios of the λ_{jk} may be known. In this case the evolutionary algorithm is easily focused on the relevant region by setting the Dirichlet parameters α_{jk} appearing in (23.9) to be in the ratio of the expected costs, with their magnitudes setting the variance in the cost ratios.

23.4.1 Illustration

Figure 23.3 shows the optimised ROC curve for a MLP with 5 hidden units and the optimised ROC curve for the logistic regressor, along with the original ROC curve of the *single* maximum likelihood MLP and logistic regressor classifiers from Figure 23.1. We emphasise that the optimised ROC curves are generally comprised of operating points for several parameter values. The ROC curve of the optimised logistic regressor is again clearly inferior to the MLP's optimised ROC – however, the optimised ROCs are clearly superior to the ROC curves for the single maximum likelihood classifier of each family. A user is thus able to select an operating point and corresponding classifier parameters from the optimised ROC curves.

[2] We adhere to the optimisation terminology for $(\mu + \lambda)$-ES, although there is a potential for confusion with the costs λ_{kj}.

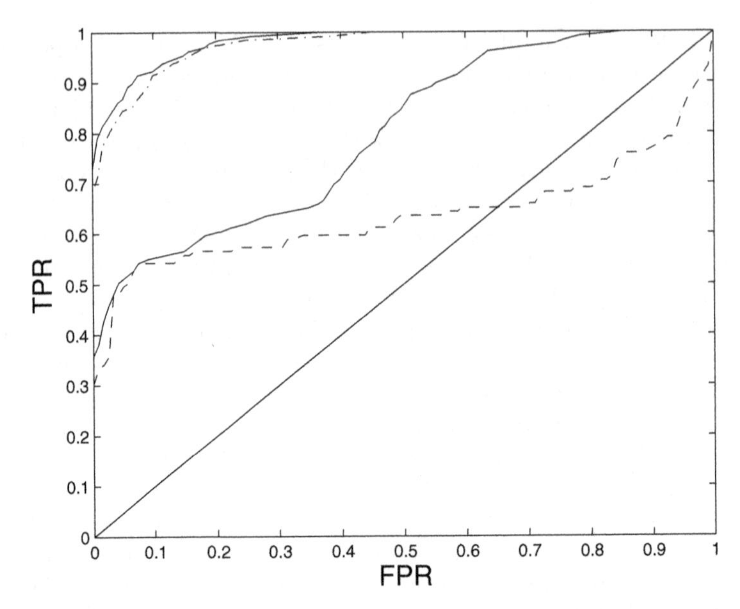

Fig. 23.3. ROC curves of maximum likelihood MLP (dashed-dotted line) and logistic regression (dashed line) classifiers, along with the optimised ROC curves for each classifier type. The ROC corresponding to random allocation is shown as the diagonal line.

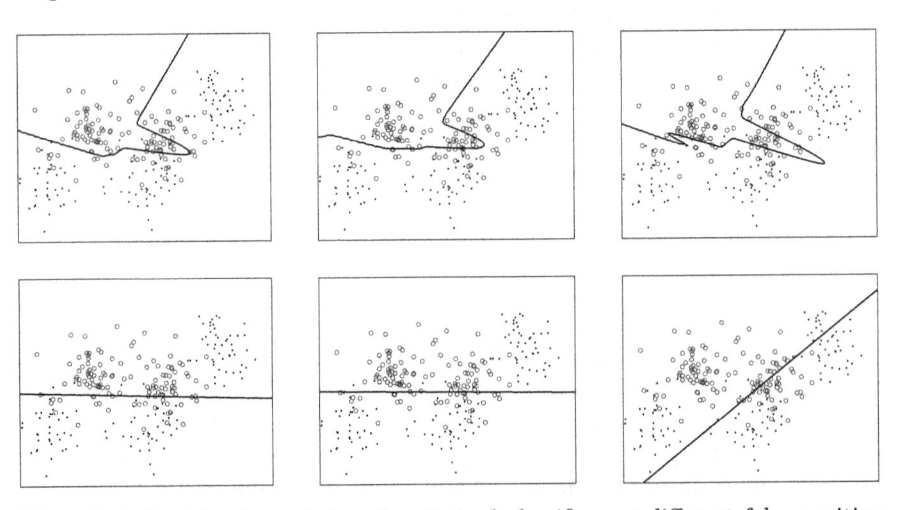

Fig. 23.4. Decision boundaries of optimised classifiers at different false positive rates. *First row:* MLPs, *Second row:* Logistic regression classifiers. *First column:* FPR=0.04, *Second column:* FPR=0.072, *Third column:* FPR=0.144.

Figure 23.4 shows the decision regions for the optimised MLPs and logistic regressors when the false positive rate C_{12} is 0.040, 0.072 and 0.144. As each point on the ROC curve may be derived from a different model parameterisation as well as corresponding to different cost matrices, the decision boundaries for the logistic regressor at different thresholds are no longer parallel to each other. The MLP decision contours also show a greater difference than those from Figure 23.2 – again demonstrating the extra flexibility available through parameter variation. Both sets of decision boundaries again show the same general trend moving along the ROC curve from left to right, with a higher true positive rate for $\mathcal{A}_1$ being achieved by moving the decision boundary so as to encompass more $\mathcal{A}_1$ observations and sacrificing the correct classification of $\mathcal{A}_2$ points.

23.4.2 Short Term Conflict Alert

As an illustration of the utility of multi-objective ROC analysis we describe its application to the optimisation of a Short Term Conflict Alert (STCA) system. This system, covering UK airspace, is responsible for providing advisory alerts to air traffic controllers if two aircraft are likely to become dangerously close. Ground-based radar is used to monitor the positions and heights of aircraft in the airspace. Having filtered out pairs of aircraft that are simply too far apart to be in any danger of collision in the next few minutes, the system makes predictions using three modules – the linear predictive filter, the current proximity filter and the manoeuvre hazard filter – whose results are combined into a final prediction. The three modules each have a number of parameters which may have different values when aircraft are in different airspace categories (for example, en route or stacked) so that $\mathbf{w}$ the vector describing the adjustable parameters has over 1500 entries.

Skilled staff of the National Air Traffic Services (NATS, the principal civil air traffic control authority for the UK) manually adjust or *tune* these parameters in order to reduce the number of false positive alerts while maintaining the true positive alerts. This tuning is performed manually on the basis of a database comprised of 170 000 aircraft pairs, containing historical and recent encounters. However, the receiver operating characteristics of the STCA system have been unknown, hampering the choice of the optimal operating point and parameters.

Figure 23.5 shows the Pareto optimal ROC front located after $T = 6000$ iterations of an MOEA which was permitted to adjust the 900 or so parameters that are routinely adjusted by NATS staff. The true and false positive rates corresponding to the manually tuned parameters $\mathbf{w}^\star$ are also marked as a cross. As the figure shows, the optimisation has located an ROC curve, several points of which dominate the manually tuned operating point. Although the ROC curve allows the choice of operating points that are a little better than the manually tuned operating point, we view as more important, however, the production of the ROC curve itself, because it reveals the true positive versus false positive rate trade-off, permitting a principled choice of the operating point to be made. In fact the current operating point $\mathbf{w}^\star$ is close to the corner

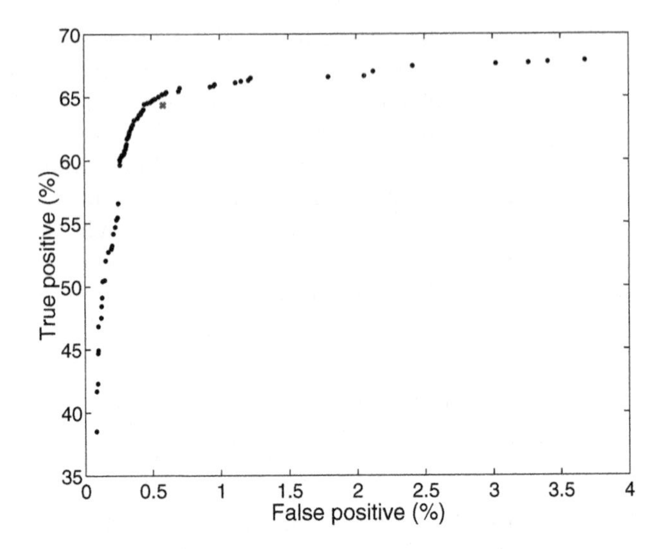

Fig. 23.5. Dots show estimates of the Pareto optimal ROC curve for STCA obtained after 6000 evaluations of a multi-objective optimiser. The cross indicates the manually tuned operating point $\mathbf{w}^{\star}$.

of the Pareto optimal curve. Choosing an operating point to the left of the corner would result in a rapidly diminishing true positive rate for little gain in the false positive rate; whereas operating points to the right of the corner provide small increases in the true positive rate at the expense of relatively large increases in the false positive rate.

Full details of the methods used and the results may be found in [13, 11], in which it is shown that the optimisation of the ROC curve can be carried out simultaneously with the optimisation of other objectives, for example the warning time of a possible conflict given to air traffic controllers.

23.5 Multi-class ROC Analysis

ROC analysis for binary classification focuses on the true and false positive rates for a particular class, although the true and false positive rates for the other class are easily derived from these. However, when discriminating between $Q > 2$ classes, focussing on a single class is likely to be misleading. We consider instead the rates at which each class is misclassified into each of the other classes. With Q classes this leads us to consider $D \equiv Q(Q-1)$ misclassification rates C_{kj} for all $j \neq k$. That is, we consider the off-diagonal elements of the confusion matrix. The true positive rates, corresponding to the diagonal elements of C, are easily determined from the off-diagonal elements since each columns sums to unity. The two objective minimisation problem for binary classification naturally generalises in the multi-class case to a D-dimensional multi-objective minimisation of the off-diagonal elements of the confusion rate matrix (23.5).

As in the binary classification case, the absolute magnitude of the misclassification costs is not important and we assume that they are normalised so that they sum to one: $\sum_{j \neq k} \lambda_{kj} = 1$. There are, therefore, $D-1 = Q(Q-1)-1$ degrees of freedom in the cost matrix, so that the Pareto front in general has dimension $D-1$ and may be thought of as a hyper-surface dividing the D-dimensional objective space of misclassification rates.

The minimisation problem (23.5) and Algorithm 4 were defined in such a way that they can be directly applied to the optimisation of ROC surfaces for multi-class problems, as we now illustrate.

23.5.1 Illustration

The synthetic data used above is simply extended to $Q = 3$ classes by augmenting it with an additional Gaussian centre, so that each of the three classes is defined by a mixture of two Gaussian densities, all with isotropic covariances $\boldsymbol{\Sigma}_j = 0.3\mathbf{I}$. All the centres are equally weighted and if $\boldsymbol{\mu}_{ji}$ for $i = 1, 2$ denotes the means of the two Gaussian components generating samples for class j, the centres are: $\boldsymbol{\mu}_{11} = (0.7, 0.3)^T$, $\boldsymbol{\mu}_{12} = (0.3, 0.3)^T$, $\boldsymbol{\mu}_{21} = (-0.7, 0.7)^T$, $\boldsymbol{\mu}_{22} = (0.4, 0.7)^T$, $\boldsymbol{\mu}_{31} = (1.0, 1.0)^T$, $\boldsymbol{\mu}_{32} = (0.0, 1.0)^T$.

We again use the MOEA to discover the Pareto optimal ROC surface for an MLP with five hidden units and softmax output units classifying 300 examples of the synthetic data. The MOEA was run for $T = 10000$ evaluations of the classifier, resulting in an estimated Pareto front or ROC surface comprising approximately 4800 mutually non-dominating parameter and cost combinations. The archive was initialised by training a single MLP using quasi-Newton optimisation of the data likelihood (see e.g. [3]) which finds a point on or near the Pareto front corresponding to equal misclassification costs; subsequent iterations of the evolutionary algorithm are therefore largely concerned with exploring the Pareto front rather than locating it.

Decision regions corresponding to various places on the Pareto optimal ROC surface are shown in the top row of Figure 23.6. The left panel shows regions that yield the smallest *total* misclassification error. This point has very similar decision regions to the Bayes optimal decision regions for equal costs (equal cost decision boundaries are shown as solid lines) as may be expected since the overlap between classes is approximately comparable and there are equal numbers in each class. Note that although no explicit measures were taken to prevent over-fitting, the decision boundaries are smooth and do not show signs of over-fitting.

By contrast with decision regions that are optimal for roughly equal costs, the middle and right panels show decision regions for imbalanced costs. The middle panel shows decision regions corresponding to minimising C_{21} and C_{23}: this, of course, can be achieved by setting λ_{21} and λ_{23} to be large, so that

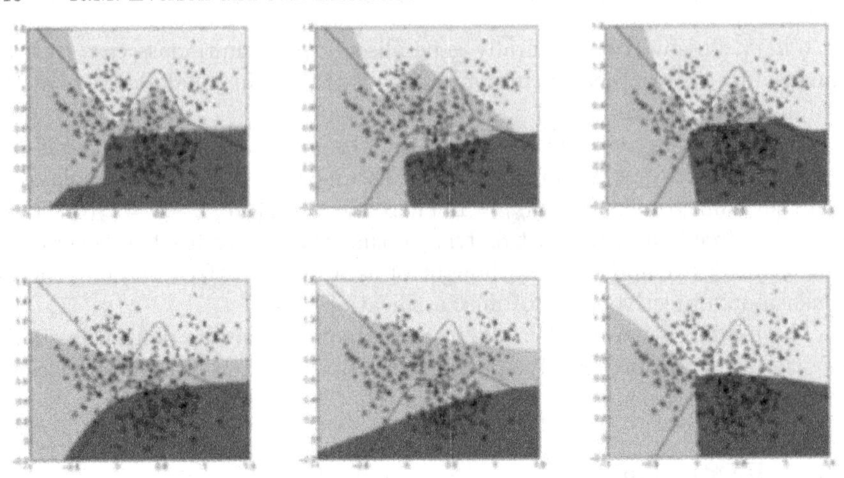

Fig. 23.6. Decision regions for various MLP *top row* and multinomial logistic *bottom row* classifiers on the multi-class ROC surface. Grey scale background shows the class to which a point would be assigned. Solid lines show the ideal equal-cost decision boundary. Symbols show actual training data. *Left column:* Parameters corresponding to minimum total misclassification error on the training data. *Middle column:* Decision regions corresponding to the minimum C_{21} and C_{23} and conditioned on this, minimum C_{31} and C_{13}. *Right column:* Decision regions corresponding to minimising C_{12} and C_{32}.

every $\mathcal{A}_2$ example (triangles) is correctly classified, no matter what the cost. For these data there are many parameterisations correctly classifying every $\mathcal{A}_2$ in the training data and we display the decision regions that also minimise C_{31} and C_{13}. For these data, it is possible to make $C_{31} = C_{13} = 0$ because $\mathcal{A}_1$ and $\mathcal{A}_3$ are adjacent only along a boundary distant from $\mathcal{A}_2$ points; such complete minimisation will not be generally possible. Of course, the penalty to be paid for minimising the $\mathcal{A}_2$ rates together with C_{31} and C_{13} is that C_{32} and C_{12} are large.

The top-right panel of Figure 23.6 shows the reverse situation: here the costs for misclassifying either $\mathcal{A}_1$ or $\mathcal{A}_3$ as C_2 are high. With these data, although not in general, it is possible to reduce C_{12} and C_{32} to zero, as shown by the decision regions which ensure that $\mathcal{A}_2$ examples (triangles) are only classified correctly when it does not result in incorrect assignment of the other two classes to $\mathcal{A}_2$. In this case the greatest misclassification rate is C_{23} (triangles as crosses).

The bottom row of Figure 23.6 shows decision boundaries for a multinomial logistic regressor(see e.g. [3]) corresponding to the same points on the Pareto optimal ROC surface as shown in the top row for the MLP. (The MOEA was run for 5000 generations resulting in an estimated Pareto front of approximately 9000 non-dominating cost and parameter combinations, although very similar results are obtained after 2000 generations.) The logistic regressor is less flexible than an MLP, having only $(d+1)Q$ parameters, where d is the dimension of the input space. On this 2-dimensional, 3-class data it therefore has only 9 adjustable parameters (compared with the 33 for the MLP), and the decision boundaries are therefore less convoluted and less well fit to the data than those for the MLP. The same trends are evident although the classification rates are lower.

The decision regions illustrated in the middle and right columns of Figure 23.6 may thought of as lying on the periphery of the Pareto surface because they correspond to one or more objectives being exactly minimised. These points are the analogues of the extreme ends of the usual two-class ROC curve where the true and false positive rates for both classes are extremised. The curvature of the ROC curve in these regions is usually small, signifying that large changes in the costs yield large changes in either the true or false positive rate, but only small changes in the other. We observe a similar behaviour here: quite large changes in the λ_{kj} in these regions yield quite small changes in the all the misclassification rates except the one which has been extremised suggesting that the curvature of the Pareto surface is low in these areas.

As we described for the STCA example, a common use of the two-class ROC curve is to locate a 'knee', a point of high curvature. The parameters at the knee are chosen as the operational parameters because the knee signifies the transition from rapid variation of true positive rate to rapid variation of false positive rate. A disadvantage of the multi-class ROC front is that its high dimension makes visualisation difficult, even for $Q = 3$ where the Pareto front is embedded in 6-dimensional space. Visualisation of these high-dimensional fronts is an area of active research; see [14] for an overview. Although, direct visualisation of the front and therefore the curvature is difficult an alternative strategy is to calculate the curvature of the manifold defined by the Pareto front and use that for selecting operating points. To date endeavours in this direction have yielded only crude approximations to the curvature even for $Q = 3$ class problems and we do not present them here. However, an alternative method of selecting a classifier in binary problems is to choose the one most distant from the diagonal of the ROC plot, and this idea can be naturally extended to the multi-class ROC surface, as we discuss in section 23.6.

As noted above, direct visualisation of the multi-class ROC surface is difficult because it is embedded in at least a 6-dimensional space. One possibility, which is explored in more depth in [10, 14], is to project the Pareto front into two or three dimensions using a data-determined nonlinear mapping such as Neuroscale [28] or the Self Organising Map [24] (see `http://www.dcs.ex.ac.uk/~reverson/research/mcroc` for examples). An alternative which we briefly discuss here is to project the Pareto front into *false positive space*. We denote by F_k the false positive rate for class k, without regard to which class the misclassification is made; thus:

$$F_k(\mathbf{w}, \boldsymbol{\lambda}) = \sum_{j \neq k} C_{kj} \qquad k = 1, \ldots, Q \tag{23.11}$$

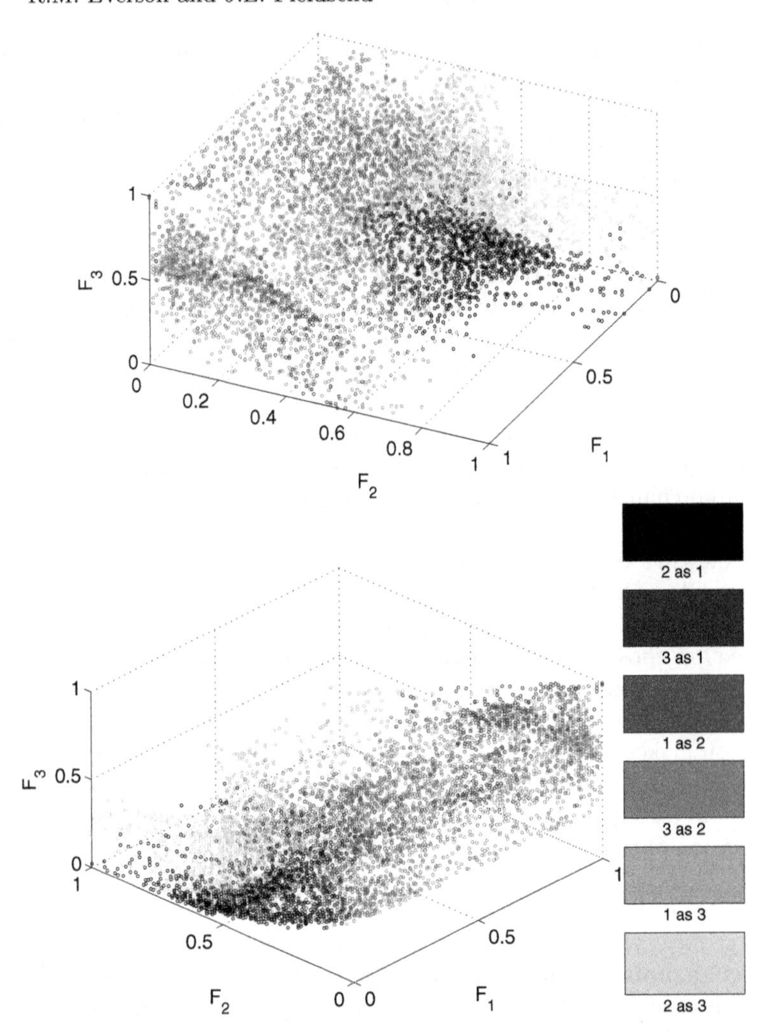

Fig. 23.7. The estimated Pareto front for synthetic data classified with a multinomial logistic regression classifier viewed in false positive space. Axes show the false positive rates for each class and different grey scales represent the class into which the greatest number of misclassifications are made. (Points better than random shown.)

where we emphasise the dependence of the false positive rates on the parameterisation and costs. Each point on the Pareto front is then plotted at the coordinates given by its Q false positive rates. This visualisation clearly loses information on *how* a point is misclassified, but colour or grey scale can be utilised to indicate the class that is most misclassified. Figure 23.7 shows the estimated Pareto front for the logistic classifier visualised in this manner.[3] Note that, along with all other projection methods, the projection into the lower dimensional false positive space does not preserve the mutual non-dominance between points on the front, which appears as a thickened cloud in three dimensions. Nonetheless this sort of visualisation can be useful for navigating the Pareto front.

23.6 Comparing Classifiers: The Gini Coefficient

In two class problems the area under the ROC curve (AUC) is often used to compare classifiers. As clearly explained by Hand and Till [18], the AUC measures a classifier's ability to separate two classes over the range of possible costs and thus be estimated using the Mann-Wilcoxon-Whitney test [18]. Unfortunately no such test is presently available for the multi-class case. The Gini coefficient is linearly related to the AUC, being twice the area between the ROC curve and the diagonal of the ROC plot. In this section we show how a natural generalisation of the Gini coefficient can be used to compare classifiers. We also draw attention to Ferri et al [12] who give another view of the volume under multi-class ROC surfaces.

By analogy with the AUC, we might use the volume of the $Q(Q-1)$-dimensional hypercube that is dominated by elements of the ROC surface for classifier A as a measure of A's performance. In binary and multi-class problems alike its maximum value is 1 when A classifies perfectly. If the classifier allocates to classes at random, without regard to the features $\mathbf{x}$, then the ROC surface is the simplex in $Q(Q-1)$-dimensional space with vertices at distance $Q-1$ along each coordinate vector. The volume of the unit hypercube dominated by this simplex is [10]:

$$\frac{1}{[Q(Q-1)]!}\left[(Q-1)^{Q(Q-1)} - Q(Q-1)(Q-2)^{Q(Q-1)}\right]. \tag{23.12}$$

When $Q = 2$ this volume (area) is just $1/2$, corresponding to the area under the diagonal in a conventional ROC plot.[4] However, when $Q > 2$ the volume not dominated by the random allocation simplex is very small; even when $Q = 3$, the volume not dominated is ≈ 0.0806. Since almost all of the unit hypercube is dominated by the random allocation simplex, we disregard this volume and instead define $G(A)$ to be the analogue of the Gini coefficient in two dimensions, namely the proportion of the volume of the $Q(Q-1)$-dimensional unit hypercube that is dominated by elements of the ROC surface,

---

[3] The Q F_k themselves may be directly minimised, but the information on how misclassifications are made is irrecoverably lost; see also Mossman [29] who equivalently maximises the true positive rates.

[4] Although binary ROC plots are usually made in terms of true positive rates versus false positive rates for one class, the false positive rate for the other class is just 1 minus the true positive rate for the other class.

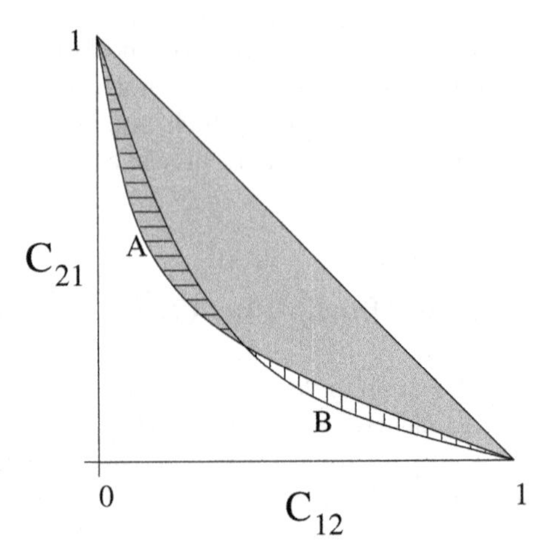

Fig. 23.8. Illustration of the G and δ measures for $Q = 2$. The shaded area corresponds to $\frac{1}{2}G(A)$, horizontal hatching indicates $\delta(A, B)$ and vertical hatching indicates $\delta(B, A)$.

but is not dominated by the random allocation simplex. This is illustrated by the shaded area in Figure 23.8 for the $Q = 2$ case. In binary classification problems this corresponds to twice the area between the ROC curve and the diagonal. In multi-class problems $G(A)$ quantifies how much better than random allocation is A. It can be simply estimated by Monte Carlo sampling of the region in the unit hypercube not dominated by the random allocation simplex.

If every point on the optimal ROC surface for classifier A is dominated by a point on the ROC surface for classifier B, then classifier B is clearly superior to classifier A. In general, however, neither ROC surface will completely dominate the other: regions of A's surface R_A will be dominated elements of R_B and vice versa; this corresponds to ROC curves that cross in binary problems. Let P denote the truncated pyramidal volume in the unit hypercube that is not dominated by the random allocation simplex. (In Figure 23.8 P is the area bounded by the origin and the points $(0, 1)$ and $(1, 0)$, but when $Q \geq 3$ note that the random allocation simplex intersects the coordinate axes at $(Q-1)$ and P is that part of the region between the simplex and the origin that also lies within the unit hypercube.) Then, to quantify the classifiers' relative performance we define $\delta(A, B)$ to be the volume of P that is dominated by elements of R_A and not by elements of R_B (marked in Figure 23.8 with horizontal lines). Note that $\delta(A, B)$ is not a metric, because although it is non-negative, it is not symmetric. Also if R_A and R_B are subsets of the same non-dominated set W (i.e., $R_A \subseteq W$ and $R_B \subseteq W$) then $\delta(A, B)$ and $\delta(B, A)$ may have a range of values depending on their precise composition [15]. Situations like this are rare in practice, however, and measures like δ have proved useful for comparing Pareto fronts.

Table 23.1. Generalised Gini coefficients and exclusively dominated volume comparisons of the logistic regression (LR) and MLP classifiers.

Q		$G(\text{LR})$	$G(\text{MLP})$	$\delta(\text{LR}, \text{MLP})$	$\delta(\text{MLP}, \text{LR})$
2	Train	0.584	0.962	0.000	0.379
	Test	0.519	0.858	0.010	0.349
3	Train	0.847	0.965	0.000	0.118
	Test	0.714	0.725	0.078	0.089

Table 23.1 shows the generalised Gini coefficient and δ measures for the multinomial logistic regressor and MLP classifiers applied to the synthetic data in both the 2 and 3 class cases. The Gini coefficients indicate that both classifiers are better than random allocation and that a substantially greater volume of P is dominated by the MLP than by the logistic regressor. The δ measures show that, using the training data, the logistic regressor does not dominate any regions that are not dominated by the MLP, although on the test set of 1000 examples the measures indicate that the logistic regressor dominates parts of the misclassification space not dominated by the MLP.

Figure 23.9 shows histograms of the distances from the random allocation simplex of points on the estimated Pareto fronts for the logistic regressor and MLP for the $Q = 3$ synthetic data. Negative distances correspond to classifiers in P, that is, closer to the origin than the random allocation simplex, while positive distances correspond to classifiers that, while non-dominated, lie beyond the random allocation simplex. These are classifiers for which a one or more misclassification rates has been allowed to become very poor in order to minimise others. As the histogram, shows the majority of classifiers are wholly better than random for both the MLP and the logistic regressor. However, the positive distances for the logistic regressor indicate that its relative inflexibility means that low misclassification rates for one class can only be achieved by sacrificing performance on others.

The distance from the random allocation simplex provides a method of selecting a single classifier from the Pareto front in the absence of other criteria on which to base the choice. Provided that the classifier lies within the unit hypercube, this criterion is equivalent to choosing the classifier which dominates the largest proportion of the region P. Figure 23.10 shows the decision regions for the logistic regressor and MLP corresponding to the most distant classifiers from which it can be seen that, in this case, the decision regions for these classifiers are quite close to the ideal equal-cost decision regions.

23.7 Conclusion

In this chapter we have considered from a multi-objective point of view the training of classifiers when the costs of misclassification are unknown. Even in

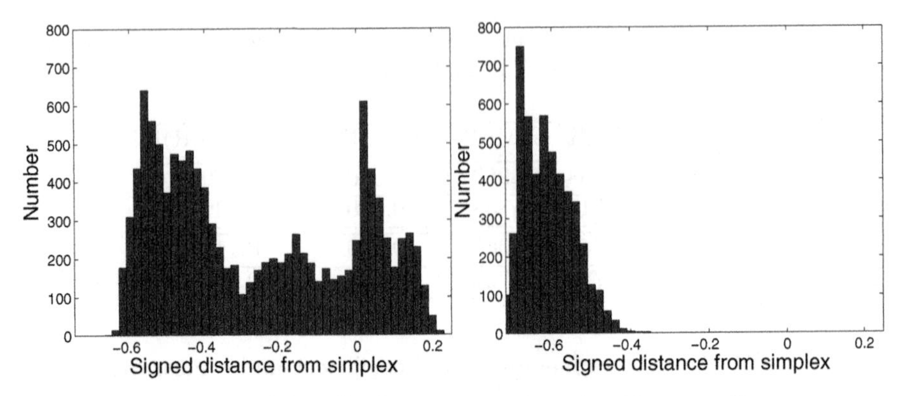

Fig. 23.9. Distances from the random classifier simplex. Negative distances correspond to models in P. *Left:* Logistic regressor; *Right:* MLP.

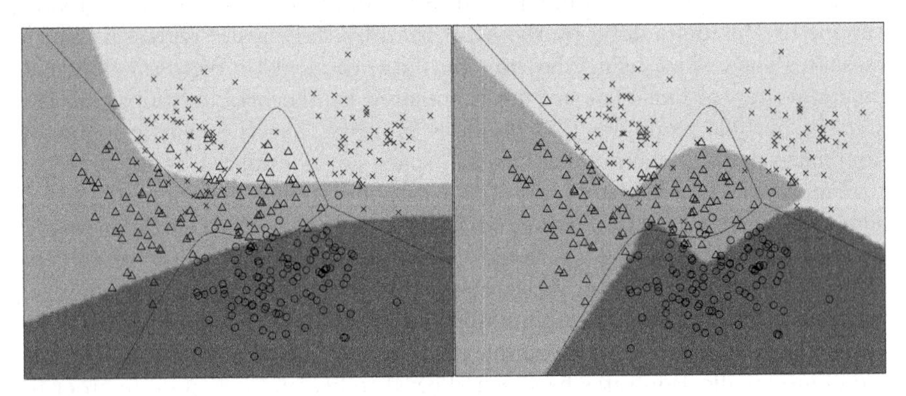

Fig. 23.10. Decision regions for the logistic regression classifier (*left*) and MLP classifier (*right*) furthest from the random allocation simplex. Solid lines show the ideal equal-cost boundaries.

classification between two classes consideration of the costs of misclassification leads naturally to a multi-objective problem which is conventionally visualised using the receiver operating characteristic curve. The multi-objective optimisation framework permits the ROC curve for $Q = 2$ classes to be naturally generalised to a surface in $Q(Q-1)$ dimensions in the general case. The resulting trade-off surface generalises the binary classification ROC curve because on it one misclassification rate cannot be improved without degrading at least one other. By viewing the classifier parameters and the misclassification costs as a single entity, we have presented a straightforward general evolutionary algorithm which is able to efficiently locate approximations to the optimal ROC surface for binary and multi-class problems alike. We remark that this algorithm is naturally able to handle other objectives (such as the warning time given to air traffic controllers in the STCA example) that the system designer must take into account.

An appealing quality of the ROC curve is that it can be plotted in two dimensions and an operating point selected from the plot. Unfortunately, the dimension of the Pareto optimal ROC surface grows as the square of the number of classes, which hampers visualisation. Projection into 'false positive space' is an effective visualisation method for 3-class problems as the false positive rates summarise the gross overall performance, allowing further analysis of exactly which classes are misclassified into which to be focused in particular regions. We regard this method as more informative than approaches which directly minimise the false positive rates [29], and therefore ignore how misclassifications are made. Nonetheless, it is likely that problems with more than three classes will require some *a priori* assessment of the important trade-offs because of the difficulty in interpreting 16 or more competing rates. Reliable calculation and visualisation of the curvature of the ROC surface is current work important for selecting operating points.

The Pareto optimal ROC surface yields a natural way of comparing classifiers in terms of the volume that the classifiers' ROC surfaces dominate. We defined and illustrated a generalisation of the Gini index for multi-class problems that quantifies the superiority of a classifier to random allocation. This naturally leads to a criterion for selecting an operating point: choose the classifier most distant from the random allocation simplex. An alternative measure for comparing classifiers in multi-class problems is the pairwise M measure described by Hand and Till [18]. However this describes the overall superiority of one classifier to another and does not permit selection of an operating point.

Finally we remark that the evaluation of the classification rates is inherently dependent on the available data. Here we have assumed that the data are of sufficient number that we can ignore any uncertainty associated with the particular data sample. Current research in this area involves bootstrapping these data in order to quantify the uncertainty in the ROC curve or surface [11] and the use of multi-objective optimisation in the presence of noise to permit reliable discovery of the Pareto optimal front with small quantities of data.

Acknowledgement

We are pleased to acknowledge support from EPSRC grant GR/R24357/01. We thank Trevor Bailey, Adolfo Hernandez, Wojtek Krzanowski, Derek Partridge, Vitaly Schetinin and Jufen Zhang for their helpful comments.

References

[1] N.M. Adams and D.J. Hand. Comparing classifiers when the misallocation costs are uncertain. *Pattern Recognition*, 32:1139–1147, 1999.

[2] M. Anastasio, M. Kupinski, and R. Nishikawa. Optimization and FROC analysis of rule-based detection schemes using a multiobjective approach. *IEEE Transactions on Medical Imaging*, 17:1089–1093, 1998.
[3] C. Bishop. *Neural Networks for Pattern Recognition.* Clarendon Press, Oxford, 1995.
[4] A.P. Bradley. The use of the area under the ROC curve in the evaluation of machine learning algorithms. *Pattern Recognition*, 30:1145–1159, 1997.
[5] C.A. Coello Coello. A Comprehensive Survey of Evolutionary-Based Multiobjective Optimization Techniques. *Knowledge and Information Systems. An International Journal*, 1(3):269–308, 1999.
[6] K. Deb. *Multi-Objective Optimization Using Evolutionary Algorithms.* Wiley, Chichester, 2001.
[7] K. Deb, A. Pratap, S. Agarwal, and T. Meyarivan. Fast and Elitist Multiobjective Genetic Algorithm: NSGA–II. *IEEE Transactions on Evolutionary Computation*, 6(2):182–197, 2002.
[8] R.O. Duda and P.E. Hart. *Pattern Classification and Scene Analysis.* Wiley, New York, 1973.
[9] C. Elkan. The foundations of cost-sensitive learning. In *IJCAI*, pages 973–978, 2001.
[10] R.M. Everson and J.E. Fieldsend. Multi-class ROC analysis from a multi-objective optimisation perspective. Technical Report 421, Department of Computer Science, University of Exeter, April 2005.
[11] R.M. Everson and J.E. Fieldsend. Multi-objective optimisation of safety related systems: An application to short term conflict alert. *IEEE Transactions on Evolutionary Computation*, 2006. (In press).
[12] C. Ferri, J. Hernández-Orallo, and M.A. Salido. Volume under the ROC surface for multi-class problems. In *ECML 2003*, pages 108–120, 2003.
[13] J.E. Fieldsend and R.M. Everson. ROC Optimisation of Safety Related Systems. In J. Hernández-Orallo, C. Ferri, N. Lachiche, and P. Flach, editors, *Proceedings of ROCAI 2004, part of the 16th European Conference on Artificial Intelligence (ECAI)*, pages 37–44, 2004.
[14] J.E. Fieldsend and R.M. Everson. Visualisation of multi-class ROC surfaces. In *Proceedings of ROCML 2005, part of the 22nd International Conference on Machine Learning (ICML 2005), forthcoming*, 2005.
[15] J.E. Fieldsend, R.M. Everson, and S. Singh. Using Unconstrained Elite Archives for Multi-Objective Optimisation. *IEEE Transactions on Evolutionary Computation*, 7(3):305–323, 2003.
[16] P. Flach, H. Blockeel, C. Ferri, J. Hernández-Orallo, and J. Struyf. Decision support for data mining: Introduction to ROC analysis and its applications. In D. Mladenic, N. Lavrac, M. Bohanec, and S. Moyle, editors, *Data Mining and Decision Support: Integration and Collaboration*, pages 81–90. Kluwer, 2003.
[17] C.M. Fonseca and P.J. Fleming. An Overview of Evolutionary Algorithms in Multiobjective Optimization. *Evolutionary Computation*, 3(1):1–16, 1995.
[18] D.J. Hand and R.J. Till. A simple generalisation of the area under the ROC curve for multiple class classification problems. *Machine Learning*, 45:171–186, 2001.
[19] J.A. Hanley and B.J. McNeil. The meaning and use of the area under a receiver operating characteristic (ROC) curve. *Radiology*, 82(143):29–36, 1982.

[20] J. Hernández-Orallo, C. Ferri, N. Lachiche, and P. Flach, editors. *ROC Analysis in Artificial Intelligence, 1st International Workshop, ROCAI-2004, Valencia, Spain*, 2004.

[21] M. T. Jensen. Reducing the Run-Time Complexity of Multiobjective EAs: The NSGA-II and Other Algorithms. *IEEE Transactions on Evolutionary Computation*, 7(5):503–515, 2003.

[22] J. Knowles and D. Corne. The Pareto Archived Evolution Strategy: A new baseline algorithm for Pareto multiobjective optimisation. In *Proceedings of the 1999 Congress on Evolutionary Computation*, pages 98–105, Piscataway, NJ, 1999. IEEE Service Center.

[23] J.D. Knowles and D. Corne. Approximating the Nondominated Front Using the Pareto Archived Evolution Strategy. *Evolutionary Computation*, 8(2):149–172, 2000.

[24] T. Kohonen. *Self-Organising Maps*. Springer, 1995.

[25] M.A. Kupinski and M.A. Anastasio. Multiobjective Genetic Optimization of Diagnostic Classifiers with Implications for Generating Receiver Operating Characterisitic Curves. *IEEE Transactions on Medical Imaging*, 18(8):675–685, 1999.

[26] M. Laumanns, L. Thiele, and E. Zitzler. Running Time Analysis of Multiobjective Evolutionary Algorithms on Pseudo-Boolean Functions. *IEEE Transactions on Evolutionary Computation*, 8(2):170–182, 2004.

[27] M. Laumanns, L. Thiele, E. Zitzler, E. Welzl, and K. Deb. Running Time Analysis of Multi-objective Evolutionary Algorithms on a Simple Discrete Optimization Problem. In J.J. Merelo Guervós, P. Adamidis, H-G Beyer, J-L Fernández-Villacañas, and H-P Schwefel, editors, *Parallel Problem Solving from Nature—PPSN VII*, Lecture Notes in Computer Science, pages 44–53. Springer-Verlag, 2002.

[28] D. Lowe and M. E. Tipping. Feed-forward neural networks and topographic mappings for exploratory data analysis. *Neural Computing and Applications*, 4:83–95, 1996.

[29] D. Mossman. Three-way ROCs. *Medical Decision Making*, 19(1):78–89, 1999.

[30] F. Provost and T. Fawcett. Analysis and visualisation of classifier performance: Comparison under imprecise class and cost distributions. In *Proceedings of the Third International Conference on Knowledge Discovery and Data Mining*, pages 43–48, Menlo Park, CA, 1997. AAAI Press.

[31] F. Provost and T. Fawcett. Robust classification systems for imprecise environments. In *Proceedings of the Fifteenth National Conference on Artificial Intelligence*, pages 706–7, Madison, WI, 1998. AAAI Press.

[32] B.D. Ripley. Neural networks and related methods for classification (with discussion). *Journal of the Royal Statistical Society Series B*, 56(3):409–456, 1994.

[33] M.J.J. Scott, M. Niranjan, and R.W. Prager. Parcel: feature subset selection in variable cost domains. Technical Report CUED/F-INFENG/TR.323, Cambridge University Engineering Department, 1998.

[34] F. Tortorella, editor. *Pattern Recognition Letters: Special Issue on ROC Analysis in Pattern Recognition*, volume 26, 2006.

[35] D. Van Veldhuizen and G. Lamont. Multiobjective Evolutionary Algorithms: Analyzing the State-of-the-Art. *Evolutionary Computation*, 8(2):125–147, 2000.

[36] X. Yao, Y. Liu, and G. Lin. Evolutionary Programming Made Faster. *IEEE Transactions on Evolutionary Computation*, 3(2):82–102, 1999.

[37] E. Zitzler and L. Thiele. Multiobjective Evolutionary Algorithms: A Comparative Case Study and the Strength Pareto Approach. *IEEE Transactions on Evolutionary Computation*, 3(4):257–271, 1999.

[38] M.H. Zweig and G. Campbell. Receiver-operating characteristic (ROC) plots: a fundamental evaluation tool in clinical medicine. *Clinical Chemistry*, 39:561–577, 1993.

24

Multi-Objective Design of Neuro-Fuzzy Controllers for Robot Behavior Coordination

Naoyuki Kubota

1-1 Minami-Osawa, Hachioji, Tokyo 192-0397, Japan
Department of System Design, Tokyo Metropolitan University

PRESTO, Japan Science and Technology Agency
kubota@comp.metro-u.ac.jp

Summary. This chapter discusses the behavioral learning of robots from the viewpoint of multiobjective design. Various coordination methods for multiple behaviors have been proposed to improve the control performance and to manage conflicting objectives. We proposed various learning methods for neuro-fuzzy controllers based on evolutionary computation and reinforcement learning. First, we introduce the supervised learning method and evolutionary learning method for multiobjective design of robot behaviors. Then, the multiobjective design of fuzzy spiking neural networks for robot behaviors is presented. The key point behind these methods is to realize the adaptability and reusability of behaviors through interactions with the environment.

24.1 Introduction

Robot learning has been discussed from various viewpoints such as industrial mechatronics, artificial intelligence, embodied cognitive science, and ecological psychology [1]-[7]. Some of robot learning is concerned with behavior acquisition. Behavior-based robotics was proposed to develop intelligent robots based on sensory-motor coordination in dynamic or unknown environments without symbolic representation and manipulation [16]-[18]. The original subsumption architecture uses control layers. A single control layer is selected alternatively according to sensory inputs, but the robot behavior is often multiobjective. Therefore, various coordination methods for multiple behaviors have been proposed to improve the control performance and to manage conflicting objectives [19]-[21].

A behavior is defined as the actions or reactions of persons or things in response to external or internal stimuli in American Heritage Dictionary. Furthermore, a behavior is also defined as a sequence of ordered action performed by a group of interacting mental and/or physical actors. Here the internal

N. Kubota: *Multi-Objective Design of Neuro-Fuzzy Controllers for Robot Behavior Coordination*, Studies in Computational Intelligence (SCI) **16**, 557–584 (2006)
www.springerlink.com

stimuli and mental states are referred in these definitions. Behavior coordination is performed according to the internal stimuli or mental states. Therefore, the behavioral learning of a robot should be also discussed with its internal state. The learning is performed in two hierarchical levels of behavior level and behavior coordination mechanism level. The coordination mechanism cannot be designed without behavior elements while each behavior should be designed acceding to the relationship with other behaviors. In general the behavioral learning can be classified into supervised learning and self-learning. The self-learning is performed by trial-and-error without exact teaching signals. The importance in the self-learning is the direct interaction with the unknown or dynamic environment, and therefore, the robot should acquire multiobjective behaviors according to the complexity of the task and the environment. Computational intelligence techniques have been applied for robot learning. Fuzzy logic and neural networks (NN) are used for robot control [28]-[35]. Furthermore, modular neural networks have been used for large size of robot control problems. Evolutionary computation methods including genetic algorithm (GA), evolutionary programming (EP), and evolution strategy (ES) have been applied for parameter tuning and structural optimization of NN and fuzzy systems (FS) [13], [22]-[3]. Furthermore, multiobjective evolutionary optimization is used in robotics [48, 49].

In this chapter, we focus on the computational intelligence techniques for multiobjective behavioral learning. Section 24.2 explains control methods for robot behaviors, and behavioral learning methodologies. Sections 24.3 and 24.4 show various methods for multi-objective design of robot behaviors and their simulation and experimental results. Section 24.5 discusses the current states of art in behavioral learning of robots, and summarizes this chapter.

24.2 Behavioral Learning in Robotics

24.2.1 Sensory-Motor Coordination

The problems of robots can be divided into two subproblems of motion planning and motion control. First, a reference trajectory is generated by solving the motion planning based on a geometric map, and the motion control for a physical robot is done according to a planned reference trajectory in classical approaches. On the other hand, sensory-motor coordination uses production rules or mapping functions represented by a pair of sensory inputs and motor outputs without geometric map building. The motion control based on sensory inputs, not reference trajectories, has an advantage of the adaptability to the facing environment. Actually, it is very difficult to update a reference trajectory in an unknown or dynamic environment. Therefore, we focus on the behavioral learning in unknown or dynamic environment. As an example, we deal with the navigation problem of a mobile robot.

A mobile robot navigation problem is one of important problems. We assume that a mobile robot has eight distance sensors to measure distances from the robot to obstacles and four light sensors to know the goal direction in unknown and dynamic environments (Fig. 24.1). Furthermore, we assume the robot doesn't know the layout and shapes of static obstacles in the unknown environment, and there exist several moving obstacles in a dynamic environment. The objective of this task is to reach a target point while avoiding collisions with any obstacles. To achieve this task, a mobile robot should have three basic behaviors of target tracing, collision avoidance, and wall following. We use simple if-then rules for the target tracing behavior that keeps the direction to the target point. The collision avoidance and wall following behaviors are designed by neural networks or fuzzy controllers. Each controller outputs the moving velocity and the steering angle of the robot.

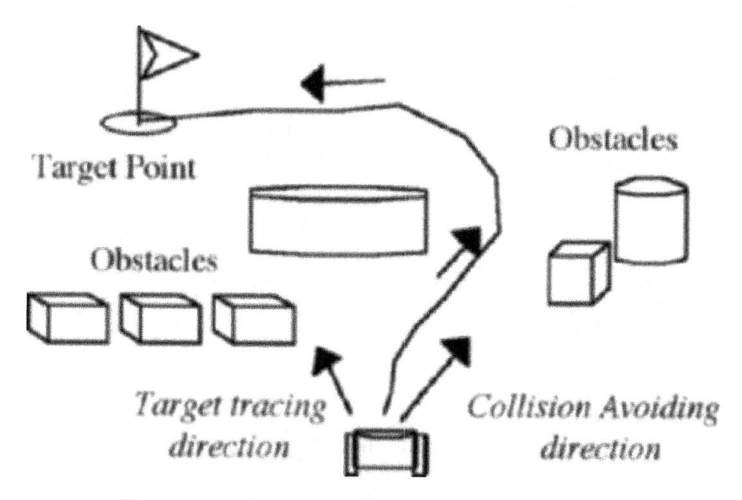

Fig. 24.1. A mobile robot navigation problem.

24.2.2 Learning Methods for Behavior Acquisition

Behavioral learning is divided into supervised learning and self-learning according to teaching signals (Fig. 24.2). Supervised learning is further divided into social learning and error-based learning. For example, least mean square algorithm is applied for behavioral learning when exact teaching signals are given to a robot. On the other hand, social learning is performed between two or more agents. Social learning is classified into imitative learning, instructive learning, and collaborative learning. Imitation is defined as the ability to recognize and reproduce others' action. The concept of imitative learning has been applied to robotics [6]-[8]. Basically, in the traditional researches of learning by observation, a motion trajectory of a human arm assembling or handling objects is measured, and the obtained data are analyzed and transformed for the motion control of a robotic manipulator. Furthermore, various neural networks have been applied to imitative learning for robots [6]-[8]. Especially, the discovery of mirror neurons is very important [5]. Each mirror neuron activates not only in performing a task, but also in observing that somebody performs the same task. While the imitative learning is basically unidirectional from a demonstrator to a learner, the instructive learning is bidirectional between an instructor and a learner. Furthermore, collaborative learning is different from the imitative learning and instructive learning, because the exact teaching data and target demonstration is not given to agents in the collaborative learning beforehand. The solution is found or searched through interaction among multiple agents. Therefore, the collaborative learning might be classified as the category of self-learning.

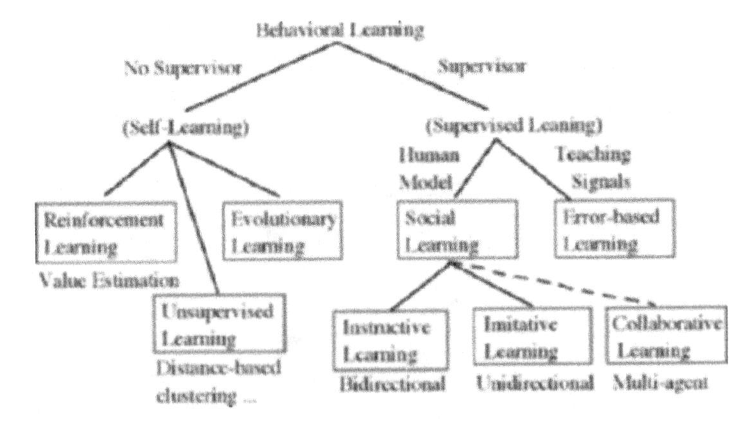

Fig. 24.2. Learning methods for robotic behaviors.

On the other hand, self-learning is divided into reinforcement learning, evolutionary learning, and unsupervised learning. Unsupervised learning does not use the external information on the learning, and performs the clustering of data according to the given criteria. This learning method is applied for the segmentation of perceptual information and data analysis in robotics. Both of reinforcement learning and evolutionary learning are used for behavior acquisition in unknown environments. Basically, a robot behavior is represented by production rules, fuzzy rules, neural networks, and others. The behavior is improved according to rewards and penalties obtained through interaction with the actual environment. A reinforcement learning method builds an agent that maximizes their expected utility only using rewards or penalties from the environment. Accordingly, the reinforcement learning does not use explicit teaching signals, but uses evaluative feedback obtained through the interaction with the environment. The reinforcement learning uses a value function used for optimizing a policy to realize a target behavior, and estimates the relationship between each state and its value in multistage decision processes based on dynamic programming [22, 23, 29]. Evolutionary learning uses a fitness function of a total cost or a total evaluation value instead of such a value function. Therefore, the evolutionary learning does not require computational memory for value functions used in reinforcement learning, but the evolutionary learning needs many episode-based search iterations.

24.2.3 Neuro-Fuzzy and Evolutionary Computing for Sensory-motor coordination

Neural networks and fuzzy systems have been applied to robotic control [29]-[35], and furthermore, evolutionary computation methods have been used for optimizing robotic controllers. For example, Floreano and Mondada [51] used evolutionary computation for optimizing a neurocontroller of a mobile robot, and Cho and Lee [52] proposed fuzzy controller for a mobile robot and its evolvability was discussed. In general, a fuzzy if-then rule is described as follows,

If x_1 is $F_{j,1}$ and x_2 is $F_{j,2}$ and ... and x_n is $F_{j,n}$ **then** y_1 is $v_{j,1}$ and y_2 is $v_{j,2}, \ldots,$ and y_o is $v_{j,o}$

where $F_{j,i}$ is a membership function for the ith input; $v_{j,i}$ is a singleton for output value of the jth rule, and n is the numbers of inputs; o is the numbers of outputs. A triangular membership function is generally described as,

$$\mu_{Fj,i}(x_i) = \begin{cases} 1 - \frac{|x_i - a_{j,i}|}{b_{j,i}}, & |x_i - a_{j,i}| \leq b_{j,i} \\ 0, & \text{otherwise} \end{cases}, \tag{24.1}$$

where $a_{j,i}$ and $b_{j,i}$ are the central value and the width of the ith membership function $F_{j,i}$ of the jth rule. Consequently, the firing strength of the jth rule (j=1, 2, ..., m) is calculated by

$$\mu_j = \prod_{i=1}^{n} \mu_{F_{i,j}}(x_i). \tag{24.2}$$

Next, we obtain resulting outputs (r=1, 2, $\ldots$, o) as follows,

$$y_k = \frac{\sum_{j=1}^{m} \mu_j v_{j,k}}{\sum_{j=1}^{m} \mu_j}, \tag{24.3}$$

where m is the number of rules. The kth output value of the hth rule is updated by the generalized delta rule,

$$\Delta v_{h,k} = -\eta \cdot \frac{\partial E_p}{\partial v_{h,k}} = \eta \cdot (Y_{p,k} - y_k) \frac{\mu_h}{\sum_{j=1}^{m} \mu_j}, \tag{24.4}$$

where $Y_{p,k}$ is the desired value for kth output of the pth pattern.

Neural networks and fuzzy rules were often applied for estimating value functions in the reinforcement learning [23]. The study of reinforcement learning is to build an intelligent agent or robot itself, because the agent of the reinforcement learning perceives the environment, makes decision according to a policy, takes actions, and updates a value function. In actor-critic methods, an agent has a value function and a policy separately. The policy for selecting an action is called the actor, while the value function is called critic. Therefore, a robot based on reinforcement learning requires a perceptual system for state recognition, an action system based on the policy for action selection and action output, and learning system for updating perceptual system and action system (Fig. 24.3).

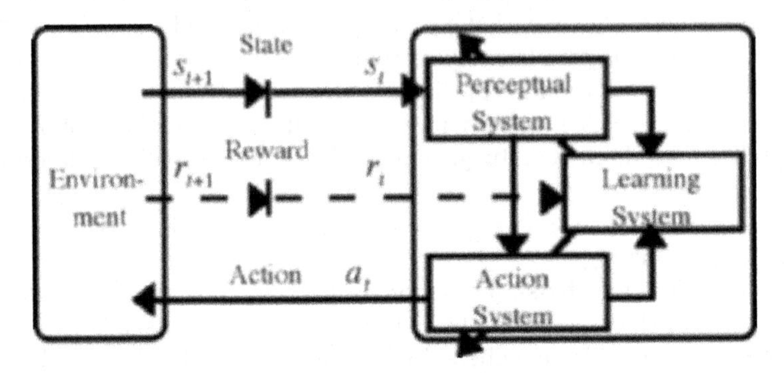

Fig. 24.3. A total architecture of a robot based on reinforcement learning.

Braitenberg proposed a thought-experiment where one builds a number of simple wheeled robots with different sensors variously connected through electrical wires to the motors [4]. In the experiment, these robots are put on a table and they display various behaviors according to the connection between sensors and motors. Some robots fall off the table, but others remain on the table. If the connection between sensors and motors of the robots fallen from the table is updated or copied from the robot remaining on the table, the robot may be able to remain on the table for long time. This kind of process can be regarded as an artificial evolution of the sensory-motor coordination on a robot. The advantage of evolution in intelligent robots is to acquire behaviors while maintaining various candidate solutions, not a single candidate solution. Evolutionary robotics is known as the research on evolutionary learning for robots [9]. The main role of evolutionary computation techniques in robotics is the optimization for modeling or problem-solving. In fact, the optimization based on evolutionary computation can be divided into three approaches of a direct use, genetic programming, and machine learning [19]. The direct use is mainly seen in applications to the numerical optimization and the combinatorial optimization for tuning control parameters and for obtaining knowledge and strategies. Genetic programming is applied to the optimization of computer programs. An individual in genetic programming can be represented as hierarchical structures and therefore the intelligent robot can obtain complex behaviors by using GP.

The machine learning is mainly used for optimizing a set of inference rules in autonomous robots. A classifier system (CS), which is well known as a genetics-based machine learning method, can learn syntactically simple string rules to guide its performance in an arbitrary environment [26]. The classifier system can be divided into *Michigan* approach and *Pittsburgh* approach. An individual is represented as a single rule is the *Michigan* approach, while an individual is represented as a set of rules in the *Pittsburgh* approach. Therefore, the *Pittsburgh* approach has a meta-level set of rule sets. Because an individual in the *Pittsburgh* approach is evaluated after an episode, a reward is given to an individual, not to a rule. Consequently, the *Pittsburgh* approach does not need to assign a reward to each rule. Therefore, discrete production rules are used in the *Michigan* approach, while neural networks and fuzzy rules are applied for representing behavior rules in the *Pittsburgh* approach. The artificial evolution of Braitenberg's thought-experiments is considered as the *Pittsburgh* approach.

24.2.4 Behavior Coordination and Behavioral Learning

A robot should acquire various behaviors through interaction with its environment, but each behavior is evaluated in the relationship with other behaviors. For example, in a navigation problem, a robot should perform collision avoidance and wall following, while performing target tracing in the actual environment. In this way, the robot must take multiple behaviors and coordinate several behaviors according to the facing environmental conditions. Computational intelligence techniques have been applied to solve this behavior coordination [17, 18]. The basic architecture for the behavior coordination is a mixture of experts [36]. In general, a complicated task is divided as a combination of behaviors based on apparent functional modularity. Each local expert is used for a divided behavior, and gating network is used for combining the outputs of local experts. Modular neural networks have the hierarchical architecture of multiple neural networks composed of local experts and their gating network. Behavior coordination can be performed by the combination of outputs from neural networks or fuzzy systems as a mixture of experts. By extending eq.(24.3), the final output is calculated by

$$z_k = \frac{\sum_{i=1}^{d} u_i y_{i,k}}{\sum_{i=1}^{d} u_i}, \quad u_i = c_i(\mathbf{x}) \tag{24.5}$$

where $\mathbf{x}$ is the sensory input $\mathbf{x} = [x_1, x_2, ..., x_n]^T$; d is the number of behaviors; $c_i(\mathbf{x})$ is the ith behavior weight function; $y_{i,k}$ is the kth output of the ith behavior. The behavior weights are updated by $c_i(\mathbf{x})$ according to the time series of perceptual information, and the robot can take a multiobjective behavior. This method can be considered as a mixture of experts, because the behavior coordination mechanism is considered as a gating network.

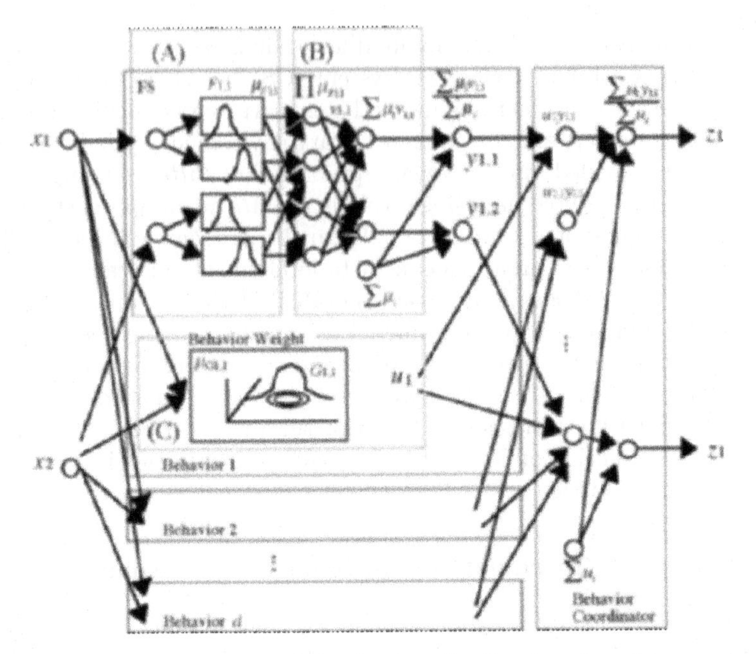

Fig. 24.4. Total architecture of behavior coordination based on fuzzy controllers.

The behavioral learning is performed in two hierarchical levels of behavior level and behavior coordination mechanism level, which are the nesting architecture because the coordination mechanism cannot be designed without behavior elements while each behavior should be designed acceding to the relationship with other behaviors. Therefore, we must consider the total architecture of behavioral learning according to available information on reward and penalty from the environment. In the following sections, we explain the detail of several methods for behavioral learning of robots.

24.3 Multi-objective Design of Robot Behaviors

Multiobjective optimization is very important in behavioral learning, because the robot should take into account not only task achievement, but also energy consumption at least. In the reinforcement learning, such an evaluation value on the task achievement is included in reinforcement signal of reward or punishment at the terminal state, but the immediate reward on energy consumption is also used for enhancing behavioral learning. On the other hand, the evolutionary learning can use not only the performance evaluation on the task achievement and immediate cost, but also the structural evaluation on the number of rules and the size of neural network. Therefore, the evolutionary learning can optimize both of rule parameters and rule structure at the same time.

24.3.1 Supervised Multi-objective Behavioral Learning

A robot can train behavior rules through the interaction with the environment, if the teaching signals are given to the robot. This learning is considered as an online behavioral learning. According to the extension of eq.(24.4), the kth output value of the hth rule in the bth behavior is updated as follows,

$$\Delta v_{b,h,k} = -\eta \frac{\partial E_p}{\partial v_{b,h,k}} = \eta (Y_{p,k} - z_k) \frac{u_b}{\sum_{i=1}^{d} u_i} \frac{\mu_{b,h}}{\sum_{j=1}^{m} \mu_{b,j}}, \tag{24.6}$$

where $Y_{p,k}$ is a desired output of the pth pattern; z_k is the output of behavior coordination. If the behaviors except the bth behavior are fixed or predefined, this learning plays the role of additional learning to improve the total performance of behavior coordination. Basically, he acquired bth behavior cannot be used separately in the decision making, because the behavior works interpolatively with other behaviors. Therefore, the objective of a behavior is not clear comparing with other behaviors.

On the other hand, if all behaviors are already designed, the robot can update the behavior weight function $c_i(\mathbf{x})$ in eq.(24.5). Here we assume the following simple function for behavior coordination,

$$u_i = c_i(\mathbf{x}) = \sum_{j=1}^{n} s_{j,i} G_{j,i}(x_j), \tag{24.7}$$

where $G_{j,i}(x_j)$ and $s_{j,i}$ are a membership function for the jth input to the ith behavior and a connection strength from the jth input to the ith behavior. Therefore, the eq.(24.5) for the final output is rewritten in the following:

$$z_k = \frac{\sum_{i=1}^{d} \left(y_{i,k} \sum_{j=1}^{n} s_{j,i} G_{j,i}(x_j) \right)}{\sum_{i=1}^{d} \sum_{j=1}^{n} s_{j,i} G_{j,i}(x_j)}. \tag{24.8}$$

The connection strength for behavior coordination is updated as follows:

$$\Delta s_{j,b} = -\eta \frac{\partial E_p}{\partial s_{j,b}} = \eta (Y_{p,k} - z_k) \frac{y_{b,k}}{\sum_{i=1}^{d} \sum_{j=1}^{n} s_{j,i} G_{j,i}(x_j)}. \tag{24.9}$$

In this way, the robot can improve the behavior rules and behavior weight functions if the suitable teaching signals are given to the robot. However, it is very difficult to prepare teaching signals in an unknown environment beforehand. Furthermore, if the suitable teaching signals are available, the robot does not need to perform behavioral learning in this problem class.

In general, since the available teaching information is much limited, the robot should acquire behaviors based on the relationship between the sensory inputs and external environmental conditions. For example, the collision avoidance behavior can be heuristically trained by using the distance measured by ultrasonic sensors or infrared sensors [42]. The moving direction

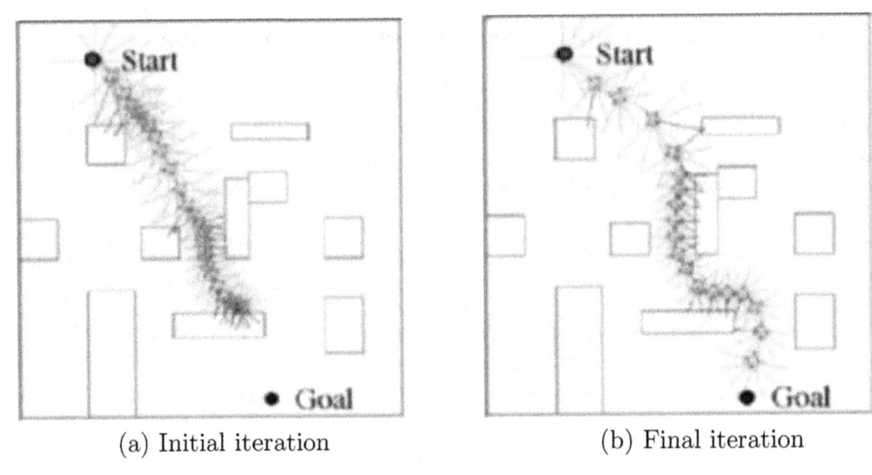

(a) Initial iteration (b) Final iteration

Fig. 24.5. Trajectories of a mobile robot at initial and final iterations.

toward the open area calculated by the measured distance. Another example is the learning of wall following behavior [42]. The robot should keep on going straight and keep the constant distance with the nearest wall in the wall following behavior.

24.3.2 Local Episode-based Learning

The robot can use local evaluation values to improve the behavior in context of the time-series of actions. Teaching signals are available in the above section, but this section discusses the learning method in case of local evaluation values instead of the teaching signals. For example, the robot can use the penalty information on collision with walls or obstacles like reinforcement learning. Since the behavior coordination does not use a value function, the robot cannot learn the relationship between a state calculated by sensory inputs and its value. Therefore, the robot should have local episode toward the success or failure, and should improve the performance of behavior coordination by updating the behavior weight functions $c_i(\mathbf{x})$. Actually, the robot can use the relationship between the mainly contributed behavior and its resultant evaluation. For example, when the robot collides with obstacles at discrete time step t, the robot can update the connection strength of collision avoidance behavior,

$$s_{j,b} \leftarrow s_{j,b} + \eta \sum_{i=0}^{T} \gamma^i G_{j,b}(x_j(t-i)), \tag{24.10}$$

where $x_j(t)$ is the jth sensory input at time step t; γ is a discount rate. This update equation is similar to profit sharing used in the genetics-based machine learning. While the profit sharing propagates the obtained reward backward to the previously selected actions, the strength connection for the

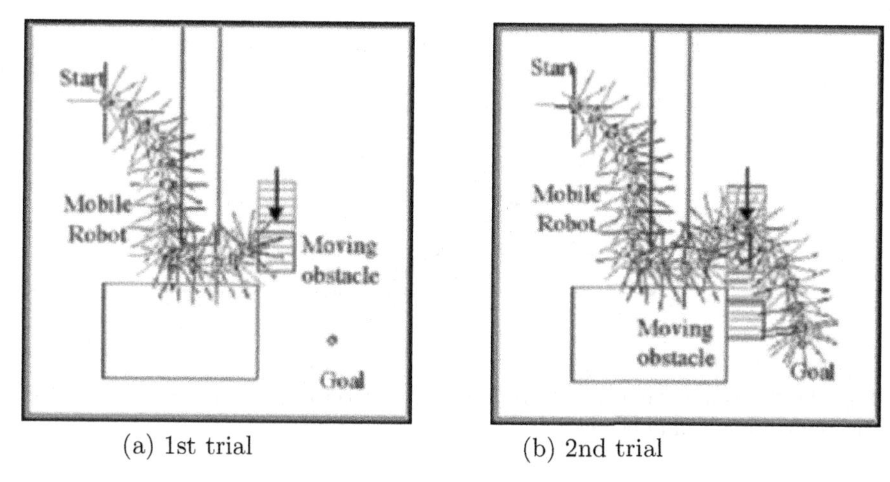

(a) 1st trial (b) 2nd trial

Fig. 24.6. Trajectories at each trial.

above method is updated by using temporally discounted membership values. Nojima et al [43] proposed a local episode-based learning method based on fuzzy inference, and applied the method for acquiring a collision avoidance behavior in a dynamic environment including several moving obstacles.

24.3.3 Evolutionary Learning for Behavior Acquisition

We explain how to apply a steady-state genetic algorithm (SSGA) for behavioral learning based on the *Pittsburgh* approach [21]. SSGA is a continuous model of generational alternation [25]. Basically, the individual selected by proportional selection scheme or "delete least fitness" scheme is eliminated, and replaced with the individual generated by crossover and mutation. SSGA should be used because it is very difficult to evaluate all candidate solutions of behavior rules in a real world. We consider the optimization problem of fuzzy controllers as numerical and combinatorial optimization problems. The decision variables are $a_{i,j}$, $b_{i,j}$, and $v_{i,k}$ described in eqs. (24.1) and (24.3) (Fig. 24.7). A candidate solution is represented in Fig. 24.8. The combination of fuzzy rules is determined by the validity parameter of each rule (r_{val}), and the combination of membership functions is determined by the validity parameter of each membership function (F_{val}). The objective is to find a fuzzy controller that minimizes errors between the target outputs and the inference results, while reducing redundant fuzzy rules and membership functions. The fitness function fit_i consists of the evaluation function G_i, the number of the membership functions M_i, and the number of fuzzy rules R_i as follows:

$$\text{fit}_i = \omega_1 \cdot G_i + \omega_2 \cdot M_i + \omega_3 \cdot R_i, \tag{24.11}$$

where ω_1, ω_2, and ω_3 are weight coefficients. The evaluation function is composed of the performance factors: time steps (P_{step}), moving length (P_{length}),

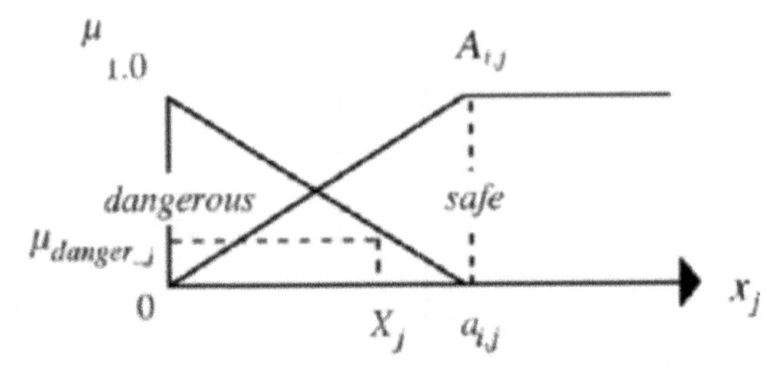

Fig. 24.7. Triangular membership functions concerning the distance x_j between the mobile robot and obstacles.

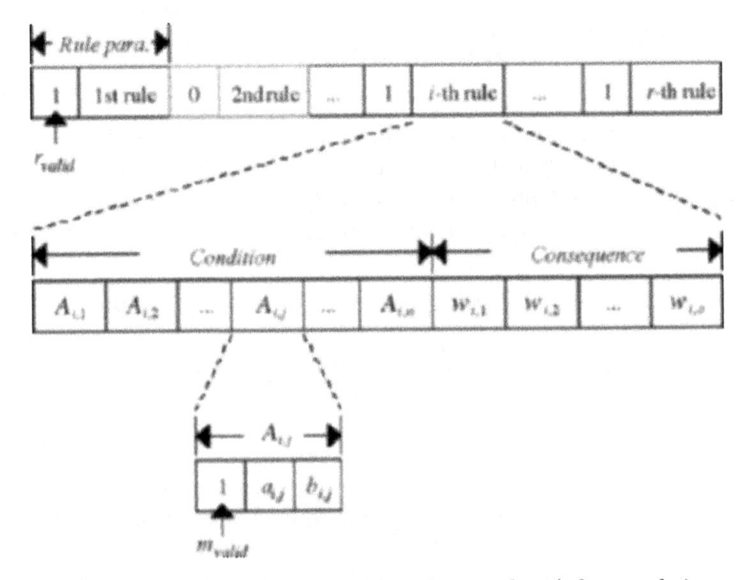

Fig. 24.8. A candidate solution representing fuzzy rules. A fuzzy rule is composed of condition and consequence parts. A membership function $A_{i,j}$ is represented by the central value $a_{i,j}$ and the width $b_{i,j}$. In the consequence part, a singleton $w_{i,j}$ is used as each output parameter.

average of the degree of danger (D_{ave}), and maximum of the degree of danger (D_{max}) on the trajectory generated by the fuzzy controller of a candidate solution as follows,

$$G_i = \omega_4 \cdot P_{\text{step}} + \omega_5 \cdot P_{\text{length}} + \omega_6 \cdot D_{\text{ave}} + \omega_7 \cdot D_{\text{max}}, \tag{24.12}$$

where ω_4, ω_5, ω_6, and ω_7 are weight coefficients.

A procedure for optimizing fuzzy controllers based on SSGA is as follows.

```
Begin
        Initialization
        Repeat
                Selection
                Crossover
                Mutation
                Evaluation (Simulation)
        Until Termination_condition = True
End
```

Fig. 24.9. The SSGA algorithm.

Initialization randomly generates and evaluates candidate solutions through the simulation of a robot. Next, Selection eliminates one individual with the worst fitness value. Crossover exchanges the combination of fuzzy rules and membership functions between two candidate solutions. Mutation changes the validity parameters and adds normal random values to the decision variables. We apply the following adaptive mutation,

$$a_{i,j} \leftarrow a_{i,j} + \left(\alpha_j \cdot \frac{\mathrm{fit}_i - \mathrm{fit}_{\mathrm{min}}}{f_{\mathrm{max}} - f_{\mathrm{min}}} + \beta_j \right) \cdot N(0,1), \tag{24.13}$$

where fit_i, $\mathrm{fit}_{\mathrm{max}}$ and $\mathrm{fit}_{\mathrm{min}}$ is the fitness values of the ith candidate solution, maximal and minimal fitness values; α_j and β_j are the coefficients for scaling and offset, respectively. The generated candidate solutions are evaluated by the simulation or experiment of the robot.

24.3.4 Perception-based Genetic Algorithm for Behavioral Learning

This section discusses the behavioral learning of a robot in a dynamic environment. A robot should be adaptive to its facing environment, if the environment changes. However, it is very difficult for the robot to perceive the environmental change by the robot itself if the robot does not have its environmental map. Furthermore, if the robot selects a different controller candidate, the dynamics of its resulting motion patterns would change. Therefore, the robot should have the perceptual information to evaluate the difference between the environmental conditions in the dynamic environment, and the robot should select a controller candidate according to the perceptual information on the facing environmental conditions. The advantage of GA is to maintain genetic diversity in a population. This means the robot can have various controller candidates adapted to different environmental conditions. Therefore, we proposed a perception-based genetic algorithm (PerGA) using the degree of the sparseness of the obstacles in the facing environment as the perceptual information, which is the average of each measured distance to obstacles in one episode [44]. The degree of sparseness of the obstacles (PCP_i) of the ith controller candidate is calculated as follows,

$$PCP_i = \frac{1}{P_{\text{step}}} \frac{1}{n} \sum_{t=1}^{P_{\text{step}}} \sum_{j=1}^{n} \frac{x_j(t)}{RNG}, \tag{24.14}$$

where $x_j(t)$ is the measured distance at discrete time step t; RNG is the sensor range; P_{step} is the number of time steps to reach a target point from the stating point used in eq.(24.8). This value increases as the robot encounters many obstacles. The selection and genetic operators are restricted in a subpopulation composed of neighboring n_{sub} individuals satisfying

$$|PCP_j - PCP_i| \leq \kappa, \tag{24.15}$$

where PCP_i is the degree of sparseness of obstacles of the ith controller candidate used at the previous episode; κ is a constant. Therefore, the controller candidate for the next episode is selected from the controller candidates used in the similar environmental conditions (Fig. 24.10). Here the individual with the worst fitness is removed in the selected subpopulation. As genetic operators, we apply multi-point crossover, simple mutation, and adaptive mutation described in the previous section.

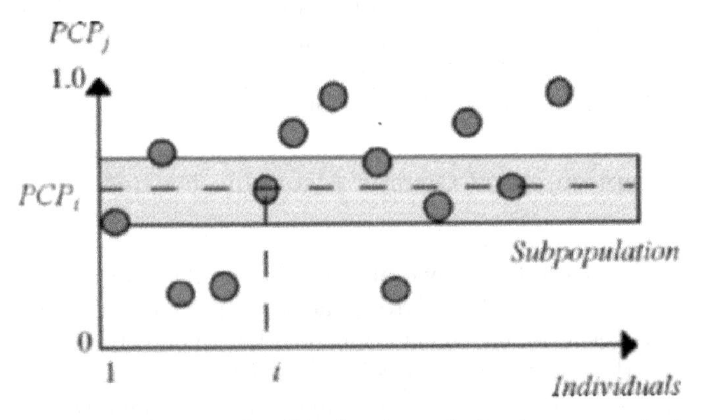

Fig. 24.10. Subpopulation in PerGA. A subpopulation is composed of n_{sub} individuals according to PCP_i. Selection and genetic operators are limited within the subpopulation to maintain various candidate controllers based on environmental conditions.

Furthermore, the performance of multiobjective design of the controller depends strongly on the weighting of the evaluation factors. Therefore, the mobile robot should update weight parameters according to the changes of the degree of sparseness and the evaluation values of P_{step}, P_{length}, D_{ave}, and D_{max}. Basically, the weight parameters of the evaluation function are updated according to the difference between the current and previous episode as follows,

$$\omega_i \leftarrow \omega_i + W \cdot \frac{1}{1+\exp\left(\frac{-EV_i(t)\cdot PCP_i}{T}\right)}, \tag{24.16}$$

$$\begin{cases} EV_1(t) = P_{\text{time}}(t) - P_{\text{time}}(t-1) \\ EV_2(t) = P_{\text{length}}(t) - P_{\text{length}}(t-1) \\ EV_3(t) = D_{\text{sum}}(t) - D_{\text{sum}}(t-1) \\ EV_4(t) = D_{\text{max}}(t) - D_{\text{max}}(t-1) \end{cases}. \tag{24.17}$$

Next, the weight parameters are normalized. Consequently, the weight parameter corresponding to the worse evaluation factor is increased. In this way, the robot can reflect the facing environmental information to the behavioral learning in the next episode.

Figure 24.11 shows the trajectories of the mobile robot with fuzzy controller based on PerGA. The environment is changed every 500 episodes and each environment is different shown as Fig. 24.11. We compared the performance of SSGA, PerGA without weight updating, and PerGA with weight updating (Fig. 24.12). Figure 24.12 shows the change of weight parameters. The weight parameter for P_{length} is increased, while the weight parameters for D_{max} and D_{ave} were decreased in Case 4. This indicates the collision avoidance behavior is rarely used. Actually it is very difficult to maintain good controllers. As a result, various good controllers might be eliminated through the evolution in Case 4. In Fig. 24.12, the fitness of PerGA is improved after the environmental change at the 2000th episode, but the SSGA takes many episodes to improve fuzzy controllers.

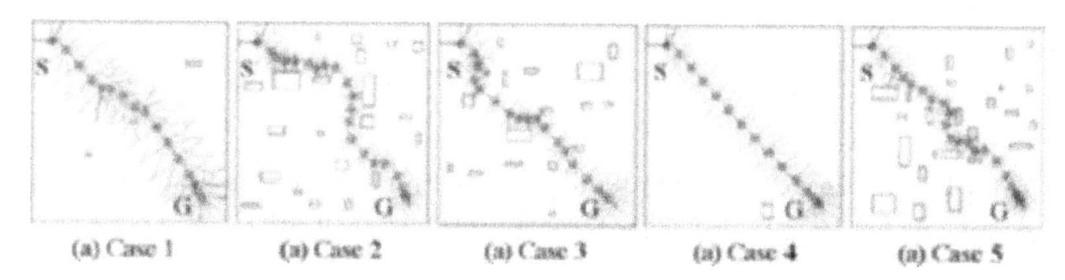

Fig. 24.11. The trajectory of the mobile robot with fuzzy controller based on PerGA. "S" and "G" indicate starting point and target point, respectively. Each environment has the different layout of obstacles.

Figure 24.14 shows the comparison results of PCP_i of SSGA and PerGA. This comparison clearly indicates PerGA can various candidate controllers. However, the individuals used in the evolution and adaptation of Per GA is less than that of SSGA. Therefore, the performance of both SSGA and PerGA is similar after many episodes.

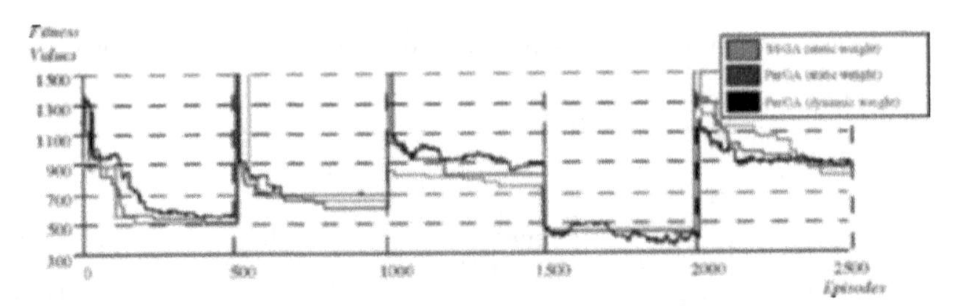

Fig. 24.12. The change of fitness values of SSGA, PerGA without weight updating, and PerGA with weight updating.

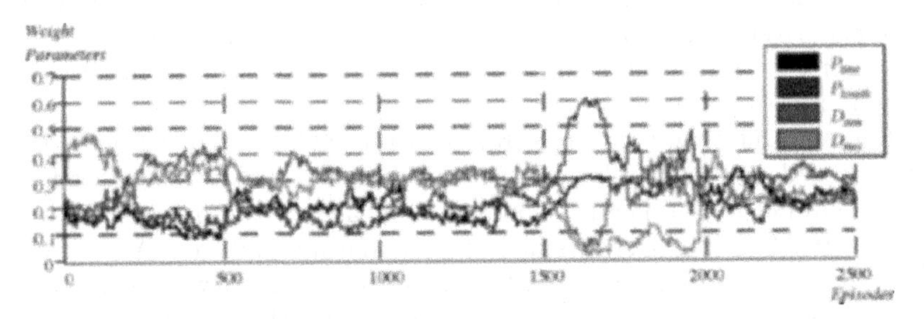

Fig. 24.13. The change of weight parameters.

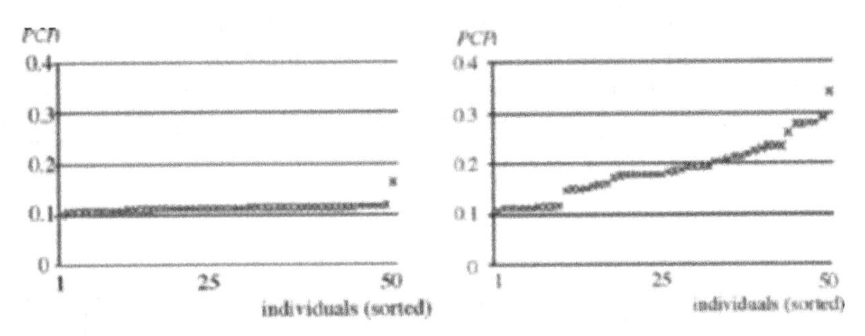

Fig. 24.14. The distribution of individuals concerning perceptual information. Left: SSGA (variance 6.15*10^{-5}); Right: PerGA (variance 3.10*10^{-3}).

24.4 Fuzzy Spiking Neural Networks and Behavior Coordination

Various types of artificial neural networks have been proposed to realize clustering, classification, nonlinear mapping, and control [13]-[22]. Basically, artificial neural networks are classified into pulse-coded neural networks and rate-coded neural networks from the viewpoint of abstraction level [31]. A pulse-coded neural network approximates the dynamics introduced the ignition phenomenon of a neuron, and the propagation mechanism of the pulse between neurons. Hodgkin-Huxley model is one of the classic neuronal spiking models with four differential equations. An integrate-and-fire model with a first-order linear differential equation is known as a neuron model of a higher abstraction level. A spike response model is slightly more general than the integrate-and-fire model, because the spike response model can choose kernels arbitrarily. On the other hand, rate-coded neural networks neglect the pulse structure, and therefore are considered as neuronal models of the higher level of abstraction. McCulloch-Pitts and Perceptron are well known as famous rate coded models [37]. One important feature of pulse-coded neural networks is the capability of temporal coding. In fact, various types of spiking neural networks (SNNs) have been applied for memorizing spatial and temporal context in robotics [37]-[41].

Floreano and his colleagues applied SNN to the control of mobile robots and blimps [38]-[40]. The contrast calculated by a Laplace filter spanning three adjacent photoreceptors is used as a sensory input for their motion control. The implementation of the SNN is simple, but its behavior is powerful. We also applied a simple spike response model, but we used a membership functions to inputs to spiking neurons, and proposed a fuzzy spiking neural network to memorize spatio-temporal pattern between sensory inputs and action outputs.

24.4.1 A Modified Simple Spike Response Model

In this section, we introduce a modified simple spike response model to reduce the computational cost. First of all, the membrane potential, or internal state $h_i(t)$ of the ith neuron at the discrete time t is given by

$$h_i(t) = \tanh(h_i^{\mathrm{syn}}(t) + h_i^{\mathrm{ref}}(t)), \tag{24.18}$$

where $h_i^{syn}(t)$ including the output pulses from the other neurons is calculated by the following equation,

$$h_i^{\mathrm{syn}}(t) = \gamma^{\mathrm{syn}} \cdot h_i(t-1) + \sum_{j+1,j!=i}^{N} w_{j,i} \cdot p_j(t-1) + h_i^{\mathrm{ext}}(t), \tag{24.19}$$

where $w_{j,i}$ is a weight coefficient from the jth to ith neuron; $p_j(t)$ is the output of jth neuron at the discrete time t; $h_i^{\mathrm{ext}}(t)$ is an input to the ith neuron from the environment; N is the number of neurons; γ^{syn} is a discount rate. Here we use a hyperbolic tangent function in order to restrict the value of an internal potential within a possible small value, because SNN has a problem of neuron burst firing. Furthermore, $h_i^{\mathrm{ref}}(t)$ is used for the refractoriness of the neuron. When the neuron is fired, R is subtracted from the refractoriness variable in the following,

$$h_i^{\mathrm{ref}} = \begin{cases} \gamma^{\mathrm{ref}} \cdot h_i^{\mathrm{ref}}(t-1) - R, & \text{if } p_i(t-1) = 1 \\ \gamma^{\mathrm{ref}} \cdot h_i^{\mathrm{ref}}(t-1), & \text{otherwise.} \end{cases} \tag{24.20}$$

where γ^{ref} is a discount rate ($0<\gamma^{\mathrm{ref}}<1.0$). When the internal state of the ith neuron is larger than the predefined threshold, a pulse output is done as follows;

$$p_i(t) = \begin{cases} 1, & \text{if } h_i(t) \geq \theta_i \\ 0, & \text{otherwise} \end{cases}, \tag{24.21}$$

where θ_i is a threshold for firing (Fig. 24.15).

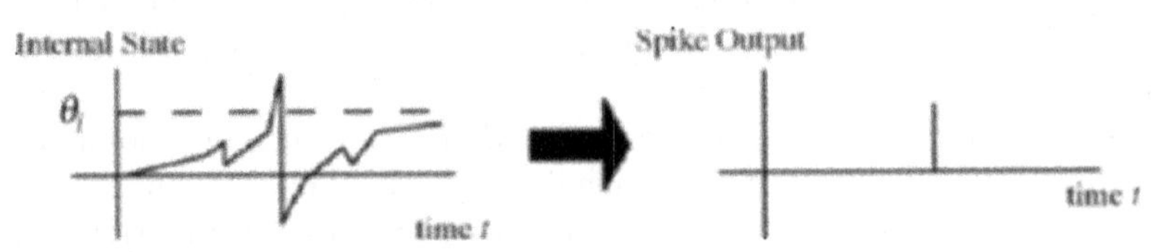

Fig. 24.15. A spiking neuron model. If the internal state of a neuron is larger than the predefined threshold, a spike output occurs.

24.4.2 Hebbian Learning of Fuzzy Spiking Neural Network

We applied a fuzzy spiking neural network (FSNN) to the motion control of a mobile robot. A neuron linking with a sensor for measuring the distance to obstacles is called an S-neuron; a neuron used for perceiving the direction to a target point is called a T-neuron; and a neuron linking with a motor is called an M-neuron. In this paper, the external input to FSNN is restricted as positive value in order to reduce computational complexity. Furthermore, the value range of different sensors is not normalized. Therefore, we apply membership functions to extract perceptual information to be used for the motion control. The membership grades (μ) are used as the external inputs to S-neurons. The FSNN is composed of input layer, S- and T- neuron layer, and M-neuron layer. We assume bidirectional connection between S-neuron and M-neuron layers. Figure 24.16 shows an example of FSNN for the mobile robot. In order to understand the relationship between a sensory input and a motor output, the robot must move in the environment. Therefore, the robot basically goes straight, and the M-neuron is used as a suppression. If a M-neuron is fired, its corresponding motor output is reduced. The kth motor output (k=1,2) of the robot is calculated by

$$v_k = 1 - \frac{\sum_{i=0}^{T} \lambda^i p_k(t-i)}{\sum_{i=0}^{T} \lambda^i}, \tag{24.22}$$

where T is the number of the referred step intervals; λ is a discount rate. In Fig. 24.16, the motor neuron is fired after the 5th and 6th sensor neurons are fired. As a result, the robot turns left.

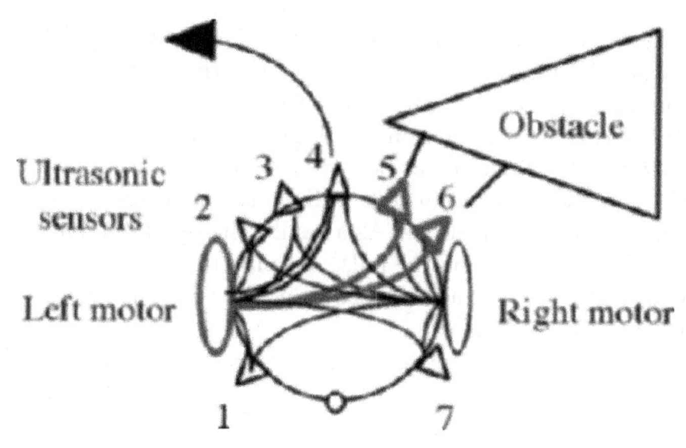

Fig. 24.16. A fuzzy spiking neural network for controlling a mobile robot. The left M-neuron is fired after the 5th and 6th S-neurons, and as a result, the robot turns right because the spike output from the M-neuron plays the role of suppression of the motor output.

The robot should learn the relationship among neurons according to its environmental condition. We use the following update rule for learning weight parameters between the S-neuron and M-neurons,

$$w_{j,i} \leftarrow \tanh(\gamma^{\mathrm{wgt}} \cdot w_{j,i} + \xi^{\mathrm{wgt}} \cdot v_i \cdot p_i(t) \cdot p_j(t-1)), \tag{24.23}$$

where v_i is the speed of the ith motor in eq.(24.22); γ^{wgt} is a discount rate; ξ^{wgt} is a learning rate. Because the output of the motor neuron plays the role of the negative effect to the motor output, the update amount of the weight connection strength increases as the speed is high. In this way, the robot can learn the collision avoiding behavior by using the simultaneous fires of the S-neuron and M-neuron. And also, the weight connection strength between T-neuron and M-neuron is updated by the above rule. Furthermore, the Hebbian learning is applied for updating the connection weights from S-neurons to T-neurons as well as from M-neurons to S-neurons. Therefore, the simultaneous firing of several neurons in the sensors can construct a spatial pattern to the M-neurons. The weight parameters are trained based on the Hebbian learning algorithm from the jth to ith neurons as follows,

$$w_{j,i} \leftarrow \tanh(\gamma^{\mathrm{wgt}} \cdot w_{j,i} + \xi^{\mathrm{wgt}} \cdot p_i(t) \cdot p_j(t-1)). \tag{24.24}$$

Here we consider two types of relationship. One is the relationship between an S-neuron and a T-neuron. If the robot moves toward the target direction near obstacles, the robot should obviously take a collision avoiding behavior, not a target tracing behavior. Therefore, when the temporally sequential firing from an S-neuron to a T-neuron occurs, its corresponding connection weight is updated into negative value. As a result, the pulse output from an S-neuron decreases the internal state of its connected T-neuron. This plays the role of suppressing the target tracing behavior, and the multiobjective behavior coordination is performed in the level of sensory inputs. The other is the relationship between the M-neuron and S-neuron. The output from M-neurons is connected with the S-neuron as positive feedback. Therefore, an S-neuron is easy to fire, because the internal state of S-neuron increases according to pulse outputs received from M-neurons. As a result, the robot can pay attention to the specific direction based on the pulse outputs from M-neurons. In this way, the robot can take various actions based on the spatio-temporal patterns of inputs and outputs by FSNN.

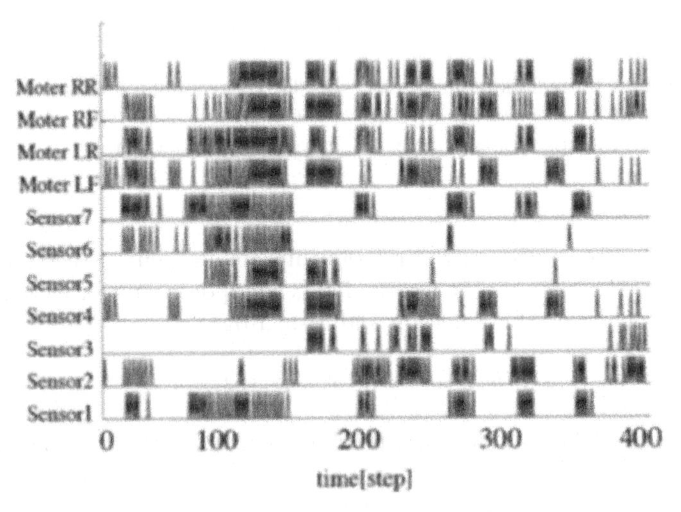

Fig. 24.17. History of spikes of each neuron.

We investigated the behavior of the FSNN in the motion control of a mobile robot. We used a partner robot called MOBiMac [43]. Basically the robot can move according to the outputs from the M-neuron. However, we used both of reverse M-neurons and forward M-neurons to realize a small sharp turn. Figure 24.17 shows the actual spikes of sensor neurons and motor neurons, and Fig. 24.18 shows the spikes from 350 to 400 time steps. For example, both of reverse M-neurons are fired after the 1st sensor neuron is fired (see Figure 24.18, (A)). This indicates the robot goes back when the robot detects obstacles in front of the robot. Both of forward M-neurons are fired immediately after the 4th sensor neuron is fired (see Figure 24.18, (B)). This indicates the robot goes forward when the robot detects obstacles before the robot. Furthermore, only the right forward M-neuron is fired after the second and third S-neurons are fired (see Figure 24.18, (C)). This indicates the robot turns right to avoid the collision with the obstacles when the robot detects obstacles at the right hand. Figure 24.19 shows trajectories of the

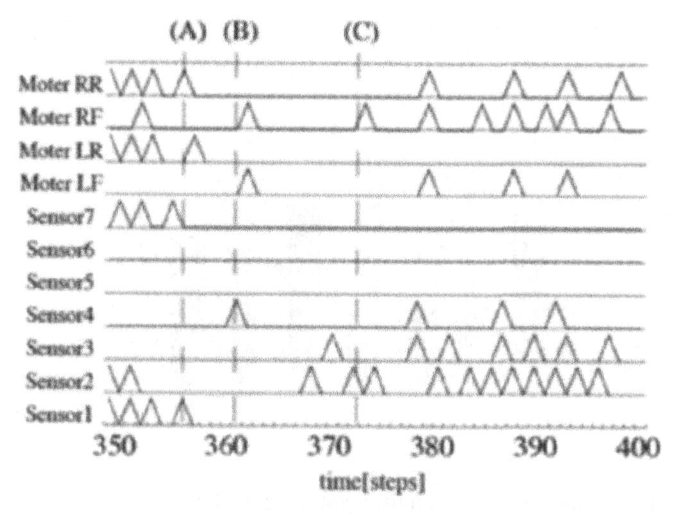

Fig. 24.18. History of spikes of each neuron from 350 to 400 time steps.

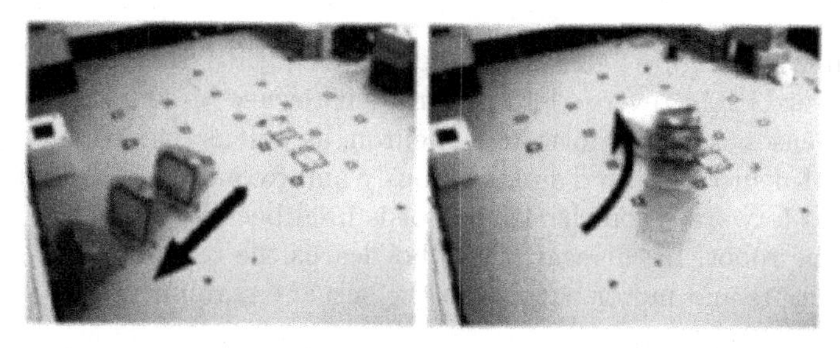

Fig. 24.19. Trajectories of the mobile robot before learning.

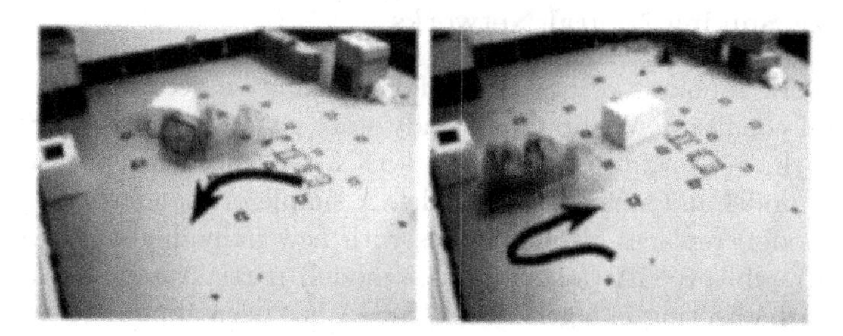

Fig. 24.20. Trajectories of the mobile robot after learning.

partner robot before learning in an experimental result. The robot collides with obstacles. As a result of collision with obstacles, the FSNN is trained gradually. Figure 24.20 shows trajectories of the robot after learning. The robot avoids collision with obstacle.

Figure 24.21 shows the comparison of simultaneously spiked times of S-neurons and M-neurons in learning process. The layout of ultrasonic sensors

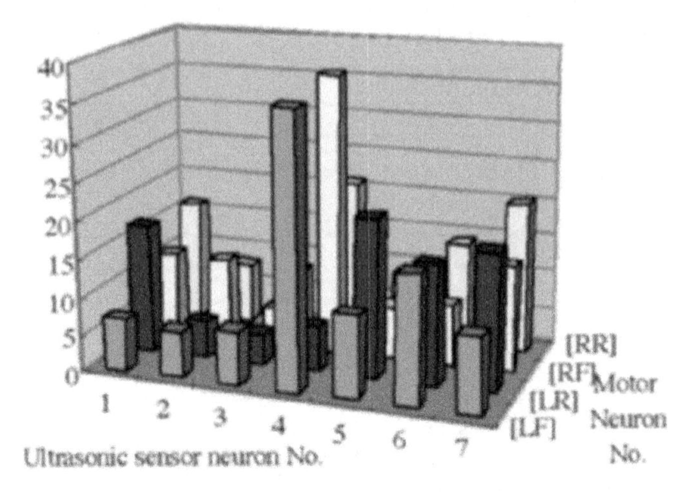

Fig. 24.21. Comparison of simultaneously spiked times of sensory inputs and motion outputs.

is also shown in Figure 24.21. The labels of [LF], [LR], [RF] and [RR] indicate right forward, right reverse, left forward, left reverse M-neurons, respectively. The 4th sensor and right forward M-neuron, as well as the 4th sensor and left forward M-neuron are fired simultaneously and frequently. This indicates the robot goes forward when the 4th sensor is fired because it is equipped at the back of the robot. In this way, the robot learns the relationship between the sensory inputs and motor outputs in the actual environment.

24.4.3 Multi-Objective Evolutionary Learning for Fuzzy Spiking Neural Networks

The network topology of FSNN should be adaptive according to the environmental condition. We apply a steady-state genetic algorithm (SSGA) for updating the network topology of the FSNN. GA can be divided into a generational model and steady-state model. A simple or standard GA (a generational model) replaces all individuals with new individuals in a generation (iteration), while SSGA (a steady-state model) partially replaces a few individuals with offspring in a generation. SSGA has been applied to incremental learning of simulation-based optimization. We also applied SSGA to various problems and showed SSGA is suitable to solve optimization problems in dynamic or changing environments. Actually, SSGA can easily maintain various best solutions obtained through environmental changes, because genetic operators are performed to only a few individuals in a generation. Therefore, we apply SSGA for solving optimization problems in this paper.

We use binary coding for weight connection. The value 1 is used if the connection exists between the S-neuron and M-neuron. Otherwise, the value is set at 0. A candidate solution is replaced by a new candidate solution generated

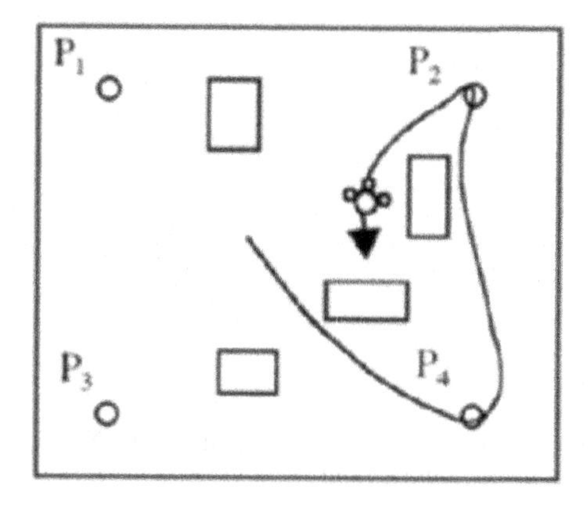

Fig. 24.22. A simulation environment including four obstacles and four target points.

by genetic operators in each generation. In this paper, the worst candidate solution is eliminated ("delete least fitness" selection) and replaced with the candidate solution generated by the crossover and mutation. We use elitist crossover and simple mutation. The elitist crossover randomly selects one individual and generates an individual by incorporating genetic information from the selected individual and best individual.

Figure 24.22 shows a simulation environment including four obstacles and four target points. A robot moves among P1, P2, P3, and P4, randomly. One trial is defined as moves from Pi to Pj. The motion of the mobile robot by the ith individual is evaluated by the following fitness function

$$f_i = f^{\mathrm{dis}} + \kappa \cdot f^{\mathrm{safe}}, \tag{24.25}$$

where $f\,dis$ is the distance between two points; $f\,safe$ is the degree of approach to obstacles; κ is a coefficient. However, the fitness value can be different according to the path between two points. Therefore, the partially inherited fitness value is used as follows;

$$\mathrm{fit}_i = (1-\alpha)\mathrm{fit}_i^{\mathrm{Pre}} + \alpha\, f_i, \tag{24.26}$$

where f_i^{Pre} is the previous fitness value; α is a coefficient. Furthermore, when the robot reaches the target point, the individual is used for the next trial according to the continuation probability. Therefore, SSGA is used as the off-line adaptation, while the Hebbian learning is used as the on-line adaption.

Figure 24.23 shows several snapshots in a simulation result of FSNN. A position of the robot is depicted every 20 steps. The sensing range is depicted as a broken line. When the mobile robot detects obstacles in the sensing range, it is depicted as a full line. Furthermore, the connection weights are depicted in the bottom of the snapshot. The depth of the color used in the line indicates weight values. A gray line indicates a small positive or negative

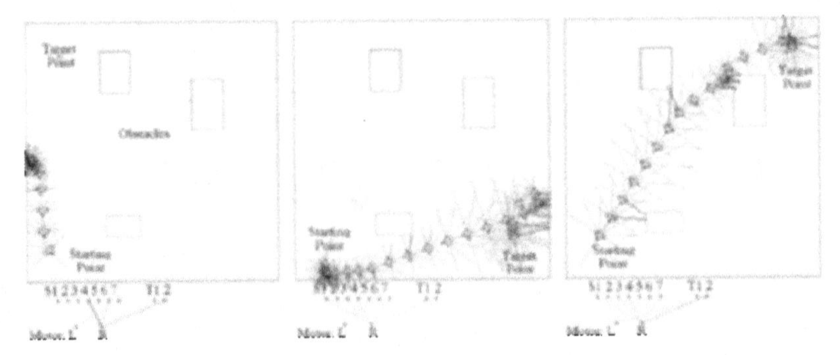

Fig. 24.23. Trajectories of a mobile robot by using FSNN in the search of SSGA. (a) 1st trial (b) 8th trial (c) 199th trial.

weight value. The lower figure shows the network connection among S-, T- and, M-neurons. The robot by the most individuals cannot reach the target point at the stage of initialization (Fig. 24.23(a)). At the 8th trial, the robot can reach the target point, because the weight connection between the T-neurons and M-neurons as well as the weight connection between the S-neurons and M-neurons are connected efficiently, but these connection is limited in the half capability of behaviors such as the left turn for collision avoidance and right turn for target tracing (Fig. 24.23(b)). Afterward, the robot finds the connection suitable to the current environment (Fig. 24.23(c)). In this way, the robot learns behaviors according to the connection pattern decided by the SSGA as the robot encounters with various situations step by step. This learning method realizes the sensory-motor coordination for multi-objective behaviors in the network structure without hidden layers.

24.5 Concluding Remarks

This chapter introduces several learning methods for multi-objective design of neuro-fuzzy controllers for robot behavior coordination. Basically, a robot must take actions based on multi-objective decision making. There are two approaches to realize the multi-objective behavioral learning. One is the direct design of a large size of total behavior rules. The other is the combination of basic behavior rules and their coordination like a mixture of experts. The robot should have behavior rules based on the mixture of experts from the viewpoints of the reusability of the acquired behaviors and life-time learning. Furthermore, each behavior used as a local expert should have explicit objective because it is easy to design the behavior coordination. If the exact teaching data for final outputs are available in the learning, supervised learning for local experts and their behavior coordination rule can be performed. However, this is not realistic, because the environmental condition of a robot is often unknown. Therefore, it is very difficult to prepare exact teaching data for behavior learning beforehand. In this situation, this chapter introduces supervised learning methods using the typical relationship between sensory inputs and action outputs and evolutionary learning methods for behavior acquisition based on steady-state genetic algorithm. The essential of the method is to realize the adaptability and reusability of behaviors through interaction with the environment.

As future works, we discuss the learnability of the proposed methods and the incremental learning for robot behavior coordination, and apply the proposed method to the behavioral learning of various mobile robots.

Acknowledgement

The author would like to thank Toshio Fukuda, Fumio Kojima, Eiji Mizutani, Yusuke Nojima, and Hironobu Sasaki for their valuable suggestions and comments.

References

[1] K. Morikawa, S. Agarwal, C. Elkan, G. Cottrell. A Taxonomy of Computational and Social Learning. In: *Proc. of Workshop on Developmental Embodied Cognition*, Edinburgh, 2001.
[2] S.J.Russell and P.Norvig. Artificial Intelligence, Prentice-Hall, Inc., 1995.
[3] R.A. Brooks, Cambrian Intelligence. The MIT Press, 1999.
[4] R.Pfeifer, C.Scheier. Understanding Intelligence, The MIT Press, 1999.
[5] M. Arbib, A. Billard, M. Iacobonni, E. Oztop. Synthetic Brain Imaging: Grasping, Mirror Neurons and Imitation, *Neural Networks*, 13:975–997, 2000
[6] A. Billard. Imitation: A Means to Enhance Learning of A Synthetic Protolanguage in An Autonomous Robot. In: K. Dautenhahn and C. L. Nehaniv (eds), Imitation in Animals and Artifacts, The MIT Press, Cambridge, pp.281–311, 2002.
[7] P. Andry, P. Gaussier, S. Moga, J.P. Banquet, J. Nadel. Learning and Communication via Imitation: An Autonomous Robot Perspective, *IEEE Trans. on Systems, Man, and Cybernetics*, 31:431–442, 2001
[8] R.P.N. Rao, A.N. Meltzoff. Imitation Learning in Infants and Robots: Towards Probabilistic Computational Models. In: *Proc. of Artificial Intelligence and Simulation of Behaviors*, 2003.
[9] S. Nolfi, D. Floreano. Evolutionary Robotics: The Biology, Intelligence, and Technology of Self-Organizing Machines, The MIT Press, 2000.
[10] G. Dudek and M. Jenkin. Computational Principles of Mobile Robotics, Cambridge University Press, 2000.
[11] J. M. Zurada, R. J. Marks II, C. J. Robinson (eds.). Computational Intelligence - Imitating Life, IEEE Press, 1994.
[12] J. Holland. Adaptation in Natural and Artificial Systems, Ann Arbor:University of Michigan Press, 1975
[13] D.B. Fogel. Evolutionary Computation, New York, IEEE Press, 1995.

[14] D. H. Ballard. An Introduction to Natural Computation, The MIT Press, 1997.
[15] D. E. Rumelhart, J. L. McClelland and the PDP Research Group. Parallel Distributed Processing Explorations in the Microstructure of Cognition. Volume 1: Foundations, The MIT Press, 1986.
[16] T. Hastie, R. Tibshirani, J. Friedman. The Elements of Statistical Learning: Data Mining, Inference, and Prediction, Springer-Verlag, New York, 2001.
[17] R.C. Arkin, Behavior-Based Robotics, The MIT Press, 1998.
[18] M. Dorigo and U. Schnepf. Genetics-based Machine Learning and Behaviour Based Robotics: A New Synthesis. *IEEE Transactions on Systems, Man, and Cybernetics*, 23(1):141–154, 1993.
[19] T. Fukuda, N. Kubota, and T. Arakawa, GA Algorithms in Intelligent Robots. In: *Fuzzy Evolutionary Computation*, Kluwer Academic Publishers, pp.81-105, 1997.
[20] T. Fukuda, N. Kubota. An intelligent robotic system based on a fuzzy approach. *Proceedings of IEEE*, 87(9):1448–1470, 1999
[21] T. Fukuda and N. Kubota Fuzzy Control Methodology: Basics and State of Art. In: *Soft Computing in Human-Related Sciences*, H.N.Teodorescu, A. Kandel, and L.C.Jain (eds.), pp.3-35, 1999.
[22] D.P. Bertsekas, J.N. Tsitsiklis. Neuro-Dynamic Programming, Athena Scientific, 1996.
[23] R.S. Sutton, A.G. Barto. Reinforcement Learning, The MIT Press, 1998.
[24] T. Kohonen. Self-organized formation of topologically correct feature maps. *Biological Cybernetics*, 43 (1982) 59-69.
[25] G.Syswerda. A Study of Reproduction in Generational and Steady-state Genetic Algorithms. In: *Foundations of Genetic Algorithms*, Morgan Kaufmann Publishers, Inc., San Mateo, 1991.
[26] D.E. Goldberg. Genetic Algorithms in Search, Optimization, and Machine Learning, Addison Welsey, 1989.
[27] G. Rudolph. Convergence Analysis of Canonical Genetic Algorithm. *IEEE Transaction on Neural Network*, 5(1):61–101, 1994.
[28] G.J. Klir, B. Yaun (ed.), Fuzzy Sets, Fuzzy Logic, and Fuzzy Systems, World Scientific Publishing, 1996.
[29] J.-S.R. Jang, C.-T. Sun, E.Mizutani. Neuro-Fuzzy and Soft Computing, Prentice-Hall, Inc., 1997.
[30] C.C. Lee. Fuzzy Logic in Control Systems: Fuzzy Logic Controller - Part I & Part II, *IEEE Trans. on Systems, Man, and Cybernetics*, Vol.20, No.2, pp.404-435, 1990.
[31] S.V. Kartalopoulos. Understanding Neural Networks and Fuzzy Logic, The IEEE Press, 1996.
[32] A. Billard. Imitation. In: M.A. Arbib (ed.), *Handbook of Brain Theory and Neural Networks*, The MIT Press, Cambridge, pp.566-569, 2002.
[33] G. Rizzolatti, M.A. Arbib. Language within our grasp, *Trends in Neuroscience*, 21, (1998), 188-194.
[34] A. Anderson, E. Rosenfeld, Neurocomputing, The MIT Press, 1988.
[35] W.T. Miller, R.S. Sutton, P.J. Werbos. Neural Networks for Control, The MIT Press, 1990.
[36] T. Caelli, L. Guan and W. Wen. Modularity in Neural Computing. *Proceedings of The IEEE*, Vol.87, No.9, pp.1497-1518, 1999.
[37] W. Maass and C.M. Bishop. Pulsed Neural Networks, The MIT Press 1999.

[38] D. Floreano, J.C. Zufferey, and C. Mattiussi. Evolving Spiking Neurons from Wheels to Wings. In: *Dynamic Systems Approach for Embodiment and Sociality*, K. Murase and T. Asakura (eds.), Magill, Advanced Knowledge International, International Series on Advanced Intelligence, Vol 6, pp. 65-70, 2003.

[39] D. Floreano, C. Mattiussi. Evolution of Spiking Neural Controllers for Autonomous Vision-based Robots. In: T. Gomi (ed.), *Evolutionary Robotics* IV, Berlin: Springer-Verlag, 2001.

[40] J. Urzelai and D. Floreano. Evolution of Adaptive Synapses: Robots with Fast Adaptive Behavior in New Environments. *Evolutionary Computation*, 9, 495-524, 2001.

[41] H. Burgsteiner. Training networks of biological realistic spiking neurons for real-time robot control. In:*Proceedings of the 9th International Conference on Engineering Applications of Neural Networks*, Elsevier Publishing, 2005.

[42] N. Kubota. Intelligent Structured Learning for A Robot Based on Perceiving-Acting Cycle. In: *Proc. of the Twelfth Yale Workshop on Adaptive and Learning Systems*, pp.199-206, 2003.

[43] Y. Nojima, F. Kojima, and N. Kubota. Local Episode-based Learning of A Mobile Robot in A Dynamic Environment. *Proc. of The Third International Symposium on Human and Artificial Intelligence Systems (HART2002)*, pp.384-388, 2002.

[44] N. Kubota, T. Morioka, F. Kojima, T. Fukuda. Learning of Mobile Robots Using Perception-Based Genetic Algorithm, *Measurement*, 29, 237-248, 2001.

[45] N. Kubota, H. Sasaki. Spiking Neural Network for Behavior Learning of A Mobile Robot.In: *Proc. of the 3rd International Symposium on Autonomous Minirobots for Research and Edutainment*, pp.267-272, 2005.

[46] N. Kubota, K. Tomoda. Behavior Coordination of A Partner Robot based on Imitation. In: *Proc. of 2nd International Conference on Autonomous Robots and Agents*, pp.164-169, 2004.

[47] N. Kubota, K. Nishida. Fuzzy Computing for Communication of A Partner Robot Based on Imitation, In: *Proc. (CD-ROM) of 2005 IEEE International Conference on Robotics and Automation*, pp.4391-4396, 2005.

[48] K. Deb. Multi-Objective Optimization using Evolutionary Algorithms, John Wiley & Sons, Chichester, UK, 2001

[49] A.M.S. Zalzala, M.C. Ang, M. Chen, A.S. Rana and Q. Wang. Evolutionary algorithms for robotic systems: principles and implementations. In: A.M.S. Zalzala and P.J. Fleming (editors), *Genetic Algorithms in Engineering Systems*, Chapter 8, pp. 161–202, IEEE Press. Control Engineering Series 55, Bath, UK, 1997.

[50] J.-Y. Donnart and J.-A. Meyer. Learning Reactive and Planning Rules in a Motivationally Autonomous Animat. *IEEE Transactions on Systems, Man and Cybernetics - Part B: Cybernetics*, 26(3), 381-395, 1996.

[51] D. Floreano and F. Mondada. Evolutionary Neurocontrollers for Autonomous Mobile Robots. *Neural Networks*, 11, 1461-1478, 1998.

[52] S.-B. Cho and G.-B. Song. Evolving CAM-Brain to control a mobile robot. *Applied Mathematics and Computation*, vol. 111, no. 2-3, pp. 147-162, 2000.

[53] S. Nolfi. Using emergent modularity to develop control systems for mobile robots. *Adaptive Behavior*, 5:343-363, 1997.

[54] I. Harvey, P. Husbands, D. Cliff, A. Thompson and N. Jakobi. Evolutionary Robotics: The Sussex Approach. In: *Robotics and Autonomous Systems*, 20:205-224, 1997.

[55] L.S. Teo, M. Khalid, and R. Yusof. Tuning of a Neuro-Fuzzy Controller designed by Genetic Algorithms. *IEEE Trans on Systems, Man and Cybernetics*, 29(2):226–236, April, 1999.

25

Fuzzy Tuning for the Docking Maneuver Controller of an Automated Guided Vehicle

J.M. Lucas, H. Martinez, and F. Jimenez

Dept. Information and Communications Engineering, University of Murcia, Murcia, Spain
jmlucas@um.es, humberto@um.es, fernan@dif.um.es

Summary. In some environments, mobile robots need to perform tasks in a precise manner. For this reason, we require obtaining good controllers in charge of these control tasks. In this work, we present a real-world application in the domain of multi-objective machine learning, which consists of an Automated Guided Vehicle (AGV), specifically, a fork-lift truck must often perform docking maneuvers to load pallets in conveyor belts. The main purpose is to improve some features of docking task as its duration, accuracy and stability, satisfying determined constraints. We propose a machine learning technique based on a multi-objective evolutionary algorithm in order to find multiple fuzzy logic controllers which optimize specific objectives and satisfy imposed constraints for docking task in charge of following up an online generated trajectory.

25.1 Introduction

In many real-world applications, mobile robots require interacting with objects in their environment, hence, they need to perform docking tasks, for instance, in order to move objects from one location to another. The performance of tasks must be suitable for the tolerances requires by the particular task. In this work, we present a real-world application in the domain of multi-objective machine learning, which consists of an Automated Guided Vehicle (AGV), specifically, a fork-lift truck must often perform docking maneuvers to load pallets in conveyor belts [21]. The main purpose is to perform docking maneuvers with specific requirements of performance. We propose a multi-objective evolutionary algorithm in order to generate and optimize fuzzy logic controllers in charge of docking control which optimize specific objectives and satisfy imposed constraints at the same time. In the evolutionary process, several objective will be taken into account and the obtained solutions must be feasible, satisfying specific constraints. In the mobile robot domain, we deal with a control problem called path tracking. The fuzzy controller, in charge of the docking task, intends to follow up an online generated trajectory until the docking point.

J.M. Lucas et al.: *Fuzzy Tuning for the Docking Maneuver Controller of an Automated Guided Vehicle*, Studies in Computational Intelligence (SCI) **16**, 585–600 (2006)
www.springerlink.com

25.1.1 Multi-Objective Evolutionary Algorithms for Constrained Optimization

Evolutionary Algorithms (EAs) are stochastic search techniques inspired by the principles of natural selection and evolution of species. EAs were initially extended to solve optimization problems with multiple objectives (often conflicting) [6, 8, 11, 15, 24, 28, 31].

EAs for multi-objective problems can be classified [11] as plain aggregating, non-Pareto and Pareto population-based approaches. The plain aggregating approach consists of a linear combination of several objectives in a single objective by means of some method such as weighted sum, goal programming and goal attainment. This approach requires the assessment of the importance of each objective by setting weights, which is very difficult for most problems and moreover, one single solution is produced at a time that may not satisfy the decision maker. Nevertheless, the simultaneous optimization of several objectives allows finding a set of alternative solutions in a single run of algorithm, therefore, decision makers have more flexibility to select suitable solutions according to their requirements. The Pareto population-based approaches allow finding several elements of the Pareto optimal set, in a single run, since they search multiple optimal solutions in parallel.

In multi-objective optimization problems, we need to find a set of trade-off solutions which are considered optimal in all the objectives to be optimized. In most real-world optimization problems, a series of constraints will arise, therefore, the feasibility of solutions must be considered in order to obtain solutions which satisfy specified constraints. The Pareto-based approach allows having a set of Pareto non-dominated and feasible solutions. Afterwards, we can choose the most satisfactory solution by applying a preference criterion.

Evolutionary algorithms have incorporated new approaches to handle constraints [5, 17, 23] and have been proved to solve multi-objective constrained optimization problems as [9, 25, 30].

25.1.2 Soft Computing Techniques for Fuzzy Control of Autonomous Robots

The fuzzy modeling of learning systems and mechanisms allows applying the fuzzy logic to the control and operation tasks of autonomous mobile robots [26, 27]. Soft computing techniques (such as EAs) are suitable to be applied to applications domains with imprecise data and/or incomplete knowledge of environment such as control problems in mobile robotics. The idea is taking advantage of an evolutionary algorithm in order to evolve the behaviors for control tasks with the aim of performing them in a more accurate and efficient manner. Several works have been proposed in the domain of the genetic design and optimization of fuzzy logic controllers (FLCs) [4, 14, 18]. Various works deal with the problem of evolving FLCs for the design of robot control systems as [3, 12, 13].

25.1.3 Organization of the Chapter

This chapter is organized as follows: Sect. 25.2 presents an application of a multi-objective evolutionary algorithm to tune a fuzzy controller in charge of the docking task for an autonomous robot. Sect. 25.3 explains the main features of the proposed multi-objective evolutionary algorithm. Sect. 25.4 describes the experiments which have been performed and the obtained results. Finally, Sect. 20.5 summarizes the most important conclusions.

25.2 Control Problem: The Docking Task

In some environments, mobile robots require docking when need to interact with objects in their environments. The final position and orientation of the robot must be suitable for the tolerances required by the particular task. In this work, we present an environment where an Automated Guided Vehicle (AGV), specifically, a fork-lift truck performs docking maneuvers to load pallets in conveyor belts [21]. The main purpose is to perform the docking task with specific requirements as duration and precision, following up an online generated trajectory. We propose a multi-objective evolutionary algorithm based partially on [16] in order to generate and optimize controllers in charge of docking control by means of tuning of parameters of fuzzy membership functions. In the evolutionary learning process, several objectives and constraints have been handled in a simultaneous manner. This process has been performed by means of the simulation system for mobile robots called *ThinkingCap-II*. New evolved controllers are simulated in that environment allowing the required data acquisition to evaluate the fitness of controller.

Docking task can be defined simply as motion from the current position to a desired position and orientation, while following a safe trajectory [2]. In fact, docking task can be performed as three distinct sub-problems [1]:

1. Moving to the required location while there is no danger of collision.
2. Moving accurately to come close to the required docking configuration.
3. Moving to the destination with the required precision, or performing some docking operation.

We consider two movements:

1. Navigation: The vehicle follows up a free curve (without constraints) towards the way-point.
2. Docking: The vehicle follows up a preplanned B-Spline curve heading to the docking point (goal point).

Fig. 25.1 shows how the fork-lift truck robot begins the docking maneuver from the waypoint *wp0* to the docking point *ra0*.

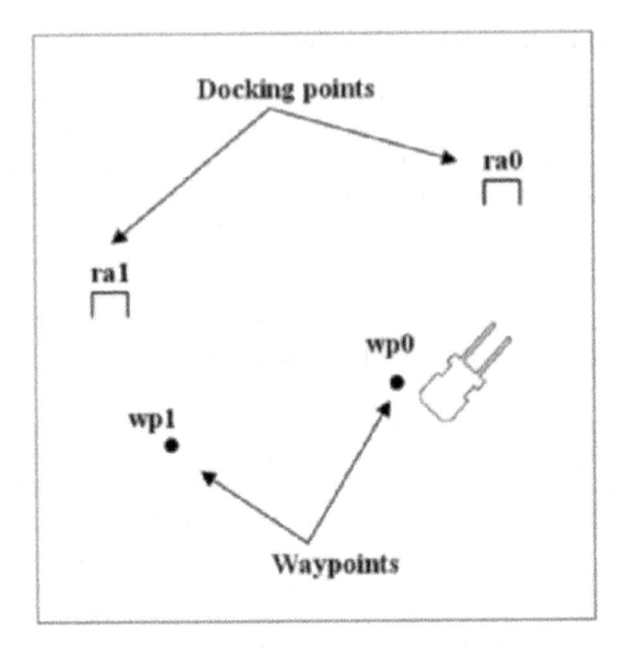

Fig. 25.1. Docking maneuver

The control problem examined in this work is a path tracking problem. Essentially, the goal is that a mobile robot or an AGV navigates properly along the desired trajectory in a two-dimensional environment. *Pure Pursuit* approach is a tracking algorithm that works by calculating the curvature that will move a vehicle from its current position to some goal position [7]. The whole point of the algorithm is to choose a goal position that is some distance ahead of the vehicle on the path. An analogy is a car driver who looks a further point in front of the car and afterwards tends to head toward that goal point gradually. In particular, *Pure Pursuit* approach has been applied in order to calculate the heading-error respect to the preplanned trajectory. The heading-error is an angle denoted by δ (see Fig. 25.2), and it is calculated by subtracting from heading the current orientation of the robot. The heading is calculated by using the look-ahead point (labeled by *"Looka"* in Fig. 25.2) in the trajectory. Fig. 25.2 depicts the fork-lift truck robot and B-Spline curve corresponding to the trajectory previously generated. Moreover, in Fig. 25.2 the goal point marked on the B-Spline curve corresponds to the docking point (the goal point).

The value of δ (heading-error) is calculated as follows:

- Dx = look-ahead point in the path.x - current position of robot.x
- Dy = look-ahead point in the path.y - current position of robot.y
- Heading = Arc tangent of Dy / Dx in the range of $-\pi$ to π
- δ = Heading - Current Orientation of Robot (where δ is normalized).

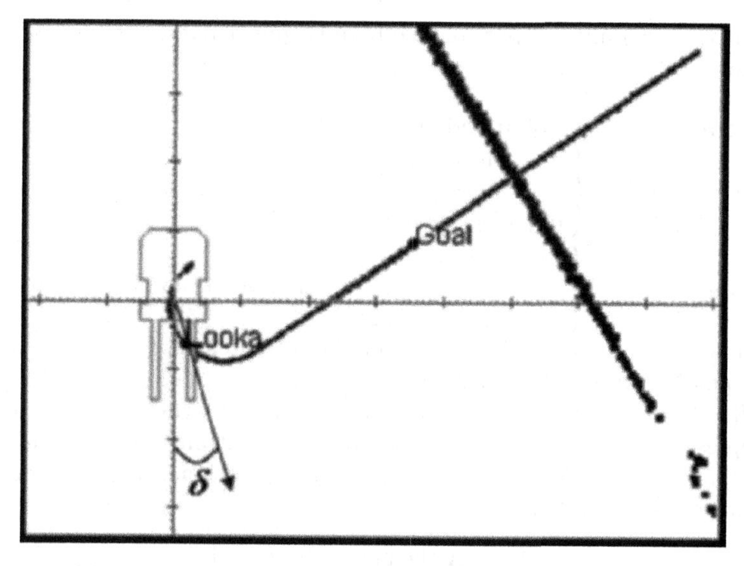

Fig. 25.2. Calculation the heading-error in the trajectory

According to the behaviors of the controllers in presence of obstacles, the navigation controller avoids the obstacles as far as possible, however during the docking maneuver, the autonomous vehicle stops immediately due to the nature of the task and security policies.

25.3 Optimization of Fuzzy Controller by Multi-objective Evolutionary Algorithms

The proposed algorithm is a Pareto-based multi-objective evolutionary algorithm (MOEA) for fuzzy tuning, i.e., it has been designed to find, in a single run, multiple non-dominated solutions according to the Pareto decision strategy. Main features of MOEA are the following:

25.3.1 Representation of Solutions

An individual in the MOEA corresponds to a fuzzy controller in charge of the docking task. A controller is a collection of N fuzzy rules in which each rule is composed by a set of n fuzzy numbers (antecedent) and m real parameters (consequent). In fact, antecedents are trapezoidal fuzzy sets that can be characterized using four parameters and the output fuzzy singletons using just one number for each one. Hence, focusing on learning rules in rule based systems, the *Pittsburgh* approach [29] has been applied. This approach is characterized by representing an entire fuzzy rule set as chromosome, although in our particular case, only one part of rule base is considered due to the controller symmetry.

The representation of an individual is the following:

$$\text{Antecedents} \Rightarrow \begin{cases} a_{11}, a_{12}, \ldots, a_{1l} \\ \cdots \\ a_{n1}, a_{n2}, \ldots, a_{nl'} \end{cases}$$

$$\text{Consequents} \Rightarrow \begin{cases} c_{11}, c_{12}, \ldots, c_{1s} \\ \cdots \\ c_{m1}, c_{m2}, \ldots, c_{ms'} \end{cases}$$

where n is the number of antecedents and m is the consequents. The value of indexes l and s depends on the number and type of fuzzy sets to code.

Afterwards, the building process of fuzzy controller rule-base from individual representation will be shown. The proposed controller is a Mamdani-type [19, 20] fuzzy controller which is in charge of the docking task. The controller contains rules of this type:

$$\text{if } \delta \in \text{A then } \omega = \text{c0}, \ \nu = \text{c1}$$

The input fuzzy set represents the heading-error δ (deg) in the trajectory (antecedent) and the two output crisp values (consequents) for angular ω (deg/s) and linear ν (m/s) velocities control. The antecedent is a trapezoidal fuzzy set and consequents are singletons in floating-point. It is assumed that the membership functions are known in advance and are fixed and the sum is 1. In order to build the controller, we use five floating-point numbers for antecedent and three and four for consequents respectively. On the one hand, the fuzzy knowledge base is composed by seven trapezoidal membership functions for the input fuzzy variable. They are labeled by NL (negative large), NM (negative medium), NS (negative small), Z (zero), PS (positive small), PM (positive medium) and PL (positive large). On the other hand, there are seven values for the angular velocity that are denoted by the same previous labels. Finally, there are four values for the linear velocity that are labeled by NZ (negative zero), NS (negative small), NM (negative medium) and NF (negative full).

The fuzzy controller is composed by seven rules which were obtained heuristically:

- Rule 1: if $\delta \in$ NL then $\omega =$ NL, $\nu =$ NZ
- Rule 2: if $\delta \in$ NM then $\omega =$ NM, $\nu =$ NS
- Rule 3: if $\delta \in$ NS then $\omega =$ NS, $\nu =$ NM
- Rule 4: if $\delta \in$ Z then $\omega =$ Z, $\nu =$ NF
- Rule 5: if $\delta \in$ PS then $\omega =$ PS, $\nu =$ NM
- Rule 6: if $\delta \in$ PM then $\omega =$ PM, $\nu =$ NS
- Rule 7: if $\delta \in$ PL then $\omega =$ PL, $\nu =$ NZ

In our case, an individual in the population is represented in a general manner as follows:

Antecedent (δ): a11, a12, a13, a14, a15
Consequent$_1$ (ω): c11, c12, c13
Consequent$_2$ (ν): c21, c22, c23, c24

From these real parameters, input and output variables sets are represented as shown Table 25.1. Immediately, the fuzzy rule-base for the before mentioned controller can be built.

Table 25.1. Representation of an individual

Input (heading-error δ)	Output$_1$ (angular velocity ω)	Output$_2$ (linear velocity ν)
NL = (-180, -180, -a_{15}, -a_{14})	NL = -c_{13}	NZ = -c_{21}
NM = (-a_{15}, -a_{14}, -a_{13}, -a_{12})	NM = -c_{12}	NS = -c_{22}
NS = (-a_{13}, -a_{12}, -a_{11}, 0)	NS = -c_{11}	NM = -c_{23}
Z = (-a_{11}, 0, 0, a_{11})	Z = 0	NF = -c_{24}
PS = (0, a_{11}, a_{12}, a_{13})	PS = c_{11}	
PM = (a_{12}, a_{13}, a_{14}, a_{15})	PM = c_{12}	
PL = (a_{14}, a_{15}, 180, 180)	PL = c_{13}	

Figs. 25.3 and 25.4 show the trapezoidal fuzzy sets for input variable and crisp values for the output variables of the initial controller.

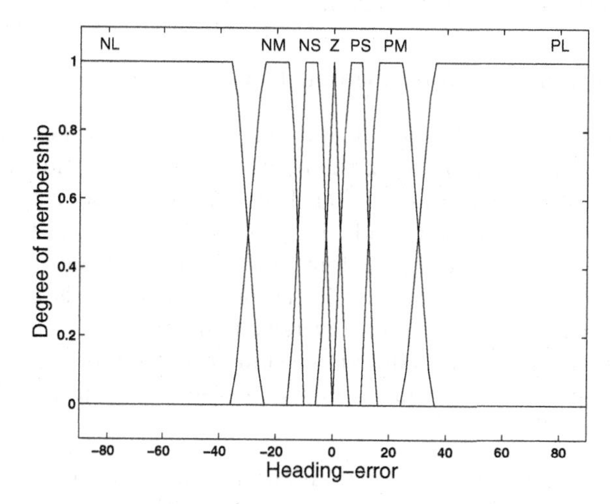

Fig. 25.3. Heading-error input δ

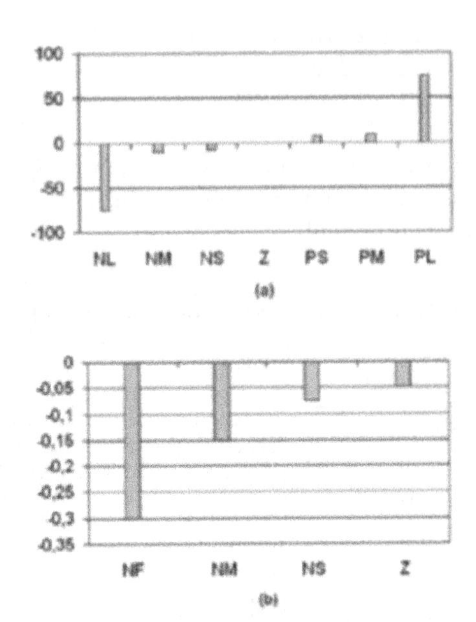

Fig. 25.4. Control outputs: (a) angular velocity ω; (b) linear velocity ν

25.3.2 Initial Population

The initial population is composed by controllers that are generated by means of random variations in their domain.

25.3.3 Selection and Generational Replacement

The selection scheme and generational replacement is based on the work presented in [16]. In each iteration of the multi-objective evolutionary algorithm, two individuals are picked at random from the population. These individuals are crossed and mutated producing two offspring. After, the best of the first offspring replaces the first parent, and the best of the second offspring replaces to the second parent only if one of the offspring is better than the parent. Note that the population diversity is maintained in populations because an offspring replaces an individual similar to itself (one of their parents). In order to determine if an individual is better than another, the following criteria are established:

- A feasible individual is better than another unfeasible one.
- One unfeasible individual x is better than another one x' if:

$$\max_{i=1,\dots,m}\left\{\frac{C_i(x)-l_i}{maxpopC_i-l_i}\right\} < \max_{i=1,\dots,m}\left\{\frac{C_i(x')-l_i}{maxpopC_i-l_i}\right\}$$

where $C_i(x) \leq l_i$, $i = 1,\ldots,m$ are constraints imposed to the fuzzy models, and $maxpopC$ is the maximum value of C_i in the current population used for normalization.

- One feasible individual is better than another one if the first one dominates the second one (Pareto dominance relation).

25.3.4 Variation Operators

In order to achieve an appropriate exploitation and exploration of the potential solutions in the search space, we apply three types of crossover operators: uniform, arithmetic and BLX-α [10] for parameter variation. Two types of the mutation operators are applied: uniform and non-uniform mutation [22].

25.3.5 Identification of Objectives and Constraints

An important issue of MOEA design is the identification of multiple objectives and constraints. Generally, we need to optimize several objectives that often conflict among them. In other words, a trade-off exists between the objectives, where improvement in one objective cannot be achieved without detriment to another. Moreover, we need to satisfy some specific constraints.

These are the objectives and/or constraints to minimize:

- **Objective/Constraint 1 - Execution time (N):** it is the required time (cycles) in order to carry out the docking maneuver. It is equivalent to the number of measures of the heading-error in the trajectory.

- **Objective/Constraint 2 - Root mean square orientation tracking error in the entire trajectory:** corresponds to the total orientation error in the entire trajectory and is calculated as follows:

$$RMSEorientationtrackingtotal = \sqrt{\sum_{1}^{N} \delta^2}$$

where δ is the heading-error in each iteration and N is the number of iterations (cycles, execution time).

- **Objective 3 - Root mean square orientation tracking error at the end of the trajectory:** corresponds to the total orientation error at the end of the trajectory and is calculated as follows:

$$RMSEorientationtrackingfinal = \sqrt{\sum_{N-L}^{N} \delta^2}$$

where L is the number of iterations that are considered at the end of the trajectory.

- **Constraint 4 - Root mean square error of control effort in the entire trajectory:** represents the variation of the angular velocity (one output of controller). The goal is to minimize the control effort by reducing the oscillations due to abrupt changes in the angular velocity. This variation is calculated as follows:

$$RMSEcontroleffort = \sqrt{\sum_{1}^{N} (\omega - \omega')^2}$$

  where ω is the current angular velocity and ω' is the previous velocity (this is the desired value in order to achieve a minimum variation).
- **Objective/Constraint 5 - Root mean square position tracking error:** it is the distance from current robot position to the reference path. This position tracking error is calculated for the entire trajectory as follows:

$$RMSEpositiontracking = \sqrt{\sum_{1}^{N} distance^2}$$

- **Objective/Constraint 6 - Orientation error goal:** it is the difference between the final orientation (angle) of robot and the goal point orientation (docking point).

Note that the objective 3 is included in the objective/constraint 2 but this one is used to avoid solutions with large heading errors in the entire trajectory. In fact, it is important that the mobile robot follows up the trajectory in an accurate way in last steps while it is reaching the docking point, so it is considered errors in final actions with the purpose of reducing them to minimum value or zero.

25.4 Experiments and Results

ThinkingCap-II (TC-II) is a framework for developing mobile robot applications. It is a joint effort between the University of Murcia (Spain) and the University of Örebro (Sweden), and it is based on previous work on *Thinking-Cap* and *BGA* architectures [26, 27]. The framework consists on a reference cognitive architecture (largely based on *ThinkingCap*) that serves as a guide for making the functional decomposition of a robotics system, a software architecture (partially based on *BGA*) that allows an uniform and reusable way of organizing software components for robotics applications, and a communication infrastructure that allows software modules to communicate in a common way independently of whether they are local or remote. In our application, the mobile robot is an Automated Guided Vehicle (AGV), in particular, it is an industrial fork-lift truck that performs docking maneuvers to load/unload pallets in conveyor belts. In TC-II platform, we work with that vehicle, defining the concrete robot corresponding to the autonomous truck and specifying a plane of the industrial plant in which the way-points and docking-points are placed. In order to perform the learning process, several simulations have been executed with the aim of evaluating the performance of evolved fuzzy controllers after docking maneuvers. During the learning process, each controller is evaluated by means of collecting the required data from several simulations in order to calculate the objectives and constraints, using the average values of all simulations.

In our experiment, we decided upon using the following parameters:

- Population size = 50 individuals
- Crossover probability = 0.6
- Mutation probability = 0.4
- Number of iterations at the end (L) = 50

Constraints limits are:

- Constraint 1 = 300
- Constraint 2 = 340
- Constraint 3 = 10
- Constraint 4 = 2.5
- Constraint 5 = 0.07

According to the variation operators, each type of operator is applied with the same probability in the corresponding category.

After learning process, we obtained a set of non-dominated and feasible solutions and subsequently, we can make a decision process in order to select one or more compromise solutions. In order to compare the solutions, we used the controller constructed manually (the initial one). The main purpose is to reduce the orientation and position tracking errors and time in the task, but it is necessary to consider the control effort to avoid unstable maneuvers by means of gradual alterations of output control velocity.

Table 25.2 summarizes the obtained results, showing the values of objectives or constraints (O/C = objective and constraint at the same time; RMSE = root mean square error) for each learned controller. These fuzzy controllers are non-dominated and feasible according to the specified constraints. In this way, they achieve better path tracking performance than the initial controller (constructed manually), performing the docking maneuvers with larger precision, but they require some more time.

Figures. 25.5 and 25.6 show the performance comparison between the initial controller and a learned controller selected from solutions which are found by the evolutionary algorithm (corresponds to the controller number 15 in Table 25.2). Each one of graphs depicts two curves: the first one with dotted line corresponds to the initial controller; and the second one with solid line corresponds to the learned (or evolved) controller. In Fig. 25.5, respects to the

Table 25.2. Results summary

Controller	O/C Time (cycles)	O/C RMSE Orientation Tracking	O RMSE Orientation Tracking Final	C RMSE Control Effort	O/C RMSE Position Tracking	O/C Orientation Error Goal
Initial	190	353.97	0.78	40.05	3.75	0.07
1	227	330.48	0.75	9.26	1.70	0.048
2	242	333.68	0.51	8.75	1.88	0.068
3	250	334.56	0.63	9.01	1.79	0.055
4	253	337.76	0.75	8.80	1.85	0.021
5	254	323.07	0.90	9.73	1.70	0.059
6	255	332.19	0.66	7.02	1.56	0.043
7	259	330.65	0.75	8.43	1.48	0.057
8	259	332.86	0.65	9.16	1.67	0.068
9	264	336.65	0.58	8.76	1.71	0.042
10	264	322.56	0.88	9.23	1.70	0.043
11	265	328.60	0.69	6.86	1.54	0.050
12	267	339.23	0.47	6.95	1.72	0.064
13	269	332.09	0.51	8.36	1.81	0.040
14	271	336.79	0.58	9.68	1.97	0.034
15	271	338.19	0.32	9.49	1.11	0.026
16	271	330.04	0.67	6.96	1.61	0.064
17	272	331.58	0.75	7.54	1.58	0.034
18	273	330.95	0.62	8.89	1.84	0.049
19	273	324.38	0.68	9.69	1.76	0.052
20	274	336.41	0.60	7.72	1.67	0.069
21	274	332.77	0.63	8.15	1.68	0.052
22	274	329.79	0.56	7.22	1.44	0.040
23	275	320.52	0.52	9.35	1.59	0.052
24	277	333.51	0.51	5.46	1.36	0.065
25	280	338.00	0.53	7.64	1.71	0.034
26	281	337.68	0.43	7.59	1.69	0.051
27	283	336.13	0.46	7.90	1.67	0.047
28	288	329.14	0.57	8.85	1.64	0.040
29	289	337.50	0.59	5.22	1.27	0.043
30	292	331.44	0.85	8.91	1.60	0.019
31	292	325.34	0.58	6.68	1.43	0.049
32	294	328.25	0.77	6.19	1.53	0.042
33	294	328.23	0.55	9.77	1.78	0.022
34	294	336.38	0.65	6.49	1.49	0.034

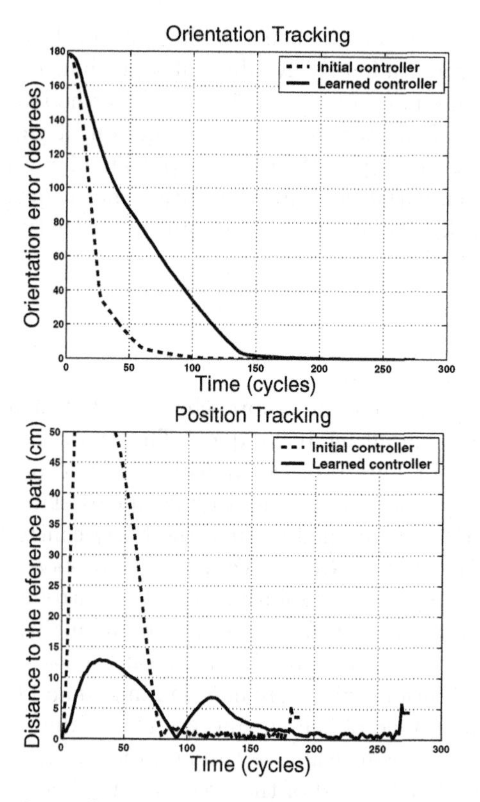

Fig. 25.5. Orientation and position tracking errors

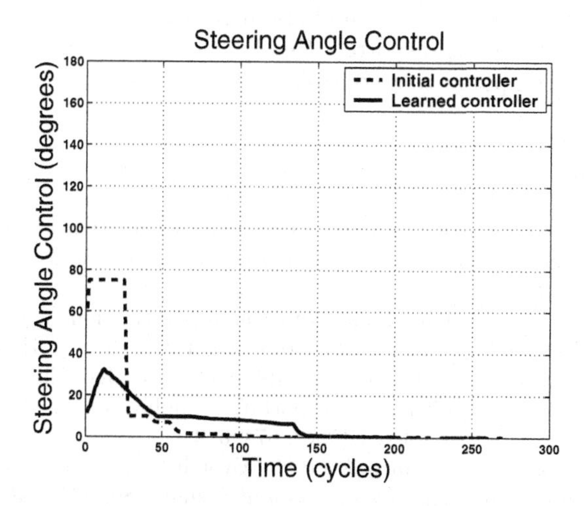

Fig. 25.6. Steering angle control

orientation tracking graph, we can see that the evolved controller reduces the orientation error gradually, avoiding abrupt changes in the control and reaching a proper final orientation of the vehicle, although a little longer time is required. In Fig. 25.6, we can see the evolution of steering angle control along the path that corresponds to the output angular velocity. The evolved controller has lower variation than the initial one, hence, the control effort to perform the task is reduced, avoiding abrupt velocity changes to achieve stable maneuvers.

25.5 Conclusions

In this chapter, we present an approach to path tracking controller tuning based on soft computing techniques. This work remarks some results in the combination of Pareto-based multi-objective evolutionary algorithms (MOEAs), handling constraints and tuning and design of fuzzy logic controllers. The MOEA was applied to tune the fuzzy controller for the docking maneuver of an Automated Ground Vehicle. The proposed evolutionary algorithm allows obtaining a set of solutions for the controller designer, reducing the hard work to design or optimize controllers manually. Moreover, multiple objectives and constraints have been handled at the same time with the purpose of obtaining solutions which are non-dominated and feasible according to specified constraints. One of the advantages is to reduce complexity because we need a single run of algorithm for tuning process. Hence, human intervention is only required at the end of the execution to choose one of the multiple solutions found by the MOEA. Objectives are often in conflict, in our case, the designer could select a compromise fuzzy controller that requires more time but with a larger accuracy and stability, for example, in cases in which fragile merchandise are transported. Finally, the performance of learned fuzzy controllers was compared to a controller obtained manually.

References

[1] Arkin RC, MacKenzie D (1994) Temporal coordination of perceptual algorithms for mobile robot navigation. *IEEE Trans. on Robotics and Automation*, 10(3): 276–286

[2] Arkin RC, Murphy RR (1990) Autonomous navigation in a manufacturing environment. *IEEE Trans. on Robotics and Automation*, 6(4): 445–454

[3] Bonissone P, Khedkar P, Chen Y (1996) Genetic algorithms for automated tuning of fuzzy controllers: a transportation application. In: *IEEE Conference on Fuzzy Systems (FUZZ-IEEE'96)*, 674–680

[4] Carse B, Fogarty T, Munro A (1996) Evolving fuzzy rule based controllers using genetic algorithms. *Fuzzy Sets and Systems*, 80: 273–293

[5] Coello CAC (2002) Theorical and numerical constraint handling techniques used with evolutionary algorithms: A survey of the state of art. *Computer Methods in Applied Mechanics and Engineering*, 191(11-12):1245–1287

[6] Coello CAC, Van Veldhuizen DA, Lamont GB (2002) Evolutionary algorithms for solving multi-Objective problems. Kluwer Academic Publishers, New York
[7] Coulter RC (1992) Implementation of the Pure Pursuit Path Tracking Algorithm. *Technical Report*, Robotics Institute, Carnegie Mellon University
[8] Deb K (2001) Multi-objective optimization using evolutionary algorithms. John Wiley & Sons
[9] Deb K, Pratap A, Meyarivan T (2001) Constrained test problems for multi-objective evolutionary optimization. *Lectures Notes in Computer Science*, 1993:284–298
[10] Eshelman LJ, Schaffer JD (1993) Real-coded genetic algorithm and interval schemata. In: *Foundations of Genetic Algorithms II*, Darrell Whitley L (eds), Morgan Kaufmann, 187–202
[11] Fonseca CM, Fleming PJ (1995) An overview of evolutionary algorithms in multiobjective optimization. *Evolutionary Computation*, 3(1):1–16
[12] Hoffmann F (2001) Evolutionary algorithms for fuzzy control system design. *Proceedings of IEEE*, 89(9): 1318–1333
[13] Homaifar A, Battle D, Tunstel E (1999) Soft computing-based design and control for mobile robot path tracking. In: *Proceedings of IEEE on Computational Intelligence in Robotics and Automation (CIRA '99)*, 35–40
[14] Homaifar A, McCormick E (1995) Simultaneous design of membership functions and rule sets for fuzzy controllers using genetic algorithms. *IEEE Trans. on Fuzzy Systems*, 3(2): 129–138
[15] Horn J, Nafpliotis N, Goldberg DE (1994) A niched Pareto genetic algorithm for multi-objective optimization. In: *Proceedings of the 1st International Conference on Evolutionary Computation*, 82–87
[16] Jiménez F, Gómez Skarmeta AF, Roubos H, Babuska R (2001) Accurate, transparent, and compact fuzzy models for function approximation and dynamic modeling through multi-objective evolutionary optimization. *The First International Conference on Evolutionary Multi-Criterion Optimization*, 653–667, Springer-Verlag
[17] Jiménez F, Verdegay JL (1999) Evolutionary techniques for constrained optimization problems. In: H.J. Zimmermann (eds), *The 7th European Congress on Intelligent Techniques and Soft Computing (EUFIT'99)*, Aachen, Germany
[18] Karr CL, Gentry EJ (1993) Fuzzy control of pH using genetic algorithms. *IEEE Trans. on Fuzzy Systems*, 1(1): 46–53
[19] Mamdani EH (1974) Applications of fuzzy algorithms for control a simple dynamic plant. *Proceedings of IEEE*, 121(12): 1585–1588
[20] Mamdani EH, Assilian S (1975) An experiment in linguistic synthesis with fuzzy logic controller. *International Journal of Man-Machine Studies*, 7: 1–13
[21] Martínez-Barberá H, Cánovas JP, Zamora MA, Gómez-Skarmeta AF (2003) i-Fork: a flexible AGV system using topological and grid maps. In: *Proceedings of IEEE Conference on Robotics and Automation*, 2147–2152
[22] Michalewicz Z (1996) Genetic Algorithms + Data Structures = Evolution Programs. 3rd eds., Springer-Verlag, London
[23] Michalewicz Z, Schoenauer M (1996) Evolutionary algorithms for constrained parameter optimization problems. *Evolutionary Computation*, 4(1):1–32

[24] Murata T, Ishibuchi H (1995) MOGA: Multi-objective genetic algorithms. In: *Proceedings of the 2nd International Conference on Evolutionary Computing*, 289–294
[25] Ray T, Kang T, Chye S (2000) An evolutionary algorithm for constrained optimization. In: Whitley D, Goldberg D, Cantú-Paz E, Spector L, Parmee I, and Beyer HG (eds) *Proceedings of the Genetic and Evolutionary Computation Conference (GECCO'2000)*, 771–777. Morgan Kaufmann, San Francisco, California
[26] Saffiotti A (1997) The use of fuzzy logic for autonomous robot navigation. *Soft Computing*, 1(4):180–197
[27] Saffiotti A, Konolige K, Ruspini EH (1995) A multivaluted-logic approach to integrating planning and control. *Artificial Intelligence*, 76(1-2):481–526
[28] Schaffer JD (1985) Multi-objective optimization with vector evaluated genetic algorithms. In: *Proceedings of the 1st International Conference on Genetic Algorithms*, 93–100
[29] Smith SF (1980) A learning System based on genetic adaptive algorithms. *Master Thesis*, Department of Computer Science, University of Pittsburgh
[30] Surry P, Radcliffe N, Boyd I (1995) A multi-objective approach to constrained optimization of gas supply networks. In: Fogarty T (eds), In: *Proceedings of the AISB-95 Workshop on Evolutionary Computing*, 993: 166–180, Springer Verlag
[31] Zitzler E, Thiele L (1999) Multiobjective evolutionary algorithms: A comparative case study and the strength pareto approach. *IEEE Transactions on Evolutionary Computation*, 3(4): 257–271

26

A Multi-Objective Genetic Algorithm for Learning Linguistic Persistent Queries in Text Retrieval Environments*

María Luque[1], Oscar Cordón[2], and Enrique Herrera-Viedma[2]

[1] Dept. of Computer Science and N.A.
University of Córdoba. 14071 - Córdoba (Spain)
mluque@uco.es

[2] Dept. of Computer Science and A.I. E.T.S. de Ingeniería Informática
University of Granada. 18071 - Granada (Spain)
{ocordon,viedma}@decsai.ugr.es

Summary. Persistent queries are a specific kind of queries used in information retrieval systems to represent a user's long-term standing information need. These queries can present many different structures, being the "bag of words" that most commonly used. They can be sometimes formulated by the user, although this task is usually difficult for him and the persistent query is then automatically derived from a set of sample documents he provides.

In this work we aim at getting persistent queries with a more representative structure for text retrieval issues. To do so, we make use of soft computing tools: linguistic information is considered for weighting the terms of Boolean queries by means of ordinal linguistic values (linguistic queries), and multiobjective evolutionary algorithms are applied to build the linguistic persistent query. Experimental results will show how using an expressive linguistic information-based query structure and a proper learning process to derive it, we can get more flexible, comprehensible and expressive user profiles.

26.1 Introduction

Persistent queries (PQs) are useful tools for information retrieval system (IRS) users having a relatively specific information need remaining fixed during a certain time period [23, 38]. By the definition of these kinds of queries, the information filtering process can be put into effect by delivering interesting information to a user, thus getting the her or him permanently updated on topics the user is interested [26].

* This work was supported by the Spanish Ministerio de Ciencia y Tecnolog under projects TIC2003-07977 and TIC2003-00877, including FEDER fundings.

M. Luque et al.: *A Multi-Objective Genetic Algorithm for Learning Linguistic Persistent Queries in Text Retrieval Environments*, Studies in Computational Intelligence (SCI) **16**, 601–627 (2006)
www.springerlink.com

Although different structures can be used to represent a PQ, it is usually difficult for a user to formulate the query independent of its structure [23, 24, 38]. Therefore, explicit PQs automatically learned from a training set of documents by means of user's relevance feedback are normally considered in information routing systems.

Soft computing tools have demonstrated to be useful in the personalization of IRSs, providing them with flexibility and some kind of "intelligence". The latter is viewed as the capability of automatically adapting to a context or service based on implicit behavior and learning instead of explicit solicitation from users [14, 22, 39].

One of the ways to add flexibility to an IRS is to make it tolerant to uncertainty and imprecision — both inherent to the user-system interaction — which can be achieved by allowing a more natural expression of users' needs [39]. For example, some flexible query languages based on the application of fuzzy set theory have been proposed which make possible simple and approximate expressions of subjective information needs [6]. In this contribution, we will deal with linguistic queries, considering them to improve the representative power of classic Boolean ones when used as PQ structures.

On the other hand, the IRS self-adaptativeness can be tackled by the machine learning perspective of soft computing, put into effect by evolutionary algorithms [1], neural networks and Bayesian networks, among others. These techniques can be hybridized with the representative power of flexible query languages to get "intelligent" IRSs [14]. In particular, evolutionary algorithms has obtained promising results in IR [15]. We will consider the use of multiobjective evolutionary algorithms [13] to automatically derive several linguistic PQs representing the user's information needs in a single run.

The aim of this contribution is to propose the use of a new, more flexible query structure —the linguistic query— to appropriately represent PQs for text retrieval and to introduce an evolutionary learning process to explicitly derive PQs of this composition. The latter will be based on a multiobjective technique able to automatically generate several PQs with a different trade-off between precision and recall in a single run.

The proposed method will be validated in a simulated text retrieval environment considering seven different information needs extracted from the classic Cranfield collection. Its efficacy will be compared with user profiles derived by one of the state-of-the-art algorithms [23, 24].

This chapter is structured as follows. Section 26.2 is devoted to introduce the preliminaries, including the PQ framework as well as the main aspects of linguistic IRSs and multigranular linguistic information. Then, Section 26.3 presents an IRS based on multigranular linguistic information that accepts linguistic weighted queries. The multiobjective GA algorithm to construct linguistic PQs is described in Section 26.4. Section 26.5 presents the experiments developed to test it and the analysis of results, while the conclusions are pointed out in Section 26.6.

26.2 Preliminaries

26.2.1 Construction of Persistent Queries

A) Information Filtering and Persistent Queries

Information filtering refers to an information seeking process where the user is assumed to be searching for information addressing a specific long-term interest [26, 38].

In an information filtering system, the user's permanent information need is represented in the form of a "profile". The most common profile structure is the "bag of words", which is based on a set of keywords representing the user's interest. Many systems assume an implicit definition of the profile by the user, although this comes with the traditional human-computer interaction "vocabulary problem", involving the difficulty for the user to select the right words to communicate with the system. This is specially important in this case as the profile can neither be too broad —as in that case the information filtering system would retrieve so many non relevant documents— nor too specific —as much valuable information can be lost.

Due to this reason, machine learning techniques have been applied to construct "implicit profiles" [24, 38]. In this case, the profile is automatically learned by the system from a training set of documents provided by the user.

Belkin and Croft suggested that IR techniques can be successfully applied to information filtering [3]. This way, the profile can be represented as a query formulated by using any IR retrieval model [2], the so called PQ [23]. Besides, IR query formulation techniques such as relevance feedback [2] or inductive query by example [12] can be applied in information filtering.

B) Flexible Persistent Queries

As different query structures from different IR retrieval models can be used to represent a PQ, the obtaining of effective retrieval results depends on the user's ability to express his information needs in the form of a query both in information filtering and in IR. It has been shown that the user often does not have a clear picture of what he is looking for and can only represent his information need in vague and imprecise terms, which results in a situation known as fuzzy-querying [37].

Flexible query languages can help to solve this problem due to their capability of personalization. A flexible query language is a language that enables a simple and approximate expression of subjective information needs [39]. For example, different linguistic IR models that use a fuzzy linguistic approach [45] to model the weights of queries and the retrieval status value of documents have been proposed in the literature [7, 8, 9, 29, 30, 31, 33], as we will see in Section 26.2.2.

Therefore, the modeling of user profiles in the form of flexible PQs (in particular, of linguistic PQs) can help us to improve both the comprehensibility and the retrieval efficacy of the obtained PQs.

C) Inductive Query by Example of Persistent Queries

Inductive Query by Example (IQBE) [12] was proposed as "a process in which searchers provide sample documents and the algorithms induce the key concepts in order to find other relevant documents". It works by taking a set of relevant (and optionally, non relevant documents) provided by a user and applying an off-line machine learning process to automatically generate a query describing the user's needs from that set. The obtained query can then be run in other IRSs to obtain more relevant documents.

Hence, IQBE techniques can be directly applied to construct PQs for information filtering, as they work in the same way as explicit profile learning methods. In this contribution, we propose a new IQBE technique, based on a multiobjective genetic algorithm, to derive several flexible PQs with different retrieval efficacy trade-offs in a single run.

26.2.2 Linguistic Information Retrieval Systems

The main activity of an IRS is to gather pertinent archived documents that best satisfy the user queries [2]. IRSs consists of three components:

1.- *A Database:* which stores the documents and the representation of their information contents (index terms).

2.- *A Query Subsystem:* which allows users to formulate their queries by means of a query language.

3.- *An Evaluation Subsystem:* which evaluates the documents for a user query obtaining a Retrieval Status Value (RSV) for each document.

The query subsystem supports the user-IRS interaction, and therefore, it should be able to account for the imprecision and vagueness typical of human communication. This aspect may be modeled by means of introducing weights in the query language. Many authors have proposed weighted IRS models using Fuzzy Set Theory [5, 8, 10, 11, 34, 42]. Usually, they assume numeric weights (values in [0,1]). However, the use of query languages based on numeric weights forces the user to quantify qualitative concepts (such as "importance"), ignoring that many users are not able to provide their information needs precisely in a quantitative form but in a qualitative one.

In fact, it seems more natural to characterize the contents of desired documents by explicitly associating a linguistic descriptor to a term in a query, like "important" or "very important", instead of a numerical value. To this end, some fuzzy linguistic IRS models [7, 33] have been proposed using a fuzzy linguistic approach [45] to model query weights and document scores.

A useful fuzzy linguistic approach which allows us to reduce the complexity of the IRS design [30] is called the ordinal fuzzy linguistic approach [27]. In this approach, the query weights and document scores are ordered linguistic terms, as we will see in the next Section.

26.2.3 Multi-Granular Linguistic Information

An ordinal fuzzy linguistic approach is an approximate technique suited for dealing with qualitative aspects of problems, defined by considering a finite and totally ordered label set $S = s_i, i \in \{0, \ldots, \mathcal{T}\}$ in the usual sense ($s_i \geq s_j$ if $i \geq j$) and with odd cardinality (7 or 9 labels). The mid-term representing an assessment of "approximately 0.5" and the rest of the terms being placed symmetrically around it [4]. The semantics of the label set is established from its ordered structure by considering that each label for the pair $(s_i, s_{\mathcal{T}-i})$ is equally informative. In some approaches [27, 29, 30], the semantics is completed by assigning fuzzy numbers defined on the [0,1] interval to the labels. These membership functions (μ_{s_i}) are described by linear trapezoidal membership functions represented by the 4-tuple $(a_i, b_i, \alpha_i, \beta_i)$ (the first two parameters indicate the interval in which the membership value is 1.0; the third and fourth parameters indicate the left and right widths of the distribution). Furthermore, we require the following operators:

1) Negation: $\text{Neg}(s_i) = s_j$, $j = \mathcal{T} - i$.
2) Maximization: $\text{MAX}(s_i, s_j) = s_i$ if $s_i \geq s_j$.
3) Minimization: $\text{MIN}(s_i, s_j) = s_i$ if $s_i \leq s_j$.

In any linguistic approach, an important parameter to be determined is the *granularity of uncertainty*, i.e., the cardinality of the label set S used to express the linguistic information. The cardinality of S must be small enough so as not to impose useless precision levels to the users, and it must be rich enough in order to allow a discrimination of the assessments in a limited number of degrees.

On the other hand, according to the uncertainty degree that a user qualifying a phenomenon has on it, the label set chosen to provide the user's knowledge will have more or less terms. When different users have different uncertainty degrees on the phenomenon, several label sets with a different granularity of uncertainty are necessary. In the latter case, we need tools for the management of multi-granular linguistic information to model these situations [28].

26.3 The Multi-Granular Linguistic IRS

In this section we review our previous work on an IRS model that accepts linguistic weighted Boolean queries and provides linguistic RSVs expressed using multi-granular linguistic information [31]. Thus, it uses multi-granular linguistic weighted queries and multi-granular linguistic RSVs. Other important property of this IRS is that it models the Boolean operators in a flexible way by means of the OWA operators [44].

Before presenting our algorithm, we provide the basic assumptions in this work. We consider a set of documents D= $\{d_1, \ldots, d_m\}$ represented by means of index terms T= $\{t_1, \ldots, t_l\}$, which describe the subject content of the documents. A numeric indexing function $F : D \times T \to [0, 1]$ is defined, called *index term weighting*. F maps a given document d_j and a given index term t_i to a numeric weight between 0 and 1. Thus, $F(d_j, t_i)$ is a numerical weight that represents the degree of significance of t_i in d_j. $F(d_j, t_i) = 0$ implies that the document d_j is not at all about the concept(s) represented by the index term t_i and $F(d_j, t_i) = 1$ implies that the document d_j is perfectly represented by the concept(s) indicated by t_i. Using the numeric values in (0,1), F can weight index terms according to their significance in describing the content of a document in order to improve the document retrieval.

26.3.1 Multi-Granular Linguistic Weighted Queries

We assume that each query is expressed as a combination of the weighted index terms which are connected by the logical operators AND ($\wedge$), OR ($\vee$), and NOT ($\neg$), and weighted with ordinal linguistic values. Each term in a query can be simultaneously weighted by means of several weights [29, 30]. Particularly, a term of a query can be weighted by means of three weights associated with different semantics. In such a way, the system is able to support the specification of user's preferences.

Each term of a query can be weighted by means of three weights associated to the following semantics:

1. *Symmetrical threshold semantics* [30]. By associating threshold weights to terms in a query, the user is asking to retrieve all documents about the topics represented by such terms. A symmetric threshold semantics is a special threshold semantics which assumes that a user may employ presence weights or absence weights in the formulation of weighted queries. Then, it is symmetrical with respect to the mid threshold value, i.e., it presents the usual behavior for the threshold values which are on the right of the mid threshold value (presence weights), and the opposite behavior for the values which are on the left (absence weights or presence weights with low values).
2. *Relative importance semantics*. This semantics defines term weights as a measure of the relative importance of each term of a query with respect to the other ones. By associating relative importance weights to terms in a query, the user is asking to see all documents whose content represents to a higher degree the concepts associated to the most important terms than to the less important ones. In practice, this means that the user requires that the computation of the RSV of a document is dominated by the more heavily weighted terms.

3. *Quantitative semantics*. This semantics defines query weights as measures of the quantity of documents that users want to consider in the computation of the final set of documents retrieved for each query term. By associating quantitative weights with the terms in a query, the user is asking to see a set of retrieved documents in which the terms with a greater quantitative weight contribute with a higher number of pertinent documents.

As in [30], we use the linguistic variable *"Importance"* to model every semantics, but with different interpretations. For example, a query term t_i with a threshold weight of value "*High*" means that the user requires documents whose content t_i should have at least a high importance value. However, the same query term t_i with a quantitative weight of value "*High*" means that the user wants a set of documents in which the term t_i contributes with a higher number of pertinent documents; and the same query term t_i with an importance weight of value "*High*" means that the user requires that the meaning of t_i must have a high importance value in the computation of the set of retrieved documents. Therefore, the problem in such a model [30] is that different linguistic weights associated with a term are assessed on the same label set, S. To solve this problem, we propose to represent the linguistic weights using multi-granular linguistic information, i.e., assuming label sets with different cardinality and/or semantics to assess the weights associated with the three semantics, called S^1, S^2 and S^3, respectively.

Then, we assume that a query is any legitimate Boolean expression whose atomic components (atoms) are 4-tuples $< t_i, c_i^1, c_i^2, c_i^3 >$ belonging to the set, $T \times S^1 \times S^2 \times S^3$; $t_i \in T$, $c_i^1 \in S^1$ is a value of the linguistic variable *"Importance"* modeling the symmetrical threshold semantics, $c_i^2 \in S^2$ is a value of the linguistic variable *"Importance"* modeling the quantitative semantics, and $c_i^3 \in S^3$ is a value of the linguistic variable *"Importance"* modeling the relative importance semantics. Therefore, the set of legitimate Boolean queries is a set of multi-granular linguistic weighted queries Q which is defined by the following syntactic rules:

1. $\forall q =< t_i, c_i^1, c_i^2, c_i^3 > \in \mathrm{T} \times S^1 \times S^2 \times S^3 \rightarrow q \in \mathrm{Q}$.
   These queries are called atoms.
2. $\forall q, p \in \mathrm{Q} \rightarrow q \wedge p \in \mathrm{Q}$.
3. $\forall q, p \in \mathrm{Q} \rightarrow q \vee p \in \mathrm{Q}$.
4. $\forall q \in \mathrm{Q} \rightarrow \neg(q) \in \mathrm{Q}$.
5. Every legitimate Boolean query $q \in \mathrm{Q}$ can only be obtained by applying rules 1-4.

26.3.2 Evaluating Multi-Granular Linguistic Weighted Queries

Usually, evaluation methods for Boolean queries work by means of a constructive bottom-up process, i.e., in the query evaluation process, the atoms are evaluated first, then the Boolean combinations of the atoms, and so forth, working in a bottom-up fashion until the whole query is evaluated. Similarly, we propose a constructive bottom-up evaluation method to process the multi-granular linguistic weighted queries. This method evaluates documents in terms of their relevance to queries by supporting the three semantics associated with the query weights simultaneously and by managing the multi-granular linguistic weights satisfactorily. Furthermore, given that the concept of relevance is different from the concept of importance, we use a label set S' to provide the relevance values of documents, which is different from those used to express the queries (S^1, S^2 and S^3).

To manage the multi-granular linguistic weights of queries, we develop a procedure based on the multi-granular linguistic information management tool defined in [28]. This procedure acts making uniform the multi-granular linguistic information before processing queries. To do so, we have to choose a label set as the uniform representation base, called *basic linguistic term set (BLTS)*, and then we have to transform (under a transformation function) all multi-granular linguistic information into that unified label set BLTS. In our case, the choice of the BLTS is easy to perform. It must be the label set used to express the output of the IRS (relevance degrees of documents), i.e., BLTS=S'.

The method to evaluate a multi-granular linguistic weighted query is composed of the following six steps:

1.- Preprocessing of the query.

The user query is preprocessed to put it into either conjunctive normal form (CNF) or disjunctive normal form (DNF), in such a way that every Boolean subexpression must have more than two atoms. Weighted single-term queries are kept in their original forms.

2.- Evaluation of atoms with respect to the symmetrical threshold semantics.

According to a symmetrical threshold semantics, a user may search for documents with a minimally acceptable presence of one term in their representations, or documents with a maximally acceptable presence of one term in their representations [29, 30]. Then, when a user asks for documents in which the concept(s) represented by a term t_i is (are) with the value *High Importance*, he/she would not reject a document with an F value greater than *High*. On the contrary, when a user asks for documents in which the concept(s) represented by a term t_i is (are) with the value *Low Importance*, he/she would not reject a document with an F value less than *Low*. Given a request $< t_i, c_i^1, c_i^2, c_i^3 >$, this means that the query weights that imply the presence of a term in a document $c_i^1 \geq s^1_{T/2}$ (e.g. *High, Very High*) must be treated differently to the query weights that imply the absence of one term in a document $c_i^1 < s^1_{T/2}$ (e.g. *Low, Very Low*). Then, if $c_i^1 \geq s^1_{T/2}$, the request $< t_i, c_i^1, c_i^2, c_i^3 >$ is synonymous with the request $< t_i, at\ least\ c_i^1, c_i^2, c_i^3 >$, which expresses the fact that the desired documents are those having F values as high as possible; and if $c_i^1 < s^1_{T/2}$, the former request is synonymous with the request $< t_i, at\ most\ c_i^1, c_i^2, c_i^3 >$, which expresses the fact that the desired documents are those having F values as low as possible. This interpretation is defined by means of a parameterized linguistic matching function

$g^1 : D\times T \times S^1 \rightarrow S^1$ [29]. Given an atom $< t_i, c_i^1, c_i^2, c_i^3 >$ and a document $d_j \in D$, g^1 obtains the linguistic RSV of d_j, called $RSV_j^{i,1}$, by measuring how well the index term weight $F(d_j, t_i)$ satisfies the request expressed by the linguistic weight c_i^1 according to the following expression:

$$RSV_j^{i,1} = g^1(d_j, t_i, c_i^1) = \begin{cases} s^1_{min\{a+\mathcal{B},T\}} & \text{if } s^1_{T/2} \leq s^1_b \leq s^1_a \\ s^1_{max\{0,a-\mathcal{B}\}} & \text{if } s^1_{T/2} \leq s^1_b \text{ and } s^1_a < s^1_b \\ Neg(s^1_{Max\{0,a-\mathcal{B}\}}) & \text{if } s^1_a \leq s^1_b < s^1_{T/2} \\ Neg(s^1_{Min\{a+\mathcal{B},T\}}) & \text{if } s^1_b < s^1_{T/2} \text{ and } s^1_b < s^1_a \end{cases} \tag{26.1}$$

such that, (i) $s_b^1 = c_i^1$; (ii) s_a^1 is the linguistic index term weight obtained as $s_a^1 = Label(F(d_j, t_i))$, being $Label : [0, 1] \rightarrow S^1$ a function that assigns a label in S^1 to a numeric value $r \in [0, 1]$ according to the following expression:

$$Label(r) = Sup_q\{s_q^1 \in S^1 : \mu_{s_q^1}(r) = Sup_v\{\mu_{s_v^1}(r)\}\}; \tag{26.2}$$

and (iii) $\mathcal{B}$ is a bonus value that rewards/penalizes the value $RSV_j^{i,1}$ for the satisfaction/dissatisfaction of request $< t_i, c_i^1, c_i^2, c_i^3 >$, which can be defined in an independent way, for example as $\mathcal{B} = 1$, or depending on the closeness between $Label(F(d_j, t_i))$ and c_i^1, for example as $\mathcal{B} = round(\frac{2(|b-a|)}{T})$.

3.- Evaluation of atoms with respect to the quantitative semantics.

In this step, documents is evaluated with regard to their relevance to individual atoms of the query, the restrictions imposed by the quantitative semantics are considered.

The linguistic quantitative weights are interpreted as follows: when a user establishes a certain number of documents for a term in the query, expressed by a linguistic quantitative weight, then the set of documents to be retrieved must have the minimum number of documents that satisfies the compatibility or the membership function associated with the meaning of the label used as linguistic quantitative weight. Furthermore, these documents must be those that better satisfy the threshold restrictions imposed on the term.

Therefore, given an atom $< t_i, c_i^1, c_i^2, c_i^3 >$ and assuming that $RSV_j^{i,1} \in S^1$ represents the evaluation according to the symmetrical threshold semantics for d_j, we model the interpretation of a quantitative semantics by means of a linguistic matching function, called g^2, which is defined between the $RSV_j^{i,1}$ and the linguistic quantitative weight $c_i^2 \in S^2$. Then, the evaluation of the atom $< t_i, c_i^1, c_i^2, c_i^3 >$ with respect to the quantitative semantics associated with c_i^2 for a document d_j, called $RSV_j^{i,1,2} \in S^1$, is obtained by means of the linguistic matching function $g^2 : D \times S^1 \times S^2 \rightarrow S^1$ as follows

$$RSV_j^{i,1,2} = g^2(RSV_j^{i,1}, c_i^2, d_j) = \begin{cases} s_0^1 & \text{if } d_j \notin \mathcal{B}^S \\ RSV_j^{i,1} & \text{if } d_j \in \mathcal{B}^S \end{cases} \tag{26.3}$$

where $\mathcal{B}^S$ is the set of documents such that $\mathcal{B}^S \subseteq Supp(\mathcal{M})$ where,

$$\mathcal{M} = \{(d_1, RSV_1^{i,1}), \ldots, (d_m, RSV_m^{i,1})\} \tag{26.4}$$

is a fuzzy subset of documents obtained according to the following algorithm:

1. $K = \#Supp(\mathcal{M})$
2. REPEAT
   $M^K = \{s_q \in S : \mu_{s_q}(K/m) = Sup_v\{\mu_{s_v}(K/m)\}\}$.
   $s^K = Sup_q\{s_q \in M^K\}$.
   $K = K - 1$.
3. UNTIL $((c_i^2 \in M^{K+1})$ OR $(c_i^2 \geq s^{K+1}))$.
4. $\mathcal{B}^S = \{d_{\sigma(1)}, \ldots, d_{\sigma(K+1)}\}$, such that $RSV^{i,1}_{\sigma(h)} \leq RSV^{i,1}_{\sigma(l)}, \forall l \leq h$.

According to g^2, the application of the quantitative semantics consists of reducing the number of documents to be considered by the evaluation subsystem for t_i in the later steps.

4.- Evaluation of subexpressions and modeling of the relative importance semantics

We argue that the relative importance semantics in a single-term query has no meaning. Then, in this step we have to evaluate the relevance of documents with respect to the subexpressions of queries composed of two atomic components.

Given a subexpression q_v with $\mathcal{I} \geq 2$ atoms, we know that each document d_j presents a partial $RSV_j^{i,1,2} \in S^1$ with respect to each atom $< t_i, c_i^1, c_i^2, c_i^3 >$ of q_v. Then, the evaluation of the relevance of a document d_j with respect to the whole subexpression q_v implies the aggregation of the partial relevance degrees $\{RSV_j^{i,1,2},\ i = 1, \ldots, \mathcal{I}\}$ weighted by means of the respective relative importance degrees $\{c_i^3 \in S^3,\ i = 1, \ldots, \mathcal{I}\}$. Therefore, as $S^1 \neq S^3$, we have to develop an aggregation procedure of multi-granular linguistic information. As said, to do so, we first choose a label set BLTS to make linguistic information uniform. In this case, BLTS=S' which is used to assess RSVs (relevance degrees of documents). Then, each linguistic information value is transformed into S' by means of the following transformation function:

Definition: *Let $A = \{l_0, \ldots, l_p\}$ and $S' = \{s'_0, \ldots, s'_m\}$ be two label sets, such that $m \geq p$. Then, a multi-granularity transformation function, $\tau_{AS'}$ is defined as $\tau_{AS'} : A \longrightarrow \mathcal{F}(S')$*

$$\tau_{AS_T}(l_i) = \{(s'_k, \alpha_k^i)\ /k \in \{0, \ldots, m\}\},$$
$$\forall l_i \in A,\ \alpha_k^i = \max_y \min\{\mu_{l_i}(y), \mu_{s'_k}(y)\}, \tag{26.5}$$

where $\mathcal{F}(S')$ is the set of fuzzy sets defined in S', and $\mu_{l_i}(y)$ and $\mu_{s'_k}(y)$ are the membership functions of the fuzzy sets associated to the terms l_i and s'_k, respectively [28].

Therefore, the result of $\tau_{AS'}$ for any linguistic value of A is a fuzzy set defined in the BLTS, S'. Using the multi-granularity transformation functions $\tau_{S^1S'}$ and $\tau_{S^3S'}$, we transform the linguistic values $\{RSV_j^{i,1,2} \in S^1,\ i = 1, \ldots, \mathcal{I}\}$ and $\{c_i^3 \in S^3,\ i = 1, \ldots, \mathcal{I}\}$ into S', respectively. Therefore, the values $RSV_j^{i,1,2}$ and c_i^3 are represented as fuzzy sets defined on S' characterized by the following expressions:

1. $\tau_{S^1S'}(RSV_j^{i,1,2}) = [(s'_0, \alpha_0^{ij}), \ldots, (s'_m, \alpha_m^{ij})]$, and
2. $\tau_{S^2S'}(c_i^3) = [(s'_0, \alpha_0^i), \ldots, (s'_m, \alpha_m^i)]$, respectively.

In each subexpression q_v, we find that the atoms can be combined using the AND or OR Boolean connectives, depending on the normal form of the user query. The restrictions imposed by the relative importance weights must be applied in the aggregation operators used to model both connectives. These aggregation operators should guarantee that the more important the query terms, the more influential they are in the determination of the RSVs. To do so, these aggregation operators must carry out two activities [27]: i) the transformation of the weighted information under the importance degrees by means of a transformation function h; and ii) the aggregation of the transformed weighted information by means of an aggregation operator of non-weighted information f. As it is known, the choice of h depends upon f. In [43], Yager discussed the effect of the importance degrees on the MAX (used to model the connective OR) and MIN (used to model the connective AND) types of aggregation and suggested a class of functions for importance transformation in both types of aggregation. For the MIN aggregation, he suggested a family of t-conorms acting on the weighted information and the negation of the importance degree, which presents the non-increasing monotonic property in these importance degrees. For the MAX aggregation, he suggested a family of t-norms acting on weighted information and the importance degree, which presents the non-decreasing monotonic property in these importance degrees.

Following the ideas shown above, we use the OWA operators ϕ^1 (with orness(W)≤ 0.5) and ϕ^2 (with orness(W)> 0.5) to model the AND and OR connectives, respectively. Hence, when $h = \phi^1$, $f = \max(Neg(weight), value)$, and when $h = \phi^2$, $f = min(weight, value)$.

Then, given a document d_j, we evaluate its relevance with respect to a subexpression q_v, called RSV_j^v, as $RSV_j^v = [(s'_0, \alpha_0^v), \ldots, (s'_m, \alpha_m^v)]$, where

1. if q_v is a conjunctive subexpression then

$$\alpha_k^v = \phi^1(max((1-\alpha_k^1), \alpha_k^{1j}), \ldots, max((1-\alpha_k^{\mathcal{I}}), \alpha_k^{\mathcal{I}j})) \qquad (26.6)$$

2. if q_v is a disjunctive subexpression then

$$\alpha_k^v = \phi^2(min(\alpha_k^1, \alpha_k^{1j}), \ldots, min(\alpha_k^{\mathcal{I}}, \alpha_k^{\mathcal{I}j})). \qquad (26.7)$$

5.- Evaluation of the whole query. In this step, the final evaluation of each document is achieved by combining their evaluations with respect to all the subexpressions using, again, the OWA operators ϕ^1 and ϕ^2 to model the AND and OR connectives, respectively.

Then, given a document d_j, we evaluate its relevance with respect to a query q as $RSV_j = \{(s'_0, \beta_0^j), \ldots, (s'_m, \beta_m^j)\}$, where $\beta_k^j = \phi^1(\alpha_k^1, \ldots, \alpha_k^{\mathcal{V}})$, if q is in CNF, and $\beta_k^j = \phi^2(\alpha_k^1, \ldots, \alpha_k^{\mathcal{V}})$, if q is in DNF, with $\mathcal{V}$ standing for the number of subexpressions in q.

Remark: *On the NOT Operator.* We should note that, if a query is in CNF or DNF form, we have to define the negation operator only at the level of single atoms. This simplifies the definition of the NOT operator. As was done in [30], the evaluation of document d_j for a negated weighted atom $< \neg(t_i), c_i^1, c_i^2, c_i^3 >$ is obtained from the negation of the index term weight $F(t_i, d_j)$. This means to calculate g^1 from the linguistic value $Label(1 - F(t_i, d_j))$.

6.- Presenting the output of the IRS

At the end of the evaluation of a user query q, each document d_j is characterized by RSV_j which is a fuzzy set defined on S'. Of course, an answer of an IRS where the relevance of each document is expressed by means of a fuzzy set is not easy to understand, and neither to manage. To overcome this problem, we present the output of our IRS by means of ordered linguistic relevance classes, as in [29, 30]. Furthermore, in each relevance class we establish a ranking of the documents using a confidence degree associated to each document.

To do so, we calculate a label $s^j \in S'$ for each document d_j, which represents its linguistic relevance class. We design an easy linguistic approximation process in S' using a similarity measure, e.g., the Euclidean distance. Each label $s'_k \in S'$ is represented as a fuzzy set defined in S', i.e., $\{(s'_0, 0), \ldots, (s'_k, 1), \ldots, (s'_m, 0)\}$. Then, we calculate s^j as

$$s^j = MAX\{s'_l | Conf(s'_l, RSV_j) = min_k\{Conf(s'_k, RSV_j)\}\}, \tag{26.8}$$

where $Conf(s'_k, RSV_j) \in [0, 1]$ is the confidence degree associated to d_j defined as

$$Conf(s'_k, RSV_j) = \sqrt{\sum_{i=0}^{k-1} (\beta_i^j)^2 + (\beta_k^j - 1)^2 + \sum_{i=k+1}^{m} (\beta_i^j)^2}. \tag{26.9}$$

26.4 A Multiobjective Genetic Algorithm to Automatically Learn Linguistic Persistent Queries

In this section we present a multiobjective Genetic Algorithm (GA) for linguistic PQ learning. The queries to be derived are legitimate queries of the multigranular linguistic information-based IRS defined in Section 26.3, thus allowing us to design expressive user profiles. The next subsections are devoted to the description of the algorithm.

26.4.1 Coding Scheme

It can be seen how the linguistic query structure considered could be represented as an expression tree, whose terminal nodes are query terms and whose inner nodes are the Boolean operators AND, OR or NOT. Besides, each terminal node has associated three ordinal linguistic values corresponding to the

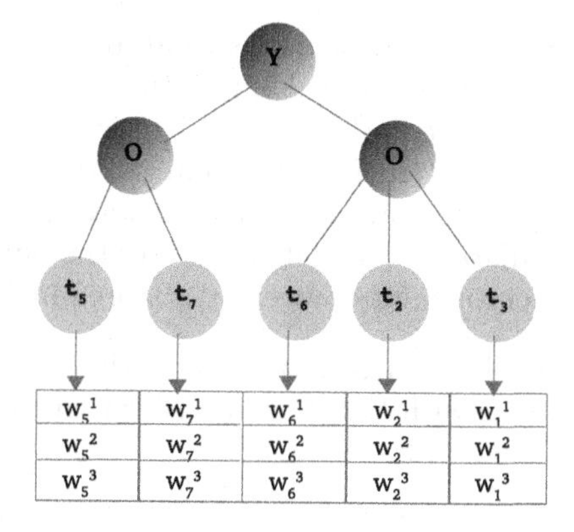

Fig. 26.1. Weighted Boolean query with linguistic weights

three semantics shown in the previous section. Figure 26.1 shows a graphical example of this kind of queries.

Hence, the natural representation would be to encode the query within a tree and to work with a Genetic Programming algorithm [32] to evolve it, as done by previous approaches devoted to the derivation of Boolean queries [16, 17, 40] or extended Boolean queries (fuzzy queries with numerical weights) [18, 19, 20, 21, 35].

However, the special characteristics of our linguistic queries allow us to deal with a much simpler representation. As seen in the previous section, the queries of our Multi-Granular Linguistic IRS are always on DNF or CNF. Hence, the query structure is not completely free but it is restricted to a disjunction of several conjunctions, or to a conjunction of several disjunctions, i.e., a fixed three-level tree structure with an OR (respectively, an AND) node in the root, several AND (respectively, OR) nodes in the second level, and the different (positive or negative) terms involved in each subexpression in the third level.

Thanks to this, we are able to design a coding scheme that represents linguistic expression trees as integers vectors, which can be represented using a usual GA chromosome. We should notice that a method of a similar coding scheme for a simpler Boolean query tree structure that does not consider term weights can be found in [25].

In our case, a chromosome C encoding a candidate linguistic PQ will be composed of two different parts, C_1 and C_2, which respectively encode the query composition (the AND-ed or OR-ed subexpressions), and the term weights. This structure presents the next features:

1. As said, every query is in CNF or DNF, so that the chromosome only encodes the subexpressions of the query (the operators are the same and there is no need to keep them in the query representation).
2. The query tree is encoded as an integer vector, where the number 0 acts a separator between subexpressions while the rest of numbers represent the

different index terms in the documentary database. Negative numbers are associated to negated terms in the query (for example, to represent the atomic expression $NOT\ t_{17}$, the number -17 is used).

3. The weights are represented as another integer vector, where each weight is encoded as its position in its label set. This way, the labels are numbered from 1 to the granularity of the label set, and the number corresponding to the selected label for the term weight is stored in the current vector.

To illustrate the coding scheme considered, the query of Figure 26.1 is encoded in a chromosome with the structure shown in Figure 26.2.

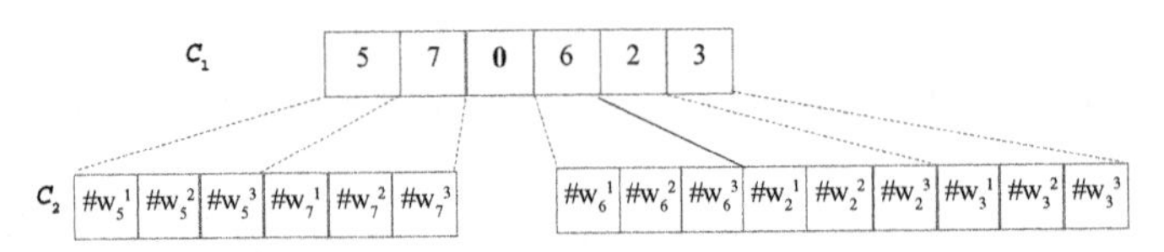

Fig. 26.2. Chromosome structure

26.4.2 Initial Gene Pool

All the individuals in the first population are randomly created, generating separately both parts of chromosomes:

1. Queries (C_1) will be composed of terms selected among those included in the set of relevant documents provided by the user, having those present in more documents a higher probability of being selected.
2. Weights (C_2) are randomly calculated, varying each gene in its respective definition interval: $\{1, ..., label\_set\_granularity\}$.

26.4.3 Fitness Function

The classical precision and recall criteria [2] —computed as shown in equation 26.10— are jointly maximized.

$$P = \frac{\sum_d r_d \cdot f_d}{\sum_d f_d} \quad ; \quad R = \frac{\sum_d r_d \cdot f_d}{\sum_d r_d} \tag{26.10}$$

where $r_d \in \{0, 1\}$ is the relevance of document d for the user, and $f_d \in \{0, 1\}$ is the retrieval of document d in the processing of the current query. Notice that both measures are defined in [0,1], where 1 is the best value that can be reached.

26.4.4 Pareto-based Multiobjective Selection Scheme

As said, the objective of our method is to automatically generate several linguistic PQs with a different trade-off between precision and recall in a single run. For this purpose, the SPEA scheme [47] has been employed as the multiobjective GA.

This algorithm introduces the elitism concept, explicitly maintaining an external population P_e. This population stores a fixed number of nondominated solutions which have been found since the start of the run. In each generation, the new nondominated solutions found are compared with the solutions in the existing external population, storing the resulting nondominated solutions on the latter. Furthermore, SPEA uses these elitist solutions, together with those in the current population, in the genetic operations, in the hope to lead the population to good areas in the search space.

The selection scheme involves the following steps:

1. The intermediate population is created from both the current population P and the external population (P_e) by means of binary tournament selection.
2. Genetic operators are used over the new individuals to get a new population (P).
3. Nondominated solutions existing in the new population are copied to the elitist population P_e.
4. The dominated and duplicated solutions are removed.

Therefore, the new elitist population is composed of the best nondominated solutions found so far, including new and old elitist solutions. To limit the growth of the elitist population, the size is restricted to a maximum number of solutions using clustering techniques (see [47] for details).

26.4.5 Genetic Operators

Due to the special nature of the chromosomes involved in this generation process (comprised by two different information levels), the design of genetic operators that is able to deal with it becomes a main task. As there exists a strong relationship between the two chromosome parts, operators working cooperatively in C_1 and C_2 are required in order to make best use of the representation considered. It can be clearly observed that the existing relationship will present several problems if not handled adequately. For example, modifications in the first chromosome part have to be automatically reflected in the second one. It makes no sense to modify the query structure, adding, deleting or changing terms and subexpressions, but continue working with the same weights. On the other hand, there is a need to develop the recombination in a correct way in order to obtain meaningful offsprings.

Taking into account these aspects, the following operators are going to be considered:

Crossover

Two different crossover operators are employed depending of the two parents' scope:

- *Crossover when both parents encode the same query (same C_1 part):* If this is the case, then the genetic search has located a promising space zone that has to be adequately exploited. This task is developed by applying a two-point crossover in C_2 (term weights), and obviously by maintaining the parent C_1 values in the offsprings.
- *Crossover when the parents encode different queries (different C_1 part):* This second case highly recommends the use of the information encoded by the parents for exploring the search space in order to discover new promising zones. In this way, an standard crossover operator is applied over both parts of the chromosomes. This operator performs as follows: a crossover point *cp* is randomly generated in C_1 for each parent and then, the genes between point *cp* and the end of C_1 are interchanged. In C_2, the crossover is developed in the same way, using the corresponding crossover points. The feasible points of crossover are the separators between subexpressions. Figure 26.3 shows an example of the crossover operator.

Mutation

Seven different operators are used, six of them acting on C_1 and one on C_2.

- *Mutation on C_1:* The mutation operators in C_1 are as follows:

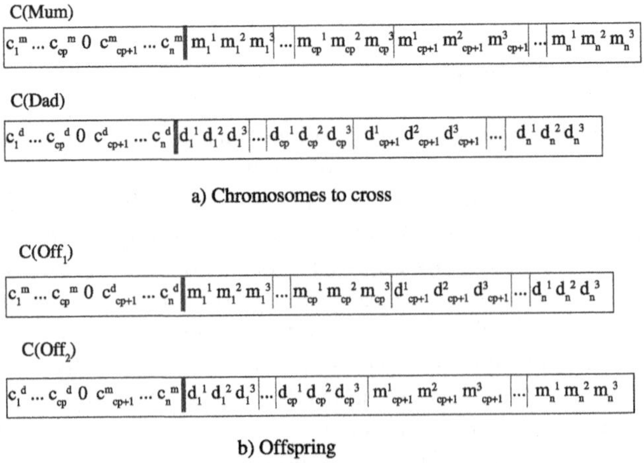

C(Mum)

$c_1^m \dots c_{cp}^m \; 0 \; c^m_{cp+1} \dots c_n^m \,|\, m_1^1 \, m_1^2 \, m_1^3 \,|\, \dots \,|\, m_{cp}^1 \, m_{cp}^2 \, m_{cp}^3 \,|\, m^1_{cp+1} \, m^2_{cp+1} \, m^3_{cp+1} \,|\, \dots \,|\, m_n^1 \, m_n^2 \, m_n^3$

C(Dad)

$c_1^d \dots c_{cp}^d \; 0 \; c^d_{cp+1} \dots c_n^d \,|\, d_1^1 \, d_1^2 \, d_1^3 \,|\, \dots \,|\, d_{cp}^1 \, d_{cp}^2 \, d_{cp}^3 \,|\, d^1_{cp+1} \, d^2_{cp+1} \, d^3_{cp+1} \,|\, \dots \,|\, d_n^1 \, d_n^2 \, d_n^3$

a) Chromosomes to cross

C(Off$_1$)

$c_1^m \dots c_{cp}^m \; 0 \; c^d_{cp+1} \dots c_n^d \,|\, m_1^1 \, m_1^2 \, m_1^3 \,|\, \dots \,|\, m_{cp}^1 \, m_{cp}^2 \, m_{cp}^3 \,|\, d^1_{cp+1} \, d^2_{cp+1} \, d^3_{cp+1} \,|\, \dots \,|\, d_n^1 \, d_n^2 \, d_n^3$

C(Off$_2$)

$c_1^d \dots c_{cp}^d \; 0 \; c^m_{cp+1} \dots c_n^m \,|\, d_1^1 \, d_1^2 \, d_1^3 \,|\, \dots \,|\, d_{cp}^1 \, d_{cp}^2 \, d_{cp}^3 \,|\, m^1_{cp+1} \, m^2_{cp+1} \, m^3_{cp+1} \,|\, \dots \,|\, m_n^1 \, m_n^2 \, m_n^3$

b) Offspring

Fig. 26.3. Explorative crossover

1. Replace selected term by a random one.
2. Negation of a term.
3. Deletion of a randomly selected separator.
4. Addition of a new separator in a valid random position.
5. Displacement of a separator.
6. Replace a subexpression by a randomly generated one with more or less terms (C_2 is automatically updated).

- *Mutation on C_2:* The mutation operator selected for C_2 is similar to the one proposed by Thrift in [41] for fuzzy rule base learning with GAs. When a mutation on a gene belonging to the second part of the chromosome is going to be performed, a local modification is developed by changing the current label to the immediately preceding or subsequent one (the decision is made at random). When the label to be changed is the first or last one in the label set, the only possible change is developed.

26.5 Experiments and Analysis of Results

26.5.1 Experiments Developed

This section is devoted to test the performance of the proposed MOGA-LPQ IQBE algorithm that automatically derives profiles represented as linguistic PQs. Since the most common profile structure is the *"bag of words"*, which is based on a set of weighted keywords representing the user's interest, we must compare our algorithm with classical methods of profile construction, in order to verify the performance. As comparison method, we have chosen one of the state-of-the-art algorithms (RSV-OKAPI) [23, 24], based on the vector space model and the probability theory [2]. We consider Robertson Selection Value (RSV) as the approach for profile learning and OKAPI BM25 as similarity function to match profiles and documents.

However, we should notice that this comparison is not fair to our algorithm due to two reasons. On the one hand, we are designing more expressive user profiles, what usually comes with a retrieval efficacy decrease (the usual interpretability-accuracy trade-off problem). On the other hand, our algorithm is able to derive several linguistic PQs (user profiles) with a different trade-off between precision and recall in a single run, thus giving more chances to the user to retrieve much/less relevant information with a larger/lesser retrieval noise at his own choice. So, the selection of a single PQ to compare it against the user profile derived by RSV-OKAPI will restrict the capabilities of our method.

The documentary database considered to design our experimental setup has been the popular *Cranfield* collection, composed of 1398 documents about Aeronautics [2]. It has been automatically indexed by first extracting the non-stop words, applying a stemming algorithm, thus obtaining a total number of 3857 different indexing terms, and then using a usual TFIDF indexing to generate the term weights in the document representations.

Among the 225 queries associated to the Cranfield collection, we have selected those presenting 20 or more relevant documents (queries 1, 2, 23, 73, 157, 220 and 225). The number of relevant documents associated to each of these seven queries are 29, 25, 33, 21, 40, 20 and 25, respectively. The relevance judgments associated to each of these selected queries have been considered to play the role of seven different user's information needs.

For each one of these queries, the documentary base has been randomly divided into two different, non overlapped, document sets, training and test, each of them composed of a fifty percent of the (previously known) relevant and irrelevant documents for the query.

MOGA-LPQ has been run five times with different initializations for each selected query during 50000 fitness function evaluations in a 2.4GHz Pentium IV computer with 1Gb of RAM. The parameter values considered are a population size of 100 individuals, an elitist population size of 25, a maximum of 10 terms by query, and 0.8 and 0.2 for the crossover and mutation probabilities (in both the C_1 and the C_2 parts). The retrieval threshold has been set to the third label of the label set.

The Pareto sets obtained in the five runs performed for each query have been put together, and the dominated solutions removed from the unified set. Then, five PQs well distributed on the Pareto front has been selected from each of the seven unified Pareto sets.

On the other hand, RSV-OKAPI has been run only one time, since it has no random components, with a profile size of 10 terms and with QTW (the term weights) equal to the RSV value.

Every selected linguistic PQ has been run on the corresponding test set once preprocessed[3] in order to evaluate their capability to retrieve relevant information for the user. The same has been done for the profile derived by RSV-OKAPI.

26.5.2 Analysis of the Pareto Sets Derived

Several quantitative metrics have been proposed in the literature to measure the quality of Pareto sets derived by multiobjective algorithms [46]. Specifi-

[3] As the index terms of the training and test documentary bases can be different, there is a need to translate training queries into test ones, removing those terms without a correspondence in the test set. Notice that, this is another source of retrieval efficacy loss for our method as two-term subexpressions where one of the terms is not present in the test document set are completely removed due to the restriction imposed on the linguistic query structure of not having single-term subexpressions (see Section 26.3). In the future, we aim at solving this problem, what would significantly improve the performance of our algorithm on the test set.

cally, we have used three different metrics; $\mathcal{M}_2^*$ and $\mathcal{M}_3^*$ and the number of nondominated solutions in the Pareto set.

Table 26.1 collects several data about the composition of the five Pareto sets generated for each query, where both the averaged value and its standard deviation are shown. From left to right, the columns contain the query number ($\#q$), the number of different non-dominated solutions obtained ($\#d$), corresponding to the number of different objective vectors (i.e., precision-recall pairs) existing among them, and the values of the two multiobjective EA metrics selected, $\mathcal{M}_2^*$ and $\mathcal{M}_3^*$, each of which is followed by their respective standard deviation values. Regarding the two later metrics, $\mathcal{M}_2^* \in [0, \#d]$ measures the diversity of the solutions found, while $\mathcal{M}_3^*$ measures the range to which the Pareto front spreads out in the objective values (in our case, the maximum possible value is $\sqrt{2} = 1.4142$). In both cases, the higher the value, the better the quality of the obtained Pareto set.

Table 26.1. Statistics of the Pareto sets obtained by the MOGA-LPQ algorithm

$\#q$	$\#d$	$\sigma_{\#d}$	M_2^*	$\sigma_{M_2^*}$	M_3^*	$\sigma_{M_3^*}$
1	7.600	0.669	3.573	0.311	1.209	0.021
2	6.200	0.522	2.836	0.251	1.175	0.031
23	10.400	0.607	4.701	0.277	**1.276**	0.008
73	5.800	0.522	2.900	0.261	1.137	0.043
157	**12.200**	0.716	**5.581**	0.307	1.229	0.020
220	5.000	0.283	2.500	0.141	1.127	0.034
225	7.000	0.566	3.163	0.244	1.201	0.016

In view of the values shown in Table 26.1, the Pareto fronts obtained are of good quality. We can see how all runs generate a number of linguistic PQs with different precision-recall trade-offs proportional to the number of relevant documents associated with the original query (for those cases where a larger number of relevant documents are provided, a larger number of different PQs are obtained in the Pareto sets); and that standard deviation values are around 0.6. The values of the $\mathcal{M}_2^*$ and $\mathcal{M}_3^*$ metrics are appropriate as well, showing a very good distribution of the Pareto fronts. We should emphasize the values of the latter, very close to 1.4142, the maximum possible value. This shows that the generated Pareto fronts cover a wide area in the space. To illustrate this fact, Figure 26.4 depicts the unified Pareto front obtained for query 157 as an example.

26.5.3 MOGA-LPQ *versus* RSV-OKAPI

To compare the considered algorithms, the average precision over eleven recall levels (P_{avg}) [2] has been taken as comparison measure as both can return the documents ordered according to their relevance. When this measure is used,

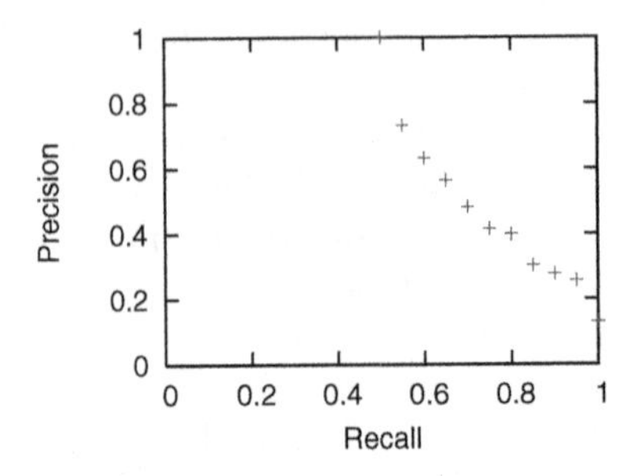

Fig. 26.4. Unified Pareto front generated for query 157

the retrieved document set is descendently ordered according to the RSV values so that the most relevant documents are at the top of the document set.

As said in Section 26.5.1, the five Pareto sets obtained by our method in the five runs performed for each of the seven queries are first unified, and then five different linguistic PQs well distributed on the resulting Pareto front are selected for each query. However, RSV-OKAPI only derives a solution per query as it is run a single time (it has no random components) and generates only one solution per run. In order to ease the comparison, we have considered the following two different performance values for each query for our algorithm:

1. Averaged results: For each Pareto set, we have chosen the PQ with the greater value of P_{avg} on the training document set and the five P_{avg} values obtained have been averaged.
2. Best result on the test set: Of the five selected PQs, we have chosen that with the greater value of P_{avg} on the test document set.

Figure 26.5 illustrates the process to obtain the best results on the test set.

Tables 26.2 and 26.3 show the retrieval efficacy for each query for MOGA-LPQ and RSV-OKAPI, respectively. In those tables, *#rel* stands for the number of relevant documents associated with that query, and *#top* for the number of relevant documents located in the *#rel* first positions of the document set (ordered by their RSV value).

The left side of Table 26.2 shows the average results. In view of them, we should notice the good results obtained by our method on the training document set: the average precision values are around 0.5, whereas the proportion of relevant documents in the first positions is around half the number of relevant documents for the query. However, on the test set, results are rather low compared to training results (we can talk about overlearning, that can be due to the preprocessing made to adapt the derived PQ to the test document collection, as mentioned before), with the P_{avg} values being around 0.1. This way, the access to new documents relevant for the user is not easy, since the user would need to examine a lot of documents before finding a useful one.

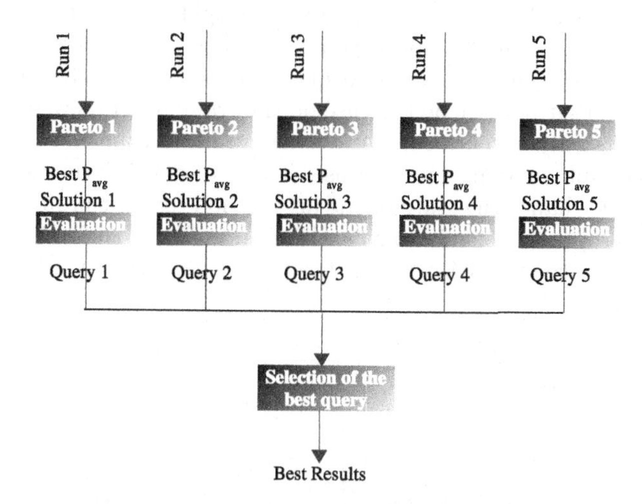

Fig. 26.5. Process to obtain the best results on the test set

Table 26.2. Retrieval Efficacy of MOGA-LPQ Linguistic PQs

Average Results							Best Results						
	Training set			Test set				Training set			Test set		
$\#q$	P_{avg}	$\#rel$	$\#top$	P_{avg}	$\#rel$	$\#top$	$\#q$	P_{avg}	$\#rel$	$\#top$	P_{avg}	$\#rel$	$\#top$
1	0.528	14	7.8	0.070	15	0.75	**1**	0.467	14	8	0.201	15	2
2	0.615	12	7.2	0.143	13	2.75	**2**	0.725	12	7	0.289	13	4
23	0.461	16	8.0	0.113	17	2.4	**23**	0.426	16	8	0.213	17	5
73	0.627	10	5.6	0.057	11	0.75	**73**	0.748	10	7	0.080	11	1
157	0.464	20	9.4	0.112	20	2.2	**157**	0.454	20	11	0.422	20	7
220	0.677	10	6.2	0.173	10	1.2	**220**	0.647	10	5	0.349	10	3
225	0.515	12	6.2	0.078	13	0.75	**225**	0.484	12	5	0.126	13	1

Table 26.3. Retrieval Efficacy of RSV-OKAPI Profiles

	Training set			Test set		
$\#q$	P_{avg}	$\#rel$	$\#top$	P_{avg}	$\#rel$	$\#top$
1	0.510	14	6	0.484	15	6
2	0.522	12	5	0.789	13	9
23	0.475	16	7	0.357	17	7
73	0.616	10	5	0.268	11	2
157	0.532	20	10	0.298	20	7
220	0.725	10	6	0.544	10	5
225	0.370	12	5	0.099	13	2

If we concentrate on the best results (the right half of Table 26.2), the training results are usually not better than those in the left side (but in queries 2 and 73). Since the selected query is the best on the test set, this does not mean that it must be the case on the training set as well. Of course, the best results fully outperforms the average results on the test set, generally around the double, making easy the user access to new information.

On the other hand, the state-of-the-art method for user profile generation shows an appropriate behavior, as can be seen in Table 26.3. P_{avg} values on the training set are over 0.4, whereas the test results are around 0.4.

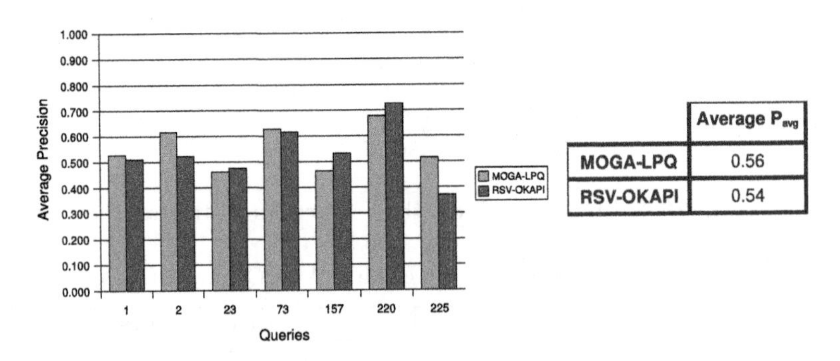

	Average P_{avg}
MOGA-LPQ	0.56
RSV-OKAPI	0.54

Fig. 26.6. Comparison between the mean P_{avg} values of MOGA-LPQ and RSV-OKAPI on the training document set

Figure 26.6 shows a comparative bar chart of the P_{avg} values got by each algorithm on the training document set (the MOGA-LPQ values correspond to the average results). Graphically, we can observe that our method outperforms RSV-OKAPI in four of the seven queries, apart from generating more expressive profiles as we shall see later. In fact, the global mean of the MOGA-LPQ P_{avg} values for the seven queries is slightly higher than that of RSV-OKAPI (0.56 in front of 0.54). Furthermore, the number of relevant documents in the first positions of the retrieved document set is very similar, as can be seen in Tables 26.2 and 26.3. However, notice again that, in each case, we are comparing the averaged results of five linguistic PQs derived by our MOGA-LPQ with that of the single user profile derived by RSV-OKAPI. Hence, if we would have chosen only the best of the five PQs in the training set performance for the comparison, it would have been much more positive for us.

Similarly, Figure 26.7 shows a comparative bar chart of the P_{avg} values got by each algorithm on the test document set (the MOGA-LPQ reported values correspond to the linguistic PQ with the best test results). In view of it, we should notice that RSV-OKAPI outperforms our IQBE method in five of the seven queries, and the differences is rather big, especially for query 2 (around 0.5). Numerically, the global mean of P_{avg} for both algorithms is 0.24

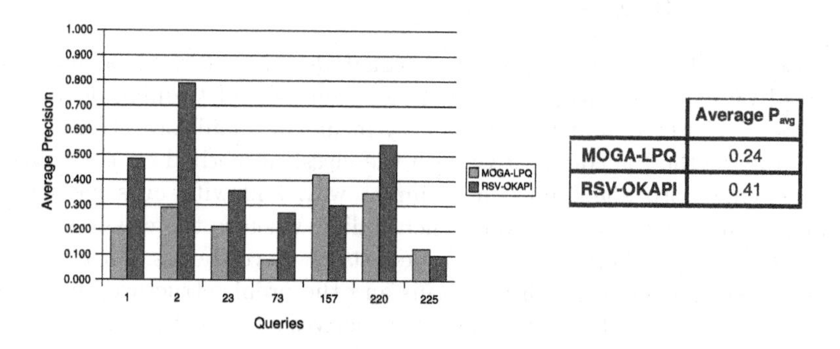

	Average P_{avg}
MOGA-LPQ	0.24
RSV-OKAPI	0.41

Fig. 26.7. Comparison between MOGA-LPQ (Best P_{avg}) and RSV-OKAPI on test document set

and 0.41 for MOGA-LPQ and RSV-OKAPI, respectively. With regards to the position of the relevant documents at the top of the document set, there is no meaningful difference between them. Therefore, the reason of the differences in the P_{avg} values is the position of the rest of the relevant documents.

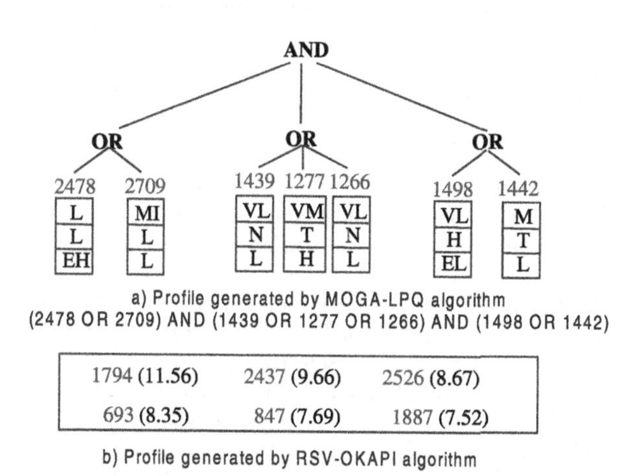

Fig. 26.8. User profiles generated by MOGA-LPQ and RSV-OKAPI for query 2

Nevertheless, although our algorithm achieves worse results than RSV-OKAPI on the test set, the modeling of user profiles as PQs clearly improves the expressivity of the user profiles when learning both the terms which is composed of the profile and its structure, instead of learning only the set of terms. As an example, Figure 26.8 shows the profiles derived by both algorithms for query 2.

26.6 Concluding Remarks

The use of soft computing tools to design PQs for text retrieval has been analyzed by constructing linguistic queries from sets of training documents extracted from the Cranfield collection by means of a multiobjective GA. As shown, our method is competitive with the classical method on the training set, behaving both algorithms in a similar way, and with ours having the advantage of a higher profile comprehensibility; whereas the classic state-of-the-art method gets better results on the test set. Nevertheless, our method is promising since it learns both the terms and the profile structure, instead of learning only the set of terms, as well as it derives a set of PQs with different precision-recall trade-off in a single run, although we must refine it with the aim of improving its capability to find new relevant information for the user.

In our opinion, many different future works arise from the present contribution. Firstly, we will search for new functions to measure the similarity between expression trees with the purpose of being able to work in the decision space[4]. On the other hand, we will try to improve our test set performance, either by designing a new preprocessing algorithm with a less aggressive way to adapt the linguistic PQs for the test document collection, or by modifying the validation process by using a division of the data set in training, validation and test document sets, as done in the last contributions proposed in the field [23, 24], instead of using the classic division in training and test sets.

References

[1] T. Bäck, D.B. Fogel, and Z. Michalewicz, editors. *Handbook of Evolutionary Computation.* IOP Publishing and Oxford University Press, 1997.

[2] R. Baeza-Yates and B. Ribeiro-Neto. *Modern Information Retrieval.* Adisson, 1999.

[3] N.J. Belkin and W.B. Croft. Information Filtering and Information Retrieval: Two Sides of the same Coin? *Communications of the ACM*, 35(12):29–38, 1992.

[4] P.P. Bonissone and K.S. Decker. Selecting Uncertainty Calculi and Granularity: An Experiment in Trading-off Precision and Complexity. In L.H. Kanal and J.F. Lemer, editors, *Uncertainty in Artificial Intelligence*, pages 217–247. North-Holland, 1986.

[5] A. Bookstein. Fuzzy Request: An Approach to Weighted Boolean Searches. *Journal of the American Society for Information Science*, 31:240–247, 1980.

[6] G. Bordogna, P. Carrara, and G. Pasi. Fuzzy Approaches to Extend Boolean Information Retrieval. In P. Bosc and J. Kacprzyk, editors, *Fuzziness in Database Management Systems*, pages 231–274. Springer-Verlag, 1995.

[4] We have worked in the objective space, but the number of solutions presenting different precision-recall values (different objective value arrays) is a little bit low with respect to the size of the elitist population. To solve this, we decided to work in the decision space, utilizing the *edit* or *Levenshtein distance* [36] to measure the similarity between expression trees, but although this measure increased the number of solutions, the run time also increased significantly.

[7] G. Bordogna and G. Pasi. A Fuzzy Linguistic Approach Generalizing Boolean Information Retrieval: A Model and its Evaluation. *Journal of the American Society for Information Science*, 44:70–82, 1993.
[8] G. Bordogna and G. Pasi. Linguistic Aggregation Operators of Selection Criteria in Fuzzy Information Retrieval. *International Journal of Intelligent Systems*, 10:233–248, 1995.
[9] G. Bordogna and G. Pasi. An Ordinal Information Retrieval Model. *International Journal of Uncertainty, Fuzziness and Knowledge-Based Systems*, 9(1):63–75, 2001.
[10] D. Buell and D.H. Kraft. A Model for a Weighted Retrieval System. *Journal of the American Society for Information Science*, 32:211–216, 1981.
[11] D. Buell and D.H. Kraft. Threshold Values and Boolean Retrieval Systems. *Information Processing & Management*, 17:127–136, 1981.
[12] H. Chen, G. Shankaranarayanan, L. She, and A. Iyer. A Machine Learning Approach to Inductive Query by Examples: An Experiment Using Relevance Feedback, ID3, Genetic Algoritms, and Simulated Annealing. *Journal of the American Society for Information Science*, 49(8):693–705, 1998.
[13] C. A. Coello, D. A. Van Veldhuizen, and G. B. Lamant. *Evolutionary Algorithms for Solving Multi-Objective Problems.* Kluwer Academic Publishers, 2002.
[14] O. Cordón and E. Herrera-Viedma. Editorial: Special Issue on Soft Computing Applications to Intelligent Information Retrieval on the Internet. *International Journal of Approximate Reasoning*, 34(2-3):89–95, 2003.
[15] O. Cordón, E. Herrera-Viedma, C. López-Pujalte, M. Luque, and C. Zarco. A Review of the Application of Evolutionary Computation to Information Retrieval. *International Journal of Approximate Reasoning*, 34:241–264, 2003.
[16] O. Cordón, E. Herrera-Viedma, and M. Luque. Evolutionary Learning of Boolean Queries by Multiobjective Genetic Programming. In *Lecture Notes in Computer Science 2439. Proc. of the PPSN-VII*, pages 710–719, Granada (Spain), 2002.
[17] O. Cordón, E. Herrera-Viedma, and M. Luque. Improving the Learning of Boolean Queries by means of a Multiobjective IQBE Evolutionary Algorithm. *Information Processing and Management*, 2005. To appear.
[18] O. Cordón, E. Herrera-Viedma, M. Luque, F. Moya, and C. Zarco. Analyzing the Performance of a Multiobjective GA-P Algorithm for Learning Fuzzy Queries in a Machine Learning Enviroment. In *Lecture Notes in Artificial Intelligence 2715. Proc. of the 10th IFSA World Congress*, pages 611–615, Istambul (Turkey), 2003.
[19] O. Cordón, F. Moya, and C. Zarco. A GA-P Algorithm to Automatically Formulate Extended Boolean Queries for a Fuzzy Information Retrieval System. *Mathware & Soft Computing*, 7(2-3):309–322, 2000.
[20] O. Cordón, F. Moya, and C. Zarco. A New Evolutionary Algorithm Combining Simulated Annealing and Genetic Programming for Relevance Feedback in Fuzzy Information Retrieval Systems. *Soft Computing*, 6(5):308–319, 2002.
[21] O. Cordón, F. Moya, and C. Zarco. Automatic Learning of Multiple Extended Boolean Queries by Multiobjective GA-P Algorithms. In V. Loia, M. Nikravesh, and L. A. Zadeh, editors, *Fuzzy Logic and the Internet*, pages 47–40. Springer, 2004.

[22] F. Crestani and G. Pasi, editors. *Soft Computing in Information Retrieval.* Physica-Verlag, 2000.

[23] W. Fan, M.D. Gordon, and P. Pathak. An Integrated Two-Stages Model for Intelligent Information Routing. *Decision Support Systems*, 2004. Submitted.

[24] W. Fan, M.D. Gordon, and P. Pathak. Effective Profiling of Consumer Information Retrieval Needs: A Unified Framework and Empirical Comparision. *Decision Support Systems*, 2005. To appear.

[25] J.L. Fernández-Villacañas and M. Shackleton. Investigation of the Importance of the Genotype-Phenotype Mapping in Information Retrieval. *Future Generation Computer Systems*, 19(1):55–68, 2003.

[26] U. Hanani, B. Shapira, and P. Shoval. Information Filtering: Overview of Issues, Research and Systems. *User Modeling and User-Adapted Interaction*, 11:203–259, 2001.

[27] F. Herrera and E. Herrera-Viedma. Aggregation Operators for Linguistic Weighted Information. *IEEE Transactions on Systems, Man and Cybernetics; Part A: Systems*, 27:646–656, 1997.

[28] F. Herrera, E. Herrera-Viedma, and L. Martínez. A Fusion Approach for Managing Multi-Granularity Linguistic Term Sets in Decision Making. *Fuzzy Sets and Systems*, 114:43–58, 2000.

[29] E. Herrera-Viedma. An Information Retrieval System with Ordinal Linguistic Weighted Queries based on Two Weighting Elements. *International Journal of Uncertainty, Fuzziness and Knowledge-Based Systems*, 9(1):77–88, 2001.

[30] E. Herrera-Viedma. Modeling the Retrieval Process for an Information Retrieval System using an Ordinal Fuzzy Linguistic Approach. *Journal of the American Society for Information Science and Technology*, 52(6):460–475, 2001.

[31] E. Herrera-Viedma, O. Cordón, M. Luque, A. G. López, and A. M. Muñoz. A Model of Fuzzy Linguistic IRS Based on Multi-Granular Linguistic Information. *International Journal of Approximate Reasoning*, 34:221–239, 2003.

[32] J. Koza. *Genetic Programming. On the Programming of Computers by Means of Natural Selection.* The MIT Press, 1992.

[33] D.H. Kraft, G. Bordogna, and G. Pasi. An Extended Fuzzy Linguistic Approach to Generalize Boolean Information Retrieval. *Information Sciences*, 2:119–134, 1994.

[34] D.H. Kraft and D.A. Buell. Fuzzy Sets and Generalized Boolean Retrieval Systems. *International Journal of Man-Machine Studies*, 19:45–56, 1983.

[35] D.H. Kraft, F.E. Petry, B.P. Buckles, and T. Sadasivan. Genetic Algorithms for Query Optimization in Information Retrieval: Relevance Feedback. In E. Sanchez, T. Shibata, and L.A. Zadeh, editors, *Genetic Algorithms and Fuzzy Logic Systems*, pages 155–173. World Scientific, 1997.

[36] V. I. Levenshtein. Binary Codes of Correcting Deletions, Insertions and Reversal. *Sov. Phys. Dokl.*, 6:705–710, 1996.

[37] M. Nikravesh, V. Loia, and B. Azvine. Fuzzy Logic and the Internet (FLINT): Internet, World Wide Web and Search Engines. *Soft Computing*, 6(4):287–299, 2002.

[38] D.W. Oard and G. Marchionini. A Conceptual Framework for Text Filtering. Technical Report CS-TR-3643, University of Maryland, College Park, 1996.

[39] G. Pasi. Intelligent Information Retrieval: Some Research Trends. In J.M. Benítez, O. Cordón, F. Hoffmann, and R. Roy, editors, *Advances in Soft Computing. Engineering Design and Manufacturing*, pages 157–171. Springer, 2003.

[40] M.P. Smith and M. Smith. The Use of Genetic Programming to Build Boolean Queries for Text Retrieval through Relevance Feedback. *Journal of Information Science*, 23(6):423–431, 1997.

[41] P. Thrift. Fuzzy Logic Synthesis with Genetic Algorithms. In *Proceedings of the Fourth International Conference on Genetic Algorithms*, pages 509–513, 1991.

[42] W.G. Waller and D.H. Kraft. A Mathematical Model of a Weighted Boolean Retrieval System. *Information Processing & Management*, 15:235–245, 1979.

[43] R.R Yager. A Note on Weighted Queries in Information Retrieval Systems. *Journal of the American Society for Information Science*, 38:23–24, 1987.

[44] R.R. Yager. On Ordered Weighted Averaging Aggregation Operators in Multicriteria Decision Making. *IEEE Transactions on Systems, Man, and Cybernetics*, 18:183–190, 1988.

[45] L.A. Zadeh. The Concept of a Linguistic Variable and its Applications to Approximate Reasoning. *Part I, II & III, Information Science*, 8:199–249, 8:301–157, 9:43–80, 1975.

[46] E. Zitzler, K. Deb, and L. Thiele. Comparison of Multiobjective Evolutionary Algorithms: Empirical Results. *Evolutionary Computation*, 8(2):173–195, 2000.

[47] E. Zitzler and L. Thiele. Multiobjective Evolutionary Algorithms: A comparative Case Study and the Strength Pareto Approach. *IEEE Transactions on Evolutionary Computation*, 3(4):257–271, 1999.

Multi-Objective Neural Network Optimization for Visual Object Detection

Stefan Roth, Alexander Gepperth and Christian Igel

Ruhr-Universität Bochum
Institut für Neuroinformatik
44780 Bochum, Germany
stefan.roth@neuroinformatik.rub.de
alexander.gepperth@neuroinformatik.rub.de
christian.igel@neuroinformatik.rub.de

In real-time computer vision, there is a need for classifiers that detect patterns fast and reliably. We apply multi-objective optimization (MOO) to the design of feed-forward neural networks for real-world object recognition tasks, where computational complexity and accuracy define partially conflicting objectives. Evolutionary structure optimization and pruning are compared for the adaptation of the network topology. In addition, the results of MOO are contrasted to those of a single-objective evolutionary algorithm. As a part of the evolutionary algorithm, the automatic adaptation of operator probabilities in MOO is described.

27.1 Introduction

When speaking of real-world object detection we usually refer to object detection tasks having some application in the commercial domain. These tasks must be solved well enough to allow their application in products, where error tolerances are usually very restrictive. In the ideal case, the detection accuracy should be comparable or superior to that of humans. When considering existing applications of real-world object detection systems it is clear that imperfect performance can lead to serious (possibly fatal, e.g., in automotive applications) problems. This imposes a tight constraint on tolerable errors rates. To make matters even more difficult, detection should not only be near-perfect but also capable of real-time operation, placing strong constraints on the complexity of the methods that are used. It is intuitively clear that these constraints will not always coexist peacefully and that methods must be developed to design and optimize systems in the presence of conflicting objectives.

S. Roth et al.: *Multi-Objective Neural Network Optimization for Visual Object Detection*, Studies in Computational Intelligence (SCI) **16**, 629–655 (2006)
www.springerlink.com

Real World Object Detection

Many commercial object detection tasks for which solutions meeting the abovementioned constraints (low classification error in real-time operation) have been proposed fall into the domains of advanced driver assistance and biometric systems. Advanced driver assistance systems typically require the detection of pedestrians [41, 44], vehicles [49, 32, 21], lane borders [13], or traffic signs [2] to ensure that the "intelligent vehicles" can construct an adequately complete representation of their surroundings and (possibly) take the appropriate actions. Within biometric systems, face recognition is of particular interest. Automatic face recognition can be found in commercial applications such as content-based image retrieval, video coding, video conferencing, automatic video surveillance of a crowd, intelligent human-computer interfaces, and identity authentication. Face detection is the inevitable first step in face recognition, aiming at localizing and extracting the regions of a video stream or an image that contain a view of a human face. Overviews of examples, problems, and approaches in the research domain of face detection can be found in [24, 51].

Thus, the problems of detecting cars and human faces are important examples of current research in visual object detection. Usually, computer vision architectures are broadly motivated by biological visual search strategies [50, 8]: a fast initial detection stage localizes likely target object locations by examining easily computable visual features, whereas a more detailed analysis is then performed on all candidate regions or object hypotheses that have been formed in the initial detection stage. The scope of this contribution is on the classification of hypotheses, that is, on the decision whether a given object hypothesis actually corresponds to a relevant object type. Based on previous work of the authors, the problems of car and face detection are discussed in-depth.

Neural Classifiers

We address the task of optimizing the weights and the structure of feed-forward neural networks (*FFNNs*) used for face and car classification in the Viisage-FaceFINDER® system and the car detection system described in [21], respectively. In both cases one goal is to increase the speed of the neural classifier, because faster classification allows for a more thorough scanning of the image, possibly leading to improved recognition. Another goal is of course enhancing the accuracy of the FFNN classifiers. It is unreasonable to expect that these two requirements can be achieved independently and simultaneously. Feed-forward neural networks have proven to be powerful tools in pattern recognition [53]. Especially (but not only) in the domain of face detection the competitiveness of FFNNs is widely accepted. As stated in a recent survey "The advantage of using neural networks for face detection is the feasibility of training a system to capture the complex class conditional density of face patterns. However, one drawback is that the network architecture has to be extensively tuned (number of layers, number of nodes, learning rates, etc.) to get exceptional performance" [51]. This drawback is addressed by variants of a hybrid optimization algorithm presented in this article.

Evolutionary Multi-objective Optimization

Given the general problem of conflicting objectives in visual object detection, ways must be devised to address and resolve it. Advanced evolutionary multi-objective optimization (EMO) considers vector-valued objective functions, where each component corresponds to one objective (e.g., speed or accuracy of a neural classifier). Such methods are capable of finding sets of trade-off solutions, none of which can be said to be superior to another without an explicit weighting of single objectives [11, 9]. From such a set one can select an appropriate compromise, which might not have been found by a single-objective approach.

There are recent studies that investigate domains of application and performance of EMO applied to FFNN design [27, 23, 16, 1, 20, 31]. It is evident from these publications that the EMO of FFNNs is under active research, and that a number of methods exist that give excellent results in FFNN structure optimization. The goal of this contribution is to show that visual object detection can profit significantly from FFNNs optimized by EMO.

Outline

In this article we summarize the results of our work in the domain of EMO for FFNN structure optimization [47, 48, 22]. We present variants of our self-adaptive, hybrid EMO and demonstrate the performance in contrast to the performance of a greedy optimization method for FFNN design known as magnitude-based pruning [38]. We evaluate our methods on an optimization problem for car classification, which will be termed *car task*, and on a face detection problem denoted *face task*.

First, in Section 27.2 we present the object detection tasks that embed the considered classification problems. We explain how the training data for classification are obtained from unprocessed ("raw") video data. In Section 27.3, the EMO framework for structure optimization of FFNNs is described. State-of-the-art methods for comparing multi-objective optimization outcomes and the experimental setup used to derive results are given in Sections 27.4 and 27.5. The results are stated in Section 27.6 and discussed in Section 27.7.

27.2 Optimization Problems

In this section, we describe the optimization problems of improving FFNNs for face and car detection. Both detection tasks were introduced in previous publications [21, 47] and share similar system architectures: a fast initial detection produces object hypotheses (so-called *regions of interest*, ROIs), which are examined by an FFNN classifier confirming or rejecting hypotheses. The initial detection stage in both cases uses heuristics designed for the fast detection of cars and faces, respectively. These heuristics are not learned but designed and are different for cars and faces. Common to all methods for initial object detection is the requirement that the "false negative" rate (the rate of disregarded true objects) be very close to zero, whereas a moderate "false positive" rate is acceptable: it is the job of the classifier to eliminate remaining false positives, if possible. This approach aims at capturing *all* objects of a class; it is accepted that sometimes an object is detected where there is none. However, usually (as shall be briefly described later) there exist additional ways of eliminating false positives beyond the scope of single-frame classification whereas there are no known methods of easily doing the reverse, that is, locating objects which are missed by the initial detection.

Input to the FFNN classifiers are ROIs produced by the initial detection step, their output is a decision whether the presented ROIs contain relevant objects or not. The decision is based purely on the image data that are contained in the ROIs; therefore, the decision can be a function of pixel values within an ROI only. The most straightforward approach would be to present the raw gray values (perhaps normalized between 0.0 and 1.0) of pixels within an ROI to an FFNN classifier for learning and online classification. However, for many problems it is profitable to perform certain transformations of the ROI data before presenting them to a classifier. Some representations facilitate classification more than others, for example, the raw data could be transformed to a representation which is more robust to image distortions or noise.

We will not discuss the initial detection stage in this contribution but emphasize the classification of ROIs instead. This binary classification, separating objects from non-objects, is achieved by estimating the true classification function from sample data. We are now going to describe the tasks of car and face classification in more detail, focusing on the process of obtaining training and online examples (also termed *feature sets*) from the raw image data, a process which is frequently termed *feature extraction*.

For both classification problems, we create four data sets of labeled examples termed D_{learn}, D_{val}, D_{test}, and D_{ext}. Labeling means assigning a class label to each example indicating whether it belongs to the "relevant object" class. For details about the data sets please see Tab. 27.1. The reason for this partitioning will become apparent when we discuss the experimental setup in Section 27.5.

27.2.1 Face Detection Data

The Viisage-FaceFINDER® video surveillance system [45] automatically identifies people by their faces in a three-step process. First, regions of the video stream that contain a face are detected, then specific face models are calculated, and finally these models are compared to a database. The final face

Table 27.1. Facts about the example data sets.

	property			usage	
data set	size(cars)	size(faces)	positives	Pruning	Evolution
D_{learn}	5000	3000	50%	Learning	Learning/Selection
D_{val}	5000	1400	50%	Crossvalidation	Crossvalidation/Selection
D_{test}	5000	2000	50%	Pareto dominance based archiving	
D_{ext}	5000	2200	50%	Estimation of generalization loss	

modeling and recognition is done using *Hierarchical Graph Matching* (*HGM*, [26]), which is an improvement of the *Elastic Graph Matching* method [34]. It is inspired by human vision and highly competitive to other techniques for face recognition [54]. To meet real-time constraints, the Viisage-FaceFINDER® requires very fast and accurate image classifiers within the detection unit for an optimal support of HGM.

Inputs to the face detection FFNN are preprocessed 20×20 pixel grayscale images, which show either frontal, upright faces or nonfaces, see Fig. 27.1. In the preprocessing step, different biologically motivated cues are fused to cluster the given images into regions of high and low significance (i.e., ROIs and background). The preprocessing comprises further rescaling, lighting correction, and histogram equalization. The fixed-size ROIs are then classified as either containing or not containing an upright frontal face by a task specific FFNN [25].

The assumption of fixed-size ROIs as input to the classifier meets the realistic application scenario for the FFNN in the Viisage-FaceFINDER® system, although in the survey on face detection by Hjelmas and Low such input patterns are regarded as unrealistic for real world face detection [24]. In Viisage-FaceFINDER® the FFNN is only a part of a sophisticated face detection

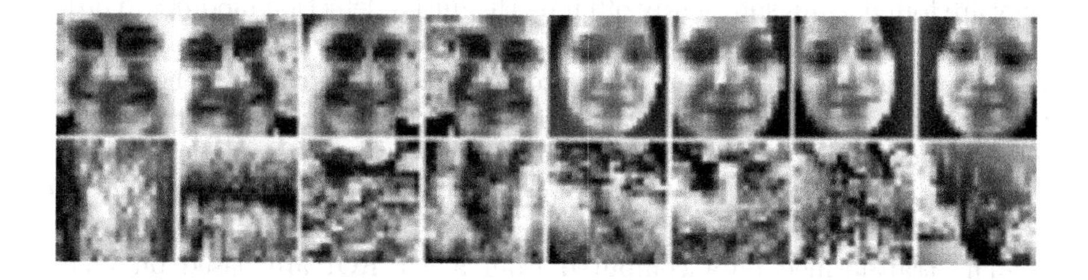

Fig. 27.1. Input to the face detection FFNN are preprocessed 20×20 pixel grayscale images showing either frontal, upright face (positive) and nonface (negative) examples. The preprocessing comprises rescaling, lighting correction, and histogram equalization.

module and its main task is to support the time consuming HGM procedure with appropriate face images.

We use the face detection problem for two slightly different test scenarios. When we refer to the *face task* in the following, we mean both of them. In the first test scenario [1], see Section 27.5.1—comparison of advanced EMO-selection vs. selection of multiple objectives via linear aggregation—the goal is to reduce the number of hidden nodes of the detection FFNN. This is because in the hardware-friendly implementation of the face detection FFNN within Viisage-FaceFINDER® the speed of the classification scales approximately linearly with the number of hidden neurons and not with the number of connections. With every hidden neuron that is saved the detection costs are reduced by approximately one percentage point. In the second test scenario [2], refer to Section 27.5.2—comparison of pruning vs. evolutionary algorithms in a multi-objective framework—, the first objective is to reduce the amount of connections in the network. This is due to another FFNN implementation where the processing speed scales approximately linearly with the number of connections as well as nodes. In both test scenarios, the second objective is identical: improving classification accuracy. A small network size and a high classification accuracy are possibly conflicting objectives (note that "the smaller the network the better the generalization" does not hold in general [3, 7]).

27.2.2 Car Detection Data

The car classification system we are going to describe here has been developed at our lab [21] as a component of a comprehensive software framework for Advanced Driver Assistance [6] containing modules responsible for lane detection, initial car detection, car tracking, traffic sign detection, and derived tasks. For the purposes of car detection, a combination of initial detection and tracking was used prior to the development of the car classifier module: initially detected objects were tracked into adjacent frames, requiring only that they are found again sufficiently often by the initial detection modules in order to be accepted as cars. It is evident that this mechanism is less-than-perfect, because it takes several frames' time to eliminate incorrect hypotheses. Furthermore, once the initial detection produces an object that is not a car but can be tracked easily, that object will be accepted as a car (example: large rectangular traffic signs). Therefore, an approach was needed that could provide an instantaneous and accurate classification of car hypotheses.

The transformed data computed from a car ROI are based on image gradients (and derived quantities) only: raw pixel values are not considered at all. Gradient information is very useful since it is usually quite robust w.r.t. changes in lighting conditions. From the two gradient images (gradients taken in x- and y-direction) denoted $G_x(x, y), G_y(x, y)$, an *energy image* is derived by the formula $E(x, y) = \sqrt{G_x^2(x, y) + G_y^2(x, y)}$. Wherever $E(x, y)$

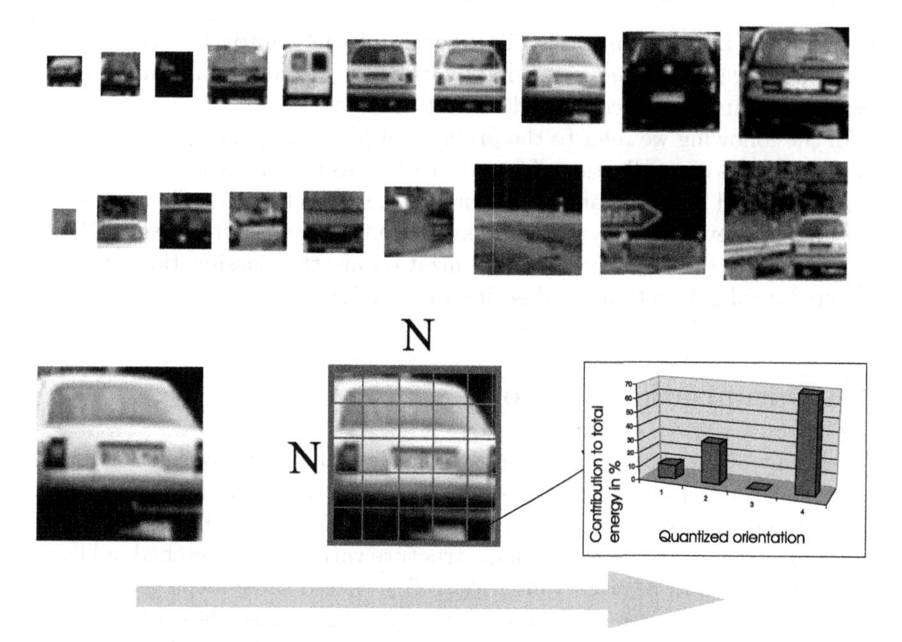

Fig. 27.2. Upper half: Positive and negative training examples for the car task. Since we are using a scale-invariant feature extraction method, examples are not rescaled and can vary significantly in size. Lower half: schematic depiction of a feature vector generated from a single training example, see Section 27.2.2 for an explanation of the quantities used here. An ROI is subdivided into N × N *receptive fields* (RF) and an identical procedure is applied to each one. Calculated orientations in each RF are quantized to k values. For each of the k possible orientation values $i \in [1,..,k]$, and RF pixels (x, y), the quantity $\nu_i = \frac{1}{E_{RF}} \sum_{(x,y)} \chi(x, y)$ is computed, where $\chi(x, y) = E(x, y)$ if $A(x, y) = i$, and 0 otherwise. E_{RF} is obtained by summing up $E(x, y)$ over the whole RF. Thus, k numbers are produced per receptive field that describe the contribution a single orientation makes to the total summed-up edge energy in a receptive field. We choose $k = 4$ and $\mathrm{N} = 7$. A feature vector thus contains N^2k values.

exceeds a certain threshold, we compute the value of the *angle image* to $A(x, y) = \arctan \frac{G_y(x,y)}{G_x(x,y)}$ and zero otherwise. Due to practical considerations, the value $A(x, y) \in [0, \pi]$ is often quantized to k orientation directions. The details of the extraction process are described in Fig. 27.2 (upper half).

Several comments are in order: The feature extraction process is designed in the described way in order to incorporate scale, translation, and lighting invariance already in the feature set, at least to a certain extent. It is a simplified implementation of orientation-selective processing, which has been shown to be abundant in biological vision systems [15]. Many successful computational models for detection and classification rely heavily on orientation-selective feature extraction (examples are wavelet representation like Gabor- or steerable filters [36, 19]), underlining the effectiveness of this approach. In addition, a favorable processing speed is ensured by using only gradient information since image gradients can be computed very efficiently.

In the following we refer to the problem of optimizing an FFNN car classifier on the basis of a collection of feature vectors as the *car task*. The car task is only considered in the second test scenario [2], see Section 27.5.2—comparison of pruning vs. evolutionary algorithms in a multi-objective framework. Hence, in the car task the objectives for optimization are the classification error and the number of connections as described in Section 27.2.1.

27.3 Optimization Methods

We present our evolutionary optimization algorithm with variants of multi-objective selection and magnitude-based pruning, respectively. Optimization is performed iteratively starting from an initial population $\mathcal{P}^{(t=0)}$ of FFNNs. An iteration t includes reproduction, structure variation, and embedded learning with some kind of cross-validation (CV). These three steps generate the offspring population $\mathcal{O}^{(t)}$. In the evolutionary algorithms, these offspring are evaluated and the individuals for the new parent population $\mathcal{P}^{(t+1)}$ are selected from $\mathcal{O}^{(t)}$ and $\mathcal{P}^{(t)}$. For every evaluation an individual of the decision space $\mathcal{X}$ (the genotype space—the space of the encoded FFNNs) is mapped to an n-dimensional vector of objective space by a quality function $\Phi : \mathcal{X} \to \mathbb{R}^n$.

27.3.1 General Properties of All Optimization Methods

In the subsequent paragraphs, each of the previously mentioned steps will be outlined, focusing on key properties common to all discussed optimization methods.

Initialization

The comparison of our results using the face task will be performed on the basis of the expert-designed 400-52-1 FFNN architecture, the *face reference topology*, proposed by Rowley et al. [40]. This FFNN has been tailored to the face detection task and has become a standard reference for FFNN based face detection [51]. No hidden neuron is fully connected to the input but to certain receptive fields, see below. The total number of connections amounts to 2905. This is in contrast to more than 21,000 in a fully connected FFNN with an equal number of hidden neurons.

In the car task, each FFNN is initially fully connected, has 196 input neurons, between 20 and 25 neurons in its hidden layer, one output neuron and all forward-shortcuts and bias connections in place. We refer to this architecture as the *car reference topology*.

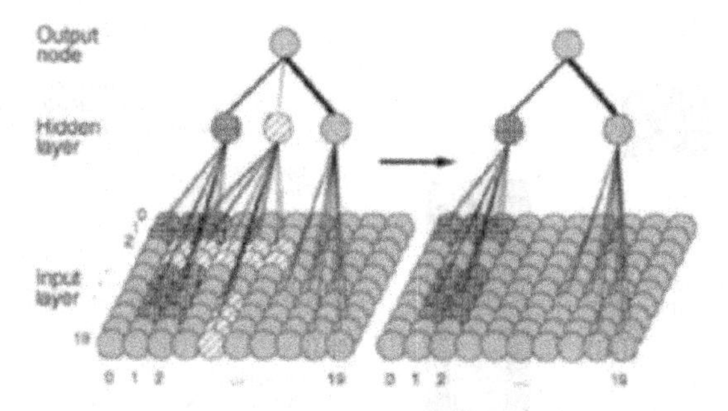

Fig. 27.3. Visualization of the *delete-node* operator within the test scenario [1]. The line widths indicate the magnitude of the corresponding weight values. The picture also visualizes the FFNN input dimension and the receptive field connectivity.

As input, the FFNN classifiers receive a feature set, representing key visual properties of the ROI which it is computed from. In the face tasks the 400 numbers in the feature set correspond to the pixels of the preprocessed image patterns, see Fig. 27.1 and Fig. 27.3, whereas in the car task the 196 numbers in the feature set encode higher-order visual properties as a result of advanced feature extraction methods, see Fig. 27.2 (lower half).

We create the parent individuals in $\mathcal{P}^{(t=0)}$ as copies of the *reference topologies*, which are initialized with different, small random weight values.

Reproduction and Variation

Each parent creates one child per generation. First, the parent is copied. The offspring is then modified by elementary variation operators. This variation process is significantly different for pruning and the evolutionary methods.

All variation operators are implemented such that their application always leads to valid FFNN graphs. An FFNN graph is considered to be *valid* if each hidden node lies on a path from an input unit to an output unit and there are no cycles. Further, the layer restriction, here set to a single hidden layer, has to be met.

Embedded Learning

Let $\mathrm{MSE}_a(D)$ and $\mathrm{CE}_a(D)$ be the mean squared error and the classification error in percent on a data set D of the FFNN represented by individual a. The weights of every newly generated offspring a are adapted by gradient-based optimization ("learning","training") of $\mathrm{MSE}_a(D_{\mathrm{train}})$. An improved version of the Rprop [28, 39] algorithm is used for at most 100 epochs of training. Finally, the weight configuration with the smallest $\mathrm{MSE}_a(D_{\mathrm{train}}) + \mathrm{MSE}_a(D_{\mathrm{val}})$ encountered during training is regarded as the outcome of the training process (i.e., some kind of CV comes in) and stored in the genome of the individual a

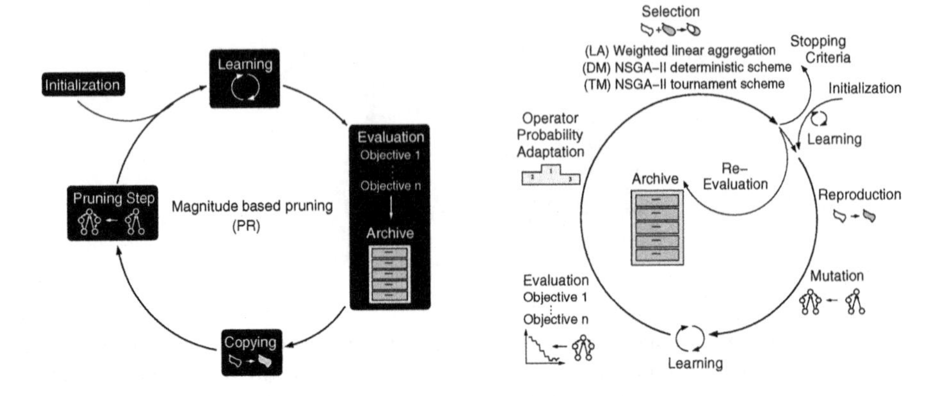

Fig. 27.4. Left, a schematic overview of the pruning method PR. Right, the hybrid evolutionary algorithm in conjunction with three different selection variants LA, DM and TM, see text.

in case of the face tasks (*Lamarckian inheritance*). In the car task, *Lamarckian inheritance* is not applied: the weights before learning are always re-initialized the same way as in the initialization of $\mathcal{P}^{(t=0)}$.

Dominance Based Archiving

For the pruning method and all evolutionary variants, another performance evaluation of the parental individuals $a \in \mathcal{P}^{(t)}$ is used to update the *external archive* $\mathcal{A}^{(t)}$ at iteration t. This performance evaluation is based on a second mapping of the individuals a to n-dimensional objective vectors $\boldsymbol{z} \in \mathbb{R}^n$. The external archive represents the outcome of a trial after its completion at $t = t_{\max}$. In the following let $n_{\text{hid}}(a)$ and $n_{\text{con}}(a)$ denote the number of hidden neurons and weights of the individual $a \in \mathcal{P}^{(t)}$, respectively.

27.3.2 Magnitude-based Network Pruning

Pruning is a well-known reductionist method for obtaining smaller FFNNs from larger ones by iteratively eliminating certain weights. Magnitude-based pruning is a simple heuristic, but has often been reported to give satisfactory results (see [38] for a review of pruning methods). In addition, it is very easy to implement and use. Preliminary experiments using more sophisticated pruning methods [38] did not yield superior results and were therefore abandoned in favor of the most simple method: magnitude-based pruning which we will refer to as method PR. The basic loop for optimization using PR is depicted in Fig. 27.4 (left). Initialization of the first population $\mathcal{P}^{(t=0)}$ is performed as described in Section 27.3.1, and reproduction simply copies the current population. Variation (here: weight elimination) is applied identically in the face and the car tasks: a percentage p of connections with the smallest absolute weight is eliminated at each iteration. Learning is performed as described in Section 27.3.1.

27.3.3 The Evolutionary Multi-objective Algorithm

Evolutionary algorithms have become established methods for the design of FFNNs, especially for adapting their topology [35, 52, 27]. They are thought to be less prone to getting stuck in local optima compared to greedy algorithms like pruning or constructive methods [38, 43].

The basic optimization loop of our hybrid evolutionary algorithm is shown in Fig. 27.4 (right). This scheme might be regarded as canonical evolutionary FFNN optimization using direct encoding and nested learning. However, there are some special features described in this section.

Initialization is performed as described in Section 27.3.1. We will sketch how offspring are created and mutated. After that, we outline the peculiarities of the nested gradient-based learning procedure within the evolutionary loop. Then we highlight the three different approaches to selection which are considered in this work. The section ends with the description of the online strategy adaptation method for adjusting operator application probabilities.

Reproduction and Variation

As mentioned in Section 27.3.1, each parent creates one child per generation; reproduction copies the parent population. The offspring population is then mutated by elementary variation operators. These are chosen randomly for each offspring from a set Ω of operators and are applied sequentially. The process of choosing and applying an operator is repeated $1 + x$ times, where x is an individual realization of a Poisson distributed random number with mean 1.

We need to distinguish the operators for the two test scenarios 1 and 2, due to the requirement of reducing the number of nodes instead of connections. In the first test scenario 1 there are 5 basic operators: *add-connection*, *delete-connection*, *add-node*, *delete-node*, and *jog-weights*:

add-connection A connection is added to the FFNN graph.

delete-connection This operator is inspired by *magnitude-based pruning*. The operator is rank-based as discussed by Braun [5]. The connections of the FFNN are sorted by the absolute value of the corresponding weights. The connection with rank number r given by

$$r := \lfloor W \cdot (\eta_{\max} - \sqrt{(\eta_{\max}^2 - 4 \cdot (\eta_{\max} - 1) \cdot u)})/(2 \cdot (\eta_{\max} - 1)) \rfloor \quad (27.1)$$

is deleted, so that connections with smaller weight have a higher probability of being removed. Here $\lfloor x \rfloor$ denotes the largest integer smaller than x,

W the number of weights, and $u \sim \mathcal{U}[0,1]$ is a random variable uniformly distributed on $[0,1]$. The parameter $1 < \eta_{\text{max}} \leq 2$ controls the influence of the rank and is set to its maximum value [46].

add-node A hidden node with bias parameter is added to the FFNN and connected to the output. For each input, a connection to the new node is added with probability $p_{\text{in}} = 1/16$.

delete-node In this rank-based node deletion operator, the hidden nodes are ordered according to their maximum output weight. The maximum output weight of a node i is given by $\max_j |w_{ji}|$, where w_{ji} is the weight of the connection from node i to node j. The nodes are selected based on eq. (27.1), such that nodes with smaller maximum output weight values have a higher probability of deletion. If node k is deleted, all connections to or from k are removed, cf. Fig. 27.3.

jog-weights This operator adds Gaussian noise to the weights in order to push the weight configuration out of local minima and thereby to allow the gradient-based learning to explore new regions of weight space. Each weight value is varied with constant probability $p_{\text{jog}} = 0.3$ by adding normally distributed noise with expectation value 0 and standard deviation $\sigma_{\text{jog}} = 0.01$.

In addition to the 5 basic operators, there are 3 task-specific mutations within scenario 1 inspired by the concept of "receptive fields", that is, dimensions of the input space that correspond to rectangular regions of the input image, cf. Fig. 27.3. The RF-operators *add-RF-connection*, *delete-RF-connection*, and *add-RF-node* behave as their basic counterparts, but act on groups of connections. They consider the topology of the image plane by taking into account that "isolated" processing of pixels is rarely useful for object detection. The RF-operators are defined as follows:

add-connection-RF A valid, not yet existing connection, say from neuron i to j, is selected uniformly at random. If the source i is not an input, the connection is directly added. Otherwise, a rectangular region of the 20×20 image plane containing between 2 and $M = 100$ pixels including the one corresponding to input i is randomly chosen. Then neuron j is connected to all the inputs corresponding to the chosen image region.

delete-connection-RF An existing connection that can be removed, say from node i to j, is selected at random. If the source i is not an input, the connection is directly deleted. Otherwise, a decision is made whether a horizontal or vertical receptive field is deleted. Assume that a horizontal field is removed. Then *delete-connection-RF*$_x(i,j)$ is applied recursively to remove the inputs from a connected pixel row:

delete-connection-RF$_x(i,j)$ Let (i_x, i_y) be the image coordinates of the pixel corresponding to the input i. The connection from i to j is deleted. If hidden node j is also connected to the input node k corresponding to pixel (i_x+1, i_y), *delete-connection-RF*$_x(k, j)$ is applied. If j is connected to node l corresponding to $(i_x - 1, i_y)$, then the operator *delete-connection-RF*$_x(l, j)$ is called.

Deletion of a vertical receptive field (i.e., a connected pixel column) is done analogously.

add-node-RF A hidden node with bias connection is added and connected to the output and a receptive field as in the *add-connection-RF* operator.

In the second test scenario [2] we use different operators for the insertion and deletion of connections than in scenario [1]. Furthermore, there is no operator for the deletion of hidden nodes; deletion of nodes happens only when nodes no longer have any ingoing or outgoing connections. We apply the basic operators *add-connection-P*, *delete-connection-P* and *add-node* for the car task. For the face task, we additionally use the operator *jog-weights* and three task-specific mutations: *add-RF-connection-P*, *delete-RF-connection-P*, and *add-RF-node*. The new mutation operators are defined as follows:

add-connection-P The operator *add-connection* is sequentially applied to the FFNN until the number of newly added connections amounts to 1% of the previously existent connections before *add-connection-P* was applied.

delete-connection-P The operator *delete-connection* is sequentially applied to the FFNN until the number of deleted connections amounts to 5% of the previously existent connections before *delete-connection-P* was applied.

add-connection-RF-P The operator *add-connection-RF* is sequentially applied to the FFNN until the number of newly added connections amounts to at least 1% of the previously existent connections before *add-connection-RF-P* was applied.

delete-connection-RF-P The operator *delete-connection-RF* is sequentially applied to the FFNN until the number of deleted connections amounts to at least 5% of the previously existent connections before the operator *delete-connection-RF-P* was applied.

Generally, weight values of new connections (produced by the operators for addition of nodes and connections) are drawn uniformly as in the first initialization of the population.

Embedded Learning

A peculiarity of the evolutionary methods is the fact that training can stop earlier due to the *generalization loss* criterion GL_α as described by Prechelt [37]. The generalization loss is computed on D_{val} for $\alpha = 5$. This is done in order to reduce the computational cost of the optimization.

Evaluations and Selection in Presence of Multiple Objectives

We are looking for sparse FFNNs with high classification accuracy. That is, we try to optimize two different objectives. There are several ways of dealing with multiple goals, and we will describe three of them in the following.

LA—Linearly Aggregated Objectives Are subject to Selection.

In the first case of the test scenario 1 the algorithm in Fig. 27.4 (right) uses a scalar fitness $a \mapsto \Phi(a) \in \mathbb{R}$ for any individual a given by the weighted linear aggregation

$$\begin{aligned}\Phi(a) := \quad & \gamma_{\mathrm{CE}} \cdot \mathrm{CE}_a^{(t)}(D_{\mathrm{train}} \cup D_{\mathrm{val}}) + \mathrm{MSE}_a^{(t)}(D_{\mathrm{train}} \cup D_{\mathrm{val}}) \\ & + \gamma_{\mathrm{hid}} \cdot n_{\mathrm{hid}}^{(t)}(a) \qquad\qquad\qquad + \gamma_{\mathrm{con}} \cdot n_{\mathrm{con}}^{(t)}(a)\end{aligned} \tag{27.2}$$

that is to be minimized. The weighting factors are chosen such that typically $\gamma_{\mathrm{CE}} \cdot \mathrm{CE}_a^{(t)}(D_{\mathrm{train}} \cup D_{\mathrm{val}}) \gg \gamma_{\mathrm{hid}} \cdot n_{\mathrm{hid}}^{(t)}(a) \approx \gamma_{\mathrm{con}} \cdot n_{\mathrm{con}}^{(t)}(a) \gg \mathrm{MSE}_a^{(t)}(D_{\mathrm{train}} \cup D_{\mathrm{val}})$ holds. Note that in test scenario 1 we tolerate an increase in the number of connections as long the number of neurons decreases.

Based on the fitness Φ, EP-style tournament selection [17] with 5 opponents is applied to determine the parents $\mathcal{P}^{(t+1)}$ for the next generation from $\mathcal{P}^{(t)} \cup \mathcal{O}^{(t)}$. We refer to the described selection method as *linearly-aggregated selection* LA and identify the complete algorithm with this selection scheme throughout this article.

DM—Vector-valued Objectives Are Subject to Deterministic Selection.

In the second case of the test scenario 1, the evolutionary algorithm in Fig. 27.4 (right) performs advanced EMO selection. It uses a selection method based on the Fast Non-Dominated Sorting Genetic Algorithm (*NSGA-II*) [12].

We first map the elements of the decision space to n-dimensional real-valued vectors $\boldsymbol{z} = (z_1, \ldots, z_n)$ of the objective space by the objective function $\Phi : \mathcal{X} \to \mathbb{R}^n$. In our case we map the individual a that has already finished training to the vector $\Phi(a) = \boldsymbol{z}_a = (n_{\mathrm{hid}}(a), \mathrm{CE}_a(D_{\mathrm{train}} \cup D_{\mathrm{val}}))$. Both objective components are subject to minimization. The elements of the objective space are partially ordered by the dominance relation $\prec$ ($\boldsymbol{z}$ dominates $\boldsymbol{z}'$) that is defined by

$$\boldsymbol{z} \prec \boldsymbol{z}' \in \mathbb{R}^n \quad \Leftrightarrow \quad \forall\, 1 \le i \le n : \; z_i \le z_i' \;\wedge\; \exists\, 1 \le j \le n : \; z_j < z_j'$$

stating that vector $\boldsymbol{z}$ performs better than $\boldsymbol{z}'$ iff $\boldsymbol{z}$ is as least as good as $\boldsymbol{z}'$ in all objectives and better with respect to at least one objective. Considering a set M of n-dimensional vectors, the subset $\mathrm{ndom}(M) \subseteq M$ consisting only of those vectors that are not dominated by any other vector of M is called the Pareto front of M. As in the *NSGA-II* (environmental) selection scheme, we first assign a rank value $\mathrm{R}^{(t)}(a)$ to each individual $a \in \mathcal{P}^{(t)} \cup \mathcal{O}^{(t)}$ based on its

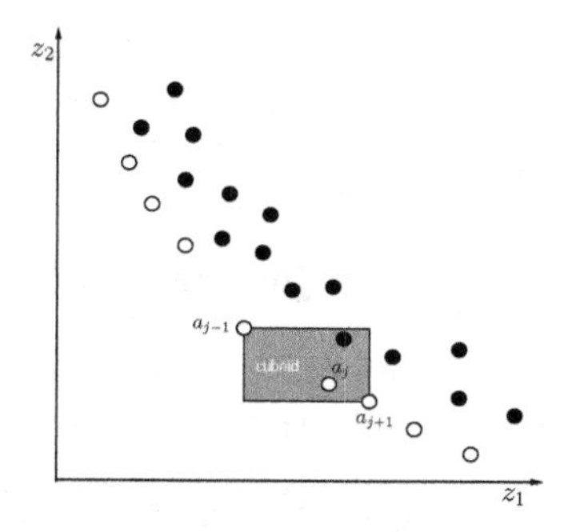

Fig. 27.5. Illustration how the crowding distance $\mathrm{C}(a_j)$ [11] is computed. The black dots are the elements of M_{i+1} and the white dots of belong to the Pareto front $\mathrm{ndom}(M_i)$.

degree of non-domination in objective space. We define the chain of subsets M_i, $i \in \mathbb{N}$, by $M_1 \supseteq M_2 := M_1 \backslash \mathrm{ndom}(M_1) \supseteq M_3 := M_2 \backslash \mathrm{ndom}(M_2) \supseteq \dots$, where $A \backslash B$ denotes the portion of set A that is not part of set B. Then the rank operator $\mathrm{R}^{(t)}(a)$ assigns to each individual $a \in \mathcal{P}^{(t)} \cup \mathcal{O}^{(t)}$ the index i of the corresponding Pareto front $\mathrm{ndom}(M_i)$ that includes the objective vector of a.

Furthermore the NSGA-II ranking takes the diversity of the population (in objective space) into account. The diversity is measured by the crowding distance $\mathrm{C}(a)$, the size of the largest cuboid (more precisely, the sum of its edge lengths) in objective space enclosing the vector $\Phi(a) = \boldsymbol{z}_a$, $a \in \mathrm{ndom}(M_i)$, but no other objective vector from $\mathrm{ndom}(M_i)$, see Fig. 27.5. Then all individuals $a \in \mathcal{P}^{(t)} \cup \mathcal{O}^{(t)}$ are sorted in ascending order according to the partial ordering relation $\leq_n$ defined by

$$\begin{aligned} a_i \leq_n a_j \Leftrightarrow & \left(\mathrm{R}^{(t)}(a_i) < \mathrm{R}^{(t)}(a_j)\right) \text{ or} \\ & \left(\mathrm{R}^{(t)}(a_i) = \mathrm{R}^{(t)}(a_j) \wedge C(a_i) \geq C(a_j)\right) \end{aligned} \tag{27.3}$$

and the first $|\mathcal{P}|$ individuals form the new parent population $\mathcal{P}^{(t+1)}$. We refer to the described selection method as *NSGA-II deterministic selection* DM throughout this article.

TM—Vector-valued Objectives Are Subject to Tournament Selection.

The selection method for the second test scenario [2] is almost completely in accordance with the *NSGA-II deterministic selection.* But it is chosen to be an EP-style tournament selection with 5 opponents to determine the parents $\mathcal{P}^{(t+1)}$ for the next generation from $\mathcal{P}^{(t)} \cup \mathcal{O}^{(t)}$, cf. selection method LA. The tournament selection is based upon the objective vector $\Phi(a) = \boldsymbol{z}_a =$

$(n_{\text{con}}(a), \text{CE}_a(D_{\text{train}} \cup D_{\text{val}}))$ and the partial relation $\leq_n$ of eq. (27.3). We refer to the described selection method as *NSGA-II tournament selection* TM throughout this article.

Search Strategy Adaptation: Adjusting Operator Probabilities

A key concept in evolutionary computation is strategy adaptation, that is, the automatic adjustment of the search strategy during the optimization process [14, 29, 42, 30]. Not all operators might be necessary at all stages of evolution. In our case, questions such as when fine-tuning becomes more important than operating on receptive fields cannot be answered in advance. Hence, the application probabilities of the variation operators are adapted using the method presented in [29, 30, 47], which is inspired by Davis' work [10]. The underlying assumption is that recent beneficial modifications are likely to be beneficial in the following generations.

The basic operators that are actually employed in a given optimization scenario are divided into groups, those adding connections, deleting connections, adding nodes, deleting nodes, and solely modifying weights. Let Ω be the set of variation operators, G the number of operator groups used in a task and $p_o^{(t)}$ the probability that $o \in \Omega$ is chosen at generation t.

The initial probabilities for operators of a single group are identical and add up to 0.2 in case of the test scenario 1 where $G = 5$. In case of the face task within test scenario 2, where no node deletion operators are used ($G = 4$), the probabilities add up to 0.25. In the car task within test scenario 2 the initial probabilities for *add-connection-P* and *add-node* are set to 0.3, and the initial probability for *delete-connection-P* it is set to 0.4. This produces a slight bias towards deleting connections.

Let $\mathcal{O}_o^{(t)}$ contain all offspring produced at generation t by an application of the operator o. The case that an offspring is produced by applying more than one operator is treated as if the offspring was generated several times, once by each operator involved. The operator probabilities are updated every τ generations. Here we set $\tau = 4$. This period is called an adaptation cycle. The average performance achieved by the operator o over an adaptation cycle is measured by

$$q_o^{(t,\tau)} := \sum_{i=0}^{\tau-1} \sum_{a \in \mathcal{O}_o^{(t-i)}} \max\left(0, \mathrm{B}^{(t)}(a)\right) \Big/ \sum_{i=0}^{\tau-1} \left|\mathcal{O}_o^{(t-i)}\right| ,$$

where $\mathrm{B}^{(t)}(a)$ represents a quality measure proportional to some kind of fitness improvement. This is for the scalar value based selection scheme, case LA,

$$\mathrm{B}^{(t)}(a) := \Phi(a) - \Phi(\text{parent}(a))$$

and for the vector-valued selection schemes DM and TM,

$$\mathrm{B}^{(t)}(a) := \mathrm{R}^{(t)}(\mathrm{parent}(a)) - \mathrm{R}^{(t)}(a) \enspace ,$$

where parent(a) denotes the parent of an offspring a. The operator probabilities $p_o^{(t+1)}$ are adjusted every τ generations according to

$$\tilde{p}_o^{(t+1)} := \begin{cases} \zeta \cdot q_o^{(t,\tau)}/q_{\mathrm{all}}^{(t,\tau)} + (1-\zeta)\cdot \tilde{p}_o^{(t)} & \text{if } q_{\mathrm{all}}^{(t,\tau)} > 0 \\ \zeta/|\Omega| \qquad\qquad + (1-\zeta)\cdot \tilde{p}_o^{(t)} & \text{otherwise} \end{cases}$$

and

$$p_o^{(t+1)} := p_{\min} \; + \; (1 - |\Omega| \cdot p_{\min})\tilde{p}_o^{(t+1)} \Big/ \sum_{o' \in \Omega} \tilde{p}_{o'}^{(t+1)} \enspace .$$

The factor $q_{\mathrm{all}}^{(t,\tau)} := \sum_{o' \in \Omega} q_{o'}^{(t,\tau)}$ is used for normalization and $\tilde{p}_o^{(t+1)}$ stores the weighted average of the quality of the operator o, where the influence of previous adaptation cycles decreases exponentially. The rate of this decay is controlled by $\zeta \in (0, 1]$, which is set to $\zeta = 0.3$ in our experiments. The operator fitness $p_o^{(t+1)}$ is computed from the weighted average $\tilde{p}_o^{(t+1)}$, such that all operator probabilities sum to one and are not lower than the bound $p_{\min} < 1/|\Omega|$. Initially, $\tilde{p}_o^{(0)} = p_o^{(0)}$ for all $o \in \Omega$.

The adaptation algorithm itself has free parameters, $p_{\min}$, τ, and ζ. However, in general the number of free parameters is reduced in comparison to the number of parameters that are adapted and the choice of the new parameters is considerably more robust. Both τ and ξ control the speed of the adaptation; a small ξ can compensate for a small τ ($\tau = 1$ may be a reasonable choice in many applications). The adaptation adds a new quality to the algorithm as the operator probabilities can vary over time. It has been empirically shown that the operator probabilities are adapted according to different phases of the optimization process and that the performance of the structure optimization benefits from this adaptation [29, 47, 30].

27.4 Evaluating Multi-objective Optimization

The performance assessment of stochastic multi-objective algorithms is in general more difficult than evaluating single-objective algorithms. The reason is that in empirical investigations, sets of sets (i.e., the non-dominated solutions evolved in multiple trials of different algorithms) have to be compared. Many ways of measuring the performance of multi-objective optimization algorithms have been proposed, here we use two unary quality indicators, the hypervolume indicator and the additive ϵ-indicator. We concisely define the performance measures used, for a detailed description of the methods we refer to the literature [33, 56, 18].

Given two sets of objective vectors $A, B \subseteq \mathbb{R}^n$ there is a common sense definition of one set being better than the other. Set A is better than B and we write $A \rhd B$ if for every element $\boldsymbol{z} \in B$ there exists an element $\boldsymbol{z}' \in A$

that is not worse than $\boldsymbol{z}$ in each objective, $\forall j \in \{1,\ldots,n\} : z'_j \leq z_j$, and $\mathrm{ndom}(A) \neq \mathrm{ndom}(B)$. Otherwise we have $A \not\triangleright B$. However, in general neither $A \triangleright B$ nor $B \triangleright A$ holds for sets A and B. Therefore, quality indicators are introduced.

An unary quality indicator assigns a real valued performance index to a set of solutions. A single indicator captures only certain aspects (preferences of a decision maker) of the evolved solutions, therefore a set of different indicators is used for the comparison of multi-objective optimization algorithms. Here, the hypervolume indicator [55] and the ϵ-indicator [56] are chosen. We use the performance assessment tools that are part of the PISA [4] software package with standard parameters.

Before the performance indicators are computed, the data are normalized. Assume we want to compare k algorithms on a particular optimization problem. For each algorithm, we have conducted T trials. For comparison, we consider the solutions in the populations after a predefined number g of evaluations. First of all, we consider the non-dominated individuals of the union of all kT populations after g evaluations. Their objective vectors are normalized such that for every objective the smallest and largest objective function value are mapped to 1 and 2, respectively, by an affine transformation. These objective vectors make up the reference set A_{ref}. The mapping to $[1,2]^n$ is fixed and applied to all objective vectors under consideration.

The hypervolume measure or $\mathcal{S}$-metric was introduced in [55] in the domain of EMO. With respect to the reference point $\boldsymbol{z}_{\mathrm{nadir}}$, it can be defined as the Lebesgue measure Λ of the union of hypercubes in objective space [9]:

$$\mathcal{S}_{\boldsymbol{z}_{\mathrm{nadir}}}(A') = \Lambda\left(\bigcup_{\boldsymbol{z}\in \mathrm{ndom}(A')} \{\boldsymbol{z}' \in \mathbb{R}^n \,|\, \boldsymbol{z} \prec \boldsymbol{z}' \prec \boldsymbol{z}_{\mathrm{nadir}}\} \right) .$$

The hypervolume indicator with respect to reference set A_{ref} is defined as

$$\mathcal{I}_{\mathcal{S},A_{\mathrm{ref}}}(A) = \mathcal{S}_{\boldsymbol{z}_{\mathrm{nadir}}}(A_{\mathrm{ref}}) - \mathcal{S}_{\boldsymbol{z}_{\mathrm{nadir}}}(A) .$$

The reference point $\boldsymbol{z}_{\mathrm{nadir}}$ is an objective vector that is worse in each objective than all individuals (here $\boldsymbol{z}_{\mathrm{nadir}} = (2.1,\ldots,2.1)$). A smaller value of $\mathcal{I}_{\mathcal{S}}$ is preferable. The additive unary ϵ-indicator is defined as

$$\mathcal{I}_{\epsilon,A_{\mathrm{ref}}}(A) = \inf\{\epsilon \in \mathbb{R} \,|\, \forall \boldsymbol{z} \in A_{\mathrm{ref}}\, \exists \boldsymbol{z}' \in A\, \forall i \in \{1,\ldots,n\} : z_i \geq z'_i - \epsilon\} .$$

Basically, the ϵ-indicator determines the smallest offset that has to be subtracted from the fitness values of the elements in A such that the resulting Pareto front covers the Pareto front of A_{ref} in objective space. A smaller value of $\mathcal{I}_{\epsilon,A_{\mathrm{ref}}}$ is preferable.

In [33] and [56], various properties of quality indicators are studied. Of particular interest is the relation to the "being better" definition given above. An unary quality indicator is $\not\triangleright$-compatible if a better indicator value for A than for B implies $B \not\triangleright A$. An indicator is $\triangleright$-complete, if $A \triangleright B$ implies a better indicator value for A than for B. Both the ϵ-indicator as well as the hypervolume indicator are $\ntrianglerighteq$-compatible and $\triangleright$-complete.

27.5 Experimental Setup

In all experiments, the FFNNs have at most one hidden layer and the activation functions are of logistic sigmoidal type. In all cases we simulate $T = 10$ trials. For each trial we set $|\mathcal{P}^{(t=0)}| = 25$.

In the car task, each FFNN in $\mathcal{P}^{(t=0)}$ is fully connected, has between 20 and 25 neurons in its hidden layer and all forward-shortcuts and bias connections in place. At each iteration t, $\mathcal{P}^{(t)}$ is initialized with small random weight values between -0.05 and 0.05. We refer to this architecture as the *car reference topology*, see Section 27.3.1. In the face tasks, $\mathcal{P}^{(t=0)}$ consists of copies of the 400-52-1 architecture of [40], the *face reference topology*, each of which is randomly initialized like the *car reference topology* at $t = 0$.

Although cross-validation is applied when training the FFNNs, the evolutionary (LA, DM, TM) or the pruning optimization (PR) may lead to overfitting, in our case they may overfit to the patterns of $D_{\text{train}} \cup D_{\text{val}}$. Hence, we additionally introduce the data set D_{test} to finally choose models that generalize well and store those in the external archive $\mathcal{A}^{(t)}$ if their objective vector $\boldsymbol{z}$ computed from D_{test} is not dominated by any member of $\mathcal{A}^{(t)}$. Members that are dominated by $\boldsymbol{z}$ are removed from the archive. That is, we use D_{test} for some kind of cross-validation of the evolutionary or pruning process. The archive $\mathcal{A}^{(t)}$ at $t = t_{\max}$ is taken to be the final outcome of an optimization trial.

The data set D_{ext}, which does not enter into the optimization at any point, is used to finally assess the performance of the members of the archive $\mathcal{A}^{(t_{\max})}$.

We train 2000 instances of the *face reference topology* and the *car reference topology* for 100 training steps using the improved Rprop learning procedure on D_{train}. From all instances and all training steps we select the networks a_{ref} (for the face and the car tasks, respectively) with the smallest classification error on $D_{\text{val}} \cup D_{\text{test}}$. When selecting the *reference topologies* a_{ref}, we decide in a similar way as in picking a solution from the evolved or pruned architectures, but taking also D_{val} into account. This is reasonable, since D_{val} has not been applied during FFNN training.

In the following figures 27.6 and 27.7, results are normalized by the performance of the *face reference topology* and *car reference topology* a_{ref}, respectively.[1] For example, the normalized classification error of an FFNN a is given by $\text{CE}'_a(D) = \text{CE}_a(D)/\text{CE}_{a_{\text{ref}}}(D)$ and the normalized number of connections and number of nodes by $n'_{\text{con}}(a) = n_{\text{con}}(a)/n_{\text{con}}(a_{\text{ref}})$ and $n'_{\text{hid}}(a) = n_{\text{hid}}(a)/n_{\text{hid}}(a_{\text{ref}})$, respectively.

[1] The reference topologies are not arbitrary, but tuned extensively by hand. They produce results that are highly competitive to other approaches in the literature.

27.5.1 Test Scenario 1

This scenario is concerned with the face task only. It has already been shown that the size of the face detection network of the Viisage-FaceFINDER® system can be successfully reduced without loss of accuracy by the scalar fitness value approach LA [48]. Here, we investigate whether we can improve these results by using the vector valued selection DM, see [47]. This is done by comparing the performance of our hybrid algorithm using either selection variant LA or DM. We strongly expect a result in favor of method DM since a single-objective method (LA) is evaluated by multi-objective performance measures which are made necessary by the nature of the problem. At the time this analysis was conducted, our goal was to challenge the popular single-objective approach to FFNN structure optimization.

We assume that the runtime of our algorithm is strongly dominated by the number of fitness evaluations (due to the efforts spent for learning) and that the number of allowed fitness evaluation is fixed. Then there is roughly no difference in runtime between the single-objective (LA) and the multi-objective (DM) approach.

All T trials of both variants LA and DM of the evolutionary algorithm described above are run for $t_{\max} = 200$ generations (i.e., 5025 fitness evaluations per trial) using only the face task. For both methods LA and DM we use the objective vector $\boldsymbol{z}_a = (n_{\text{hid}}, \text{CE}(D_{\text{test}}))$ of every individual $a \in \mathcal{P}^{(t)}$ to iteratively update the external archive.

27.5.2 Test Scenario 2

In this scenario we aim to show that the significantly higher effort of implementing and applying advanced EMO by method TM can be worthwhile compared to the simpler pruning method PR [22]. Both the car and the face task are considered for this investigation: methods are compared within tasks, and the results of the within-task comparisons are contrasted. All trials T of method PR are performed for $t_{\max} = 90$ iterations at $p = 10\%$.[2] All TM trials are performed for $t_{\max} = 200$ generations. For both methods PR and TM we compute the objective vector $\boldsymbol{z}_a = (n_{\text{con}}, \text{CE}(D_{\text{test}}))$ of every individual $a \in \mathcal{P}^{(t)}$ to update the external archive.

27.6 Results

The normalized results of test scenario 1 are shown in Fig. 27.6. Indicator results confirm what is evident from visual inspection of the two Pareto fronts: method DM performs considerably better than method LA.

[2] No regular FFNNs were ever produced afterwards. Reducing the pruning percentage p to the point where valid FFNNs could be produced for 200 generations did not change results.

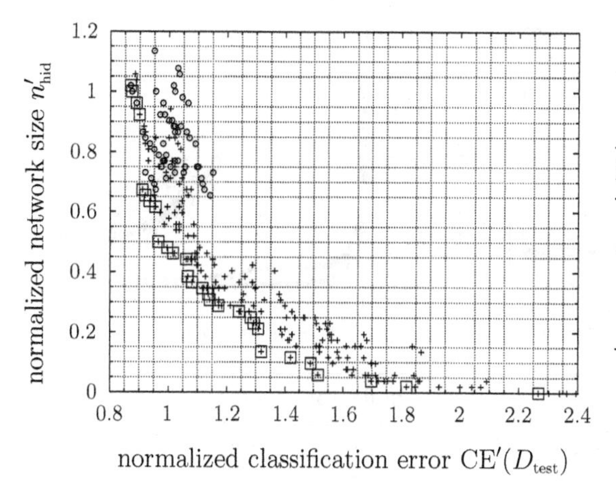

indicator	method	value
$\mathcal{I}_\epsilon$	LA	0.745
$\mathcal{I}_\epsilon$	DM	0.138
$\mathcal{I}_\mathcal{S}$	LA	0.673
$\mathcal{I}_\mathcal{S}$	DM	0.137

Fig. 27.6. Evolved solutions by selection variants LA and DM in test scenario [1]. The left plot shows the two objectives, the normalized classification error $\text{CE}'(D_{\text{test}})$ and the normalized number of hidden neurons n'_{hid}, for all FFNNs of all Pareto fronts of all trials. The circles represent the outcomes of selection method LA, and the crosses the results of the variant DM. The non-dominated FFNNs from the union of all trial outcomes from both methods (the "meta Pareto front") are highlighted. The right table shows the results of performance indicators explained in the text. The distributions over all trials of indicator results for methods LA and DM differ in a statistically significant way when compared by the Wilcoxon rank sum test ($p < 0.005$).

The normalized results of test scenario [2] are shown in Fig. 27.7. One perceives the surprisingly similar performance of the two methods when applied to cars. While the evolutionary method performs better, the differences are small and the errors of the generated FFNNs are similar in similar regimes of n_{con}. In contrast, the differences between the two methods are quite pronounced when applied to the face task: here, EMO is clearly superior. In both tasks, the distributions of the ϵ- and hypervolume indicators differ in a statistically significant way. All results persist when considering the vectors $(n_{\text{hid}}, \text{CE}(D_{\text{ext}}))$ and $(n_{\text{con}}, \text{CE}(D_{\text{ext}}))$ instead of $(n_{\text{hid}}, \text{CE}(D_{\text{test}}))$ and $(n_{\text{con}}, \text{CE}(D_{\text{test}}))$, respectively. This shows that no significant overfitting has occurred.

27.7 Discussion

When considering scenario [1], it is evident that method DM performs better, but this is not surprising since we evaluated a single-objective method in a multi-objective setting. Because a prior decision about the relative weighting of objectives was taken before optimization when using method LA, it can

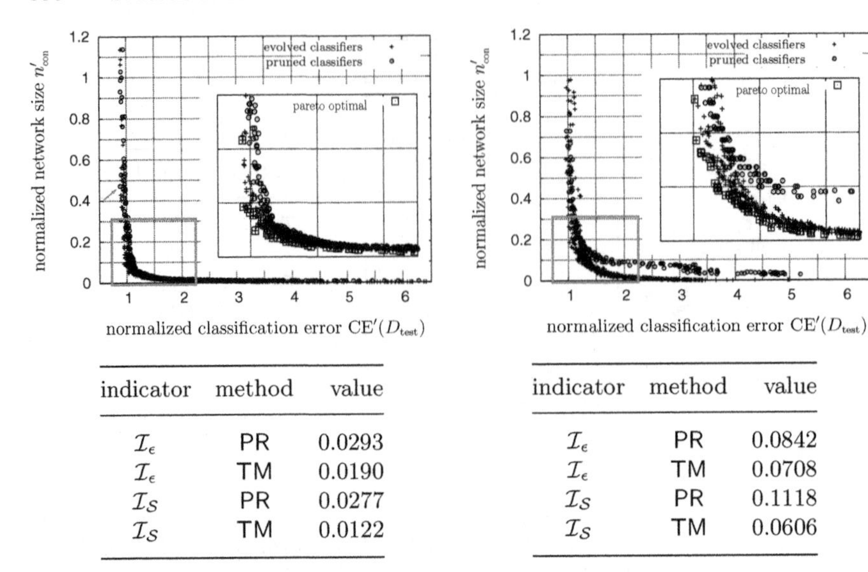

indicator	method	value
$\mathcal{I}_\epsilon$	PR	0.0293
$\mathcal{I}_\epsilon$	TM	0.0190
$\mathcal{I}_S$	PR	0.0277
$\mathcal{I}_S$	TM	0.0122

indicator	method	value
$\mathcal{I}_\epsilon$	PR	0.0842
$\mathcal{I}_\epsilon$	TM	0.0708
$\mathcal{I}_S$	PR	0.1118
$\mathcal{I}_S$	TM	0.0606

Fig. 27.7. Evolved solutions by selection variants TM and PR in the second scenario [2]. Left: results from the car task. Right: results from the face task. Shown on top are the unions of all trial outcomes; members of the meta Pareto fronts are shown in the magnifications. The only pruned FFNN in the meta Pareto front of the car task is indicated by an arrow. The performance indicators (tables at the bottom) are explained in Section 27.4. The distributions over all trials of indicator results for methods PR and TM differ in a statistically significant way when compared by the Wilcoxon rank sum test ($p < 0.005$).

be expected to give adequate results in the corresponding region of objective space, whereas solutions in other regions are less likely to be selected. In contrast, methods like DM and TM can select solutions from all regions of objective space: when using multi-objective performance indicators, which measure performance over all of objective space, this leads to the significant differences in performance that are observed. Single-objective methods like LA are useful whenever the desired trade-off between objectives is known a priori. However, this will not usually be the case, and therefore multi-objective optimization is clearly the strategy of choice.

Considering test scenario [2], we interpret the result of the car task as an indication that the problem class is intrinsically easier (w.r.t. the magnitude of the classification error of the best conceivable FFNN) than the face task since both the ϵ- and the hypervolume indicator results of both methods lie more closely together. This suggests similar performance.[3] We assert that the

[3] The absolute values of the performance indicators depend on reference sets. When comparing two performance indicator values for one task, one should therefore look at their difference, which is independent of the reference set, and not at their quotient.

simpler optimization method can yield competitive performance on simpler tasks. For the more difficult face task, a sophisticated (here: evolutionary) optimization strategy is clearly favorable.

For the support of our interpretation about the difficulty of both tasks in scenario 2, we observe that the (absolute) error $\mathrm{CE}(D_{\mathrm{test}})$ of the best trained car reference topology is 3.5 times smaller than $\mathrm{CE}(D_{\mathrm{test}})$ of the corresponding face reference topology (see Section 27.5). As the results plainly show, optimization is unable to improve classification accuracy greatly compared to the reference topologies, which constitute approximate optima in this respect. Therefore this difference in classification performance should be considered meaningful. Furthermore, optimization in the car task produced FFNNs without a hidden layer, which nevertheless had an (absolute) classification accuracy of about 80%. We take this as a hint that the problem is almost linearly separable and therefore can be considered "easy".

Finally, we want to compare the performance of the methods PR and LA. Both are essentially single-objective methods, since they compute a scalar fitness value from a weighted average of objectives. In the case of LA, this is trivially fulfilled by construction, see eq. (27.2), whereas pruning takes the coefficient of one objective (classification error) to be zero and focuses on the other one (number of connections) only. However, pruning does not perform selection at all since it simply copies the current population into the next. Thus, the success of structure adaptation is not taken into account. Therefore, there is no bias towards certain trade-off solutions in selection. In contrast, method LA uses a scalar fitness value to select the individuals of the next parental population. Hence, only individuals optimized for a single fixed trade-off between objectives are archived. The comparison between these single-objective methods is possible by contrasting them to their advanced multi-objective counterparts TM and DM and comparing the results. It turns out that the performance differences between methods LA and DM are much more pronounced considering any of the given performance indicators than the corresponding differences between methods PR and TM. This supports our previous explanations and shows that pruning, although an amazingly primitive method by itself, can significantly profit from multi-objective performance evaluation. To conclude, when using optimization methods in a multi-objective setup, great care must be taken not to introduce a bias restricting search to certain regions of objective space.

Acknowledgment

Christian Igel acknowledges support from Viisage Technology AG under contract "SecureFaceCheck".

References

[1] Hussein A. Abbass. Speeding up backpropagation using multiobjective evolutionary algorithms. *Neural Computation*, 15(11):2705–2726, 2003.

[2] C Bahlmann, Y Zhu, V Ramesh, M Pellkofer, and T Koehler. A system for traffic sign detection, tracking, and recognition using color, shape and motion information. In *Proceedings of the IEEE Symposium on Intelligent Vehicles*, pages 255–260, 2005.

[3] P. L. Bartlett. The sample complexity of pattern classification with neural networks: The size of the weights is more important than the size of the network. *IEEE Transactions on Information Theory*, 44(2):525–536, 1998.

[4] Stefan Bleuler, Marco Laumanns, Lothar Thiele, and Eckart Zitzler. Pisa – A platform and programming language independent interface for search algorithms. In Carlos M. Fonseca, Peter J. Fleming, Eckart Zitzler, Kalyanmoy Deb, and Lothar Thiele, editors, *Evolutionary Multi-Criterion Optimization (EMO 2003)*, volume 2632 of *LNCS*, pages 494 – 508. Springer-Verlag, 2003.

[5] H. Braun. *Neurale Netze: Optimierung durch Lernen und Evolution.* Springer-Verlag, 1997.

[6] T Bücher, C Curio, H Edelbrunner, C Igel, D Kastrup, I Leefken, G Lorenz, A Steinhage, and W von Seelen. Image processing and behaviour planning for intelligent vehicles. *IEEE Transactions on Industrial electronics*, 90(1):62–75, 2003.

[7] Rich Caruana, Steve Lawrence, and C. Lee Giles. Overfitting in neural networks: Backpropagation, Conjugate Gradient, and Early Stopping. In *Advances in Neural Information Processing Systems*, volume 13, pages 402–408. MIT Press, 2001.

[8] MM Chun and JM Wolfe. Visual attention. In EB Goldstein, editor, *Blackwell's Handbook of Perception*, chapter 9, pages 272–310. Oxford, UK: Blackwell, 2001.

[9] C. A. Coello Coello, D. A. Van Veldhuizen, and G. B. Lamont. *Evolutionary Algorithms for Solving Multi-objective Problems.* Kluwer Academic Publishers, New York, 2002.

[10] Lawrence Davis. Adapting operator probabilities in genetic algorithms. In J. David Schaffer, editor, *Proceedings of the Third International Conference on Genetic Algorithms, ICGA'89*, pages 61–69. Morgan Kaufmann, 1989.

[11] Kalyanmoy Deb. *Multi-Objective Optimization using Evolutionary Algorithms.* John Wiley & Sons, 2001.

[12] Kalyanmoy Deb, Samir Agrawal, Amrit Pratap, and T. Meyarivan. A fast and elitist multiobjective genetic algorithm: NSGA-II. *IEEE Transactions on Evolutionary Computation*, 6(2):182–197, 2002.

[13] E.D. Dickmanns and B.D. Mysliwetz. Recursive 3-D road and relative ego-state recognition. *IEEE Transactions on Pattern Analysis and Machine Intelligence*, 14(2):199–213, 1992.

[14] Agoston Endre Eiben, Robert Hinterding, and Zbigniew Michalewicz. Parameter control in evolutionary algorithms. *IEEE Transactions on Evolutionary Computation*, 3(2):124–141, 1999.

[15] D Ferster and K.D. Miller. Neural mechanisms of orientation selectivity in the visual cortex. *Annual Review of Neuroscience*, 23:441–471, 2000.

[16] J. E. Fieldsend and S. Singh. Pareto evolutionary neural networks. *IEEE Transactions on Neural Networks*, 16(2):338–354, 2005.

[17] David B. Fogel. *Evolutionary Computation: Toward a New Philosophy of Machine Intelligence*. IEEE Press, Piscataway, NJ, USA, 1995.
[18] C. M. Fonseca, J. D. Knowles, L. Thiele, and E. Zitzler. A tutorial on the performance assessment of stochastic multiobjective optimizers. Presented at the Third International Conference on Evolutionary Multi-Criterion Optimization (EMO 2005), 2005.
[19] W Freeman and E Adelson. The design and use of steerable filters. *IEEE Transactions on Pattern analysis and machine intelligence*, 13(9):891–906, 1991.
[20] N. Garcia-Pedrajas, C. Hervas-Martinez, and J. Munos-Perez. Multi-objective cooperative coevolution of artificial neural networks (multi-objective cooperative networks). *Neural Networks*, 15:1259–1278, 2002.
[21] A Gepperth, J Edelbrunner, and T Bücher. Real-time detection and classification of cars in video sequences. In *Proceedings of the IEEE Symposium on Intelligent Vehicles*, pages 625–631, 2005.
[22] A. Gepperth and S. Roth. Applications of multi-objective structure optimization. In M. Verleysen, editor, *5th European Symposium on Artificial Neural Networks (ESANN 2005)*, pages 279–284. d-side Publications, 2005.
[23] J. Gonzalez, I. Rojas, J. Ortega, H. Pomares, J. Fernandez, and A. Diaz. Multi-objective evolutionary optimization of the size, shape, and position parameters of radial basis function networks for function approximation. *IEEE Transactions on Neural Networks*, 14(6):1478– 1495, November 2003.
[24] Eric Hjelmas and Boon Kee Low. Face detection: A survey. *Computer Vision and Image Understanding*, 83:236–274, 2001.
[25] H. Martin Hunke. Locating and tracking of human faces with neural networks. Master's thesis, University of Karlsruhe, Germany, 1994.
[26] Michael Hüsken, Michael Brauckmann, Stefan Gehlen, Kazunori Okada, and Christoph von der Malsburg. Evaluation of implicit 3D modeling for pose invariant face recognition. In A. K. Jain and N. K. Ratha, editors, *Defense and Security Symposium 2004: Biometric Technology for Human Identification*, volume 5404 of *Proceedings of SPIE*. The International Society for Optical Engineering, 2004.
[27] C. Igel and B. Sendhoff. Synergies between evolutionary and neural computation. In M. Verleysen, editor, *13th European Symposium on Artificial Neural Networks (ESANN 2005)*, pages 241–252. d-side Publications, 2005.
[28] Christian Igel and Michael Hüsken. Empirical evaluation of the improved Rprop learning algorithm. *Neurocomputing*, 50(C):105–123, 2003.
[29] Christian Igel and Martin Kreutz. Operator adaptation in evolutionary computation and its application to structure optimization of neural networks. *Neurocomputing*, 55(1–2):347–361, 2003.
[30] Christian Igel, Stefan Wiegand, and Frauke Friedrichs. Evolutionary optimization of neural systems: The use of self-adptation. In M. G. de Bruin, D. H. Mache, and J. Szabados, editors, *Trends and Applications in Constructive Approximation*, number 151 in International Series of Numerical Mathematics, pages 103–123. Birkhäuser Verlag, 2005.
[31] Yaochu Jin, Tatsuya Okabe, and Bernhard Sendhoff. Neural network regularization and ensembling using multi-objective evolutionary algorithms. In *Proceedings of the Congress on Evolutionary Computation (CEC 2004)*, pages 1–8. IEEE Press, 2004.

[32] J Kaszubiak, M Tornow, RW Kuhn, B Michaelis, and C Knoeppel. Real-time vehicle and lane detection with embedded hardware. In *Proceedings of the IEEE Symposium on Intelligent Vehicles*, pages 619–624, 2005.

[33] J. D. Knowles and D. W. Corne. On metrics for comparing non-dominated sets. In *Congress on Evolutionary Computation Conference (CEC 2002)*, pages 711–716. IEEE Press, 2002.

[34] M. Lades, J. C. Vorbrüggen, J. Buhmann, J. Lange, C. von der Malsburg, R. P. Würtz, and W. Konen. Distortion invariant object recognition in the dynamic link architecture. *IEEE Transactions on Computers*, 42:301–311, 1993.

[35] S. Nolfi. Evolution and learning in neural networks. In M. A. Arbib, editor, *The Handbook of Brain Theory and Neural Networks*, pages 415–418. MIT Press, 2nd edition, 2002.

[36] M Oren, C Papageorgiou, P Sinha, T Osuna, and T Poggio. Pedestrian detection using wavelet templates. In *Proc. Computer Vision and Pattern Recognition, Puerto Rico*, pages pp. 193–199, 1997.

[37] Lutz Prechelt. Early stopping – but when? In Genevieve B. Orr and Klaus-Robert Müller, editors, *Neural Networks: Tricks of the Trade*, volume 1524 of *LNCS*, chapter 2, pages 57–69. Springer-Verlag, 1999.

[38] Russell D. Reed and Robert J. Marks II. *Neural Smithing*. MIT Press, 1999.

[39] Martin Riedmiller. Advanced supervised learning in multi-layer perceptrons – From backpropagation to adaptive learning algorithms. *Computer Standards and Interfaces*, 16(5):265–278, 1994.

[40] Henry A. Rowley, Shumeet Baluja, and Takeo Kanade. Neural network-based face detection. *IEEE Transactions on Pattern Analysis and Machine Intelligence*, 20(1):23–38, 1998.

[41] A Shashua, Y Gdalyahu, and G Hayun. Pedestrian detection for driving assistance systems: Single-frame classification and system level performance. In *Proceedings of the IEEE Symposium on Intelligent Vehicles*, pages 1–6, 2004.

[42] James Edward Smith and Terence C. Fogarty. Operator and parameter adaptation in genetic algorithms. *Soft Computing*, 1(2):81–87, 1997.

[43] Achim Stahlberger and Martin Riedmiller. Fast network pruning and feature extraction by using the unit-OBS algorithm. In Michael C. Mozer, Michael I. Jordan, and Thomas Petsche, editors, *Advances in Neural Information Processing Systems*, volume 9, pages 655–661. The MIT Press, 1997.

[44] M Szarvas, A Yoshizawa, M Yamamoto, and J Ogata. Pedestrian detection using convolutional neural networks. In *Proceedings of the IEEE Symposium on Intelligent Vehicles*, pages 224–229, 2005.

[45] Viisage Technology AG. `http://www.viisage.com`.

[46] L. Darrell Whitley. The GENITOR algorithm and selection pressure: Why rank-based allocation of reproductive trials is best. In James David Schaffer, editor, *Proceedings of the Third International Conference on Genetic Algorithms, ICGA'89*, pages 116–121. Morgan Kaufmann, 1989.

[47] S. Wiegand, C. Igel, and U. Handmann. Evolutionary multi-objective optimisation of neural networks for face detection. *International Journal of Computational Intelligence and Applications*, 4(3):237–253, 2004.

[48] S. Wiegand, C. Igel, and U. Handmann. Evolutionary optimization of neural networks for face detection. In M. Verleysen, editor, *12th European Symposium on Artificial Neural Networks (ESANN 2004)*, pages 139–144. Evere, Belgium: d-side publications, 2004.

[49] C Wöhler and J. K. Anlauf. Real-time object recognition on image sequences with the adaptable time delay neural network algorithm - applications for autonomous vehicles. *Image and Vision Computing*, 19(9-10):593–618, 2001.
[50] J. M. Wolfe. Visual search. In H. D. Pashler, editor, *Attention*. London UK: University College London Press, 1998.
[51] M.-H. Yang, D. J. Kriegman, and N. Ahuja. Detecting faces in images: A survey. *IEEE Transactions on Pattern Analysis and Machine Intelligence*, 24(1):34–58, 2002.
[52] X. Yao. Evolving artificial neural networks. *Proceedings of the IEEE*, 87(9):1423–1447, 1999.
[53] G. P. Zhang. Neural Networks for Classification: A Survey. *IEEE Transactions on System, Man, and Cybernetics - Part C*, 30(4):451 – 462, 2000.
[54] W. Zhao, R. Chellappa, P. Phillips, and A. Rosenfeld. Face recognition: A literature survey. *ACM Computing Surveys (CSUR)*, 35(4):399 – 458, 2003.
[55] Eckart Zitzler and Lothar Thiele. Multiobjective optimization using evolutionary algorithms — a comparative case study. In Agoston E. Eiben, Thomas Bäck, Marc Schoenauer, and Hans-Paul Schwefel, editors, *Fifth International Conference on Parallel Problem Solving from Nature (PPSN V)*, pages 292–301. Springer-Verlag, 1998.
[56] Eckart Zitzler, Lothar Thiele, Marco Laumanns, Carlos M. Fonseca, and Viviane Grunert da Fonseca. Performance assessment of multiobjective optimizers: An analysis and review. *IEEE Transactions on Evolutionary Computation*, 7(2):117–132, 2003.

Index